MALDI Mass Spectrometry Imaging
From Fundamentals to Spatial Omics

New Developments in Mass Spectrometry

Editor-in-chief:
Frank Sobott, *University of Leeds, UK*

Series editors:
Juan F. Garcia-Reyes, *Universidad de Jaén, Spain*
Marek Domin, *Boston College, USA*
Zhongping Yao, *The Hong Kong Polytechnic University, China*

Titles in the Series:
1: Quantitative Proteomics
2: Ambient Ionization Mass Spectrometry
3: Sector Field Mass Spectrometry for Elemental and Isotopic Analysis
4: Tandem Mass Spectrometry of Lipids: Molecular Analysis of Complex Lipids
5: Proteome Informatics
6: Capillary Electrophoresis–Mass Spectrometry for Metabolomics
7: Lipidomics: Current and Emerging Techniques
8: Processing Metabolomics and Proteomics Data with Open Software: A Practical Guide
9: Advanced Fragmentation Methods in Biomolecular Mass Spectrometry: Probing Primary and Higher-order Structure with Electrons, Photons and Surfaces
10: Advanced Mass Spectrometry-based Analytical Separation Techniques for Probing the Polar Metabolome
11: Ion Mobility–Mass Spectrometry: Fundamentals and Applications
12: MALDI Mass Spectrometry Imaging: From Fundamentals to Spatial Omics

How to obtain future titles on publication:
A standing order plan is available for this series. A standing order will bring delivery of each new volume immediately on publication.

For further information please contact:
Book Sales Department, Royal Society of Chemistry, Thomas Graham House, Science Park, Milton Road, Cambridge, CB4 0WF, UK
Telephone: +44 (0)1223 420066, Fax: +44 (0)1223 420247
Email: booksales@rsc.org
Visit our website at books.rsc.org

MALDI Mass Spectrometry Imaging
From Fundamentals to Spatial Omics

Edited by

Tiffany Siegel
Maastricht University, The Netherlands
Boehringer Ingelheim Pharma GmbH & Co. KG, Germany
Email: tiffany.porta@gmail.com

ROYAL SOCIETY
OF **CHEMISTRY**

New Developments in Mass Spectrometry No. 12

Print ISBN: 978-1-83916-241-1
PDF ISBN: 978-1-83916-519-1
EPUB ISBN: 978-1-83916-520-7
Print ISSN: 2045-7545
Electronic ISSN: 2045-7553

A catalogue record for this book is available from the British Library

The Royal Society of Chemistry is a charity, registered in England and Wales, Number 207890, and a company incorporated in England by Royal Charter (Registered No. RC000524), registered office: Burlington House, Piccadilly, London W1J 0BA, UK, Telephone: +44 (0) 20 7437 8656.

For further information see our website at www.rsc.org

For general enquiries, please contact books@rsc.org

For EU product safety enquiries, please email books@rsc.org or contact Royal Society of Chemistry Worldwide (Germany) GmbH, Römischer Hof, Unter den Linden 10, 10117 Berlin.

Dedication

To Marty R.L. Paine.

Who taught me that true friendship does not know distance.

New Developments in Mass Spectrometry No. 12
MALDI Mass Spectrometry Imaging: From Fundamentals to Spatial Omics
Edited by Tiffany Siegel

Published by the Royal Society of Chemistry, www.rsc.org

Preface

The story of this project began in early 2020 when the Royal Society of Chemistry contacted me with the idea of an up-to-date book on mass spectrometry imaging (MSI). I was pleased by the opportunity to provide our community with a comprehensive update. And, luckily, a circle of renowned expert authors and contributors joined the endeavor.

In editing this book my focus was on providing an instructive and comprehensive resource for students and scientists not previously exposed to MSI. I believe that this goal was reached. Furthermore, the book can also serve as an update to the more seasoned MSI experts, providing additional insights and inspiration.

Our focus is exclusively on matrix-assisted laser desorption/ionization (MALDI). MALDI is by far the most widely used ionization technique for biological surface analysis and, in past decades, has matured into an indispensable technical ally in life sciences and biological research. The basis for this success story is the remarkable methodological and technical advances that have boosted MALDI's sensitivity, specificity and potential for disease/(bio)marker identification. Many of the considerations relevant for MALDI can be translated to other ionization methods.

The scope of this book thus covers MALDI-MSI in its entirety, from the basics of the MALDI ionization process, to different source geometries and detection systems, to the latest research and applications in metabolomics, lipidomics, or proteomics. Also included is a practical guide for each class of molecules that can be analyzed by MALDI-MSI. Finally, insights into new innovative applications beyond -omics are provided.

The first section of this book (Chapters 1–6) starts with the fundamentals of MALDI and reflects on implications for imaging. Chapter 2 is a historical view of how MALDI-MSI has developed in the past 20 years since the beginning of the century, explaining its current popularity in various fields.

New Developments in Mass Spectrometry No. 12
MALDI Mass Spectrometry Imaging: From Fundamentals to Spatial Omics
Edited by Tiffany Siegel

Published by the Royal Society of Chemistry, www.rsc.org

Chapters 3 and 4 offer a detailed picture of the instrumentation used for MALDI-MSI, namely the ionization sources and different designs on one hand, and the detection systems used on the other hand. Chapter 5 is a practical guide for sample handling and preparation for MALDI-MSI, which may serve as a handbook for beginners as well as experts who want to extend their knowledge. Chapter 6 reflects on the current challenges that the community is facing while considering quantitative aspects of MALDI-MSI.

The second section (Chapters 7–12) depicts the most recent developments and applications of MALDI-MSI in the omics landscape, namely in metabolomics (Chapter 7), lipidomics (Chapter 8), (neuro)peptidomics (Chapter 9), glycomics (Chapter 10), and proteomics (Chapter 11). This section ends with a critical insight into the crucial role bioinformatics plays in MALDI-MS imaging-based spatial omics (Chapter 12).

In the third section we go beyond omics and look more broadly at other applications of MALDI-MSI, namely those in microbiology (Chapter 13), drug discovery and development (Chapter 14), forensics (Chapter 15), environmental pollutants (Chapter 16), and clinics (Chapter 17). In Chapter 18 we discuss advances that allowed investigation of the finest structures of the elegant organ that is the eye. Finally, Chapter 19 touches upon the combination and integration of MALDI-MSI with other modalities, often appreciated to maximize information that can be obtained from a single sample/tissue section.

I would like to express my sincere gratitude to all authors who have contributed to this project. Your dedication, despite the many restrictions imposed by the COVID-19 pandemic, has been outstanding. I feel privileged by the opportunity to work with you.

My thanks go out to, by order of appearance in the book: Richard Knochenmuss, Pierre Chaurand, Jens Soltwisch, Boone Prentice, Jeff Spraggins, Berta Cillero Pastor, Eva Cuypers, Anahit Campbell, Jonathan Stauber, Hanna Bednarz, Karsten Niehaus, Shadrack Mutuku, Shane Ellis, Jörg Hanrieder, Calvin Blaschke, Rick Drake, Eduardo Chicano-Gálvez, Michel Salzet, Isabelle Fournier, Benjamin Balluff, Jacob Porter, Laura Sanchez, Jeremy Barry, Peter Marshall, Reid Groseclose, Robert Bradshaw, Simona Francesce, Mélanie Lagarrigue, Mira Merdas, Charles Pineau, Kristina Schwamborn, Kevin Schey, Gus Grey, and Martina Marchetti-Deschmann.

I would furthermore like to thank our reviewers for critically assessing and improving the content. They are, in alphabetical order: Theodore Alexandrov (EMBL, Germany), Chris Anderton (Pacific Northwest National Laboratory, USA), Ian Anthony (M4i, Maastricht University, Netherlands), Michael Becker (Boehringer Ingelheim, Germany), Eva Cuypers (M4i, Maastricht University, Netherlands), Gregory Hamm (AstraZeneca, UK), Peter Hoffmann (University of South Australia, Australia), Lieke Lamont (Leiden University, Netherlands), Lingjun Li (University of Wisconsin-Madison, USA), Marcel Niehaus (Bruker, Germany), Michelle Reid (Cristal Therapeutics, Netherlands), Alison Scott (University of Maryland School of Medicine, USA), David Siegel (Boehringer Ingelheim, Germany), Andrew Smith (University of

Milano-Bicocca, Italy), Sara Tortorella (Molecular Horizon, Italy), Mark Towers (Waters Corporation, UK), Michiel Vandenbosch (M4i, Maastricht University, Netherlands), and Rob Vreeken (Janssen Pharmaceutica, Belgium). Special personal thanks to Britt Claes (M4i, Maastricht University, Netherlands) for following the evolution of the book with me and her encouragement; and to my husband David, for continually believing in me and encouraging me to bring this book to completion. I am forever grateful to all of you, and I am glad that we could make this happen together as members of our great MSI community.

I would furthermore like to thank the team at the Royal Society of Chemistry for their continued support. I am particularly grateful to Janet Bridgewater and Liv Towers from the Royal Society of Chemistry editorial office for making the compilation of this book an exceptionally smooth experience. A big thank you to the graphic designers for the wonderful book cover.

I am immensely proud of the contributions to this book and I am thus confident that it will serve as a valuable resource for many adventures in the exciting field of MSI.

Tiffany Siegel
Maastricht University, The Netherlands
Boehringer Ingelheim Pharma GmbH & Co. KG, Germany

Biographies

Biography of the Editor

Tiffany Siegel, Ph.D., started her journey with mass spectrometry imaging in Geneva in 2008 when she joined the group of Prof. Gerard Hopfgartner. In Geneva, she worked on developing new approaches for pharmaceuticals, forensics, and plant biology employing MALDI-MSI and other related surface sampling techniques. She then moved to the group of Prof. Ron Heeren to continue her academic career as a postdoctoral researcher at the AMOLF institute in Amsterdam. With the relocation of the group to Maastricht, Tiffany became a member of the Maastricht MultiModal Molecular Imaging Institute (M4i) in the role of Tenured Assistant Professor at the Department of Imaging Mass Spectrometry of M4i (2016–2021). Her research group focused on the translation between the development of innovative MS-based instrumentation and direct application to intraoperative mass spectrometry and clinical diagnostics. In 2021 she joined the drug development efforts of Boehringer Ingelheim applying MALD-MSI to the field of drug metabolism and pharmacokinetics. Tiffany furthermore serves as Communication Officer on the executive board of the Mass Spectrometry Imaging Society MSIS and as Vice-Chair of the European Scientific Committee of MSACL, the Association for Mass Spectrometry and Advances to the Clinical Lab.

New Developments in Mass Spectrometry No. 12
MALDI Mass Spectrometry Imaging: From Fundamentals to Spatial Omics
Edited by Tiffany Siegel

Published by the Royal Society of Chemistry, www.rsc.org

Biographies of the Authors

Benjamin Balluff studied bioinformatics at the Weihenstephan University of Applied Sciences, Germany. Afterward he did his Ph.D. at the Institute of Pathology, Helmholtz Zentrum Muenchen, Germany, to search for biomarkers in gastric cancer using mass spectrometry imaging (MSI). During his PostDoc time at Leiden University Medical Center, he focused on the analysis of MSI data to describe intra-tumor heterogeneity. In 2015, he became an Assistant Professor for Imaging Bioinformatics at Maastricht University where he strives to advance MSI in terms of single cell analyses and integration with other omics technologies.

Jeremy Barry is a Scientific Investigator at GlaxoSmithKline. He earned B.S. degrees in chemistry and forensic science from the University of New Haven. His M.S. was obtained from the University of North Carolina at Greensboro where he studied pseudo-stationary phase capillary electrophoresis. At North Carolina State University he received his Ph.D. under Dr David Muddiman aiding in the design and characterization of matrix-assisted laser desorption electrospray ionization (MALDESI) for applications in direct analysis and mass spectrometry imaging. He started at GSK in 2014 performing small molecule structural analysis and in 2016 moved to the MS Bioimaging team. His primary focus has been on improving the validity and reproducibility of quantitative mass spectrometry imaging especially in the realm of drug distribution.

Hanna Bednarz is currently working as a researcher and lecturer at the Medical School OWL of the University of Bielefeld. Her main research interest is the correlation between localized pathological events, such as cancer, tumor development, or inflammation and specific changes in cellular metabolism, with the aim of elucidating the causes, basics, and consequences of these perturbations. She obtained her Ph.D. in Biology in 2012 and a Diploma degree in Molecular Biotechnology in the area of molecular techniques, omics analysis, bacterial communication, host–microbe interactions, and microscopy. In recent years, her postdoctoral research has focused on developing novel methods for interdisciplinary research to bridge the fields of biology, microbiology, and molecular medicine with health research. By combining light microscopy, electron microscopy, and metabolomics studies with the new capabilities of imaging mass spectrometry, she aims to advance knowledge of specialized physiological adaptations of tissues over space and time.

Calvin Blaschke is a Ph.D. candidate in Dr Drake's lab at the Medical University of South Carolina (MUSC). He aims to make the detection of glycan biomarkers a more relevant clinical laboratory tool by utilizing mass spectrometry imaging to observe glycan compositions with high-throughput and sensitive methods. His undergraduate research ranged from studying the genetics of pathogenic fungi, developing chromatin remodeling assays, and mining RNA-Seq data for insights on the diet of bats at Coker University, the University of Chicago, and Queen Mary University of London, respectively. At MUSC, Calvin has developed a novel serum and plasma glycan profiling method. He is currently applying this rapid workflow to large cohorts for robust glycan biomarker discovery and adapting its use for additional clinical biofluids.

Robert Bradshaw is a Senior Lecturer in Analytical Science at Sheffield Hallam University (SHU). He obtained a B.Sc. (Hons) in Forensic Bioscience from SHU in 2009 and then went on to complete a Ph.D. on a project entitled "Towards the integration of MALDI-MSI into the current Home Office Fingermark Examination Workflow" under the supervision of Prof. Simona Francese. During his Ph.D. and a Postdoctoral Research position, he worked closely with the UK Home Office and West Yorkshire Police to further develop this novel approach for the analysis of fingermarks obtained in criminal investigations. He has co-authored a total of 16 of the papers published in this niche research area. Within his current role, Robert is a Course leader for M.Sc. Analytical Chemistry, M.Sc. Pharmaceutical analysis and PGCert Mass spectrometry. His research is predominantly within environmental chemistry and forensic science. From an environmental chemistry perspective, he aims to detect and quantify potential harmful species derived from plastic food packaging/containers using a range of mass spectrometry (both profiling and imaging) and chromatography-based methodologies. This will inform the safety of the materials that are currently being employed and the extent of leaching that may occur in different scenarios. Within his forensic research, he is interested in the use of modern analytical tools (mass spectrometry and spectroscopy) to produce forensically relevant methods to detect and identify counterfeit products. This will be of interest to both law enforcement agencies and industry leaders.

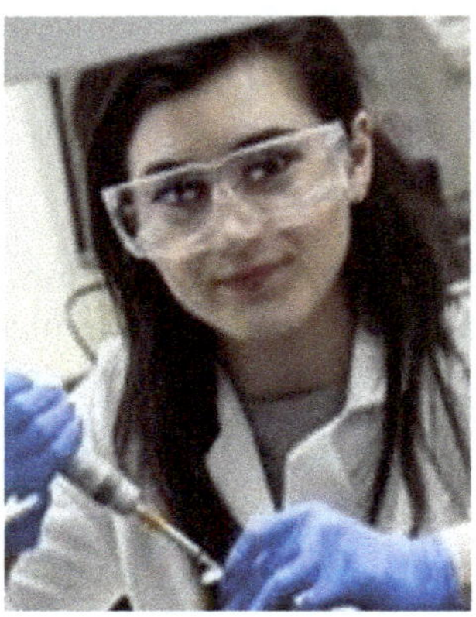

Anahit Campbell holds a M.S. degree in Bioanalytical Chemistry from the University of Wisconsin – Milwaukee where she studied mass spectrometry characterization of novel alkylating agents under Dr Xiaohua Peng. She joined the ImaBiotech team in the summer of 2020 where she is currently a Senior Associate Scientist.

Pierre Chaurand (Ph.D. 1994, Université Paris Sud, Orsay, France) is a Professor of Chemistry at the Université de Montréal (2009–present). His background is in fundamental and analytical mass spectrometry (MS). With over 29 years of expertise, he is one of the pioneers of the mass spectrometry imaging (MSI) technology. His research interests are focused on the development of new strategies and methods to improve the specificity and sensitivity of MSI with applications in clinical biology and forensic science. Dr Chaurand has over 100 peer reviewed publications and book chapters in the field of MS, with over 70 in the specific field of MSI.

Eduardo Chicano-Gálvez is the Coordinator of Proteomics Unit at Maimónides Biomedical Research Institute of Córdoba (IMIBIC; Spain) where he is in charge of optimizations and analysis of mass spectrometry imaging techniques and the analysis of quantitative proteomics data acquired by LC–MS (DIA, DDA, PRM). Eduardo received his Ph.D. from the University of Córdoba (Spain) and a Bioinformatics Master's degree from the International University of Andalusia (UNIA) before continuing to work at IMIBIC Proteomics Unit as a Senior Research Technician where he uses proteomics analysis to investigate a wide range of projects. He has active collaborations with many researchers around the world, focusing mainly in the biomedical area but also in environmental, agricultural, and other basic research areas. He is currently working on the finding of new prognosis/diagnosis biomarkers both in fluid and tissue, and developing new approaches combining MSI data and/or LC–MS data using AI, neural networks, classical machine learning methods, and systems biology.

Berta Cillero Pastor studied Molecular Biology and Biochemistry at the Universidad Autónoma de Madrid. She obtained her Ph.D. at the INIBIC Insitute of La Coruña (Spain) on the effect of cytokines in the field of osteoarthritis by employing proteomics techniques. She then moved to Amsterdam with her own fellowship to work at AMOLF to employ Mass Spectrometry Imaging technology in biomedical applications. Currently she is the CORE lab leader of the Division of Imaging Mass Spectrometry at M4i, Maastricht University.

Eva Cuypers is a Master in Pharmaceutical Sciences and obtained her Ph.D. in Toxicology in 2008 at KU Leuven (Belgium). She worked for more than 15 years as a research expert and principal investigator in an analytical toxicology laboratory (KU Leuven) where she gained practical expertise in analytical sciences including method development, validation, and valorization. During a 6 month internship at the lab of Prof. Ron Heeren in AMOLF (Amsterdam), she developed a method to perform imaging mass spectrometry on cut open single hairs for the detection of drugs. For this highly innovative research, she received multiple awards, including the once-in-a-lifetime Young Scientist Award of The International Society of Forensic Toxicologists. In 2019, she joined the group of Prof. Heeren (M4i, Maastricht) to push the boundaries of imaging mass spectrometry toward single cell research. Recently, she became the principal investigator of her own group at M4i with research focusing on hypothesis-driven translational imaging mass spectrometry to empower clinical diagnostics.

Richard Drake is a Professor in the Department of Cell and Molecular Pharmacology and Experimental Therapeutics at the Medical University of South Carolina and SmartState Endowed Chair in Proteomics. He is also Director of the MUSC Proteomics Center, a mass spectrometry-centric facility with state-of-the-art resources for imaging mass spectrometry, glycobiology, and protein PTM applications. He is an experienced protein biochemist and glycobiologist, with particular expertise in MS imaging of tumor tissues and biomarker discovery from clinical fluids. Current efforts are in the large-scale application of *N*-glycan MS imaging methods developed in his laboratory to liver, prostate, pancreas and breast tumors. Method development of MS imaging approaches for biofluids, cells, and other types of cellular glycan classes is ongoing, as well as other extracellular matrix targeted assays. The goal is to develop glycan biomarkers associated with tumor, stroma, and immune regions of tumors, in two and three dimensions. Furthermore, these glycan maps are being used to identify the localized glycoprotein carriers and develop new approaches to effectively isolate and characterize tissue glycopeptides.

Shane R. Ellis obtained his Ph.D. from the University of Wollongong (UOW) in 2013 and is currently an Australian Research Council Future Fellow at UOW where he leads the Mass Spectrometry Imaging Group within the Molecular Horizons Institute. From 2014–2019 he was a Tenured Assistant Professor at the Maastricht MultiModal Molecular Imaging Institute (M4i) at Maastricht University and led the Instrumentation and Application Development Group within the Division for Imaging Mass Spectrometry. Preceding this was a two-year postdoctoral research position FOM-AMOLF in Amsterdam where he worked on the development of Timepix active pixel detectors for various ion imaging applications. His research focuses on both the continued development of MSI technologies and their application to studying region-specific metabolism in tissues and cells, with a particular focus on lipidomics.

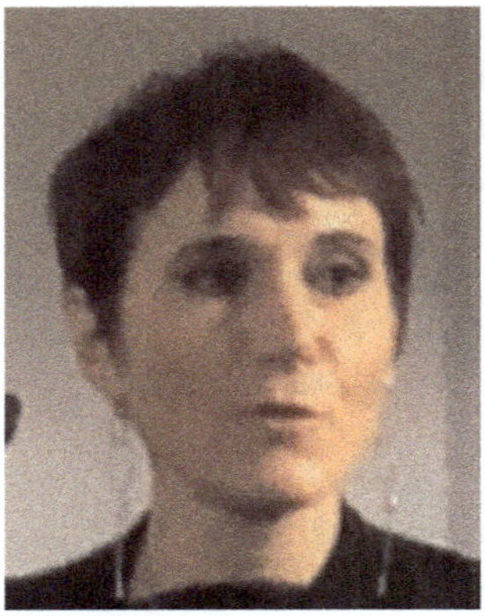

Isabelle Fournier is a Distinguished Professor at the University of Lille and held a position as co-director of the Proteomics, Inflammatory Response & Mass Spectrometry Lab Inserm U1192. Prof. Fournier is an analytical chemist specialized in Mass Spectrometry applied to biology and clinics. She started her research career in 1996 during a Ph.D. at University Pierre & Marie Curie in Paris working on the fundamental mechanisms associated with MALDI. She continued her work on fundamental MALDI for 2 years in the group of Prof. Michael Karas in Frankfurt. In 2002, she started to develop MALDI-MS Imaging at the University of Lille. In 2004, after obtaining an associate professor position, she established her own group on MALDI-MS Imaging and contributed to the field with several developments including new MALDI matrices, strategies for imaging of FFPE tissues and proteins, and the development of tagged-probes for specific imaging of mRNA and antigens. In 2009, she was hired as a Full Professor and awarded a Junior position at Institut Universitaire de France. In 2010 she confounded the Imabiotech Company which provides services in MALDI-MSI. Since 2012, she has been interested in the development of Spatially-Resolved Proteomics in combination with MALDI-MS Imaging for clinical applications in the field of oncology. For 5 years, she has also been interested in the development of *in vivo* MS as a tool for guided surgery and intraoperative analysis. She has authored 177 publications (h-index 43, 6442 citations), 10 patents, 12 book chapters and presented at 120 conferences. In 2017, she was distinguished for her contribution in clinical mass spectrometry by the international distinguish award from MSACL. She was a board member of the COST Action dedicated to MS Imaging and was involved in the foundation of the Mass Spectrometry Imaging Society. For 5 years she has been Vice Chair on the H2020 FET-Open Programs. In 2019 she was awarded a senior position at Institut Universitaire de France. Since 2021 she has been part of the scientific council of the French Network IBiSA which coordinates the national policy of accreditation and support for biology, health, and agronomy infrastructures.

Simona Francese is a Professor of Forensic and Bioanalytical Mass Spectrometry at Sheffield Hallam University, UK. She holds a degree in Chemistry obtained from the University of Salerno, Italy, and a Ph.D. in Chemical Sciences awarded in 2004 by the same University with a one-year period spent at the University of Leeds, UK, thanks to a Marie Curie fellowship. She was a post-doctoral fellow at the University of Leeds in 2004 and 2006, and a research associate at the University of Florence, Italy (2006–2008). She took on a lectureship in biomedical sciences at Sheffield Hallam University in 2008, becoming Reader in 2014. In her current role, she is the Head of the Bioanalytical Research Group and has been the Impact Lead for the UK REF 2021 submission while also being heavily engaged in learning and teaching activities. She is an expert in the development of MALDI-MS Imaging applications and has pioneered its development for the analysis of latent fingermarks to profile offenders. More recently she has engaged in the development of MALDI-MS-based methods for the robust detection, provenance, and mapping of blood in stains and marks, publishing 30 papers (out of a total of 70) on these topics. Her research has been implemented in police casework in the UK and Europe and has been partly funded by the Home Office, West Yorkshire Police, and The Defence Science and Technology Laboratory, UK. She is one of the 60 invited members of the International Fingermark research Group and a member of the European Network of Forensic Sciences. Since March 2017, she has been the Chair of the EU COST Action CA16101 MULTi-modal imaging of FOREnsic SciEnce Evidence (MULTI-FORESEE)-tools for Forensic Science'. She engages public dissemination at all levels and one recent endeavor was the delivery of a TED talk in Vancouver on molecular fingerprinting.

Gus Grey is a Senior Research Fellow in the Department of Physiology at the University of Auckland. He serves as the current Academic Director of the Biomedical Imaging Research Unit, a Core imaging facility within the School of Medical Sciences. His research interests lie in ocular lens function and the development of novel therapeutic strategies to treat lens cataracts. In addition, he collaborates with many groups in Auckland to apply mass spectrometry imaging to study neurodegeneration, cancer treatment, osteoarthritis, diabetes, and other major human diseases. Dr Grey received his B.Tech. (Biomedical Science) and Ph.D. (Physiology) from the University of Auckland, before undertaking post-doctoral training at the Medical University of South Carolina and Vanderbilt University. He then returned to Auckland to establish his own research lab and mass spectrometry imaging platform, which is part of the University of Auckland Mass Spectrometry Hub, a cross-faculty network of instrumentation and expertise that he coordinates.

Reid Groseclose is a Director and GSK Fellow in Bioimaging at GlaxoSmithKline. He earned his B.S. in chemistry from the University of Tennessee in 2004 and his Ph.D. in chemistry from Vanderbilt University in 2009 in the lab of Richard Caprioli, where his work focused on exploring the utility of imaging mass spectrometry and on-tissue proteomics for the analysis and classification of cancerous tissues. He joined GlaxoSmithKline in 2009, where his research interests have centered on the application of mass spectrometry imaging to assess the tissue distribution of drugs and their metabolites. His group is particularly interested in using MSI to quantify the localized tissue concentration of drug candidates and investigating mechanisms of pharmacology and safety.

Jörg Hanrieder obtained his M.Sc. in Chemistry from the University of Leipzig (2005) and his Ph.D. in Analytical Neurochemistry from Uppsala University, Sweden, in 2010 under supervision of Jonas Bergquist and Malin Andersson. He continued with a Postdoc within SIMS and MALDI imaging in Andy Ewing's lab at Chalmers University in Gothenburg, Sweden. In 2015, he started his own group as an Assistant Professor at the University of Gothenburg Sahlgrenska Academy Med School as well as at the Queen Square Inst. of Neurology, University College London, where he holds an appointment as senior PI. He is now an Associate Professor of Neurobiology and his current group includes 4 Ph.D. students and 3 Postdocs (www.neurochem.se).

Richard Knochenmuss received his Ph.D. in Chemical Physics at Washington State University in 1984. He has worked in both academia and industry in several fields, including spectroscopy and intermolecular interactions, fundamentals of laser desorption/ionization methods, mass spectrometer design, and metabolomics/biomarkers. Currently he is consulting in mass spectrometry and related methods, especially ion mobility.

Mélanie Lagarrigue, Ph.D., obtained her Ph.D. in analytical chemistry in 2007 from the Pierre and Marie Curie University of Paris on capillary electrophoresis coupled to mass spectrometry for the detection of chemical warfare agent degradation products in biological and environmental matrices. She then moved to Rennes to launch MALDI-MSI in the research group of Dr Charles Pineau. Her post-doctoral research focused on MALDI-MSI of proteins and proteomics, particularly in the field of reproductive biology. She contributed to the evaluation of laser technologies and statistical tools dedicated to MALDI-MSI and to the use of top-down proteomics for the identification of proteins detected by MALDI-MSI. Her academic research then focused on the development of quantitative MALDI-MSI to investigate the tissular distribution of chlordecone, an organochlorine pesticide responsible for persistent water and soil pollution in the French West Indies. In 2016, she was appointed as a lecturer in biochemistry at the University of Rennes 1 in the research group of Dr Charles Pineau. She continues the investigation of the mechanisms of action of chemical pollutants using the MALDI FT-ICR facility and proteomics strategies to identify partner proteins and molecular responses associated with the exposure to these molecules.

Martina Marchetti-Deschmann obtained a Ph.D. in Chemistry at the University of Vienna. She is a Full Professor for Mass Spectrometric Methods/ Analytical Chemistry at the TU Wien (Vienna University of Technology), where she is heading the Research Group of Mass Spectrometric Bio- and Polymer Analysis in the Research Division of Imaging and Instrumental Analytical Chemistry at the CTA. Martina has been working in the field of Instrumental Analytical Chemistry for over 20 years focusing on the development and application of state-of-the-art and high-end mass spectrometry for molecular bulk analysis and imaging approaches, with the goal to translate analytical and technological advances to industrial applications. Her research is documented in more than 100 publications and was awarded the Beynon Prize from the Journal "Rapid Communications in Mass Spectrometry" in 2007 and the Fritz Feigl Prize in 2013 by the Austrian Society of Analytical Chemistry (ASAC). Martina's group was the first lab combining molecular and elemental imaging in a multi-instrument approach and developing sample handling strategies for MS imaging which allow the use of the very same sample at different instruments to maintain local analyte information down to the cellular level in a modern correlated, multimodal manner.

Peter Marshall graduated from the University of Hull in 1984 and afterward completed a Ph.D. at the University of Reading in collaboration with ICI Plant Protection Division (now Syngenta) at Jealott's Hill, Berkshire. He joined Glaxo (now GlaxoSmithKline) in February 1988 as a Natural Product Extraction Chemist before moving into the mass spectrometry group in 1993. His many roles included the development of methodologies for analyzing compounds from single resin beads as part of the combinatorial chemistry era. This work also included the use of Inductively Coupled Plasma mass spectrometry (ICP-MS) which led us into MALDI-MS imaging about 18 years ago. Peter pioneered the introduction of MALDI-MS Imaging into GSK and was an active member of the European Union Cooperation in Science and Technology (COST) Action for MALDI-MS Imaging co-representing the UK. Peter is currently serving on the board of the newly formed Mass Spectrometry Imaging Society. Peter's use of MALDI-MS has recently morphed into Ultra High Throughput Screening by MALDI and through a collaboration between GSK, Bruker, and Analytik Jena/Cybio has led to the development of a fully Automated system that has the potential to spot and analyze a full diversity 2 million compound assay in ~8 days.

Mira Merdas will complete her Ph.D. degree in pharmaceutical sciences in 2021 from the University of Rennes I – France in collaboration with the international pharmaceutical company "Servier" and the research group of Dr Charles Pineau. She received her M.Sc. in analytical chemistry in 2017 and a second M.Sc. in molecular chemistry in 2018 from the University of Rennes 1. She acquired experience in analytical chemistry, especially in MALDI-MSI using FT-ICR instrumentation. Her current research area focuses on the optimization of new approaches of on-tissue chemical derivatization to study the distribution of drugs and chemical pollutants and their metabolites.

Shadrack Mutuku is currently a Postdoctoral Research Fellow within Molecular Horizons at The University of Wollongong. He completed a Doctor of Philosophy degree from the University of Adelaide in Australia in 2020. His research focused on spatial mapping of lipid profiles of clinical prostate tissues by matrix-assisted laser desorption/ionization mass spectrometry imaging (MALDI-MSI) to identify markers of prostate cancer aggressiveness and androgen receptor-targeted treatment response as well as bioanalytical characterization and visualization of enzalutamide uptake in patient-derived explant cultures of prostate cancer. His Ph.D. thesis was highly regarded to be of exceptional quality resulting in the award of a Dean's Commendation for Doctoral Thesis Excellence. Prior to Adelaide, Dr Shadrack M. Mutuku attended the University of Nottingham, England where he pursued a M.Sc. Degree in Drug Discovery and Pharmaceutical Sciences. Here, his Master's project involved using an immobilized artificial membrane as a lipid bilayer model to study its effects on drug–receptor kinetics. He also holds a first-class honors degree in B.Sc. Biochemistry and Molecular Biology from the Jomo Kenyatta University of Agriculture and Technology, Kenya.

Karsten Niehaus studied biology at the Cologne and Bielefeld universities. His research focused on the analysis of the molecular aspects of plant–microbe interactions. Besides the genetic analysis of bacterial and plant genes, controlling these interactions and the general concepts of cell–cell interactions were analyzed. He has been the head of the Department for Proteome and Metabolome Research at Bielefeld University since 2005. The Department concentrates mainly on interdisciplinary approaches in the field of bacterial metabolism, molecular plant microbe interactions, and for about 10 years has investigated the physiological functions associated with human disease. The group contributed to the installment of the Technology Platform at the Centre for Biotechnology (CeBiTec) at the University of Bielefeld. Presently seven mass spectrometry systems have been installed to support proteome and metabolome analysis. He is responsible for the initiation of the master's programs in "Genome Based Systems Biology" and "Interdisciplinary Biomedicine". Imaging methods such as light microscopy, confocal laser scanning microscopy, and electron microscopy were a constant accompaniment, supporting his research. Merging classical imaging methods, with the analytical power of mass spectrometry, was a consequent extension of this interest.

Charles Pineau has been working in the field of reproductive biology over the past thirty years and is a renowned specialist in spermatogenesis. He has heavily invested in the development of proteomics, integrative proteomics, and imaging mass spectrometry technologies and has applying these to answer biological and clinically relevant problems in the field of testicular pathophysiology and reproductive toxicology. Charles Pineau currently serves as Research Director at IRSET (www.irset.org; Inserm unit 1085, Rennes, France). He is also Director of BIOSIT, a technological institute which brings together 14 core facilities (https://biosit.univ-rennes1.fr) among which is the PROTIM Core Facility (www.protim.eu). Charles's team focuses on the deciphering of the testicular proteome in mammals with a priority given to the identification of novel germ cell proteins that play a key role in the normal and pathological spermatogenesis. His team also uses MALDI Imaging mass spectrometry to help better understand the deleterious effects of small molecules (environmental toxicants, pharmaceutical molecules) on the male and female reproductive spheres and gain new insights into the molecular mechanisms underlying their cellular toxicity. Charles Pineau serves as a recognized scientific expert in the field of male reproduction, proteomics, and reproductive toxicology for national and international funding agencies. He is a scientific consultant for several pharmaceutical companies. Of note is that Charles is one of the past Presidents of the French Proteomics Society. Currently, he is PI of the French initiative (Chromosome 14 team) in the C-HPP and he is the new chair-elect of the Human Proteome Project (https://hupo.org).

Boone M. Prentice received his B.S. degree in Chemistry with Honors and Distinction from Longwood University (Farmville, VA) in 2008 and his Ph.D. degree in Chemistry from Purdue University (West Lafayette, IN) in 2013. His Ph.D. research under the mentorship of Professor Scott A. McLuckey focused on ion trap mass spectrometry (MS) instrumentation development as well as gas-phase ion/ion and ion/molecule reactions involving biopolymers. Dr Prentice then worked as a postdoctoral research fellow in biochemistry in Professor Richard Caprioli's laboratory at Vanderbilt University (Nashville, TN) where he studied matrix-assisted laser desorption/ionization (MALDI) imaging mass spectrometry instrumentation and applications. Specifically, he worked on Fourier transform ion cyclotron resonance (FT-ICR) and time-of-flight (TOF) MS systems and applied imaging mass spectrometry to the study of diabetes, cancer, drug delivery, and infectious disease. Dr Prentice joined the Department of Chemistry at the University of Florida as an Assistant Professor in the fall of 2018. His research focuses on developing next-generation bioanalytical imaging mass spectrometry technologies using novel gas-phase ion chemistry methods and instrumentation to better understand the molecular basis of health and disease.

Jacob Porter graduated from the University of East Anglia in 2015, studying forensic chemistry. His master's thesis, under the supervision of Dr Andrew Mayes, focused on characterizing microplastics with fluorescent dyes. He completed his Ph.D. at Florida International University in 2020 with Dr Francisco Fernandez-Lima, studying biomolecular structural characterization with high-resolution mass spectrometry and mobility spectrometry. His current research topics as a postdoctoral scholar with Dr Laura Sanchez include using imaging mass spectrometry to study small molecule signaling in high grade serous ovarian cancer and cross-microbial interactions. While disparate, his areas of academic interest thus far have been unified by research that he feels aligns with topics of societal and ethical import.

Michel Salzet has been a professor since 1998 and is the director of the U1192 Inserm, PRISM (http://www.laboratoire-prism.fr) unit at the University of Lille (ULille). Pr Salzet received his Ph.D. in Biochemistry at ULille in 1993, where he was a Chair of Immunology in 1997. After his Ph.D., Pr Salzet worked as an Associate Researcher at the Institute of Neurosciences at the College at Old Westbury (USA) and was subsequently nominated as Senior Research Scientist at the Beth Israel, Mind Body Institute of Harvard Medical School (1998). He was a member of the Institut Universitaire de France from 1998 until 2003. Pr Salzet is an expert in the field of neuroimmunology and proteomics. He has trained 28 Ph.D. students and 26 Masters students including 8 resident physicians. He has participated in 33 thesis committees (11 international) and 14 HDR. He has published 362 articles, 31 book chapters (Coordinator of Book: 10, Patents: 16, h-Index: 62, Google scholar), Conferences: 194 (102 on invitation). Pr Salzet is also co-founder of the Start-Up IMABIOTECH started in 2009 and Clic-Imaging platform (2013), Yes-Tec (2020). He creates the Proteomics Master's Degree Program of Lille (2000), Bioinformatics and Omics (2020), became the director of the Doctoral School of Biology and Health of Lille (1998–2002), was responsible for the Genopole program for University of Lille 1 (1998–2003) and from 2013 to 2017, was responsible for the international relationship of Ulille. At the national level, he was a Member of the Governing Board of DVS Biomarkers and Companion Tests, AVIESAN (2015–2017), Scientific Delegate at CNRS (2009–2012), Member of the Scientific Council of the French Natural History Museum Paris (2009–2012) and since 1999, he has been a Member of the National Council of University section 68 and expert for the AERES and HCERES. At the international level, he was a Member of the Governing Board of European Graduates School of Neurosciences Euron (2011–2016), a Member of the Managing group for personalized medicine for European citizens (2009–2012) and a Member of the Core Group of LESC of European Sciences Foundation (2009–2012). Since 2021 he has been in the EU Beating Cancer Plan Stakeholder Contact Group. As an international expert, he was the Remote Evaluator (2016) then Vice Chair (2017), H2020 RIA FET-OPEN, Panel B: Biology. Since 2002 he has been the Remote Evaluator, International: 5th PRDC, DFG, LECMA-AFI-ISAO, FWO, NWO, Austrian Science Fund, NSF, MRC, Health and Medical Research Fund, ARC, FNRS, ISF, IBSF, Marsden Found of Science, CRSNG, FRSQ, FCI (Board member), NIH, NIDA. He is presently a distinguished Professor (PRCE2). He received in 1993 the Wicart Hagelstein Medaillon and in 2003 the "Grand Prix des Sciences" by the Lille Science Academy of the Sciences. He received the Stefanick Award (2020), MATWIN (2015), Altran (2014) innovation prize, the Quebec Sciences prize (2013) with Pr X. Roucou on the Ghost Proteins and the INPI (Nord of France) (2008). He obtained more than 30 grants from national (CNRS, Inserm, INCA, National Valorisation Organization) and international (NIH, NSF & NIDA (USA); FRSQ & IRSC (Canada); FWO (Belgium)) agencies and 11 industrial contracts.

Laura Sanchez earned a Ph.D. in Chemistry at the University of California, Santa Cruz, with Roger Linington as an NSF graduate research fellow. She joined Pieter Dorrestein's lab at UC San Diego as an NIH IRACDA Fellow and her research helped establish methods for probing and characterizing metabolic exchanges in polymicrobial communities, specifically those associated with cheese rinds. She has been at UC Santa Cruz since January 2021 as an Associate Professor of Chemistry and Biochemistry and her NIH and NSF funded research program utilizes imaging mass spectrometry techniques to probe how cells and microbes use chemistry to coordinate activities in a variety of biological systems.

Kevin Schey is currently a Professor of Biochemistry, Chemistry and Ophthalmology & Visual Sciences and a senior member of the Mass Spectrometry Research Center at Vanderbilt University. He also serves as Deputy Director of Research for the NIH Center on Imaging Mass Spectrometry. His research interests lie in the areas of proteomics of aging in ocular tissues, exosome proteomics, and imaging mass spectrometry. He is particularly interested in method development for IMS and applications to diseases of ocular tissues. Dr Schey received his B.S. in Chemistry from Muhlenberg College in Allentown, PA and his Ph.D. in Analytical Chemistry from Purdue University working under the mentorship of Dr Graham Cooks. After post-doctoral training at the University of Chicago he began his academic career at the Medical University of South Carolina in Charleston, SC. After 18 years at MUSC, he moved to Vanderbilt University where he directs the MSRC Core laboratories in Imaging MS, Proteomics, and small molecule MS.

Kristina Schwamborn, M.D., Ph.D., received her M.D. from Heinrich-Heine-University in Dusseldorf, Germany, where she did her thesis work on human herpes virus type 8 transmission. Starting her residency, she joined the Institute of Pathology at the RWTH Aachen University in Germany. She received her Ph.D. from RWTH Aachen University for proteomic studies focusing on prostate and bladder cancer biomarkers. After a two-year postdoctoral fellowship at the Mass Spectrometry Research Center at Vanderbilt University in Nashville, TN, USA she continued her residency at the Institute of Pathology at the Technical University of Munich, Germany. Currently, she is working as a consultant at the Institute of Pathology at the Technical University of Munich, Germany. Her current research focus centers on the application of imaging mass spectrometry in pathology diagnosis and risk stratification.

Jens Soltwisch is a research associate at the Institute of Hygiene at the University of Münster in Germany. He studied physics in Münster and after his finishing his degree in 2008 continued at the same University to obtain his Ph.D. in 2012. After a brief postdoctoral fellowship at the AMOLF institute in Amsterdam in 2013, he returned to his alma mater. He is currently working in the group "Biomedical Mass Spectrometry" at the Institute of Hygiene as a research associate. Since his undergraduate studies, he has been involved in fundamental research and method development centering on MALDI mass spectrometry. After investigating the influence of laser parameters on the MALDI-process, in recent years his attention has shifted toward method development in MALDI-MSI imaging as well as post-ionization modalities. He was the driving force behind the development of MALDI-2 and its early implementation for MALDI-MSI. He has contributed to close to 50 publications on MALDI-MS and MALDI-MSI and presented his work at numerous international congresses. In 2021 he was awarded the Mattauch-Herzog-Prize of the German Society for Mass Spectrometry.

Jeffrey M. Spraggins is a faculty member in the departments of Cell and Developmental Biology, Biochemistry, and Chemistry at Vanderbilt University in Nashville, TN. He is also Director of the Biomolecular Multimodal Imaging Center (BIOMIC) and is a member of the Mass Spectrometry Research Center. Dr Spraggins received his B.A. in Chemistry from the College of Wooster and his Ph.D. in Analytical Chemistry from the University of Delaware (2009), where he studied gas-phase fragmentation mechanisms of modified biomolecules and metal–sulfide cluster ion–molecule reactions. Broadly, Dr Spraggins' research falls into two categories: (1) Developing novel mass spectrometry technologies to maximize imaging performance enabling molecular histology at cellular resolution and (2) combining imaging mass spectrometry with a variety of other biomedical imaging technologies to create new integrated modalities capable of providing a systems biology view of tissue at cellular resolution. Recently, he has begun to apply these advanced imaging technologies to the construction and analysis of 3-D multimodal molecular atlases. As part of large-scale mapping initiatives such as the Human Biomolecular Atlas Program (HuBMAP, NIH) and the Kidney Precision Medicine Project (KPMP, NIH), he is working on bringing together 3-D imaging mass spectrometry with a variety of *in vivo* and *ex vivo* imaging technologies to understand the cellular and molecular architecture of human tissues and how the tissue microenvironment is altered during aging and disease.

Jonathan Stauber, Ph.D. is the founder and Chief Scientific Officer who created ImaBiotech in 2009 using Mass Spectrometry Imaging routinely to better assess drug efficacy. He developed different methods for measuring protein expression and metabolism in disease tissues and drug development programs. He developed a tissue digestion and quantification approach resulting in 47 granted patents and 30 publications and book chapters on the development and applications of Mass Spectrometry Imaging (MSI). Dr Stauber received his Ph.D. in proteomics and molecular biology from the University of California (San Diego, USA) and the University of Technology of Lille (France), and a Post-Doc integrating a multimodal imaging platform with MSI and MRI techniques at AMOLF institute (Amsterdam, The Netherlands).

Contents

Section 1

Chapter 1 An Introduction to MALDI Ionization Mechanisms for Users of Mass Spectrometry Imaging 3
Richard Knochenmuss

1.1 Introduction 3
1.2 Laser Ablation in MALDI 4
1.2.1 Laser Pulse Dependence 5
1.2.2 Plume Pressure, Temperature and Velocity 5
1.2.3 Laser Spot Size 6
1.2.4 Local Thermal Equilibrium 6
1.3 Primary Ionization 7
1.3.1 Preformed Ions and "Lucky Survivors" 7
1.3.2 Mechanical Charge Separation 7
1.3.3 Thermal Ionization 8
1.3.4 Non-thermal Ionization 9
1.3.5 Metal Surfaces 9
1.4 Secondary Ionization 10
1.4.1 Matrix Suppression Effect 11
1.4.2 Analyte Suppression Effect 12
1.5 MALDI-2 15
1.6 Conclusions 16
References 17

New Developments in Mass Spectrometry No. 12
MALDI Mass Spectrometry Imaging: From Fundamentals to Spatial Omics
Edited by Tiffany Siegel

Published by the Royal Society of Chemistry, www.rsc.org

Chapter 2 MALDI Mass Spectrometry Imaging – Past, Present, and Future Challenges **20**
Pierre Chaurand

2.1 Opening Statement 20
2.2 The Beginning 21
2.3 The Last Two Decades 23
2.3.1 MSI Data Acquisition and Instrumentation 24
2.3.2 Automated Matrix Deposition 25
2.3.3 On-tissue Chemistries 25
2.3.4 Imaging of Xenobiotics 27
2.4 Future Challenges 28
2.4.1 The Push for Higher Spatial Resolution 28
2.4.2 Standard Protocols and Methods 29
2.4.3 Increasing Sensitivity and Specificity 29
2.5 Concluding Remarks 31
References 31

Chapter 3 Instrumentation for MALDI-MSI – Part I Ionization Sources and Design **39**
Jens Soltwisch

3.1 Introduction 39
3.2 Pressure Regime 40
3.2.1 Plume Development 40
3.2.2 Vacuum Ion Sources 41
3.2.3 Intermediate Pressure Ion Sources 42
3.2.4 Atmospheric Pressure Ion Sources 44
3.3 Spatial Resolution in MALDI-MSI 44
3.4 Optical Setups 47
3.4.1 Front Side Illumination 48
3.4.2 Transmission Mode Illumination 50
3.5 Post-ionization 51
3.5.1 MALDI-2 52
3.5.2 MALDESI/LAESI 53
3.5.3 Alternative Approaches 53
3.6 Conclusions and Future Directions 54
References 55

Chapter 4 Instrumentation for MALDI-MSI – Part II Detection Systems **59**
Boone M. Prentice and Jeffrey M. Spraggins

4.1 Introduction 59

4.2 Mass Analyzers and Their Performance for MALDI Imaging 60
4.2.1 Sector Analyzers 60
4.2.2 Quadrupole Mass Filters 62
4.2.3 Time-of-flight 63
4.2.4 QTOF 65
4.2.5 Ion Traps 66
4.2.6 Triple Quadrupoles 68
4.2.7 FT-ICR 68
4.2.8 Orbitrap 71
4.3 Ion Mobility Spectrometry for Improved MALDI-MSI Performance 72
4.3.1 Drift Tube Ion Mobility Spectrometry 72
4.3.2 Traveling Wave Ion Mobility Spectrometry 74
4.3.3 Trapped Ion Mobility Spectrometry 75
4.3.4 High-field Asymmetric Waveform Ion Mobility 77
4.4 Conclusion and Future Perspectives 78
Abbreviations 78
Acknowledgements 79
References 80

Chapter 5 Sample Preparation of Biological Tissues for MALDI-MSI 87
Berta Cillero-Pastor and Eva Cuypers

5.1 Introduction 87
5.2 Sample Collection 88
5.3 Embedding and Storage 89
5.4 Sectioning 90
5.4.1 Frozen Material 90
5.4.2 FFPE Material 91
5.4.3 Important Considerations for Tissue Sectioning 91
5.5 Washing 92
5.5.1 Metabolites 92
5.5.2 Lipids 92
5.5.3 Proteins, Glycans and Peptides 92
5.6 Enzymatic Digestion 93
5.6.1 Proteins 94
5.6.2 Glycans 94
5.7 Derivatization 94
5.7.1 Reagent Selection 95
5.7.2 Deposition Method of Derivatization Reagents 95

5.8 Matrix Selection and Application Methods 95
5.8.1 Matrix Application and Recrystallization 97
5.9 General Considerations, Conclusion and Future Directions 98
References 99

Chapter 6 Quantitative MSI: A Review of Biomarker and Drug Quantifications Using Mass Spectrometry Imaging 105
Anahit Campbell and Jonathan Stauber

6.1 Introduction 105
6.2 History and Progress 106
6.3 Method Workflow 107
6.3.1 Method Development (MD) 108
6.3.2 Method Validation (MV) 114
6.3.3 Data Analysis 117
6.4 Versatility of QMSI 121
6.4.1 Lipids 122
6.4.2 Peptides and Amines 122
6.4.3 Polymers and Polyimides 124
6.4.4 Alcohol and Phenols 124
6.4.5 Nanoparticles 125
6.5 Current Challenges 125
6.6 Conclusion and Future Directions 126
References 127

Section 2

Chapter 7 MALDI Mass Spectrometry Imaging and Metabolomics – A Glimpse into the Local Metabolic State of Cells and Tissues 133
Hannah Bednarz and Karsten Niehaus

7.1 Introduction 133
7.2 You Can't Have it All – Sample Retrieval, Preparation and Matrix Deposition – Crucial Steps of the MSI Experiment for Metabolic Analysis 134
7.2.1 Metabolic Quenching 135
7.2.2 Preparation of Flash-frozen Tissue Samples 135
7.2.3 If Matrix Chemistry Fails to Ionize the Target Metabolite or Metabolite Group – Derivatization 138
7.2.4 Selection of the Right Matrix for the MALDI-ionization Process – Targeting the Peak of the Iceberg in Metabolomics Research 139

7.3 Most Used Instrumentation with Strengths and Current Limitations 142
7.3.1 Resolution Power, Exact Mass and Isotope Distribution 142
7.3.2 Tandem Mass Spectrometry and Ion Mobility 143
7.4 Strategies for Structural Analysis/Identification 144
7.5 Applications 146
7.5.1 MALDI-MSI of Plant Metabolites 146
7.5.2 Application of Metabolite Imaging to Animal Tissues 147
7.5.3 MALDI-MSI in Pharmacology and Toxicology 148
7.5.4 MALDI-MSI of Metabolites in Animal Disease Models 148
7.5.5 Spatial Analysis of Metabolites in Human Tissues 148
7.5.6 Imaging of Metabolites with High Lateral Resolution – Emerging Applications 149
7.6 Conclusions and Future Directions 150
Acknowledgements 151
References 151

Chapter 8 MALDI Mass Spectrometry Imaging in Lipidomics 157
Shadrack M. Mutuku and Shane R. Ellis

8.1 Introduction 157
8.2 Sample Preparation for Lipid MSI 158
8.2.1 Tissue Fixation 158
8.2.2 Tissue Washing 159
8.2.3 MALDI Matrices 159
8.2.4 Matrix Application 160
8.3 Lipid Identification in MSI 161
8.3.1 Identification of Lipids Detected with MALDI-MSI 161
8.4 Selected Applications of MALDI-MSI of Lipids 167
8.4.1 Cancer Research 167
8.4.2 Brain Injury and Disease 170
8.4.3 Liver Disease 173
8.5 Recent Developments in MALDI-MSI of Lipids 173
8.5.1 Strategies to Improve the Ionization Efficiency of Lipids 173
8.5.2 Lipid Analysis and Imaging from Single Cells 175
8.5.3 Imaging Stable Isotope Labeled Lipids 177

8.6 Conclusions and Future Directions 178
Abbreviations 178
Acknowledgements 179
References 179

Chapter 9 MALDI Mass Spectrometry Imaging of Neuropeptides 188
Jörg Hanrieder

9.1 Introduction 188
9.2 Methods 189
9.2.1 Sample Preparation for MALDI-MS Imaging of Neuropeptides 189
9.2.2 Data Acquisition and Data Analysis 193
9.2.3 Molecular Identification 195
9.3 MALDI MS-based Neuropeptide Imaging in Biological Research 196
9.3.1 Neuropeptide Imaging in Cells and Animal Models 196
9.3.2 MALDI-MSI of Neuropeptides in Neurodegenerative Diseases 197
9.4 Conclusions and Future Directions 202
Acknowledgements 202
References 202

Chapter 10 MALDI Mass Spectrometry Imaging and Glycomics 207
Calvin R. K. Blaschke and Richard R. Drake

10.1 Introduction 207
10.2 *N*-glycans 208
10.2.1 *N*-glycan Method Development and Optimization 210
10.2.2 *N*-glycan Biological Applications 213
10.3 Glycosphingolipids 216
10.3.1 Glycosphingolipids Method Development and Optimization 217
10.3.2 Biological Application of Glycosphingolipids 218
10.4 Glycosaminoglycans 222
10.5 Glycosides 223
10.6 Non-conjugated Saccharides 224
10.7 Conclusions and Future Directions 226
Acknowledgements 227
References 227

Chapter 11 MALDI Mass Spectrometry Imaging and Spatially-resolved Proteomics **234**
Eduardo Chicano-Galvez, Michel Salzet and Isabelle Fournier

11.1 From Protein MS Imaging to Spatially-resolved Proteomics: Combining Deep Proteome Identification with Spatial Information 234
11.2 MALDI-MSI of Proteins 238
11.2.1 MALDI-MSI of Intact Proteins or Top-down pMSI 238
11.2.2 MALDI-MSI of Digested Proteins or Bottom-up pMSI 241
11.2.3 Specific MALDI of Proteins Using Tagged Probes 244
11.3 Protein Identification in MALDI-MSI and Spatially-resolved Proteomics 245
11.3.1 *In Situ* Identification 246
11.3.2 Off-tissue Identification and Spatially-resolved Proteomics 248
11.3.3 MALDI-MSI Guided Spatially-resolved Proteomics 251
11.4 Conclusions and Future Directions 251
Abbreviations 252
Acknowledgements 253
References 253

Chapter 12 The Role of Informatics and Data Analysis in MALDI Mass Spectrometry Imaging **262**
Benjamin Balluff

12.1 Introduction – The Role of Informatics for the Success of Mass Spectrometry Imaging in Biomedical Research 262
12.2 Software for MSI Data Creation and Handling 263
12.2.1 Software for MSI Data Acquisition 263
12.2.2 MSI Data Representation and Handling 264
12.3 Data Processing 265
12.3.1 Baseline Subtraction 265
12.3.2 Smoothing 268
12.3.3 Spectral Resampling 268
12.3.4 Spectral Alignment and Recalibration 269
12.3.5 Normalization 269

12.3.6 Peak Picking and Peak Filtering 270
12.3.7 Image Processing and Visualization 271
12.4 Data Analysis 273
12.4.1 Finding Discriminatory *m/z* Values 273
12.4.2 Supervised Machine Learning 275
12.4.3 Unsupervised Projection and Clustering Methods 277
12.4.4 Identification of *m/z* Values 279
12.4.5 Data Integration 280
12.5 Image Registration 280
12.5.1 Integration with *Ex Vivo* Techniques 280
12.5.2 Integration with *In Vivo* Techniques 281
12.6 Outlook – MSI Software in the AI Age 282
References 283

Section 3

Chapter 13 MALDI Mass Spectrometry Imaging for Microbiology 293
Jacob E. Porter and Laura M. Sanchez

13.1 Introduction to Microbiology as a Field of Study for MSI Development 293
13.2 Sample Preparation for Microbial MSI 295
13.3 Applications for Imaging Microbial Monocultures 299
13.4 Applications for Imaging Microbial Co-cultures 303
13.5 Applications for Imaging Host–Microbe Interactions 305
13.6 Compound Identification 306
13.7 Conclusions and Future Directions 309
Abbreviations 310
Acknowledgements 311
References 311

Chapter 14 MALDI Mass Spectrometry Imaging in Drug Discovery and Development 321
Jeremy A. Barry, Peter S. Marshall and M. Reid Groseclose

14.1 Introduction 321
14.1.1 Mass Spectrometry Imaging 322
14.1.2 The Drug Discovery and Development Pipeline 323
14.2 Early Discovery 325
14.2.1 High-throughput Screening 326
14.2.2 Complex *In Vitro* Models 326

14.3 Preclinical Phase of Development 327
14.3.1 Drug Disposition and Pharmacokinetics 328
14.3.2 Targeted Drug Delivery 334
14.3.3 Pharmacodynamics 335
14.3.4 Disease Pathogenesis 336
14.3.5 Toxicology and Safety Investigations 338
14.4 Clinical and Beyond 340
14.5 Conclusion and Future Directions 342
References 343

Chapter 15 MALDI Mass Spectrometry Imaging in the Forensic Arena 359
Robert Bradshaw and Simona Francese

15.1 Introduction 359
15.2 MALDI-MSI Application in Forensics 360
15.2.1 Fingermarks 360
15.2.2 Biological Tissues (*Post-mortem* Analysis) 365
15.3 MALDI-MSI Application in Forensic Toxicology 366
15.3.1 MALDI-MSI of Drugs and Metabolites in Hair 366
15.3.2 Potential use of MALDI-MSI of Drugs and Metabolites in Tissues 372
15.4 Prospective Applications of MALDI-MSI in the Forensic Arena 373
15.4.1 Forensic Entomology 373
15.4.2 Microbial Forensics 375
15.4.3 Forensic Botany 376
15.4.4 Food Forensics 377
15.4.5 Forensic Analysis of Fibers 378
15.4.6 Ink/Documents 379
15.5 Conclusions 380
References 381

Chapter 16 MALDI Mass Spectrometry Imaging for Toxicological Evaluation of Environmental Pollutants 387
Mélanie Lagarrigue, Mira Merdas and Charles Pineau

16.1 Introduction 387
16.2 Sample Preparation for the Study of Environmental Pollutant Distribution 389
16.2.1 Tissue Collection 389
16.2.2 Quantification (Optional) 392
16.2.3 On-tissue Chemical Derivatization (Optional) 392

16.2.4 Matrix Deposition 393
16.2.5 Acquisition 393
16.3 Application of MALDI-MSI for the Study of Environmental Pollutant Distribution 393
16.3.1 Quantitative MALDI-MSI of Environmental Pollutants 393
16.3.2 MALDI-MSI of Pesticides in Plants 396
16.3.3 MALDI-MSI of Nanoparticles and Plastics 396
16.4 Combination of MALDI-MSI and LC–MS/MS for the Identification of Responses Induced by the Exposure to Environmental Pollutants 397
16.4.1 Alteration of Energy and Lipid Metabolisms after Exposure to Bisphenol S 398
16.4.2 Metabolic Responses in Plants 400
16.5 Conclusions and Future Directions 402
Abbreviations 403
References 403

Chapter 17 MALDI Mass Spectrometry Imaging in the Clinical Landscape 408
Kristina Schwamborn

17.1 Introduction 408
17.2 Possibilities and Advantages of MALDI-MSI in Pathology Research and Practice 410
17.3 Tumor and Disease Detection 411
17.4 Tumor and Disease Subtyping 412
17.5 Outcome and Response Prediction 415
17.6 Challenges and Considerations 416
17.7 Conclusions and Future Directions 420
References 421

Chapter 18 MALDI Mass Spectometry Imaging for Eye Diseases 432
Angus C. Grey and Kevin L. Schey

18.1 Ocular Structure, Function, and Common Diseases 432
18.1.1 General Introduction to the Eye 432
18.1.2 Anterior Eye – Cornea, Lens, and Cataract 434
18.1.3 Posterior Eye – Retina, Macular Degeneration and Optic Nerve 435
18.2 MALDI-MSI in the Anterior Eye 436
18.2.1 Corneal Structure, Corneal Dystrophies and Keratoconus 436

18.2.2 The Lens in Aging and Cataracts 436
18.3 MALDI-MSI in the Posterior Eye 441
18.3.1 Retina and Macular Degeneration 441
18.3.2 Optic Nerve, Trauma and Glioma 444
18.4 MALDI-MSI for Ocular Therapeutics Detection 445
18.4.1 Advantages and Applications in Pre-clinical Settings 445
18.4.2 Drug Detection in the Eye for Off-target Effects 448
18.4.3 On-target Ocular Drug Detection 449
18.5 Conclusions and Future Directions 450
Acknowledgements 451
References 451

Chapter 19 Correlative Multimodal Mass Spectometry Imaging – Imaging Across the Scales 457
Martina Marchetti-Deschmann

19.1 Introduction 457
19.2 Correlative, Hybrid, Multisensor, Multimodal Imaging – Is There a Difference? 458
19.3 MSI in the Context of Correlative Multimodal Imaging 460
19.4 Combination of MS Techniques 463
19.5 Combination of MS with Other Modalities 465
19.5.1 *In Vivo* and *Ex Vivo* Imaging 465
19.5.2 X-Ray Fluorescence Imaging 467
19.5.3 Optical Imaging 468
19.5.4 Electron Microscopy 470
19.5.5 Atomic Force Microscopy 471
19.6 Challenges for Multimodal Imaging 472
19.7 Conclusions and Future Directions 473
References 473

Subject Index 477

Section 1

CHAPTER 1

An Introduction to MALDI Ionization Mechanisms for Users of Mass Spectrometry Imaging

RICHARD KNOCHENMUSS

RKResearch GmbH, Switzerland
Email: rkrcontact@rknochenmuss.ch

1.1 Introduction

Matrix-Assisted Laser Desorption/Ionization (MALDI) was a rather surprising phenomenon when it appeared in the late 1980s, prompting immediate questions about the underlying mechanisms. The topic continues to generate significant interest and controversy more than 30 years later. In particular, generation of the first (primary) ions remains hotly debated. Reviews[1,2] can be consulted for a more detailed discussion, this contribution is intended to give a brief overview, with an emphasis on those aspects which are particularly relevant for MALDI mass spectrometry imaging (MALDI-MSI) applications.

Starting with a solid sample and ending with free ions, MALDI is a convolution of a series of chemical and physical processes. This makes it considerably more complicated than some mass spectrometry ionization techniques. Compare, for example, electron impact, in which a gas phase analyte is ionized in an isolated, well-defined event. The "D" (desorption)

New Developments in Mass Spectrometry No. 12
MALDI Mass Spectrometry Imaging: From Fundamentals to Spatial Omics
Edited by Tiffany Siegel

Published by the Royal Society of Chemistry, www.rsc.org

and "I" (ionization) aspects are not independent in MALDI, it is necessary to consider them together. The laser interaction with the solid sample determines not only what ions are initially created, but sets the conditions for the subsequent sample ejection and expansion, or so-called plume. Early ions continue to react in the plume as long as collisions occur, which is orders of magnitude longer than the laser pulse. At the end of the process, the observed mass spectra rarely, if ever, reflect the true composition of the original sample. Much useful information is obtained, of course, but it is important to keep in mind the ways in which the results reflect the processes inherent in the method.

We begin with the desorption/ablation event, then consider primary and secondary ionization reactions, which include suppression phenomena with considerable importance for MS imaging.

1.2 Laser Ablation in MALDI

As is apparent from the "desorption" in the MALDI name, the early picture of the role of the laser was to gently warm the surface such that some molecules would smoothly evaporate. This was supported by the observation, which made MALDI so useful, that large, fragile molecules were found in the mass spectrum. Since then, much more has been learned about this aspect of MALDI,[3,4] and it has become clear that the general term "ablation" is more appropriate, since the event is not necessarily very smooth. This has been directly visualized by, for example, Leisner *et al.*[5] and others[4] (Figure 1.1).

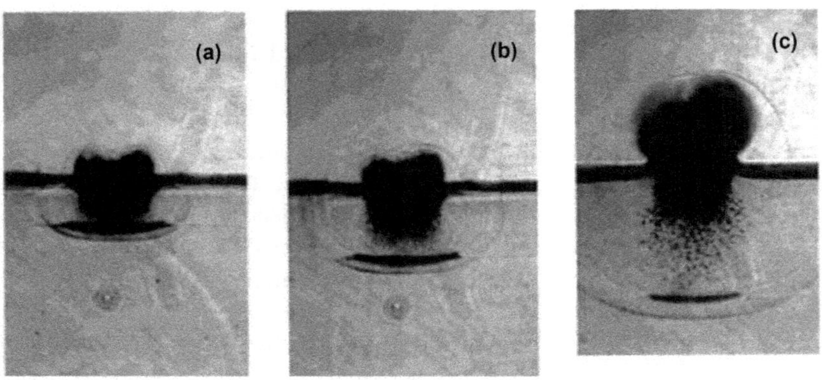

Figure 1.1 Ablation of aqueous copper sulfate with 8 ns, 1064 nm laser pulses. Delay times after the laser pulse: (a) 300 ns; (b) 550 ns; (c) 1000 ns. While not a MALDI sample, this example shows some of the phenomena that can occur both above and below the surface during ablation. Shock waves, nucleation regions, and dense ejecta plumes are readily apparent. The ablated material is not uniformly vaporized, condensed material ranging from clusters of a few molecules to macroscopic particles is found, the latter have been collected and characterized.[6]
Reproduced from ref. 4 with permission from American Chemical Society, Copyright 2003.

1.2.1 Laser Pulse Dependence

The characteristics of the plume vary considerably with laser pulse duration and energy. This contributes to the substantial changes in mass spectral quality and content *vs* laser intensity that all users are familiar with. Molecular dynamics simulations of MALDI[7–9] provide detailed insight as well as visualization of the differences between long (several nanosecond), low energy (~10 mJ cm^{-2}) pulses and short (pico- or femtosecond), powerful (100s of mJ cm^{-2}) ones. The former are not so different from the original picture of soft desorption, but do not eject much analytically useful material. The latter cause large scale material disintegration (phase explosion) with creation of big pieces of condensed material (up to macroscopic) that may be too cold to fully vaporize later. Such an event is also of limited analytical value since the dense plume hinders ions from being properly collected and focused in the mass spectrometer.[10] Also, one or a few shots may ablate most of the available material, but provide proportionately few ion counts.

These effects have consequences for spatial resolution in imaging. Even if the optical focus is the same, a weaker laser pulse typically ablates a spatially smaller region than a more powerful one. In some instruments this is largely due to the beam profile, which may be gaussian or similar peaked function. In that case, only the center of the illuminated region will ablate substantially. This is obviously not the case for top-hat or other more homogeneous profiles. There is also a significant mechanical component. A deeper and more violent ablation event will affect a considerable region around the illuminated region, blasting some of it away. For these reasons, lower laser fluences may be preferable for achieving the best spatial resolution.

1.2.2 Plume Pressure, Temperature and Velocity

Because considerable material is heated above the boiling or sublimation point in a few nanoseconds (for example by a Nd:YAG or nitrogen laser), the initial pressure in the ablation plume may reach tens of atmospheres. This is not gentle evaporation from the surface. The pressurized mix of gas and condensed material does not expand uniformly. The top layers are energized most by the laser and are not hindered in their expansion. They reach high velocities, around 1000 m s^{-1}. Deeper layers are cooler (the laser is attenuated by the upper layers) and may be ejected with velocities as low as 100 m s^{-1}.[11] This velocity pattern means that there is little downstream mixing of sample layers in the plume, as long as it can expand freely. Typical peak plume temperatures are calculated to reach over 1000 K, consistent with previous experiments.[12]

If the ablation is performed in a background gas, rather than in a vacuum, the plume cannot expand as much before it stagnates and mixes with the background. However, given that the starting conditions are much higher in pressure and density than even ambient air, a significant portion of the expansion is essentially identical in all MALDI events.

1.2.3 Laser Spot Size

The rather harsh initial plume conditions decay with the plume expansion, which is adiabatic (insignificant heat flow to/from the surroundings). This is an important fact, since it leads to a MALDI spot size effect: even at the same fluence (laser energy per area), changing the laser focus leads to different results.[13–16] This has recently been investigated in oversampled imaging, as well.[17,18] In particular, a higher fluence is typically needed to obtain a good signal from a smaller spot. In MALDI imaging this tradeoff can be important, since higher spatial resolution is often desired. The effect arises because the plume expands not only perpendicular to the irradiated surface, but also laterally. For a small spot, lateral expansion is somewhat faster than for a wide spot.[19] This fast radial expansion means less time for ion–molecule reactions in the plume, so there are fewer analyte ions created.

1.2.4 Local Thermal Equilibrium

The high pressures and temperatures in the early plume imply many intermolecular collisions, raising the question whether the material in the plume reaches or approaches thermal equilibrium. Some models of MALDI ionization are based on this assumption.[20–22] As noted above, the plume is in fact very inhomogeneous, especially perpendicular to the surface, and changes considerably with time. This means that at most a transient local thermal equilibrium (LTE) might be established. This is most likely during the early hot, dense period, since forward and backward reaction rates are highest. As the plume expands and cools, equilibria must shift, but reaction rates also decrease, so even if LTE is lost at some point, relative ion populations may remain near their values from earlier times. The topic of LTE in the plume is not easily explored experimentally because it is very difficult to sample the full plume, not least due to the wide velocity range noted above. In nearly all such studies, the acquired MALDI mass spectra are a narrow slice out of the whole picture, typically of the fastest, topmost ions, which are the easiest to collect.

As is not surprising from the above considerations, some data are consistent with LTE, others are not, and some show a transition between the two situations. For example, it was found in one case that the matrix : analyte ion ratios corresponded to the thermodynamically expected equilibria for a tripeptide in two different matrices.[23] However, this required a relatively high laser fluence. At low fluence other ratios were observed. This is consistent with the idea that plume LTE needs a sufficiently long period of high temperature and density, along with a large quantity of matrix primary ions *vs.* the quantity of analyte. Other examples of such thermodynamically predicted ratios have been reported.[24,25] As will be discussed below, models incorporating full plume physics and chemical kinetics also exist and are a superset of those that assume equilibrium.

1.3 Primary Ionization

Primary ionization is the formation (or separation, in the case of "preformed" ion pairs) of the first ions in a MALDI event. These may or may not react further in secondary reactions with neutral molecules or with other ions (especially neutralization with opposite charged ions). Despite many years of investigation, it has proven difficult to definitively elucidate primary ionization mechanisms in MALDI. It is possible or even probable that multiple mechanisms exist, and more than one may be active for a given matrix/analyte/laser combination.

A difficulty which has repeatedly arisen in studies of MALDI mechanisms is devising experiments and methods which can clearly differentiate between primary ionization mechanisms. This is largely a consequence of the fact that measured MALDI results are never purely dependent on primary ionization, but also on the complex ablation event and secondary reactions. Results which might initially seem simple and clear tend to be far less conclusive on closer inspection. In the discussion below we will briefly mention a few of the complications that occur.

1.3.1 Preformed Ions and "Lucky Survivors"

An early proposal for primary ionizations is the "Lucky Survivor" preformed ion model of Karas *et al.*[26,27] Free ions of matrix and analyte obviously exist in the preparation solution, and the model suggests that they remain free in the dried sample. They are then simply released by ablation. Most may not be set free, but some escape – the lucky survivors. A nice demonstration of the existence of candidate ions in the solid state was provided by using pH indicator dyes as analytes in MALDI preparations. The dried matrix crystals retain the colors of the solution.[28]

However, unpaired ions are unfavorable in solid materials, and ions in the dried sample are expected to be stoichiometrically coordinated with an oppositely-charged counterion. But releasing an ion pair in the ablation is not helpful, since it has no net charge. Separating these ion pairs is not trivial, they are bound by strong electrostatic forces. In addition, the model also postulates most ions are entrained in neutral chunks. Although these certainly evolve and release some ions,[29,30] simulations and data suggest that matrix clusters and chunks mostly remain intact,[6,11] so few free ions are released.

One can certainly imagine that the Lucky Survivors mechanism contributes some ions in MALDI. After all, salt crystals yield ions by laser ablation. But it makes no quantitative predictions and seems to rely on processes of very low ion yield.

1.3.2 Mechanical Charge Separation

Rather than separating preformed ions as in the previous model, charge separation by mechanical fracturing of matrix crystals (triboelectricity) is a possibility. This occurs due to the unequal free energies of the newly formed surfaces. Further physical separation occurs as the crystal fragments fly apart. However, triboelectricity has not been found to be necessary for an effective MALDI matrix, so this role of this mechanism seems limited in traditional MALDI. It may be much more relevant in "laserspray" or "inlet" methods, in which crystal fracture during rapid sublimation has been associated with ion production.[31]

Other surface-based mechanisms involve ion fractionation in ablated droplets. Ions of one polarity may be enriched on the surface compared to the interior. Final mechanical ion separation occurs by bursting of bubbles or other mechanical instabilities of the droplets.[32] The yield of such processes is quite low, and the ions are again released largely in clusters and chunks of neutrals.[33]

1.3.3 Thermal Ionization

Ions can be created from neutrals by heat, for example by electron emission from molecules, atoms and surfaces. Some early mass spectrometers used hot wires as ion sources. Gaseous organic molecules were ionized by thermally activated electron transfer to the wire. While a recent miniaturized hot wire source has shown some limited results in MALDI-like applications,[34] this method fails for many biomolecules which are analyzed by MALDI, because they fragment too easily at higher temperature. In the late 90s it was then proposed that in the hot early MALDI plume the matrix itself could be acting as a solvent to facilitate ion separation. Since most, but not all, matrix materials are polar molecules, this was called the Polar Fluid Model (PFM).[35,36] There continues to be significant interest in the concept.[20–22]

In the PFM the hot matrix fluid reduces the energetic requirements for charge separation by solvating the nascent ions or ion pairs, just as in a beaker of water. Instead of 500–700 kJ mol^{-1} to separate matrix ion pairs (for example $2M \rightarrow M^+$ and M^-, or $2M \rightarrow MH^+ + (M-H)^-$), the energy required could be reduced to 100–200 kJ mol^{-1}. If these reactions approach equilibrium in the hot plume, this would generate enough ions to be easily measurable.

More detailed theoretical investigation shows that there are fundamental reasons why this model is very unlikely to be relevant for MALDI.[37] For example, at the temperatures and pressures in the early plume, organic substances have low dielectric constants.[37,38] Even the dielectric constant of water is near 1 under such conditions[39] *vs.* 80 at lab conditions. This means ion solvation is nearly negligible.

Comparison of measured MALDI results with thermal predictions has also shown poor correlation in several cases.[40–42] Other results were initially interpreted as evidence for thermal mechanisms, but later shown to be equally compatible with non-thermal models.[43,44] Again, secondary reactions in the plume complicated the interpretation of these data.

1.3.4 Non-thermal Ionization

MALDI can be performed with a wide range of laser wavelengths, but here we consider only ultraviolet excitation as it is most widespread. The laser energy must be efficiently deposited, so the absorption spectrum of the matrix in the solid state is a key factor in determining MALDI efficacy.[45–48] This spectrum is typically much broader and shifted to the red compared to the spectrum in solution. This reflects interactions between neighboring matrix molecules and delocalized electronic states. The energy needed to ionize matrix molecules is on the order of a few (2–3) quanta of typical ultraviolet MALDI laser light (337 to 355 nm). These facts suggest ionization mechanisms involving a small number of electronically excited matrix ions.

A model based on electronic excitation has been proposed for some matrices, involving migration of excitations in the matrix solid and pooling reactions of two excitations.[19] The model also includes secondary ionization, as well as full plume physics. It is therefore called the Coupled Chemical and Physical Dynamics (CPCD) model of MALDI.[49] This remains the only comprehensive model of MALDI to date.

There is considerable evidence for exciton-based mechanisms in some matrix materials, such as dihydroxybenzoic acids (DHB).[50] Based on the photophysical properties relevant for this mechanism, the CPCD has correctly predicted the relative MALDI performance of these isomers.[42] Previously there was no explanation for the fact that they do not perform equally well. These results were also not compatible with thermal models.

The CPCD model has also successfully reproduced extensive data for wavelength- and fluence-dependent performance of DHB, alpha-cyano-4-hydroxycinnamic acid (CHCA), and alpha-cyano-4-chlorocinnamic acid (ClCCA) matrices.[40,41,51,52] The laser wavelength studies led to an important result: Rather than being fluence dependent,[53] MALDI ion yield is actually a function of the number of laser photons deposited. At a constant wavelength, these are functionally equivalent, but the underlying mechanisms are apparently quantized. This is also reproduced by the CPCD.

1.3.5 Metal Surfaces

A phenomenon which can be particularly relevant for MALDI imaging involves the sample support. If the sample is either sufficiently thin that some laser light reaches the support (a few 100 nm), or the sample is ablated down to the support, the support material can significantly affect the ion yield.[54,55] It was found that a stainless-steel support led to dramatically better signal than a gold one, for several matrices. This effect was proposed to involve interaction of the matrix excited states with the conduction band of the metal, leading to more efficient matrix ionization. It is useful to be aware of this effect since gold supports may otherwise be preferred due to their chemical inertness. The effect can also be taken advantage of, since the surface enhancement for a thin sample can be substantial, giving much better signal than a thicker sample. However, the sample does need to be quite thin. In the experiments noted above, this was achieved by rapidly moving an electrospray sample applicator over the substrate. There may be a physically similar enhancement when using thin (sputtered) metal over coatings on a MALDI sample (to suppress surface charging), but it appears to be less dramatic.

1.3.5.1 *Types of Primary Ions*

Regardless of the mechanism by which the initial ions are created, only a few types of ions are typically observed. These are defined by the charge carrier which is involved. If electrons are transferred, radical cations and anions are formed. Considering a sample of pure matrix:

$$2M \rightarrow M^{+} \text{ and } M^{-}, \text{ where M is a matrix molecule}$$

In most common matrix molecules these are radical ions.

Alternatively, a proton disproportionation might be involved:

$$2M \rightarrow MH^{+} \text{ and } (M\text{–}H)^{-}$$

Proton transfer ions tend to be energetically preferred for many of the classical matrix materials (DHB, SA, CHCA, *etc*) which have acidic and basic sites. However, electron transfer ions can be dominant for other matrices (DCTB, etc[56]), and both types may appear in the same spectrum. Which ones are dominant depends on the relative ionization energies, electron and proton affinities, and basicities of the species involved. For many matrices, interconversion reactions (between electron and proton transfer ion pairs) are rather low in energy, a few tens of kJ mol^{-1}, and are therefore facile in the early plume.

Metal adduct ions are common in positive ion mode, since the commercial matrix (or the sample) often contains considerable salt, and can be enhanced by addition of metal salts (for example, silver ions are particularly useful for some synthetic polymers). The larger the cation, the lower the binding energy with the matrix or analyte. Protons (H^{+}) bind the strongest by far, with affinities up to 900 kJ mol^{-1}. Sodium affinities are around 150 kJ mol^{-1}, for K^{+} they are near 100 kJ mol^{-1}. Metal adducts may form directly from the corresponding matrix salt, or by interconversion reactions. Anion adducts are rare because anions are large and diffuse, leading to low binding energies with neutral molecules.

Assuming the laser energy is directly absorbed, corresponding electron or proton transfer ions of analytes can also be formed, but the matrix is typically in excess and most primary ions will be of the types mentioned above.

1.4 Secondary Ionization

Reactions of the primary ions with neutrals (or other ions) in the plume are probably the most important process for MALDI imaging. Regardless of how primary ions are formed, and which ions they are, they are typically embedded in a plume composed mostly of neutral matrix molecules and clusters. Secondary reactions involving matrix species are therefore typically the most important.

1.4.1 Matrix Suppression Effect

One of the most straightforward methods of investigating these reactions involves changing the relative amounts of matrix and analyte. It was observed that increasing the analyte mole fraction in MALDI preparations could lead to a strong decrease, or even complete disappearance, of matrix ions in the mass spectrum.[57,58] This has been named the Matrix Suppression Effect (MSE). The MSE is not only dependent on the matrix : analyte (M : A) ratio, but also on laser fluence. At a given M : A, there is less MSE at higher fluence.[58] There is also an effect of analyte molecular weight – larger analytes cause MSE at lower M : A.[59] Finally, MSE also depends on the laser spot size – it requires less analyte for a larger spot.[49]

The MSE was initially surprising, but is readily understood in terms of ion–molecule reactions in the expanding plume. As noted above, the number of initially generated primary ions is limited by the number of laser photons, so stronger pulses generate more. The sample is ablated, and the ion/neutral mixture begins to expand. Also, as noted above, the initial state of this fluid is dense and hot, so collisions are frequent and there is sufficient activation energy for many reactions. The reaction of matrix ions with analytes is normally energetically favorable (or else the matrix is inappropriate for the analyte), so reactions like the following begin to create analyte ions:

$$MH^+ + A \leftarrow\rightarrow M + AH^+$$

Similar reactions can occur for electron or cation transfer. This reaction may be kinetically limited or move toward LTE depending on the reaction activation energy, local plume conditions, and the evolution of the local plume over time.[23–25] The net ion mole fraction is not high, perhaps in the order of 10^{-5}, maybe as low as 10^{-9}.[18] It is higher early in the plume before recombination of ion pairs reduces it,[60] but the mole fraction of analyte for MSE may be 1–10%. Therefore, the above reaction is pushed far to the right. With enough analyte, matrix ions are so depleted that they become negligible in the mass spectrum. On the other hand, if more MH^+ is provided (higher laser fluence), more neutral analyte is needed for full depletion. Of course, matrix ion depletion and MSE occur in negative polarity as well.[59]

The MSE is reproduced in the CPCD model because it includes the full forward and reverse kinetics of such reactions, and how they are modulated by the plume density and temperature. The reaction rates use a linear free energy approximation: the activation energy, and hence rate constant, is assumed to be inversely related to the reaction free energy (a more negative ΔG implies a lower activation energy), as is common for many organic reaction mechanisms. The extent of the reaction is affected by how the speed of the plume expansion compares to the secondary reaction rates. This in turn depends on the laser spot size, as noted above. In imaging, a small spot is often preferred since spatial resolution is so important. Unfortunately, this means faster expansion and decreased analyte ion formation. It may sometimes be possible to qualitatively evaluate this by the degree of MSE observed.

The CPCD includes ions of both polarities, which make it possible to examine how ion–molecule reactions in one polarity affect those in the other. This is an aspect which is seldom considered, since only one polarity is observed in the mass spectrum. They are connected *via* neutrals, especially analyte, which can also be a limiting reagent. For example, if the above reaction to form a protonated analyte is energetically highly favorable (analyte more basic than the matrix), the corresponding reaction in the opposite polarity is probably less favorable:

$$(\mathrm{M{-}H})^- + \mathrm{A} \rightarrow \mathrm{M} + (\mathrm{A{-}H})^-$$

This means the positive matrix ions will react faster with neutral analyte and consume more of it. This reduces the negative analyte yield compared to what it would be otherwise. The ratio of positive to negative ions in the corresponding mass spectra will then not be one, for either matrix or analyte.[61] To put it another way, if the analyte signal is poor, it might be useful to look at the opposite polarity. The reaction thermodynamics and kinetics might actually favor those products, even if the original polarity "should" give good results.

An important MSE observation is that all ion types are affected at the same time. For example, matrix ions of different types often appear in positive mode. Protonated species are often dominant, but radical cations and alkali adducts are frequently also strong. Adding enough analyte to induce MSE will suppress all of these ions, even if the analyte only appears as, for example, AH^+. This is again a consequence of extensive secondary ion molecule reactions, which allow the system to proceed toward the kinetically and thermodynamically most favorable products. It implies electron transfer reactions to interconvert the ion types.

The key condition for MSE is that enough analyte is present to consume all the available primary ions. This must happen before the plume has expanded enough that reactions become too slow. This kinetic aspect is the reason why larger analytes induce MSE at lower concentration than smaller ones. An analyte of 10 000 Da is physically much larger than one of 100 Da, with a correspondingly larger collision cross section. It is therefore more efficient in reaction with primary ions.

1.4.2 Analyte Suppression Effect

We are now in a position to consider the situation, which is nearly always encountered in MALDI, and especially in imaging: more than one analyte. It should be clear from the discussion of the MSE that ion–molecule reactions with all analytes will occur in parallel, which causes the analyte signals to be coupled. Taking the example of proton transfer reactions again, the matrix–analyte reactions with analytes A and B are:

$$MH^+ + A \leftarrow\rightarrow M + AH^+$$

$$MH^+ + B \leftarrow\rightarrow M + BH^+$$

In addition, direct analyte–analyte reactions can occur:

$$AH^+ + B \leftarrow\rightarrow A + BH^+$$

Because the matrix is in excess, even under MSE conditions, A and B are most strongly coupled by the reversibility of the matrix reactions:

$$AH^+ + M \leftarrow\rightarrow A + MH^+$$

$$MH^+ + B \leftarrow\rightarrow M + BH^+$$

giving a net reaction:

$$AH^+ + B \leftarrow\rightarrow A + BH^+$$

As for the MSE, it should not be forgotten that the opposite polarity reactions are also taking place, affecting the results in the observed polarity. The most important consequence of these reactions is that it is difficult to obtain a mass spectrum which even approximately reflects the composition of the sample. In the extreme case, called the Analyte Suppression Effect (ASE), some analytes can completely suppress others.[62]

The conditions for ASE are, not surprisingly, similar to those for MSE: sufficient analyte concentration, both compared to primary ions (laser fluence) and to other analytes. The analyte which is favored is the one which forms the energetically most favored ions. In the examples above, it would be the analyte with the highest proton affinity. Of course, this assumes that the reaction kinetics of the two analytes are not very different. The only case where a general kinetic limitation might arise is for electron transfer reactions, in the so-called Marcus inverted region. At low ΔG, rates of electron transfer reactions increase with ΔG, but can slow at high ΔG due to poor overlap of vibronic wavefunctions.[63]

Both ASE and MSE are illustrated in Figure 1.2 by a series of measurements of an equimolar mixture of 5 analytes.[64] The M : A ratio was changed over 4 orders of magnitude. At the highest analyte concentration (lowest M : A mole ratio, 77), the thermodynamically most favored species is dominant in the spectrum, while only very weak signals of 2 other analytes were observed. No matrix ions are visible. At higher M : A = 690, the other analytes increase in relative intensity, though the least favorable species is still not visible. Matrix ions begin to appear only at M : A = 6900, as does the last component of the mixture. Even at this dilution, relative intensities in the spectrum must be interpreted with caution.

Both MSE and ASE were investigated by MALDI imaging of dried droplet samples.[65] As seen in Figure 1.3, these results illustrate how sample

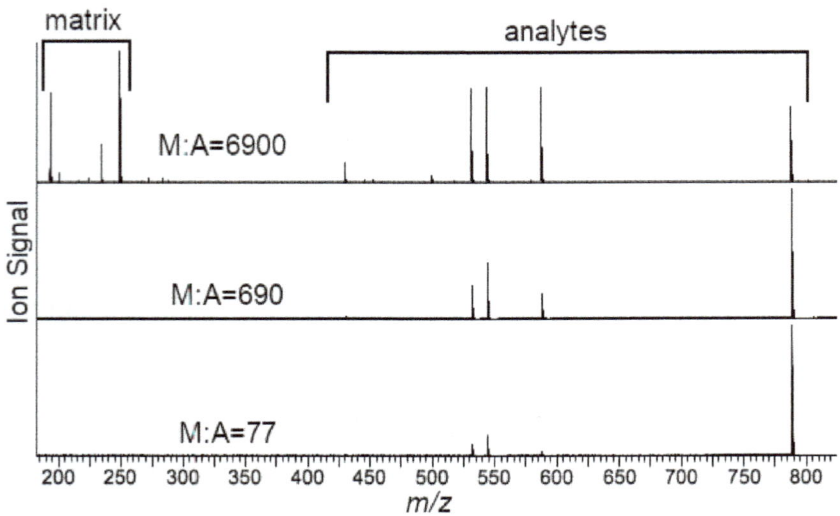

Figure 1.2 Illustration of MSE and ASE using an equimolar mixture of 5 analytes in DCTB matrix. The matrix : analyte mixing ratios are indicated.[64] See the text for more discussion.

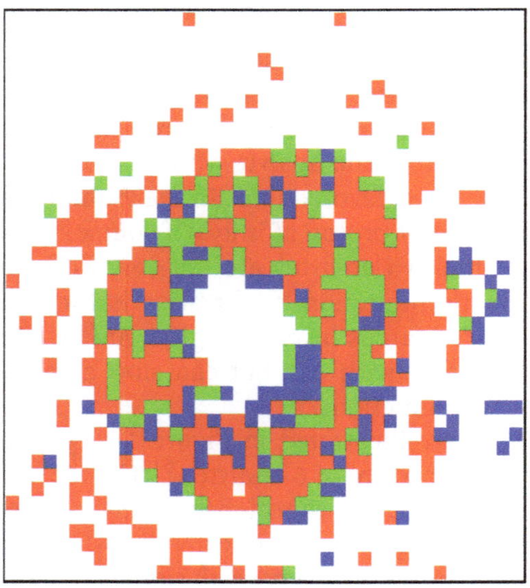

Figure 1.3 MALDI images of yohimbine and caffeine in a DHB dried drop sample. The molar ratios were 4 : 1 : 36 (yohimbine : caffeine : matrix). In green pixels the yohimbine : caffeine signal ratio was correct (4 : 1) within a factor of 2. Red indicates suppression of the caffeine signal, while blue shows suppression of yohimbine. Suppression of caffeine is most common, appearing in 61% of pixels.
Adapted from ref. 65 with permission from American Chemical Society, Copyright 2004.

inhomogeneity at the scale of the pixels, 100 μm, is reflected in the suppression effects. Even when the M : A ratio in the preparation solution is low enough, not all pixels exhibit suppression, though most do. The inverse can also occur, some pixels may exhibit suppression even if it is not typical in the sample. The best results might then be obtained by filtering pixels (or spectra summed in one pixel) based on the relative magnitudes of matrix peaks, since these give some indication of suppression. Of course, a different, more uniform, sample preparation (other than dried drop) would likely provide a far more consistent result.

Strong ASE is a common problem in MALDI imaging of lipids. Boskamp and Soltwisch investigated suppression of different glycerophospholipid classes in both positive and negative modes.[66] In positive-ion mode, phosphatidylcholines (PC) suppressed phosphatidylethanolamines (PE) or phosphatidylserines (PS), in a concentration-dependent manner. As could be expected from the considerations noted above, in negative polarity the situation was reversed, and PC was difficult to observe. Interestingly, a cross-polarity effect was also observed, the presence of PC enhanced the signal of other lipids in negative mode. This was attributed to the reactions leading to PC positive ions.

Methods to control suppression effects generally involve modified sample preparations. An example is ensuring that the matrix and analyte are not uniformly co-crystallized, by applying them separately.[67] This is roughly the situation in MALDI tissue imaging, but works against sensitivity, which is also an issue in imaging. Another option is derivatization by applying an appropriate reagent on top of the rest of the preparation.[68] This changes the chemical nature of some sample components, with corresponding modification of the suppression patterns. Alternatively, expected sample constituents can be spiked either to act as reference suppressors,[69] or as internal standards.[70]

Tissue sections may contain considerable salt, which reduces sensitivity. Desalting[71] may restore signal and thereby ameliorate nonspecific suppression effects, but does not decrease analyte–analyte suppression. In an interesting reversal of strategy, Popkova and Schiller proposed saturating the matrix solution with cesium iodide.[72] Most of the ions in the resulting spectra were Cs^+ adducts, which gave better lipid coverage. ASE is much reduced because analyte affinities for large alkali cations are low. As an indirect result, there are small absolute differences in the reaction thermodynamics for cationization of the various analytes, reducing competition between analytes.

1.5 MALDI-2

An important recent extension of MALDI is the so-called MALDI-2 technique.[73] Earlier efforts to increase sensitivity by "post-ionization" of neutral molecules in the MALDI plume met with limited success. The key insight in the MALDI-2 method was to introduce a few hundred Pa of buffer gas to slow and compress the plume. Crossing the compressed plume with a time-delayed second laser about 0.5 mm above the sample was found to increase signals of many analytes by up to 2 orders of magnitude. In the context of imaging, this obviously makes small ablation spot sizes and high spatial resolution much more feasible.

As shown by systematic variation of parameters,[74] much of the enhanced signal is generated in a primary + secondary reaction cascade (denoted Mechanism 1), similar, but not identical to normal MALDI. Neutral matrix molecules (not large aggregates) are 2-photon ionized, generating radical primary ions. These react in the compressed plume to form proton transfer matrix species, which continue to react with analytes as collisions allow. This mechanism requires the second laser to be of sufficiently short wavelength that the energy of two photons is above the matrix ionization potential. This is typically shorter than the wavelength of the ablation laser. A Mechanism 2 is also proposed to also be active, involving conventional MALDI of the clusters and aggregates entrained in the plume. This mechanism does not require a short wavelength second laser. Finally, some analytes may be directly 2-photon ionized by the second laser, if they have suitable chromophores.

Given that very similar ion–molecule reactions are involved, one could imagine that suppression effects are also an issue in MALDI-2, just as in MALDI. However, even though it is compressed by the background gas, the MALDI-2 plume is still much less dense than the early stages of the MALDI expansion. This has the advantage that secondary reactions have less opportunity to proceed toward LTE. Also, the method is based on greatly increasing the number of primary ions, which also prevents suppression effects in normal MALDI. Consistent with these considerations, one study has found no correlation between analyte proton affinity and MALDI-2 signal, in contrast to normal MALDI.[75] This in turn means reduced ASE, as was found in images of pharmaceuticals in tissues. It seems possible that the worst suppression effects in standard MALDI can be reduced or avoided using MALDI-2, but questions still remain about response factors and relative sensitivities.[76]

1.6 Conclusions

Although experimentally very simple, MALDI is a superposition of multiple physical and chemical processes, and it is found to be rather complex when examined in detail. In spite of this, many important MALDI phenomenon, especially those involving suppression effects and relative sensitivities can be qualitatively and, in some cases quantitatively, understood in terms of reactions of primary matrix ions with analytes in the expanding plume. Many of these effects are particularly pronounced in MALDI imaging because tissue samples are complex and sample preparation options are limited. The framework presented here can guide experimental design, and especially aid in interpretation of results. The "M" in MALDI does not stand for "Magic", even if many questions remain.

References

1. R. Knochenmuss, *Analyst*, 2006, **131**, 966–986.
2. C. N. McEwen and B. S. Larsen, *Int. J. Mass Spectrom.*, 2015, **377**, 515–531.
3. K. Dreisewerd, *Chem. Rev.*, 2003, **103**, 395–426.
4. G. Paltauf and P. E. Dyer, *Chem. Rev.*, 2003, **103**, 487–518.
5. A. Leisner, A. Rohlfing, U. Röhling, K. Dreisewerd and F. Hillenkamp, *J. Phys. Chem. B*, 2005, **109**, 11661–11666.
6. S. Alves, M. Kalberer and R. Zenobi, *Rapid Commun. Mass Spectrom.*, 2003, **17**, 2034–2038.
7. L. V. Zhigilei, P. B. S. Kodali and B. J. Garrison, *J. Phys. Chem. B*, 1998, **102**, 2845–2853.
8. L. V. Zhigilei and B. J. Garrison, *Appl. Phys. Lett.*, 1999, **74**, 1341–1343.
9. R. Knochenmuss and L. V. Zhigilei, *J. Phys. Chem. B*, 2005, **109**, 22947–229957.
10. G. R. Kinsel, M. E. Gimon-Kinsel, K. J. Gillig and D. H. Russell, *J. Mass Spectrom.*, 1999, **34**, 684–690.
11. R. Knochenmuss and L. V. Zhigilei, *J. Mass Spectrom.*, 2010, **45**, 333–346.
12. A. Koubenakis, V. Frankevich, J. Zhang and R. Zenobi, *J. Phys. Chem. A*, 2004, **108**, 2405–2410.
13. K. Dreisewerd, M. Schürenberg, M. Karas and F. Hillenkamp, *Int. J. Mass Spectrom. Ion Processes*, 1995, **141**, 127–148.
14. D. Feldhaus, C. Menzel, S. Berkenkamp and F. Hillenkamp, *J. Mass Spectrom.*, 2000, **35**, 1320–1328.
15. S. Guenther, M. Koestler, O. Schulz and B. Spengler, *Int. J. Mass Spectrom.*, 2010, **294**, 7–15.
16. M. Niehaus and J. Soltwisch, *Sci. Rep.*, 2018, **8**, 1–10.
17. M. Wiegelmann, K. Dreisewerd and J. Soltwisch, *J. Am. Soc. Mass Spectrom.*, 2016, **27**, 1952–1964.
18. R. T. Steven, A. M. Race and J. Bunch, *J. Am. Soc. Mass Spectrom.*, 2016, **27**, 1419–1428.
19. R. Knochenmuss, *J. Mass Spectrom.*, 2002, **37**, 867–877.
20. Y. J. Bae, J. C. Choe, J. H. Moon and M. S. Kim, *J. Am. Soc. Mass Spectrom.*, 2013, **24**, 1807–1815.
21. Y. J. Bae and M. S. Kim, *Annu. Rev. Anal. Chem.*, 2015, **8**, 41–60.
22. I.-C. Lu, C. H. Lee, Y.-T. Lee and C.-K. Ni, *Annu. Rev. Anal. Chem.*, 2015, **8**, 21–39.
23. K. Breuker, R. Knochenmuss, J. Zhang, A. Stortelder and R. Zenobi, *Int. J. Mass Spectrom.*, 2003, **226**, 211–222.
24. G. R. Kinsel, D. Yao, F. H. Yassin and D. S. Marynick, *Eur. J. Mass Spectrom.*, 2006, **12**, 359–367.
25. E. A. Lee, M. E. Kinsel and G. R. Kinsel, *J. Am. Soc. Mass Spectrom.*, 2020, **31**, 2321–2327.
26. T. W. Jaskolla and M. Karas, *J. Am. Soc. Mass Spectrom.*, 2011, **22**, 976–988.

27. M. Karas, M. Glückmann and J. Schäfer, *J. Mass Spectrom.*, 2000, **35**, 1–12.
28. R. Krüger, A. Pfenninger, I. Fournier, M. Glückmann and M. Karas, *Anal. Chem.*, 2001, **73**, 5812–5821.
29. I. Fournier, A. Brunot, J.-C. Tabet and G. Bolbach, *Int. J. Mass Spectrom.*, 2002, **213**, 203–215.
30. I. Fournier, A. Brunot, J.-C. Tabet and G. Bolbach, *J. Mass Spectrom.*, 2005, **40**, 50–59.
31. C. N. McEwen, V. S. Pagnotti, E. D. Inutan and S. Trimpin, *Anal. Chem.*, 2010, **82**, 9164–9168.
32. L. Zilch, J. T. Maze, J. W. Smith, G. E. Ewing and M. F. Jarrold, *J. Phys. Chem. A*, 2008, **112**, 13352–13363.
33. I. Bhattacharyya, J. T. Maze, G. E. Ewing and M. F. Jarrold, *J. Phys. Chem. A*, 2011, **115**, 5723–5728.
34. X. Luo, P.-T. Tue, K. Sugiyama and Y. Takamura, *Sci. Rep.*, 2017, **7**, 15170.
35. X. Chen, J. A. Carroll and R. C. Beavis, *J. Am. Soc. Mass Spectrom.*, 1998, **9**, 885–891.
36. S. Niu, W. Zhang and B. T. Chait, *J. Am. Soc. Mass Spectrom.*, 1998, **9**, 1–7.
37. R. Knochenmuss, *J. Am. Soc. Mass Spectrom.*, 2014, **25**, 1521–1527.
38. A. A. Maryott and E. R. Smith, *Natl. Bur. Stand. Circ. 514*, 1951.
39. M. Uematsu and E. U. Franck, *J. Phys. Chem. Ref. Data*, 1980, **9**, 1291–1306.
40. R. Knochenmuss, *Analyst*, 2014, **139**, 147–156.
41. R. Knochenmuss, *J. Am. Soc. Mass Spectrom*, 2015, **26**, 1645–1648.
42. K. M. Kirmess, R. Knochenmuss, G. J. Blanchard and G. R. Kinsel, *J. Mass Spectrom.*, 2016, **51**, 79–85.
43. S. H. Ahn, K. M. Park, Y. J. Bae and M. S. Kim, *J. Mass Spectrom.*, 2013, **48**, 299–305.
44. R. Knochenmuss, *J. Mass Spectrom.*, 2013, **48**, 998–1004.
45. K. N. Robinson, R. T. Steven and J. Bunch, *J. Am. Soc. Mass Spectrom.*, 2018, **29**, 501–511.
46. D. A. Allwood, R. W. Dreyfus, I. K. Perera and P. E. Dyer, *Rapid Commun. Mass Spectrom.*, 1996, **10**, 1575–1578.
47. D. A. Allwood, R. W. Dreyfus, I. K. Perera and P. E. Dyer, *Appl. Surf. Sci.*, 1997, **110**, 154–157.
48. M. Wiegelmann, J. Soltwisch, T. W. Jaskolla and K. Dreisewerd, *Anal. Bioanal. Chem.*, 2013, **405**, 6925–6932.
49. R. Knochenmuss, *Annu. Rev. Anal. Chem.*, 2016, **9**, 365–385.
50. P. Setz and R. Knochenmuss, *J. Phys. Chem. A*, 2005, **109**, 4030–4037.
51. J. Soltwisch, T. W. Jaskolla and K. Dreisewerd, *J. Am. Soc. Mass Spectrom.*, 2013, **24**, 1477–1488.
52. J. Soltwisch, T. W. Jaskolla, F. Hillenkamp, M. Karas and K. Dreisewerd, *Anal. Chem.*, 2012, **84**, 6567–6576.
53. P. Demirev, A. Westman, C. T. Reimann, P. Håkansson, D. Barofsky, B. U. R. Sundqvist, Y. D. Cheng, W. Seibt and K. Siegbahn, *Rapid Commun. Mass Spectrom.*, 1992, **6**, 187–191.
54. G. McCombie and R. Knochenmuss, *J. Am. Soc. Mass Spectrom.*, 2006, **17**, 737–745.

55. R. Knochenmuss, G. McCombie and M. Faderl, *J. Phys. Chem. A*, 2006, **110**, 12728–12733.
56. T. D. McCarley, R. L. McCarley and P. A. Limbach, *Anal. Chem.*, 1998, **70**, 4376–4379.
57. T.-W. D. Chan, A. W. Colburn and P. J. Derrick, *Org. Mass Spectrom.*, 1991, **26**, 342.
58. R. Knochenmuss, F. Dubois, M. J. Dale and R. Zenobi, *Rapid Commun. Mass Spectrom.*, 1996, **10**, 871–877.
59. R. Knochenmuss, V. Karbach, U. Wiesli, K. Breuker and R. Zenobi, *Rapid Commun. Mass Spectrom.*, 1998, **12**, 529–534.
60. R. Knochenmuss and L. V. Zhigilei, *Anal. Bioanal. Chem.*, 2012, **402**, 2511–2519.
61. R. Knochenmuss, *Int. J. Mass Spectrom.*, 2009, **285**, 105–113.
62. R. Knochenmuss, A. Stortelder, K. Breuker and R. Zenobi, *J. Mass Spectrom.*, 2000, **35**, 1237–1245.
63. R. A. Marcus, *Pure Appl. Chem.*, 1997, **69**, 13–29.
64. A. J. Hoteling, W. F. Nichols, D. J. Giesen, J. R. Lenhard and R. Knochenmuss, *Eur. J. Mass Spectrom.*, 2006, **12**, 345–358.
65. G. McCombie and R. Knochenmuss, *Anal. Chem.*, 2005, **76**, 4990.
66. M. S. Boskamp and J. Soltwisch, *Anal. Chem.*, 2020, **90**, 5222–5230.
67. X. Lou, J. L. J. van Dongen, L.-G. Milroy and E. W. Meijer, *Rapid Commun. Mass Spectrom.*, 2016, **30**, 2628–2634.
68. Q. Wu, T. J. Comi, B. Li, S. S. Rubakhin and J. V. Sweedler, *Anal. Chem.*, 2016, **88**, 5988–5995.
69. L. Jadoul, N. Smargiasso, F. Pamelard, D. Alberts, N. Agnes, E. DePauw and R. Longuespée, *OMICS*, 2016, **20**, 110–121.
70. D. A. Pirman, A. Kiss, R. M. A. Heeren and R. A. Yost, *Anal. Chem.*, 2013, **85**, 1090–1096.
71. H.-Y. Wang, C. B. Liu and H.-W. Wu, *J. Lipid Res.*, 2011, **52**, 840–849.
72. Y. Popkova and J. Schiller, *Rapid Commun. Mass Spectrom.*, 2017, **31**, 411–418.
73. J. Soltwisch, H. Kettling, S. Vens-Cappell, M. Wiegelmann, J. Müthing and K. Dreisewerd, *Science*, 2015, **348**, 211–215.
74. A. Potthoff, K. Dreisewerd and J. Soltwisch, *J. Am. Soc. Mass Spectrom.*, 2020, **31**, 1844–1853.
75. F. P. Y. Barré, M. R. L. Paine, B. Flinders, A. J. Trevitt, P. D. Kelly, R. Ait-Belkacem, J. P. Garcia, L. B. Creemers, J. Stauber, R. J. Vreeken, B. Cillero-Pastor, S. R. Ellis and R. M. A. Heeren, *Anal. Chem.*, 2019, **91**, 10840–10848.
76. F. B. Eiersbrock, J. M. Orthen and J. Soltwisch, *Anal. Bioanal. Chem.*, 2020, **412**, 6875–6886.

CHAPTER 2

MALDI Mass Spectrometry Imaging – Past, Present, and Future Challenges

PIERRE CHAURAND

Dept. of Chemistry, Université de Montréal, Montreal, QC, Canada
Email: pierre.chaurand@umontreal.ca

2.1 Opening Statement

In the past two decades, mass spectrometry imaging (MSI) has taken the MS community (and beyond!) by storm. This is exemplified by the steadily growing number of research groups with MSI activities, the increasing abundance of publications in the field,[1,2] the strong interest of MS manufacturers in designing or adapting MS platforms to MSI, and the current number of events, meetings, and conferences that include or are dedicated to MSI. This extraordinary explosion is in large part due to the seminal body of work of Richard Caprioli and colleagues at Vanderbilt University (Nashville TN, USA) in early 2000, which demonstrated that using matrix-assisted laser desorption/ionization (MALDI)-MS, intact proteins could be detected from thinly cut tissue sections and precisely mapped to the underlying histology. From that point on, (MALDI)-MSI transitioned from a MS curiosity to a full-blown technology capable of addressing key biological questions in a wide range of domains ranging from developmental biology, pharmaceutical science, plant science, forensics, and, possibly the most exciting, the understanding of disease onset and progression. Whereas MALDI-MSI was initially focused on the imaging of intact proteins, today, a wide range of very

New Developments in Mass Spectrometry No. 12
MALDI Mass Spectrometry Imaging: From Fundamentals to Spatial Omics
Edited by Tiffany Siegel

Published by the Royal Society of Chemistry, www.rsc.org

diverse biomolecules (*i.e.* metabolites, lipids, peptides, proteins) can be studied. This is due to the abundance of novel sample preparation protocols and methods that have been developed and optimized by numerous groups around the world. Even if MALDI-MSI is still by far the most used technology to investigate the molecular content of tissue sections, numerous novel atmospheric pressure desorption ionization technologies have emerged and are used today for a wide range of applications. The early days of MSI were exclusively dominated by the natural coupling of MALDI to time-of-flight (TOF) mass analyzers. Progressively, other MS platforms including QTOFs, ion traps, FT/ICRs, orbitraps, and ion mobility have been coupled to MALDI ion sources. These offer collision cross section/ion mobility separation, exact mass measurements, and MS/MS abilities surpassing those of TOFs and have also been extensively used for MSI of lower molecular weight (MW) compounds.

The beginning of my professional career (1994) as a research associate at the Institute for Laser Medicine at the University of Dusseldorf (Germany) under the supervision of the late Pr. Raimund Kaufmann and Pr. Bernhard Spengler (currently Professor at the University of Giessen, Germany), followed by my years as Research Faculty in the Caprioli laboratory at Vanderbilt University (1998–2009), to my current position as Professor of Chemistry at the Université de Montréal, have allowed me to witness the precursors and contribute to the very beginning and the continuous evolution of MALDI-MSI. In this chapter, I will give my perspective of events that led to the advent of modern MALDI-MSI, list the key technological advances from the past two decades, and present my vision of the future outlining areas where MALDI-MSI technology needs to progress.

2.2 The Beginning

As for most technological breakthroughs, MSI sits on the shoulders of giants. The concept of imaging surfaces using MS was first introduced in 1962 in a seminal article by R. Castaing and G. Slodzian published (in French) in the Journal de Microscopie[3] where they cite: "*The principle consists in using this 'secondary ion emission' to form an image of the surface, then, by way of mass spectrometry, to isolate, amongst all the emitted ions that contribute to the image, those that are characteristic of a particular element and therefore, carry with them the distribution map of this element. The initial total ion image is thus successively processed to yield as many 'elemental images' as there are elements and isotopes of the same elements in the bombarded surface.*" This extraordinary vision led to the advent of ion microscopes and subsequently to imaging SIMS.[4] In 1982, Chandra *et al.* reported, for the first time, imaging of ions from a tissue section.[5] They used an ion microscope to demonstrate that calcium ions are present in the amyloplasts of root-cap cells of corn, pea, and lettuce. However, it was only in the late 90s that SIMS imaging was applied to acquire molecular images of biological materials.[6,7] In the early 2000, the possibilities of imaging SIMS were further improved with the use of polyatomic ions as primary projectiles allowing the analyzable MW range to be expanded up to ~1 kDa with spatial resolutions in the 1 μm range.[8]

In parallel to SIMS, the first laser microprobes coupled to TOF mass analyzers were developed in the mid-70s by Franz Hillenkamp and co-workers and used for the analysis of biomaterials including tissue sections.[9,10] With the development of laser microprobes, the extraordinary foresight of Hillenkamp and co-workers has led to the advent of modern MALDI-MS.[11,12] In 1994, Spengler and co-workers presented the first examples of laser desorption ionization (LDI)-MSI of biological tissue sections using a homebuilt LDI-TOF instrument equipped with a co-axial laser irradiation ion source at the 1994 ASMS conference.[13] In particular, stunning LDI-MSI ion images of K^+ ions acquired from erythrocytes with a spatial resolution of 0.50 μm were shown. This poster also presented impressive peptide ion images acquired from dry-droplet MALDI prepared samples using a 2,5-DHB matrix. Substance P ion images from the rim crystals were acquired with a spatial resolution of 1 μm. These results and others demonstrating the potential of imaging tissue sections with a laser microprobe MS instrument were ultimately published in 2002.[14] I started my postdoctoral fellowship with Bernhard Spengler in the fall of 1994 and was privileged to see first-hand the potential of this instrument. The development of software to pilot automated data acquisition and display the selected ion images was also of interest. In 1995, Gusev *et al.* published an interesting paper where they imaged by MALDI-MS peptides and dyes previously separated by thin layer chromatography (TLC).[15] A α-CHCA/fructose matrix mixture was in this case initially prepared on a second slide, which when in contact with the wet TLC slide allowed for homogeneous matrix transfer. MALDI-MSI data acquisition was then manually performed at a spatial resolution of 0.5 mm on a laser microprobe TOF-MS instrument. 3D ion images from the TLC plate were reconstructed and the optimized protocol allowed for minimal analyte planar diffusion (−0.5 mm) and resulted in an absolute detection limit in the picogram range.

In 1997, Caprioli and co-workers (while still at the University of Texas Medical School, Houston TX, USA) published their ground-breaking article describing the potential of MALDI-MSI for proteomic analysis of thinly cut tissue sections.[16] This is today one of the most cited papers in the field of MSI. The key experiment was the analysis of human buccal mucosa epithelial cells (cheek cells) directly deposited on a target plate. A matrix (α-cyano-hydroxycinnamic acid, α-CHCA) was applied by electrospray deposition before MALDI-MSI analysis, performed at a spatial resolution of 25 μm. Numerous protein signals with MWs ranging from 3000 to 12 000 Da were successfully detected and mapped to the buccal aggregated cells. In a second similar experiment, rat pancreatic sections were analyzed after α-CHCA electrospray deposition. The signal for insulin at *m/z* 5802 was found to be clearly associated with the pancreatic islet. In 1998, Richard Caprioli moved to Vanderbilt University, and that same year I joined his lab as Research Faculty to continue the development of MALDI-MSI technology.

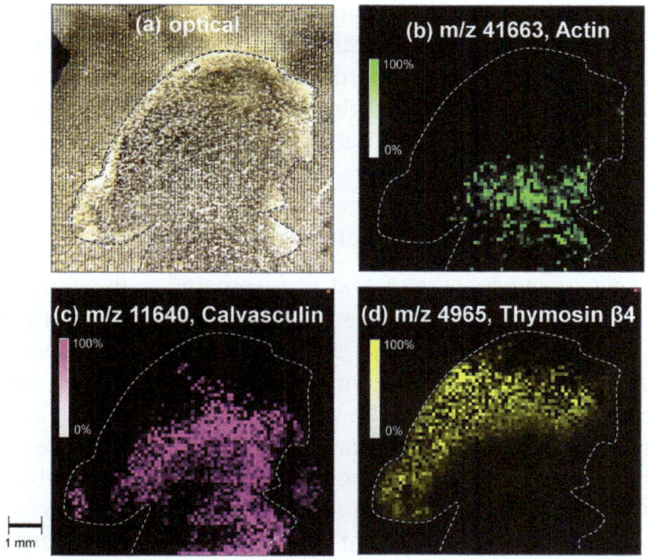

Figure 2.1 Selected protein images from a human glioblastoma section. (a) Section mounted on a metal plate, coated with matrix (the lines are from ablation of the matrix with the laser). (b–d) Mass spectrometric images of proteins showing high concentration in the proliferating area of the tumor (d) and other proteins present specifically in the ischemic and necrotic areas (b and c).
Reproduced from ref. 17 with permission from Springer Nature, Copyright 2001.

One of the first key papers to be published (in Nature Medicine) from the Caprioli lab is entitled "*Imaging mass spectrometry: A new technology for the analysis of protein expression in mammalian tissues*".[17] Here, the authors displayed several protein ion images acquired by MALDI-MSI from a full mouse coronal brain section and of a human glioma tumor sample (Figure 2.1), demonstrating the uniqueness of signal expression with exact correlation to the underlying histology. This work is in my view the first example that direct correlations between molecular expression and tissue histology can be made. It also introduced the potential of the technology for the study of disease including cancer, and foresees its usefulness in the clinical setting for diagnostic and, in the context of cancer, for tumor margin assessment.

2.3 The Last Two Decades

The past two decades have seen a phenomenal explosion of knowledge in MSI. Protocols and methods for sample preparation for the analysis of a wide range of biomolecules have been proposed. Numerous software for automated data acquisition, visualization, processing, and statistical analyses have been developed. MS instrumentation has been rethought for or adapted to imaging applications. Today, automated sample preparation systems are commercially available, and most vendors of MALDI-MS instrumentation propose solutions for MSI. In this section, are listed what I consider to be some of the major technological advances and breakthroughs that have propelled MALDI-MSI to a state-of-the-art molecular technology.

2.3.1 MSI Data Acquisition and Instrumentation

One of the keys to the success of MSI is automated data acquisition. While the early researchers in the field were patient enough to manually acquire their data, automated data acquisition had to be addressed and implemented. An early example of a homemade MSI acquisition software was proposed by Stoeckli *et al.*[18] It allowed instrument control (a MALDI-TOF mass spectrometer), data acquisition, and data analysis. Later acquisition and analysis software solutions were in large part developed by MALDI-MS instrument vendors. Besides instrument control software, precise control of the sample position and movement with micrometer accuracy is also indispensable. Sample movement control using high accuracy stepper motors has also been implemented in modern MALDI-MS instruments.

MSI data acquisition speed has also leap-frogged in the past two decades. Whereas the first MALDI-MSI acquisitions were performed with compact nitrogen N_2 lasers operating at repetition rates of a few Hz,[17] modern MALDI-MSI systems operate with solid state lasers capable of data MSI acquisition at repetition rates of up to 20 kHz.[19,20] MSI data acquisition times were literally cut from a few days to a few minutes! The use of solid-state lasers also allows to better define the laser focus on-target, which today can be in the order of a few micrometers.[19,21] This evolution in performance now allows the acquisition of megapixel ion images (Figure 2.2)[22] and opens the door to high throughput 3D and whole-body MSI.[23]

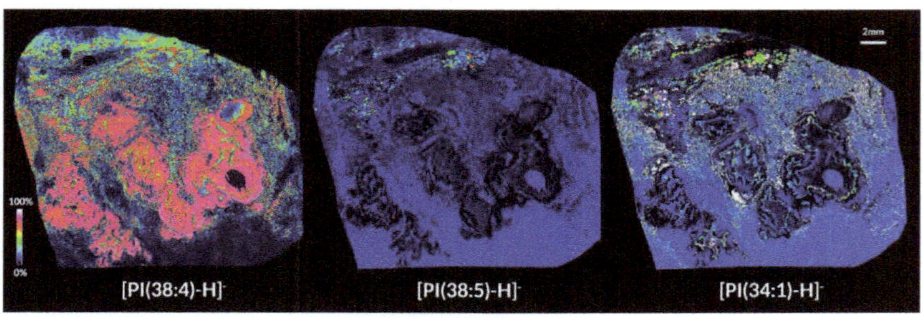

Figure 2.2 Ultra-high speed MALDI-TOF MSI data (20 μm) of human lung tissue from a cystic fibrosis patient consisting of over 1 million total pixels collected in ~6 h. Highlighted are three ion images of lipids observed throughout the course of data acquisition. Image credit: Jeffrey Spraggins, Vanderbilt University (Nashville TN, USA).
Adapted from ref. 22 with permission from John Wiley & Sons, Copyright © 2016 WILEY-VCH Verlag GmbH & Co. KGaA, Weinheim.

Since MALDI-MS is a pulsed technique, it was historically coupled to time-of-flight (TOF) mass analyzers. The first MALDI-MSI experiments were therefore performed on TOF-MS instruments. With growing interest in MSI, QTOFs, ion traps, QIT-TOFs, FT/ICRs, Orbitraps, and ion mobility TOF MS systems have progressively been equipped with MALDI ion sources opening new analytical possibilities.[24–31] Whereas QTOF and ion trap systems allow for MS/MS imaging of selected compounds,[24,32] FT/ICR and Orbitrap instruments allow exact mass MSI.[33,34]

2.3.2 Automated Matrix Deposition

One of the prerequisites of MALDI-MSI is homogeneous matrix application on the tissue sections while minimizing analyte delocalization. In the context of MSI, homogeneous matrix deposition is extremely important allowing consistent pixel-to-pixel molecular desorption/ionization throughout the sample surface. Early attempts were made by dragging a ~400 μl volume of matrix over the tissue section in a cold room.[17] While some success was obtained, reproducibility was very limited. Electrospray matrix deposition was also used[16] but ultimately the resulting dry matrix deposition proved detrimental for protein MSI. Handheld pneumatic artist paintbrushes were then adapted for on-tissue matrix application offering homogeneous matrix deposition control including sample wettability and matrix thickness. By the mid-2000s, the first automated pneumatic matrix deposition systems were developed and commercialized.[35,36] These allowed precise control of the multiple parameters allowing homogeneous matrix deposition. In parallel to matrix spray deposition systems, automated chemical printing systems were also developed and commercialized for matrix deposition.[37,38] Matrix coating is achieved by droplet printing in a Cartesian array over the surface of the tissue section. The downside of this approach is that the attainable imaging spatial resolution is limited to the center-to-center distance between two adjacent matrix spots, at best in the range of 200 μm. Beyond matrix spray and droplet deposition, matrix sublimation systems were also developed[39] and recently commercialized.[40] These allow for very homogeneous solvent-free matrix deposition and have for the most part been used for lipidomics studies[41,42] in some cases at cellular scale spatial resolutions.[43,44] Recent method developments now allow for proteomics studies.[45,46]

The main advantage of automated sample preparation for MALDI-MSI is a higher degree of reproducibility. Consistency and reproducibility in matrix deposition are both very critical parameters to consider when performing 3D-MSI or longitudinal (clinical) studies (*i.e.* implying a large cohort), where all samples cannot be simultaneously analyzed.

2.3.3 On-tissue Chemistries

In 2007, Groseclose *et al.* proposed a novel method for on-tissue identification of proteins in spatially discrete regions using tryptic digestion followed by MALDI-MSI and MS/MS analysis.[38] In this workflow, trypsin was initially homogeneously deposited over the tissue section using an automated chemical printer. The resulting tryptic peptides were then analyzed by MSI after matrix deposition. This approach was subsequently adapted for the proteomics analysis of formalin-fixed paraffin embedded (FFPE) tissue sections, opening the potential of retroactively analyzing millions of fully annotated clinical samples stored around the world for the past decades in hospital biobanks.[47] This was a major milestone to allow for MSI to step closer to the histopathology and clinical communities because FFPE is the classical process by which most of the clinical biopsies are preserved. More recently, other enzymes have also been explored such as *N*-glycosidase F (PNGaseF) to liberate and analyze N-linked glycan species from proteins,[48] and metalloproteinases to digest and analyze extracellular matrix proteins (*i.e.* collagens and elastin).[49] These high-throughput technological advances are applicable to any tissue regardless of type, origin, or disease status and are thus relevant to all research: basic, translational, or clinical.

Beyond proteolytic enzymes, the automated matrix deposition systems can also be used for homogeneous deposition of molecular reactive tags for *in-situ* analyte derivatization prior to MSI. *In situ* chemistries are typically performed in a targeted manner toward a class of compounds to enhance detection sensitivity and specificity by MALDI-MS. The potential of reactive mass-tags was beautifully illustrated in an impressive article by Lemaire *et al.* where they developed specific targeting probes for either protein (antibody probe) or mRNA (oligonucleotide probe) MALDI-MSI.[50] This approach is based on probes modified with a photocleavable linker coupled to a tag cleaved and detected using mass spectrometry. The liberated mass-tag is then used as a reporter ion for the *in-situ* presence of the targeted analyte. This strategy was recently used by Yagnik *et al.* where a multiplex assay of twelve antibody probes was developed.[51]

Several reactive tags containing a fixed charge have been developed to enhance the detection sensitivity of targeted analytes. For example, Cobice *et al.* have derivatized ketone steroids *in situ* with the Girard-T reagent prior to MALDI-MSI analyses.[52] In this case, derivatization permitted detection and imaging of poorly ionizable endogenous steroid hormones. More recently, Shariatgorji and co-workers developed fluoromethylpyridinium-based reactive matrices that facilitate the covalent charge-tagging of molecules containing phenolic hydroxyl and/or primary or secondary amine groups.[53] These reactive matrices react rapidly at ambient temperature and preserve the localization of endogenous compounds for subsequent MALDI-MSI analysis. Moreover, the derivatization agent crystallizes on the surface of the tissue in an MS-compatible solvent and no additional matrix is needed. The authors used these reactive tags to perform a comprehensive mapping of neurotransmitter networks in specific brain regions (Figure 2.3). This is in my view one of the most elegant studies ever performed with MALDI-MSI.

The reactive chemistries one can design to specifically target analytes are endless. I foresee here continuous close collaborative efforts between analytical

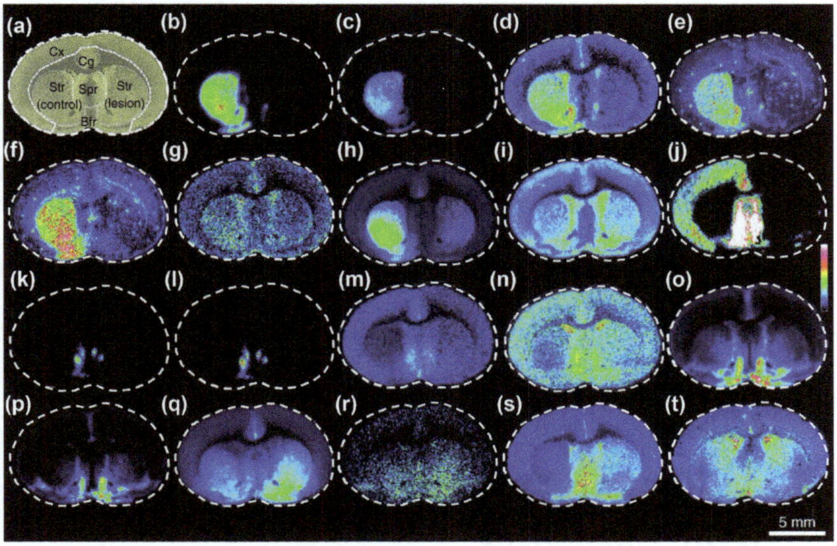

Figure 2.3 MALDI FT/ICR MS images of neurotransmitters and metabolites acquired from a tissue section of a unilateral 6-hydroxydopamine (6-OHDA) lesioned Parkinson's disease model rat treated with 3,4-dihydroxyphenylalanine (l-DOPA). (a) Optical image of a coronal tissue section from a rat brain. Imaging experiments were conducted on a tissue section from a unilateral 6-OHDA-lesioned animal that had been treated with subchronic daily l-DOPA for 3 weeks and received a final dose of (deuterated) l-DOPA-d3. (b–t) Images showing the distributions of endogenous dopamine (b), 3-methoxytyramine (c), 3,4-dihydroxyphenylacetaldehyde (DOPAL) (d), DOPAL (e; double derivatized), 3,4-dihydroxyphenylacetic acid (f; double derivatized), 3-methoxy-4-hydroxyphenylacetaldehyde (g), homovanillic acid (h), l-DOPA metabolite 3-*O*-methyldopa (i), norepinephrine (j), epinephrine (k), epinephrine (l; double derivatized), dihydroxyphenylethylene glycol (m), 3-methoxy-4-hydroxyphenylethyleneglycol (n), serotonin (o), 5-hydroxyindolealdehyde (p), 5-hydroxyindoleacetic acid (q), 5-hydroxytryptophol (r), γ-aminobutyric acid (s) and lysine (t) facilitated by reactive matrix. Bfr, basal forebrain region; Cx, cortex; Cg, cingulate cortex; Str, striatum; Spr, septal region. MALDI-MSI experiments were performed four times independently from different animals with similar results. Data are shown using a rainbow scale (representing ion intensity scale) for best visualization. Lateral resolution, 80 μm.
Reproduced from ref. 53 with permission from Springer Nature, Copyright 2019.

and organic chemists for the design of clever reactive tags offering high specificity and sensitivity imaging of targeted compounds from tissue sections.[54]

2.3.4 Imaging of Xenobiotics

In 2003, Reyzer *et al.* published a landmark paper which demonstrated that administered pharmaceutical compounds could be detected, imaged and quantified from thin tissue sections by MALD-MS.[55] They showed that an anti-tumor drug could be unambiguously determined in mouse tumor tissue using a MALDI–QqTOF-MS instrument by monitoring the dissociation of the protonated drug to its predominant fragment ion. They also showed that the drug MALDI-MS response was proportional to the amount of drug in a tissue with good correlation with that obtained by LC–MS/MS quantitation. These extraordinary findings opened the door for (MALDI-)MSI to become today an indispensable tool for many researchers and for the pharmaceutical industry. The knowledge of tissue drug distribution and local abundance may have a direct impact on a number of areas in drug development, including pharmacology, pharmacokinetics, safety, drug–drug interactions, transport, and metabolism.[56–62] Its adoption by the pharmaceutical industry is in my view one of the major successes of MSI technology. Beyond pharmaceutical compounds, MALDI-MSI has also been employed in forensic applications to monitor the presence of substances of abuse in hair samples[63,64] and in fingermarks.[65,66]

2.4 Future Challenges

2.4.1 The Push for Higher Spatial Resolution

While cellular spatial resolution (*i.e.* 5–15 μm) is today attainable with most commercially available MALDI-MS systems,[22,42] the push for subcellular MALDI-MSI is inexorable.[44,67] From experience, the demand (and the need) from multiple collaborators to push MSI technology in the sub-micrometer realm is there to study single cells in their natural tissue habitat.[68,69] Prototype ion sources with front[21] or transmission[70,71] laser illumination geometries have been designed and have demonstrated (sub)micrometer MSI capabilities. Beyond laser ion optics, other challenges exist:

1) One of the consequences of irradiating sub-micrometer sample surfaces is the production of lower numbers of ions requiring more sensitive MS systems. This consequence may be in large part alleviated with the recent development of the laser post-ionization MALDI-2 technology,[72,73] especially when coupled to state-of-the-art ion mobility MSI instrumentation.[30]
2) Another aspect is the ability to generate ‘perfect’ tissue sections, especially from fresh-frozen tissue specimens, without (micro)tearing or cracking. Fresh-frozen sections are typically cut with thicknesses in the 10 μm-range. Subcellular MSI may require thinner sections. For subcellular imaging, cryofracture methodologies abundantly used by the imaging SIMS community[74] may need to be adapted to MALDI-MSI.
3) The currently available automated matrix dispensing systems can generate (sub)micrometer scale MALDI crystals. To image cellular features such as the nucleus or mitochondria, crystal dimensions will need to be significantly smaller and perfectly organized on samples. Some progress in matrix deposition is necessary to better control on-tissue crystal formation and dimension.

Ultimately, to successfully image tissues at sub-micrometer dimensions by MALDI-MS, the spatial resolution needs to be at minimum 1/10th of the cell dimension – in the order of 500 nm (or less). Laser spot diameter on a sample will therefore also need to be perfectly controlled correcting for small sample height variations. This can already be achieved when accurately controlling the laser focal distance.[75] Considering our current collective knowledge, I believe this is attainable within the next few years.

2.4.2 Standard Protocols and Methods

MSI has the inherent capability of tissue classification based on molecular phenotypes. This has been widely exploited within the MSI community for the detection and grading of disease. For MSI technology to be used in the clinical setting, it will not be sufficient to show that biological differences can be correctly classified. Rather it is necessary that by using strict quality control procedures established and accepted by the MSI community, identical classifications can be consistently made across sites, operators, and instruments. This will require that sample processing protocols, data acquisition methods (and even possibly the MSI data acquisition platform), as well as data analysis be standardized. This is especially vital for the analysis of FFPE tissue blocks, which is a standard for histopathology worldwide. Consistency and reproducibility will need to be assessed *via* multicenter validation studies. Such initiatives have already taken place within the (MALDI)-MSI community.[76,77]

2.4.3 Increasing Sensitivity and Specificity

MALDI-2 technology has already had a significant impart toward increasing analytical sensitivity up to one hundred-fold for a wide range of compounds including administered pharmaceuticals,[78] oligosaccharides,[79] and steroids,[80] the latter being globally unobserved by conventional MALDI-MS.

Classical (MALDI) MSI tissue sample preparation approaches can only broadly analyze specific classes of biomolecules. Based on the matrix employed and its mode of deposition, different classes of biomolecules (*i.e.* proteins, peptides, glycans, lipids or metabolites) can be targeted for analysis. Furthermore, most MALDI-MSI methods are developed to maximize the amount of information recovered therefore limiting sensitivity. Strategies to improve analytical specificity and sensitivity in MSI and open the technique for the analysis of novel classes of endogenous molecules need to be developed. To incorporate some degree of molecular specificity, several approaches can be proposed:

1) Alternative desorption/ionization agents, namely nanoparticles and metals, have been developed. These surface-assisted laser desorption ionization (SALDI)-based approaches can be employed for MSI of tissue sections and often allow the specific detection of subclasses of lipids

and metabolites.[2] For example, my lab has recently developed two metal-based sample preparation approaches which allow the detection of cholesterol[81] and triglycerides[82] with very high specificity and sensitivity. These methods have been used to study Alzheimer's disease[83] and non-alcohol fatty liver disease (Figure 2.4).[84] More recently, Chen *et al.* used TiO_2 and ZnO nanoparticles to image low MW molecules including neurotransmitters from mouse brain tissue sections.[85]

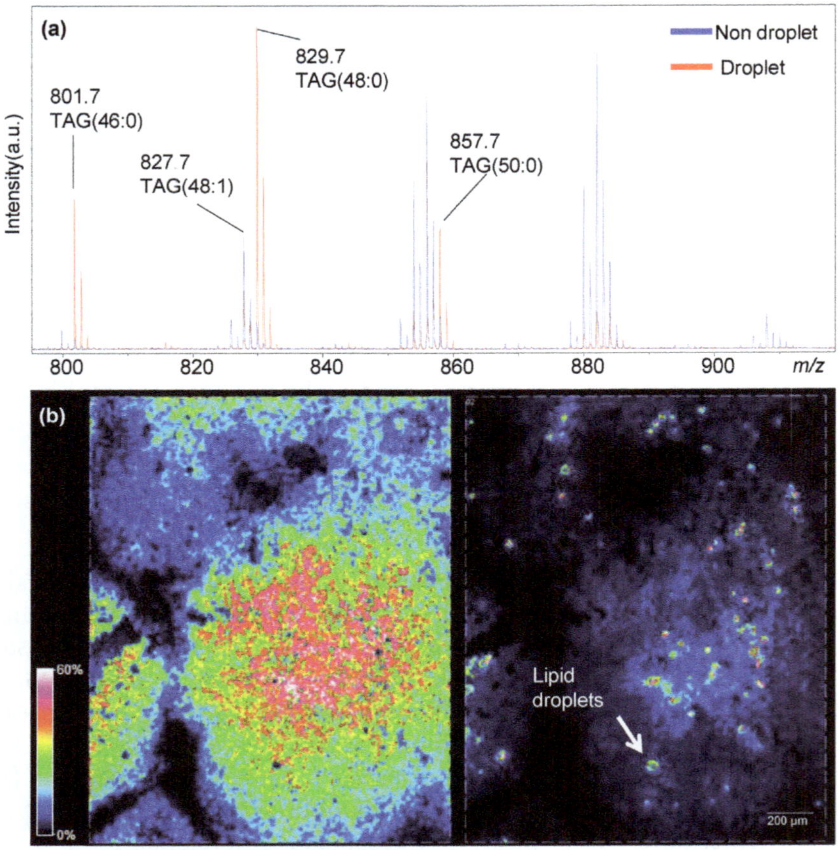

Figure 2.4 (a) Typical gold-assisted LDI MS profiles acquired from a non-alcoholic steatohepatitis (NASH) liver section from droplet and non-droplet regions showing a clear differentiation across histologies. Lipid droplets tend to contain a higher abundance of short chain saturated TGs (≤50 carbon). (b) Gold-assisted LDI MSI acquired from a NASH liver section from droplet and non-droplet regions at 10 µm of spatial resolution showing clear distribution differences for TG(48 : 1) and TG(48 : 0). TG(48 : 0) was found clearly observed sequestered within the lipid droplets.
Adapted from ref. 84 with permission from Springer Nature, Copyright 2019.

2) Tailored surfaces for the specific capture of selected analytes can be developed and used for MSI. Early on, Caprioli and co-workers demonstrated that the rat pituitary could be in contact with a C-18 coated bead layer for the specific transfer of peptides for subsequent MALDI-MSI analysis.[16] We have also demonstrated that proteins could be transferred from thin tissue sections to nitrocellulose membranes while maintaining their spatial organization.[86] More recently, Wu *et al.* demonstrated that imprint MSI can be used to analyze lipids in rat brain tissue samples.[87] Here, tissues were imprinted on slides coated with a dopamine-modified TiO_2 monolith resulting in the adsorption of endogenous analytes onto the coating and found that many lipids, including cholesterol, ceramides, phosphatidylethanolamines, and phosphatidylcholines, can be detected and imaged using this approach. The surface chemistries one can design for the specific capture of biomolecules are endless. It can be envisioned that tailored chemistries with high affinity for targeted subclasses of analytes will continue to emerge conferring high specificity and high sensitivity to (MALDI)-MSI.

2.5 Concluding Remarks

After two decades of collective efforts by a growing number of laboratories world-wide, MSI has become a mature analytical method to investigate the organization of a wide range of biomolecules in various tissue specimens. As evidenced, the body of literature around MSI is growing at phenomenal rates. Based on the enormous potential of MSI technology, the continually active methodology and technology developments which are ongoing world-wide and the growing interest that we have witnessed in the past two decades, I foresee that in the decades to come MSI will have a bright future with continuous expansion in diverse fields of application. It is today recognized as a major and essential analytical tool in our ongoing quest to understand life and disease, but its analytical possibilities are still in their infancy.

References

1. P. Chaurand, Imaging mass spectrometry of thin tissue sections: a decade of collective efforts, *J. Proteomics*, 2012, **75**(16), 4883–4892.
2. W. H. Müller, A. Verdin, E. De Pauw, C. Malherbe and G. Eppe, Surface-assisted laser desorption/ionization mass spectrometry imaging: A review, *Mass Spectrom. Rev.*, 2020, DOI: 10.1002/mas.21670. Epub ahead of print. PMID: 33174287.
3. R. Castaing and G. Slodzian, Microanalyse par emission ionique secondaire, *J. Microsc.*, 1962, **1**, 395–410.
4. A. Benninghoven, Surface investigation of solids by the statical method of secondary ion mass spectroscopy (SIMS), *Surf. Sci.*, 1973, **35**, 427–457.

5. S. Chandra, J. F. Chabot, G. H. Morrison and A. C. Leopold, Localization of calcium in amyloplasts of root-cap cells using ion microscopy, *Science*, 1982, **216**(4551), 1221–1223.
6. P. J. Todd, J. M. McMahon, R. T. Short and C. A. McCandlish, Organic SIMS of Biologic Tissue, *Anal. Chem.*, 1997, **69**(17), 529A–535A.
7. P. J. Todd, T. G. Schaaff, P. Chaurand and R. M. Caprioli, Organic ion imaging of biological tissue with secondary ion mass spectrometry and matrix-assisted laser desorption/ionization, *J. Mass Spectrom.*, 2001, **36**(4), 355–369.
8. D. Touboul, F. Kollmer, E. Niehuis, A. Brunelle and O. Laprévote, Improvement of biological time-of-flight-secondary ion mass spectrometry imaging with a bismuth cluster ion source, *J. Am. Soc. Mass Spectrom.*, 2005, **16**(10), 1608–1618.
9. F. Hillenkamp, E. Unsold, R. Kaufmann and R. Nitsche, Laser microprobe mass analysis of organic materials, *Nature*, 1975, **256**, 119–120.
10. F. Hillenkamp, E. Unsöld, R. Kaufmann and R. Nitsche, A high-sensitivity laser microprobe mass analyzer, *Appl. Phys.*, 1975, **8**(4), 341–348.
11. M. Karas, D. Bachmann and F. Hillenkamp, Influence of the wavelength in high-irradiance ultraviolet laser desorption mass spectrometry of organic molecules, *Anal. Chem.*, 1985, **57**(14), 2935–2939.
12. F. Hillenkamp, M. Karas, R. C. Beavis and B. T. Chait, Matrix-Assisted Laser Desorption/Ionization Mass Spectrometry of Biopolymers, *Anal. Chem.*, 1991, **63**(24), 1193A–1203A.
13. B. Spengler, M. Hubert and R. Kaufmann, MALDI ion imaging and biological ion scanning with a new scanning UV-laser microprobe, in *42nd ASMS Conference on Mass Spectrometry and Allied Topics*, Chicago, 1994, May 29–June 3.
14. B. Spengler and M. Hubert, Scanning microprobe matrix-assisted laser desorption ionization (SMALDI) mass spectrometry: Instrumentation for sub-micrometer resolved LDI and MALDI surface analysis, *J. Am. Soc. Mass Spectrom.*, 2002, **13**(6), 735–748.
15. A. I. Gusev, O. J. Vasseur, A. Proctor, A. G. Sharkey and D. M. Hercules, Imaging of thin-layer chromatograms using matrix-assisted laser desorption/ionization mass spectrometry, *Anal. Chem.*, 1995, **67**(24), 4565–4570.
16. R. M. Caprioli, T. B. Farmer and J. Gile, Molecular imaging of biological samples: localization of peptides and proteins using MALDI-TOF MS, *Anal. Chem.*, 1997, **69**(23), 4751–4760.
17. M. Stoeckli, P. Chaurand, D. E. Hallahan and R. M. Caprioli, Imaging mass spectrometry: a new technology for the analysis of protein expression in mammalian tissues, *Nat. Med.*, 2001, **7**(4), 493–496.
18. M. Stoeckli, T. B. Farmer and R. M. Caprioli, Automated mass spectrometry imaging with a matrix-assisted laser desorption ionization time-of-flight instrument, *J. Am. Soc. Mass Spectrom.*, 1999, **10**(1), 67–71.
19. N. Ogrinc Potočnik, T. Porta, M. Becker, R. M. A. Heeren and S. R. Ellis, Use of advantageous, volatile matrices enabled by next-generation high-speed matrix-assisted laser desorption/ionization time-of-flight imaging

employing a scanning laser beam, *Rapid Commun. Mass Spectrom.*, 2015, **29**(23), 2195–2203.

20. J. M. Spraggins and R. M. Caprioli, High-speed MALDI-TOF imaging mass spectrometry: rapid ion image acquisition and considerations for next generation instrumentation, *J. Am. Soc. Mass Spectrom.*, 2011, **22**(6), 1022–1031.
21. M. Kompauer, S. Heiles and B. Spengler, Atmospheric pressure MALDI mass spectrometry imaging of tissues and cells at 1.4-μm lateral resolution, *Nat. Methods*, 2017, **14**(1), 90–96.
22. J. M. Spraggins, D. G. Rizzo, J. L. Moore, M. J. Noto, E. P. Skaar and R. M. Caprioli, Next-generation technologies for spatial proteomics: Integrating ultra-high speed MALDI-TOF and high mass resolution MALDI FTICR imaging mass spectrometry for protein analysis, *Proteomics*, 2016, **16**(11–12), 1678–1689.
23. E. H. Seeley and R. M. Caprioli, 3D Imaging by Mass Spectrometry: A New Frontier, *Anal. Chem.*, 2012, **84**(5), 2105–2110.
24. M. L. Manier, M. L. Reyzer, A. Goh, V. Dartois, L. E. Via, C. E. Barry and R. M. Caprioli, Reagent Precoated Targets for Rapid In-Tissue Derivatization of the Anti-Tuberculosis Drug Isoniazid Followed by MALDI Imaging Mass Spectrometry, *J. Am. Soc. Mass Spectrom.*, 2011, **22**(8), 1409–1419.
25. D. M. Drexler, T. J. Garrett, J. L. Cantone, R. W. Diters, J. G. Mitroka, M. C. Prieto Conaway, S. P. Adams, R. A. Yost and M. Sanders, Utility of imaging mass spectrometry (IMS) by matrix-assisted laser desorption ionization (MALDI) on an ion trap mass spectrometer in the analysis of drugs and metabolites in biological tissues, *J. Pharmacol. Toxicol. Methods*, 2007, **55**(3), 279–288.
26. S. Shimma, Y. Sugiura, T. Hayasaka, N. Zaima, M. Matsumoto and M. Setou, Mass imaging and identification of biomolecules with MALDI-QIT-TOF-based system, *Anal. Chem.*, 2008, **80**(3), 878–885.
27. D. S. Cornett, S. L. Frappier and R. M. Caprioli, MALDI-FTICR imaging mass spectrometry of drugs and metabolites in tissue, *Anal. Chem.*, 2008, **80**(14), 5648–5653.
28. S. M. Khalil, A. Römpp, J. Pretzel, K. Becker and B. Spengler, Phospholipid Topography of Whole-Body Sections of the Anopheles stephensi Mosquito, Characterized by High-Resolution Atmospheric-Pressure Scanning Microprobe Matrix-Assisted Laser Desorption/Ionization Mass Spectrometry Imaging, *Anal. Chem.*, 2015, **87**(22), 11309–11316.
29. M. C. Djidja, E. Claude, M. F. Snel, P. Scriven, S. Francese, V. Carolan and M. R. Clench, MALDI-ion mobility separation-mass spectrometry imaging of glucose-regulated protein 78 kDa (Grp78) in human formalin-fixed, paraffin-embedded pancreatic adenocarcinoma tissue sections, *J. Proteome Res.*, 2009, **8**(10), 4876–4884.
30. K. V. Djambazova, D. R. Klein, L. G. Migas, E. K. Neumann, E. S. Rivera, R. Van de Plas, R. M. Caprioli and J. M. Spraggins, Resolving the Complexity of Spatial Lipidomics Using MALDI TIMS Imaging Mass Spectrometry, *Anal. Chem.*, 2020, **92**(19), 13290–13297.

31. E. S. Rivera, K. V. Djambazova, E. K. Neumann, R. M. Caprioli and J. M. Spraggins, Integrating ion mobility and imaging mass spectrometry for comprehensive analysis of biological tissues: A brief review and perspective, *J. Mass Spectrom.*, 2020, **55**(12), e4614.
32. S. Khatib-Shahidi, M. Andersson, J. L. Herman, T. A. Gillespie and R. M. Caprioli, Direct molecular analysis of whole-body animal tissue sections by imaging MALDI mass spectrometry, *Anal. Chem.*, 2006, **78**(18), 6448–6456.
33. A. Römpp and B. Spengler, Mass spectrometry imaging with high resolution in mass and space, *Histochem. Cell Biol.*, 2013, **139**(6), 759–783.
34. A. Buck, A. Ly, B. Balluff, N. Sun, K. Gorzolka, A. Feuchtinger, K.-P. Janssen, P. J. Kuppen, C. J. van de Velde, G. Weirich, F. Erlmeier, R. Langer, M. Aubele, H. Zitzelsberger, M. Aichler and A. Walch, High-resolution MALDI-FT-ICR MS imaging for the analysis of metabolites from formalin-fixed, paraffin-embedded clinical tissue samples, *J. Pathol.*, 2015, **237**(1), 123–132.
35. W. P. Mounfield and T. J. Garrett, Automated MALDI matrix coating system for multiple tissue samples for imaging mass spectrometry, *J. Am. Soc. Mass Spectrom.*, 2012, **23**(3), 563–569.
36. M. Lagarrigue, M. Becker, R. Lavigne, S. O. Deininger, A. Walch, F. Aubry, D. Suckau and C. Pineau, Revisiting rat spermatogenesis with MALDI imaging at 20-microm resolution, *Mol. Cell. Proteomics*, 2011, **10**(3), M110.005991.
37. H.-R. Aerni, D. S. Cornett and R. M. Caprioli, Automated Acoustic Matrix Deposition for MALDI Sample Preparation, *Anal. Chem.*, 2006, **78**(3), 827–834.
38. M. R. Groseclose, M. Andersson, W. M. Hardesty and R. M. Caprioli, Identification of proteins directly from tissue: in situ tryptic digestions coupled with imaging mass spectrometry, *J. Mass Spectrom.*, 2007, **42**(2), 254–262.
39. J. A. Hankin, R. M. Barkley and R. C. Murphy, Sublimation as a method of matrix application for mass spectrometric imaging, *J. Am. Soc. Mass Spectrom.*, 2007, **18**(9), 1646–1652.
40. A. P. Wisman, Y. Tamada, S. Hirohata, K. Gomi, E. Fukusaki and S. Shimma, Mapping haze-komi on rice koji grains using β-glucuronidase expressing Aspergillus oryzae and mass spectrometry imaging, *J. Biosci. Bioeng.*, 2020, **129**(3), 296–301.
41. J. A. Hankin, S. E. Farias, R. M. Barkley, K. Heidenreich, L. C. Frey, K. Hamazaki, H. Y. Kim and R. C. Murphy, MALDI mass spectrometric imaging of lipids in rat brain injury models, *J. Am. Soc. Mass Spectrom.*, 2011, **22**(6), 1014–1021.
42. A. Thomas, J. L. Charbonneau, E. Fournaise and P. Chaurand, Sublimation of New Matrix Candidates for High Spatial Resolution Imaging Mass Spectrometry of Lipids: Enhanced Information in Both Positive and Negative Polarities after 1,5-Diaminonapthalene Deposition, *Anal. Chem.*, 2012, **84**(4), 2048–2054.

43. M. Dufresne, D. Guneysu, N. H. Patterson, M. M. Marcinkiewicz, A. Regina, M. Demeule and P. Chaurand, Multimodal detection of GM2 and GM3 lipid species in the brain of mucopolysaccharidosis type II mouse by serial imaging mass spectrometry and immunohistochemistry, *Anal. Bioanal. Chem.*, 2017, **409**(5), 1425–1433.
44. A. Maimó-Barceló, J. Garate, J. Bestard-Escalas, R. Fernández, L. Berthold, D. H. Lopez, J. A. Fernández and G. Barceló-Coblijn, Confirmation of sub-cellular resolution using oversampling imaging mass spectrometry, *Anal. Bioanal. Chem.*, 2019, **411**(30), 7935–7941.
45. L. E. Lin, P. R. Su, H. Y. Wu and C. C. Hsu, A Simple Sonication Improves Protein Signal in Matrix-Assisted Laser Desorption Ionization Imaging, *J. Am. Soc. Mass Spectrom.*, 2018, **29**(4), 796–799.
46. J. Yang and R. M. Caprioli, Matrix Sublimation/Recrystallization for Imaging Proteins by Mass Spectrometry at High Spatial Resolution, *Anal. Chem.*, 2011, **83**(14), 5728–5734.
47. M. R. Groseclose, P. P. Massion, P. Chaurand and R. M. Caprioli, High-throughput proteomic analysis of formalin-fixed paraffin-embedded tissue microarrays using MALDI imaging mass spectrometry, *Proteomics*, 2008, **8**(18), 3715–3724.
48. T. W. Powers, B. A. Neely, Y. Shao, H. Tang, D. A. Troyer, A. S. Mehta, B. B. Haab and R. R. Drake, MALDI imaging mass spectrometry profiling of N-glycans in formalin-fixed paraffin embedded clinical tissue blocks and tissue microarrays, *PLoS One*, 2014, **9**(9), e106255.
49. P. M. Angel, S. Comte-Walters, L. E. Ball, K. Talbot, A. Mehta, K. G. M. Brockbank and R. R. Drake, Mapping Extracellular Matrix Proteins in Formalin-Fixed, Paraffin-Embedded Tissues by MALDI Imaging Mass Spectrometry, *J. Proteome Res.*, 2018, **17**(1), 635–646.
50. R. Lemaire, J. Stauber, M. Wisztorski, C. Van Camp, A. Desmons, M. Deschamps, G. Proess, I. Rudlof, A. S. Woods, R. Day, M. Salzet and I. Fournier, Tag-mass: specific molecular imaging of transcriptome and proteome by mass spectrometry based on photocleavable tag, *J. Proteome Res.*, 2007, **6**(6), 2057–2067.
51. G. Yagnik, Z. Liu, K. J. Rothschild and M. J. Lim, Highly Multiplexed Immunohistochemical MALDI-MS Imaging of Biomarkers in Tissues, *J. Am. Soc. Mass Spectrom.*, 2021, **32**(4), 977–988.
52. D. F. Cobice, C. L. Mackay, R. J. A. Goodwin, A. McBride, P. R. Langridge-Smith, S. P. Webster, B. R. Walker and R. Andrew, Mass Spectrometry Imaging for Dissecting Steroid Intracrinology within Target Tissues, *Anal. Chem.*, 2013, **85**(23), 11576–11584.
53. M. Shariatgorji, A. Nilsson, E. Fridjonsdottir, T. Vallianatou, P. Källback, L. Katan, J. Sävmarker, I. Mantas, X. Zhang, E. Bezard, P. Svenningsson, L. R. Odell and P. E. Andrén, Comprehensive mapping of neurotransmitter networks by MALDI–MS imaging, *Nat. Methods*, 2019, **16**(10), 1021–1028.
54. P. Chaurand, Introducing specificity and sensitivity in imaging MS, *Bioanalysis*, 2015, **7**(18), 2279–2281.

55. M. L. Reyzer, Y. Hsieh, K. Ng, W. A. Korfmacher and R. M. Caprioli, Direct analysis of drug candidates in tissue by matrix-assisted laser desorption/ionization mass spectrometry, *J. Mass Spectrom.*, 2003, **38**(10), 1081–1092.
56. M. Stoeckli, D. Staab and A. Schweitzer, Compound and metabolite distribution measured by MALDI mass spectrometric imaging in whole-body tissue sections, *Int. J. Mass Spectrom.*, 2007, **260**(2), 195–202.
57. S. Castellino, M. R. Groseclose and D. Wagner, MALDI imaging mass spectrometry: bridging biology and chemistry in drug development, *Bioanalysis*, 2011, **3**(21), 2427–2441.
58. A. Nilsson, T. E. Fehniger, L. Gustavsson, M. Andersson, K. Kenne, G. Marko-Varga and P. E. Andrén, Fine mapping the spatial distribution and concentration of unlabeled drugs within tissue micro-compartments using imaging mass spectrometry, *PLoS One*, 2010, **5**(7), e11411.
59. N. Sun and A. Walch, Qualitative and quantitative mass spectrometry imaging of drugs and metabolites in tissue at therapeutic levels, *Histochem. Cell Biol.*, 2013, **140**(2), 93–104.
60. A. B. McEwen, C. M. Henson and S. G. Wood, Quantitative whole-body autoradiography, LC-MS/MS and MALDI for drug-distribution studies in biological samples: the ultimate matrix trilogy, *Bioanalysis*, 2014, **6**(3), 377–391.
61. R. J. A. Goodwin, Z. Takats and J. Bunch, A Critical and Concise Review of Mass Spectrometry Applied to Imaging in Drug Discovery, *SLAS Discovery*, 2020, **25**(9), 963–976.
62. F. Tobias and A. B. Hummon, Considerations for MALDI-Based Quantitative Mass Spectrometry Imaging Studies, *J. Proteome Res.*, 2020, **19**(9), 3620–3630.
63. T. Porta, C. Grivet, T. Kraemer, E. Varesio and G. Hopfgartner, Single Hair Cocaine Consumption Monitoring by Mass Spectrometric Imaging, *Anal. Chem.*, 2011, **83**(11), 4266–4272.
64. B. Flinders, E. Beasley, R. M. Verlaan, E. Cuypers, S. Francese, T. Bassindale, M. R. Clench and R. M. A. Heeren, Optimization of Sample Preparation and Instrumental Parameters for the Rapid Analysis of Drugs of Abuse in Hair samples by MALDI-MS/MS Imaging, *J. Am. Soc. Mass Spectrom.*, 2017, **28**(11), 2462–2468.
65. N. Lauzon and P. Chaurand, Detection of exogenous substances in latent fingermarks by silver-assisted LDI imaging MS: perspectives in forensic sciences, *Analyst*, 2018, **143**(15), 3586–3594.
66. G. Groeneveld, M. de Puit, S. Bleay, R. Bradshaw and S. Francese, Detection and mapping of illicit drugs and their metabolites in fingermarks by MALDI MS and compatibility with forensic techniques, *Sci. Rep.*, 2015, **5**, 11716.
67. K. Ščupáková, B. Balluff, C. Tressler, T. Adelaja, R. M. A. Heeren, K. Glunde and G. Ertaylan, Cellular resolution in clinical MALDI mass spectrometry imaging: the latest advancements and current challenges, *Clin. Chem. Lab. Med.*, 2020, **58**(6), 914–929.

68. K. Ščupáková, F. Dewez, A. K. Walch, R. M. A. Heeren and B. Balluff, Morphometric Cell Classification for Single-Cell MALDI-Mass Spectrometry Imaging, *Angew. Chem. Int. Ed.*, 2020, **59**(40), 17447–17450.
69. I. S. Gilmore, S. Heiles and C. L. Pieterse, Metabolic Imaging at the Single-Cell Scale: Recent Advances in Mass Spectrometry Imaging, *Annu. Rev. Anal. Chem.*, 2019, **12**(1), 201–224.
70. A. Zavalin, J. Yang, K. Hayden, M. Vestal and R. M. Caprioli, Tissue protein imaging at 1 mum laser spot diameter for high spatial resolution and high imaging speed using transmission geometry MALDI TOF MS, *Anal. Bioanal. Chem.*, 2015, **407**(8), 2337–2342.
71. M. Niehaus, J. Soltwisch, M. E. Belov and K. Dreisewerd, Transmission-mode MALDI-2 mass spectrometry imaging of cells and tissues at subcellular resolution, *Nat. Methods*, 2019, **16**(9), 925–931.
72. J. Soltwisch, H. Kettling, S. Vens-Cappell, M. Wiegelmann, J. Müthing and K. Dreisewerd, Mass spectrometry imaging with laser-induced postionization, *Science*, 2015, **348**(6231), 211–215.
73. J. Soltwisch, B. Heijs, A. Koch, S. Vens-Cappell, J. Höhndorf and K. Dreisewerd, MALDI-2 on a Trapped Ion Mobility Quadrupole Time-of-Flight Instrument for Rapid Mass Spectrometry Imaging and Ion Mobility Separation of Complex Lipid Profiles, *Anal. Chem.*, 2020, **92**(13), 8697–8703.
74. T. P. Roddy, D. M. Cannon, S. G. Ostrowski, N. Winograd and A. G. Ewing, Identification of Cellular Sections with Imaging Mass Spectrometry Following Freeze Fracture, *Anal. Chem.*, 2002, **74**(16), 4020–4026.
75. M. Kompauer, S. Heiles and B. Spengler, Autofocusing MALDI mass spectrometry imaging of tissue sections and 3D chemical topography of nonflat surfaces, *Nat. Methods*, 2017, **14**(12), 1156–1158.
76. J. A. Barry, R. Ait-Belkacem, W. M. Hardesty, L. Benakli, C. Andonian, H. Licea-Perez, J. Stauber and S. Castellino, Multicenter Validation Study of Quantitative Imaging Mass Spectrometry, *Anal. Chem.*, 2019, **91**(9), 6266–6274.
77. A. Römpp, J. P. Both, A. Brunelle, R. M. Heeren, O. Laprévote, B. Prideaux, A. Seyer, B. Spengler, M. Stoeckli and D. F. Smith, Mass spectrometry imaging of biological tissue: an approach for multicenter studies, *Anal. Bioanal. Chem.*, 2015, **407**(8), 2329–2335.
78. F. P. Y. Barré, M. R. L. Paine, B. Flinders, A. J. Trevitt, P. D. Kelly, R. Ait-Belkacem, J. P. Garcia, L. B. Creemers, J. Stauber, R. J. Vreeken, B. Cillero-Pastor, S. R. Ellis and R. M. A. Heeren, Enhanced Sensitivity Using MALDI Imaging Coupled with Laser Postionization (MALDI-2) for Pharmaceutical Research, *Anal. Chem.*, 2019, **91**(16), 10840–10848.
79. B. Heijs, A. Potthoff, J. Soltwisch and K. Dreisewerd, MALDI-2 for the Enhanced Analysis of N-Linked Glycans by Mass Spectrometry Imaging, *Anal. Chem.*, 2020, **92**(20), 13904–13911.
80. T. Bien, E. A. Hambleton, K. Dreisewerd and J. Soltwisch, Molecular insights into symbiosis—mapping sterols in a marine flatworm-algae-system

using high spatial resolution MALDI-2-MS imaging with ion mobility separation, *Anal. Bioanal. Chem.*, 2021, **413**(10), 2767–2777.
81. M. Dufresne, A. Thomas, J. Breault-Turcot, J.-F. Masson and P. Chaurand, Silver-Assisted Laser Desorption Ionization For High Spatial Resolution Imaging Mass Spectrometry of Olefins from Thin Tissue Sections, *Anal. Chem.*, 2013, **85**(6), 3318–3324.
82. M. Dufresne, J. F. Masson and P. Chaurand, Sodium-Doped Gold-Assisted Laser Desorption Ionization for Enhanced Imaging Mass Spectrometry of Triacylglycerols from Thin Tissue Sections, *Anal. Chem.*, 2016, **88**(11), 6018–6025.
83. L. K. Hamilton, M. Dufresne, Joppé, E. Sandra, S. Petryszyn, A. Aumont, F. Calon, F. Barnabé-Heider, A. Furtos, M. Parent, P. Chaurand, Fernandes and J. L. Karl, Aberrant Lipid Metabolism in the Forebrain Niche Suppresses Adult Neural Stem Cell Proliferation in an Animal Model of Alzheimer's Disease, *Cell Stem Cell*, 2015, **17**(4), 397–411.
84. H. Alamri, N. H. Patterson, E. Yang, P. Zoroquiain, A. Lazaris, P. Chaurand and P. Metrakos, Mapping the triglyceride distribution in NAFLD human liver by MALDI imaging mass spectrometry reveals molecular differences in micro and macro steatosis, *Anal. Bioanal. Chem.*, 2019, **411**(4), 885–894.
85. C. Chen, S. R. Laviolette, S. N. Whitehead, J. B. Renaud and K. K. C. Yeung, Imaging of Neurotransmitters and Small Molecules in Brain Tissues Using Laser Desorption/Ionization Mass Spectrometry Assisted with Zinc Oxide Nanoparticles, *J. Am. Soc. Mass Spectrom.*, 2021, **32**(4), 1065–1079.
86. E. Fournaise and P. Chaurand, Increasing specificity in imaging mass spectrometry: high spatial fidelity transfer of proteins from tissue sections to functionalized surfaces, *Anal. Bioanal. Chem.*, 2015, **407**(8), 2159–2166.
87. Q. Wu, S. S. Rubakhin and J. V. Sweedler, Quantitative Imprint Mass Spectrometry Imaging of Endogenous Ceramides in Rat Brain Tissue with Kinetic Calibration, *Anal. Chem.*, 2020, **92**(9), 6613–6621.

CHAPTER 3

Instrumentation for MALDI-MSI – Part I Ionization Sources and Design

JENS SOLTWISCH

Institute of Hygiene, Westfälische Wilhelms-Universität Münster, Germany
Email: jenssol@uni-muenster.de

3.1 Introduction

Matrix-assisted laser desorption/ionization mass spectrometry imaging (MALDI-MSI) as a general technique is a relatively quick method, providing high mass accuracy and spatial resolution and enabling orthogonal techniques for structural elucidation. Additionally, it can provide capabilities for a large number of analyte classes, ranging from small molecules (*e.g.* metabolites and pharmaceuticals) to the detection of large molecules (*e.g.* proteins). Obviously, however, no single MALDI-MSI ion source can provide all these features, but its design has to be adapted to the targeted class of molecules and application, or to answer a specific research question.

In general, MALDI-MSI ion sources can be divided into three pressure regimes of operation: high vacuum, intermediate pressure between 0.5 and 10 mbar, and atmospheric pressure. All three regimes implicate specific demands to ion optics to ensure effective ion transfer as well as sensitive and accurate mass measurement. This chapter introduces the general concept of MALDI-MSI ion sources for all three regimes and compares and discusses their advantages and drawbacks. Regarding laser irradiation of the sample, this chapter will give an overview over the most common laser optic

New Developments in Mass Spectrometry No. 12
MALDI Mass Spectrometry Imaging: From Fundamentals to Spatial Omics
Edited by Tiffany Siegel

Published by the Royal Society of Chemistry, www.rsc.org

configurations for standard MALDI-MSI and introduce some special-purpose solutions for high spatial resolution. With growing demand for higher spatial resolution, sensitive MSI analysis from decreasing pixel size becomes increasingly challenging. Different post-ionization modalities (*e.g.* laser post-ionization or plasma-based techniques) but also electrospray-assisted methods have therefore been implemented in MALDI-MSI ion sources to effectively increase ion yields and will be introduced in this chapter.

3.2 Pressure Regime

3.2.1 Plume Development

After the absorption of energy from the laser pulse in a MALDI event, the material is ejected from the sample surface in what is often described as the MALDI plume. This plume consists of a dense molecular vapor from matrix and analyte and most often also contains larger clusters and aggregates. Thermodynamics, spatial development, and gas phase chemistry of the highly excited and dynamic volume of this MALDI plume have a key influence on the outcome of every MALDI experiment. In contrast to the initial processes of material ejection and primary ionization, however, it is not solely determined by the laser irradiation parameters and matrix morphology but also by the pressure prevalent in its surrounding. Under high vacuum conditions (usually $p < 10^{-5}$ mbar), ejected ions and neutrals expand into the ion source unhindered. Velocities for single molecules under these conditions are reported to be between 350 and 1000 m s^{-1} for thermal desorption conditions and can be estimated to be on the order of 10 and 100 m s^{-1} for larger clusters presumably produced by ablation.[1,2] At the other extreme, *i.e.* at atmospheric pressure (AP) conditions, plume expansion is heavily restricted by the surrounding gas. Rapid build-up of pressure leads to a strong containment of the molecular plume under these conditions and a quick deceleration.[3] A third pressure regime often utilized in MALDI ion sources can be found between about 0.5 and 10 mbar. Under these so-called fine vacuum or intermediate pressure conditions, the molecular particles and small clusters of the ejected plume are considerably affected by the residual gas in the ion source. With a mean free path of roughly 10 to 200 μm in the background gas, molecules on the leading edge of the plume are locally compressed and the expansion of the plume is slowed down in a process described as a "snow plough" effect.[4] Containment is however considerably lower than under atmospheric pressure conditions leading to a less dense plume.

Next to a cloud of single molecules and small molecular aggregates, other parts of the plume are ejected as larger clusters of various sizes ranging from hundreds to millions of matrix and analyte molecules. These clusters leave the surface with much slower initial velocities than desorbed molecules. Due to their increased inertia, they are much less affected by the background pressure. Even if the general MALDI mechanism for ionization can be considered largely independent of the surrounding pressure in the source, key features for MSI such as molecule fragmentation, cluster-formation, charge distribution, and ion yield of the experiment can differ starkly for the different regimes.

3.2.2 Vacuum Ion Sources

Historically, (axial) time of flight (TOF) analyzers have been the method of choice for MALDI-MS. In the classical setup, the MALDI ion source is integrated into the high vacuum of the TOF instrument. The laser spot on the sample thus constitutes the starting position, and the sharp laser pulse is used as the starting time for the TOF measurement. An individual mass spectrum is collected for each laser pulse by accelerating the ions toward the field free drift tube and detector in a constant electric field, either instantly or after a delayed extraction of up to hundreds of nanoseconds. After the initial MALDI event, dedicated ion optics in the first part of the ion path effectively catch the radially expanding cloud of ions and focus a wide range of *m*/*z*-ratios onto the detector.[5] Typically, high voltages in the order of tens of kilovolts are used for extraction and acceleration of the molecular ions. To prevent overly high field strength and inhomogeneity of the field, sharp edges or pointy ends have to be avoided in the design of the ion source. Only minor losses during the collection and flight time of ions make axial MALDI–TOF-MS an extremely sensitive method.

Due to the direct ion acceleration in high vacuum sources, ions reach the detector within a few microseconds after the MALDI event. Because of this rapid detection, thermally excited ions have relatively little time to undergo fragmentation. Increased thermal load, caused by *e.g.* excessively high laser fluences can however lead to thermal in-source or post source decay (ISD: during or before acceleration; PSD; in the field free path).[6] Independent of the thermal excitation, ISD can also be caused by rapid radical driven gas phase reactions inside the early plume on a timescale of a few tens of ns.[7]

It is noteworthy that the MALDI process takes place inside the acceleration region of the TOF measurement. Ions generated *via* MALDI in this region show a broad range of initial starting velocities induced by the initial plume expansion. These deviations and distributions in the starting velocity introduced during material ejection are reflected in the recorded flight times and lead to peak shifting (*i.e.* deviation in mean starting velocity) and peak broadening (*i.e.* starting velocity distribution). While techniques like delayed extraction and the use of reflectrons can reduce their influence, these effects inherently limit the resolving power and mass accuracy of axial TOF-MS.

Next to the MALDI generated starting velocity, minute changes in the extracting electric fields can also lead to deviations in the uptake of kinetic energy and compromise mass accuracy.

In axial MALDI–TOF-MS imaging of, for example, tissue sections of biological surfaces, samples must therefore be mounted on a conductive surface to assure for homogenous extraction conditions across the complete sample and to minimize detrimental surface charging effects. Translucent substrates such as indium tin oxide (ITO) coated glass slides are regularly used to allow for the light microscopic analysis of the samples but also coating with a thin layer of gold has been reported to provide sufficient conductivity.[8] Especially during the investigation of larger areas, additional changes in the electric fields are often caused by deviations in the distance between the starting position and the extracting ion optics. Variations in acceleration distance translate into differences in kinetic energy and therefore strongly compromise mass accuracy across the imaged area. In modern axial MALDI–TOF imaging instrumentation used sample stages therefore often include measures to adjust for changes in sample thickness and to counteract a possible tilt of the sample stage on the fly.[9] As described later in this chapter, this type of adjustment of sample height is also helpful in ensuring a homogenous laser spot size on the sample also in intermediate pressure and AP-MALDI ion sources. Changes in sample height introduced by the topography of the sample itself, however, are in all cases much harder to account for.[10]

In conclusion, unmatched speed and sensitivity of measurement are great assets for axial MALDI–TOF-MSI. The resolving power and mass accuracy of the technique, however, are limited by the measuring principle, especially for large areas or long measurement times. Very regular calibration of the instrument using external and if possible internal calibrants is therefore highly advisable in any axial MALDI–TOF-MSI measurement.

3.2.3 Intermediate Pressure Ion Sources

To increase mass accuracy and resolution of mass spectra acquired for MALDI-MSI, the mass measurement has to be energetically decoupled from the MALDI process. Consequently, MALDI ions have to be produced in a dedicated ion source and transported to the mass spectrometer using ion optics. While this is generally possible in high vacuum, long transfer times, as compared to the almost instant TOF measurement, lead to increased thermal fragmentation of MALDI generated ions before analysis.[11] To reduce these effects, ions can be generated at elevated pressures between 0.5 and 10 mbar. Under these intermediate pressure conditions MALDI generated ions undergo a number of low energy collisions with the background gas molecules and are effectively slowed down. Next to thermalizing the macroscopic forward motion of the molecular ions, ro-vibronic excitation within the molecular structure is also effectively drained into the bath gas. This process, dubbed collisional cooling, effectively reduces thermal fragmentation and opens up the opportunity to transfer intact molecular ions into virtually any type of mass analyzer.[12,13]

Because of the pressure difference between the ion source and the high vacuum of the analyzer, differential pumping and ion transfer along the pressure gradient become necessary. For this, ions are collected in multipole setups and guided toward an aperture using collisional focusing.[14]

Alternatively, one or multiple ion funnel systems can be used to collect and focus the ions.[15] In contrast to high vacuum ion sources, much lower voltages in the order of a few hundred volts are used for extraction. Because pressure conditions between 0.5 and 10 mbar are especially susceptible to gas discharges, however, care has to be taken in the design of the ion optics.

Regarding MALDI-MSI, the complete energetic decoupling of the MALDI process and MS measurement relaxes the requirements in MALDI ion source design and choice of experimental parameters. While effective extraction of the ions into the ion optics is still of paramount importance, small deviations in the starting velocity or the extracting potential are not reflected in the mass spectrometric performance. For example, adjustment of the laser power of the focal spot size has no effect on the calibration of the mass spectrometer. Because ions are thermalized, the detrimental effects of surface charging on ion velocities and overall ion extraction are strongly reduced. This allows for the use of non-conductive sample substrates such as histological glass slides.

The increase in mass accuracy and resolving power especially provided by FT-MS mass analyzers (*i.e.* FT-ICR and orbitrap-MS) in comparison to time-of-flight instruments enables a more accurate analysis of highly complex mass spectra generated from tissue samples.[16] This is especially true for the annotation of molecular species based on accurate mass. The decoupling of ion generation and mass analysis also allows for straightforward implementation of tandem MS and ion mobility modalities into MALDI-MSI workflows,[17,18] *i.e.* in comparison to (axial) time-of-flight instruments.

On the downside, ion transfer and differential pumping increase the risk of ion losses, possibly compromising the sensitivity of the instrument. Often the transfer of ions can only be optimized for a certain window of *m/z* ratios. Next to preventing excessive fragmentation of desired analyte ions, collisional cooling also stabilizes oligo-molecular clusters and small complexes generated during the MALDI process. These largely undefined ions contribute to what is commonly known as "chemical noise" and may reduce the signal-to-noise (*S/N*) ratio of the measurement.[19]

The energetic decoupling can significantly slow the acquisition speed and sample rates. Ion transfer from the ion source to the mass spectrometer adds additional time to the analysis. Additionally, ion packages generated from single laser pulses disperse due to interaction with the background gas. Especially for high repetition rate lasers, this can lead to a mixing of ions from separate MALDI events and the formation of a constant stream of ions that reaches the mass analyzer, much like the result of a constant ion source such as electrospray ionization (ESI). Measures to prevent the intermixing of ion packages from different pixels further limit the duty cycle of intermediate pressure ion sources in MALDI-MSI as compared to the direct analysis in a high vacuum.

In most cases, FT-MS or orthogonal extraction TOF (oTOF) instruments are coupled to intermediate ion sources. While duty cycles in oTOF MS are limited by the transportation times of ions to the analyzer, in FT-MS, mass measurement is most often the time limiting factor.

3.2.4 Atmospheric Pressure Ion Sources

Similar to intermediate pressure, the MALDI process can be effectively decoupled from mass measurement using AP ion sources. Here, the MALDI process takes place under ambient conditions and the MALDI-plume is heavily contained by the surrounding gas. Again, the ablated material is effectively thermalized in terms of kinetic and internal energy resulting in relatively stable ion species. For transportation of ions from atmospheric pressure to the high vacuum of the mass analyzer, different differential pumping systems have been developed. The general design of AP-MALDI ion sources often leans on ion optics developed for ESI or other ambient ionization mechanisms and is based on the interplay of gas flow and applied voltages. Their layout can include (heated) capillaries, skimmers or ion funnel systems.[20,21] This way, AP-MALDI sources can be coupled to virtually any type of mass analyzer. Analogous to intermediate pressure ion sources this enables the use of instruments that provide high-mass accuracy and resolving power, largely independent of any MALDI parameter or sample property. The increased containment caused by the high pressure of the surrounding gas leads to a much denser plume and more rapid cooling processes. This can benefit the survival of larger clusters ejected during the MALDI process or favor their formation by condensation inside the dense plume and increase the level of chemical noise or unwanted adduct peaks in the resulting mass spectrum.[20]

For MSI, ambient conditions inside the ion source simplify sample preparation and sample transfer. They also enable the analysis of volatile analytes and the use of volatile matrix compounds. In MS imaging, the gradual volatilization of such compounds in a vacuum over the course of a measurement can easily compromise the correct display of the distribution of analyte or prevent the analysis all together in the case of a fully sublimated matrix. In infrared (IR)-MALDI-MSI applications, this is especially true for water, often used as an intrinsic matrix that is highly volatile in a vacuum. On the downside, constituents of ambient air like oxygen and water, but also trace contaminations can induce unwanted side reactions (*e.g.* analyte oxidation) during the MALDI process or contribute to chemical noise.

3.3 Spatial Resolution in MALDI-MSI

Somewhat unsatisfactorily, the term spatial resolution in MALDI-MSI lacks a solid general definition and even more so standardized ways to be measured. In a broader sense, it describes the smallest spatial feature that can be resolved in a given MALDI-MSI experiment. Spatial resolution can be limited by the quality of the sample and its preparation on the one side, and by the pixel size as defined in the experimental parameters of the MSI method on the other. Here, only the latter will be discussed. As a rule of thumb, the spatial resolution of a MALDI-MSI experiment of an appropriately prepared sample is about a factor of two to three larger than the employed pixel size.

In general, two fundamentally different approaches to MALDI-MSI are possible: i) microscope, and ii) microprobe or scanning mode MALDI-MSI. Analogous to bright field microscopy, in the microscope mode a large area of sample is evenly irradiated with a laser beam to trigger a MALDI event. Ions generated on or near the sample surface are ion-optically imaged onto a pixelated detector analogous to the CCD chip of a digital camera.[22] Using the time-of-flight principle, all pixels of the detector simultaneously and independently record mass spectra, each specifically connected to a defined starting position on the sample. Pixel size in microscope mode imaging is independent of laser irradiation parameters but is defined by the magnification of the employed ion optics in combination with the size of each microsensor of the detector. Spatial resolution is therefore, in principle, only restricted by the limit of diffraction of the ion optical setup that is potentially much lower than that of its light optical counterpart. Despite of apparent advantages of the microscope mode in terms of acquisition speed and spatial resolution, technical challenges in realizing a dedicated mass spectrometric setup have so far hampered wide spread application of the technique.[23] Here, especially the pixelated ion detection system that provides the necessary resolving power and mass accuracy for MALDI-MSI has proven to be challenging.

Typically, therefore, MALDI imaging experiments are conducted in the microprobe mode. Here, the absorption laser is finely focused onto a predefined sample area and a mass spectrum is generated. In this approach, pixel size is defined by the distance between neighboring raster spots. To ensure that all information included in the generated mass spectrum exclusively represents the sample volume linked to the respective pixel, the size of the ablated area has to be smaller than the pixel size of the image generated. In other words, the volume irradiated and removed by the laser has to originate from within the pixel boundaries.

In this context, it is important to differentiate between the (optical) laser spot size, defined by, *e.g.* the full width at half maximum (FWHM) of the laser intensity profile, and the "MALDI-active" volume of the sample. While the laser spot size is independent of the employed laser power, the "MALDI-active" spot size is not. Because MALDI is a threshold process, the effective spot size $d_{eff}(E)$ changes with laser pulse energy E, especially for Gaussian beams, and may also differ greatly between different matrices at the same laser setting (Figure 3.1A). The size of the ablation crater is often a good indication for the size of the "MALDI-active spot" and should be used to evaluate the limits of an actual MALDI-MSI setup rather than the (optical) laser spot size.

With regard to spatial resolution, it is worthwhile to examine the optical system of MALDI-MSI ion source in more detail. Because the size of the ablation crater and the laser spot size are inherently linked, pixel size in a scanning approach is ultimately limited by the limits of diffraction of the employed optical system. For a focused laser beam, the spot diameter d_0 is described by:

$$d_0 = \frac{2\lambda}{\pi\varphi} M^2 \tag{3.1}$$

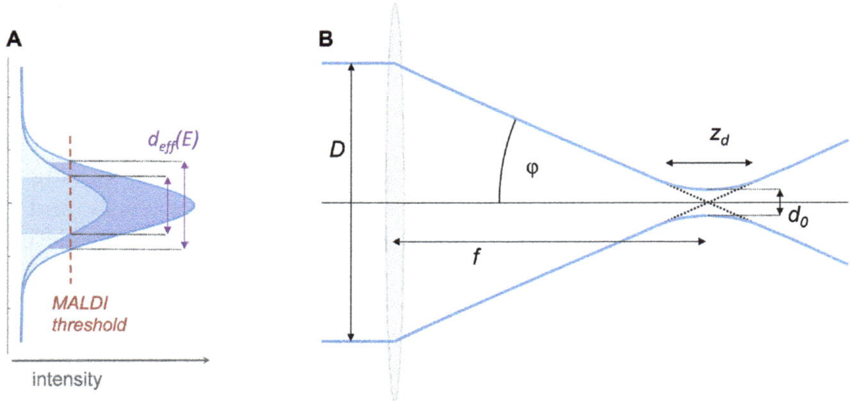

Figure 3.1 (A) Gaussian beam profile at different laser pulse energies (E) with schematic depiction of two different effective spot sizes $d_{\text{eff}}(E)$; (B) Schematics of focusing a Gaussian beam D: diameter of incoming beam, f: focal length, φ: angle of divergence, d_0: laser spot diameter in focus, z_d: depth of focus.

where λ is the wavelength of the laser light, φ the divergence angle (half angle) and M^2 a beam quality factor (also see Figure 3.1).

Particularly a small laser spot size can be achieved by increasing the divergence angle. From a practical point of view, this requires a large beam diameter of the incoming laser beam D and a short focal length f of the focusing lens (Figure 3.1B). Implications of these theoretical considerations for the design of MALDI-MSI ion sources will be discussed in more detail below in Section 3.4. With regard to spot size, eqn (3.1) reveals large differences between UV- and IR-MALDI. Using comparable optical systems, typical IR wavelengths used for MALDI around 3 μm produce a laser spot size roughly one order of magnitude larger than UV-MALDI with wavelengths around 337–355 nm. Also apparent from eqn (3.1), the laser beam quality has a large influence on laser spot size. While an ideal Gaussian beam has a beam quality of $M^2 = 1$, laser systems employed for MALDI can vary between $M^2 = 1.1$ and $M^2 = 1.5$ for solid state laser systems (Nd:YAG, Nd:YLF) and easily surpass a value of 5 for N_2-Lasers in UV MALDI. M^2 values in the range of 1.5 are typical for various IR systems (Er:YAG, OPO).

In an ideal MALDI-MSI experiment, ablation crater and desired pixel size are matched in a way that the complete pixel is sampled without interaction with neighboring pixels. Therefore, minute laser spot sizes are not always desirable because they can only sample a fraction of large pixel areas. The easiest way to adjust the laser spot size to the desired pixel size is by defocusing the laser spot on the sample surface. For this, one of the lenses of the optical setup is placed on a translational stage, moved along the optical axis and the irradiated area on the sample surface changes in size. Especially for laser systems with a low beam quality, defocusing, however, leads to strong changes of the laser beam profile and a less defined crater.

Alternatively, d_0 can be changed while leaving the optical system in focus by variation of the divergence angle φ (compare eqn (3.1)).[24] For this, typically different beam expanders are employed to change the parameter D of the initial beam. In a different approach, the optical system remains unchanged, and a small laser spot is used to scan the complete pixel area. Here different systems based on movement of the laser beam,[25,26] or rapid scanning of the stage,[27] have been described.

To circumvent the dogma of "pixel size should be larger than laser spot size", oversampling methods can be employed. In this approach, each pixel is irradiated until all matrix is removed and no new ions are produced from its volume.[28] Moving to the next pixel with a raster step size smaller than the produced "dead spot", only freshly irradiated areas contribute to the mass spectrum resulting in an effective MALDI spot size smaller than the employed laser spot. Similarly, the stop-motion scheme of recording pixels can be replaced by continuously moving the stage at a constant speed and recording mass spectra into a pre-defined time bin.[29] The mass spectrum in each time bin is directly connected to a specific area on the sample irradiated by the "moving" laser. In this acquisition scheme often coined "continuous raster mode", pixel size is defined by the sample stage velocity and scan time per bin.

Apart from the laser irradiation parameters, the employed sample stage can be a limiting factor, especially for high spatial resolution. Here, pixel size often approaches the minimum step size of the employed stage motors, often leading to detrimental aliasing effects and distorted images. Additionally, precision and play of moving parts such as lead screws can influence the point stability of the intended raster.

3.4 Optical Setups

Generally, the optical setup of all MALDI-MSI ion sources consists of a laser, optics for beam shaping and delivery, and focusing optics. While historically N_2-Lasers (337 nm) have often been employed for MALDI-MS, the third harmonic of diode pumped solid state (DPSS) lasers (Nd:YAG:355 nm, Nd:YLF:349 nm) are now commonly used for MALDI-MSI. In contrast to N_2 Lasers (repetition rate typically $<$100 Hz), the DPSS technology allows for repetition rates up to 10 kHz and has a longer lifetime. Additionally, DPSS lasers produce near-Gaussian beams with M^2 values down to 1.1, beneficial to produce small laser spot sizes.[30,31] Less common in application, IR–MALDI setups often employ Er:YAG lasers (2.94 μm) or OPO systems with a tunable wavelength around 3 μm.

For beam shaping, beam expanders are commonly used to tailor the beam diameter in order to achieve the desired divergence angle.[32] To increase beam quality, mode filtering can also be applied in the beam shaping unit, *e.g.* by the use of an aperture.[30] In some cases, more complex beam shaping optics have been used to produce structured or flat-top beam profiles.[24,31]

The focusing optics consist of one or more focusing lenses. In the case of high and intermediate pressure ion sources, the main focusing lens is often placed inside the vacuum of the ion source to minimize focal length and produce a large divergence angle. In these cases, adjusting the laser focus on the sample can be done by moving a second correction lens along the optical path outside the vacuum. Alternatively, sample height can be adjusted by a suitably equipped stage.[10] Because focal depth decreases with laser spot size, proper focusing and height adjustment across the complete sample are of major importance for successful MALDI-MSI using small laser spots. While at a spot size of 20 μm in diameter, focal depth is well above 1 mm, for a 5 μm spot it drops to only 80 μm (also see Figure 3.1). For any pixels on the sample with a height outside this focal depth, laser spot size will be considerably larger than in focus, leading to a decrease in laser fluence and a drop in signal intensity.

3.4.1 Front Side Illumination

As described in Section 3.2, ion optics are an essential part of every MALDI ion source. Their design is typically based on a rotational symmetry around an axis perpendicular to the sample plate originating from the laser spot on the sample (Figure 3.2). In these ion sources, the laser beam is usually introduced through a gap or an opening in the ion optics and hits the sample under an angle, often between 45° and 70° with respect to the sample surface. Because the MALDI process is largely unaffected by the angle of

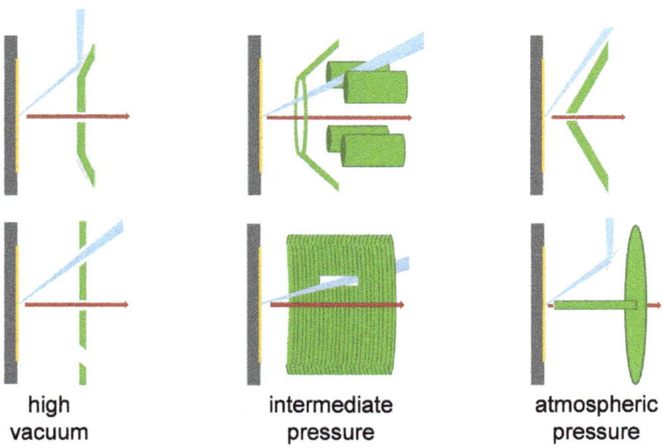

Figure 3.2 Schematics of common MALDI ion source designs for high vacuum (right), multipole and ion funnel-based systems for intermediate pressure between 0.5 and 10 mbar (middle) and skimmer and capillary based systems for atmospheric pressure (left). Ion optics are presented in green, laser path in blue, and the path of the ions toward the mass spectrometer in red.

incidence,[33] this robust design allows for a sensitive and homogenous extraction of ions. The angled illumination is therefore the standard found in most MALDI ion sources.

In high vacuum MALDI instruments, the laser beam is typically delivered either parallel to the sample surface and reflected onto the sample by a small mirror mounted on the extracting ion optics or through a small hole in the extraction plate. In intermediate pressure ion sources based on multipole ion guides, the laser beam is often guided through the gap between the rods of the multipole. In ion funnel setups, a dedicated hole in the funnel wall is commonly used. In atmospheric pressure ion sources, commonly a mirror is placed close to the capillary or onto the skimmer setup and is used to reflect the laser onto the sample surface.

For MALDI-MSI at high spatial resolution, angular illumination has some inherent limitations. Hitting the sample under an angle, the laser spot is stretched in one dimension, leading to an elliptical shape on the sample surface. This limits pixel size to the semi-major axis of the resulting oval. Also, because the sample is illuminated under an angle, any difference in sample height moves the position of the laser spot on the sample surface in a lateral direction. This can distort the resulting ion image and impede a true pixel-by-pixel image fusion with optical modalities.

Most severely, however, the production of small laser spots in front side (angular) illumination is limited by the ion optics that dominate the space close to the sample surface. As apparent from eqn (3.1), laser spot size is inversely proportional to the angle of divergence. This angle is ultimately determined by the size of the gap, opening or mirror and its distance to the sample surface (Figure 3.2).

Sensitive ion transfer is paramount to generate meaningful data, especially for small pixel sizes. The design of MALDI-MSI ion sources therefore prioritizes on effective ion optics and finds as much space for laser delivery as possible in a second step. Switching these priorities, the use of larger holes, gaps or mirrors, will either compromise ion extraction by interfering with ion optics, or lead to a shallower angle of illumination, thereby increasing the ellipticity of the laser spot. In common MALDI-MSI ion sources with front side illumination, this results in very shallow angles of divergence to values below 5° (half angle) and limits the laser spot size to minimum values between 25 μm and 5 μm.

The group of Spengler has presented an exception to this rule. By integrating the ion transfer capillary of an AP-MALDI ion source into the illumination optics (Figure 3.3A), they were able to realize an angle of divergence as large as 60° (half angle) and an angle of incidence of 90° relative to the sample surface.[34] Because the extracting capillary blocks a small part of the central portion of the laser profile, beam quality is somewhat deteriorated in this setup and leads to an increase in the effective M^2 factor (see eqn (3.1)). Effectively, this setup enabled MALDI-MSI measurements with ablation craters as small as 1.4 μm.[34]

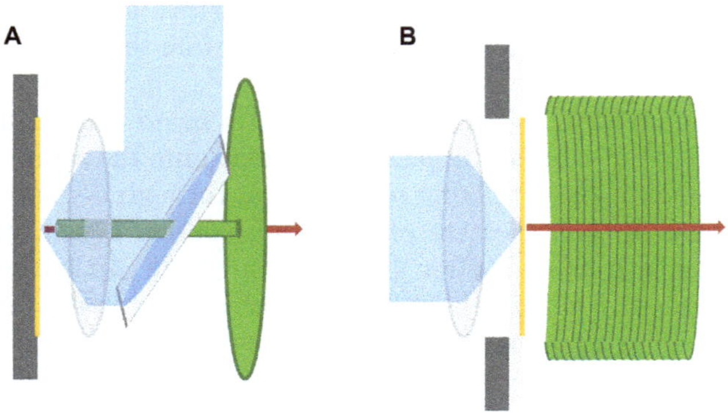

Figure 3.3 (A) Schematics of laser beam delivery in an AP-MALDI setup dedicated to high spatial resolution. (B) Schematics of laser beam delivery in transmission mode; ion collection and transfer using an ion funnel at intermediate pressure was chosen as an example. Ion optics are presented in green, laser path in blue and the path of the ions toward the mass spectrometer in red.

3.4.2 Transmission Mode Illumination

In an alternative approach, laser light can also be delivered to the sample from its backside (Figure 3.3B). This technique dubbed transmission or backside illumination was already employed in early laser desorption ionization instruments and some of the first MALDI instruments in the 1980s.[35,36] For transmission mode MALDI-MS (t-MALDI-MS), the sample and matrix layer both have to be relatively thin and mounted on a transparent substrate. The light passes through the low absorbing sample and deposits its energy into the thin matrix layer. In processes similar to conventional MALDI, material is ejected from the surface of the sample, and matrix and analyte molecules are partially ionized. Ion yields for transmission mode MALDI, however, have been described to be lower than for front side illumination.[37]

For t-MALDI-MSI, the sample stage is designed in a "picture frame" type setup. This allows for unobstructed optical access from the back side similar to a bright-field microscope. Because no ion optics interfere with the light optical setup behind the sample, dedicated microscope objectives with high numerical aperture, or in other words large angle of divergence, can be employed.[38] This way, angles between 25° and 30° (half angle) have been realized in transmission mode illumination. With no ion optics obstructing parts of the laser profile, Gaussian beams with a low effective M^2 factor can be employed leading to ablation craters as small as 1.2 μm in diameter.[39]

While the generation of very small ablation craters makes transmission mode MALDI-MSI a prime candidate for high spatial resolution, reduced ion yields connected to this illumination geometry have long prevented the further development of its application. With only minute sample volumes ablated per pixel at high spatial resolution, insufficient ionization limits its analytical range to highly abundant and easily ionizable analyte species. With the advent of new post-ionization strategies that can help to overcome some of these limitations, transmission mode illumination has lately seen somewhat of a renaissance and was successfully combined with laser post-ionization (MALDI-2) for MALDI-MSI at pixel sizes down to 600 nm with oversampling.[39–41]

3.5 Post-ionization

Inherently, MALDI-MS suffers from a low overall ion yield. Depending on the analyte and matrix, it has been estimated that only one out 10^5 to 10^7 molecules that are ejected from the sample surface during a UV–MALDI event is also ionized.[42] Positively speaking, the apparent giant reservoir of desorbed neutrals harbors great potential to further increase the sensitivity of the technique by orders of magnitude. Consequently, different approaches have been pursued to ionize desorbed molecules in a secondary process most often called post-ionization.[43] Ideally, such a technique is aimed to produce many ions for a broad range of molecules in an untargeted and unbiased fashion.

To successfully post-ionize desorbed neutrals inside a MALDI plume, sufficient energy has to be deposited into their electronic system. In photoionization, this energy is directly delivered to the analyte molecule by the absorption of photons. For this, the MALDI plume is irradiated by a second laser pulse, or in some cases by incoherent light sources delivered parallel to the sample surface. In untargeted approaches, single photon ionization (SPI) using VUV light as well as non-resonant multi-photon ionization can be used.[44–46] Here, energy delivered to each molecular species is not tuned to its respective ionization energy. Consequently, molecules with a high ionization threshold may not be ionized, while excess thermal load may be significantly increased for analyte molecules with a low ionization energy. In more targeted approaches, specific molecular species or a particular molecular class of interest can be ionized by selective multi-photon processes (Resonance-Enhanced Multi-Photon Ionization, REMPI).[47] Advantageously, in these approaches, the photon-energy delivered to each molecule is specifically tailored to match the electronic structure of the respective analyte species. While these techniques are highly effective, their benefit is limited in non-targeted applications such as most MALDI imaging experiments.

In an alternative approach, indirect techniques take a detour through chemical (post-)ionization.[43] In a multistep process, a suitable chemical intermediary is initially excited or ionized and transfers its charge or excitation energy to the analyte in a secondary process. Beneficially, primary excitation can be adjusted to match the ionization or excitation energy of this intermediary. Subsequent distribution of charge in secondary processes allows for the ionization of a much broader group of analyte species than the direct approach. To utilize any type of this chemical ionization, ejected material and reaction partners have to undergo sizeable interaction by collisions in the gas phase. Consequently, this largely omits its use in high vacuum. For intermediate and atmospheric pressure ion sources, however, different techniques successfully exploit this concept for post-ionization in MALDI-MSI. It should be noted that in AP-MALDI ion sources components of the laboratory atmosphere like oxygen and water, but also traces of chemical softeners or other volatile compounds, can lead to unwanted chemical reactions within the MALDI plume and with the excited or ionized intermediary. The same is true for intermediated ion sources fed with air from the laboratory environment. These unwanted compounds may produce sizable background signal especially when post-ionization strategies are employed.

3.5.1 MALDI-2

As described in Section 3.2, the MALDI plume expansion is restricted and confined under intermediate and atmospheric pressure conditions. This leads to the temporary formation of a dense cloud largely consisting of gaseous matrix molecules and smaller aggregates intermixed with a smaller number of co-ejected neutral analyte species. In this situation, the matrix can be exploited as the chemical intermediary introduced above. In a technique coined MALDI-2, a second pulsed laser is focused into the MALDI plume about 500 μm above the sample surface at a typical delay of 5–15 μs after the start of the initial MALDI event.[48,49] The beam path is aligned to be parallel to the sample surface to avoid interaction of the laser light with the sample. For irradiation, nanosecond long pulses of a wavelength below 280 nm and pulse energies of 200–800 μJ are used. Under these conditions, a large number of matrix molecules are selectively ionized in resonance enhanced two-photon processes. At the same time, most analyte species remain largely unaffected due to a low absorptivity. Constricted by the background gas, collisions inside the MALDI-plume between the abundant newly formed matrix ions and neutral analyte species now induce charge transfer processes that result in an effective post-ionization of a wide range of analyte classes. Optimal conditions for the formation of this "temporary reaction vessel" are essential for the success of a MALDI-2 experiment and largely depend on the interplay of pressure conditions and timing between the two laser pulses. Initially developed for intermediate pressure ion sources,[18,48,50] MALDI-2 was recently also implemented for an atmospheric pressure setup.[51]

A relatively young technique, MALDI-2 has already revealed great potential for MALDI-MSI for the imaging of different lipid classes.[52–54] Here, the technique helps to overcome ion suppression effects and significantly increases the number of detected lipid species and lipid classes.[55] Especially for small pixel sizes, the crucial increase in signal intensity helps to record information-rich MSI data at high spatial resolution. Apart from lipids, MALDI-2 was shown to increase signal intensities for a number of exogenous (pharmaceutical) compounds as well as for glycans in the negative ion mode.[56,57]

3.5.2 MALDESI/LAESI

IR–MALDI-MSI can exploit water as an endogenous matrix, ubiquitous to almost all biological samples.[58] The omission of an exogenous matrix allows for the analysis of samples that are difficult to coat with matrix and the absence of matrix signals simplifies data interpretation for the analysis of small molecules and metabolites. Ion yields in IR–MALDI-MS, however, are extremely low.[59] To increase ion yields, closely related techniques called MALDESI (matrix assisted laser desorption electrospray ionization) and LAESI (laser ablation electrospray ionization) have been developed for use in dedicated AP ion sources.[60–62] In these approaches, material ejected during the MALDI process is combined with an electrospray that is positioned to spray parallel to the sample surface. Neutral molecules ejected from the sample surface are caught in droplets of the electro-sprayed solvent and subsequently ionized in a classical ESI process. Generated ions are transferred to the mass spectrometer using a suitable interface. Again, the use of a two-step process allows for the soft yet effective post-ionization of a wide range of molecular classes ranging from lipids and small metabolites to large proteins.[63–66] Somewhat limited in spatial resolution by the typically used wavelength of around 3 μm (see eqn (3.1)), it was successfully employed in the analysis of "challenging" samples such as plant material with a pixel size as low as 150 μm in the front side and 60 μm in transmission mode.[65,67] These samples are usually very heterogeneous in morphology and are often difficult to section. Additionally, matrix application and analyte extraction are often hampered by their physico-chemical properties. In the analysis of intact proteins, ESI assisted techniques produce higher charge states as compared to standard MALDI, which is beneficial for structural elucidation.[63] In standard imaging applications, however, signal intensity and molecular coverage typically lack behind the values for conventional UV–MALDI and MALDI-2 methods.

3.5.3 Alternative Approaches

Next to ESI, alternative methods for post-ionization can be combined with AP-MALDI-MSI. Here, excitation and ionization of the intermediary are achieved in gas discharge or photoionization. While most of these techniques based on either atmospheric pressure photo ionization (APPI) or atmospheric pressure chemical ionization (APCI) have not been applied to MALDI-MSI, they are most often combined with laser desorption ionization (LDI). Naturally a combination with MALDI-MSI could be an obvious extension of this work.[43,46,68] In these techniques, excitation and ionization of the intermediary can either take place in a separate chamber prior to mixing with the desorption plume or after the mixing inside the transfer ion optics. In a first successful combination with MALDI-MSI and plasma post-ionization, ejected neutrals from the plume are transported into the instrument using a gas flow and are mixed with charged reaction partners in a dedicated chamber on their way to the mass spectrometer.[40] In this approach, reaction partners are produced using a cold plasma in a dedicated chamber, fully integrated with the ion transfer system that allows for stable and reliable operation during long imaging runs.

3.6 Conclusions and Future Directions

Starting with the experimental conversion of ion sources designed for regular dried droplet MALDI for imaging purposes more than 20 years ago, in recent years the design and development of ion sources for MALDI-MSI have branched and produced a large number of dedicated solutions. Application of MALDI-MSI now ranges from rapid screening of lipid fingerprints for clinical applications (*e.g.* distinguishing healthy from tumorous tissue) over ultra-high mass resolution and tandem MS experiments for structural elucidation to ever improving spatial resolution rapidly approaching the cellular scale.

Roughly, ion sources for MALDI-MSI can be separated into two major groups: vacuum MALDI on the one side and intermediate and atmospheric pressure MALDI on the other. In high vacuum, mass analysis is energetically coupled to the MALDI process. While high vacuum MALDI-MSI is unrivaled in acquisition speed and can provide a high sensitivity over a large m/z range, this coupling inherently limits resolving power and mass accuracy. Any changes to MALDI parameters like laser power, spot size, but also sample height or position of the laser spot are directly reflected in shifting m/z values. Very regular external, and if possible internal calibration, is therefore highly important.

In intermediate pressure and AP-MALDI ion sources on the other hand, the MALDI process is energetically decoupled from mass analysis. For this, ions are effectively thermalized in collisions with the background gas. This allows for the use of any type of mass analyzer, on the condition that suitable ion optics and vacuum system are provided. Accuracy and resolving power are largely independent of MALDI parameters. Also, tandem MS as well as ion mobility separation prior to MS analysis become readily available. The necessary transport of ions from the ion source to the analyzer, however, decreases acquisition speed and may lead to loss of ions. Rapid cooling of the expanding MALDI plume can lead to increased formation of clusters and aggregates, manifesting as an increase in chemical noise in the acquired mass spectra. In AP-MALDI sources this might even necessitate the use of thermal "desolvation" using heated capillaries or a heated sheath gas. Additionally, contaminations in ambient air may also contribute to an increase of chemical noise.

Spatial resolution in MALDI-MSI depends on a number of factors that include sample quality and sample preparation on the one side but also optical capabilities of the ion source on the other. Rapidly growing demand for fast and high throughput measurements at increasingly high spatial resolution requires for dedicated and optimized design considerations. Next to a suitable sample stage that provides the necessary step size, accuracy, and point stability, laser optical as well as ion optical setups offer a wide range of technical challenges. Desirably, a large angle of divergence that enables small laser spot sizes on the sample has to be combined with thorough extraction of the produced ions to ensure high sensitivity. At the same time, routine and high throughput application call for a reliable "plug-and-play" design. In the future, these seemingly conflicting tasks may increase the already apparent trend toward rugged and fast MALDI-imaging "work horses", with a pixel size down to 10–15 μm and acceptable MS performance on the one hand and more dedicated specialized equipment on the other. While the first group may rely on proven "classical" MALDI technology, the second will have to explore innovative strategies to achieve increased sensitivity and high specificity in molecular annotation as well as increasingly small pixel size of 2 μm and below. With regard to molecular annotation and identification in MALDI-MSI, innovative techniques based on in-source gas phase chemistry like derivatization or specific fragmentation are currently on the advent and may be flanked by increased molecular coverage introduced by post-ionization.[69,70] Developments like transmission mode illumination, highly effective ion transfer optics like ion funnels as well as different post-ionization strategies like MALDI-2 or plasma post-ionization currently open the door toward cellular or even sub-cellular resolution and sub-micron pixel sizes and may become important pieces to the puzzle to that end of the ongoing progress in the design of ion sources for MALDI-MSI.[15,39,41,48]

References

1. K. Dreisewerd, *Chem. Rev.*, 2003, **103**, 395–425.
2. M. Niehaus and J. Soltwisch, *Sci. Rep.*, 2018, **8**, 7755.
3. T. A. Schmitz, J. Koch, D. Günther and R. Zenobi, *Appl. Phys. B*, 2010, **100**, 521–533.
4. S. Amoruso, A. Sambri and X. Wang, *J. Appl. Phys.*, 2006, **100**, 013302.
5. S. R. Ellis, J. Soltwisch and R. M. A. Heeren, *J. Am. Soc. Mass Spectrom.*, 2014, **25**, 809–819.
6. R. Kaufmann, B. Spengler and F. Lutzenkirchen, *Rapid Commun. Mass Spectrom.*, 1993, **7**, 902–910.
7. T. Kocher, A. Engstrom and R. A. Zubarev, *Anal. Chem.*, 2005, **77**, 172–177.
8. A. Scherl, C. G. Zimmermann-Ivol, J. D. Dio, A. R. Vaezzadeh, P.-A. Binz, M. Amez-Droz, R. Cochard, J.-C. Sanchez, M. Glückmann and D. F. Hochstrasser, *Rapid Commun. Mass Spectrom.*, 2005, **19**, 605–610.
9. M. Piotrowski, B. Malys and K. G. Owens, *J. Am. Soc. Mass Spectrom.*, 2019, **30**, 489–500.
10. M. Kompauer, S. Heiles and B. Spengler, *Nat. Methods*, 2017, **14**, 1156–1158.
11. D. Papanastasiou, O. Belgacem, M. Sudakov and E. Raptakis, *Rev. Sci. Instrum.*, 2008, **79**, 055103.
12. A. N. Krutchinsky, A. V. Loboda, V. L. Spicer, R. Dworschak, W. Ens and K. G. Standing, *Rapid Commun. Mass Spectrom.*, 1998, **12**, 508–518.

13. J. Soltwisch, J. Souady, S. Berkenkamp and K. Dreisewerd, *Anal. Chem.*, 2009, **81**, 2921–2934.
14. A. V. Loboda, S. Ackloo and I. V. Chernushevich, *Rapid Commun. Mass Spectrom.*, 2003, **17**, 2508–2516.
15. M. E. Belov, S. R. Ellis, M. Dilillo, M. R. L. Paine, W. F. Danielson, G. A. Anderson, E. L. de Graaf, G. B. Eijkel, R. M. A. Heeren and L. A. McDonnell, *Anal. Chem.*, 2017, **89**, 7493–7501.
16. A. Römpp and B. Spengler, *Histochem. Cell Biol.*, 2013, **139**, 759–783.
17. S. R. Ellis, M. R. L. Paine, G. B. Eijkel, J. K. Pauling, P. Husen, M. W. Jervelund, M. Hermansson, C. S. Ejsing and R. M. A. Heeren, *Nat. Methods*, 2018, **15**, 515–518.
18. J. Soltwisch, B. Heijs, A. Koch, S. Vens-Cappell, J. Höhndorf and K. Dreisewerd, *Anal. Chem.*, 2020, **92**, 8697–8703.
19. A. V. Loboda and I. V. Chernushevich, *Int. J. Mass Spectrom.*, 2005, **240**, 101–105.
20. V. M. Doroshenko, V. V. Laiko, N. I. Taranenko, V. D. Berkout and H. S. Lee, *Int. J. Mass Spectrom.*, 2002, **221**, 39–58.
21. S. G. Moyer and R. J. Cotter, *Anal. Chem.*, 2002, **74**, 468A–476A.
22. L. A. Klerk, A. F. M. Altelaar, M. Froesch, L. A. McDonnell and R. M. A. Heeren, *Int. J. Mass Spectrom.*, 2009, **285**, 19–25.
23. A. N. Verenchikov and M. I. Yavor, *Int. J. Mass Spectrom.*, 2021, **463**, 116547.
24. M. Wiegelmann, K. Dreisewerd and J. Soltwisch, *J. Am. Soc. Mass Spectrom.*, 2016, **27**, 1952–1964.
25. N. O. Potocnik, T. Porta, M. Becker, R. M. A. Heeren and S. R. Ellis, *Rapid Commun. Mass Spectrom.*, 2015, **29**, 2195–2203.
26. A. Bednařík, M. Machálková, E. Moskovets, K. Coufalíková, P. Krásenský, P. Houška, J. Kroupa, J. Navrátilová, J. Šmarda and J. Preisler, *J. Am. Soc. Mass Spectrom*, 2019, **30**, 289–298.
27. M. A. Müller, M. Kompauer, K. Strupat, S. Heiles and B. Spengler, *J. Am. Soc. Mass Spectrom.*, 2021, **32**, 465–472.
28. M. F. Snel and M. Fuller, *Anal. Chem.*, 2010, **82**, 3664–3670.
29. F. Barré, B. Rocha, F. Dewez, M. Towers, P. Murray, E. Claude, B. Cillero-Pastor, R. Heeren and T. Porta Siegel, *Int. J. Mass Spectrom.*, 2019, **437**, 38–48.
30. A. Zavalin, J. Yang and R. Caprioli, *J. Am. Soc. Mass Spectrom.*, 2013, **24**, 1153–1156.
31. A. Holle, A. Haase, M. Kayser and J. Hoehndorf, *J. Mass Spectrom.*, 2006, **41**, 705–716.
32. H. Kettling, S. Vens-Cappell, J. Soltwisch, A. Pirkl, J. Haier, J. Müthing and K. Dreisewerd, *Anal. Chem.*, 2014, **86**, 7798–7805.
33. A. Westman, T. Huth-Fehre, P. Demirev, J. Bielawski, N. Medina, B. U. R. Sundqvist and M. Karas, *Rapid Commun. Mass Spectrom.*, 1994, **8**, 388–393.
34. M. Kompauer, S. Heiles and B. Spengler, *Nat. Methods*, 2017, **14**, 90.
35. D. W. Lorch and H. Schäfer, *Fresenius Z. Anal. Chem.*, 1981, **308**, 246–248.

36. U. Seydel and B. Lindner, *Int. J. Quantum Chem.*, 1981, **20**, 505–512.
37. M. Schürenberg, T. Schulz, K. Dreisewerd and F. Hillenkamp, *Rapid Commun. Mass Spectrom.*, 1996, **10**, 1873–1880.
38. A. Zavalin, E. M. Todd, P. D. Rawhouser, J. Yang, J. L. Norris and R. M. Caprioli, *J. Mass Spectrom.*, 2012, **47**, 1473–1481.
39. M. Niehaus, J. Soltwisch, M. E. Belov and K. Dreisewerd, *Nat. Methods*, 2019, **16**, 925–931.
40. E. A. Elia, M. Niehaus, R. T. Steven, J.-C. Wolf and J. Bunch, *Anal. Chem.*, 2020, **92**, 15285–15290.
41. T. Bien, S. Bessler, K. Dreisewerd and J. Soltwisch, *Anal. Chem.*, 2021, **93**, 4513–4520.
42. K. N. Robinson, R. T. Steven, A. M. Race and J. Bunch, *J. Am. Soc. Mass Spectrom.*, 2019, **30**, 1284–1293.
43. X. Ding, K. Liu and Z. Shi, *Mass Spectrom. Rev.*, 2021, **40**, 566–605.
44. K. Dreisewerd, M. Schürenberg, M. Karas and F. Hillenkamp, *Int. J. Mass Spectrom. Ion Processes*, 1995, **141**, 127–148.
45. A. S. Woods, S. N. Jackson, E. K. Lewis, T. Egan, L. Muller, J.-C. Tabet and J. A. Schultz, *J. Proteome Res.*, 2013, **12**, 1668–1677.
46. A. Akhmetov, J. F. Moore, G. L. Gasper, P. J. Koin and L. Hanley, *J. Mass Spectrom.*, 2010, **45**, 137–145.
47. L. Hanley, R. Wickramasinghe and Y. P. Yung, *Annu. Rev. Anal. Chem.*, 2019, **12**, 225–245.
48. J. Soltwisch, H. Kettling, S. Vens-Cappell, M. Wiegelmann, J. Müthing and K. Dreisewerd, *Science*, 2015, **348**, 211–215.
49. A. Potthoff, K. Dreisewerd and J. Soltwisch, *J. Am. Soc. Mass Spectrom.*, 2020, **31**, 1844–1853.
50. S. R. Ellis, J. Soltwisch, M. R. L. Paine, K. Dreisewerd and R. M. A. Heeren, *Chem. Commun.*, 2017, **53**, 7246–7249.
51. M. Niehaus, K. N. Robinson, T. Murta, E. A. Elia, A. M. Race, B. Yan, R. T. Steven and J. Bunch, *J. Am. Soc. Mass Spectrom.*, 2020, **31**, 2287–2295.
52. A. P. Bowman, J. F. J. Bogie, J. J. A. Hendriks, M. Haidar, M. Belov, R. M. A. Heeren and S. R. Ellis, *Anal. Bioanal. Chem.*, 2019, **412**, 2277–2289.
53. T. Bien, M. Perl, A. C. Machmüller, U. Nitsche, A. Conrad, L. Johannes, J. Müthing, J. Soltwisch, K.-P. Janssen and K. Dreisewerd, *Anal. Chem.*, 2020, **92**, 7096–7105.
54. T. Bien, E. A. Hambleton, K. Dreisewerd and J. Soltwisch, *Anal. Bioanal. Chem.*, 2021, **413**, 2767–2777.
55. M. S. Boskamp and J. Soltwisch, *Anal. Chem.*, 2020, **92**, 5222–5230.
56. B. Heijs, A. Potthoff, J. Soltwisch and K. Dreisewerd, *Anal. Chem.*, 2020, **92**, 13904–13911.
57. F. P. Y. Barré, M. R. L. Paine, B. Flinders, A. J. Trevitt, P. D. Kelly, R. Ait-Belkacem, J. P. Garcia, L. B. Creemers, J. Stauber, R. J. Vreeken, B. Cillero-Pastor, S. R. Ellis and R. M. A. Heeren, *Anal. Chem.*, 2019, **91**, 10840–10848.

58. Y. Li, B. Shrestha and A. Vertes, *Anal. Chem.*, 2007, **79**, 523–532.
59. F. Hillenkamp, T. W. Jaskolla and M. Karas, in *MALDI MS*, John Wiley & Sons, 2013, Ltd, pp. 1–40.
60. M. Nazari and D. C. Muddiman, in *Advances in MALDI and Laser-Induced Soft Ionization Mass Spectrometry*, ed. R. Cramer, Springer International Publishing, Cham, 2016, pp. 169–182.
61. P. Nemes and A. Vertes, *Anal. Chem.*, 2007, **79**, 8098–8106.
62. Y. H. Rezenom, J. Dong and K. K. Murray, *Analyst*, 2008, **133**, 226–232.
63. A. Kiss, D. F. Smith, B. R. Reschke, M. J. Powell and R. M. A. Heeren, *Proteomics*, 2014, **14**, 1283–1289.
64. J. S. Sampson, A. M. Hawkridge and D. C. Muddiman, *J. Am. Soc. Mass Spectrom.*, 2006, **17**, 1712–1716.
65. M. C. Bagley, C. L. Pace, M. Ekelöf and D. C. Muddiman, *Analyst*, 2020, **145**, 5516–5523.
66. P. Nemes, A. S. Woods and A. Vertes, *Anal. Chem.*, 2010, **82**, 982–988.
67. J.-P. Hieta, A. Vaikkinen, S. Auno, H. Räikkönen, M. Haapala, G. Scotti, J. Kopra, P. Piepponen and T. J. Kauppila, *J. Am. Soc. Mass Spectrom.*, 2017, **28**, 1060–1065.
68. M. M. Nudnova, L. Zhu and R. Zenobi, *Rapid Commun. Mass Spectrom.*, 2012, **26**, 1447–1452.
69. C. Bookmeyer, J. Soltwisch, U. Röhling and K. Dreisewerd, *ChemPlusChem*, 2020, **85**, 1559–1563.
70. F. Wäldchen, B. Spengler and S. Heiles, *J. Am. Chem. Soc.*, 2019, **141**, 11816–11820.

CHAPTER 4

Instrumentation for MALDI-MSI – Part II Detection Systems

BOONE M. PRENTICE[a] AND JEFFREY M. SPRAGGINS*[b]

[a] Department of Chemistry, University of Florida, Gainesville, Florida 32611, USA; [b] Department of Cell and Developmental Biology and Mass Spectrometry Research Center, Vanderbilt University School of Medicine, Nashville, Tennessee 37240, USA
*Email: jeff.spraggins@vanderbilt.edu

4.1 Introduction

Modern instrumentation and computing capabilities have enabled researchers to move beyond reductionist biology and, instead, probe how the components of biological entities (*e.g.* molecules, cells, and tissues) interact to reveal the underlying biology of disease. This systems biology approach has been accelerated by advancements in high-throughput 'omics' technologies, however, genetic and molecular information are only part of the story. The challenge lies in understanding how these parts interact and how perturbations to the system relate to disease. Imaging effectively offers a 'blueprint' as to how biological components work together by providing spatial context to molecular information. From the advent of complex microscopes in the late 1600s to modern modalities such as magnetic resonance, positron emission tomography, and advanced microscopy, imaging technologies have always been at the forefront of our understanding of biochemistry and biology. However, relative to the new omics technologies,

New Developments in Mass Spectrometry No. 12
MALDI Mass Spectrometry Imaging: From Fundamentals to Spatial Omics
Edited by Tiffany Siegel

Published by the Royal Society of Chemistry, www.rsc.org

these classical biomedical imaging technologies have limited molecular specificity. Because of its unique combination of spectral performance and spatial context, matrix-assisted laser desorption/ionization (MALDI) mass spectrometry imaging (MSI, also commonly referred to as imaging mass spectrometry, IMS) now finds itself uniquely positioned to bridge the gap between the information-rich genomics, proteomics, and metabolomics approaches and biomedical imaging technologies. MALDI imaging combines the sensitivity and specificity of mass spectrometry with the spatial insight of classic histology. As such, imaging performance is directly tied to the analytical performance of the mass spectrometer. Herein, we provide a review of the history, operating principles, and performance characteristics of modern mass analyzers and ion mobility mass spectrometry (IM-MS) platforms as they relate to the MALDI imaging experiment.

4.2 Mass Analyzers and Their Performance for MALDI Imaging

A wide variety of modern mass spectrometers are now capable of performing imaging mass spectrometry experiments. When selecting a platform for use, the mass analyzer is an important component of the instrument to consider, as each option has advantages and disadvantages associated with its performance (see Table 4.1). It is vital to understand the relative figures of merit of each mass analyzer so that users may intelligently decide which platform is best suited for their desired experiment. Additionally, a well-informed practitioner will understand the limitations of the acquired data and thereby accurately interpret the results of the imaging experiment. Although this chapter largely focuses on MALDI MSI, we start with a brief discussion of secondary ion mass spectrometry (SIMS) imaging to recognize the important role SIMS has played in defining the selection of the mass analyzer for early experiments. As imaging setups became more sophisticated and commercial MALDI sources became widespread, the selection of mass analyzers evolved to meet the needs of new experiments.

4.2.1 Sector Analyzers

The first spatially-resolved measurements acquired using mass spectrometers were performed in the 1960s using SIMS ionization sources.[1] These early instruments were termed "ion microscopes" and consisted of ion optic collection systems that maintained the spatial positions of the desorbed ions from the sample surface through the mass analyzer and to the detector.[2] Magnetic sector mass analyzers common to the time were well suited to transmitting a beam of ions generated from the SIMS source to a detector in this stigmatic manner. A few decades later, ions desorbed from biological samples were studied, mainly *via* spatial profiling experiments as opposed to microscopic imaging.[3,4] In either case, high-energy SIMS sources

Table 4.1 Approximate figures of merit for mass analyzers that are commonly used in imaging mass spectrometry experiments.

	Upper mass range	Resolving power ($m/\Delta m_{FWHM}$)	Mass accuracy	Dynamic range	Acquisition rate	Duty cycle[a]	MS/MS
Double-focusing sector	10 000	100–100 000	Single digit ppm	10^7–10^{12}	10 kHz	Low	keV, single collision
Quadrupole mass filter	10 000	1000–5000	50–100 of ppm	10^5	20 kHZ	Low	No
Time-of-flight	150 000	10 000–50 000	10–100 ppm	10^4–10^6	20 kHz	High	keV, single collision
QqTOF	10 000	20 000–50 000	1–10 ppm	10^5	20 kHz	High	Beam-type CID
Quadrupole ion trap	8000	1000–10 000	20–100 ppm	10^4	10 Hz	High	Ion trap CID, MS^n
Linear ion trap	8000	1000–10 000	20–100 ppm	10^5	10 Hz	High	Ion trap CID, MS^n
Triple quadrupole	10 000	1000–5000	50–100 of ppm	10^6	20 kHZ	Low	Beam-type CID
FT-ICR	25 000	100 000–10 000 000	<1 ppm	10^4	0.5 to 2 Hz	High	SORI CID, MS^n
Orbitrap	20 000	100 000–500 000	<1 ppm	10^4	0.5 to 2 Hz	High	No

[a]Relative duty cycle listed for a scan of the complete mass range using a pulsed MALDI source.

of the day were prone to atomization and fragmentation, so studies with the first ion microscopes were generally limited to studying metal ions sputtered from the surfaces of alloys and to tracking elemental distributions in manufactured catalysts.[5] The elemental ions produced by SIMS sources necessitated the ability to analyze low *m/z* values, did not require an extensive mass range, and operated at high vacuum, matching well the performance characteristics of magnetic sectors.

A single-focusing magnetic mass analyzer uses a single wedge-shaped (*i.e.* sector) magnetic field to separate ions by momentum dispersion.[6,7] As ions are accelerated into the magnetic field using an applied electric field, their arc through the sector represents a balance between the Lorentz force pulling inward toward the center of the magnet and the centrifugal force pulling outward. Under user-selected operating conditions of energetic stability, these forces are balanced, and ions of a defined momentum (and thus selected *m/z*) are transmitted through the device. Most magnetic sectors require the magnetic field to be scanned to sequentially focus ions of select *m/z* at the exit slit of the sector and to the detector. Thus, at any single point during the scan, all *m/z* values except the *m/z* of interest are discarded, resulting in low efficiency and poor duty cycle for the mass analyzer. However, double-focusing sector instruments, whereby an electric sector (E) is used to correct for kinetic energy dispersions and improve the mass resolution of momentum analysis performed in the magnetic sector (B), are still used today.

Since their first invention independently by Aston and Dempster, which was inspired by J. J. Thomson's work,[6,7] magnetic sectors have been noted for their ability to accurately measure the relative abundances of isotopes.[8] Additionally, the transmission of a beam of ions minimizes the effects of space charge, relative to the ion trapping mass analyzers discussed below, and thus offers a wide dynamic range (*e.g.* modern sector instruments can provide dynamic ranges of 10^{12} when employing multi-detector modes).[9,10] This abundance accuracy and wide dynamic range along with fairly high resolving power (10 000 to 100 000) (note that the resolving power here is defined as $m/\Delta m$, where Δm is measured at full width at half maximum [FWHM]) and good sensitivity lends itself nicely to modern-day laser ablation inductively coupled plasma (LA-ICP) elemental imaging experiments.[11–13] Commercial platforms are also available with multi-detection systems, where ions of dispersed *m/z* values emanating from the sector are focused to a plane containing an array of multiple detectors (*e.g.*, ten individual Faraday cups), in part compensating for the low duty cycle.[14] In addition to measuring elemental content in biological tissues, LA-ICP-EB instruments are widely used in geochemical communities for isotope-ratio measurements of substrates such as fossils, minerals, magmatic rocks, and other sediments.[15]

4.2.2 Quadrupole Mass Filters

High energy SIMS sources produce ions with a wide range of initial kinetic energies. Momentum analysis in magnetic sector mass analyzers is sensitive to dispersions in velocity, so high-energy SIMS experiments in the early days of imaging often resulted in lower mass resolution. This accelerated the adoption of quadrupole mass filters (QMFs, also called quadrupole mass analyzers [QMAs]), which were much less sensitive to the kinetic energy of secondary ions produced by SIMS.[16] QMFs consist of an array of four rods with precisely machined hyperbolic surfaces.[17] Opposite phases of a radio frequency (RF) potential are applied to the two-rod pairs, creating a radial quadrupolar trapping field within the device.[17] Ions placed into the quadrupolar field adopt oscillating frequencies termed "secular frequencies" that are dependent on their *m/z* ratios.[18] Radial divergence of the ion beam is partially compensated by the electrodynamic focusing achieved in the quadrupolar field of the QMF, resulting in high transmission efficiency through the device. However, like sectors, QMFs also suffer from poor duty cycle, as the RF potential and quadrupolar direct current (DC) potential are scanned to transmit ions of select *m/z* values through the device. As lower-energy cluster primary ion beams replaced high-energy atomic primary ion beams in SIMS sources, intact ions of larger molecular weight could be produced from surfaces.[19] The limited mass range of both sectors and QMFs would eventually be incompatible with higher *m/z* molecular imaging experiments. Additionally, the mass resolving power of QMFs is relatively low (typically 1000 to 5000, and commonly of unit-mass resolution). Still, the robustness, stability, and high transmission efficiency of QMFs make them attractive devices. The most widespread and useful adoption of QMFs in recent years has been in so-called hybrid instruments, where multiple mass analyzers are combined in a single instrument platform (*vide infra*).

4.2.3 Time-of-flight

The production of larger *m/z* value secondary ions from cluster primary ion beams in SIMS necessitated a mass analyzer with a larger mass range and better mass resolution.[16] Time-of-flight mass analyzers offer improvements in both of these figures of merit and operate by the pulsed acceleration of an ion packet into a field-free drift region. All ions in the source are accelerated to roughly the same kinetic energy, so once they enter the drift region, ions of smaller *m/z* ratios possess greater velocities and traverse the drift region more quickly, thereby arriving at the detector earlier in time than larger *m/z* ions.[20] Early TOF mass analyzers were plagued by ion spread in the source (*e.g.* velocity, position, and directional spreads). This was especially problematic for gas-phase ionization sources (*e.g.* electron ionization). However, a dedicated community of users steadily improved TOF mass resolution with innovations such as delayed extraction (also called time-lag focusing) to combat position or velocity spread and the reflectron to correct for kinetic energy dispersions.[21–25] Modern TOF instruments offer resolving powers over 50 000.[26] Unlike sectors and QMFs that analyze only a single mass at a time, there is no need to scan the TOF and all ions are detected sequentially with differing flight times as they arrive at the detector. In theory, the mass range of the TOF analyzer is unlimited, as larger molecular weight ions simply require longer flight times to reach the detector. Improvements in the speed and stability of high voltage electronics have allowed for the acquisition of an entire mass spectrum across a broad mass range in only a few hundred microseconds. This multiplexed advantage, coupled with high transmission efficiency, offered much improved detection efficiency compared to sectors and QMFs.

The pulsed nature of TOF required SIMS ionization sources to also be pulsed.[27] However, the development of a pulsed ionization source, matrix-assisted laser desorption/ionization (MALDI), independently by both Tanaka[28] as well as Karas and Hillenkamp[29] in the mid-1980s offered a perfect combination with TOF mass analyzers. The first MALDI experiments were performed at high vacuum, which matched well with the high vacuum requirement of TOF (*i.e.* to prevent scattering collisions in the field-free drift region of the analyzer). The pulsed nature of MALDI lends itself nicely to microprobe image acquisition, where a raster of the tissue surface is performed with the laser and mass spectra from discrete spots are sequentially acquired. MALDI offered a "soft" ionization technique that could readily produce large molecular weight, intact ions for the first time from biological samples. Indeed, the first MALDI images were acquired for peptide analytes.[30–32] Modern imaging experiments have enabled the detection of analytes over 100 kDa, though roughly 70 kDa remains a practical mass limit.[33] MALDI also offered a solid surface from which to produce ions, largely eliminating directional spread (also known as the turn-around problem) in TOF. However, as TOF ion separation depends on the energy of ions emanating from the source, the surface of the sample is still an important consideration. For example, the sample should be flat and parallel to the entrance to the TOF mass analyzer to avoid position spread that can degrade mass resolution. Similarly, conductive substrates such as indium tin oxide-coated glass microscope slides should be used for MSI experiments to ensure a uniform electric field within the source region.[34] This high voltage electric field (*i.e.* several kV) is used for pulsed ion acceleration into the mass analyzer. Insulating or non-conductive substrates can disrupt the uniformity of this field, resulting in unequal ion acceleration and diminished mass resolution. It should be noted that this discussion here is largely limited to axial TOF mass analyzers, whereby the sampling of ions is performed in the same axis as the ion flight path. While TOF instruments using orthogonal acceleration (oa) involve similar considerations, the merits of oaTOF MS are discussed separately in the QqTOF section below.

Advances in laser technology have provided for state-of-the-art MALDI TOF systems with superb acquisition rates up to 20 kHz.[35] Rapid acquisition is useful in instances where a large number of pixels are required to sample the tissue, such as in experiments involving many replicated tissue sections or in instances where large areas of tissue must be sampled at high spatial resolution.[36] For example, investigators have recently reported the use of high-speed MALDI TOF lipid imaging to acquire a one million pixel, 20 μm-spatial resolution image of human lung tissue from a cystic fibrosis patient in under six hours.[37] By comparison, previous instrument configurations that only sampled one pixel per second would have required almost two weeks of instrument time! While most instruments perform microprobe imaging, select instruments are also capable of microscope imaging modes.[38] TOF analyzers equipped with sensitive multichannel plate (MCP) detectors can achieve single-ion detection, demonstrating excellent sensitivity.[39] This sensitivity has been put to good use, as sophisticated laser optics in several research labs have been used to achieve near sub-micrometer MALDI laser spot diameters on TOF platforms, resulting in only a small amount of material ablated for mass analysis and ion detection.[40–46] Commercial instruments are now available that offer spatial resolutions of less than 10 μm.[37,47] MALDI TOF platforms are widely available in both linear (ideal for high sensitivity and broad mass range detection of, for example, proteins) and reflectron (with an improved mass resolution for more abundant small molecules such as lipids) configurations. These instruments are also noted for their lower cost and simplicity of operation.

4.2.4 QTOF

Although the mass resolution of TOF analyzers represented a significant improvement over that of QMAs, the unambiguous identification of analytes of interest remained challenging. The desire to use tandem mass spectrometry (MS/MS or MS^n) to identify ions of interest more confidently in MSI experiments quickly became widespread. This desire, along with the fact that hybrid mass analyzers equipped with MALDI sources became increasingly common in the early 2000s, led to the use of QTOF instruments for MSI.[48–50] Tandem MS involves two stages of mass analysis separated by an ion transformation step. Most commonly, an MS/MS experiment involves the selection of a precursor ion of interest by a mass analyzer, the fragmentation of this selected ion by energetic collisions with a neutral gas target *via* a process termed collision-induced dissociation (CID),[51] and the subsequent mass analysis of the resulting fragment ions by a mass analyzer. Tandem-in-space instruments perform these three steps in different regions of the instrument. With a QqTOF mass spectrometer, mass selection of the precursor ion is first achieved with a quadrupole mass filter (Q) *via* appropriate selection of RF and DC potentials.[52] The *m/z* value of interest is then passed through an RF-only ion guide (q), which acts only as a transmission device (*i.e.* it does not perform ion isolation or mass selection). The RF-only ion guide is often termed a collision cell and is filled with a few millitorr of inert gas such as helium, nitrogen, or argon. Ions are accelerated into the collision cell and undergo several energetic collisions (*i.e.* tens of electron volts) as they travel through the device. This ion guide is frequently a quadrupole device, but can also be another electrodynamic multipole (*e.g.* hexapole) or other ion optic device (*e.g.* traveling wave ion guide). As the ions are activated as they traverse through the cell, this is called a "beam-type" CID experiment. The RF-only quadrupole can also be replaced with another multipole device, such as a hexapole (h), which provides for a higher charge capacity and improved ion transmission efficiency.[53] The resulting product ions are then analyzed in the TOF mass analyzer. The fragment ions produced *via* a tandem mass spectrometry experiment can allow for the structural elucidation of the original parent ion and provide for a high level of molecular specificity. It is worth noting that MS/MS experiments can also be performed on hybrid sectors as well as TOF/TOF platforms.[54–56] The collision energies on these instruments are typically much higher (*i.e.* involving single, keV energy collisions) and can result in significantly different fragmentation mechanisms (*e.g.* charge remote fragmentation pathways).[57] However, these higher collision energies can also result in increased ion scattering and lower fragmentation efficiency.

The QTOF configuration can also improve mass resolution and instrument duty cycle. In this configuration, ions are injected orthogonally into the TOF mass analyzer perpendicular to the direction of ion extraction.[58] This eliminates the need to accelerate ions directly from the source into the TOF mass analyzer, so the electric potential applied to the sample surface can be much lower than with MALDI TOF platforms (*i.e.* a few hundred volts instead of a few thousand volts). As the ion source is now "decoupled" from the mass analyzer, the uniformity of the electric field within the MALDI source region no longer directly affects mass resolution. Position spread due to the surface of the sample is no longer a concern, meaning that there is no need for delayed extraction and the use of conductive substrates is no longer required. Instead, the velocity and directional distributions in the radial dimension become important, which can be minimized *via* collisional cooling in the quadrupole and using appropriately tuned ion optics at the entrance to the TOF mass analyzer. This can allow for a resolving power of 50 000 at *m/z* 1000 and mass accuracies of 1 to 10 parts per million (ppm).[58] The instrument duty cycle is improved by matching the lateral fill time into the ion extraction region with the total time of flight within the mass analyzer. In this way, once the first packet of ions has reached the detector at the end of the flight tube, the second packet of ions has filled the entrance of the TOF analyzer and is ready to be pulsed into the field-free drift region, maximizing the fraction of ions produced from the ion source that is analyzed.[59] This timing may be particularly useful with continuous ionization sources such as desorption electrospray ionization (DESI). The beam-type transmission on QqTOF platforms enables high-speed data acquisition, similar to that achieved with TOF platforms.[60] However, the presence of a quadrupole mass analyzer and other RF multipoles can limit the transmission of high *m/z* ions to the TOF mass analyzer, decreasing the effective mass range during MSI.[61]

4.2.5 Ion Traps

The desire to perform tandem mass spectrometry during MSI experiments led to the use of ion trap mass analyzers.[62–65] MALDI sources capable of operating at intermediate pressures, rather than at high vacuum, were developed for these platforms. For example, Garrett and Yost successfully coupled a MALDI source that operated at 0.17 torr to a linear ion trap mass analyzer.[63] DESI sources, which sampled ions at ambient pressure,[66,67] also emerged around the same time as linear ion trap mass analyzers and were well suited to these platforms, which operated at much higher pressures than sector and TOF mass analyzers. Like the QqTOF instruments mentioned above, ion trap platforms utilize sources that are decoupled from the mass analyzer, eliminating the need for high source voltages and the use of conductive substrates. Both two-dimensional (2-D) linear ion trap (LIT)[68,69] and three-dimensional (3-D) quadrupole ion trap (QIT)[70] platforms are attractive because of their ability to accumulate ions and perform highly efficient tandem mass spectrometry. Tandem MS performed at each pixel during the imaging experiment allows for the spatial mapping of fragment ions. Multiple isomers present in the precursor ion population can be differentiated *via* this approach (so long as each isomer produces a diagnostic fragment ion), thereby improving the molecular specificity of the imaging experiment. For example, "tandem mass spectrometry imaging" has been used to differentiate structural isomers of phosphatidylethanolamine (PE) lipids in developing mouse embryos.[71] Ion trap platforms tend to be simple, robust, and inexpensive. Although LITs have a higher charge capacity than QITs, both have limited dynamic ranges due to space charge effects that arise from confining ions within a finite trapping volume.[72] Ion traps possess intermediate resolving power (1000 to 10 000), intermediate mass accuracy (50–100 ppm), and low mass ranges (commercially up to roughly 8000, though research-grade instruments have achieved detection of analytes of over 100 kDa[73,74]). However, like QMFs, ion traps are readily incorporated in hybrid instrumentation.

In contrast to QqTOF platforms, ion trap platforms are tandem-in-time instruments, which perform precursor ion isolation, ion activation, and fragment ion mass analysis sequentially in the same physical location (*i.e.* the ion trap) at different time steps. Ion isolation can be achieved similar to a quadrupole mass filter (*e.g. via* the appropriate selection of RF and DC potentials applied to the quadrupole electrodes).[75] Ion activation is achieved *via* the application of a supplemental alternating current (AC) potential to the trap electrodes.[76–79] The frequency of this supplemental waveform is selected to match the secular frequency of the ion of interest, offering a second level of selectivity following the initial precursor ion isolation step. The collision energies in this "ion trap CID" experiment are lower than those achieved in beam-type activation and are typically limited to a few electronvolts.[80] Fragment ions are then analyzed by scanning the RF potential of the ion trap, causing ions of increasing m/z values to achieve unstable secular frequencies and resulting in their ejection from the ion trap into an external detector (typically an electron multiplier).[81]

The ability to trap ions also opens opportunities to perform more exotic tandem MS experiments. Multiple stages of tandem MS can be performed. For example, first-generation product ions can be isolated and then subjected to CID before analyzing the second-generation fragment ions (*i.e.* an MS/MS/MS or MS^3 experiment).[75] Instead of the energetic collisions with neutral gas commonly employed in CID, ion dissociation can be achieved using a variety of other methods and can result in improved or complementary structural information. For example, photodissociation (PD) can be performed by aligning an ultraviolet (UV) or infrared (IR) laser into the ion trap to irradiate a mass-selected ion population.[82] Ion/electron reactions have been performed using "chimera" or ICR ion traps to achieve high energy fragmentation.[83,84] Other investigators have used ion/molecule reactions to perform gas-phase chemical derivatization of the ion of interest (*e.g.* ozone-induced dissociation).[85–87] Recently, gas-phase charge inversion ion/ion reactions have been used to alter the ion type subjected to CID.[88,89] Ion/ion reactions have been used to access more favorable ion dissociation chemistry in order to identify fatty acyl *sn*-chain positions in phosphatidylcholines in rat brain tissue.[90] Gas-phase reactions are particularly attractive in MSI because they can be performed rapidly (*e.g.* in less than 1 second) and, unlike analogous condensed phase tissue derivatization approaches, do not irrevocably alter the original sample.[91] As MSI analyses of lipids and metabolites in biological tissues have become more popular, the importance of tandem MS for accurate analyte identification has become vital.

4.2.6 Triple Quadrupoles

A downside to the use of ion trap instrumentation is that tandem-in-time MS/MS experiments take longer to perform than tandem-in-space MS/MS experiments. Similar to QqTOF mass spectrometers, triple quadrupole (QqQ) instruments offer another "beam-type" ion activation platform with high acquisition speeds in imaging mass spectrometry experiments.[92] In these instruments, an initial quadrupole mass filter (Q) is used to perform precursor ion isolation, followed by CID in an RF-only quadrupole (q) and subsequent product ion analysis in a second quadrupole mass filter (Q). Initially described by Yost and Enke,[93] QqQ platforms are capable of many useful analysis scans designed to improve molecular specificity and signal-to-noise ratio (S/N), including product ion scans, precursor ion scans, neutral loss scans, and selected reaction monitoring.[94] While the rapid acquisition speed of the triple quadrupole is attractive, the necessity of scanning one or both quadrupole mass filters results in a low instrument duty cycle. Improvements in the speed of ion trap MS/MS experiments and the availability of QqTOF platforms with better mass resolution have greatly reduced the ubiquity of imaging experiments performed with QqQ instruments, though the sensitivity of QqQ platforms can still be attractive, particularly in drug and metabolite MSI.[95]

4.2.7 FT-ICR

The rise in hybrid instrumentation led to the facile coupling of external MALDI, SIMS, and DESI ion sources to Fourier transform ion cyclotron resonance (FT-ICR) mass analyzers.[96–98] As with ion traps, the ion source is decoupled from the mass analyzer for both FT-ICR and Orbitrap (*vide infra*) mass analyzers, allowing for ionization and mass analysis steps to be independently optimized. Although FT-ICR had been used previously to study ions generated from internal MALDI sources (*i.e. via* probes admitted in the ultrahigh vacuum region of the ICR cell), sample admission was cumbersome.[99–101] In FT-ICR, ions placed in a magnetic field adopt cyclotron frequencies dependent on their m/z values.[102] The balance of Lorenz and centrifugal forces, along with electrostatic trapping plates, allows for ions to be trapped within the ICR cell. Ion detection is performed by exciting ions to larger cyclotron radii. This excitation process also brings the ions in phase with one another and results in coherent motion of the ion packets. The coherent motion of the ion packets induces an image current between two oppositely placed, capacitively-coupled electrodes in the ICR cell. The resulting image current is a superposition of the cyclotron frequencies of the various m/z ions in the cell. This image current decays over a few seconds as ions de-phase and relax back to the center of the ICR cell. Individual frequencies can be extracted by performing a Fourier transform of the time-domain transient signal.

The cyclotron frequency of an ion can be measured very accurately, giving FT-ICR mass analyzers unsurpassed resolving powers (over 1 000 000) and mass accuracies (less than 1 ppm).[103] High-resolution accurate mass (HRAM) measurements allow for the facile identification of ion elemental compositions, providing for tentative analyte identifications that can later be confirmed using tandem mass spectrometry.[104] For example, several databases have been created to allow for accurate mass values to be searched against theoretical m/z values for many common lipids and metabolites.[105] The extreme resolving power of FT-ICR can even allow for isotopic fine structure (IFS) analysis, giving even greater confidence in assigning empirical chemical formulas.[106] This high resolving power also offers improved peak capacity, which is critical in MSI experiments. For example, investigators at the National High Magnetic Field Laboratory have used a 21 Tesla MALDI FT-ICR instrument to resolve lipid m/z differences in rat brain tissue as small as 1.79 mDa and with root-mean-square mass accuracies below 100 ppb.[103] Resolving powers of at least 300 000 at m/z 800 are generally needed in lipid imaging experiments to adequately separate all isobaric ions (see Figure 4.1). However, this exquisite resolving power comes at the cost of scan speed. Transient lengths of 0.5 to 4 seconds are commonly used in imaging experiments, resulting in spectral acquisition rates of only 0.25 to 2 Hz, much slower than the 20 kHz speeds available on modern TOF systems. Larger m/z ions have lower frequencies and thus require longer transient lengths to achieve the same resolution.[107,108]

The mass range of FT-ICR analyzers is in part determined by the size of the magnet and can also be limited by the transmission efficiency of the ion optics before the ICR cell. The use of gated trapping and modification of

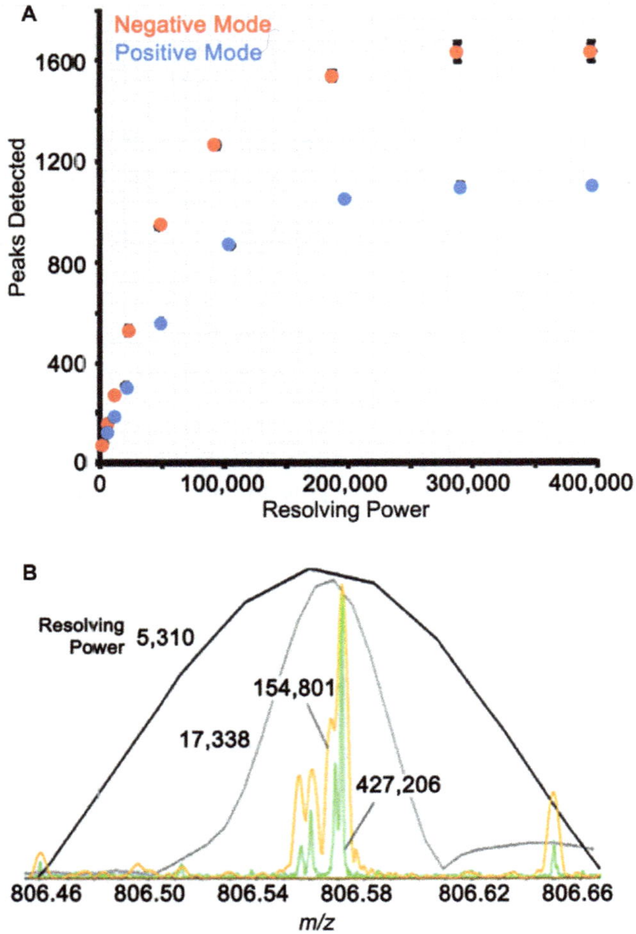

Figure 4.1 Number of peaks detected as a function of resolving power. (A) Data provided in both positive and negative ionization modes from a lipid extract of mouse brain tissue showing that peak capacity is not maximized until ~300 000 resolving power. (B) Spectral data are provided for this experiment with increasing resolving power over the mass range of *m/z* 806.46–806.66 highlighting the molecular complexity of direct tissue analysis. Data provided by Dr Jeffrey M. Spraggins of Vanderbilt University.

the ion optics on a 15T system recently allowed for the detection of protein ions up to *m/z* 22 000 in MALDI imaging experiments.[109] Like ion traps, space charge effects typically limit the dynamic range of FT-ICR analyzers to less than that of transmission mass analyzers such as sectors, QMFs, and TOF instruments. However, the hybrid instrumentation of modern FT-ICR platforms has recently been used to selectively enrich for ions of interest during MALDI-MSI.[110] These investigators used the mass-resolving quadrupole to isolate select *m/z* values for accumulation in a downstream hexapole ion trap prior to their admission to the ICR cell, demonstrating roughly 40-fold improvements in image sensitivity and dynamic range compared to the full scan experiments.[110] FT-ICR platforms also necessitate the use of a cryomagnetic, resulting in a large instrument footprint that is both expensive to purchase and to maintain using liquid helium. However, advances in magnet technology are reducing or eliminating the need to refill ICR magnets with liquid helium, reducing instrument size and operating costs.

4.2.8 Orbitrap

The turn of the century saw the emergence of another FT mass analyzer capable of high mass resolution and high mass accuracy measurements without the need for a cryomagnet, the Orbitrap.[111] This mass analyzer operates on the principle of orbital trapping, in which ions are injected into a hollow device and circulate around a central spindle electrode.[112] An attractive potential applied to the spindle bends ions toward this central electrode and balances the opposing centrifugal force, allowing for stable radial trapping. Due to the shapes of the electrodes, ions oscillate axially across the spindle as coherent packets or disks of ions. This axial oscillation is *m/z* dependent and can be detected as an image current as the disks oscillate between the two halves of the analyzer. As with FT-ICR mass analyzers, the resulting image current is a superposition of the cyclotron frequencies of the various *m/z* ions in the cell and individual frequencies can be extracted by performing a Fourier transform of the time-domain transient signal. As ions are focused to a disk instead of a central point, the charge capacity of Orbitrap analyzers is higher than on most FT-ICR instruments. However, the dynamic range of both systems is still low compared to other analyzers. Though the top-end resolution may not be quite as high as with FT-ICR analyzers, Orbitrap platforms still provide excellent mass resolution and accuracy (<2 ppm) in MSI experiments.[41,113–116] Newer "high-field" Orbitrap instruments[117] offer nearly 500 000 resolving power at *m/z* 400 and the resolving power of the Orbitrap decreases more slowly at increasing *m/z* compared to FT-ICR analyzers, which may be beneficial for higher *m/z* protein imaging experiments. "Extended mass range" (EMR) Orbitrap instruments now offer mass ranges up to *m/z* 20 000.[118] Most Orbitrap analyzers are available on hybrid instrument platforms. Multiple recent reports have leveraged hybrid LIT-Orbitrap instruments to perform high mass resolution precursor ion measurements (MS^1) in the Orbitrap analyzer while simultaneously performing tandem mass spectrometry analysis on select ions of interest (MS^2) in the LIT analyzer.[119,120] This provides the molecular specificity and structural information of a classical tandem mass spectrometry experiment while minimizing the loss in instrument duty cycle and still obtaining full mass range data during image acquisition.

4.3 Ion Mobility Spectrometry for Improved MALDI-MSI Performance

Ion mobility-MS provides 2-D separations on the basis of molecular structure (apparent surface area, Å^2) and *m/z*.[121–125] While there are multiple ion mobility technologies (see Figure 4.2), all achieve molecular separation through the interplay between collisional forces with a carrier buffer gas and an opposing electric field. Interactions between analytes and the inert gas are defined differently for each ion mobility technique,[131] but in all cases the molecular size-to-charge ratio of an ion determines its mobility and causes ions to exit from the ion mobility cell at differing times (*i.e.* drift time).[132,133] The elution time from the ion mobility device plus the time it takes for the population of ions to reach the detector is often referred to as the arrival time distribution.[131] Similar to the retention times in chromatography, these arrival times can distinguish molecules of similar mass and can be used to calculate the collision cross section (CCS) for improved specificity and to assist in structural identification.[125,134,135] Coupling ion mobility and mass spectrometry imaging provides a means to improve analytical detection characteristics such as sensitivity, specificity, *S*/*N*, and dynamic range enabling greater molecular coverage, improved structural characterization, and high spatial resolution capabilities for MALDI imaging.

4.3.1 Drift Tube Ion Mobility Spectrometry

Drift tube ion mobility spectrometry (DTIMS) is the simplest of the gas-phase separation technologies with its origin dating back to the 1950s and 1960s with the foundational theoretical work of McDaniel[136–138] and Mason and Schamp[139] and the instrumental development efforts of McDaniel[136–138] laying the groundwork for the modern ion mobility systems of today. DTIMS is a time-dispersive separation device that functions by driving ions through an inert buffer gas using an electrostatic field (*i.e.* DC potential). In modern DTIMS systems, ions are typically focused using electrodynamic, RF potentials applied to stacked ring electrodes (see Figure 4.2A). Ions with larger CCSs encounter more drift gas molecules compared to ions with smaller CCS values, allowing separation as a function of mobility.[121] DTIMS has been demonstrated in varying pressure regimes, but reduced pressure (<5 Torr)[123,138,140] provides greater ion transmission efficiency through the ion mobility cell making it more advantageous for molecular imaging experiments where sensitivity is paramount for high spatial resolution experiments. Commercial DTIMS-MS platforms can routinely achieve moderate-to-high mobility resolving powers (∼60 (CCS/ΔCCS))[121,141] and produce consistent CCS measurements.[135] This has made DTIMS invaluable for measuring absolute CCS values in structural studies[121,130] and for creating unified experimental and

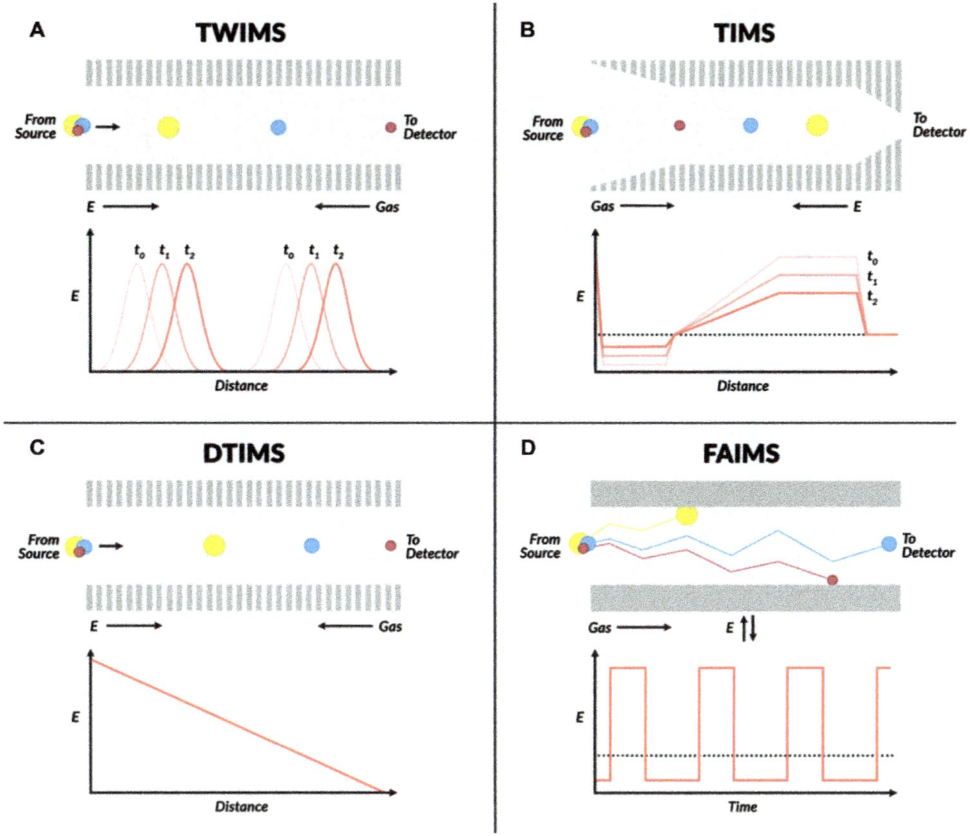

Figure 4.2 Overview of common ion mobility technologies. (A) Drift tube ion mobility spectrometry (DTIMS), (B) traveling wave ion mobility spectrometry (TWIMS), (C) trapped ion mobility spectrometry (TIMS), and (D) high-field asymmetric waveform ion mobility (FAIMS).

theoretical CCS value databases.[142–145] Drift tube measurements will certainly play a key role in analyte identification for MALDI MSI in the future as DTIMS reference data are what other ion mobility methods are calibrated to. Despite some of the earliest examples of ion mobility-based MALDI imaging utilizing DTIMS,[146–150] there have been a limited number of recent examples due to a lack of a commercially available MALDI DTIMS instrument. Still, DTIMS has been coupled to other ion sources for MSI including infrared matrix-assisted laser desorption electrospray ionization (IR-MALDESI) demonstrating its utility as a gas-phase separation technology for improving peak capacity and improving *S*/*N* in imaging experiments.[151]

4.3.2 Traveling Wave Ion Mobility Spectrometry

Traveling wave ion mobility spectrometry (TWIMS) is perhaps the most common form of ion mobility coupled with MALDI imaging as it was the first commercial ion mobility imaging platform.[126,152–155] TWIMS was first described by Giles *et al.* in 2004[156] and was released commercially in 2006 as an electrospray ionization (ESI) system. Then in 2008, it became available with MALDI capabilities. TWIMS is performed by transmitting ions through an inert buffer gas using a series of ring electrodes that radially confine the ions with an RF voltage and propel them through the cell with a DC voltage that is applied to each electrode in succession (see Figure 4.2B). As the DC voltage traverses the length of the TWIMS cell, it creates a traveling electric field wave that propels ions through the buffer gas with a velocity that is dependent on mobility.[156,157] TWIMS commonly yields a resolving power of ~40 (CCS/ΔCCS),[125] which is sufficient for resolving many isobaric ion species[126,158] and some isomeric ions, often with the aid of chemical derivatization[129] or metal adduction.[159,160] CCS values for many analyte classes are also frequently calculated using TWIMS-based analyses, leading to the curation of several databases for species such as proteins and steroids.[134,161]

The Clench research group has been one of the primary drivers of MALDI TWIMS imaging development.[162–164] In one of the early TIMS imaging reports, they demonstrated the capabilities of the technology by imaging the anticancer drug, vinblastine, in whole-body rat tissues.[164] TWIMS was used to filter interfering isobaric chemical noise improving the sensitivity of the intact molecular ion of vinblastine and the transition of *m*/*z* 811 to *m*/*z* 751 in MS/MS imaging experiments. Following up on this work, Djidja and co-workers were able to both map and fragment tryptic peptides in pancreatic tumor tissues.[162] Using TWIMS, they imaged peptides of glucose-regulated protein 78 kDa (Grp78) showing the protein to have higher expression within tumor regions, as confirmed by immunohistochemical staining. TWIMS was able to differentiate isobaric peptides in this analysis and they were able to perform *in situ* MS/MS identification, a challenging experiment, particularly for singly charged peptides. The Heeren group also used MALDI TWIMS to study cancer. They used ion mobility to differentially image and fragment stearoylcarnitine and its close isobars in a breast tumor xenograft model.[165] Through co-registration with optical microscopy images, they were able to uncover specific localizations of acylcarnitine, phosphatidylcholine, and sphingomyelin species to viable, necrotic, and hypoxic tumor sub-regions. These early ion mobility studies touched on an important issue for MALDI imaging, where potentially confounding factors such as spectral complexity and molecular specificity present challenges in accurately determining spatial distributions of endogenous molecules, and how gas-phase separations offer a potential solution to the problem.

4.3.3 Trapped Ion Mobility Spectrometry

Trapped ion mobility spectrometry (TIMS) is a high resolving power, field-dispersive ion mobility technology that was introduced by Fernandez-Lima and co-workers in 2011.[166,167] TIMS separations are carried out in an augmented ion funnel located in the first vacuum stage of the mass spectrometer. The TIMS funnel consists of an entrance funnel, TIMS tunnel, and exit funnel (see Figure 4.2C) with the most recent designs including a dual funnel design that enables simultaneous accumulation and ion mobility analysis, dramatically improving the throughput and duty cycle of the experiment.[166,167] Unlike other ion mobility devices that drive ions through a 'stationary' buffer gas, TIMS separates ions in space using an axial electric field gradient (EFG) to hold ions stationary against a gas flow that is passing through the funnel. They are trapped at specific locations within the TIMS tunnel where the force of the gas collisions is in equilibrium with the force exerted on the ions by the electric field gradient. The electric field gradient is then gradually reduced/scanned eluting ions with ascending mobilities. Note, this is inverse to how elution is performed with other ion mobility technologies that transmit ions in descending order. The ion mobility resolving power is proportional to the electric field gradient scan rate, which is defined by the scan time (typically 50–500 ms) and voltage range and can achieve resolving powers >200.[168]

For molecular imaging, TIMS was first integrated with a MALDI QqTOF platform by Spraggins and co-workers in 2019.[127] Subsequently, TIMS has been utilized for lipid and metabolite imaging with both MALDI and MALDI-2.[127,128,169–173] In the initial MALDI TIMS work, high-resolution ion mobility was used for *in situ* separation of closely isobaric lipid species.[127] A common challenge in MALDI imaging experiments is the differentiation of overlapping isotopic distributions of lipids with varying adduction (*e.g.* H^+, Na^+, and K^+). TIMS has also been used for high-resolution gas-phase separations to image lipid isomers using MALDI.[128,170,172] As an example, Figure 4.3 highlights MALDI TIMS data showing the separation and mapping of isomeric lipids in human kidney tissue. With a TIMS scan time of 400 ms and $1/K_0$ range of 0.8–1.7 V s cm^{-2} (EFG scan rate = 0.32 V s^{-1}),

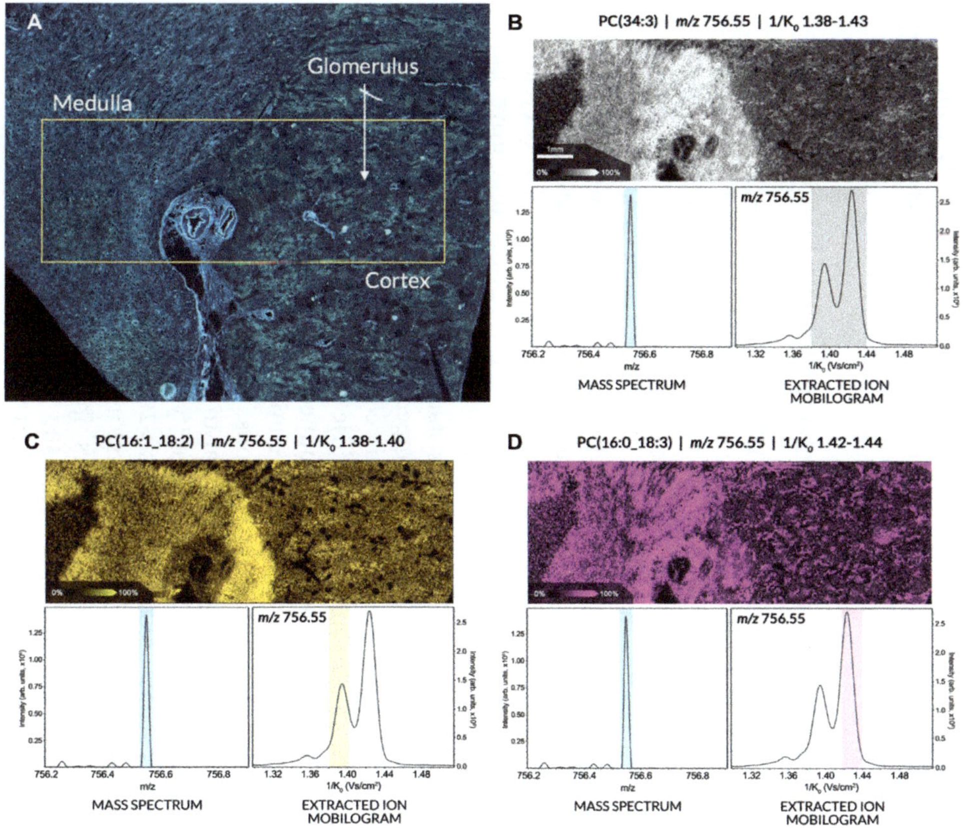

Figure 4.3 MALDI-MSI data generated from human kidney tissue collected at 20 μm-spatial resolution using trapped ion mobility spectrometry (TIMS). (A) Complimentary autofluorescence microscopy provides histological context highlighting the medulla and cortex regions of the tissue. (B–D) Each panel highlights positive ion mode images from selected mobility ranges for *m*/*z* 756.55. Although only one peak is detected in the selected mass window (blue), ion mobility is able to distinguish multiple isomeric species. Data provided by Katerina V. Djambazova and Dr Jeffrey M. Spraggins of Vanderbilt University.

multiple lipid species were observed at *m/z* 756.55. The image of the average ion intensity covering both species ($1/K_0$ 1.38–1.44, see Figure 4.3B) shows a greater intensity in the medulla of the kidney (left) with a sparse distribution in the cortex (right). However, visualizing each TIMS resolved species separately improves image contrast and reveals differences in their spatial distributions, particularly in the cortex of the kidney (see Figure 4.3C and D). These species were determined to be isomeric phosphatidylcholine (16:1_18:2) and (16:0_18:3), respectively. This capability is critically important for MALDI MSI as isomeric species cannot be distinguished based on mass accuracy alone, no matter the resolving power. Yet, ion mobility can differentiate these species without the need for performing targeted MS/MS experiments allowing for both high specificity and broad molecular coverage imaging, maintaining the multiplexed advantage of MALDI MSI.

4.3.4 High-field Asymmetric Waveform Ion Mobility

High-field asymmetric waveform ion mobility (FAIMS), or differential mobility spectrometry (DMS), is a spatially-dispersive form of ion mobility that has seen recent application to liquid droplet-based MSI technologies such as desorption electrospray ionization (DESI)[174] and direct liquid droplet sampling technologies[175–177] but has yet to be utilized for MALDI MSI. When coupled to a mass spectrometer, the FAIMS device is typically positioned between an atmospheric pressure ion source and the inlet of the mass spectrometer.[178–180] Ions are carried by an inert buffer gas through closely spaced electrodes upon which a high-frequency asymmetric waveform of alternating strong and weak electric fields is applied perpendicular to the flow of the carrier gas, commonly referred to as a dispersion electric field (see Figure 4.2D).[178,179,181] Gas-phase separation of ions results from differences in ion mobility in the strong and weak electric fields, causing deviation from their initial trajectory. Scanning the secondary weak DC field, known as the compensation voltage, selectively transmits ions within specific mobility windows. Although the earliest forms of the FAIMS device reported in the early 1990s were flat plate designs, cylindrical instruments were developed shortly thereafter. Most modern instruments incorporate cylindrical FAIMS[178,179,181] or more recently, chip-based FAIMS platforms that minimize separation times, enable higher dispersion electric fields, and greatly reduce the size of the device.[182,183] FAIMS is not typically used to determine structural characteristics of the molecule like CCS but is instead used to improve analytical figures of merit including *S/N*, sensitivity, and dynamic range.[178] Because FAIMS operates at atmospheric pressure, it has yet to be utilized for MALDI imaging, however, because the technical challenges of integrating FAIMS with atmospheric pressure MALDI (AP-MALDI)[184–187] would be minimal, this is likely a coming development in the field.

4.4 Conclusion and Future Perspectives

As detailed above, mass spectrometry technologies have rapidly advanced in the areas of sensitivity, specificity, molecular coverage, and dynamic range enabling dramatically improved MALDI imaging capabilities and opening the door to previously unforeseen application areas. As MALDI source technologies and sample preparation methods push the field toward higher spatial resolution experiments, it is critical that the development of mass analyzers and gas-phase enrichment technologies keep pace. Next-generation MALDI MSI platforms will need to overcome the challenges of direct sampling of complex chemical mixtures from ever decreasing pixel sizes while, at the same time, improving sensitivity to enable deeper molecular characterization of specific tissue foci and cell-types within the tissue microenvironment. Some obvious technologies are already beginning to emerge, and it is just a matter of time before they are utilized for MALDI-MSI. For example, extended mass range FTMS platforms are likely to play a large role in the future of spatial proteomics and ultra-high resolving power ion mobility platforms such as cyclic ion mobility (cIM)[188] and structures for lossless manipulations (SLIM)[189–191] will certainly enable improved specificity and peak capacity for tissue imaging. These advanced MALDI MSI capabilities will enable researchers to describe the interplay of biomolecular species more fully within cellular environments. Our goal has been to provide users and those interested in MALDI MSI with an understanding of how MALDI imaging instrumentation has developed over time, their operating principles, and their performance characteristics. Primed with this information we hope that users can more effectively assess the advantages and disadvantages of each platform and tailor MALDI MSI experiments to better fit their needs.

Abbreviations

MS	mass spectrometry
MALDI	matrix-assisted laser desorption/ionization
MSI	mass spectrometry imaging
IMS	imaging mass spectrometry
ESI	electrospray ionization
AP-MALDI	atmospheric pressure matrix-assisted laser desorption/ionization
IR-MALDESI	infrared matrix-assisted laser desorption electrospray ionization
DESI	desorption electrospray ionization
SIMS	secondary imaging mass spectrometry
LA-ICP	laser ablation inductively coupled plasma
LC	liquid chromatography
GC	gas chromatography

FTMS	Fourier transform mass spectrometry
FT-ICR	Fourier transform ion cyclotron resonance
Q	quadrupole mass filter
q	rf-only quadrupole
QqQ	triple quadrupole
TOF	time-of-flight
LIT	linear ion trap
h	hexapole
E	electric sector
B	magnetic sector
QMA	quadrupole mass analyzer
QMF	quadrupole mass filter
CASI	continuous accumulation of selected ions
IFS	isotopic fine structure
HRAM	high-resolution accurate mass
DC	direct current
RF	radio frequency
CID	collision-induced dissociation
OzID	ozone-induced dissociation
IR	infrared
UV	ultraviolet
PD	photodissociation
EFG	electric field gradient
CCS	collisional cross section
DTIMS	drift tube ion mobility spectrometry
TWIMS	traveling wave ion mobility spectrometry
TIMS	trapped ion mobility spectrometry
FAIMS	high-field asymmetric waveform ion mobility
SLIM	structures for losses manipulations
cIM	cyclic ion mobility
FFPE	formalin-fixed, paraffin-embedded

Acknowledgements

The authors would like to thank Katerina V. Djambazova (Vanderbilt University) for generating the human kidney MALDI TIMS dataset and Drs. Lukasz Migas and Raf Van de Plas (TU Delft) for their assistance in processing the TIMS data. Support for these studies was provided by NIH Common Fund and National Eye Institute (U54EY032442 awarded to J.M.S.) and by the NIH Common Fund and National Institute of Diabetes and Digestive and Kidney Diseases (U54DK120058 awarded to J.M.S.), and by the NIH National Institute of General Medical Sciences (R01GM138660 awarded to B.M.P.). The prototype MALDI timsTOF MS used to generate these data was developed as part of the National Science Foundation Major Research Instrument Program (CBET – 1828299 awarded to J.M.S.). Resolving power data were generated using a Bruker 15T solariX FT-ICR MS at Vanderbilt

University that was acquired through the NIH Shared Instrumentation Grant Program (1S10OD012359 awarded to Richard Caprioli).

References

1. R. Castaing and G. Slodzian, *J. Microsc.*, 1962, **1**, 395–410.
2. C. Colliex, *C. R. Phys.*, 2019, **20**, 746–755.
3. S. Chandra, J. F. Chabot, G. H. Morrison and A. C. Leopold, *Science*, 1982, **216**, 1221.
4. R. Day, S. Unger and R. Cooks, *Anal. Chem.*, 1980, **52**, 557–572.
5. S. Järås, *Appl. Catal.*, 1982, **2**, 207–218.
6. A. J. Dempster, *Phys. Rev.*, 1918, **11**, 316–325.
7. F. W. Aston, *Philos. Mag.*, 1919, **38**, 707–714.
8. J. S. Becker and H.-J. Dietze, *Fresenius J. Anal. Chem.*, 2000, **368**, 23–30.
9. Thermo Scientific ELEMENT XR: extended dynamic range high resolution ICP-MS, Application Note: 30064, https://assets.thermofisher.com/TFS-Assets/LSG/Application-Notes/AN-30064-ICP-MS-ELEMENT-XR-AN30064-EN.pdf.
10. S. C. Wilschefski and M. R. Baxter, *Clin. Biochem. Rev*, 2019, **40**, 115–133.
11. J. S. Becker, M. Zoriy, J. S. Becker, J. Dobrowolska and A. Matusch, *J. Anal. At. Spectrom.*, 2007, **22**, 736–744.
12. J. S. Becker, *et al.*, *Mass Spectrom. Rev.*, 2010, **29**, 156–175.
13. A. Matusch, *et al.*, *Anal. Chem.*, 2012, **84**, 3170–3178.
14. Neoma multicollector ICP-MS - illuminating the edge of discovery, Thermo Scientific Product Brochure, https://assets.thermofisher.com/TFS-Assets/CMD/brochures/br-30600-neoma-mc-icp-ms-br30600-en.pdf.
15. J. S. Becker, H. Sela, J. Dobrowolska, M. Zoriy and J. S. Becker, *Int. J. Mass Spectrom.*, 2008, **270**, 1–7.
16. M. Pacholski and N. Winograd, *Chem. Rev.*, 1999, **99**, 2977–3006.
17. R. E. March, *J. Mass Spectrom.*, 1997, **32**, 351–369.
18. H. G. Dehmelt, Radiofrequency spectroscopy of stored ions I: storage, in *Advances in Atomic and Molecular Physics*, ed. D. R. Bates and I. Estermann, Academic Press, New York, 1967; vol. 3, pp. 53–72.
19. N. Winograd, The development of secondary ion mass spectrometry (SIMS) for imaging, in *The Encyclopedia of Mass Spectrometry*, ed. M. L. Gross and R. M. Caprioli, Elsevier, Boston, 2016, pp. 103–112.
20. W. C. Wiley and I. H. McLaren, *Rev. Sci. Instrum.*, 1955, **26**, 1150–1157.
21. B. A. Mamyrin, V. I. Karataev, D. V. Shmikk and V. A. Zagulin, *Zh. Eksperim. i Teor. Fiz.*, 1973, **64**, 82–89.
22. S. M. Colby, T. B. King, J. P. Reilly and D. Lubman, *Rapid Commun. Mass Spectrom.*, 1994, **8**, 865–868.
23. R. S. Brown and J. J. Lennon, *Anal. Chem.*, 1995, **67**, 1998–2003.
24. M. Karas, *et al.*, *Anal. Chem.*, 1995, **67**, 675–679.
25. M. Vestal, P. Juhasz and S. Martin, *Rapid Commun. Mass Spectrom.*, 1995, **9**, 1044–1050.
26. M. Vestal, *J. Mass Spectrom.*, 2009, **44**, 303–317.

27. J. Schwieters, *et al.*, *J. Vac. Sci. Technol. A*, 1991, **9**, 2864–2871.
28. K. Tanaka, *et al.*, *Rapid Commun. Mass Spectrom.*, 1988, **2**, 151–153.
29. M. Karas and F. Hillenkamp, *Anal. Chem.*, 1988, **60**, 2299–2301.
30. R. M. Caprioli, T. B. Farmer and J. Gile, *Anal. Chem.*, 1997, **69**, 4751–4760.
31. L. Li, R. W. Garden and J. V. Sweedler, *Trends Biotechnol.*, 2000, **18**, 151–160.
32. B. Spengler and M. Hubert, *J. Am. Soc. Mass Spectrom.*, 2002, **13**, 735–748.
33. A. van Remoortere, *et al.*, *J. Am. Soc. Mass Spectrom.*, 2010, **21**, 1922–1929.
34. P. Chaurand, *et al.*, *Anal. Chem.*, 2004, **76**, 1145–1155.
35. J. M. Spraggins and R. M. Caprioli, *J. Am. Soc. Mass Spectrom.*, 2011, **22**, 1022–1031.
36. B. M. Prentice and R. M. Caprioli, *J. Postdoctoral Res.*, 2016, **4**, 3–13.
37. J. Spraggins, *et al.*, *Proteomics*, 2016, **16**, 1678–1689.
38. J. H. Jungmann, *et al.*, *J. Am. Soc. Mass Spectrom.*, 2012, **23**, 1679–1688.
39. M. E. Kurczy, T. R. Northen, S. A. Trauger and G. Siuzdak, *Methods Mol. Biol.*, 2015, **1203**, 141–149.
40. A. Zavalin, *et al.*, *J. Mass Spectrom.*, 2012, **47**, 1473–1481.
41. A. Römpp and B. Spengler, *Histochem. Cell Biol.*, 2013, **139**, 759–783.
42. A. Zavalin, J. Yang and R. Caprioli, *J. Am. Soc. Mass Spectrom.*, 2013, **24**, 1153–1156.
43. A. D. Feenstra, M. E. Dueñas and Y. J. Lee, *J. Am. Soc. Mass Spectrom.*, 2017, **28**, 434–442.
44. M. Kompauer, S. Heiles and B. Spengler, *Nat. Methods*, 2017, **14**, 90–96.
45. M. Niehaus, J. Soltwisch, M. E. Belov and K. Dreisewerd, *Nat. Methods*, 2019, **16**, 925–931.
46. E. C. Spivey, J. C. McMillen, D. J. Ryan, J. M. Spraggins and R. M. Caprioli, *J. Mass Spectrom.*, 2019, **54**, 366–370.
47. A. Zavalin, J. Yang, K. Hayden, M. Vestal and R. M. Caprioli, *Anal. Bioanal. Chem.*, 2015, 407.
48. M. L. Reyzer, Y. Hsieh, K. Ng, W. A. Korfmacher and R. M. Caprioli, *J. Mass Spectrom.*, 2003, **38**, 1081–1092.
49. J. Stauber, *et al.*, *J. Am. Soc. Mass Spectrom.*, 2010, **21**, 338–347.
50. P. J. Trim, *et al.*, *Anal. Bioanal. Chem.*, 2010, **397**, 3409–3419.
51. S. A. Mcluckey, *J. Am. Soc. Mass Spectrom.*, 1992, **3**, 599–614.
52. I. V. Chernushevich, A. V. Loboda and B. A. Thomson, *J. Mass Spectrom.*, 2001, **36**, 849–865.
53. K. Sannes-Lowery, R. H. Griffey, G. H. Kruppa, J. P. Speir and S. A. Hofstadler, *Rapid Commun. Mass Spectrom.*, 1998, **12**, 1957–1961.
54. R. Kaufmann, P. Chaurand, D. Kirsch and B. Spengler, *Rapid Commun. Mass Spectrom.*, 1996, **10**, 1199–1208.
55. D. Suckau, *et al.*, *Anal. Bioanal. Chem.*, 2003, **376**, 952–965.

56. B. M. Prentice, C. W. Chumbly and R. M. Caprioli, *J. Mass Spectrom.*, 2015, **50**, 703–710.
57. C. Cheng and M. L. Gross, *Mass Spectrom. Rev.*, 2000, **19**, 398–420.
58. M. Guilhaus, D. Selby and V. Mlynski, *Mass Spectrom. Rev.*, 2000, **19**, 65–107.
59. J. Coles and M. Guilhaus, *TrAC, Trends Anal. Chem.*, 1993, **12**, 203–213.
60. F. Barré, *et al.*, *Int. J. Mass Spectrom.*, 2019, **437**, 38–48.
61. E. D. Inutan, J. Wager-Miller, K. Mackie and S. Trimpin, *Anal. Chem.*, 2012, **84**, 9079–9084.
62. F. J. Troendle, C. D. Reddick and R. A. Yost, *J. Am. Soc. Mass Spectrom.*, 1999, **10**, 1315–1321.
63. T. J. Garrett and R. A. Yost, *Anal. Chem.*, 2006, **78**, 2465–2469.
64. T. J. Garrett, *et al.*, *Int. J. Mass Spectrom.*, 2007, **260**, 166–176.
65. K. E. Burnum, *et al.*, *Endocrinology*, 2008, **149**, 3274–3278.
66. J. M. Wiseman, S. M. Puolitaival, Z. Takáts, R. G. Cooks and R. M. Caprioli, *Angew. Chem., Int. Ed.*, 2005, **44**, 7094–7097.
67. J. M. Wiseman, D. R. Ifa, Q. Song and R. G. Cooks, *Angew. Chem., Int. Ed.*, 2006, **45**, 7188–7192.
68. J. C. Schwartz, M. W. Senko and J. E. P. Syka, *J. Am. Soc. Mass Spectrom.*, 2002, **13**, 659–669.
69. J. W. Hager, *Rapid Commun. Mass Spectrom.*, 2002, **16**, 512–526.
70. W. Paul and H. Steinwedel, Apparatus for separating charged particles of different specific charges, *U.S. Pat.*, 2939952, 1960.
71. K. E. Burnum, *et al.*, *J. Lipid Res.*, 2009, **50**, 2290–2298.
72. P. S. H. Wong and R. G. Cooks, *Ion Trap Mass Spectrometry*, ed. Purdue University, West Lafayette, IN.
73. B. M. Prentice and S. A. McLuckey, *Anal. Chem.*, 2012, **84**, 7562–7569.
74. U. P. Schlunegger, M. Stoeckli and R. M. Caprioli, *Rapid Commun. Mass Spectrom.*, 1999, **13**, 1792–1796.
75. J. N. Louris, *et al.*, *Int. J. Mass Spectrom. Ion Processes*, 1990, **96**, 117–137.
76. R. E. March, *et al.*, *Int. J. Mass Spectrom. Ion Processes*, 1989, **95**, 119–156.
77. R. E. March, *et al.*, *Int. J. Mass Spectrom. Ion Processes*, 1990, **99**, 109–124.
78. R. E. March, *et al.*, *Int. J. Mass Spectrom. Ion Processes*, 1991, **110**, 159–178.
79. R. L. Alfred, F. A. Londry and R. E. March, *Int. J. Mass Spectrom. Ion Processes*, 1993, **125**, 171–185.
80. J. M. Wells and S. A. McLuckey, *Methods Enzymol.*, 2005, **402**, 148–185.
81. J. E. Fulford, *et al.*, *J. Vac. Sci. Technol.*, 1980, **17**, 829–835.
82. D. R. Klein, *et al.*, *Anal. Chem.*, 2018, **90**, 10100–10104.
83. M.-E. N. Born and B. M. Prentice, *Int. J. Mass Spectrom.*, 2020, **452**, 116338.

84. T. Baba, J. L. Campbell, J. C. Y. Le Blanc, P. R. S. Baker and K. Ikeda, *J. Lipid Res.*, 2018, **59**, 910–919.
85. M. C. Thomas, *et al.*, *Anal. Chem.*, 2008, **80**, 303–311.
86. B. L. Poad, *et al.*, *J. Am. Soc. Mass Spectrom.*, 2010, **21**, 1989–1999.
87. M. R. L. Paine, *et al.*, *Angew. Chem., Int. Ed.*, 2018, **57**, 10530–10534.
88. B. M. Prentice and S. A. McLuckey, *Chem. Commun.*, 2013, **49**, 947–965.
89. D. J. Foreman and S. A. McLuckey, *Anal. Chem.*, 2020, **92**, 252–266.
90. J. T. Specker, S. L. V. Orden, M. E. Ridgeway and B. M. Prentice, *Anal. Chem.*, 2020, **92**, 13192–13201.
91. J. R. Bonney and B. M. Prentice, *Anal. Chem.*, 2021, **93**, 6311–6322.
92. G. Hopfgartner, E. Varesio and M. Stoeckli, *Rapid Commun. Mass Spectrom.*, 2009, **23**, 733–736.
93. R. A. Yost and C. G. Enke, *Anal. Chem.*, 1979, **51**, 1251A–1264A.
94. J. C. Schwartz, A. P. Wade, C. G. Enke and R. G. Cooks, *Anal. Chem.*, 1990, **62**, 1809–1818.
95. L. Lamont, *et al.*, *Anal. Chem.*, 2018, **90**, 13229–13235.
96. K. K. Kutz, J. J. Schmidt and L. Li, *Anal. Chem.*, 2004, **76**, 5630–5640.
97. I. M. Taban, *et al.*, *J. Am. Soc. Mass Spectrom.*, 2007, **18**, 145–151.
98. C. R. Anderton and L. J. Gamble, *Microsc. Today*, 2016, **24**, 24–31.
99. C. Koster, J. A. Castoro and C. L. Wilkins, *J. Am. Chem. Soc.*, 1992, **114**, 7572–7574.
100. M. L. Easterling, T. H. Mize and I. J. Amster, *Anal. Chem.*, 1999, **71**, 624–632.
101. J. T. Brenna, W. R. Creasy, W. McBain and C. Soria, *Rev. Sci. Instrum.*, 1988, **59**, 873–879.
102. A. G. Marshall, C. L. Hendrickson and G. S. Jackson, *Mass Spectrom. Rev.*, 1998, **17**, 1–35.
103. A. P. Bowman, *et al.*, *Anal. Chem.*, 2020, **92**, 3133–3142.
104. D. S. Cornett, S. L. Frappier and R. M. Caprioli, *Anal. Chem.*, 2008, **80**, 5648–5653.
105. M. Sud, *et al.*, *Nucleic Acids Res.*, 2007, **35**, D527–D532.
106. Y. I. Kostyukevich, G. N. Vladimirov and E. N. Nikolaev, *J. Am. Soc. Mass Spectrom.*, 2012, **23**, 2198–2207.
107. J. M. R. Spraggins, G. David, Moore, L. Jessica, Rose, L. Kristie, Hammer, D. Neal, Skaar, P. Eric, Caprioli and M. Richard, *J. Am. Soc. Mass Spectrom.*, 2015, **26**, 974–985.
108. M. Dilillo, *et al.*, *Sci. Rep.*, 2017, **7**, 603.
109. B. M. Prentice, D. J. Ryan, R. Van de Plas, R. M. Caprioli and J. M. Spraggins, *Anal. Chem.*, 2018, **90**, 5090–5099.
110. B. M. Prentice, *et al.*, *Anal. Chem.*, 2020, **92**, 13092–13100.
111. A. Makarov, *et al.*, *Anal. Chem.*, 2006, **78**, 2113–2120.
112. R. A. Zubarev and A. Makarov, *Anal. Chem.*, 2013, **85**, 5288–5296.
113. K. Strupat, *et al.*, *J. Am. Soc. Mass Spectrom.*, 2009, **20**, 1451–1463.
114. H. Ye, J. Wang, T. Greer, K. Strupat and L. Li, *ACS Chem. Neurosci.*, 2013, **4**, 1049–1056.

115. B. Li, D. R. Bhandari, C. Janfelt, A. Römpp and B. Spengler, *Plant J.*, 2014, **80**, 161–171.
116. B. Spengler, *Anal. Chem.*, 2015, **87**, 64–82.
117. A. Michalski, *et al.*, *Mol. Cell. Proteomics*, 2012, **11**, O111.013698.
118. Z. L. VanAernum, *et al.*, *Anal. Chem.*, 2019, **91**, 3611–3618.
119. D. C. Perdian and Y. J. Lee, *Anal. Chem.*, 2010, **82**, 9393–9400.
120. A. D. Feenstra, R. L. Hansen and Y. J. Lee, *Analyst*, 2015, **140**, 7293–7304.
121. J. C. May and J. A. McLean, *Anal. Chem.*, 2015, **87**, 1422–1436.
122. G. F. Verbeck, B. T. Ruotolo, H. A. Sawyer, K. J. Gillig and D. H. Russell, *J. Biomol. Tech.*, 2002, **13**, 56–61.
123. A. B. Kanu, P. Dwivedi, M. Tam, L. Matz and H. H. Hill, *J. Mass Spectrom.*, 2008, **43**, 1–22.
124. M. A. Ewing, M. S. Glover and D. E. Clemmer, *J. Chromatogr. A*, 2016, **1439**, 3–25.
125. J. N. Dodds, J. C. May and J. A. McLean, *Anal. Chem.*, 2017, **89**, 12176–12184.
126. F. Barré, *et al.*, *Int. J. Mass Spectrom.*, 2019, **437**, 38–48.
127. J. M. Spraggins, *et al.*, *Anal. Chem.*, 2019, **91**, 14552–14560.
128. K. V. Djambazova, *et al.*, *Anal. Chem.*, 2020, **92**, 13290–13297.
129. L. Ahonen, *et al.*, *J. Chromatogr. A*, 2013, **1310**, 133–137.
130. F. Lanucara, S. W. Holman, C. J. Gray and C. E. Eyers, *Nat. Chem.*, 2014, **6**, 281–294.
131. V. Gabelica, *et al.*, *Mass Spectrom. Rev.*, 2019, **38**, 291–320.
132. J. Hofmann, H. S. Hahm, P. H. Seeberger and K. Pagel, *Nature*, 2015, **526**, 241–244.
133. T. Wyttenbach, P. R. Kemper and M. T. Bowers, *Int. J. Mass Spectrom.*, 2001, **212**, 13–23.
134. M. F. Bush, *et al.*, *Anal. Chem.*, 2010, **82**, 9557–9565.
135. S. M. Stow, *et al.*, *Anal. Chem.*, 2017, **89**, 9048–9055.
136. W. S. Barnes, D. W. Martin and E. W. McDaniel, *Phys. Rev. Lett.*, 1961, **6**, 110–111.
137. G. E. Keller, D. W. Martin and E. W. McDaniel, *Phys. Rev.*, 1965, **140**, A1535–A1546.
138. E. W. McDaniel, D. W. Martin and W. S. Barnes, *Rev. Sci. Instrum.*, 1962, **33**, 2–7.
139. E. A. Mason and H. W. Schamp, *Ann. Phys.*, 1958, **4**, 233–270.
140. J. T. Moseley, R. M. Snuggs, D. W. Martin and E. W. McDaniel, *Am. Phys. Soc.*, 1969, **178**, 240–248.
141. Y. M. Ibrahim, *et al.*, *Int. J. Mass Spectrom.*, 2015, **377**, 655–662.
142. J. M. Dilger, M. S. Glover and D. E. Clemmer, *J. Am. Soc. Mass Spectrom*, 2017, **28**, 1293–1303.
143. J. M. Dilger, S. J. Valentine, M. S. Glover and D. E. Clemmer, *J. Am. Soc. Mass Spectrom.*, 2013, **24**, 768–779.
144. J. A. Picache, *et al.*, *Chem. Sci.*, 2019, **10**, 983–993.
145. X. Zheng, *et al.*, *Chem. Sci.*, 2017, **8**, 7724–7736.

146. J. A. McLean, W. B. Ridenour and R. M. Caprioli, *J. Mass Spectrom.*, 2007, **42**, 1099–1105.
147. P. Dwivedi, *et al.*, *J. Mass Spectrom.*, 2010, **45**, 1383–1393.
148. S. N. Jackson, *et al.*, *J. Mass Spectrom.*, 2007, **42**, 1093–1098.
149. S. N. Jackson, *et al.*, *J. Am. Soc. Mass Spectrom.*, 2008, **19**, 1655–1662.
150. A. S. Woods, *et al.*, *J. Proteome Res.*, 2006, **5**, 1484–1487.
151. M. Ekelof, *et al.*, *J. Am. Soc. Mass Spectrom.*, 2020, **31**, 642–650.
152. D. J. Harvey, *et al.*, *J. Am. Soc. Mass Spectrom.*, 2012, **23**, 1955–1966.
153. W. B. Ridenour, M. Kliman, J. A. McLean and R. M. Caprioli, *Anal. Chem.*, 2010, **82**, 1881–1889.
154. K. Skraskova, *et al.*, *Methods*, 2016, **104**, 69–78.
155. L. Xu, *et al.*, *J. Am. Soc. Mass Spectrom.*, 2015, **26**, 924–933.
156. K. Giles, *et al.*, *Rapid Commun. Mass Spectrom.*, 2004, **18**, 2401–2414.
157. S. D. Pringle, *et al.*, *Int. J. Mass Spectrom.*, 2007, **261**, 1–12.
158. J. Stauber, *et al.*, *J. Am. Soc. Mass Spectrom.*, 2010, **21**, 338–347.
159. M. Groessl, S. Graf and R. Knochenmuss, *Analyst*, 2015, **140**, 6904–6911.
160. A. L. Rister, T. L. Martin and E. D. Dodds, *J. Mass Spectrom.*, 2019, **54**, 429–436.
161. M. Hernandez-Mesa, B. Le Bizec, F. Monteau, A. M. Garcia-Campana and G. Dervilly-Pinel, *Anal. Chem.*, 2018, **90**, 4616–4625.
162. M. C. Djidja, *et al.*, *J. Proteome Res.*, 2009, **8**, 4876–4884.
163. M. C. Djidja, *et al.*, *Proteomics*, 2009, **9**, 2750–2763.
164. P. J. Trim, *et al.*, *Anal. Chem.*, 2008, **80**, 8628–8634.
165. M. Bot, *et al.*, *J. Lipid Res.*, 2013, **54**, 1265–1274.
166. F. Fernandez-Lima, D. A. Kaplan, J. Suetering and M. A. Park, *Int. J. Ion Mobil. Spectrom.*, 2011, **14**, 93–98.
167. F. A. Fernandez-Lima, D. A. Kaplan and M. A. Park, *Rev. Sci. Instrum.*, 2011, **82**, 126106.
168. K. Michelmann, J. A. Silveira, M. E. Ridgeway and M. A. Park, *J. Am. Soc. Mass Spectrom*, 2015, **26**, 14–24.
169. T. Bien, E. A. Hambleton, K. Dreisewerd and J. Soltwisch, *Anal. Bioanal. Chem.*, 2021, **413**, 2767–2777.
170. T. Fu, *et al.*, *J. Mass Spectrom.*, 2020, **55**, e4531.
171. E. K. Neumann, *et al.*, *Anal. Chem.*, 2020, **92**, 13084–13091.
172. J. Soltwisch, *et al.*, *Anal. Chem.*, 2020, **92**, 8697–8703.
173. T. Bien, E. A. Hambleton, K. Dreisewerd and J. Soltwisch, *Anal. Bioanal. Chem.*, 2021, **413**, 2767–2777.
174. K. Y. Garza, *et al.*, *Anal. Chem.*, 2018, **90**, 7785–7789.
175. C. L. Feider, N. Elizondo and L. S. Eberlin, *Anal. Chem.*, 2016, **88**, 11533–11541.
176. R. L. Griffiths, *et al.*, *Anal. Chem.*, 2020, **92**, 2885–2890.
177. T. Porta, E. Varesio and G. Hopfgartner, *Anal. Chem.*, 2013, **85**, 11771–11779.
178. R. Guevremont, *J. Chromatogr. A*, 2004, **1058**, 3–19.
179. B. M. Kolakowski and Z. Mester, *Analyst*, 2007, **132**, 842–864.

180. R. W. Purves and R. Guevremont, *Anal. Chem.*, 1999, **71**, 2346–2357.
181. E. V. Krylov, *Int. J. Mass Spectrom.*, 2003, **225**, 39–51.
182. A. A. Shvartsburg, Y. M. Ibrahim and R. D. Smith, *J. Am. Soc. Mass Spectrom.*, 2014, **25**, 480–489.
183. A. A. Shvartsburg, *et al.*, *Anal. Chem.*, 2009, **81**, 8048–8053.
184. D. R. Bhandari, M. Schott, A. Rompp, A. Vilcinskas and B. Spengler, *Anal. Bioanal. Chem.*, 2015, **407**, 2189–2201.
185. B. Li, D. R. Bhandari, C. Janfelt, A. Rompp and B. Spengler, *Plant J.*, 2014, **80**, 161–171.
186. V. V. Laiko, S. C. Moyer and R. J. Cotter, *Anal. Chem.*, 2000, **72**, 5239–5243.
187. V. V. Laiko, M. A. Baldwin and A. L. Burlingame, *Anal. Chem.*, 2000, **72**, 652–657.
188. K. Giles, *et al.*, *Anal. Chem.*, 2019, **91**, 8564–8573.
189. S. V. Garimella, *et al.*, *J. Am. Soc. Mass Spectrom.*, 2014, **25**, 1890–1896.
190. A. V. Tolmachev, *et al.*, *Anal. Chem.*, 2014, **86**, 9162–9168.
191. I. K. Webb, *et al.*, *Anal. Chem.*, 2014, **86**, 9169–9176.

CHAPTER 5

Sample Preparation of Biological Tissues for MALDI-MSI

BERTA CILLERO-PASTOR* AND EVA CUYPERS

Division of Imaging Mass Spectrometry, Maastricht MultiModal Molecular Imaging (M4i) Institute, 6229 ER Maastricht, Netherlands
*Email: b.cilleropastor@maastrichtuniversity.nl

5.1 Introduction

Mass spectrometry imaging (MSI) is a powerful technique to visualize the spatial distribution of molecular compounds in tissue sections. Sample preparation is a crucial step in the MSI workflow since it can introduce artifacts and inappropriate handling can cause degradation or delocalization of the analytes, possibly leading to incorrect interpretations and conclusions. Therefore, careful treatment of the tissue material immediately after surgical removal and/or animal sacrifice is essential. This includes temperature and storage, but also tissue sectioning and the use/choice of embedding and fixation media. In this chapter, we will focus more specifically on sample preparation for matrix-assisted laser desorption/ionization (MALDI), as it is by far the most used ionization source used for MSI. The different steps and complexity of sample preparation vary and depend on the type of tissue to be analyzed. For fresh frozen material, in some applications, no sample preparation apart from sectioning and matrix application is needed, while for formalin-fixed paraffin-embedded (FFPE) tissue, a multi-step process is required, which usually

New Developments in Mass Spectrometry No. 12
MALDI Mass Spectrometry Imaging: From Fundamentals to Spatial Omics
Edited by Tiffany Siegel

Published by the Royal Society of Chemistry, www.rsc.org

involves at least deparaffinization. Antigen retrieval, enzymatic digestion or additional chemical treatments may also be necessary depending on the molecular class to be analyzed. The last, but essential, step in the MALDI-MSI sample preparation is matrix application, which must be done regardless of any previous sample preparation.[1] In this chapter, the different sample preparation steps from organ collection to matrix application will be described and discussed.

5.2 Sample Collection

Sample collection is the first step in any MALDI-MSI experiment. Molecular preservation is of utmost importance. Therefore, when retrieving an organ from a living organism the process must be as fast as possible immediately after the animal is sacrificed, the surgical resection or the biopsy excision to prevent possible molecular degradation.

The preferred method is snap freezing of the sample, also referred to as flash-freezing. Organs are typically frozen by placing the specimen in an aluminum foil boat on top of an acetone/isopentane bath cooled with dry ice or liquid nitrogen. Fragile specimens such as kidney are prone to fracture. Snap freezing in dry ice-chilled isopropanol followed by a wash in dry ice-chilled isopentane to remove excess isopropanol has been shown to be an alternative method to maintain the molecular and tissue integrity.[2] The specimens can be then stored at −80 °C prior to cryosectioning.[3] Variations of this protocol consist of the incorporation of a perfusion step with phosphate-buffered saline (PBS) to avoid any possible blood contamination, followed by snap-freezing in liquid nitrogen.[4] Other organs such as the heart can be perfused with oxygenated Krebs–Henseleit buffer to avoid blood contamination and spectral interference.[5] However, this washing step can modify the molecular signature as it introduces salts and therefore might also create ion suppression or introduce adducts. Therefore, perfusion or washing the tissue with PBS or other buffers is not recommended, unless the contamination of blood or body fluid produces an important ion suppression of the molecules of interest.[6]

In some cases, and to preserve the morphology and structural integrity of the organ, such as in the case of lungs, inflation is often used.[7] Formalin or cryoprotectants such as optimal cutting temperature medium (OCT), sucrose or dimethyl sulfoxide among others are commonly used by pathologists to maintain the tissue architecture, however it is advised to test different inflation media depending on the MALDI-MSI application of interest. Moreover, extensive freezing times (~60 seconds up to 90 seconds maximum for larger specimens) should be avoided as this will lead to artifacts during cryosectioning such as tissue fragmentation.

To avoid molecular degradation, samples should always be transported on dry ice (before and after sectioning). It is preferred to store the samples as retrieved from the animal at −80 °C and section them right before proceeding to MSI. The tissue can be stored at −80 °C for up to several years.

Other methods to preserve post-mortem molecular degradation include a heat stabilization step at 95 °C (*e.g.* Stabilizor T1, Denator, Mölndal, Sweden) prior to freezing. This is particularly interesting for metabolite and neuropeptide analyses that are prone to quick degradation.[8]

Alternatively to snap freezing, samples can be fixed in formalin. However, formalin fixation is usually followed by embedding in paraffin. FFPE offers an interesting alternative to overcome molecular degradation issues and it is routinely used in both clinical diagnostics and research and can be easily stored at room temperature.[9]

PAX fixation offers an incredibly powerful alternative to formalin fixation since cross-linking of proteins is avoided in this manner while also maintaining a comparably good preservation of the histomorphology of the tissue.

5.3 Embedding and Storage

It is important to keep in mind that preparing sections for any MSI-based measurement is similar to the procedures used in the preparation of frozen sections for immunostaining or dye staining, *i.e.*, section thicknesses ranging from 5 to 20 μm, depending on sample density properties. However, extra attention should be paid to the embedding media since they may interfere or hamper the quality of the obtained mass spectra. Therefore, cryo-sectioning from snap-frozen blocks of tissue is preferred. Ideally, snap freezing is carried out without any embedding material. However, small samples or tissues with very different densities (bone-cartilage material) need an embedding media to facilitate the sectioning. Embedding allows these samples to retain a good shape and facilitates the cutting process. The use of OCT should be avoided at any time to cover the whole tissue since the polymer molecules cause severe ionization suppression with possibly decreasing molecular sensitivity.[10] Alternative embedding compounds are described in the literature: pre-cooled semi-liquid gel of 2% sodium carboxymethylcellulose (CMC) for rodent whole-body sections;[11] 10% solution (W/V) of gelatin for the embedding of hair samples;[12] a combination of 20% gelatin and 7.5% CMC for embedding of bone samples[13] and poly-L-lysine-based tissue embedding for dry plants.[14] However, gelatin-embedded material should not be exposed above its melting point (35 °C) since melting will introduce delocalization and ion suppression. Paraffin, often combined with formalin fixation and mentioned in the previous section as FFPE, is often used by pathology laboratories and can be used for glycan, metabolite, protein, and peptide analysis.[15] However, the detection of certain lipids and some metabolites is usually compromised. An exception to the prescribed methods is the longitudinal sectioning of hair. It was described by Flinders *et al.* that longitudinal sections of hair can be made without embedding medium and at room temperature.[16]

For each MSI application and tissue of interest, the proper embedding medium needs to be carefully optimized. Crumbling, tissue cracks should be avoided while tissue morphology, structural integrity, and shape should be maintained. A recent study described an MSI, polymer-based embedding medium, poly[*N*-(2-hydroxypropyl)methacrylamide] (pHPMA), capable of supporting delicate tissues without the typical ion suppression of OCT. However, although ideal, for some applications this medium synthesized by the authors is not commercially available.[17] For example, nine media have been evaluated by Nelson *et al.* as substrates for embedding whole zebra fish compatible with MALDI-MSI analysis.[17]

Dannhorn *et al.* demonstrated that a combination of different materials such as hydroxypropyl methylcellulose (HPMC) and polyvinylpyrrolidone (PVP) media preserves the histology of a tissue, avoids any melting due to the low-temperature properties, and is also affordable compared to HPMA.[2]

Finally, embedding media serve as a mold where large sample cohorts can be processed at the same reducing time and batch effects.

In summary, costs, ease of preparation, and commercial availability are also key aspects to consider.

5.4 Sectioning

During sample preparation workflows, it is important to be aware of the related batch variations. Indeed, batch effects, a phenomenon already known from other omics fields, are a potential systematic source of technical variation. It significantly impedes the reliable interpretation and comparison of results, in particular from large-scale studies performed in translational clinical research.[18–20] In MALDI-MSI, batch effects can appear at different levels: sectioning, matrix application or instrument robustness among other steps. For example, it has been shown that differences in section thickness can be related to different spatial intensities.[21] Moreover, the time between sectioning until further processing can cause noticeable variation related to molecular degradation.[22] Furthermore, fixation as well as storage type and time might cause systematic variation to individual or sets of samples.[22] These factors should all be considered as section variations. Slide variations might appear since sections at different positions on the slide can also be exposed differently to batch effects of the measurement (*e.g.*, due to increasing matrix sublimation in the source or fading detector sensitivity). Taking this into account, it is advised to randomize sections on slides to minimize this effect. The next two sections will address specific recommendations related to frozen and FFPE material, respectively.

5.4.1 Frozen Material

Cryostats are typically used for sectioning of frozen material. The thickness of the section and the cutting temperature depend on the application (type of tissue and embedding), but generally samples are cut between 7 and 20 μm.[23]

Once the section is made, the transfer of the tissue section to the glass slide is the next challenge. The preferred method is transferring the tissue by gently picking it up with an artist's brush and positioning it on a cold glass slide (kept in a cryostat chamber at −20 °C). The tissue is then thaw-mounted onto the slide by placing a finger at the backside of the slide, thus heating up the tissue section (until the white trace created by humidity disappears). Care should be taken during the transfer to avoid deforming the tissue.[24] Thicker sections will be more difficult to adhere compared to thin sections. The use of conductive tapes can also facilitate the transfer of the tissue to the slide and improve the ionization process in some cases.[25]

In order to preserve the molecular integrity, a drying step using a nitrogen/air flow just after cryosectioning followed by a vacuum packing step can be incorporated. The drying step is used to avoid bleaching of the tissue in the cryostat. In addition, UV treatment prior to MSI is a good alternative to decontaminate tissues from clinically relevant pathogens such as herpesviridae, papovaviridae human immunodeficiency virus, or SARS-CoV-2, which may be present in human tissue samples while preserving the biodistributions of analytes within the tissue.[26,27]

5.4.2 FFPE Material

FFPE material is sectioned using a microtome and generally cut between 2–7 μm.

The FFPE sections are usually placed in a water bath at about 40–45 °C, followed by mounting the sections on glass slides and storing at room temperature afterwards. Poly-lysine coated slides can be used to prevent tissue loss caused by extensive washing steps as well as detaching when introduced in the mass spectrometer.[28] Tissue microarrays (TMAs) are the typical case to illustrate this since antigen retrieval and extensive washing steps can cause the loss of the relatively small tissue cores.

5.4.3 Important Considerations for Tissue Sectioning

- Select the correct embedding media depending on your tissue and analytes of interest.[14]
- Fresh-frozen tissue is less stable and more fragile than FFPE tissue and thus results in more variation and more artifacts during sectioning than FFPE tissue sectioning.
- Conventional TOF instruments require the use of special conductive indium tin oxide (ITO) slides to ensure sufficient ion extraction.
- To compare and overlay the imaged tissue with the stained histological section and to minimize batch effects, the tissue shape should be retained, and deformities, scratches, and rolled-up edges should be avoided.
- Sample preparation steps, such as temperature variations and time from harvesting to measurement, should be as standardized as possible for all samples within a study to limit variability introduced due to technical deviations. Moreover, samples should be randomized on slides to avoid slide batch effects.[29]

5.5 Washing

Washing steps are critical to remove interference molecules and improve the extraction and ionization of the analytes of interest. The length of the washing steps differs between applications as well as by the method used to apply the solvent. The most common methods include pipetting or the use of a beaker. A wet tissue in contact with the samples can be used for very delicate sections to enable local washing.[30]

Below are some of the more often used washing methods classified per molecular class:

5.5.1 Metabolites

Frozen tissue sections: Although not commonly used it has been shown that a simple wash with an organic solvent such as chloroform for 15 seconds at −20 °C improves the detection of small molecules (below *m/z* 500). This step significantly removes most glycerophospholipids. Consequentially, ATP-related energy metabolites, amino acids and derivatives, glucose derivatives, and glycolysis pathway metabolites can be visualized with enhanced ion intensity and good reproducibility.[31]

FFPE samples: To detect metabolites from FFPE, a xylene washing step to remove paraffin needs to be performed prior to matrix application.[32]

5.5.2 Lipids

Highly abundant endogenous potassium and sodium ions might create artifacts and favor the ionization of some lipids; therefore, desalting samples prior to matrix application is often used for lipid analysis. Isotonic solutions such as ammonium acetate can improve the detection of protonated lipids, as sodiated and potassiated species are removed by the washing step. An advantage is the maintenance of the tissue osmolarity with this washing step and therefore the preservation of tissue morphology. Ammonium formate solution has a similar effect in the removal of alkali metal ions and it is useful to resolve the isobaric species caused by these endogenous ions.[33] Lipids are typically measured in frozen samples since the removal of paraffin washes off many lipids. Specifics of sample preparation for the detection of lipids can be found in Chapter 8 of this book.

5.5.3 Proteins, Glycans and Peptides

Frozen tissue sections: Typical washing steps for peptide analysis include organic solvents for the removal of lipids, salts, and other interferences in the low *m/z* range. Isopropanol, glacial acetic acid, toluene, or chloroform washings have been successfully used in frozen sections to remove salts and lipids from the tissue.[34–37] Most commonly used protocols include several washing steps such as 30 seconds in 70% EtOH, 30 seconds in 100% EtOH, 30 seconds in

Table 5.1 Commonly used protocols as washing steps for FFPE samples for protein and peptide analysis.

70% EtOH	100% EtOH	Carnoy's fluid	100% EtOH	H_2O with 0.2% TFA
30 s	30 s	2 min	30 s	30 s

deionized water, and again 30 seconds in 70% EtOH and 30 seconds in 100% EtOH. Other protocols include Carnoy's fluid (EtOH–chloroform–acetic acid, 6/3/1, v/v/v) and several dip washing steps in water 0.2% TFA.[38]

Other acetonitrile-based solvents, *e.g.* 50 : 50 (v/v) acetonitrile/water, have been shown to improve protein intensities. This protocol has been used for lens sections consisting of repeated washing steps of 3×30 seconds, and a drying step with a nitrogen stream after the third wash.[39]

FFPE tissue sections: the removal of paraffin is crucial, and samples must be heated up at 60 degrees for about 1 hour followed by washing in toluene or xylene (typically 2×2 min). Afterward, tissue sections are typically washed using different variations of ethanol and Carnoy's fluid followed by a rehydration step in a graded ethanol series.[40,41] These steps are followed by enzymatic digestion (for digested proteins or glycans) or matrix application for intact protein analysis. An example of this protocol can be found in Table 5.1.

Each of the steps is a potential source of analyte delocalization or degradation. Delocalization is a phenomenon where molecules migrate/diffuse across and away from the tissue. Hydrophilic molecules are in general more prone to delocalization. Consequently, extensive sample prep protocols with several washing steps suffer more from these artifacts.[42]

5.6 Enzymatic Digestion

Enzymatic digestion is typically performed for protein or glycan analysis. Most of the studies of MALDI-MSI for protein detection need a bottom-up approach since large species are difficult to ionize. In addition, identification of intact proteins is challenging (top down) due to the tissue complexity and the lack of methods incorporating protein isolation and therefore protein yield. Bottom-up approaches consist of the direct visualization of peptides resulting from enzymatic digestion, followed by tandem mass spectrometry (MS/MS) fragmentation. Introduction of chromatographic separation coupled to MALDI-MSI increases the number of identifications. In the case of FFPE samples, antigen retrieval and protein digestion help to overcome some of the limitations caused by formalin fixation such as protein cross-linking: after paraffin removal, an antigen retrieval step is introduced for epitope retrieval at a temperature of 95 °C for approx. 25 minutes. Buffers for this step vary from 10 mM citraconic anhydride buffer, to 10 mM sodium citrate and 10 mM tris.[43] After this step, the enzyme is typically applied to automatic robotic systems where the number of layers, flow rate, or even humidity and temperature conditions can be controlled.[44]

5.6.1 Proteins

Similar to classical proteomics, the most employed enzyme is trypsin (cleaves C-terminally after lysine and arginine). The addition of MS-compatible detergents like octyl-β-glucoside and surfactants such as Triton or SDS can enhance the number of peptides and proteins detected.[45,46] Concentration of trypsin and time of incubation are factors that need to be optimized for every tissue application. In general, trypsin is applied between 30 min to 18h at a starting concentration of 3.2 ng mm^{-2}. Other enzymes such as chymotrypsin, elastase, pepsin, or collagenase can also be used.[47–49] However, these enzymes are only useful for tissues rich in collagens or elastin, such as skin. Recent advances in proteomics coupled to laser microdissection with workflows such as nanopots have shown an enormous increase in protein identification numbers. The methods analyze predefined tissue voxels, generating quantitative cell-type-specific images for >2000 proteins with 100 μm spatial resolution.[50]

Due to the extensive sample treatment protocols for protein digestion, the inclusion of quality control standards and evaluation of batch effects is recommended, more importantly for clinical applications.[51,52]

5.6.2 Glycans

Amidase PNGase F is an enzyme commonly used in the analyses of glycoproteins. Tissues are typically incubated overnight (18 h) at 37 °C in a saturated humid environment.[53] Enzyme combinations for glycan and protein digestion allow the detection of glycans and peptides on the same slide. Critical steps for a successful enzymatic digestion include a clean and saturated humid environment for instance in a saturated atmosphere of MeOH/H_2O (1/1 v/v). More details on sample preparation specific to glycans can be found in Chapter 10 of this book.

5.7 Derivatization

An often-raised challenge in MALDI-MSI is sensitivity. Ion suppression, low concentrations, and methodological (*i.e.* matrix interferences) as well as sample-related (*i.e.* endogenous) interferences can challenge the visualization of certain molecules. A viable solution can be chemical derivatization, a method of chemically tagging the target analyte with a larger, pre-charged moiety that increases ionization efficiency. Derivatization is extensively used in gas and liquid chromatography, where the derivatization reaction mostly takes place in solution during the sample preparation stage. For MALDI-MSI, on-tissue derivatization is required, making it even more challenging. In recent years, there has been an increasing number of research papers describing promising results in targeting key biological molecules and taking advantage of derivatization strategy: *N*-glycans,[54] amino acids, fatty acids,[55] neurotransmitters and vitamin metabolites.[56] Other successful applications in drug localization were shown by Chacon,[57] Flinders[58] and Barré *et al.*[59] Although derivatization can significantly increase the overall sensitivity, there are also possible drawbacks that should be considered: 1) the derivatization reagents should be selected based on the target functional group meaning that the target analyte must be selected before sample preparation; 2) there is a significantly higher risk of analyte delocalization; 3) spatial resolution might be jeopardized, especially when derivatization and matrix are deposited in different stages; 4) when increased temperatures are involved, tissue integrity and stability may be compromised. Moreover, parameters such as co-crystallization with matrix, deposition method, solvent composition, and reagent concentration should be taken into account.[60]

5.7.1 Reagent Selection

As mentioned earlier, the choice of the derivatization reagent is based on the functional group to be targeted. Table 5.2 gives a brief overview of the most frequently targeted functional groups and possibly used derivatization reagents.

5.7.2 Deposition Method of Derivatization Reagents

Multiple matrix deposition methods for derivatization reagents are described in Table 5.3; they all have specific advantages and disadvantages. Depending on the required spatial resolution, extraction efficiency, and experiment time, the correct application method should be chosen.[60]

5.8 Matrix Selection and Application Methods

MALDI is a soft ionization technique, and therefore ions of larger mass ranges can be measured and identification can be achieved with tandem MS (MS^2). In brief, MALDI is carried out by co-crystallizing the matrix and analyte. Ionization is triggered by irradiating the laser to the co-crystal. Given the importance of the matrix in the ionization process, it is crucial to select the appropriate matrix for successful imaging measurement. The choice of matrix depends upon the type of analyte involved (see Table 5.4). In general, relative proton affinity of matrix and analytes of interest are key players. However, for most matrices, their suitability in playing the role of matrix for certain compounds is determined empirically. Below you find an overview of the currently most common matrices. However, it should be mentioned, this is a non-limitative list and new matrices and alternatives to organic matrices are developed on a regular basis.

Table 5.2 Derivatization reagents and targets.

Functional group	Target	Derivatization reagent	Reference
Aliphatic alcohols, phenols, and thiols	3-Methoxysalicylamine	1,1′-Thiocarbonyldiimidazole (TCDI)	57
	Cannabinoids	2-Fluoro-1-methylpyridinium *p*-toluenesulfonate (FMPTS)	61
Carboxylic acids	Free fatty acids and phospholipids	2-Picolamine in the presence of 2,2-dipyridyl disulfide (DPDS) and triphenylphosphine (TPP)	
	Free fatty acids and phospholipids	*N,N*-Dimethylpiperazine iodide (DMPI)	62
	Tricarboxylic acid cycle intermediates, fatty acids, and bile acids and their metabolites	*N,N,N*-Trimethyl-2-(piperazin-1-yl)ethan-1-aminium iodide (TMPA)	63
Amines	Peptides	(*N*-Succinimidyloxycarbonylmethyl)tris(2,4,6-trimethoxyphenyl)phosphonium bromide (TMPP)	64
	Peptides, amino acids and neurotransmitters	4-Hydroxy-3-methoxycinnamaldehyde (CA)	65,66
	Amine acids	*p-N,N,N* Trimethylammonioanilyl *N*-hydroxysuccinimidyl carbamate iodide (TAHS)	67–69
	Neurotransmitters	2,4-Diphenyl-pyranylium (DPP) and 2,4-diphenylpyranylium tetrafluoroborate (DPP-TFB), 2-fluoro-1-methyl pyridinium (FMP-10)	66,70–74
	GABA	mTRAQ® reagents	75
Ketones/Aldehydes	Androgens	Girard's T reagent (GT)	
	Triamcinolone acetonide	Girard's T reagent (GT)	59
	Sterols and oxysterols	Girard's P reagent (GP)	76
	Corticosteroids	2,4-Dinitrophenylhydrazine (DNPH)	58

Table 5.3 Derivatization reaction deposition techniques.

Technique	Advantages	Disadvantages
Acoustic multi-spotter	Uniform, fast, good reproducibility	Droplet application
Electrospray deposition	Homogenous	Limited time for analyte–matrix interaction
Pneumatic sprayer	Homogenous	Droplet size not constant
ImagePrep™	Controlled condition, automated, homogenous	Droplet size not constant, expensive
Dry-coating	Cheap, high-purity matrix	Poor analyte–matrix extraction
Sublimation	Homogenous, reproducible, fast	Poor analyte–matrix extraction
Desktop inkjet printer	Uniform droplets (multichannel)	Slow, solvent compatibility, clogging

Table 5.4 Matrix selection based on analytes of interest.

Analyte class	Matrix		Polarity	Reference
Drugs	CHCA	α-Cyano-4-hydroxycinnamic acid	+	77–80
	SA	Sinapinic acid	+	79,81
	9-AA	9-Aminoacridine	−	77,82
	DAN	Diaminonaphthalene	+/−	82
	DHB	2,5-Dihydroxybenzoic acid	+	77–79
(Glycero)Phospholipids	DHB	2,5-Dihydroxybenzoic acid	+	83,84
	MBT	2-Mercaptobenzothiazole	+/−	85
	NOR	Norharmane	+/−	
Cardiolipins	DHA	2,6-Dihydroxyacetophenone	−	86
Endotoxins	NOR	Norharmane	+/−	87
Sugars/phosphorylated metabolites	9-AA	9-Aminoacridine	+	88
Nucleotides	9-AA	9-Aminoacridine	−	89
Proteins/peptides	CHCA	α-Cyano-4-hydroxycinnamic acid	+	13
	DHAP	2,5-dihydroxyacetophenone	−	90
	SA	Sinapinic acid	+	91
Proteins/lipids	2,5-cDHA	(*E*)-4-(2,5-dihydroxyphenyl)-but-3-en-2-one	+/−	92
	COOH-NHMe	2-(Methylamino)benzoic acid	+/−	93
Glucose, nucleotide, Antioxidant	NEDC	*N*-(1-Naphthyl)-ethylenediamine dihydrochloride	−	94–97

5.8.1 Matrix Application and Recrystallization

The matrix application procedure is one of the most critical steps of the sample preparation procedure. Indeed, the quality of the mass spectra and the subsequent imaging results, including spatial resolution, will depend on

Table 5.5 Matrix application methods with pros (+) and cons (−).

Application method	Extraction efficiency	Spatial resolution	Reproducibility
TLC sprayer (manual)	+	−	−
Air-brush (manual)	+	+/−	−
Automatic sprayer	+	+/−	+
Sublimation	−	++	+

the co-crystallization of the matrix–analyte complex formed on the tissue surface. Several matrix application methods are described for MALDI-MSI.[87,91,98,99] Manual spraying with nebulizer sprayer (TLC sprayer or air-brush), automatic deposition of small droplets with robotic devices, and sublimation are the most commonly used ones. The extraction efficiency and image spatial resolution that can be acquired will depend on the matrix application method,[99] as depicted Table 5.5.

As indicated in Table 5.5, spatial resolution that can be acquired with sublimation is by far the best, meaning that the crystal size when applying the matrix by sublimation is small enough to reach subcellular spatial resolution. However, the disadvantage of very small matrix crystals is generally that the method has less extraction efficiency. The main factor in the extraction efficiency is the wetness of the method. Sprayer applications in general are wetter since solvents are used to dissolve the matrix leading to better extraction efficiencies. The disadvantage of more wet methods is the possibility of delocalization of the compounds of interest. Finding the balance between extraction efficiency and maintaining spatial localization is always a crucial factor in imaging mass spectrometry. Matrix recrystallization can be a possible option for combining the advantages of sublimation, *i.e.* small crystal size, and improving the extraction efficiency. This was shown for example using 9-AA based sublimation combined with recrystallization for the detection of endogenous metabolites.[100] However, it is not always successful and will depend on the matrix, the recrystallization conditions (water- or organic-based) and the analytes of interest.[101]

5.9 General Considerations, Conclusion and Future Directions

This chapter highlights how sample preparation for MSI, and MALDI-based imaging in particular, is not a cookbook recipe. It is highly dependent on many factors such as sampling and sectioning methods, analytes of interest, sample type, embedding media, spatial resolution, or matrix deposition. Moreover, the first and critical step, *i.e.* sample preparation, will determine the success rate of the MSI experiments. If the sample preparation is not adequate, the target ion signals might not be detected, or artifacts will affect the resulting images. Therefore, it is advised to include quality controls if possible. Standards can be spotted next to or on the sample in order to check matrix extraction efficiency and/or derivatization performance. Controlling or checking delocalization of compounds however is much more challenging. Depending on the target analyte, H&E or immunohistochemistry staining overlays may in some cases give an idea about possible analyte delocalization. Ideally, full sample preparation should be tested and optimized for each application and sample type before moving on to full studies including large cohorts of (clinical) samples. Since the MSI field is more and more moving towards single cell analysis, high spatial resolution imaging is required. This also means an extra challenge in sample preparation for MALDI-MSI since the crystal size of the MALDI-matrix is currently one of the main limiting factors. Although spraying devices have made great progress within the last few years, the crystal size of the matrix is in most cases not small enough for single cell analysis. Therefore, sublimation is currently the only matrix deposition method that reaches the required crystal size. As discussed previously, this comes with certain disadvantages, including less sensitivity. Moreover, sublimation is not possible for all matrices, in particular salts are very difficult to sublime.

In general, it can be concluded that sample preparation methods and the development thereof need to keep up with the fast evolution of MALDI-MSI devices otherwise sample preparation will be the limiting factor in pushing the sensitivity and spatial resolution boundaries.

References

1. K. Ščupáková, B. Balluff, C. Tressler, T. Adelaja, R. M. A. Heeren, K. Glunde and G. Ertaylan, *Clin. Chem. Lab. Med.*, 2020, **58**, 914–929.
2. A. Dannhorn, E. Kazanc, S. Ling, C. Nikula, E. Karali, M. P. Serra, J.-L. Vorng, P. Inglese, G. Maglennon, G. Hamm, J. Swales, N. Strittmatter, S. T. Barry, O. J. Sansom, G. Poulogiannis, J. Bunch, R. J. A. Goodwin and Z. Takats, *Anal. Chem.*, 2020, **92**, 11080–11088.
3. M. R. L. Paine, J. Liu, D. Huang, S. R. Ellis, D. Trede, J. H. Kobarg, R. M. A. Heeren, F. M. Fernandez and T. J. MacDonald, *Sci. Rep.*, 2019, **9**, 2205.
4. T. C. Rohner, D. Staab and M. Stoeckli, *Mech. Ageing Dev.*, 2005, **126**, 177–185.
5. A. Aljakna Khan, N. Bararpour, M. Gorka, T. Joye, S. Morel, C. A. Montessuit, S. Grabherr, T. Fracasso, M. Augsburger, B. R. Kwak, A. Thomas and S. Sabatasso, *Sci. Rep.*, 2021, **11**, 5135.
6. K. Veerasammy, Y. X. Chen, S. Sauma, M. Pruvost, D. K. Dansu, T. Choetso, T. Zhong, D. Marechal, P. Casaccia, R. Abzalimov and Y. He, *J. Vis. Exp.*, 2020 Dec 22, 166.
7. S. Baijnath, A. Shobo, L. A. Bester, S. D. Singh, G. Kruger, T. Naicker and T. Govender, *J. Mol. Histol.*, 2016, **47**, 213–219.
8. C. Esteve, E. A. Tolner, R. Shyti, A. M. J. M. van den Maagdenberg and L. A. McDonnell, *Metabolomics*, 2016, **12**, 30.

9. J. Hermann, H. Noels, W. Theelen, M. Lellig, S. Orth-Alampour, P. Boor, V. Jankowski and J. Jankowski, *Anal. Bioanal. Chem.*, 2020, **412**, 1263–1275.
10. L. A. McDonnell and R. M. Heeren, *Mass Spectrom. Rev.*, 2007, **26**, 606–643.
11. P. J. Trim, *Methods Mol. Biol.*, 2017, **1618**, 175–189.
12. B. Flinders, E. Cuypers, T. Porta, E. Varesio, G. Hopfgartner and R. M. A. Heeren, *Methods Mol. Biol.*, 2017, **1618**, 137–147.
13. M. Vandenbosch, S. P. Nauta, A. Svirkova, M. Poeze, R. M. A. Heeren, T. P. Siegel, E. Cuypers and M. Marchetti-Deschmann, *Anal. Bioanal. Chem.*, 2020, **413**(10), 2683–2694.
14. Y. He, W. Guo, K. Luo, Q. Sun, Z. Lin and Z. Cai, *J. Chromatogr. A*, 2019, **1608**, 460389.
15. M. B. O'Rourke, M. P. Padula, C. Smith, P. Youssef, S. Cordwell, P. Witting, G. Sutherland and B. Crossett, *J. Vis. Exp.*, 2018, **16**(131), 56778.
16. B. Flinders, E. Cuypers, H. Zeijlemaker, J. Tytgat and R. M. Heeren, *Drug Test. Anal.*, 2015, **7**, 859–865.
17. K. A. Nelson, G. J. Daniels, J. W. Fournie and M. J. Hemmer, *J. Biomol. Tech.*, 2013, **24**, 119–127.
18. B. Balluff, C. Hopf, T. Porta Siegel, H. I. Grabsch and R. M. A. Heeren, *J. Am. Soc. Mass Spectrom.*, 2021, **32**(3), 628–635.
19. W. W. B. Goh, W. Wang and L. Wong, *Trends Biotechnol.*, 2017, **35**, 498–507.
20. J. Kuligowski, D. Perez-Guaita, I. Lliso, J. Escobar, Z. Leon, L. Gombau, R. Solberg, O. D. Saugstad, M. Vento and G. Quintas, *Talanta*, 2014, **130**, 442–448.
21. R. Longuespee, K. Kriegsmann, M. Cremer, C. Zgorzelski, R. Casadonte, D. Kazdal, J. Kriegsmann, W. Weichert, K. Schwamborn, M. Fresnais, P. Schirmacher and M. Kriegsmann, *Proteomics: Clin. Appl.*, 2019, **13**, e1800074.
22. R. J. A. Goodwin, *J. Proteomics*, 2012, **75**, 4893–4911.
23. R. Longuespée, K. Kriegsmann, M. Cremer, C. Zgorzelski, R. Casadonte, D. Kazdal, J. Kriegsmann, W. Weichert, K. Schwamborn, M. Fresnais, P. Schirmacher and M. Kriegsmann, *Proteomics: Clin. Appl.*, 2019, **13**, e1800074.
24. S. A. Schwartz, M. L. Reyzer and R. M. Caprioli, *J. Mass Spectrom.*, 2003, **38**, 699–708.
25. A. Svirkova, A. Turyanskaya, L. Perneczky, C. Streli and M. Marchetti-Deschmann, *Analyst*, 2018, **143**, 2587–2595.
26. J. G. Swales, A. Dexter, G. Hamm, A. Nilsson, N. Strittmatter, F. Michopoulos, C. Hardy, P. Morentin-Gutierrez, M. Mellor, P. E. Andren, M. R. Clench, J. Bunch, S. E. Critchlow and R. J. A. Goodwin, *Anal. Chem.*, 2018, **90**, 6051–6058.
27. A. Dannhorn, S. Ling, S. Powell, E. McCall, G. Maglennon, G. N. Jones, A. J. Pierce, N. Strittmatter, G. Hamm, S. T. Barry, J. Bunch, R. J. A. Goodwin and Z. Takats, *Anal. Chem.*, 2021, **93**, 2767–2775.

28. A. Zavalin, E. M. Todd, P. D. Rawhouser, J. Yang, J. L. Norris and R. M. Caprioli, *J. Mass Spectrom.*, 2012, **47**, i.
29. B. Balluff, C. Hopf, T. Porta Siegel, H. I. Grabsch and R. M. A. Heeren, *J. Am. Soc. Mass Spectrom.*, 2021, **32**, 628–635.
30. E. R. van Hove, D. F. Smith, L. Fornai, K. Glunde and R. M. Heeren, *J. Am. Soc. Mass Spectrom.*, 2011, **22**, 1885–1890.
31. H. Yang, W. Ji, M. Guan, S. Li, Y. Zhang, Z. Zhao and L. Mao, *Metabolomics*, 2018, **14**, 50.
32. B. Balluff and L. A. McDonnell, *Methods Mol. Biol.*, 2018, **1730**, 345–357.
33. H. Y. Wang, C. B. Liu and H. W. Wu, *J. Lipid Res.*, 2011, **52**, 840–849.
34. Y. Schober, T. Schramm, B. Spengler and A. Römpp, *Rapid Commun. Mass Spectrom.*, 2011, **25**, 2475–2483.
35. K. K. Kutz, J. J. Schmidt and L. Li, *Anal. Chem.*, 2004, **76**, 5630–5640.
36. J. Stauber, L. MacAleese, J. Franck, E. Claude, M. Snel, B. K. Kaletas, I. M. Wiel, M. Wisztorski, I. Fournier and R. M. Heeren, *J. Am. Soc. Mass Spectrom.*, 2010, **21**, 338–347.
37. M. Dilillo, R. Ait-Belkacem, C. Esteve, D. Pellegrini, S. Nicolardi, M. Costa, E. Vannini, E. L. D. Graaf, M. Caleo and L. A. McDonnell, *Sci. Rep.*, 2017, **7**, 603.
38. B. Enthaler, T. Bussmann, J. K. Pruns, C. Rapp, M. Fischer and J.-P. Vietzke, *Rapid Commun. Mass Spectrom.*, 2013, **27**, 878–884.
39. D. M. Anderson, M. G. Nye-Wood, K. L. Rose, P. J. Donaldson, A. C. Grey and K. L. Schey, *J. Mass Spectrom.*, 2020, **55**, e4473.
40. C. T. McDowell, Z. Klamer, J. Hall, C. A. West, L. Wisniewski, T. W. Powers, P. M. Angel, A. S. Mehta, D. N. Lewin, B. B. Haab and R. R. Drake, *Mol. Cell. Proteomics*, 2021, **20**, 100012.
41. J. Oetjen, D. Lachmund, A. Palmer, T. Alexandrov, M. Becker, T. Boskamp and P. Maass, *Anal. Bioanal. Chem.*, 2016, **408**, 6729–6740.
42. K. Ščupáková, B. Balluff, C. Tressler, T. Adelaja, R. M. A. Heeren, K. Glunde and G. Ertaylan, *Clin. Chem. Lab. Med.*, 2020, **58**, 914–929.
43. P. M. Angel, S. Comte-Walters, L. E. Ball, K. Talbot, A. Mehta, K. G. M. Brockbank and R. R. Drake, *J. Proteome Res.*, 2018, **17**, 635–646.
44. A. Blutke, N. Sun, Z. Xu, A. Buck, L. Harrison, S. C. Schriever, P. T. Pfluger, D. Wiles, T. Kunzke, K. Huber, J. Schlegel, M. Aichler, A. Feuchtinger, K. Matiasek, S. M. Hauck and A. Walch, *Sci. Rep.*, 2020, **10**, 14461.
45. B. Cillero-Pastor, G. B. Eijkel, A. Kiss, F. J. Blanco and R. M. A. Heeren, *Arthritis Rheum.*, 2013, **65**, 710–720.
46. B. Banstola, E. T. Grodner, F. Cao, F. Donnarumma and K. K. Murray, *Anal. Chim. Acta*, 2017, **963**, 76–82.
47. B. Enthaler, M. Trusch, M. Fischer, C. Rapp, J. K. Pruns and J. P. Vietzke, *Anal. Bioanal. Chem.*, 2013, **405**, 1159–1170.
48. P. M. Angel, S. Comte-Walters, L. E. Ball, K. Talbot, A. Mehta, K. G. M. Brockbank and R. R. Drake, *J. Proteome Res.*, 2018, **17**, 635–646.
49. K. R. Tucker, L. A. Serebryannyy, T. A. Zimmerman, S. S. Rubakhin and J. V. Sweedler, *Chem. Sci.*, 2011, **2**, 785–795.

50. P. D. Piehowski, Y. Zhu, L. M. Bramer, K. G. Stratton, R. Zhao, D. J. Orton, R. J. Moore, J. Yuan, H. D. Mitchell, Y. Gao, B.-J. M. Webb-Robertson, S. K. Dey, R. T. Kelly and K. E. Burnum-Johnson, *Nat. Commun.*, 2020, **11**, 8.
51. A. M. Judd, D. B. Gutierrez, J. L. Moore, N. H. Patterson, J. Yang, C. E. Romer, J. L. Norris and R. M. Caprioli, *J. Mass Spectrom.*, 2019, **54**, 716–727.
52. B. Balluff, C. Hopf, T. Porta Siegel, H. I. Grabsch and R. M. A. Heeren, *J. Am. Soc. Mass Spectrom.*, 2021, **32**, 628–635.
53. B. Heijs, A. Potthoff, J. Soltwisch and K. Dreisewerd, *Anal. Chem.*, 2020, **92**, 13904–13911.
54. S. Holst, B. Heijs, N. de Haan, R. J. van Zeijl, I. H. Briaire-de Bruijn, G. W. van Pelt, A. S. Mehta, P. M. Angel, W. E. Mesker, R. A. Tollenaar, R. R. Drake, J. V. Bovee, L. A. McDonnell and M. Wuhrer, *Anal. Chem.*, 2016, **88**, 5904–5913.
55. Q. Wu, T. J. Comi, B. Li, S. S. Rubakhin and J. V. Sweedler, *Anal. Chem.*, 2016, **88**, 5988–5995.
56. K. W. Smith, B. Flinders, P. D. Thompson, F. L. Cruickshank, C. L. Mackay, R. M. A. Heeren and D. F. Cobice, *ACS Omega*, 2020, **5**, 13430–13437.
57. A. Chacon, I. Zagol-Ikapitte, V. Amarnath, M. L. Reyzer, J. A. Oates, R. M. Caprioli and O. Boutaud, *J. Mass Spectrom.*, 2011, **46**, 840–846.
58. B. Flinders, J. Morrell, P. S. Marshall, L. E. Ranshaw and M. R. Clench, *Anal. Bioanal. Chem.*, 2015, **407**, 2085–2094.
59. F. P. Barre, B. Flinders, J. P. Garcia, I. Jansen, L. R. Huizing, T. Porta, L. B. Creemers, R. M. Heeren and B. Cillero-Pastor, *Anal. Chem.*, 2016, **88**, 12051–12059.
60. C. Harkin, K. W. Smith, F. L. Cruickshank, C. Logan Mackay, B. Flinders, R. M. A. Heeren, T. Moore, S. Brockbank and D. F. Cobice, *Mass Spectrom. Rev.*, 2021, Epub ahead of print.
61. E. Beasley, S. Francese and T. Bassindale, *Anal. Chem.*, 2016, **88**, 10328–10334.
62. S. S. Wang, Y. J. Wang, J. Zhang, T. Q. Sun and Y. L. Guo, *Anal. Chem.*, 2019, **91**, 4070–4076.
63. C. Sun, W. Liu, Y. Geng and X. Wang, *Anal. Chem.*, 2020, **92**, 12126–12131.
64. J. Franck, K. Arafah, A. Barnes, M. Wisztorski, M. Salzet and I. Fournier, *Anal. Chem.*, 2009, **81**, 8193–8202.
65. M. E. Duenas, E. A. Larson and Y. J. Lee, *Front. Plant Sci.*, 2019, **10**, 860.
66. C. Esteve, E. A. Tolner, R. Shyti, A. M. van den Maagdenberg and L. A. McDonnell, *Metabolomics*, 2016, **12**, 30.
67. S. Toue, Y. Sugiura, A. Kubo, M. Ohmura, S. Karakawa, T. Mizukoshi, J. Yoneda, H. Miyano, Y. Noguchi, T. Kobayashi, Y. Kabe and M. Suematsu, *Proteomics*, 2014, **14**, 810–819.
68. M. Arts, Z. Soons, S. R. Ellis, K. A. Pierzchalski, B. Balluff, G. B. Eijkel, L. J. Dubois, N. G. Lieuwes, S. M. Agten, T. M. Hackeng,

L. J. C. van Loon, R. M. A. Heeren and S. W. M. Olde Damink, *Angew. Chem., Int. Ed. Engl.*, 2017, **56**, 7146–7150.
69. E. Takeo, Y. Sugiura, T. Uemura, K. Nishimoto, M. Yasuda, E. Sugiyama, S. Ohtsuki, T. Higashi, T. Nishikawa, M. Suematsu, E. Fukusaki and S. Shimma, *Anal. Chem.*, 2019, **91**, 8918–8925.
70. Q. Cao, Y. Wang, B. Chen, F. Ma, L. Hao, G. Li, C. Ouyang and L. Li, *ACS Chem. Neurosci.*, 2019, **10**, 1222–1229.
71. A. S. Davison, N. Strittmatter, H. Sutherland, A. T. Hughes, J. Hughes, G. Bou-Gharios, A. M. Milan, R. J. A. Goodwin, L. R. Ranganath and J. A. Gallagher, *Metabolomics*, 2019, **15**, 68.
72. E. Sugiyama, M. M. Guerrini, K. Honda, Y. Hattori, M. Abe, P. Kallback, P. E. Andren, K. F. Tanaka, M. Setou, S. Fagarasan, M. Suematsu and Y. Sugiura, *iScience*, 2019, **20**, 359–372.
73. M. Shariatgorji, A. Nilsson, R. J. Goodwin, P. Kallback, N. Schintu, X. Zhang, A. R. Crossman, E. Bezard, P. Svenningsson and P. E. Andren, *Neuron*, 2014, **84**, 697–707.
74. M. Shariatgorji, A. Nilsson, E. Fridjonsdottir, T. Vallianatou, P. Kallback, L. Katan, J. Savmarker, I. Mantas, X. Zhang, E. Bezard, P. Svenningsson, L. R. Odell and P. E. Andren, *Nat. Methods*, 2019, **16**, 1021–1028.
75. T. Ito and M. Hiramoto, *Anal. Bioanal. Chem.*, 2019, **411**, 6847–6856.
76. E. Yutuc, R. Angelini, M. Baumert, N. Mast, I. Pikuleva, J. Newton, M. R. Clench, D. O. F. Skibinski, O. W. Howell, Y. Wang and W. J. Griffiths, *Proc. Natl. Acad. Sci. U. S. A.*, 2020, **117**, 5749–5760.
77. M. Nishidate, M. Hayashi, H. Aikawa, K. Tanaka, N. Nakada, S. I. Miura, S. Ryu, T. Higashi, Y. Ikarashi, Y. Fujiwara and A. Hamada, *Drug Metab. Pharmacokinet.*, 2019, **34**, 209–216.
78. I. Rzagalinski, B. Kovačević, N. Hainz, C. Meier, T. Tschernig and D. A. Volmer, *Anal. Chem.*, 2018, **90**, 12592–12600.
79. T. Greer, R. Sturm and L. Li, *J. Proteomics*, 2011, **74**, 2617–2631.
80. B. Flinders, E. Beasley, R. M. Verlaan, E. Cuypers, S. Francese, T. Bassindale, M. R. Clench and R. M. A. Heeren, *J. Am. Soc. Mass Spectrom.*, 2017, **28**, 2462–2468.
81. S. Ntshangase, S. Mdanda, S. D. Singh, T. Naicker, H. G. Kruger, S. Baijnath and T. Govender, *ACS Omega*, 2019, **4**, 21169–21177.
82. J. Dong, W. Ning, D. J. Mans and J. D. Mans, *Anal. Methods*, 2018, **10**, 572–578.
83. J. Schiller, R. Suss, B. Fuchs, M. Muller, M. Petkovic, O. Zschornig and H. Waschipky, *Eur. Biophys. J.*, 2007, **36**, 517–527.
84. Y. Sugiura, Y. Konishi, N. Zaima, S. Kajihara, H. Nakanishi, R. Taguchi and M. Setou, *J. Lipid Res.*, 2009, **50**, 1776–1788.
85. E. Astigarraga, G. Barreda-Gomez, L. Lombardero, O. Fresnedo, F. Castano, M. T. Giralt, B. Ochoa, R. Rodriguez-Puertas and J. A. Fernandez, *Anal. Chem.*, 2008, **80**, 9105–9114.
86. H.-Y. J. Wang, S. N. Jackson and A. S. Woods, *J. Am. Soc. Mass Spectrom.*, 2007, **18**, 567–577.

87. A. J. Scott, B. Flinders, J. Cappell, T. Liang, R. S. Pelc, B. Tran, D. P. Kilgour, R. M. Heeren, D. R. Goodlett and R. K. Ernst, *Pathog. Dis.*, 2016, **74**.
88. R. L. Vermillion-Salsbury and D. M. Hercules, *Rapid Commun. Mass Spectrom.*, 2002, **16**, 1575–1581.
89. S. R. Fagerer, S. Nielsen, A. Ibáñez and R. Zenobi, *Eur. J. Mass Spectrom.*, 2013, **19**, 39–47.
90. T. Wenzel, K. Sparbier, T. Mieruch and M. Kostrzewa, *Rapid Commun. Mass Spectrom.*, 2006, **20**, 785–789.
91. M. L. Salum, S. L. Giudicessi, T. Schmidt De León, S. A. Camperi and R. Erra-Balsells, *J. Mass Spectrom.*, 2017, **52**, 182–186.
92. J. Yang, J. L. Norris and R. Caprioli, *J. Mass Spectrom.*, 2018, **53**, 1005–1012.
93. P. Huang, C. Y. Huang, T. C. Lin, L. E. Lin, E. Yang, C. Lee, C. C. Hsu and P. T. Chou, *Anal. Chem.*, 2020, **92**, 7139–7145.
94. F. P. Y. Barre, B. S. R. Claes, F. Dewez, C. Peutz-Kootstra, H. F. Munch-Petersen, K. Gronbaek, A. H. Lund, R. M. A. Heeren, C. Come and B. Cillero-Pastor, *Anal. Chem.*, 2018, **90**, 14198–14206.
95. R. Chen, W. Xu, C. Xiong, X. Zhou, S. Xiong, Z. Nie, L. Mao, Y. Chen and H. C. Chang, *Anal. Chem.*, 2012, **84**, 465–469.
96. J. Hou, S. Chen, N. Zhang, H. Liu, J. Wang, Q. He, J. Wang, S. Xiong and Z. Nie, *Analyst*, 2014, **139**, 3469–3475.
97. J. Wang, S. Qiu, S. Chen, C. Xiong, H. Liu, J. Wang, N. Zhang, J. Hou, Q. He and Z. Nie, *Anal. Chem.*, 2015, **87**, 422–430.
98. F. J. Tsai, S. Y. Chen, Y. C. Liu, H. Y. Liao and C. J. Chen, *PLoS One*, 2017, **12**, e0181423.
99. E. Gemperline, S. Rawson and L. Li, *Anal. Chem.*, 2014, **86**, 10030–10035.
100. T. Morikawa-Ichinose, Y. Fujimura, F. Murayama, Y. Yamazaki, T. Yamamoto, H. Wariishi and D. Miura, *J. Am. Soc. Mass Spectrom.*, 2019, **30**, 1512–1520.
101. J. Yang and R. M. Caprioli, *Anal. Chem.*, 2011, **83**, 5728–5734.

CHAPTER 6

Quantitative MSI: A Review of Biomarker and Drug Quantifications Using Mass Spectrometry Imaging

ANAHIT CAMPBELL AND JONATHAN STAUBER*

ImaBiotech Corp, 50 Milk Street, Boston, MA 02109, USA
*Email: jstauber@imabiotech.com

6.1 Introduction

QMSI, or quantitative mass spectrometry imaging, has undergone developments in the last decade that have widened its application, particularly to support biomarker validation and drug biodistribution studies.[1] We have entered a realm of imaging merged with quantification. Industries such as pharmaceutical, clinical, chemical, biological, and even industrial manufacturing have undoubtedly benefited from the ability to spatially quantify a label-free compound. Although a number of additional methods (*e.g.*, LC–MS and LC–MS/MS, ELISA, and gene expression assays) provide sensitive quantitative data, QMSI has seen some serious improvements over the last decade. There are several types of MSI ionization techniques including SIMS (secondary ion mass spectrometry), DESI (desorption electrospray ionization), MALDI (matrix-assisted laser desorption/ionization), among others, all offering different capabilities. This chapter focuses on quantitative MALDI-MSI in particular, as it is the most commonly used

New Developments in Mass Spectrometry No. 12
MALDI Mass Spectrometry Imaging: From Fundamentals to Spatial Omics
Edited by Tiffany Siegel

Published by the Royal Society of Chemistry, www.rsc.org

technique for spatially quantifying endogenous or exogenous molecules within tissue samples. Ellis *et al.* emphasize the next development to solidify QMSI is method validation,[2] the process of which is described in detail in this chapter. This chapter will not only evaluate the step-by-step approach of method selection and method development (MD), method validation (MV) and finally quantitative sample analysis, but also establish future guidelines for the QMSI community. Although not an exhaustive list of all contributing parameters, this chapter will assess the role and impact of the following parameters that contribute to a robust QMSI method, including sample storage, sample handling, sample preparation, normalization methods and experimental execution, matrix deposition, instrumental parameters, and data processing strategies.

6.2 History and Progress

QMSI is a powerful method by which to determine not only the spatial distribution of the drug, but also the relative or absolute concentrations of this distribution. Absolute MSI quantification was first introduced in 2011 using MALDI instruments[5–7] and since then much progress has been forged. This included some of the first work to suggest solutions to some of the greatest challenges in QMSI including those of reproducibility and ion suppression, and in addition it led to applications of internal standard (IS) normalization.[6,7] One such advancement is the development of the mimetic tissue method by Castellino *et al.* in 2013.[8] Despite QMSI still being a non-GLP (Good Laboratory Practice) technique, recently QMSI results have been added as supplementary information to support GLP studies in toxicology programs,[9] drug distribution studies in preclinical or clinical stages,[3,5,10–16] and biomarker monitoring.[6] Other categories of compounds that have been studied with QMSI and are discussed in detail further in this chapter include lipids, proteomics, polymers, and alcohols.

Although QMSI is not the only technique to perform quantitative imaging, it is arguably an exceedingly advantageous technique to identify and measure biomarkers and drug compounds within tissue as compared to other methods of quantitative distribution analysis in tissues. For example, autoradiography, despite its high sensitivity and accuracy, involves the use of radioactive material which requires difficult handling and waste removal. Additionally, the lack of specificity between drug compound and possible metabolites makes it deficient as compared to QMSI.[3] On the other hand, LC–MS/MS is a more well-established quantitative method. However, the preservation of spatial information is at least partially lost and the microdissection that would be required to preserve this spatial information is difficult and time consuming. A detailed evaluation of the two methods is presented later in this text in Section 6.5 'Current Challenges'.

In that regard, QMSI is advantageous as it is a label-free technique that presents fully spatial information of both drugs and metabolites.[4] Despite the many favorable features of QMSI, the method still holds limitations, most significantly, ion-suppression effects. This is due to the lack of method of separation in MSI overall, therefore resulting in the ionization and analysis of thousands of molecules at the same time. The source of the ion-suppression can occur from a variety of sources through the tissue histology and QMSI sample preparation process. This creates the challenges of detecting species that are in low abundance with QMSI. Ion-suppression effects also impact analytical factors such as reproducibility, accuracy, precision, *etc.* This chapter will discuss these challenges in further detail and present solutions and examples of such solutions, as a means to illustrate the progress made not only in MSI, but also in QMSI especially in the last decade.

6.3 Method Workflow

The complexity of QMSI lies in the attainability and preservation of analytical parameters despite the multivariate optimization of the many steps included in sample preparation, instrument method optimization, and overall workflow. A workflow for qMALDI or even other QMSI involves a feedback loop of three critical processes including method development (MD), method validation (MV), and tissue quantification (including QC). A diagram of the workflow can be found in Figure 6.1. MD allows setting up

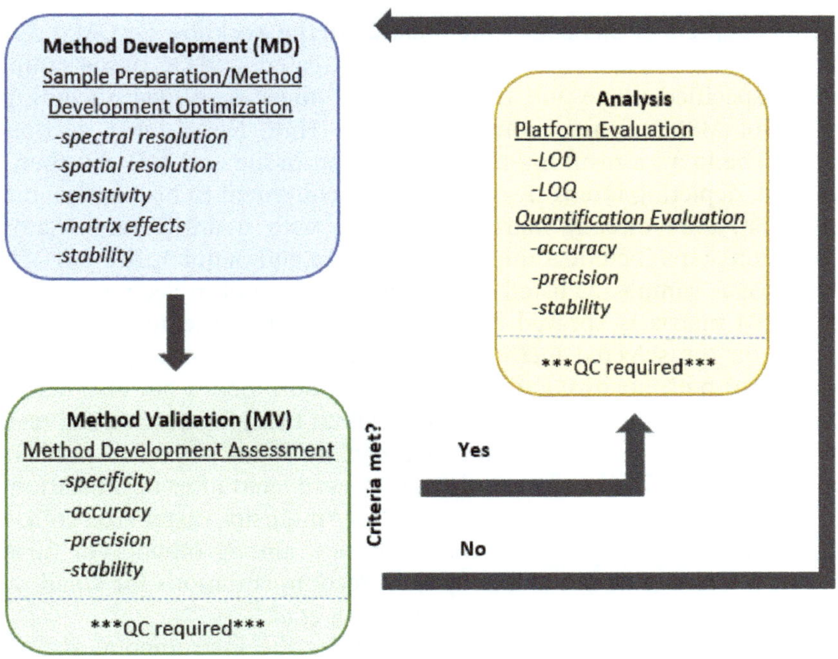

Figure 6.1 A diagram of the workflow for QMSI is presented introducing the method development, method validation, and quantification of samples with validation steps as a feedback loop.

the optimal condition for every step with minimal delocalization and optimized sensitivity. MV characterizes the performance of the analysis. Seven parameters for evaluating analysis have been compiled from the literature:[14,16–20]

- stability (of the compound) (*the ability of the method to control for degradation and fragmentation of the compound of interest*);
- spectral resolution (*the ability of the method to selectively distinguish the peak of interest of the compound of interest*);
- spatial resolution (*the ability of the method to distinguish the compound in the relevant tissue section*);
- accuracy (*the proximity of the concentration to the true value*);
- precision (*the proximity of the concentration replicates to one another*);
- specificity (*the ability of the method to distinguish between the compound of interest and the matrix*);
- repeatability (*the statistical proximity of intra-experimental runs to another*);
- reproducibility (*the statistical proximity of inter-experimental runs to one another*);
- evaluation of the IS (purity and fragmentation).

6.3.1 Method Development (MD)

As other chapters of this book have illustrated, the workflow for QMSI tissue analysis includes cryosectioning the control, treated, and QC tissue samples along a specified plane and then mounting them on slides, including a section for MSI and a section for staining (H&E, Nissel, *etc.*), sectioned serially. The images resulting from the stained tissue and MSI can then be overlayed, depicting intensities spatially and congruent to histological areas of interest, thus directly marrying imaging with quantification. Sample preparation steps include a dilution series that allows quantification of the treated tissue sample, as listed and described in detail in Section 6.3. Next, the MALDI matrix is sprayed onto the sample slide (including the IS depending on the selected method). This chapter lists not only a variety of matrices for both positive and negative ionization modes but also derivatization agents and their application in the QMSI field, for which a full review and listing of their selection and use can be found in other published studies.[21–24] And finally, the acquisition is conducted after optimization of several instrumental parameters such as laser intensity, raster size, collision voltage, octupole settings, and transfer optics, among others. For further reading, a series of method development recommendations for small molecules analyzed by MSI are included in Wang *et al.*[25]

The purpose of the MD stage is to optimize the instrumental and analytical parameters of the overall sample preparation and acquisition steps described previously. Specifically, for tissue samples this includes any necessary tissue pre-treatment (*e.g.* washing to get rid of ion suppression effects) and optimization of the matrix agent and solvents used.[22] Although not the focus of this chapter, the influence of the matrix and spraying conditions are significant to QMSI in that they aid in the extraction of the compound by bringing it toward the surface of the tissue. Therefore, it is an important step to optimize, as mentioned before, to avoid delocalization and degradation of the compound of interest.[21]

To accurately determine the concentration of a particular compound in treated tissue, a calibration curve must be established. In the case of an exogenous compound quantification (drug or xenobiotic), the compound itself is used to build the calibration curve on control tissue. If the analyte of interest is an endogenous marker (present in the tissue of interest) to the tissue, a labeled compound (typically the deuterated analog of the compound) is required to build the calibration curve for quantification of the endogenous molecule.

To build a linear standard curve, in most cases, the following formula can be applied:

$$y = mx + b \tag{6.1}$$

where y is the measured intensity (ratio) of the compound and IS, m is a multiple, x is the concentration, and b is the y-intercept.[25] Higher concentrations skew variance, therefore to correct for this, curve weighing, the formula for which is $1/x$ would be applied where x is the concentration. Further reading on calibration curve validation can be found in a work by Stauffer.[26] Calibration curves should include 6–8 standard points (and $<75\%$ rejection of points, if the user respects the LC–MS guidelines) in addition to a blank point. After the rejection of outliers,[27] the calibration curve should result in an R^2 of 0.9500–0.9999.[24] This step also includes confirmation of the detection of the molecule of interest. This is achieved through the determination of the limit of detection (LOD) the formula for which is:

$$\text{LOD} = \frac{3s}{m} \tag{6.2}$$

where s is the standard deviation of the signal intensity and m is the slope of the calibration curve, and the LOQ (limit of quantification), the formula for which is:

$$\text{LOQ} = \frac{10s}{m} \tag{6.3}$$

It should be noted that the multiple "3" for the LOD and "10" for the LOQ are accepted to achieve a confidence interval of 95%.[28] We have observed that the LOD and LOQ can vary depending on the MALDI instrument. For example, because the signal to noise ratio (S/N) of a MALDI–FT-ICR instrument is high, without baseline correction, the resulting LOD (calculated by the formula above) will be falsely high. To further expand on the topic, it is important to consider the definition of the LOD. As an analytical metric, the LOD is defined as the lowest concentration of a compound that can be reliably detected within a given analytical method from a standard solution. "Reliably," in this case, means the capacity to discriminate the detection of the lowest concentration from a blank tissue in at least 95.5% of cases.[29] In the case of MSI, since the measurement involves replicate positions, the LOD is dependent on the number of raster position spots on the ROI (region of interest). For this reason, we observe that for most MALDI or high-resolution QMSI methods, 300–400 positions or pixels per region of interest are necessary to achieve a sufficient average spectrum. This is due to matrix deposition and the subsequent crystal size that impact the variability of the detection, which can be remedied by inclusion of an IS compound. Other minor acquisition parameters such as the raster window and grid implement can allow for more ion transfer without sacrificing for spatial resolution, as in the case with a wider diameter laser raster. By extension, the LLOQ (lower limit of quantification) and the ULOQ (upper limit of quantification) can be used to define the LDR (linear dynamic range)[28] which have been found for QMSI to generally fall within a 2.5–3 log range from a 5.0 log dilution range, as observed experimentally.

6.3.1.1 *Internal Standard (IS)*

To determine the concentration of a compound in tissue, a control tissue that is from the same or similar animal and of the dosed tissue is required in addition to an internal standard (IS). The signal from the internal standard allows for the correction of suppression effects from tissue samples and the impact of a heterogeneous tissue sample. Ideally, this IS would be a deuterated or isotopic form of the compound, however, if such an IS is not available, an analog would be sufficient but with the carried risk that the analog may fragment differently than the target compound or even interfere with the detection of the compound of interest or have differing ionization efficiencies. Lastly it should be noted that an endogenous molecule and even matrix can be utilized as an IS, provided that the concentration is homogeneous *in*, in the case of endogenous molecules, and *on*, in the case of matrix, the tissue.[30] Nevertheless, the calculated ratio for the molecule as compared to the IS is calculated as:

$$\text{ratio} = \frac{\text{Response of the compound of interest in the tissue}}{\text{Internal standard response on the tissue}} \tag{6.4}$$

where $0.1 < \text{ratio} < 10$ is to be considered a valid linear range.[11] The IS is added to the MALDI matrix (after optimizing the concentration) and sprayed with the matrix on the entire slide. The calibration curve is spotted on control tissue; hence the method is referred to as "on-tissue." An advantage of the "on-tissue" method is that the IS compound, and therefore the normalization, is far more representative of the compound of interest than the

TEC method described below. In addition, it is also a single step, thus increasing throughput. Figure 6.2 shows images of spotted tissue and the subsequent signal intensity ratio obtained from the compound (COC) and its deuterated form COC-d_3.[11]

6.3.1.2 *Tissue Extinction Coefficient (TEC)*

There is a second approach called TEC which is defined as:

$$\text{TEC} = \frac{\text{Standard response on tissue}}{\text{Standard response on the slide}} \tag{6.5}$$

where the TEC value <1.[31] The advantage of this method is that TEC acts like an IS so it allows for QMSI even without access to IS compounds and can also be employed when it is not possible to apply the standard onto the tissue, by obtaining a TEC value for correction in the data analysis stage from the calibration curve from the ITO plate directly. Similarly, to the process of quantification using an IS, the steps of TEC preparation include dissolving the compound of interest in the matrix solution and then spraying this matrix solution, containing the compound, onto the entire ITO slide. A calibration curve is included but rather than spotting it on control tissue, it is spotted instead on the ITO slide directly and therefore is referred to as "in-solution." Figure 6.3 shows the image and quantification results of propranolol ($[M+H]+$ ion; *m*/*z* 260.16) and olanzapine ($[M+H]+$ ion; *m*/*z* 313.15) used in two separate experiments.[31] It should be noted that one disadvantage of TEC experiments is that they result in an additional image per region during the acquisition step as there is not only the images of the treated tissue sections, but also an additional image of a control that has a deposit of the compound of interest. Also, while an IS normalizes each individual pixel, TEC normalizes based on the ROI selected. Although it is possible to select several distinct regions to better normalize the tissue, it still does not account for each pixel, as with the IS method.

6.3.1.3 *Mimetic Tissue*

Here, the control tissue is homogenized with a spiked concentration of the compound standard. This method is referred to as the "in-tissue" method. Advantages include excellent measure of extraction efficiency of the analyte, which is critical for the correction of matrix effects and ion suppression related to potential binding to proteins. It is noteworthy that the homogenization preparation and spiking of the homogenate often involves a higher temperature and therefore potentially degrades the tissue or compound, or causes cell lysis (release of intracellular molecules into the extracellular content). This in turn may provide a higher ion suppression effect. Additionally, compared to the method of spotting a dilution range, spiking the compound into the homogenate is tedious in preparation, and

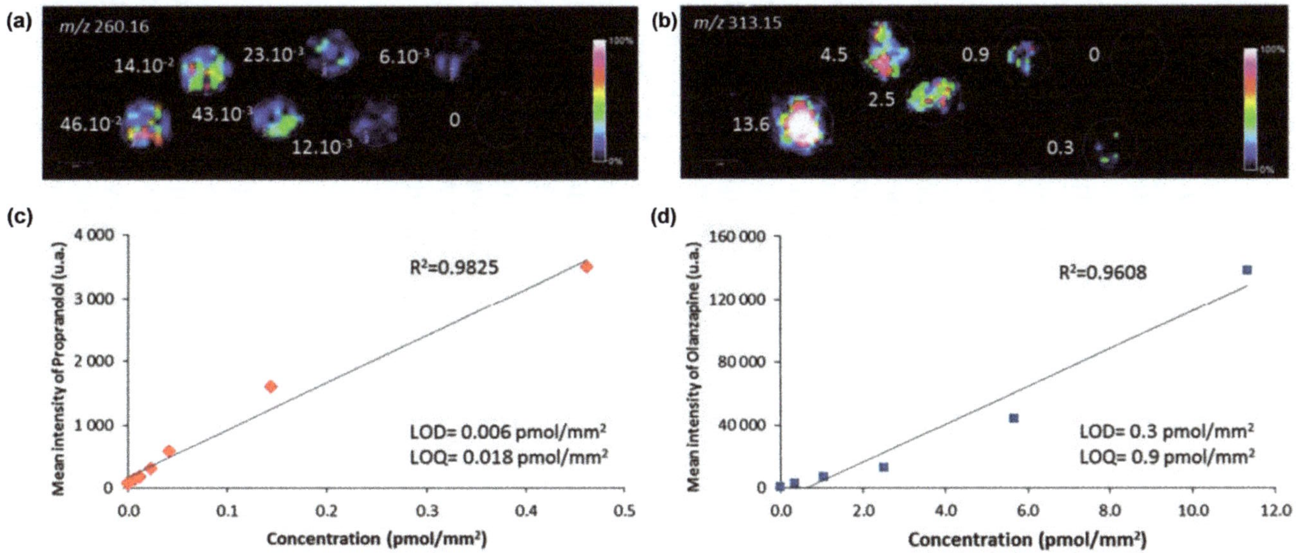

Figure 6.2 Images of MS calibration curve (on ITO) above and the respective standard calibration curve, below, using the TEC method, with propranolol ([M + H]+ ion, *m/z* 260.16) (left) and olanzapine ([M + H]+ ion, *m/z* 313.15) (right). The intensities obtained from the MSI experiment were then plotted against the respective concentration for both compounds.
Reproduced from ref. 31 with permission from Elsevier, Copyright 2012.

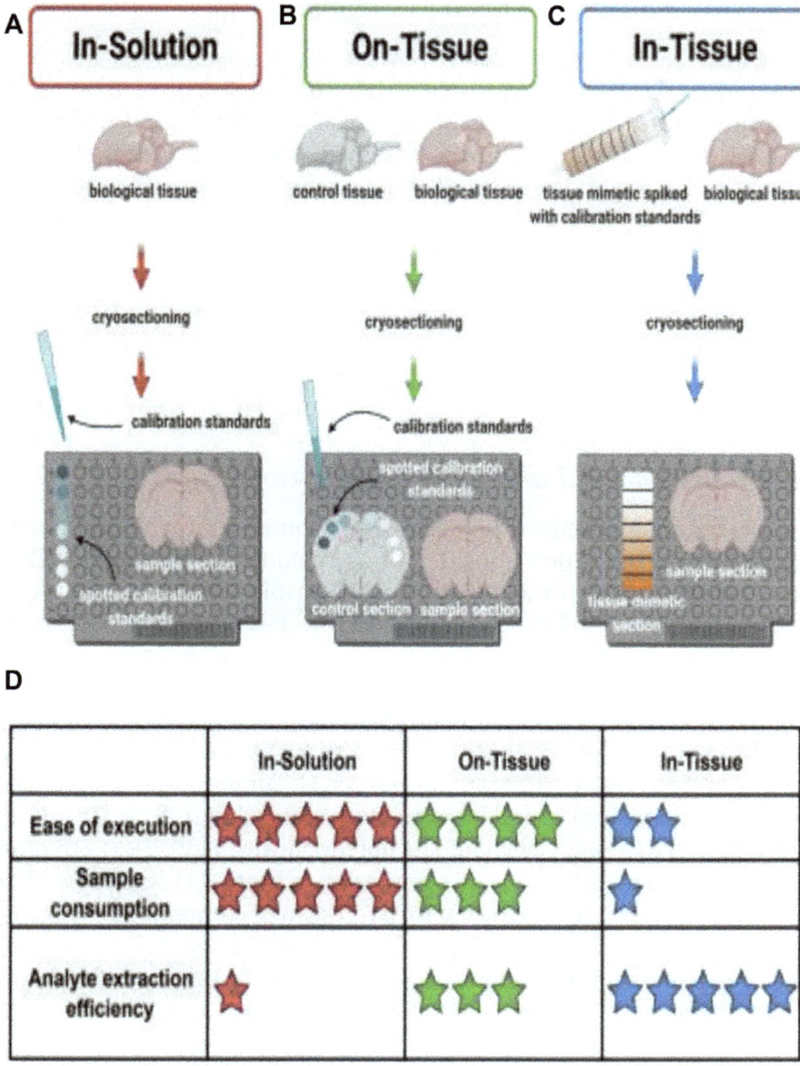

Figure 6.3 A diagram comparing the three most common approaches to normalization in QMSI, including "in-solution," "on tissue", and "in-tissue." Each method includes the material used, process, plate diagram, and finally a brief evaluation summary of the method.
Reproduced from ref. 21 with permission from American Chemical Society, Copyright 2020. Adapted from ref. 68 with permission from Springer Nature, Copyright 2014.

difficult to modify, and not to mention requires a high consumption of tissue. Lastly, although theoretically the mimetic "in-tissue" method preserves the integrity of the micro-environment of the tissue sufficiently, few if any studies have compared factors such as ion suppression and compound tissue extraction with the three methods mentioned above.[8,32]

6.3.1.4 Normalization Method Comparison

Tobias and Hummon describe and compare the approaches of "in-solution" and "on-tissue" (*i.e.* IS, TEC), both of which have been described above, and "in tissue" (mimetic tissue), for the establishment of a calibration curve, as shown in Figure 6.3.[21] There are other manners of applying standards other than the three most widely used methods described above, including depositing the standard beneath the tissue, "standard beneath tissue," sandwiching the standard between the tissue (*i.e.* half volume beneath, half volume on top), "standard sandwich", and mixing the standard with the matrix and depositing the combination as a drop, "standard pre-mix." Chumbley *et al.* concluded that these additional methods displayed no statistical difference at the 95% confidence interval between the results of the standard beneath, standard sandwich, standard pre-mixed, and the standard on-tissue (described prior) methods.[14]

6.3.1.5 Instrumental and Experimental Conditions

Lastly, it is critical to discuss other factors that contribute to the QMSI method development process. Such considerations include the type of MALDI-MS modes, including TOF (time-of-flight), QQQ (triple quadrupole mass spectrometry), IMS (ion mobility separation), FTICR (Fourier-transform ion cyclotron resonance), among others. Lamont *et al.* evaluate the first three mentioned modes to determine the analytical performance. From the study, it was determined that when comparing MS runs, TOF mode produced optimal LOD at two orders of magnitude higher sensitivity than that of the other modes and produced the best R^2. However, upon application of combining QQQ with MRM (multiple reaction monitoring), this mode produced significantly improved results to that QQQ alone.[32] It should be noted that although MRM can provide for an improved dynamic range, the limitation is that QQQ is a targeted technique and is not versatile in allowing for a full scan of compounds. So, it can be concluded that the analyzer mode must be carefully selected for the objective. From the conclusion of this stage, in the context of QMSI, the sensitivity, the specificity, spectral resolution, spatial resolution, and stability (of the compound)[17] can be measured and optimized and improved prior to proceeding to the next stage. What is significant in summarizing this portion of the discussion is that the many parameters introduced in this section impact the overall analysis and each holds their role in the analytical robustness of the experiment and the capacity for quantification. For further information on improving measures that impact MSI overall, refer to Chapter 3 (for information regarding ionization source parameters), Chapter 4 (for information regarding the different mass analyzers), and Chapter 5 (for information regarding sample preparation in more detail).

6.3.2 Method Validation (MV)

While the MD stage optimizes the condition of a method, the MV stage validates the conditions in terms of stability, accuracy, precision, and repeatability of the analysis with certain criteria. MD and MV include conditions and results that are required to validate the SOP (Standard Operating Protocol) including inter-day and intra-day repeatability using quality control over multiple days of analysis. First, to assess inter-day validation, duplicate or ideally triplicate quality controls of various concentrations are prepared and analyzed during the extent of the experiment. External analytical methods (LCMS, LC–MS/MS) are also applicable in order to verify the same quantitative results for a solution sample, if the sample is homogeneous.[33] Nevertheless, a detailed discussion of method validation is presented below.

6.3.2.1 *Quality Controls*

The QC stage is a continuous process that ensures the continued validation and integrity of the MD stage. To determine the quality of the quantification with regard to the specificity, accuracy, precision, reproducibility, and stability, the inclusion of a wide distribution of concentration range (*i.e.* that consists of a low quality control (LQC) (~10 nM), middle quality control (MQC) (~100 nM), and high quality control (HQC) (~1000 nM)) can be employed.[18] These QC materials require stability in their storage (freezer), over the bench, and that they are homogeneous. If the QC is found to vary plate to plate, then as required, the results of the entire plate are not acceptable, thereby eliminating the need to make a plate-to-plate comparison of the test material. The integration of a robust QC is often viewed as an optional step of analysis, for example for PK studies or toxicity studies.[34] Arguably, however, considering the complex QMSI workflow from sample preparation to acquisition and the low LOD required for most applications of QMSI, and fine-drawn results, a robust QC stage for a quantitative study of any sort should not be optional.

The QC sample sections consist of homogenized tissue of interest containing a standard compound (*i.e.* compound of interest) and its preparation follows a similar standardized process to the other tissue sections, rather than matching the mode of the quantification strategy (*i.e.* "in-solution," *etc.*). QC replicates must be prepared and be analyzed under optimized method conditions from the MD stage.[17] This process includes sectioning and mounting the spiked homogenate tissue section to the same slide and on the same day as the control, treated, and other control tissue sections so that the stability of the tissue and its handling is identical for the contents of the entire slide. Care should be taken to sustain the stability of all materials used by minimizing contamination with other materials or degradation under UV and high temperature.[35] To ensure stability, old and new stock batches can be compared in a concentration to ensure stability,[18] and QC samples can be sectioned prior to other tissue sections, *etc.* Additionally, the plane and seriality of the sectioning should be preserved to reduce variation.

Such requirements in uniformity can be extended to the dilution range and the replicate tissue sections in that they are all mounted ideally on a single slide to remain consistent with the matrix application, storage conditions, and acquisition. In the case that the material cannot be mounted on a single slide, the QC can serve as a metric of reproducibility amongst the slides to ensure a proper comparison of data. Additionally, the control tissue used for the reference control and the control tissue on which the dilution series is spotted must be of identical tissue type and sectioning plane as the treated tissue. In the case of a heterogeneous tissue organ, the tissue can be homogenized for the purposes of creating a homogeneous section for spotting a calibration curve.[20] To determine if substantial heterogeneity of a tissue section is present, the profile of the tissue can be analyzed and compared, statistically. In the case of a whole-body section, either a more rigorous approach of selecting several control tissues from the various representative organs of interest for both the control standard and control tissue for dilution, or a simpler option of using a single tissue (with liver tissue being the preferred choice) that is most representative of the collective whole-body organs, can be utilized.

The metrics discussed above can be evaluated both for inter- and intra-experimental runs. Within ImaBiotech, the acceptance criteria for QC precision is ≤25% for projects that last a few weeks or more. A reference diagram of the tissue and solutions found on a QMSI slide including the discussed QC tissue sections and the sections and solutions described previously can be found in Figure 6.4.

To determine the accuracy and precision, (sample) standard deviation (s) and relative standard deviation (RSD) can be calculated. The standard deviation can be calculated using the following formula:

$$s = \sqrt{\frac{\sum_{i=1}^{N} (x_i - \bar{x})^2}{N - 1}} \tag{6.6}$$

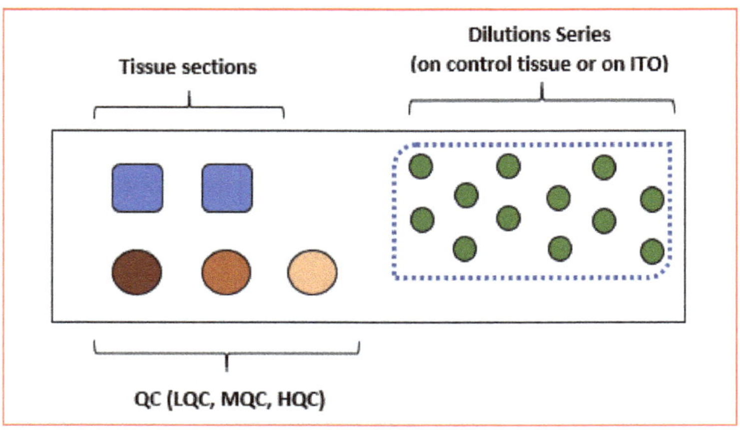

Figure 6.4 A diagram of the various components of a QMSI experiment, on a slide. This includes the tissue sections to be analyzed, dilution spots to determine concentration, and finally the QC to determine the analytical metrics of the experiment.

where s is the sample standard deviation, N is the number of observations, x_i is an observed value, and $\bar{x}$ is the mean of the observed value. After determining s, the relative standard deviation can be determined by:

$$\mathrm{RSD} = \frac{s}{|x|} * 100\,\% \quad (6.7)$$

Finally, s for accuracy and RSD for precision (that can be used to evaluate the experimental repeatability and reproducibility) can be converted to a percentage to evaluate the metric criteria. Additionally, statistical treatments such as "student's t test" can be applied to comparatively evaluate the standard deviations.[28]

6.3.2.2 *Quality Control Studies in QMSI*

A thorough and robust study was conducted jointly by two separate entities to evaluate various standard methods of development (mimetic model *vs.* dilution series) in terms of their analytical parameters of mean concentration determination, repeatability, reproducibility, and accuracy. This study specifically focused on the variability found in analysts, site, and instruments, and the variability was measured by deviations in the QC signal. In addition, the groups evaluated the variability of the method by means of cross-instrumental validation. The results concluded a multivariable association. The most interesting finding perhaps was that the mimetic model and dilution series both resulted in comparable and statistically acceptable quantification values. Therefore, as the authors concluded, the decision on method development should be determined by the pros and cons of each method.[19]

Another study with perhaps one of the most rigorous quantitative approaches to the validation methodology is from Källback *et al.* In the study, the authors included a validation method confirming the selectivity (expressed as a ratio), accuracy (expressed as standard deviation), precision (expressed as standard deviation), and recovery in both intra-experimental as well as cross-instrumental (MALDI–TOF and MALDI–FTICR) methods (both expressed as percentages). Six replicates of dosed mice were analyzed to determine the dose mean, standard deviation, and p-value. Lastly, the authors included cross-instrumental validation with both LC–MS/MS-MRM (see Table 6.1).[18] These and several other studies evaluated inter-instrumental orthogonal validation and concluded no significant difference at the 95% confidence interval for MALDI and HPLC–MS/MS results for the parameters examined,[11,14,18–20] except for the ionization efficiency between LC–MS/MS and MALDI.[16,18] However, a comprehensive normalization method can correct for this variability.[16]

6.3.3 Data Analysis

In terms of data analysis, the minimum peak threshold must be established as the intensity for which would vary depending on the experiment. To

Table 6.1 A summary of the results of the analytical parameters evaluated for the quantification of Citalopram analyzed by MALDI–TOF-MSI and MALDI–FTICR-MSI and LC – MS/MS (reference method). The comparison evaluated the selectivity, accuracy, precision, and recovery. The results of analysis amongst six replicates of animals were also obtained with the mean dose, standard deviation, and p-value obtained. Reproduced from ref. 18, https://doi.org/10.1021/acs.analchem.0c03203, under the terms of the CC BY 4.0 license, https://creativecommons.org/licenses/by/4.0/.

	MALDI–TOF-MSI			MALDO–FTICR-MSI			
Parameters	Slide V1	Slide V2	Slide V3	Slide V4	Slide V5	Slide V6	LC–MS/MS
Selectivity radio (LLOQ/blank sample) (≥5)	11.1			33.0			5.6
Accuracy ($n=6$)							
LLOQ (≤±20%) [%]	−5.1	17.0	6.3	−14.2	13.0	−2.4	−8.7
LQC (≤±15%) [%]	4.0	−0.7	−8.9	−6.3	13.6	5.7	−2.1
MQC (≤±15%) [%]	5.9	−1.1	4.5	6.1	−12.0	5.7	1.4
HQC (≤±15%) [%]	5.6	−2.2	14.9	3.0	9.2	5.7	7.2
Precision ($n=6$)							
LLOQ (≤20%) [%]	14.8	10.1	9.9	13.1	16.4	3.8	10.8
LQC (≤15%) [%]	10.0	6.9	14.5	9.8	10.2	8.5	1.6
MQC (≤15%) [%]	10.6	11.0	7.3	12.6	10.8	9.6	2.3
HQC (≤15%) [%]	14.2	10.8	13.8	5.9	10.6	12.4	2.6
Coefficient of determination							
R^2	0.990	0.977	0.984	0.996	0.989	0.998	0.999
Recovery (analyte)							
LQC [%]	20	23	17	32	28	31	62
MQC [%]	35	32	26	34	39	38	82
HQC [%]	45	40	44	42	55	56	88
Recovery (Internal Standard)							
LQC [%]	33	49	40	37	34	31	60
MQC [%]	33	47	43	36	45	34	87
HQC [%]	43	49	50	42	49	37	89
Quantitation		MALDI–TOF-MSI ($n=3$)		MALDI–FTICR-MSI ($n=3$)			LC–MS/MS ($n=6$)
Animal D1 [pmol mg^{-1}]		54.1 ± 8.1, $p=0.739$		60.7 ± 4.8, $p=0.218$			55.9 ± 3.3
Animal D2 [pmol mg^{-1}]		50.8 ± 6.8, $p=0.156$		57.8 ± 5.2, $p=0.635$			59.5 ± 0.8
Animal D3 [pmol mg^{-1}]		53.2 ± 4.3, $p=0.177$		59.2 ± 1.6, $p=0.409$			58.3 ± 1.1
Animal D4 [pmol mg^{-1}]		51.5 ± 7.3, $p=0.593$		59.6 ± 5.4, $p=0.221$			54.1 ± 0.7
Animal D5 [pmol mg^{-1}]		52.9 ± 5.1, $p=0.372$		56.1 ± 8.3, $p=0.971$			56.3 ± 1.1
Animal D6 [pmol mg^{-1}]		42.4 ± 4.6, $p=0.346$		45.0 ± 2.3, $p=0.695$			45.7 ± 1.1
Average concentration [pmol mg^{-1}]		51.1 ± 4.4, $p=0.153$		56.9 ± 6.0, $p=0.648$			55.0 ± 4.9

ensure reproducibility, the threshold selection must be standardized. This is demonstrated by Bowman *et al.* to be 6σ above the baseline noise.[36] Although this threshold can be adjusted, if set too low the resulting data will be riddled with noise, and if set too high, low intensity peaks of interest may be falsely excluded. Regardless, the LOQ must be above the threshold value in order to fall within a range to determine quantification.

There are several types of quantification that can be completed with QMSI. Apart from qualitative analysis, the quantitative data of tissue sections includes relative or absolute quantification, that is comparative quantification between tissue regions or quantification of an exact concentration of the analyte of interest. This also applies to specific tissue analysis such as depth profiling, which will be described later. Regardless of the quantification type, it is important to translate between different units of quantification ($mol\,g^{-1}$ or $mol\,L^{-1}$); this is achieved by first determining the concentration of the surface area (mm^2) of the tissue by representing, in software, the tissue intensity from a particular *x,y* coordinate as pixels that comprise a tissue image. Specifically, the ROIs (which must include ≥400 positions (or pixels) to collect sufficient data, as an independent variable from that of the raster size) are extracted intensities from the mass spectra. Furthermore, the IS ratio discussed above is applied to the intensity determined in this stage to account for suppression effects of the tissue. Upon determination of the surface area, accounting for IS correction, and determining the concentration of the compound within the tissue from the constructed calibration curve, we can combine concentration and the surface as $mol\,mm^{-2}$. As a last step, histological areas of interest can be delineated to assign the quantitative concentrations to an area.

To summarize, working from the following equation for a linear calibration curve, from eqn (6.1):

$$y = mx + b$$

where m is the slope and b is the y-intercept, can be written to solve for x:

$$x = \frac{y - b}{a} = \frac{\text{mol}}{\text{mm}^2} \tag{6.8}$$

To convert this to a concentration in the units of mass, we use the following calculation to apply the molecular weight of the target molecule:

$$\frac{\text{mol}}{\text{mm}^2} * \frac{\text{g}}{\text{mol}} = \frac{\text{g}}{\text{mm}^2} \tag{6.9}$$

Lastly, we apply the following calculation to apply the density of the tissue:

$$\frac{\text{g}}{\text{mm}^2} * \frac{1}{d*h} = \frac{\text{g}}{\text{g}} \text{ of tissue} \tag{6.10}$$

where h is the height of the tissue and d is the density of the tissue. The thickness and thus density of the sectioned tissue are calculated as a "wet" section prior to placing the tissue in the desiccator to eliminate moisture. It should be noted that the dry tissue amounts to ~80% of the measurements of the "wet" tissue section thickness, as determined experimentally.

Employing the calculations above using data garnered from the pixel intensity, the tissue distribution profiles of a compound can be quantified. During data analysis, the response of the platform can be evaluated by metrics of LOD and LOQ, meanwhile the response of the quantification can be evaluated by the accuracy, precision, and inevitably, the stability of the experimental compounds. Again, the minimum number of pixels for an ROI must be evaluated for quantification.

The process of this determination is outlined by Bonnel *et al.* First, the molecules in the skin tissue sample of the study were subtracted after imaging (not shown below). Next, the intensity of the compounds was imaged and determined (see Figure 6.5). This was then spatially attributed to their pixel acquisition (including contamination subtraction, raw data extraction, pixel alignment) (see Figure 6.5) and the correlations of the pixel to the depth of the skin tissue. Finally, the graphed data (concentration *vs* depth) can be plotted as shown in Figure 6.5. It should be noted that the larger error bars, especially found in the epidermis layer, are due to the cryosectioning of the later from the dermis layer, where smudging or other contamination can occur.[12]

Additionally, histological delineation of the drug/compound distribution was demonstrated by Huizing *et al.* who evaluated colon tissue samples and extended this work to make conclusions about the drugs penetration and pharmacokinetics.[13] Another capability of QMSI is for the determination of a 3D tissue concentration distribution and the visual reconstruction. This technique requires a series of well-distributed consecutive sections of a tissue that are imaged, reconstructed together, and any missing tissue concentration extrapolated. Further recommended reading of 3D QMSI studies includes an initial publication on the technique by Seeley and Caprioli,[37] data mining strategies for 3D construction,[38] and a robust study on the localization of long-chain acylcarnitines in a spinal cord injury.[39] Although there are other techniques for 3D quantification, such as quantitative autoradiography (QWBA), qMALDI bypasses many of the associated challenges of QWBA such as the use of radioactive materials, loss of metabolite information, and the varied molecular behavior of a labeled *vs.* a non-labeled molecule.[40]

Although this chapter will not describe the hardware and software requirements for a QMSI experiment in detail, further reading on the topics of, for example data storage and software capabilities, can be found in the comprehensive works of Veselkov *et al.*,[41] Picard de Muller *et al.*,[42] and Källback *et al.*[43] As for QMSI software platforms, there are a variety of available resources currently available, including Multimaging (ImaBiotech, France), a multimodal imaging platform that combines QMSI with microscopy.[44,45] Other examples include MSiReader (North Carolina State University, Raleigh, NC), available as an open-source customizable Matlab application for MSI data and a software package.[46,47]

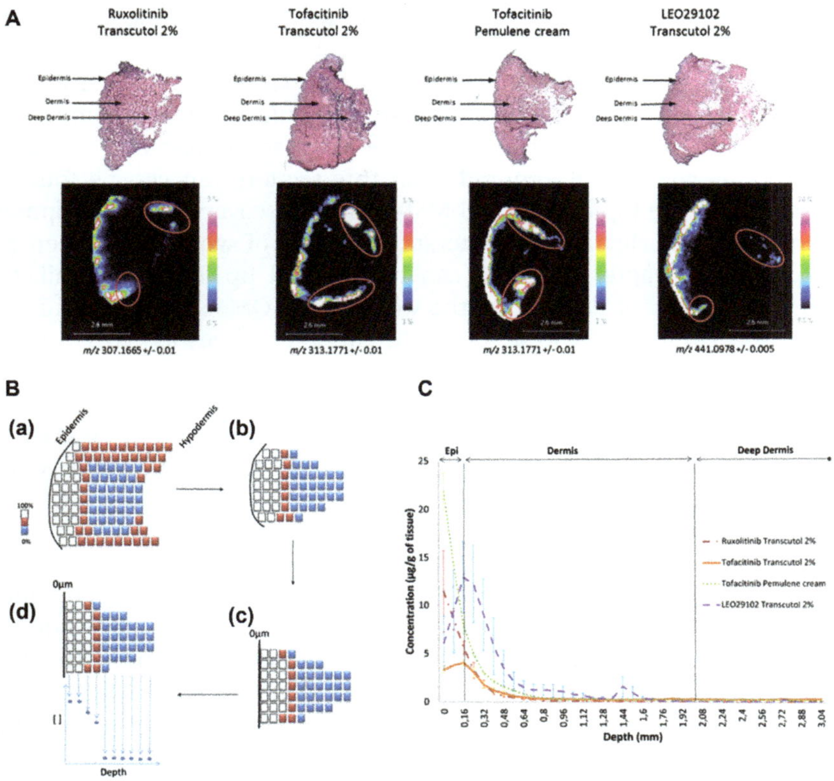

Figure 6.5 (A) Staining and MSI images for skin tissue, depicting the intensity determined for the analytes. In conjunction with histological information obtained from the H&E stained images, the MSI data shows that the epidermis of the skin sample has the highest concentration of the drug compound. (B) Pixel acquisition and alignment for the data of Figure 6.5A. The figure depicts first the raw data (a), the data after processing (contaminated data pixels removed) (b), alignment processing of the data (c), and finally calculation of the drug compound penetration depth. (C) A graph depicting the information of Figure 6.5b with the various drug compounds/formulations.
Reproduced from ref. 12 with permission from Springer Nature, Copyright 2018.

6.4 Versatility of QMSI

The following section not only provides a display of the versatility of molecules that can be quantitatively analyzed by QMSI, but also an overview of the basic considerations required for various categories of molecules. Such a preliminary assessment is important to determine the drug pharmacokinetic ADME properties (such as the solubility, polarity, and binding affinity) through the detection and quantification of the drug compound in tissue. Furthermore, the discussion below addresses common issues to QMSI including limiting molecular weight compounds, matrix and ion suppression issues, poor spectral resolution, and poor sensitivity. Although there are categories of compounds beyond the few evaluated below, the chosen categories are represented due to the unique approaches the literature review offers. Common categories such as small molecules (which encompass the majority of the pharmaceutical field) are of great interest but due to the focus of this chapter are omitted from this section, but can be found in Chapter 14 on the topic of MALDI-MSI in drug discovery and development.

Further summaries and considerations, many of which have been presented in this chapter on QMSI can be followed up in more detail. One particular strategy of illustrating the workflow of QMSI can be found from Tobias and Hummon, who describe the different stages as "metro lines," and the different steps as stops. What is key in this form of presentation is the larger branched multidirectional enterprise that comprises QMSI.[21] A QMSI review by Ellis *et al.* details the MD process not only for MALDI but other QMSI techniques such as DESI, SIMS, and LA-ICP-MS[2] (Figure 6.6).

6.4.1 Lipids

Lipids are of interest in biological fields as lipid species (such as ceramide-1-phosphate and sphingosine-1-phosphate) affect biological pathways. The pharmaceutical field also holds lipids in interest particularly in the context of topical treatments as the composition of the skin and its endogenous and exogenous lipid layers affect the drug compound's interaction with the skin.[15] In a study by Landgraf *et al.*, the authors quantify endogenous phospholipids in nerve tissue (using phosphatidylcholine (PCho) as the IS due to the similar ionization properties and extraction efficiencies to that of the analyte) with consideration for determining the analytical variability as a result of the MD stage. Specifically, the study included an evaluation of signal variability (decrease), normalization (IS evaluation and potential spatial and spectral interference), and challenges with quantification.[5] Tissue lipidomics is also of rising interest for biomarker discovery. The variability due to ion suppression effects in the quantification of phosphocholines in tissue was evaluated by Jadoul *et al.* The authors propose the approach of depositing standard solutions by means of spraying onto heterogeneous tissue to better quantify the analyte in various environments, therefore preventing solvent effects (which for example cannot be achieved by the alternative method of a spiked homogenized sample).[48]

6.4.2 Peptides and Amines

Although proteins are difficult to detect using MALDI due to their high molecular weight, attempting to detect peptides in a bottom-up approach makes protein analysis indirectly compatible with MALDI. However, as Waliczek *et al.* highlight, polymer solutions are often at trace concentrations, creating a challenge for the ionization of these molecules. The authors employed salt derivatives 2,4,6-triphenylpyridinium and 2,4,6-trimethylpyridinium, salts

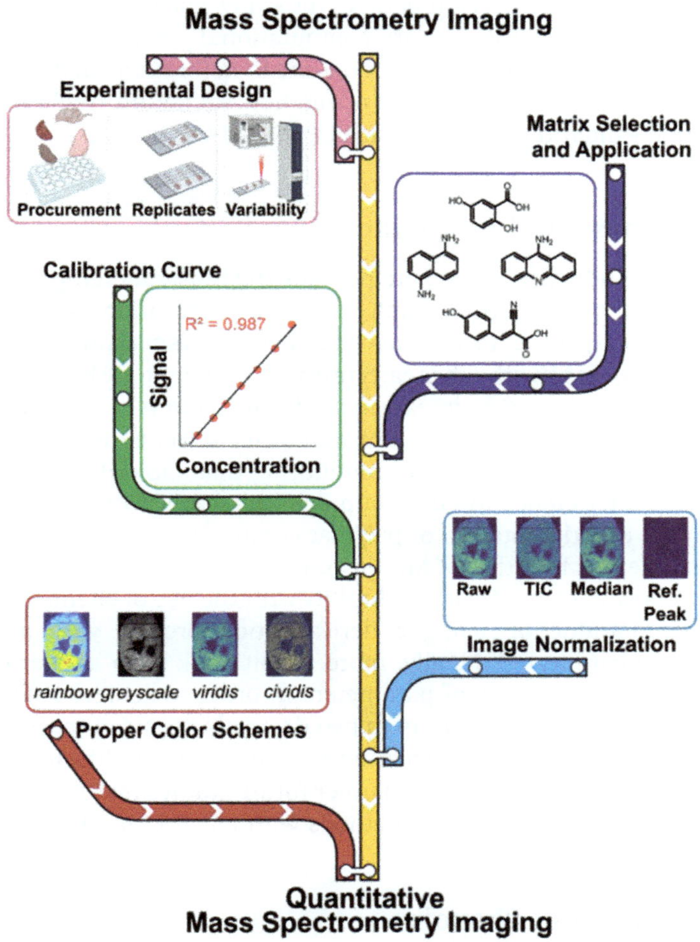

Figure 6.6 A schematic summary illustrating the interconnectedness of the QMSI method development. This includes starting with the experimental design and decision making, towards the acquisition, and finally the post-acquisition processing and data interpretation.
Reproduced from ref. 21 with permission from American Chemical Society, Copyright 2020.

that are both inexpensive and accessible. These highly nucleophilic derivatization agents resulted in detection of the amine group at 1×10^{-18} levels.[49] In a novel validated method by Ferreira *et al.*, melatonin was quantified by the approach of pixel intensity in the gray scale, while verifying analytical parameters such as inter- and intra-day precision. The MALDI results achieved LODs in the 1×10^{-12} pM range, confirming the high sensitivity of the QMSI method as compared to other instrumental techniques.[50] In another work, a demonstration of the quantification of the tyrosine kinase inhibitor chemotherapy drug compound Imatinib by Sammour *et al.* exhibits a systematic approach to the method development for the study. For example, rather than a manual ROI selection, the authors computationally defined the selection, which was optimized using a novel non-linear calibration mode. The study also included a rigorous execution of method validation (including inter-instrumental validation of MALDI with UPLC–ESI-QTOF-MS).[51]

6.4.3 Polymers and Polyimides

Polyimides, including polymers, are perhaps the most emerging category for QMSI. As mentioned in this chapter, sample preparation (solubility) and stability are key aspects of method development, and especially so for polymer and polyimides, along with other categories of compounds. These issues have been addressed with studies in early work from Zagorevski *et al.* and Trimin *et al.* who evaluated polymers in TFA and other solvents. Furthermore, the authors also describe the fragment monitoring that allowed them to display the potential interference in end-group determination, which poses an obstacle in QMSI experimentation of polymer compounds.[52,53]

Other studies such as that of Murgasova *et al.* and Thieme *et al.* overcame issues with polymers and their insolubility with the use of chemical reaction (ring-opening) induction using a derivatization process with a reagent, *N*-methylethaolamine.[54,56] Finally, more recent work such as that of Payne *et al.* presents optimization of parameters in order to achieve resolution of repeating end group units with up to two decimals points of mass accuracy. This allows enabling the confirmation of end group modifications with a high level of confidence.[55] In this manner future studies may aim to not only improve QMSI but also enhance methods with reaction monitoring.

6.4.4 Alcohol and Phenols

As the analysis and quantification of low polarity compounds such as alcohols and phenols with MALDI are not possible to achieve, the conversion of the compound (into either a polar compound or one that will result in identifiable spectral features) and then the subsequent analysis does allow for the investigation. This is possible through the derivatization of the analyte of interest.[57,58] There are many demonstrations of various derivative agents being employed to aid in the ionization efficiency for phenolic hydroxyl among many other categories of compounds, proving the derivatization method an effective strategy for difficult to detect species.[24] Particularly interesting is the study by Kaya *et al.* who analyzed a compound with the complications of poor detection and ionization, ion suppression, analyte fragmentation, and a host of matrix effects. Their method of derivatization using 4-(*N*-methyl)-pyridinium boronic acid proves viable for a host of diol containing groups.[59] Other quantitative studies of alcohols and phenols include the assessment of phenolic compounds in chocolate[60] and ethoxylated alcohols in friction reducers.[57]

6.4.5 Nanoparticles

In the context of pharmaceutical compounds, nanoparticle drug development has become increasingly popular in the treatment of brain disease and conditions as the permeability of the nanoparticles allows for the penetration of the blood brain barrier (BBB). One such study of such an application of QMSI is from Barré *et al.* in which the team reviewed the quantification of nanoparticles in brain tissue sections using the different QMSI approaches of DESI, SIMS, and MALDI.[61] Other studies such as that by Luo *et al.* present quantification of nanoparticle morphology from the MALDI spectrum that is then treated with the Monte Carlo simulation method.[62]

6.5 Current Challenges

As illustrated by this chapter, QMSI is a versatile analytical method that offers the unique advantage of label-free spatial quantitative distribution at a sensitive level. As described in the literature in this chapter, it is possible to detect 100 nM (or 100 ng g^{-1}) concentration in tissue with a pixel size of 50 microns. This represents a detection limit at the femtogram level.[63] Theoretically, this is also found to be in the fact that QMSI requires a ng scale of tissue which by volume is 1 µL for a single pixel which equates to femtogram level concentrations while LC–MS requires a g scale of tissue which by volume is 1 mL for analysis which equates to picogram level concentrations, thereby making QMSI an intrinsically more sensitive method in the context of volume of tissue required for analysis.[60] Furthermore, if a single isolated pixel is found for example, to have a femtogram level concentration, this concentration would be far below the level of detection for LC–MS capabilities considering that the instrument requires mL levels of prepared sample. This is especially critical to toxicity studies and other pharmacological investigations. Additionally, this method can be used to analyze an expansive variety of compounds, as summarized in detail in the previous section of the chapter. Therefore, it can be concluded that for the quantity of ablated tissue with QMSI *vs.* LC–MS, QMSI can be considered more sensitive due to its intrinsic characteristics. However, it should be noted that in practice for some cases, if sensitivity is lacking for the detection of analytes by MALDI-MSI, one would consider performing a tissue extract and run an LC–MS(/MS), which could provide better sensitivity in this case due to sample clean-up (*i.e.* reduced ion suppression due to liquid chromatography separation of analytes).

Although the focus of this chapter has been qMALDI, there are increasingly more instrumental techniques and capabilities developed to increase the variety of applications for QMSI. However, MSI and even more so QMSI can fall victim to various issues, many that arise during the MD stage. These issues include contamination, delocalization, reproducibility, low throughput, high cost, and a high required skill level (knowledge in sample sectioning, MS parameters and large data analysis). Beginning with the broad issue of contamination, it is perhaps the most problematic as it can arise in almost every step of MD and its origin may be difficult to define and thus correct. Tissue contamination can arise from the harvesting of the tissue, during any event of tissue handling, during cryosectioning, mounting, and even during scanning. The spraying stage can also introduce matrix contamination if care is not taken to thoroughly flush the lines. However, with a rigorous adherence to a well-developed SOP, the risk of contamination can be avoided. Delocalization is another issue that greatly affects the quantification results for tissue samples. Delocalization occurs when the compound is delocalized from its origin in the tissue during the matrix deposition process. If the matrix solution does not include the correct ratio of aqueous and organic solvent and/or the spraying conditions are not optimized in consideration with the matrix solvent, there is a risk of delocalization of the compound and therefore an inconsistent quantitative measurement of the spatial intensity. Although the issue of delocalization can be remedied by matrix deposition by sublimation, the method of sublimation is found to hold poor reproducibility, especially regarding producing uniform matrix crystal sizes, and for this reason, has been omitted as an option for matrix deposition for the purposes of QMSI.[64]

For the reasons above, as well as the exhaustive considerations required during the sample preparation and acquisition steps, QMSI can suffer from irreproducible results. As described previously, reproducibility can be monitored and, if necessary, corrections can be made with the application of MV and QC. Lastly, as presented in this chapter, QMSI has low throughput due to the multistage process of MD and the cost and maintenance of the instrument. Therefore, the process of QMSI, in most cases, also becomes a costly and time-consuming process, and as such, is a process that requires a high level of skill and training.

6.6 Conclusion and Future Directions

Having described and emphasized the importance of the stages of MD and its outcome on the subsequent stages of MV and QC, QMSI would benefit from improved MD which again includes sample preparation and acquisition. In particular, bench tasks such as sectioning, pipetting, measuring, and weighing have the potential to greatly impact the proceeding steps. Therefore, reducing systematic error by incorporating automation, care in handling and execution, and frequent calibration, can lead to better results further down the pipeline.

There has been and is a continuing need for improved strategies for the detection of additional compounds and categories of compounds. For example, although there are solutions for large molecular weight compounds, there are pitfalls such as diminished throughput when analyzing these molecules. For example, the average sensitivity for proteins after trypsin digestion is about 100 μM, which only allows the detection and quantification of abundant proteins.

Perhaps the most exciting area of QMSI (and bioinformatics in general) is the prospect of integrating machine-learning and deep-learning platforms to ever increase the analytical parameters of accuracy, precision, reproducibility, *etc.* For this reason, the advancement of QMSI most likely lies in the development and application of automated data processing platforms. A recent published work by Race *et al.* aimed to integrate histological and MSI data to allow for simple back and forth annotation transfer and multimodal registration workflows.[65]

In conclusion, this chapter has illustrated the workflow for the QMSI (MALDI) process and emphasized the specific considerations for the technique. This review also described the application of MV and QC, which are often lacking from the current QMSI literature. Further guidelines from governing agencies should be referenced to ensure proper adherence[66,67] on MD and MV. It has been illustrated that when a robust analytical approach is applied to the three phases of workflow outlined (MD, MV, QC), QMSI results are not only analytically sound, but QMSI also becomes an unmatched versatile option for the spatial quantification of a growing pool of compounds.

References

1. S. Schulz, M. Becker, M. R. Groseclose, S. Schadt and C. Hopf, *Curr. Opin. Biotechnol.*, 2019, **55**, 51.
2. S. R. Ellis, A. L. Bruinen and R. M. Heeren, *Anal. Bioanal. Chem.*, 2014, **406**, 1275.
3. N. Sun and A. Walch, *Histochem. Cell Biol.*, 2013, **140**, 93.
4. S. Castellino, M. R. Groseclose and D. Wagner, *Bioanalysis*, 2011, **3**, 2427.
5. R. R. Landgraf, T. J. Garrett, M. C. Prieto Conaway, N. A. Calcutt, P. W. Stacpoole and R. A. Yost, *Rapid Commun. Mass Spectrom.*, 2011, **25**, 3178.
6. J. Stauber, *Bioanalysis*, 2013, **4**, 2095.
7. D. Bonnel, R. Legouffe, N. Willand, A. Baulard, G. Hamm, B. Deprez and J. Stauber, *Bioanalysis*, 2011, 1399.
8. M. R. Groseclose and S. Castellino, *Anal. Chem.*, 2013, **85**, 10099.
9. M. R. Groseclose, S. B. Laffan, K. S. Frazier, A. Hughes-Earle and S. Castellino, *J. Am. Soc. Mass Spectrom.*, 2015, **26**, 887.
10. M. L. Manier, J. M. Spraggins, M. L. Reyzer, J. L. Norris and R. M. Caprioli, *J. Mass Spectrom.*, 2014, **49**, 665.
11. D. A. Pirman, R. F. Reich, A. Kiss, R. M. A. Heeren and R. A. Yost, *Anal. Chem.*, 2013, **85**, 1081.
12. D. Bonnel, R. Legouffe, A. H. Eriksson, R. W. Mortensen, F. Pamelard, J. Stauber and K. T. Nielsen, *Anal. Bioanal. Chem.*, 2018, **410**, 2815.
13. L. R. S. Huizing, J. McDuffie, F. Cuyckens, M. van Heerden, T. Koudriakova, R. M. A. Heeren and R. J. Vreeken, *Anal. Chem.*, 2021, **93**, 2144.
14. C. W. Chumbley, M. L. Reyzer, J. L. Allen, G. A. Marriner, L. E. Via, C. E. Barry and R. M. Caprioli, *Anal. Chem.*, 2016, **88**, 2392.

15. P. J. Hart and M. R. Clench, MALDI-MSI of Lipids in Human Skin, *Methods Mol. Biol.*, 2017, **1618**, 29.
16. N. Takai, Y. Tanaka, A. Watanabe and H. Saji, *Bioanalysis*, 2013, **5**, 603.
17. D. A. Skoog, D. M. West and F. J. Holler, *Skoog and West's Fundamentals of Analytical Chemistry*, Cengage Learning, Belmont (Calif.), 6th edn, 2014.
18. P. Källback, T. Theodosia Vallianatou, A. Nilsson, R. Shariatgorji, N. Schintu, M. Pereira, F. Barré, H. Wadensten, P. Svenningsson and P. E. Andrén, *Anal. Chem.*, 2020, **92**, 14676.
19. J. Barry, R. Ait-Belkacem, W. M. Hardesty, L. Benakli, C. Andonian, H. Licea-Perez, J. Stauber and S. Castellino, *Anal. Chem.*, 2019, **91**, 6266.
20. S. L. Koeniger, N. Talaty, Y. Luo, D. Ready, M. Voorbach, T. Seifert, S. Cepa, J. A. Fagerland, J. Bouska, W. Buck, R. W. Johnson and S. Spanton, *Rapid Commun. Mass Spectrom.*, 2011, **25**, 503.
21. F. Tobias and A. B. Hummon, *J. Proteome Res.*, 2018, **19**, 3620.
22. J. Leopold, Y. Popkova, K. Engel and J. Schiller, *Biomolecules*, 2018, **8**, 173.
23. C. Harkin, K. W. Smith, F. L. Cruickshank, C. L. Mackay, B. Flinders, R. M. A. Heeren, T. Moore, S. Brockbank and D. F. Cobice, *Mass Spectrom. Rev.*, 2021, 1.
24. Q. Zhou, A. Fülöp and C. Hopf, *Anal. Bioanal. Chem.*, 2020, **413**, 2599.
25. P. Wang and R. W. Giese, *J. Chromatogr. A*, 2017, **1486**, 35.
26. M. T. Stauffer, *Calibration and Validation of Analytical Methods: A Sampling of Current Approaches*, IntechOpen, London, 2018.
27. A. P. R. Zabell, F. E. Lytle and R. K. Julian, *Clin. Mass Spectrom.*, 2016, **16**, 25.
28. D. C. Harris, *Exploring Chemical Analysis*, W.H. Freeman, New York, 2013.
29. D. A. Armbruster and T. Pry, *Clin. Biochem. Rev.*, 2008, S49.
30. S. D. Turker, W. B. Dunn and J. Wilkie, *Appl. Spectrosc. Rev.*, 2016, **52**, 73.
31. G. Hamm, D. Bonnel, R. Legouffe, F. Pamelard, J.-M. Delbos, F. Bouzom and J. Stauber, *J. Proteomics*, 2012, **75**, 4952.
32. L. Lamont, D. Hadavi, B. Viehmann, B. Flinders, R. M. Heeren, R. J. Vreeken and T. Porta Siegel, *Anal. Bioanal. Chem.*, 2021, **413**, 2779.
33. J. A. Barry, M. R. Groseclose and S. Castellino, *Bioanalysis*, 2019, **11**, 1099.
34. Y.-J. Xue, H. Gao, Q. C. Ji, Z. Lam, X. Fang, Z. Lin, M. Hoffman, D. SchulzJander and N. Weng, *Bioanalysis*, 2012, **4**, 2637.
35. M. A. Reeve and A. G. Buddie, *Plant Methods*, 2018, **14**, 90.
36. A. P. Bowman, G. T. Blakney, C. L. Hendrickson, S. R. Ellis, R. M. Heeren and D. F. Smith, *Anal. Chem.*, 2020, **92**, 3133.
37. E. H. Seeley and R. M. Caprioli, *Anal. Chem.*, 2012, **84**, 2105.
38. H. Thiele, S. Heldmann, D. Trede, J. Strehlow, S. Wirtz, W. Dreher, J. Berger, J. Oetjen, J. H. Kobarg, B. Fischer and P. Maass, *Biochim. Biophys. Acta, Proteins Proteomics*, 2014, **1844**, 117.

39. J. Quanico, L. Hauberg-Lotte, S. Devaux, Z. Laouby, C. Meriaux, A. Raffo-Romero, M. Rose, L. Westerheide, J. Vehmeyer, F. Rodet, P. Maass, D. Cizkova, N. Zilka, V. Cubinkova, I. Fournier and M. Salzet, *Sci. Rep.*, 2018, **8**, 16083.
40. E. G. Solon, A. Schweitzer, M. Stoeckli and B. Prideaux, *AAPS J.*, 2009, **12**, 11.
41. K. Veselkov, J. Sleeman, E. Claude, J. P. Vissers, D. Galea, A. Mroz, I. Laponogov, M. Towers, R. Tonge, R. Mirnezami, Z. Takats, J. K. Nicholson and J. I. Langridge, *Sci. Rep.*, 2018, **8**, 4053.
42. G. Picard de Muller, R. Ait-Belkacem, D. Bonnel, R. Longuespée and J. Stauber, *J. Am. Soc. Mass Spectrom.*, 2017, **28**, 2635.
43. P. Källback, A. Nilsson, M. Shariatgorji and P. E. Andrén, *Anal. Chem.*, 2016, **88**, 4346.
44. Multimaging™ Solutions. https://www.imabiotech.com/multimodal-imaging/.
45. R. Ait-Belkacem, V. Bol, G. Hamm, F. Schramme, B. Van Den Eynde, L. Poncelet, F. Pamelard, J. Stauber and B. Gomes, *SLAS Discovery*, 2017, **22**, 1182.
46. MSiReader, https://msireader.wordpress.ncsu.edu/.
47. M. T. Bokhart, M. Nazari, K. P. Garrard and D. C. Muddiman, *J. Am. Soc. Mass Spectrom.*, 2017, **29**, 8.
48. L. Jadoul, N. Smargiasso, F. Pamelard, D. Alberts, A. Noël, E. De Pauw and R. Longuespée, *OMICS: J. Integr. Biol.*, 2016, **20**, 110.
49. M. Waliczek, M. Kijewska, M. Rudowska, B. Setner, P. Stefanowicz and Z. Szewczuk, *Sci. Rep.*, 2016, **6**, 37720.
50. M. S. Ferreira, D. N. de Oliveira, C. C. Mesquita, A. P. de Lima Barbosa, G. F. Anhê and R. R. Catharino, *J. Anal. Sci. Technol.*, 2016, **7**, 28.
51. D. Abu Sammour, C. Marsching, A. Geisel, K. Erich, S. Schulz, C. Ramallo Guevara, J.-H. Rabe, A. Marx, P. Findeisen, P. Hohenberger and C. Hopf, *Sci. Rep.*, 2019, **9**, 10698.
52. D. V. Zagorevskii, M. J. Nasrullah, V. Raghunadh and B. C. Benicewicz, *Rapid Commun. Mass Spectrom.*, 2005, **20**, 178.
53. S. Trimpin, S. Keune, H. J. Räder and K. Müllen, *J. Am. Soc. Mass Spectrom.*, 2006, **17**, 661.
54. R. Murgasova, D. M. Hercules and J. R. Edman, *Macromolecules*, 2004, **37**, 5732.
55. D. Thieme, H. Sachs and M. Thevis, *J. Mass Spectrom.*, 2008, **43**, 974.
56. M. E. Payne and S. M. Grayson, *J. Visualized Exp.*, 2018, 136.
57. J. L. York, R. H. Magnuson, D. Camdzic and K. A. Schug, *Rapid Commun. Mass Spectrom.*, 2019, **33**, 1286.
58. R. S. Borisov, N. Y. Polovkov, D. I. Zhilyaev and V. G. Zaikin, *J. Anal. Chem.*, 2015, **70**, 1542.
59. I. Kaya, S. M. Brülls, J. Dunevall, E. Jennische, S. Lange, J. Mårtensson, A. G. Ewing, P. Malmberg and J. S. Fletcher, *Anal. Chem.*, 2018, **90**, 13580.
60. D. N. de Oliveira, A. C. B. Camargo, C. F. O. R. Melo and R. R. Catharino, *Food Res. Int.*, 2018, **103**, 8.

61. F. Barré, R. M. A. Heeren and N. Ogrinc, *Curr. Pharm. Des.*, 2017, **23**, 174.
62. Z. Luo, Y. Zhao, T. Darwish, Y. Wang, J. Hou and F. Stellacci, *Nat. Commun.*, 2018, **9**, 1343.
63. R. Caprioli, M. Andersson and P. Andren, *Neuroproteomics*, 2009, 115.
64. T. Morikawa-Ichinose, Y. Fujimura, F. Murayama, Y. Yamazaki, T. Yamamoto, H. Wariishi and D. Miura, *J. Am. Soc. Mass Spectrom.*, 2019, **30**, 1512.
65. A. M. Race, D. Sutton, G. Hamm, G. Maglennon, J. P. Morton, N. Strittmatter, A. Campbell, O. J. Sansom, Y. Wang, S. T. Barry, Z. Takáts, R. J. Goodwin and J. Bunch, *Anal. Chem.*, 2021, **93**, 3061.
66. Food and Drug Administration, *Analytical Procedures and Methods. Methods Validation for Drugs and Biologics*, 2015, https://www.fda.gov/files/drugs/published/Analytical-Procedures-and-Methods-Validation-for-Drugs-and-Biologics.pdf.
67. National Institute of Standard and Technology, http://www.nist.com/.
68. T. Porta, A. Lesur, E. Varesio and G. Hopfgartner, *Anal. Bioanal. Chem.*, 2015, **407**, 2177.

Section 2

CHAPTER 7

MALDI Mass Spectrometry Imaging and Metabolomics – A Glimpse into the Local Metabolic State of Cells and Tissues

HANNAH BEDNARZ[a] AND KARSTEN NIEHAUS*[b]

[a] Bielefeld University, Medical School OWL, Sustainable Environmental Health Sciences, 33615, Bielefeld, Germany; [b] Bielefeld University, Proteome and Metabolome Research, Center for Biotechnology (CeBiTec) & Faculty of Biology, 33615, Bielefeld, Germany
*Email: kniehaus@cebitec.uni-bielefeld.de

7.1 Introduction

In 1971 Horning and Horning[1] published an article termed "Metabolic profiles: Gas-phase methods for analysis of metabolites". This publication can be seen as a pre-phase of recent metabolomics studies. The word metabolome appeared first in 1998, in a review on the post-genome functional studies of yeast and within a metabolite pool analysis of *Escherichia coli*. Within a few years, metabolomics became a valuable tool not only in the analysis of microorganisms. It has become a part of biotechnology, biomarker-discovery, biomedicine, food technology, forensics, systems biology, synthetic biology, and other fields.[2]

The metabolism defines the highly dynamic conversion of organic substances within organisms. This conversion is conducted by a network of enzymes and is organized in pathways which aim either toward anabolism,

New Developments in Mass Spectrometry No. 12
MALDI Mass Spectrometry Imaging: From Fundamentals to Spatial Omics
Edited by Tiffany Siegel

Published by the Royal Society of Chemistry, www.rsc.org

the biosynthesis of complex molecules from smaller ones; or catabolism, the breaking down of complex substrates, concurrently producing the energy that fuels the anabolism. While in plants or microorganisms metabolism is typically divided into primary (fundamental) and secondary (beneficial, not essential) metabolism, in vertebrates usually the terms endogenous, from own origin, and exogenous, of external origin, are applicable.

Metabolites are the intermediates of metabolism that are rather small molecules (usually up to 1500 Da in size), with the most chemically complex characteristics within the group of OMICS biomolecules. The term refers to lipids, carbohydrates, amino acids, nucleic acids, but also cofactors, vitamins, and others. Consequentially, they are highly diverse in biochemical characteristics, as they can contain either polar or nonpolar groups, show hydrophilic or hydrophobic properties and can also be volatile at room temperature.

Next to biosynthesis and energy production, the intermediates of the metabolism are involved in a multitude of processes, including among others intra- and extracellular cell signaling, modification of protein function and activity, and the regulation of gene expression. They are being passed through the network of enzymes with a specific flux and their biotransformation includes a variety of reactions, like oxidation, reduction, and hydrolysis.

Just as diverse as the family of metabolites, as broad is the spectrum of methods and instrumentation available for the purpose of metabolic analysis, which is furthermore under constant technical development. While the definition of the metabolome refers to the whole set of metabolites within an organism at a given time point, metabolomic analyses can either aim for the identification of the whole interacting system or for some selected single metabolites or for a specific functional subgroup of interest. As mentioned before, metabolism is highly dynamic and it can alter within seconds in response to environmental changes, or any other triggering event. A glimpse at this current metabolic state reflects the energy and health constitution of an organism and is therefore of great value for analysis.

Metabolic conversions in multi-cellular organisms are moreover unique to specialized cells and tissues, and every cell is in dynamic metabolic interchange with its microenvironment. As metabolites are usually not recognized by antibodies for immunohistochemistry or for visualization with specific stains, methods for visualization of their localization are sparse. Therefore, MSI presents a possibility to gain an insight into the local abundance and distribution of metabolites in cells and in tissues entities, organs, and even in whole organisms.

7.2 You Can't Have it All – Sample Retrieval, Preparation and Matrix Deposition – Crucial Steps of the MSI Experiment for Metabolic Analysis

The focus of this chapter is to point out technical requirements that are specific for MSI of small molecules. The general sample processing prior to MSI analysis of metabolites includes sample retrieval, metabolic quenching, embedding and tissue sectioning, optional deparaffination and/or derivatization steps, and finally matrix application (see Figure 7.1). Based on the depicted variety of metabolites and their structural and chemical characteristics, a multitude of methods exists for the analysis of metabolites, unlike for the analysis of other building blocks of life. All these protocols, either aiming at sample preparation or the choice of the right matrix for the given question cannot be entirely covered in detail within this section. For further information on the topic of sample preparation please refer to Chapter 5 of this book.

7.2.1 Metabolic Quenching

Highly dynamic processes of adaptation or perturbations of the metabolism upon slight changes of the environment or upon any triggering event, make the quenching step one of the most essential elements of metabolomics experiments. Metabolism quenching implements methods intending to immediately terminate enzymatic activity and to halt the metabolic turnover. Any delay between tissue extraction and the quenching step can lead to false results, which can originate from enzymatic conversions, or from the chemical transformation reactions, *e.g.* the oxidation of lipids. Lability of specific compounds was reported especially from neurology studies in the brain, where rapid phosphocreatine depletion or ATP to ADP metabolism can occur during postmortem ischemia. Some methods able to overcome these conversions were reported, which include fast heat-stabilization of ex-vivo tissues by the inactivation of enzymes at 95 °C[3] or by *in-situ* focused microwave irradiation (FMW), or the *in-situ*-freezing or funnel-freezing of animals under anesthesia.[4] Some of these methods, however, are limited to specimens from model organisms and cannot be adapted to human samples.

In most frequent cases, the user can decide to either rapidly freeze the extracted sample or chemically fix the tissue in order to stop the enzymatic conversions.[5] The decision for further analysis has to be made at this stage. Taking into account that mass spectrometry is a very sensitive analytical technique, and the computational data analysis reveals any slight difference between measurements, it is crucial to keep the chosen and trained methods during the whole experiment. The best way is to define standard operating procedures (SOPs).

7.2.2 Preparation of Flash-frozen Tissue Samples

In general, fresh and snap-frozen sample material is considered to be the best choice for the recovery and further analysis of metabolites. This involves a minimum amount of processing steps and therefore, both the higher number and the more varied spectrum of the metabolites are conserved and can be characterized. On the other hand, since the sample has to be maintained under frozen conditions during further steps, any freeze–thaw cycles should be avoided. Moreover, the preservation of the internal structures is

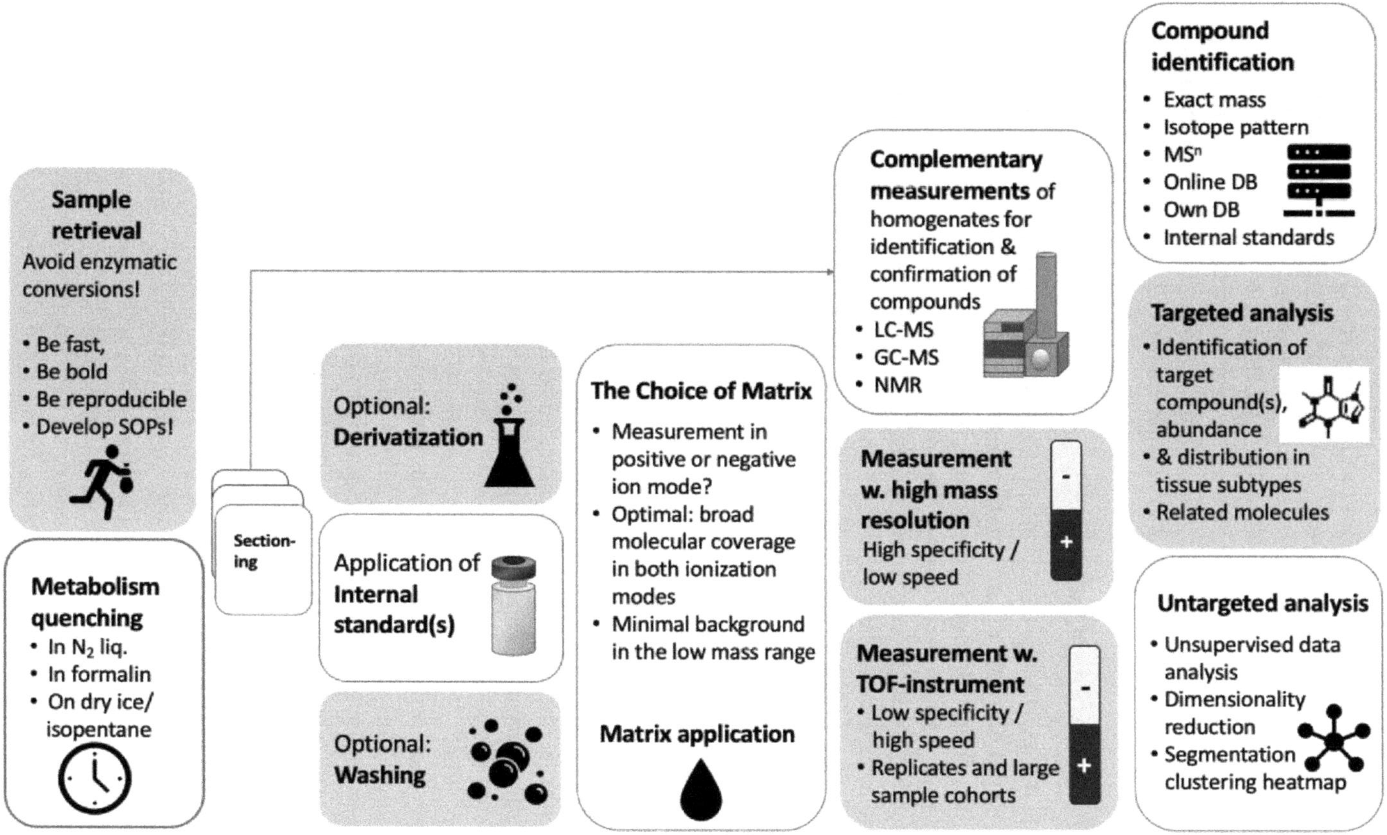

Figure 7.1 Crucial steps in the workflow for MALDI-MSI measurement of metabolites.

not optimal in flash-frozen tissues and artifact fissures are likely to appear during sectioning.

Rapid freezing can be achieved by the placement of the extracted tissues in the optimal-sized container, like a PE tube, while holding it with a pair of forceps and quickly immersing it in liquid nitrogen. Direct immersion of tissues in liquid nitrogen should be avoided, as it is more damaging. Isopentane, cooled by dry ice or liquid nitrogen tends to be less damaging to tissue samples, however it is volatile, flammable and irritating to skin, eyes and the respiratory tract, so it should be handled with care. In order to obtain better tissue integrity and to avoid cracks or fissures, smaller samples can also be frozen on top of a dry ice layer, placed on aluminum foil. Once frozen, the sample has to be kept at least at -80 °C, better at -150 °C or below for long term storage. The sectioning has to be performed in *cryo*-microtomes and the samples can be frozen-embedded in diverse materials, in order to facilitate the sectioning procedure. Therefore, pure water, chilled to just above the freezing point can be subsequently used to place and embed the sample on the cutting target. Alternatively, carboxymethyl cellulose (CMC, 4–5%) can also be used for the freeze-embedding of samples prior to sectioning. The standard embedding in optimal cutting temperature (OCT) compounds can result in the smearing of the ingredients during sectioning and the later appearance of the component masses in the mass spectrum at around 700–1000 Da. This can lead to some ionic suppression of the overall signal intensity. Moreover, the peaks resulting from the OCT can overlay with tissue-specific signals, which can sometimes be overcome by using instruments with high resolving power only.

The depletion of lipids by acetone,[6] ethanol,[7] or cold chloroform[8] can minimize the ionic suppression in lipid-rich specimens and lead to signal enhancement for some metabolites and other small molecules.

As fresh-frozen tissue sections should not be exposed to any freeze–thaw cycles, it is possible to either store the frozen sections until the measurement or to dehydrate the sample as fast as possible, in order to eliminate enzymatic conversions. This step can be performed by placing object slides with frozen sections in a desiccator with a connected and pre-warmed vacuum pump and by applying rapid vacuum to the sections.

Vast amounts of specimens, that are archived in collections from clinical or medical studies, as well as originating from other fields of research, however, are conserved in the form of chemically fixed tissues, mostly as formalin-fixed, paraffin-embedded (FFPE) samples. Therefore, tissues are conserved in formalin, preferably directly after excision, followed by dehydration in ethanol, clearance in xylene, and embedding in paraffin blocks. Sections of these blocks are being gained upon requirement; the paraffin, however, has to be removed ahead of any analysis. These samples were formerly not used for metabolite analysis on a routine basis, as the prerequisite of more subsequent processing cycles was considered to wash out, to modify, and to delocalize small molecules. However, several studies have proved to obtain a considerable amount of analytes from FFPE specimens, both in LCMS-based analysis in homogenized samples,[9] as well as by means of MALDI-MSI along with distinctive localization in tissue sections.[10,11] Comprehensive comparison of metabolites, that can be obtained from tissue specimens and from the corresponding cell tissue cultures, both as frozen and FFPE material, showed however significantly lower values for FFPE material.[12] Around 55% of metabolites could be detected in FFPE cell cultures and 40% in human tumor tissues in comparison to their frozen counterparts. The approaches usually tend to observe a significant loss of metabolites, while using FFPE tissue, however with varying results in dependence upon the respective substance class. The substantial losses mostly refer to lipids, fatty acid esters, steroids, peptides, amino acids, or carbohydrates. Moreover, the reactions of formaldehyde with amine groups introduce methylene cross-links between proteins, leading to fixation of the perfused tissues as a result.[13] Cross-links between proteins and nucleic acids are also likely to be formed. Beyond these proven modifications, more reactions of the highly polar formaldehyde with other tissue components are likely to take place, probably also affecting to some extent the chemically diverse family of metabolites. These modifications have not been systematically studied so far. Possible changes to compound structures and shifts of *m/z* values due to chemical modification have therefore to be considered, when FFPE tissues are chosen for metabolic analysis.

Nevertheless, where freshly frozen material is difficult to obtain or a highly conserved ultrastructure is a prerequisite, purposeful metabolomic analysis can also be performed on the basis of FFPE sample materials.

7.2.3 If Matrix Chemistry Fails to Ionize the Target Metabolite or Metabolite Group – Derivatization

The vast amount of different compounds being continuously processed through metabolic pathways in any living organism results in a wide spectrum of metabolites at different concentrations at the chosen time point of analysis. Some can appear in low abundance and close to the detection limit or the peaks can become overlaid by peaks of matrix fragments or by other interfering signals. Depending on the chosen ionization mode, metabolites can be hard to ionize in *e.g.* a MALDI source. Multiple methods of on-tissue chemical derivatization (OTCD) were adapted from related analytical methods or were developed exclusively for MALDI-MSI of metabolites. These methods aim to increase the overall coverage of metabolites, the enhancement of ionization efficiency, and thus that of signal intensity or the increase of *m/z* value of the target molecule in order to perform a shift in the mass range. Numerous methods of derivatization in metabolome studies have been summarized in current and comprehensive reviews.[14,15]

In this context, particular mention should be made of 4-hydroxy-3-methoxycinnamaldehyde (CA), *N,N,N*-trimethylammonioanilyl *N*-hydroxysuccinimidyl carbamate iodide (TAHS) or 2,3-diphenyl-pyranylium tetrafluoroborate (DPP-TFB), which are widely in use for the on-tissue derivatization of amine groups, leading to enhanced analysis of amino acids or neurotransmitters. CA in combination with 4-hydroxy-3-methoxycinnamic acid (ferulic acid, FA) as a matrix was shown to efficiently enhance the detection of endogenous amines, even when the targets were pre-coated in advance with CA/FA mixture and the fresh-frozen tissues of rat brain or pig adrenal glands were placed on top for subsequent derivatization.[16] Carbonyl groups and steroid substances can be enhanced by Girard's reagent T (GirT) and carboxyl groups can be targeted by 2-picolylamine (2PA).[17]

It is obvious that any derivatization step will increase the *m*/*z* ratio of compounds, which has to be considered during data analysis, when the identification of compounds is performed by their *m*/*z* values or by their MS^n fragment pattern, which will also be altered. In order to verify any potential identification, the substances should be purchased as pure chemical standards, followed by their derivatization, fragmentation, and comparison of the generated patterns.

7.2.4 Selection of the Right Matrix for the MALDI-ionization Process – Targeting the Peak of the Iceberg in Metabolomics Research

A myriad of metabolite structures has been described. The Human Metabolome Database 4.0 (www.hmdb.ca) contains 114 265 metabolites. About 30 000 pathways for small molecules can currently be accessed in the associated SMPD-database for the human organism only. The spectrum of secondary metabolites being synthesized by various plant species is estimated to be even higher.[18] The average mass of detected and quantified metabolites within the human metabolome database (HMDB) shows a binominal distribution with a broad peak around 150 to 350 Da and another at around 800 Da (see Figure 7.2). The latter one is mainly formed by different types of lipids, that are discussed in more detail in Chapter 8 of this book. A minimal cell requires around 300 metabolites forming central-, amino acid, nucleotide-, lipid- and co-factor-metabolism.[19] Central metabolites such as glucose (180 Da), glucose phosphate (260 Da), glycerol-3-phosphate (172 Da), pyruvate (88 Da), methionine (149 Da) and adenine (135 Da) show low molecular weights. ATP (507 Da), NADP (744 Da), and co-enzyme A (767 Da), as other co-factors, show higher molecular weights, but only about 1% of all detected and quantified metabolites of the HMDB exceed the mass of 1000 Da.

Nevertheless, due to the chemical variety and complexity of metabolite species, there does not exist a one and only ultimate analytical method of choice for metabolite analysis. Even under optimally chosen conditions, only a subset of metabolites will be detected. This limitation applies to any mass spectrometry-based method, as their analysis spectrum strongly depends on ionization efficiency and the effectiveness of transfer of positive or negative electric charge to the target molecules. In order to increase this efficiency,

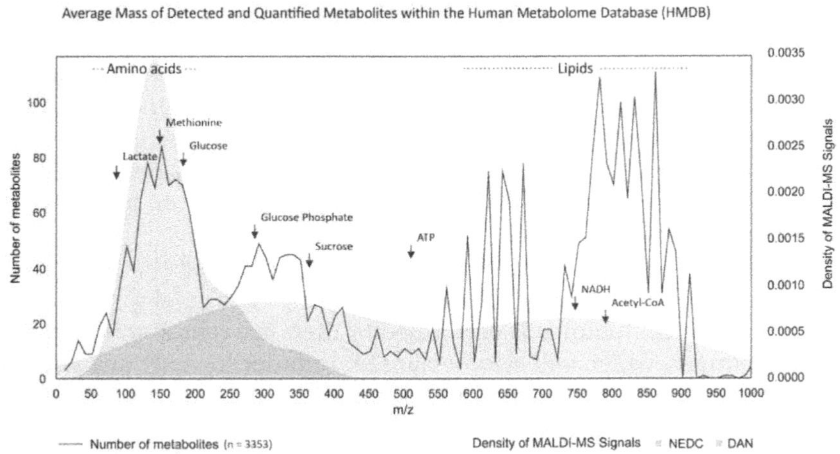

Figure 7.2 Average mass and distribution of identified and quantified metabolites within the human metabolome database (black curve) and detection of these metabolites in MSI experiments in human brain tumors and adenocarcinoma tissues by using *N*-(1-naphthyl)ethylenediamine dihydrochloride (NEDC, olive shape) or 1,5-diaminonaphthalene (DAN, blue shape) as the matrix. Several key metabolites within the range of *m/z* 0–1000 are indicated.

the MALDI process utilizes the application of substances, called matrices, which are supposed to extract the analytes from the sample, co-crystallize, absorb the energy of the MALDI laser (mostly in the UV range), and efficiently transform the energy toward ionization effects. Accordingly, in the MS imaging process, the matrix is being applied to the tissue section by methods of spraying,[20] sublimation,[21] and even immersion that can be an alternative in some cases,[22] or other coating or printing methods (refer to Chapter 5).

Many matrix substances for the analysis of metabolites are known, some have been adapted from the dried droplet MALDI analysis mode and many more are being developed by rational design. The optimal matrix substance should fulfill several needs:

A) It shall cover a broad range of analytes and therefore a broad mass range;
B) Optimally it shall create both positive and negatively-charged ions, thus enhancing the spectrum of metabolite detection from one sample;
C) Background signals of the matrix itself shall be preferably minimal, especially in the low mass range <1000 Da, where many amino acids and other metabolites of the central metabolism are detected as singly charged ions (see Figure 7.2);
D) The size of the analyte–matrix cocrystals shall be small and their distribution even, optimally producing a surface rather than individually

dispersed crystals. The matrix on-tissue deposition should be tested microscopically, by scanning electron microscopy at best, in order to verify the selected sample preparation procedures; especially in cases when high lateral resolution imaging of metabolites in substructures or even at subcellular resolution shall be achieved;

E) Neither the matrix itself shall fragment under UV-excitation, nor shall it enhance fragmentation of the analytes, as both would contribute to additional signals in the low mass range of metabolites;

F) The matrix shall be stable under vacuum of the MALDI source during the whole period of the measurement, which can take from minutes up to several hours, depending on the instrument, the sample size, and the chosen resolution. To our knowledge, the existing matrices currently fulfil only a part of these "ultimate" requirements for the analysis of metabolites.[23]

Well established and widely applied MALDI-MSI matrices, 2,5-dihydroxybenzoic acid, (2,5-DHB) and alpha-cyano-4-hydroxycinnamic acid (CHCA) have both been used for the detection of metabolites in tissues;[24–26] nevertheless, they tend to contribute to multiple noise signals in the low mass range and novel matrix substances have shown to perform better in the field of metabolic analyses. In the last few years, several matrix substances have been either designed or established as MALDI matrices especially for the detection of small metabolic compounds in a broad spectrum of tissue specimens, such as 9-aminoacridine (9-AA),[11,27] 1,5-diaminonaphthalene hydrochloride (DAN-HCL),[7,28–31] *N*-(1-naphthyl)ethylenediamine dihydrochloride (NEDC),[32,33] or 4,5-(bis(dimethylamino)naphthalen-1-yl)furan-2,5-dione (4-maleicanhydrido proton sponge, MAPS),[34] all applicable for negative ion detection. Unfortunately, there is not one matrix that suits all needs. *N*-(1-Naphthyl)ethylenediamine dihydrochloride (NEDC) works well for small metabolites, but fails to detect molecules with a higher mass that are better detected by 1,5-diaminonaphthalene (DAN) (see Figure 7.2).

Conjugated polymers are useful matrices for dual-mode measurements, while lacking any background signals in the low mass spectrum.[35] An up-to-date overview of novel matrices for MALDI-MSI was currently presented by Zhou and co-workers.[15]

Enhanced detection of metabolites in MALDI-MS analysis was reported by combining matrix substances.[36] This principle can also be transferred to MS imaging.[37] This possibility broadens the spectrum of methods, which can be optimized in preliminary experiments with regard to the research issue. The other possibility is to use automated inkjet printing to apply a fine grid of different matrices and to use the same pattern for the MSI measurement.[38] In this way, the advantageous characteristics of multiple matrix substances can be jointly used within one tissue section, however taking into account the decreased lateral resolution, which is limited by the size of the spotted matrix droplets.

Mapping of metabolites in tissues can be performed even without any application of a matrix; the detection will nevertheless be narrowed to a limited number of molecules, those that most likely absorb the energy of the UV laser without fragmentation. Especially in regard to plants specimens, compounds of the flavonoid family, or small secondary metabolites specific to lichens can be detected by laser induced ionization (LDI-MSI).[39] The good performance of flavonoids or coumarins in LDI makes the derivates suitable for the application as MALDI matrices (CHCA, 3-Hydroxycoumarin).

The focus of this book lies on methods of MALDI-MSI. However, it should be mentioned that in the related methods of SALDI (surface-assisted laser desorption/ionization)[40] or nanostructure-assisted laser desorption/ionization (NALDI) approaches complement the classic MALDI in the field of modern bioanalytics.[41] Replacing the classical matrix with nanomaterials like nanoparticles or nanolayers of metal or metal oxides (*e.g.* Au, Ag, TiO_2 *etc.*)[42] or other nanosubstrates (graphene oxide,[43] carbon dots,[44] and carbon nanotubes[45]) have successfully been applied in the field of spatial analysis of metabolites in tissues. By applying gold nanoparticles on MALDI targets (AuNPET), MSI of tissue imprints to these surfaces could be performed for metabolite distributions in garlic cloves.[46] Taken together, these matrix-free methods allow for improvement of the selectivity and sensitivity of the spectrometric measurement of biologically important small molecules. Nevertheless, the alternatives are currently not a universal substitution for classical small molecule matrices in imaging of the metabolome.

7.3 Most Used Instrumentation with Strengths and Current Limitations

7.3.1 Resolution Power, Exact Mass and Isotope Distribution

Classical metabolome analysis includes a separation step, *e.g.* liquid chromatography, coupled *via* electrospray ionization to a mass analyzer. In MSI experiments no separation of the analytes takes place before they enter the mass analyzer. For this reason, the resolution power and mass accuracy of the mass analyzer are of special importance in the MSI of metabolites.

Fourier transform ion cyclotron resonance (FT-ICR) mass spectrometry provides the best mass resolution power, in recent instruments better than 1.6×10^6 FWHM at m/z 400, and a mass accuracy better than 0.1 ppm error, that also allows for the analysis of the isotope fine structure at a resolution of 2 mDa.[47] The exact mass, in combination with the isotope distribution, allows for the calculation of the sum formula of an analyte of interest.[48] Even nearly isobaric metabolites can be differentiated, identified and localized with a good dynamic range for (semi-) quantification.[47] Unfortunately, FT-ICR instruments are expensive in their acquisition and their maintenance, not easy to operate, and slow in sample scanning, limiting their general use.

MALDI-Orbitrap FTMS has a mass resolution power $>0.5\times10^6$ FWHM at 200 m/z (Orbitrap Fusion), and an accuracy between 1–5 ppm. These well-established instruments are convenient to use, and data files can be analyzed using a number of software tools. A further advantage is the

MS^n -potential of these instruments, which adds a further analytical dimension for identification. Orbitrap mass analyzers are often combined with the Spectroglyph-[49] or Transmit-MALDI source.[50] The Spectroglyph source offers the possibility to switch between an ESI- and MALDI-ionization without any change in the hardware. The Transmit-MALDI source allows a lateral resolution of 1.4 μm, the best resolution of all commercial instruments.[51]

MALDI-Time of Flight (TOF) instruments are the most common instruments used for MSI. The advantage of these instruments is the robustness, ease of use, and high speed of measurement. The rapifleX (Bruker) operates at 40 true pixels per second, supported by a 10 kHz 3D laser that has a diameter of 5 μm enabling MSI at 10 μm resolution. Unfortunately, the resolution power is only about 40 000 FWHM with a mass accuracy of about 5 ppm for small molecules. A direct comparison of one sample analyzed by MALDI FT-ICR- and MALDI–TOF-MS was presented by Buck and co-workers.[10] Spectral analysis revealed that about one-third of the detected peaks remained unresolved by MALDI–TOF, which led to a 3–5 times lower number of *m*/*z* features compared to FTICR measurements. Nevertheless, MALDI–TOF allowed the reproduction and verification of individual markers identified previously by MALDI–FTICR MSI, indicating that high resolution measurements can help metabolite identification as validation in MALDI–TOF-MS high throughput experiments.

From this point of view, nearly all MALDI instruments are generally suited for MSI of metabolites, since all offer a mass resolution that separates the isotopes within the relevant mass range of $m/z < 1000$.

7.3.2 Tandem Mass Spectrometry and Ion Mobility

With currently available instruments, the users have to decide between throughput and ease of use, *versus* high resolution and high accuracy of the masses.[52] Tandem mass spectrometry can enable a compromise. In contrast to peptides, metabolites cannot be sequenced by MS/MS experiments. Nevertheless, a fairly good mass accuracy, in combination with tandem mass spectrometry, can significantly improve the identification of small molecules.[48] Adding a further dimension of separation by including ion mobility can further help to identify more metabolites within one sample. Metabolite identification confidence is simultaneously enhanced by incorporating orthogonal collision cross section (CCS) data matching, as recently shown for 146 plant natural products in a chromatography-based analysis.[53]

High lateral resolution is realized by a very small laser focus, which in turn ablates only small amounts of metabolites, resulting in a low sensitivity. To some extent, this limitation can be overcome by laser-induced post-ionization (MALDI-2). MALDI-2 produces an increase in the number of detected features as well as in ion signal intensities.[54] Combining MALDI-2 with trapped ion mobility spectrometry (TIMS) resulted in a further improvement in signal separation and sensitivity.[55] In addition, the use of a second separation dimension such as trapped ion mobility spectrometry strengthens the confidence of the identification based on the collision cross section (CCS). This combination is also beneficial for small isobaric metabolites, especially for lipids.[56]

7.4 Strategies for Structural Analysis/Identification

Metabolomics pool measurements can give a first clue on the adaptation of a cell, tissue, or organ to a changing environment. Unfortunately, metabolite pools alone can be misleading since a high flux through a pathway can result in higher pools but rising pools can also be explained by a block in the same pathway. Visualization of a metabolic pathway for a tissue or organ can be the first step to analyze local functions, or hints to the consequences of a disease (see Figure 7.3). Getting information on the metabolic flux would be more relevant. To analyze metabolic fluxes heavy isotopes, most frequently ^{13}C- and ^{15}N-labeled nutrients, are commonly used. Dynamic labeling, done

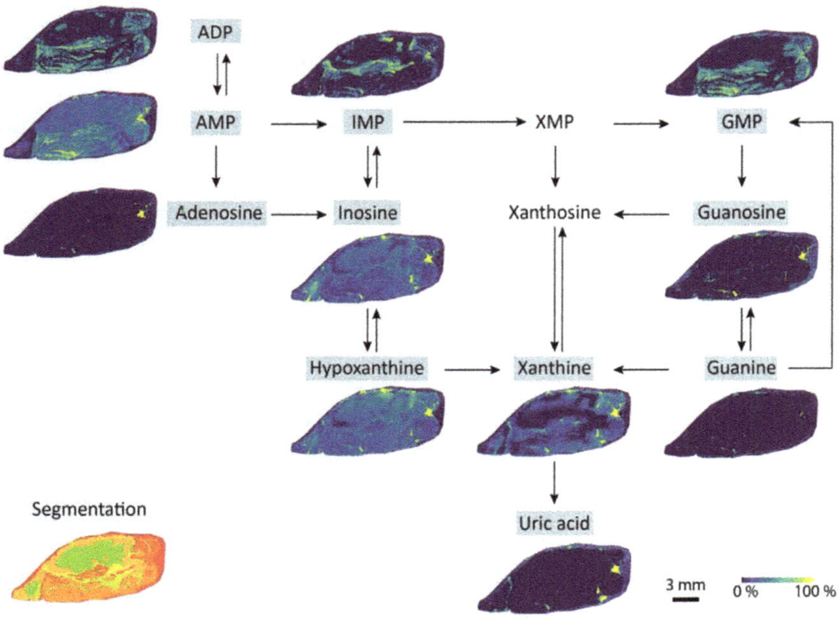

Figure 7.3 MALDI-MSI imaging of metabolic pools in rat brains with DAN-HCL as the matrix. The distribution of any metabolite species represents an insight into the localization of the pool of this metabolite at the time of sample retrieval and metabolism quenching. The interpretation of these metabolite pools has to be done carefully. On the one hand, a high flux through a pathway can result in higher pools, but rising pools can also be explained by a block in the same pathway. Segmentation of all results with the selected algorithm can show chemical similarities within tissue areas and support the identification of functional entities. Figure courtesy of Judith Neumann, unpublished results.

in a pulse-chase experiment, follows the wave of the heavy isotopes through the metabolic pathway. Steady-state labeling requires an equilibrium of labeling and the analysis of the distribution of the heavy isotopes in certain metabolites.[57] Both approaches require intensive computational data integration and a good knowledge of the biochemical pathways within the analyzed organism. Extracellular metabolite concentrations, isotopic labeling, quantification of the isotopomers, and data integration with the help of metabolic flux models are used to better understand the metabolic adaptations of cancer cells.[58] Since metabolism is coupled to specific cells, tissues and organs, imaging of metabolic fluxes would be extremely helpful to comprehend the interaction of cancer cells with its neighboring tissues. Louie and co-workers[59] labeled metabolites of a tumor-bearing mouse with $^{2}H_2O$-enriched water to incorporate deuterium into tissues, as a result of active metabolism. The deuterium was rapidly incorporated into the newly synthesized lipids, generating kinetic images that distinguish metabolic active regions from regions with lower fluxes. A more recent analysis using ^{13}C- and ^{15}N-labeling of metabolites, in combination with high-resolution gas cluster ion beam secondary ion mass spectrometry (GCIB-SIMS), allowed the localization of a channeled de-novo purine synthesis at a sub-cellular level in HeLa cells.[60]

In GC–MS analysis metabolite identification is well supported by databases such as the Golm metabolite database (GMD) or the NIST database. In addition, molecular formulas can be filtered by exact mass, isotope distribution, elemental composition, H/C-ratio, and other heuristic rules.[61] Unfortunately, in MSI experiments the quality of mass spectra is influenced by a number of variables that are not easy to control. The 3D-topology and uniformity of sample and the applied matrix have a significant impact on mass resolution and accuracy. To some extent, careful sample preparation and mass shift correction can overcome this problem.[62] As in chromatography-MS-based metabolomics, a combination of high resolution, exact mass, isotope pattern, and fragmentation are needed to ensure a confident identification of metabolites. This approach is further supported by databases such as the human metabolome database, the METASPACE project,[63] and others. Combining ion mobility, tandem mass spectrometry, good mass accuracy, and a user-defined metabolite database for a particular instrumental setup will result in increased success in metabolite identification.[55,64]

Besides the database itself, reporting of metabolite data should be standardized.[65,66] Due to the different sample preparations, ionization techniques and mass spectrometers, MSI-data cannot be compared to a database as for GC–MS data sets. One approach focused on MSI data is the METASPACE project.[63] Finally, the software used to analyze metabolome data in MSI approaches has an impact on the quality of identification and quantification. The biochemical interpretation of the MSI data depends on the correct processing of the raw data. The data analysis pipeline generally includes peak picking, normalization, deisotoping, peak identification by several rules, image reconstruction, quantification, and in a growing number of projects also comparison with multimodal image data. Especially in medicinal projects histopathology and image analysis by software approaches such as QuPath become increasingly important.[67] During the last few years, a number of vendor-dependent as well as vendor-independent software packages have been developed. A recent approach to combine the essential functions in one pipeline is LipostarMSI.[68] This vendor-neutral software allows an automated molecular identification, based beside other criteria, on MS/MS-data. MS/MS spectra can either be acquired in parallel with the MSI acquisition or *via* separate LC – MS/MS experiments, where identifications are then linked with MSI data *via* precursor ion matching. Nevertheless, a functionality to analyze flux data from heavy isotope labeling MSI experiments is missing.

7.5 Applications

MALDI-MSI builds bridges between analytic modalities of histochemistry, cytochemistry, and those of the compound analysis and identification.

Especially regarding metabolites, this method offers one of the first possibilities to visualize a greater band-width of these compounds directly in tissues, therefore enabling analysis of the compounds in their spatial relationship to each other and within the context to morphological and functional structures. Whereas peptides and proteins can be accessed by specific antibodies or the conversion of chromogenic substrates by localized, tissue-specific enzymes, histological structures such as nuclei, cytoplasm, fibers, collagen, or lipids reach areas that can be exposed by specific stains. Neither the usage of antibodies nor the utility of staining can generally be applied to the visualization and detection of metabolites and hence MSI offers an exciting novel possibility. While method development is still an ongoing task and MALDI-MSI is gaining robustness, sensitivity and spatial resolution with every methodological improvement, multiple studies begin to tackle open questions of anatomy, physiology, biochemistry, or immunology or define new ones, induced by the novel analytical technique. Still, despite the possibility to extract ionic maps of compound distributions of many specimens, MALDI-MSI analysis of metabolites can be quite laborious. Therefore, while setting-up an experimental setup it should always be considered whether and in which way should the results of the MALDI-MS imaging experiments help to answer the existing questions, such as does the underlying problem have a spatial component? Otherwise, other analytical methods for metabolomic analysis, such as HPLC–MS, NMR, or GC–MS tend to be more robust, sensitive, quantitative, and available for high-throughput analysis and therefore should still be preferred.

7.5.1 MALDI-MSI of Plant Metabolites

In plants, as in all multicellular organisms, physiological functions are connected to specific tissues and allow them to carry out functions that support the general fitness of the entire organism. The metabolome of leaves, flowers, seeds, stems, and roots is for obvious reasons different from each other, since these organs realize very different functions. But also, an organ such as a leaf or a root is not physiologically homogeneous. Even substructures such as vascular bundles and their surrounding cells are morphologically as well as physiologically differentiated. Only the interplay of cellular form and physiological function, on the level of the single cell, will explain the performance of the entire individual. In this context MSI can contribute to an understanding of the plants physiology at the level of systems biology.[69] MSI offers a novel approach to analyze the distribution of a huge diversity of plant metabolites over time and space.[70]

Plant leaves,[71] petals,[72] roots,[73] and other small or flat parts of the plant can be harvested, mounted on a glass slide, dried, and analyzed by MALDI-MS after matrix application. The advantage of this approach is that it is fast and therefore artifact-minimizing. Plant waxes of the cuticula can be removed by dipping the sample in chloroform, resulting in a better and more sensitive detection of primary and secondary metabolites of the tissues beneath the epidermal cells.[74] In such sample preparations the topology of the sample can interfere with the quality of the recorded mass spectra. This problem can be overcome to some extent, by autofocusing MALDI-MS,[50] but even so, this problem is not solved in general and for all instrumentation.

Otherwise, sample preparation for MSI of plants often starts with snap-frozen materials and cryo-sectioning. Due to a high-water content and rigid cell walls, cryo-sectioning of plant materials can be unsatisfactory. Careful snap freezing, embedding of materials in carboxy-methyl cellulose, and adaptation of the temperature during sectioning can overcome the problem to some extent.[75] Dry plant materials can tend to crumble during cryosection, a problem that can be solved by gluing the sections to an adhesive tape while sectioning.[76] In most studies DHB is the matrix of choice, covering a wide range of metabolites from amino acids to sugars and secondary metabolites.[77]

MSI has been used to study the primary metabolism of plants,[78] adaptation to biotic[71] and abiotic stresses,[79] symbiosis,[80] developmental biology,[76] environmental biology[81] and to study the broad range of secondary plant metabolites.[24] It is estimated that there are more than 200 000 secondary metabolites in plants.[82] This huge amount of different secondary metabolites is a characteristic challenge in analyzing plant samples. In liquid chromatography–mass spectrometry analysis the huge number of different metabolites can be addressed by setting up a user-specific library.[53] A very similar approach could be useful in MSI as well. MSI analyses of the metabolic flux in plant tissues by stable isotope labeling are currently missing.

7.5.2 Application of Metabolite Imaging to Animal Tissues

The mouse brain is perhaps the most often analyzed sample in MSI development.[83] Besides the technological development, animal samples play an outstanding role in MSI for disease models, toxicology, pharmacology, and developmental biology. As in all other analyses, sample preparation is the most critical step of the entire analytical process. Often snap-frozen tissues are used to analyze the metabolome of animal tissues. Freezing can be carried out in liquid nitrogen, but freezing samples on dry ice seems to result in better thin sections.[84] The ATP/ADP-ratio can be used, besides other labile metabolites, to judge the effectiveness in sample preparation for animal metabolomics.[85] As well as cryo-sectioning, formalin-fixed paraffin-embedded (FFPE) animal tissues are analyzed by MSI. A careful comparison of fresh-frozen to FFPE tissues revealed that about 72% of the metabolites could be detected by MSI in both samples.[11] In addition, pretreatment of the section can effectively increase the sensitivity for the detection of several analytes.[86]

7.5.3 MALDI-MSI in Pharmacology and Toxicology

MSI is very well suited for marker-free visualization of pharmacological compounds in animal tissues.[87] It is possible to detect administered drug and metabolite distributions in a whole rat sagittal tissue section by a single MSI analysis.[88] In contrast to radioactive labeling of the drug and autoradiography, MSI is easier, cost efficient, and a metabolic conversion of the drug is detected. Further on, changes in the metabolism of the tissues, induced by the pharmacological compound, become visible.[87] In time-series experiments pharmacokinetic data can be obtained.[89]

As well as the precise localization of a drug in entire animals, organs, or tissues, MSI is used to characterize toxicological side effects of drugs. MSI is used for safety evaluation of novel compounds, with a special focus on metabolic alterations in distinct localizations.[90] MSI is suited to also follow very local alterations, that are potentially overseen by classical metabolomics approaches.[91]

7.5.4 MALDI-MSI of Metabolites in Animal Disease Models

Animal models that represent a human disease are widely used. Animal models are crucial tools for the elucidation and understanding of gene functions, exploring pathogenesis of human diseases, and developing new drugs. Models for stroke, ischemia, fatty liver disease, fractures, and many others have been established. CRISPR-Cas9, and other genome editing technologies broaden this approach to genetically induced diseases.[92]

MSI can be used to analyze atherosclerosis,[93] diseases of the central nervous system,[94] cancer,[95] and other diseases.

7.5.5 Spatial Analysis of Metabolites in Human Tissues

MALDI-MSI experiments in human tissues are in most cases performed in relation to clinical studies, as the most common specimens are biopsies or tissue resections. The open questions for metabolomic research are those regarding the diagnostic analyses in pathological alterations, in basic research of physiological and molecular mechanisms of disease development, and in forensic analysis.

Metabolomic profiling in clinical research of cancer development often includes the visualization of metabolic hallmarks of cancer cells and tissues, *e.g.* the Warburg effect and motility of cancer cells. Metabolite-biomarkers support tumor-segmentation and clustering approaches, the definition of necrotic, mucoid or stromal structures, or the definition of tumor margins.

Evident alterations to carnitine and to fatty-acid metabolism were imaged in breast cancer tissues, which led to the verification of the increased activity of enzymes involved in beta-oxidation in these cancerous tissue regions.[7] Cancer associated metabolic pathways of energy metabolism, Krebs cycle or antioxidative reactions, as well as zinc-related metabolism were visualized within the tumor tissues in prostate cancer.[96,97] In addition to the increased abundance of metabolites of the central metabolism, gain-of-function mutation in isocitrate dehydrogenase (IDH) that leads to the production of oncometabolite 2-hydroxyglutarate could be visualized in patient-derived samples of glioblastoma brain tumors.[30,34] This can be used for the rapid (<5 min) identification of the IDH-mutation status, which supports the required classification of diffuse gliomas, and tumor diagnosis.[98]

Imaging of responses, that are induced in patient derived cells or tissues for *e.g.* in tumor organoids by specific tumor therapeutics, can provide precious information on the anticipated outcome of the therapy and support decisions for precise treatments, as could be shown in the case of colorectal tumor organoids[99] or in liver metastases of gastrointestinal stromal tumors.[100]

7.5.6 Imaging of Metabolites with High Lateral Resolution – Emerging Applications

The recent development of instrumentation (AP-SMALDI, MALDI-2) has led to a significant increase in lateral resolution and sensitivity in MALDI-MSI. Furthermore, improvements of sample preparation methods and the development of novel matrices with homogenous deposition and minimal cocrystal-forming properties are fueling the expansion of application methods, also in the context of metabolite imaging.

In recent years, MALDI-MSI contributed widely to the analysis of the physiology and metabolism of endogenous but also exogenous compounds in insects, a topic which was comprehensively reviewed by Yang and co-workers.[101] Insects tend to be challenging specimens, due to their small size, their complex morphology, and fragile organs and structures, which can collapse during sectioning. The outcomes of the studies however contribute to knowledge in insect biology, as well as to ecological and agricultural fields of research.

3D-cell culture models can be performed in small-scale experiments without the need for animal or human studies. These multicellular structures, *e.g.* tumor spheroids, can offer the optimal opportunity for pharmaceutical or toxicological studies. Despite the small size, they implement effects of a tissue environment as well as metabolic gradients and diffusion processes.[102,103] The highest challenges in high-resolution MSI of metabolites address single cell heterogeneity, in accordance with results of other omics-modalities, as single cell genomics, transcriptomics, and proteomics.

Exciting novel insights are also being gained in the field of metabolic analysis of microbial colonies and communities.[104–106] The dynamics of chemical communication (quorum sensing) within colonies of a single species, induction of biofilm formation or swarming activity and interaction with other bacterial communities, and the therefore locally induced metabolic alterations can optimally be addressed by MALDI-MSI.

7.6 Conclusions and Future Directions

The technique of MSI offers the first possibility for *in-situ* detection of a broad spectrum of metabolites and their visualization within the spatial tissue context. Nevertheless, despite the steady technological and methodological improvement in the field of MALDI-MSI, still the detection is limited to only a subset of possible metabolites. Also, some other challenges have to be faced, but they bear the potential to significantly improve the technique.

One of the key challenges is the enhancement of detection sensitivity. The aim is to improve the extraction of metabolites from tissues without significant delocalization, paired with enhanced ionization of metabolites to nearly cover up the whole spectrum of both more, as well as less abundant metabolites in the given tissue at a given timepoint. Therefore, it is crucial to improve our understanding of the reactions, which take place in the MALDI ionization process, in order to gain full control of the parameters and to shift the analysis either toward a broad untargeted analysis or toward a more specific analysis of target chemical groups or compounds.

Further improvements in the MALDI matrices and/or of chemical derivatization are needed, toward the establishment of the most universal procedure for the analysis of metabolites, instead of choosing from the myriad of currently existing methods. In addition, this step should also improve the analysis of tissues with different characteristics. Irrespective of whether dense, lipid-rich, structured tissues (as *e.g.* in plants) are to be analyzed – the extraction and detection of metabolites should be comparable and unbiased. This would also include more knowledge and security in the analysis of fixed tissues, which are more convenient in use than fresh, snapped-frozen specimens and give the best structure-preservation. In summary, the method of MSI for metabolites offers the opportunity to address both new aspects and unanswered questions regarding the spatial organization of metabolomic networks in organisms. The method may currently be limited to the detection of only a subset of metabolites – but this glimpse is exciting and promising for further future applications.

Acknowledgements

The authors thank Judith Martha Neumann for the support in the preparations of figures and the supply of exemplary MSI results.

References

1. E. C. Horning and M. G. Horning, *Clin. Chem.*, 1971, **17**, 802–809.
2. T. Cajka and O. Fiehn, *Anal. Chem.*, 2016, **88**, 524–545.
3. E. Q. Blatherwick, G. J. Van Berkel, K. Pickup, M. K. Johansson, M. E. Beaudoin, R. O. Cole, J. M. Day, S. Iverson, I. D. Wilson, J. H. Scrivens and D. J. Weston, *Xenobiotica*, 2011, **41**, 720–734.
4. I. A. Mulder, C. Esteve, M. J. Wermer, M. Hoehn, E. A. Tolner, A. M. van den Maagdenberg and L. A. McDonnell, *Proteomics*, 2016, **16**, 1652–1659.
5. G. A. Dienel, *J. Neurochem.*, 2021, **158**, 1007–1031.
6. C. Sun, Z. Li, C. Ma, Q. Zang, J. Li, W. Liu, H. Zhao and X. Wang, *J. Pharm. Biomed. Anal.*, 2019, **176**, 112797.
7. C. Sun, F. Wang, Y. Zhang, J. Yu and X. Wang, *Theranostics*, 2020, **10**, 7070–7082.
8. H. Yang, W. Ji, M. Guan, S. Li, Y. Zhang, Z. Zhao and L. Mao, *Metabolomics*, 2018, **14**, 50.
9. A. D. Kelly, S. B. Breitkopf, M. Yuan, J. Goldsmith, D. Spentzos and J. M. Asara, *PLoS One*, 2011, **6**, e25357.
10. A. Buck, A. Ly, B. Balluff, N. Sun, K. Gorzolka, A. Feuchtinger, K. P. Janssen, P. J. Kuppen, C. J. van de Velde, G. Weirich, F. Erlmeier, R. Langer, M. Aubele, H. Zitzelsberger, M. Aichler and A. Walch, *J. Pathol.*, 2015, **237**, 123–132.
11. A. Ly, A. Buck, B. Balluff, N. Sun, K. Gorzolka, A. Feuchtinger, K. P. Janssen, P. J. Kuppen, C. J. van de Velde, G. Weirich, F. Erlmeier, R. Langer, M. Aubele, H. Zitzelsberger, L. McDonnell, M. Aichler and A. Walch, *Nat. Protoc.*, 2016, **11**, 1428–1443.
12. S. Cacciatore, G. Zadra, C. Bango, K. L. Penney, S. Tyekucheva, O. Yanes and M. Loda, *Mol. Cancer Res.*, 2017, **15**, 439–447.
13. S. Magdeldin and T. Yamamoto, *Proteomics*, 2012, **12**, 1045–1058.
14. C. Harkin, K. W. Smith, F. L. Cruickshank, C. Logan Mackay, B. Flinders, R. M. A. Heeren, T. Moore, S. Brockbank and D. F. Cobice, *Mass Spectrom. Rev.*, 2021, 1–33.
15. Q. Zhou, A. Fulop and C. Hopf, *Anal. Bioanal. Chem.*, 2020, **413**, 2599–2617.
16. M. L. Manier, J. M. Spraggins, M. L. Reyzer, J. L. Norris and R. M. Caprioli, *J. Mass Spectrom.*, 2014, **49**, 665–673.
17. M. E. Duenas, E. A. Larson and Y. J. Lee, *Front. Plant Sci.*, 2019, **10**, 860.
18. O. Fiehn, *Methods Mol. Biol.*, 2006, **323**, 439–447.
19. M. Breuer, T. M. Earnest, C. Merryman, K. S. Wise, L. Sun, M. R. Lynott, C. A. Hutchison, H. O. Smith, J. D. Lapek, D. J. Gonzalez,

V. de Crecy-Lagard, D. Haas, A. D. Hanson, P. Labhsetwar, J. I. Glass and Z. Luthey-Schulten, *eLife*, 2019, **8**, e36842.
20. D. Velickovic, G. Zhang, D. Bezbradica, A. Bhattacharjee, L. Pasa-Tolic, K. Sharma, T. Alexandrov, C. R. Anderton and K. Consortium, *J. Am. Soc. Mass Spectrom.*, 2020, **31**, 508–516.
21. M. Dufresne, N. H. Patterson, J. L. Norris and R. M. Caprioli, *Anal. Chem.*, 2019, **91**, 12928–12934.
22. J. Schafermann, G. Kliewer, J. Losch, H. Bednarz, M. Giampa and K. Niehaus, *J. Mass Spectrom.*, 2020, **55**, e4488.
23. C. D. Calvano, A. Monopoli, T. R. I. Cataldi and F. Palmisano, *Anal. Bioanal. Chem.*, 2018, **410**, 4015–4038.
24. H. Bednarz, N. Roloff and K. Niehaus, *J. Agric. Food Chem.*, 2019, **67**, 13470–13477.
25. C. J. Earnshaw, V. A. Carolan, D. S. Richards and M. R. Clench, *Rapid Commun. Mass Spectrom.*, 2010, **24**, 1665–1672.
26. R. T. Steven and J. Bunch, *Anal. Bioanal. Chem.*, 2013, **405**, 4719–4728.
27. T. Morikawa-Ichinose, Y. Fujimura, F. Murayama, Y. Yamazaki, T. Yamamoto, H. Wariishi and D. Miura, *J. Am. Soc. Mass Spectrom.*, 2019, **30**, 1512–1520.
28. A. R. Korte and Y. J. Lee, *J. Mass Spectrom.*, 2014, **49**, 737–741.
29. H. Liu, R. Chen, J. Wang, S. Chen, C. Xiong, J. Wang, J. Hou, Q. He, N. Zhang, Z. Nie and L. Mao, *Anal. Chem.*, 2014, **86**, 10114–10121.
30. J. M. Kampa, U. Kellner, C. Marsching, C. Ramallo Guevara, U. J. Knappe, M. Sahin, M. Giampa, K. Niehaus and H. Bednarz, *Neuropathology*, 2020, **40**, 546–558.
31. M. Sahin, J. M. Neumann, B. Riefke, H. Bednarz, K. Gutberlet, M. Giampa, K. Niehaus and A. Fatangare, *Toxicol. Lett.*, 2020, **334**, 36–43.
32. F. P. Y. Barre, B. S. R. Claes, F. Dewez, C. Peutz-Kootstra, H. F. Munch-Petersen, K. Gronbaek, A. H. Lund, R. M. A. Heeren, C. Come and B. Cillero-Pastor, *Anal. Chem.*, 2018, **90**, 14198–14206.
33. J. Wang, S. Qiu, S. Chen, C. Xiong, H. Liu, J. Wang, N. Zhang, J. Hou, Q. He and Z. Nie, *Anal. Chem.*, 2015, **87**, 422–430.
34. M. Giampa, M. B. Lissel, T. Patschkowski, J. Fuchser, V. H. Hans, O. Gembruch, H. Bednarz and K. Niehaus, *Chem. Commun.*, 2016, **52**, 9801–9804.
35. K. Horatz, M. Giampa, Y. Karpov, K. Sahre, H. Bednarz, A. Kiriy, B. Voit, K. Niehaus, N. Hadjichristidis, D. L. Michels and F. Lissel, *J. Am. Chem. Soc.*, 2018, **140**, 11416–11423.
36. J. Dong, W. Ning, D. J. Mans and J. D. Mans, *Anal. Methods*, 2018, **10**, 572–578.
37. S. R. Shanta, T. Y. Kim, J. H. Hong, J. H. Lee, C. Y. Shin, K. H. Kim, Y. H. Kim, S. K. Kim and K. P. Kim, *Analyst*, 2012, **137**, 5757–5762.
38. A. Urbanek, S. Holzer, K. Knop, U. S. Schubert and F. von Eggeling, *Anal. Bioanal. Chem.*, 2016, **408**, 3769–3781.

39. P. Le Pogam, B. Legouin, A. Geairon, H. Rogniaux, F. Lohezic-Le Devehat, W. Obermayer, J. Boustie and A. C. Le Lamer, *Sci. Rep.*, 2016, **6**, 37807.
40. W. H. Muller, A. Verdin, E. De Pauw, C. Malherbe and G. Eppe, *Mass Spectrom. Rev.*, 2020, 1–48.
41. P. Pomastowski and B. Buszewski, *Nanomaterials*, 2019, **9**, 260.
42. P. Rafols, D. Vilalta, S. Torres, R. Calavia, B. Heijs, L. A. McDonnell, J. Brezmes, E. Del Castillo, O. Yanes, N. Ramirez and X. Correig, *PLoS One*, 2018, **13**, e0208908.
43. Y. K. Kim and D. H. Min, *Langmuir*, 2014, **30**, 12675–12683.
44. S. N. Baker and G. A. Baker, *Angew. Chem., Int. Ed. Engl.*, 2010, **49**, 6726–6744.
45. S. Xu, Y. Li, H. Zou, J. Qiu, Z. Guo and B. Guo, *Anal. Chem.*, 2003, **75**, 6191–6195.
46. M. Misiorek, J. Sekula and T. Ruman, *Phytochem. Anal.*, 2017, **28**, 479–486.
47. A. P. Bowman, G. T. Blakney, C. L. Hendrickson, S. R. Ellis, R. M. A. Heeren and D. F. Smith, *Anal. Chem.*, 2020, **92**, 3133–3142.
48. T. Kind and O. Fiehn, *BMC Bioinf.*, 2006, **7**, 234.
49. M. E. Belov, S. R. Ellis, M. Dilillo, M. R. L. Paine, W. F. Danielson, G. A. Anderson, E. L. de Graaf, G. B. Eijkel, R. M. A. Heeren and L. A. McDonnell, *Anal. Chem.*, 2017, **89**, 7493–7501.
50. M. Kompauer, S. Heiles and B. Spengler, *Nat. Methods*, 2017, **14**, 1156–1158.
51. M. Kompauer, S. Heiles and B. Spengler, *Nat. Methods*, 2017, **14**, 90–96.
52. S. Schulz, M. Becker, M. R. Groseclose, S. Schadt and C. Hopf, *Curr. Opin. Biotechnol*, 2019, **55**, 51–59.
53. M. Schroeder, S. W. Meyer, H. M. Heyman, A. Barsch and L. W. Sumner, *Metabolites*, 2019, **10**, 13.
54. E. C. Spivey, J. C. McMillen, D. J. Ryan, J. M. Spraggins and R. M. Caprioli, *J. Mass Spectrom.*, 2019, **54**, 366–370.
55. J. Soltwisch, B. Heijs, A. Koch, S. Vens-Cappell, J. Hohndorf and K. Dreisewerd, *Anal. Chem.*, 2020, **92**, 8697–8703.
56. P. O. Helmer, I. D. Nordhorn, A. Korf, A. Behrens, R. Buchholz, F. Zubeil, U. Karst and H. Hayen, *Anal. Chem.*, 2021, **93**, 2135–2143.
57. J. M. Buescher, M. R. Antoniewicz, L. G. Boros, S. C. Burgess, H. Brunengraber, C. B. Clish, R. J. DeBerardinis, O. Feron, C. Frezza, B. Ghesquiere, E. Gottlieb, K. Hiller, R. G. Jones, J. J. Kamphorst, R. G. Kibbey, A. C. Kimmelman, J. W. Locasale, S. Y. Lunt, O. D. Maddocks, C. Malloy, C. M. Metallo, E. J. Meuillet, J. Munger, K. Noh, J. D. Rabinowitz, M. Ralser, U. Sauer, G. Stephanopoulos, J. St-Pierre, D. A. Tennant, C. Wittmann, M. G. Vander Heiden, A. Vazquez, K. Vousden, J. D. Young, N. Zamboni and S. M. Fendt, *Curr. Opin. Biotechnol*, 2015, **34**, 189–201.
58. M. R. Antoniewicz, *Exp. Mol. Med.*, 2018, **50**, 1–13.

59. K. B. Louie, B. P. Bowen, S. McAlhany, Y. Huang, J. C. Price, J. H. Mao, M. Hellerstein and T. R. Northen, *Sci. Rep.*, 2013, **3**, 1656.
60. V. Pareek, H. Tian, N. Winograd and S. J. Benkovic, *Science*, 2020, **368**, 283–290.
61. T. Kind and O. Fiehn, *BMC Bioinf.*, 2007, **8**, 105.
62. A. McCann, S. Rappe, R. La Rocca, M. Tiquet, L. Quinton, G. Eppe, J. Far, E. De Pauw and C. Kune, *Anal. Bioanal. Chem.*, 2021, **413**, 2831–2844.
63. A. Palmer, P. Phapale, I. Chernyavsky, R. Lavigne, D. Fay, A. Tarasov, V. Kovalev, J. Fuchser, S. Nikolenko, C. Pineau, M. Becker and T. Alexandrov, *Nat. Methods*, 2017, **14**, 57–60.
64. J. M. Spraggins, K. V. Djambazova, E. S. Rivera, L. G. Migas, E. K. Neumann, A. Fuetterer, J. Suetering, N. Goedecke, A. Ly, R. Van de Plas and R. M. Caprioli, *Anal. Chem.*, 2019, **91**, 14552–14560.
65. A. R. Fernie, A. Aharoni, L. Willmitzer, M. Stitt, T. Tohge, J. Kopka, A. J. Carroll, K. Saito, P. D. Fraser and V. DeLuca, *Plant Cell*, 2011, **23**, 2477–2482.
66. L. W. Sumner, A. Amberg, D. Barrett, M. H. Beale, R. Beger, C. A. Daykin, T. W. Fan, O. Fiehn, R. Goodacre, J. L. Griffin, T. Hankemeier, N. Hardy, J. Harnly, R. Higashi, J. Kopka, A. N. Lane, J. C. Lindon, P. Marriott, A. W. Nicholls, M. D. Reily, J. J. Thaden and M. R. Viant, *Metabolomics*, 2007, **3**, 211–221.
67. P. Bankhead, M. B. Loughrey, J. A. Fernandez, Y. Dombrowski, D. G. McArt, P. D. Dunne, S. McQuaid, R. T. Gray, L. J. Murray, H. G. Coleman, J. A. James, M. Salto-Tellez and P. W. Hamilton, *Sci. Rep.*, 2017, **7**, 16878.
68. S. Tortorella, P. Tiberi, A. P. Bowman, B. S. R. Claes, K. Scupakova, R. M. A. Heeren, S. R. Ellis and G. Cruciani, *J. Am. Soc. Mass Spectrom.*, 2020, **31**, 155–163.
69. L. P. de Souza, M. Borghi and A. Fernie, *Int. J. Mol. Sci.*, 2020, **21**, 8987.
70. B. A. Boughton, D. Thinagaran, D. Sarabia, A. Bacic and U. Roessner, *Phytochem. Rev.*, 2016, **15**, 445–488.
71. L. Becker, S. Bellow, V. Carre, G. Latouche, A. Poutaraud, D. Merdinoglu, S. C. Brown, Z. G. Cerovic and P. Chaimbault, *Anal. Chem.*, 2017, **89**, 7099–7106.
72. D. Holscher, R. Shroff, K. Knop, M. Gottschaldt, A. Crecelius, B. Schneider, D. G. Heckel, U. S. Schubert and A. Svatos, *Plant J.*, 2009, **60**, 907–918.
73. K. Takahashi, T. Kozuka, A. Anegawa, A. Nagatani and T. Mimura, *Plant Cell Physiol.*, 2015, **56**, 1329–1338.
74. S. Cha, H. Zhang, H. I. Ilarslan, E. S. Wurtele, L. Brachova, B. J. Nikolau and E. S. Yeung, *Plant J.*, 2008, **55**, 348–360.
75. B. A. Boughton and D. Thinagaran, *Methods Mol. Biol.*, 2018, **1778**, 241–252.
76. K. Gorzolka, J. Kolling, T. W. Nattkemper and K. Niehaus, *PLoS One*, 2016, **11**, e0150208.
77. D. Sturtevant, Y. J. Lee and K. D. Chapman, *Curr. Opin. Biotechnol*, 2016, **37**, 53–60.

78. A. R. Korte, M. D. Yandeau-Nelson, B. J. Nikolau and Y. J. Lee, *Anal. Bioanal. Chem.*, 2015, **407**, 2301–2309.
79. L. D. Sarabia, B. A. Boughton, T. Rupasinghe, A. M. L. van de Meene, D. L. Callahan, C. B. Hill and U. Roessner, *Metabolomics*, 2018, **14**, 63.
80. D. Velickovic, H. L. Liao, R. Vilgalys, R. K. Chu and C. R. Anderton, *J. Nat. Prod.*, 2019, **82**, 1382–1386.
81. A. K. Mullen, M. R. Clench, S. Crosland and K. R. Sharples, *Rapid Commun. Mass Spectrom.*, 2005, **19**, 2507–2516.
82. O. Fiehn, *Plant Mol. Biol.*, 2002, **48**, 155–171.
83. G. Xu and J. Li, *J. Comp. Neurol.*, 2019, **527**, 2158–2169.
84. K. Veerasammy, Y. X. Chen, S. Sauma, M. Pruvost, D. K. Dansu, T. Choetso, T. Zhong, D. Marechal, P. Casaccia, R. Abzalimov and Y. He, *J. Visualized Exp.*, 2020, **166**, e62008.
85. G. A. Dienel, *Neurochem. Res.*, 2020, **45**, 2586–2606.
86. Y. Chen, W. Tang, A. Gordon and B. Li, *J. Am. Soc. Mass Spectrom.*, 2020, **31**, 1066–1073.
87. E. Davoli, M. Zucchetti, C. Matteo, P. Ubezio, M. D'Incalci and L. Morosi, *Mass Spectrom. Rev.*, 2020, **40**, 201–214.
88. S. Khatib-Shahidi, M. Andersson, J. L. Herman, T. A. Gillespie and R. M. Caprioli, *Anal. Chem.*, 2006, **78**, 6448–6456.
89. M. Nishidate, M. Hayashi, H. Aikawa, K. Tanaka, N. Nakada, S. I. Miura, S. Ryu, T. Higashi, Y. Ikarashi, Y. Fujiwara and A. Hamada, *Drug Metab. Pharmacokinet.*, 2019, **34**, 209–216.
90. D. Bonnel and J. Stauber, *Methods Mol. Biol.*, 2017, **1641**, 129–140.
91. J. M. Kampa, M. Sahin, M. Slopianka, M. Giampa, H. Bednarz, R. Ernst, B. Riefke, K. Niehaus and A. Fatangare, *Toxicol. Lett.*, 2020, **325**, 43–50.
92. Y. Xu and Z. Li, *Comput. Struct. Biotechnol. J.*, 2020, **18**, 2401–2415.
93. M. Martin-Lorenzo, G. Alvarez-Llamas, L. A. McDonnell and F. Vivanco, *Expert Rev. Proteomics*, 2016, **13**, 69–81.
94. Z. Dai, Y. Sun, X. Zhao and X. Pu, *Cell Tissue Res.*, 2020, **380**, 415–424.
95. G. Arentz, P. Mittal, C. Zhang, Y. Y. Ho, M. Briggs, L. Winderbaum, M. K. Hoffmann and P. Hoffmann, *Adv. Cancer Res.*, 2017, **134**, 27–66.
96. M. K. Andersen, S. Krossa, T. S. Hoiem, R. Buchholz, B. S. R. Claes, B. Balluff, S. R. Ellis, E. Richardsen, H. Bertilsson, R. M. A. Heeren, T. F. Bathen, U. Karst, G. F. Giskeodegard and M. B. Tessem, *Anal. Chem.*, 2020, **92**, 3171–3179.
97. M. K. Andersen, T. S. Hoiem, B. S. R. Claes, B. Balluff, M. Martin-Lorenzo, E. Richardsen, S. Krossa, H. Bertilsson, R. M. A. Heeren, M. B. Rye, G. F. Giskeodegard, T. F. Bathen and M. B. Tessem, *Cancer Metab.*, 2021, **9**, 9.
98. R. Longuespee, A. K. Wefers, E. De Vita, A. K. Miller, D. E. Reuss, W. Wick, C. Herold-Mende, M. Kriegsmann, P. Schirmacher, A. von Deimling and S. Pusch, *Acta Neuropathol. Commun.*, 2018, **6**, 21.
99. X. Liu, C. Flinders, S. M. Mumenthaler and A. B. Hummon, *J. Am. Soc. Mass Spectrom.*, 2018, **29**, 516–526.

100. D. Abu Sammour, C. Marsching, A. Geisel, K. Erich, S. Schulz, C. Ramallo Guevara, J. H. Rabe, A. Marx, P. Findeisen, P. Hohenberger and C. Hopf, *Sci. Rep.*, 2019, **9**, 10698.
101. F. Y. Yang, J. H. Chen, Q. Q. Ruan, H. S. A. Saqib, W. Y. He and M. S. You, *Arch. Insect Biochem. Physiol.*, 2020, **103**, e21643.
102. L. H. Tucker, G. R. Hamm, R. J. E. Sargeant, R. J. A. Goodwin, C. L. Mackay, C. J. Campbell and D. J. Clarke, *Anal. Chem.*, 2019, **91**, 9522–9529.
103. I. Palubeckaite, L. Crooks, D. P. Smith, L. M. Cole, H. Bram, C. Le Maitre, M. R. Clench and N. A. Cross, *J. Mass Spectrom.*, 2020, **55**, e4461.
104. S. J. Dunham, J. F. Ellis, B. Li and J. V. Sweedler, *Acc. Chem. Res.*, 2017, **50**, 96–104.
105. E. U. Brockmann, D. Steil, A. Bauwens, J. Soltwisch and K. Dreisewerd, *Anal. Chem.*, 2019, **91**, 15081–15089.
106. R. C. Pessotti, B. L. Hansen, V. M. Zacharia, D. Polyakov and M. F. Traxler, *Anal. Chem.*, 2019, **91**, 14818–14823.

CHAPTER 8

MALDI Mass Spectrometry Imaging in Lipidomics

SHADRACK M. MUTUKU[a] AND SHANE R. ELLIS*[a,b]

[a] Molecular Horizons and School of Chemistry and Molecular Bioscience, University of Wollongong, Wollongong, NSW, Australia; [b] Illawarra Health and Medical Research Institute, Wollongong, NSW, Australia
*Email: sellis@uow.edu.au

8.1 Introduction

Lipids encompass a broad family of generally hydrophobic compounds containing hydrogen-substituted carbons often with polar headgroups, which are generally soluble in organic solvents. According to the International Lipid Classification and Nomenclature Committee[1] there are eight major recognized categories of lipids: fatty acids, glycerides, glycerophospholipids, sphingolipids, sterols, prenols, saccharolipids, and polyketides[1,2] with each containing multiple sub-classes.[3,4] This rich structural diversity reflects the vast and complex biochemical functions lipids have in many biological processes in addition to their well-known roles of being primary stores of energy and key structural scaffolds of cell membranes.[3,5–9] Lipids are made of generally simple molecular templates that combine to give rise to the many complex lipid structures found throughout biology.[10] This is realized through a series of biosynthetic pathways involving numerous possible metabolic building blocks.[11] For instance, glycerol is a fundamental backbone for synthesizing glycerophospholipids, where substitution at the *sn*-3 position with a variety of headgroups including ethanolamine, choline, serine, inositol, and glycerol gives rise to different glycerophospholipid classes. Additionally, different fatty

New Developments in Mass Spectrometry No. 12
MALDI Mass Spectrometry Imaging: From Fundamentals to Spatial Omics
Edited by Tiffany Siegel

Published by the Royal Society of Chemistry, www.rsc.org

acyl ester and/or ether substituents at the *sn*-1 and *sn*-2 positions can vary in chain length as well as the number and position of double bonds leading to hundreds of possible phospholipid species in each subclass.[1] Another example is sphingoid bases for the generation of various sphingolipids.[12] Consequently, there can be >100 000 possible lipid structures in nature[13] and a complete lipidome – a term that describes the comprehensive composition of all lipids found within a biological system such as a cell or tissue – is yet to be fully elucidated. The development of analytical technologies to better illuminate the diversity and composition of lipids is thus an ever-growing area of research.

The biological significance of lipids combined with their vast structural diversity has meant their detailed quantitative and qualitative analysis is well-suited for mass spectrometry, spawning the field of lipidomics.[14–17] Lipids are arguably the most commonly studied analyte using mass spectrometry imaging (MSI) because: (i) many common lipid species such as phospholipids and sphingolipids readily ionize using conventional ionization methods such as matrix-assisted laser desorption/ionization (MALDI)[18] and electrospray ionization (ESI);[19] and (ii) lipid analysis requires arguably the most simple preparation procedures (often only requiring sectioning and matrix application) making their analyses generally robust and reproducible. Indeed, in almost every MSI study to date focusing on lipids heterogeneous lipid distributions have been observed that reflect their multifaceted roles in driving many cellular and physiological processes. Lipids have also proven to be a powerful molecular family for the classification of diseased tissues[20,21] leading to localized lipid profiles even being used for real-time classification of tissues during surgery.[22,23]

8.2 Sample Preparation for Lipid MSI

MALDI-MSI of lipids requires arguably the most straightforward sample preparation of all analyte classes (*i.e.* no enzymatic digestion or washing is generally needed, although in some cases these can be advantageous[24,25]). Analysis of other analyte classes, such as proteins, requires the removal of lipids to aid their detection. Many of the sample preparation details covered in Chapter 5 of this book are applicable to lipid analysis. Below we address pertinent aspects specific to lipid analysis.

8.2.1 Tissue Fixation

While MALDI-MSI of lipids is almost exclusively performed on freshly frozen tissue, given the widespread use of formalin fixation for tissue presentation and storage, it has become necessary to understand the effect of formalin fixation on lipid MSI. The process of formalin fixation involves the chemical cross-linking of amine-containing molecules, mostly proteins. As a result, depleted signals for amine-containing lipids are expected in formalin-fixed tissues compared to freshly frozen tissues. Nonetheless, studies have demonstrated many lipid classes, such as phospholipids and sphingolipids, that lack available amine groups can be detected and imaged with minimal delocalization following formalin fixation in both positive- and negative-ion mode, including from brain,[26] lung[28] and kidney tissues.[29] Formalin fixation can also alter the types of adducts detected depending on the composition of the fixation solution.[26,27]

Formalin fixation is often combined with paraffin embedding (FFPE) for long-term tissue preservation and storage. The process of paraffin embedding requires organic solvent washes that can result in the removal and re-distribution of most lipid species, in addition to losses caused by fixation itself (see above). Surprisingly though, several lipid species and numerous metabolites have been detected from FFPE tissues with histologically distinct localizations, suggesting the distribution of at least a small number of lipid species are unexpectedly maintained in FFPE tissue.[28]

8.2.2 Tissue Washing

Many lipids are detected as multiple cationized adducts ($[M+H]^+$, $[M+Na]^+$, and $[M+K]^+$) that arise due to the presence of endogenous salts in the tissue. This leads to rather complex mass spectra, which can complicate data interpretation and therefore lipid identification (see Section 8.3). The most common means to coalesce lipid signals to the protonated form is to wash the tissue section with aqueous ammonium formate[29] or ammonium acetate.[25,30]

Another approach is washing the tissue with a specific salt additive to favor the formation of certain alkali adducts, such as lithium adducts of phospholipids.[30–32] This allows one to take advantage of the richer MS/MS data obtained when fragmenting lithium adducts, for example, better identifying acyl chains from phosphatidylcholine (PC) lipids in the positive-ion mode.[33–36]

8.2.3 MALDI Matrices

As with any MALDI experiment, use of a suitable matrix is essential for successful detection of lipids. While determining the matrix composition and concentration is a rather empirical process, several general rules can be applied for MALDI-MSI of lipids to ascertain a reasonable starting point. Firstly, neutral and basic lipids are often best detected in the positive-ion mode using acidic matrices, including the traditional MALDI matrices α-cyano-4-hydroxycinnamic acid (CHCA) and 2,5-dihydroxybenzoic acid (DHB).[37–39] In contrast, acidic lipids are often best detected in the negative-ion mode using basic matrices favoring lipid deprotonation. It is important to emphasize that these rules are not universal. For example, negative-ion mode analysis of lipids using DHB will still lead to the detection of abundant acidic lipids such as phosphatidylinositols (PI) and sulfatides (ST). Speaking to the robustness of sample preparation for MALDI-MS of lipids, the vast majority of MALDI matrices will generate some lipid-related ions. For a summary of some popular matrices used for the MALDI analysis of lipids the reader is referred to refs. 40 and 41.

Much work in recent years has searched for alternative matrices that allow for a more efficient ionization of lipids. Recently, Perry *et al.* have compared a variety of matrices (*i.e.* 9-aminoacridine (9AA), 5-chloro-2-mercaptobenzothiazole (CMBT), 1,5-diaminonaphthalene (DAN), 2,5-dihydroxyacetophenone (DHA), and DHB) in terms of the lipid-related signals they generate in both positive- and negative-ion mode.[42] In positive-ion mode, 9-AA and DAN were found to favor the formation of protonated ions compared to CMBT, DHA, and DHB, which favored the formation of sodiated and potassiated ions from liver tissue. In the negative-ion mode, DAN and 9-AA produced the highest number of deprotonated lipid signals from liver tissue.[42]

It can be desirable to obtain complementary positive-ion and negative-ion mode data from the same tissue section to obtain a more comprehensive overview of lipid distributions. Both DAN[43] and norharmane[44] have been shown to perform well for lipid imaging in both polarities, among others.[42] One approach used a checkerboard acquisition pattern to allow consecutive acquisition of positive- and negative-ion MSI data from the same tissue sample.[44] This was applied to both a Zebra Finch bird brain and tumor xenograft tissue and combined with probabilistic latent semantic analysis (pLSA) combining data from both polarities. When applied to a tumor xenograft, different pLSA components could distinguish tissue regions corresponding to viable, early-necrotic, and necrotic tissue regions. An alternate method for acquiring dual polarity lipid spectra from the same tissue using DAN matrix was reported by Kaya *et al.* whereby careful optimization of laser energy allowed for consecutive positive- and negative-ion mode analysis from the same sampling positions.[45] When applied to rat small intestine mucosa, the combined analysis allowed imaging of many lipid species including PC and lysophosphatidylcholines (LPC) in positive-ion mode and PI, phosphatidylethanolamines (PE), phosphatidic acids (PA), ceramide 1-phosphate (CerP), cholesterol sulfate, bile acids, and free fatty acids in the negative-ion mode using DAN matrix.[45]

Another requirement for matrix choice especially relevant when using axial-TOF instruments operating under high vacuum is the vacuum stability of the matrix. Under such conditions, common matrices with relatively high vapor pressure such as DHA can sublimate in less than an hour, leading to a reduction in signal during an experiment. Such effects may become significant for larger area-imaging or high-resolution imaging that necessitate long acquisition times. One approach to overcome this is the use of vacuum stable matrices such as those containing ketone functionalities, including (*E*)-4-(2,5-dihydroxyphenyl)but-3-en-2-one.[46] Alternatively, with recent improvements in high speed, MALDI-TOF technology data acquisition may be completed before significant sublimation of the matrix has occurred.[47–49] Such approaches can enable the use of semi-volatile matrices that perform well for lipid analysis.

8.2.4 Matrix Application

Both spray-coating and sublimation are widely used for MALDI-MSI of lipids. Lipids appear to be the most suitable analyte class for analysis using sublimation.[43,50] Recent studies have investigated the roles of class-specific diffusion of lipids from the tissue into the matrix layer produced with sublimation.[51]

Matrix recrystallization combined with sublimation is also increasingly applied to lipids and can reduce matrix-related signals and improve signal intensity for many lipids.[52–55] This entails brief incubation of a matrix-coated slide in a sealed container with an aqueous-organic solvent mixture. Recrystallization by freezing of glass slides after sublimation with DAN matrix has been used for imaging gangliosides (GM) such as GM1(d36 : 1), GM1(d38 : 1), GM3(d36 : 1), and GM3(d38 : 1) throughout rat brains.[56] We wish to emphasize again that such improvements may not be universal and may be heavily dependent on the choice of matrix, tissue, and ion source, among other parameters.

Besides the matrix itself, the choice of solvent used to dissolve the matrix for spray-based disposition is another key parameter. However, it should be noted that high-quality lipid data using MALDI-MSI have been achieved using a variety of solvents, including methanol/water, ethanol/water, acetonitrile/water, and mixtures of organic solvents such as methanol and chloroform that are chosen to replicate standard solvents used for lipid extraction from tissues and cells.[57]

Beyond matrix-based techniques, direct laser desorption/ionization (LDI) of lipids from active substrates where the surface itself acts as the desorption and ionization promoting medium can also yield rich lipid spectra. Silicon nanospot arrays (NAPA)[58] and nanostructure-initiator mass spectrometry (NIMS)[59] are some variants of surface-assisted LDI platforms. Although matrix-free LDI methods can offer a benefit for low molecular weight (LMW) compound detection without background interference from matrix-related ions, their application in lipid studies is not as widespread as for MALDI. Silver-coated NIMS has been used for brain imaging in Smith–Lemli–Opitz syndrome where cholesterol and 7-dehydrocholesterol were detected as $[M+Ag]^+$ ions at a resolution of 75 μm.[60] Additionally, NIMS has been deployed to measure isotopic enrichment of lipids to define dynamic changes in tumor metabolism in a Trp-53 (tumor suppressor gene) deficient mouse model.[61] A comparison of MALDI-MS and NAPA-MS found differential and complementary detection of 178 assigned metabolites in hepatocyte extract and urine samples on an identical instrument. While NAPA-MS detected more of the LMW compounds. MALDI-MS was better at measuring labile and higher molecular weight species.[62] NAPA has been shown to offer substantial improvements in signal for triacylglycerides (TGs) compared to MALDI, with up to 100-fold gain observed when analyzing mouse lung and human skin tissue.[63]

8.3 Lipid Identification in MSI

8.3.1 Identification of Lipids Detected with MALDI-MSI

MALDI-MSI is capable of detecting many lipid-related signals corresponding to a wide variety of lipid species. While many MSI applications (*e.g.* those using multivariate data analysis to classify different tissue regions based on spectral similarities and differences) do not necessarily require knowledge of the exact lipids detected, interpreting the data in terms of different lipid-related biological processes occurring throughout the sample does require identification of lipids. Lipid identification in MSI presents several challenges, and users must be aware of the level of lipid identification possible (and the associated ambiguity) given the exact nature of the mass spectrometry data acquired.[2,64]

8.3.1.1 *Lipid-related Ions Generated by MALDI-MSI*

Lipid spectra generated by MALDI analysis of biological material often leads to multiple ions observed for the same lipid species. For example, in positive-ion mode many lipid species (*e.g.* PC, PE, TG and glycosphingolipids (GSL)) can be simultaneously detected as $[M+H]^+$, $[M+Na]^+$, and $[M+K]^+$ ions, with the alkali adducts predominately arising from endogenous salt present within the sample. While care must be taken to ensure the distributions of different adducts are not mistakenly interpreted as changing lipid distributions, in some cases, the spatial distributions of different adducts can provide insight into varying Na^+/K^+ ratios throughout the tissue. For example, decreased levels of potassiated PC ions and concomitant increase in the corresponding sodiated ions have been associated with brain regions following traumatic brain injury.[65] Similar observations have been reported following ischemic damage caused by stroke.[66] In the latter case, the origin of the altered adduct distributions was further confirmed with secondary ion mass spectrometry-MSI (SIMS-MSI) that enabled direct MSI analysis of endogenous Na^+ and K^+. Washing of tissue samples with aqueous buffer solutions containing aqueous ammonium formate or ammonium acetate prior to matrix application can remove most endogenous salts and favor the formation of $[M+H]^+$ ions (see above).[25] However, care and optimization are needed to ensure the washing process does not lead to lipid delocalization.

In negative-ion mode, the majority of lipid ions are observed as $[M{-}H]^-$ anions that can simplify spectral interpretation. For some lipids such as GMs and cardiolipins, alkali adducted species may also be observed in the form of $[M{-}2H+Na]^-$ and $[M{-}2H+K]^-$ ions. For some lipid species such as ceramides (Cer) $[M+Cl]^-$ ions can also be observed.

8.3.1.2 *In-source Fragmentation of Lipids during MALDI-MSI*

MALDI analysis of lipids can result in some in-source fragmentation that must be considered to avoid erroneous lipid identification. In-source fragmentation is typically increased with increasing laser energy, the use of so-called "hot" matrices and is dependent on ion transmission conditions and the extent of collisional cooling occurring during ion transport through the mass spectrometer. While this topic cannot be covered in detail here, several of the most common in-source fragmentation processes observed during MALDI-MSI analysis of lipids are outlined below. Further information can be found in ref. 67.

1. The dissociation of phospholipid headgroups can readily occur during MALDI. For example, the formation of an ion at *m/z* 184.073 in positive-ion mode can indicate loss of the phosphocholine headgroup from PC and SM lipids (typically from their protonated ions) and may overlap with the signal from endogenous phosphocholine. An equally common fragmentation process is the neutral losses of $N(CH_3)_3$ (−59.073) and phosphocholine (−183.066) from PC and SM lipids. This process occurs to a greater extent from alkali adducted species and may lead to incorrect lipid identification. For example, loss of $N(CH_3)_3$ from $[PC(34:1)+K]^+$ forms a fragment ion with an *m/z* value with the same mass as $[PA(36:2)+K]^+$ (a lipid unlikely to form abundant signals in positive-ion mode). Alternatively, neutral loss of the phosphocholine headgroup of PC (and possibly other headgroups) will form ions with the same *m/z* value as dehydrated diacylglyceride (DG)s. Such fragments have sometimes been mistakenly attributed to endogenous DGs. Similar processes can be observed in the negative-ion mode. For example, loss of the serine headgroup (−87.032 Da) from PS lipids will form a PA lipid with the same acyl chain composition.
2. While headgroup losses are the most common source of in-source fragments it may also lead to the formation of free fatty acid anions resulting from the dissociation of esterified fatty acids from phospholipids. Care must be taken to ensure the assignment of fatty acid signals does not arise from phospholipid fragmentation (*e.g.* by analyzing phospholipid standards under identical analysis conditions).

8.3.1.3 Identification of Lipids Using Accurate Mass Measurements

Identification of lipids to the sum-composition level using accurate mass measurements can be performed using the same methods widely used for shotgun lipidomics.[68] This level of identification allows identification of lipid class and subclass (*e.g.* a PE lipid) and the combined number of carbons and double bonds in both acyl chains (*e.g.* PE(40:6) defines a PE lipid with 40 carbons and 6 double bonds across both acyl chains). Such identifications typically require a mass accuracy of ≤3 ppm that is readily obtained on Orbitrap, Fourier-Transform Ion Cyclotron Resonance (FTICR) and modern orthogonal-TOF mass spectrometers. If using database matching to aid in identifications, manual curation of an identified lipid is often needed to minimize the risk of false positives. For example, it is essential to consider whether identified lipids will form logical ions in a given polarity based on their molecular structure and if such lipids are consistent with lipid biosynthesis processes. An essential requirement of confident identification based on accurate mass is the ability to resolve nearby isobaric interferences that may lead to mass shifts in the centroid *m/z* value of a peak if not completely resolved. While mass resolutions ≥150 000 @ *m/z* 750 (as available on Orbitrap and FTICR) can resolve most isobaric interferences, given the complex spectra generated by MALDI-MSI analysis, baseline resolution of all detected ions is not always achieved even at this resolving power.[69] Using a 21 T-FTICR for MALDI-MSI of rat brain tissue, it was demonstrated that there are numerous peak pairs detected that are separated by less than 2 mDa (Figure 8.1).[70] These require a mass resolving power of ∼400 000 @ *m/z* 750 to resolve. Nonetheless, identification based on

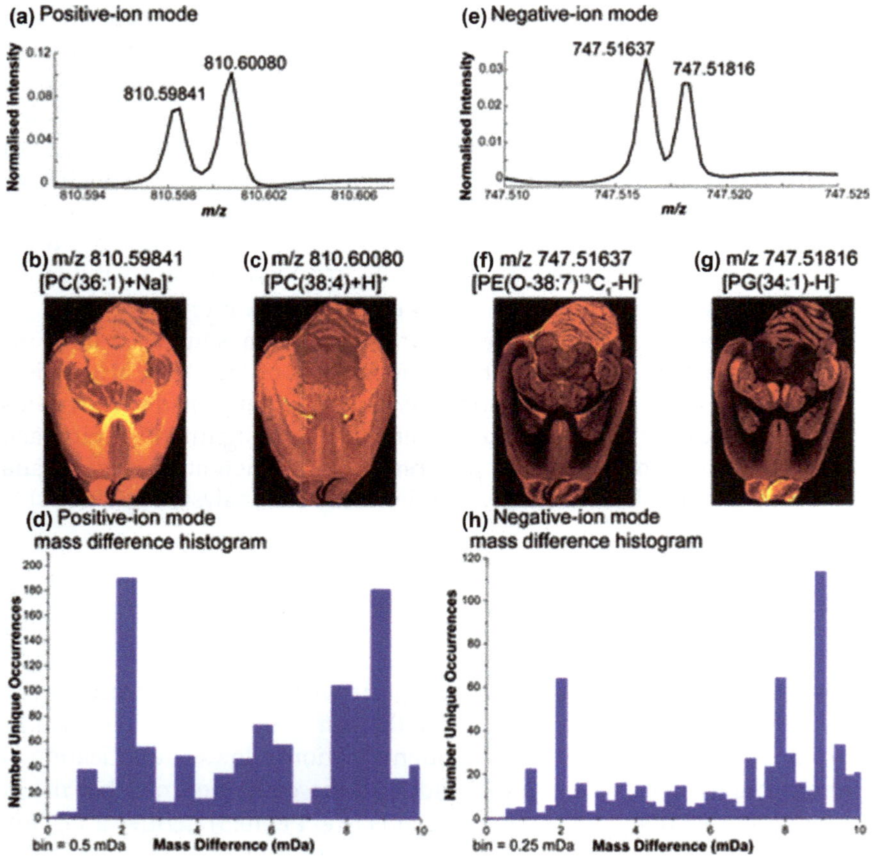

Figure 8.1 (a) Example of two lipids detected in positive-ion mode MALDI-MSI analysis of rat brain tissue with a mass difference of 2.4 mDa. (b and c) Ion images and sum-composition identifications of the two lipids shown in (a) identified as $[PC(36:1)+Na]^+$ and $[PC(38:4)+H]^+$. (d) Histograms showing the mass differences between adjacent peaks in the positive-ion mode from rat brain tissue. (e) Example of two lipids detected in the negative-ion mode with a mass difference of 1.79 mDa. (f and g) Ion images and sum-composition identifications of the two lipids shown in (e) identified as the isotopologues $[PE(O\text{-}38:7)^{13}C_1–H]^-$ and $[PG(34:1)–H]^-$. (d) Histograms showing the mass differences between adjacent peaks in the negative-ion mode from rat brain tissue. Reproduced from ref. 70, https://doi.org/10.1021/acs.analchem.9b04768, with permission from American Chemical Society, Copyright 2020.

accurate mass is a powerful approach for sum-composition identifications of the many lipids. Importantly, accurate mass data is provided simultaneously with MSI data if performed on a suitable mass spectrometer.

8.3.1.4 Identification of Lipids Using Tandem Mass Spectrometry

Tandem mass spectrometry (MS/MS) is the gold standard for lipid identification and the same fragmentation rules used for LC–MS/MS and shotgun lipidomics can be applied to MALDI-MSI data. Conventional collision-induced dissociation (CID) can identify many lipids to the molecular lipid species level by detection of both head group and fatty acyl-related fragment ions. Both lipid class as well as the length and degree of unsaturation of each acyl chain are defined at this identification level. In the negative-ion mode, it is common to observe fragment ions corresponding to fatty acyl anions and the neutral loss of head groups from phospholipids. Fatty acyl neutral losses can also be observed in positive-ion mode (*e.g.* from protonated PC species) but are typically low in abundance. As another example, in the positive-ion mode analysis of GSLs MS/MS can identify the structure of the sphingoid base.

Identification of lipids detected during MALDI-MSI experiments can be performed using both on-tissue MALDI-MS/MS or by comparison of LC–MS/MS data produced by analysis of a comparable sample extract. While on-tissue MS/MS has the advantage that precursor ions are generated under the same conditions as MALDI-MSI it is typically deployed in a targeted manner on selected precursor ions as acquisitions are often performed manually. Several multiplexing strategies have been developed using hybrid ion trap-Orbitrap instruments that allow the automated acquisition of multiple MS/MS spectra.[71–73] Using a MALDI-enabled Orbitrap Elite and a method based on data-dependent acquisition (DDA), it has been shown that ion trap MS/MS can be acquired in parallel with a 240 000 resolution (@*m/z* 400) Orbitrap scan. In this approach, full scan (MS1) Orbitrap data is acquired at every second position across a sample and encodes both the MSI data and accurate mass measurements of all detected ions.[73,74] At the same time ion trap MS/MS are acquired under DDA conditions that ensure almost every nominal *m/z* value is selected multiple times for MS/MS throughout the acquisition to allow for signal averaging of the MS/MS spectra obtained from each precursor window. This method was combined with Analysis of Lipid Experiments software[75] to allow the automated identification of lipids using both MS1 accurate mass and ion trap MS/MS. Applied to mouse brain tissue and combining data from both positive- and negative-ion mode, acquisitions for over 100 lipids at the molecular lipid species level could be identified, including the resolution for several isomeric species (*i.e.* PS 18 : 0_20 : 1 and PS 18 : 1_20 : 0). An alternative mode for multiplexed MS/MS acquisitions has been reported by Prentice *et al.* using a MALDI-TOF/TOF instrument. Here, up to three precursor ions produced from the same laser shots could be isolated using their different arrival times at the precision timed ion selector provided their arrival times differed by more than 6–7%.[76] Fragment ions produced by post-source decay of each precursor produced overlapping MS/MS spectra. Due to the temporal separation of precursor ions it was possible to resolve signals from identical fragments produced from different precursors, for example, the *m/z* 163 fragment that is characteristic of potassiated PC and SM lipids. In addition to solely identifying lipids, MS/MS-imaging can also be used to differentially image isobaric or isomeric lipids that are indistinguishable at the MS1 level.[77]

Identification using LC–MS/MS of sample extracts has the advantage of typically higher sensitivity and the possibility of differentiating lipids in the chromatographic dimension that may be difficult to resolve using MS/MS alone (*e.g.* glucosylceramide (GluCer) and galactosylceramide (GalCer) lipids). However, one must be aware that due to the higher sensitivity it is possible that additional lipids are observed that are not detected in the MALDI-MSI data. Correlation of MALDI-MSI and LC–MS/MS datasets is typically done using accurate mass matching of precursor ions. As one example, Moerman *et al.* have recently used this approach to aid in the identification of lipid species detected during the MALDI-MSI analysis of human carotid atherosclerosis tissue.[78]

8.3.1.5 Isomer-resolved MALDI-MSI of Lipids

A limitation of conventional MS/MS using CID is the inability to distinguish many isomeric lipids such as those that vary only in fatty acyl double bond (db) positions and *sn*-position of fatty acyls on the glycerol backbone.[15] The vast majority of lipid MSI data may thus represent the combined distributions of multiple isomeric lipids that remain unresolved. In recent years, there has been a significant drive to develop MSI methods that allow differential imaging of db and/or *sn*-positional isomers to address this challenge. These methods include ozone-induced dissociation,[74] Paternò–Büchi reactions,[79–81] electron-induced dissocation,[82] and charge inversion using ion–ion reactions.[83] Other methods such as ultraviolet photodissociation[84] and epoxidation[85] have been used for isomer-resolved imaging desorption electrospray ionization (DESI)-MSI but have not yet been adapted for MALDI. All these methods are MS/MS (MS2) or MS3-based approaches meaning they are targeted experiments typically applied to a small number of selected lipid species in a given MSI experiment. For further information on the individual methods, please refer to the relevant references. It is expected that isomer-resolved MSI will continue to develop rapidly as it shows much promise for revealing alterations in lipid metabolism at a level of molecular specificity not possible with other MSI approaches.

8.3.1.6 Resolving Lipids Using Ion Mobility

Ion mobility also enables the resolution of some isobaric and isomeric lipid signals during both MALDI-MSI experiments[86–91] and those using other MSI ion sources.[92–94] As one example demonstrating the power of ion mobility in resolving complex lipid spectra, trapped ion mobility spectrometry (TIMS) combined with MALDI-MSI has been used to resolve isobaric signals arising from isobaric $[PC(36:4)+K]^+$ and the M + 2 isotope of $[PC(36:5)+K]^+$ that would otherwise require a mass resolving power of ~100 000 to resolve.[86] Ion mobility can also significantly increase the peak capacity of MSI experiments by providing an additional dimension of separation. Using TIMS combined with a Q-TOF mass spectrometer, an increased peak capacity of 250% has been demonstrated.[86] However, identification of ion mobility separated lipids beyond that possible with *m/z* values is a challenge and typically requires data from synthetic standards to match collision cross-sections with those observed from tissue. This has so far limited the use of ion mobility as a means for broadband identification. The development of collision cross-section libraries offers one means by which this challenge is being addressed.[88,95,96]

8.4 Selected Applications of MALDI-MSI of Lipids

8.4.1 Cancer Research

Cells display an altered energetic hallmark during neoplastic transformation.[97,98] Often, there is a switch to greater glycolytic metabolism for aerobic conversion of pyruvate to lactate, an unusual energetic inefficient process for ATP generation but favorable for cancer proliferation, the so-called "Warburg effect".[99,100] In this cellular dyshomeostasis, the heightened flux of carbon-containing substrates ultimately leads to changes in lipid metabolism.[101] Solid cancers particularly display metabolic reprogramming characterized by increased extracellular fatty acid uptake from surrounding adipose tissue and overexpression of various *de novo* lipogenesis enzymes such as fatty acid synthase, ATP citrate lyase, and acetyl-CoA carboxylase.[101–103] The characterization of tumor margins and migrating tumor cells is important in routine clinical pathological assessment to gain a prognostic outlook. MALDI-MSI is particularly advantageous in showing biochemical profiles of heterogenous solid tumors bearing multiple cell types and its utilization in spatial lipidomics of cancer is a rapidly growing area of research.

For many colorectal cancer patients, diagnosable symptoms present at the late stage of disease that results in half of the cases developing hard-to-treat metastasis.[104] Hence a better insight into the biochemical processes in the colorectal tumor microenvironment is desirable. Using a MALDI Q-TOF system with CHCA matrix, spatial lipidomic profiling of tumor-adjacent areas *versus* remote areas was used to explore the field-effects of tumors.[105] Elevated levels of PC(16:0_18:1), LPC(16:0), and LPC(18:1) were detected in cancerous regions, and so, are suggestive of their biomarker potential.[105] In another human colon cancer lipidome investigation using an LTQ-Orbitrap analyzer with a 2-mercaptobenzothiazole matrix, the proportion of K^+ adducts of PC(34:1), PC(36:1), PC(36:3), and SM(d34:1) in relation to Na^+ and H^+ adducts was found to be significantly increased between adenomatous compared to healthy mucosa. This observation MSI was further underpinned by survival analysis that showed poor survival outcomes for *KCNAB2* genes that encode for regulatory subunits of potassium voltage-gated channels.[106] A recent report that applied laser-post ionization MALDI-MSI (MALDI-2) studied the distribution of complex GSL species at different stages of colorectal cancer. Although this was a small study of a few patients, d18 : 1 based globotriaosylceramide (Gb3Cer/Gb4Cer) species of varying fatty acyl chain lengths and degree of unsaturation were specifically mapped to the glandular epithelium but not the stroma. Various fucosylated lipoforms of GSL were also observed indicative of heterogeneity in morphological profiles within and across tissues (Figure 8.2).[107] In the context of colon cancer drug design, GSL research is critical for chemotherapeutic target delivery by means of antibody/toxin recognition.

Prostate cancer is hormonally driven by aberrant androgen receptor signaling with a phenotype of multifocal and lipogenic tumors.[108–110] Clinically the disease is diagnosed by serum prostate-specific antigen (PSA) testing and the Gleason score (GS) system to pathologically grade tumors based on the level of cell differentiation.[111–113] However, both methods still have an imprecise ability to reliably distinguish men with indolent tumors from aggressive cases,[114] and so, spatial lipidomics has been leveraged to identify more sensitive tumor lipid classifiers. Two studies have detected distinct regional abundances of LPC in epithelial tissue. Positive-ion mode MALDI-TOF imaging uncovered higher amounts of LPC(16 : 0) in benign epithelium compared to cancer using sublimated 9-AA in a cohort of 31 patients.[115] Recently, Andersen *et al.* used NEDC and DHB to measure lipids and metabolites in 15 patients and also observed reduced LPC(16 : 0) levels in tumors.[116] This phenomenon was suggested to be a predictor of biochemical recurrence after radical prostatectomy. Furthermore, unsupervised principal component analysis (PCA) and orthogonal partial least squares discriminant analysis models of dual polarity data delineated spectra of histopathological defined normal, stroma, and tumor prostate compartments.[116] A MALDI-MSI metabolomics study also using quercetin and 9-AA independently for dual-polarity imaging on a FTICR-MSI system demonstrated that DG(34 : 1), TG(52 : 3), TG(56 : 3), and SM (d36 : 1) were localized in non-cancerous regions while PC(16 : 0) and PE(34 : 1) were high in tumors.[117] An increased ratio of ATP to ADP and AMP was also detected and interpreted as an increased energy demand to support tumor cell growth.[117] Another investigation combining MALDI-MSI with liquid surface extraction analysis was able to distinguish GS (4 + 3) tumors from GS (3 + 4) tumors based on cardiolipin distribution.[118] This study employed segmentation by way of bisecting k-means clustering method to map tumor, normal, normal tissue with macrophages, adipose, and atrophic regions.[118]

As mentioned above, spatial segmentation of comprehensive lipid spectral data can detect patterns or differences in region-specific lipid metabolism alterations. Downstream statistical analysis approaches rely on regions of interest that can be manually selected areas based on tissue pathology or be

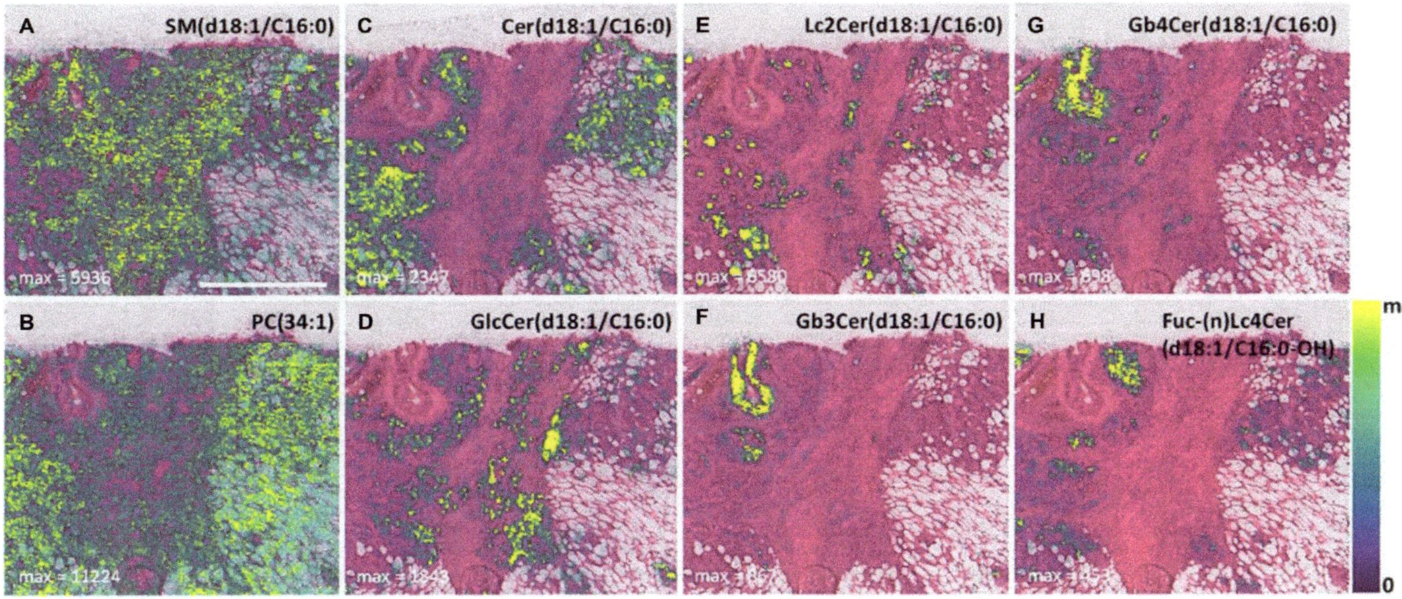

Figure 8.2 Spatial distribution of multiple lipid classes having a 34 : 1 sum-composition in human colorectal cancer tissue by MALDI-2-MSI. (A and B) SM(d34 : 1) and PC(34 : 1) exhibit similar ubiquitous abundance across the tissue; (C and D) Cer(d34 : 1) and GlcCer(d34 : 1) are localized to the tumor microenvironment; (E) Lc2Cer(d34 : 1) is more specific to glandular structures; (F and G) Gb3Cer(d34 : 1) and Gb4Cer(d34 : 1) are observed to have region-specific distribution in non-stromal tissue compartments; and (H) Fuc-(n)Lc4Cer is observed in potentially dysplastic cells.
Reproduced from ref. 107 with permission from American Chemical Society, Copyright 2020.

entirely unbiased using morphologically-driven lipid features. Clustering analysis has also been used to define head and neck carcinoma tissue, parotid glands and adenolymphoma specimens.[119] The wide application of these univariate and multivariate statistical methods[120,121] can potentially feed into deep learning and artificial intelligence pipelines to aid tissue classification based on multidimensional MSI data.[122]

In lung cancer, phospholipid profiling by MALDI-MSI showed fatty acyl chain elongation of phospholipids in both mouse and human lungs to be mostly regulated by ELOVL6, an elongase enzyme.[123] Hence, providing a therapeutic rationale for targeting this key part of phospholipid metabolism. MALDI-MSI has also been used to identify lipid markers of radiation-induced lung injury, a limiting factor of curative cancer treatment regimens. Following application of DHB matrix and imaging on a TOF/TOF system, PC(14 : 0_16 : 0), PC(16 : 0_16 : 0), PC(16 : 0_16 : 1), and PC(36 : 5)/PC(38 : 8) were detected in reduced abundance in irradiated animal tissues compared to naïve. Localized increases in ether-linked PC and polyunsaturated diacyl-PC ions were observed after radiation treatment.[124]

Medulloblastoma radiation treatment is debilitating and incurable in at least one-third of patients. Three-dimensional MALDI-MSI experiments in a whole mouse brain murine model delineated lipid alterations between metastatic and non-metastatic primary tumors.[125] This study used a norharmane matrix for negative-ion mode analysis on a MALDI-TOF system and accurate MS/MS lipid identification conducted on an Orbitrap Elite mass spectrometer. By way of area under the curve analysis of semi-supervised bisecting k-means spatial segments, PA(16 : 0_18 : 1), PA(18.0_20.4), PE(*P*-18 : 1_18 : 1), PE(16 : 0_20 : 4), PE(18 : 1/18 : 1), and PS(16 : 0_18 : 1) were decreased in metastatic tissue while PS(18 : 0_22 : 6), PI(18 : 0_20 : 4), phosphatidylinositol phosphate (PIP)(18 : 0_20 : 4), and phosphatidylinositol 4,5-bisphosphate (PIP_2)(18 : 0_20 : 4) were increased in metastatic tissue.[125]

8.4.2 Brain Injury and Disease

The brain has among the highest proportion of lipid content of any tissue with neuronal and glial cell membranes rich in sphingolipids and phospholipids[126] and is one of the most studied tissues using MSI. Lipid changes associated with traumatic brain injury (TBI) have been studied by several groups using MALDI-MSI. To study TBI-induced lipid changes, the controlled cortical impact (CCI) mouse model is often used and designed to simulate concussion.[127,128] Using TBI models, various lipid species have been shown to be elevated at the injury site. Mallah *et al.* demonstrated the elevation of a positive ion at *m/z* 856.59 at the site of TBI that was assigned as $[PC(42:9)+H]^+$ (but could also be $[PC(40:6)+Na]^+$).[129] A later study by the same group also observed elevated signals for acylcarnitines in injured brain lesions.[130] In particular, the palmitoylcarnitine signal was highest in the *substantia nigra* 3 days post-injury before signals decreased. Alterations in sphingolipids such as Cer have also been observed during TBI.[131]

Region-specific changes associated with neurodegenerative diseases have also been studied using MALDI-MSI. A central tenet in neurodegenerative disorders is the dysfunction of the endosomal–lysosomal system[132,133] leading to accumulation of sphingolipids[134] or unesterified cholesterol.[135] MALDI-MSI has studied fatty acid chain length-dependent gangliosides in Niemann–Pick disease type C (NPC), a fatal neurodegenerative condition caused by genetic mutations of the NPC1 or cholesterol-binding glycoprotein (NPC2).[136] Accumulation of GM2(d36 : 1) and GM2(d38 : 1) species was detected in lobule X of the Purkinje layer of mutant animals (*Npc*1$^{-/-}$) compared to control animals (*Npc*1$^{+/+}$).[136]

Amyloid β-peptide plaques drive Alzheimer's disease (AD) research. MALDI-MSI has been used to provide new understanding regarding the roles of lipids in AD. A novel approach described by Kaya and colleagues augmented dual-polarity MALDI-MSI with complementary imaging modalities,[137] where conserved tissue integrity and morphology were compatible with subsequent immunohistochemistry (IHC) and multiplexed immunofluorescence. Staining of glial cells, cytoskeleton, and cell nuclei by anti-glial fibrillary acidic protein, anti-βIII-tubulin and 4,6-diamidino-2-phenylindole, respectively, correlated with varied spatial lipid signatures of the molecular layer, granular layer, and white matter regions of a transgenic (tgSwe) AD model.[137] This group later developed a multimodal MSI strategy that involved consecutive protein MSI (for visualizing amyloid-plaque associated peptides) and dual-polarity lipid MSI on the same mouse cerebellum section. As an example, at 10 μm spatial resolution, MALDI-MS images of Cer(d18 : 1_18 : 0), LPC(16 : 0), and LPC(18 : 0) corresponded with fluorescence staining of amyloid plaques of transgenic (tgArcSwe) AD mouse brain cortical regions (Figure 8.3).[138] LPC(16 : 0), PC(32 : 0), PC(33 : 1), PC(34 : 1), and PC(35 : 3) were co-localized to molecular and granular layers.[138] In negative-ion mode CerP(18 : 0), PG(34 : 1), PE(38 : 4), PS(40 : 6), PI(36 : 4), and PI(38 : 4) were localized to molecular and/or granular layers while SM(36 : 1) and ST lipids were unique to white matter.[138] A decrease of various sulfatide species, ST(d36 : 1), ST(d38 : 1), ST(d40 : 1(OH)), ST(d42 : 1), and ST(d42 : 1(OH)) together with elevated levels of lysophospholipids (LPA (16 : 0), LPA(18 : 1), LPA(18 : 0), LPE(16 : 0), LPE(18 : 0), LPC(16 : 0), LPC(18 : 0), and LPI(18 : 0)) within amyloid plaques was argued to be linked with degeneration of myelin lipids in the tgArcSwe AD mouse model.[138,139] LPC abundance was implicated in promoting amyloid β plaque oligomerization and fibrillization consistent with induction of neurotoxicity[139] through recruitment of apolipoprotein E.[140] In a further separate study, Kaya *et al.* found that long-chain sphingosine (>18C) base, monosialogangliosides (GM1, GM2, and GM3) were co-located with amyloid plaques in the hippocampus and cortical regions of an AD (5xFAD) mouse brain.[141]

Glucocerebrosidase (GCase) is responsible for the breakdown of GluCer and its mutation can lead to accumulation of GluCer.[142] MALDI FTICR in combination with LC–ESI-MS/MS[143] has been applied to study Gaucher disease (GD) pathology in a 4L/PS-NA mouse model with mutant GCase. Using MALDI-MSI,

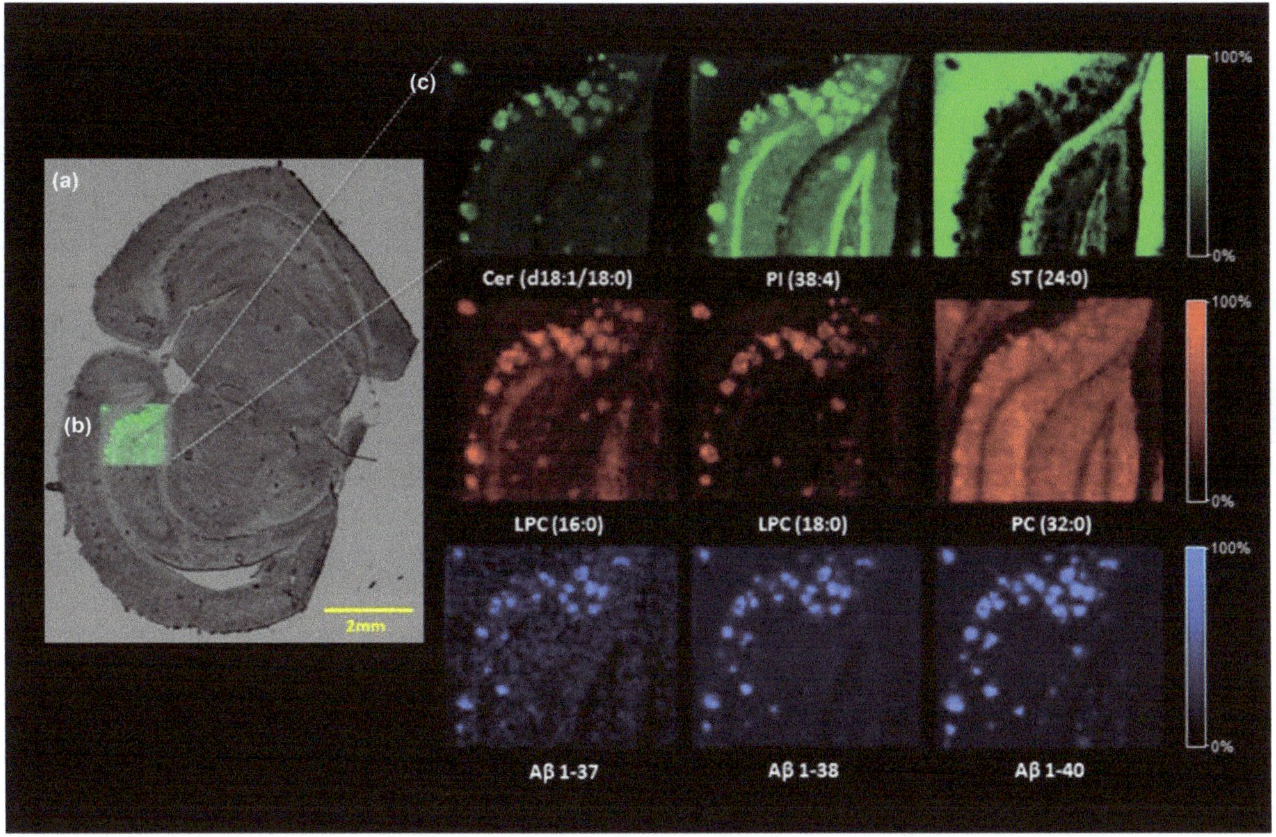

Figure 8.3 Combined dual polarity MALD-MSI of amyloid-plaque associated lipids and peptides. (a) Brightfield image of the coronal mouse brain section of a transgenic (tgArcSwe) mouse model. (b) Hippocampal region analyzed by MSI highlighted in green. (c) Selected lipid ion distributions in the negative-ion mode, Cer(d18 : 1/18 : 0), PI(38 : 4) and ST(42 : 1); positive-ion mode, LPC(16 : 0), LPC(18 : 0), and PC(32 : 0); and the amyloid-β peptides acquired consecutively to dual-polarity lipid MSI; Aβ 1-37, Aβ 1-38, and Aβ 1-40.
Reproduced from ref. 138 with permission from American Chemical Society, Copyright 2017.

the region-specific accumulation of a variety of hexosylceramide (HexCer) was observed compared to wild type tissue, likely reflecting the upregulation of specific Cer synthases. Based on LC–ESI-MS/MS data, a large proportion of these hexosyl sugars were not GalCer but GluCer species such as GluCer(d18 : 1_18 : 0) and GluCer(d18 : 1_22 : 0). The accumulation of these GSL species (including lactosylceramide) was suggested to impede degradation of key lysosomal proteins, amyloid precursor protein, and α-synuclein.[126]

8.4.3 Liver Disease

Alongside being the major detoxification organ, the liver is a central site of numerous other biochemical processes, including lipid metabolism. In recent years, region-specific and disease-state-specific lipid metabolism changes associated with the progression of non-alcoholic fatty liver disease (NAFLD) have been performed. Hall *et al.* applied MALDI-MSI to study lipid zonation and remodeling associated with the progression of NAFLD in both human and mouse liver tissue.[144] PC(32 : 0) and SM(40 : 1) were found to be distributed primarily in the central and portal veins. In both normal and NAFL tissue PC(34 : 2) and PC(36 : 4) were associated with zone 1 and PC(34 : 1) and PC(36 : 1) with zone 3. As the disease progressed to non-alcoholic steatohepatitis (NASH), loss of lipid zonation was observed. The presence of arachidonic acid-containing lipids was associated with the enzyme lysophosphatidylcholine acyltransferase 2 using IHC. Combined with the observation of pro-inflammatory eicosanoids that are produced following hydrolysis of arachidonic acid from phospholipids by phospholipase A2, this study adds to the breadth of knowledge suggesting that arachidonic acid-containing lipids are key drivers of inflammation and oxidative damage in NASH. In another study, MALDI-MSI using fresh frozen biopsies from 23 patients following bariatric surgery demonstrated that localized lipid profiles combined with multivariate data analysis could categorize both the heterogeneity of disease progression within individual biopsies and between bioposies.[145] Good agreement was found between MSI-based classification of non-steatotic *vs* steatotic regions compared with histology. In addition, arachidonic acid-containing lipids such as PE and PI were decreased in steatotic regions. This again points to a major role of lipid distribution driving localized inflammation *via* eicosanoid cascade. TGs in NAFLD have also been studied using sodium-doped gold-assisted LDI MSI at a spatial resolution of 10 μm.[146] TG(48 : 2), TG(48 : 1), and TG(48 : 0) showed spatial changes in simple steatosis compared to cirrhosis, where FA desaturation was more distinctly observed with disease progression.

8.5 Recent Developments in MALDI-MSI of Lipids

8.5.1 Strategies to Improve the Ionization Efficiency of Lipids

Common to all contemporary ionization methods is the typically low and variable ionization efficiency observed for many molecules meaning only one in 10^5–10^7 of desorbed molecules may be ionized.[147] Due to the high structural diversity of the lipidome, ionization efficiencies among many lipid species vary dramatically, and a given MSI experiment only detects a subset of the lipid species present.

8.5.1.1 Laser Post-ionization (MALDI-2)

In efforts to increase the overall sensitivity and breadth of lipid species detectable in MALDI-MSI, several approaches that aim to improve overall ionization efficiencies have been reported in recent years. Laser post-ionization (MALDI-2) was reported in 2015[148] and involves irradiation of the MALDI plume with a second UV laser with a wavelength below the two-photon ionization threshold of the matrix. When performed under elevated pressure conditions (source pressures >3 mbar), MALDI-2 induces a second MALDI-like event within the confined plume that can ionize many lipid species not ionized by the initial MALDI event. Further details on the principle of MALDI-2 are provided in Chapters 1 and 3 of this book. MALDI-2 can improve the signal intensities for many lipid species by up to 100-fold and enable the detection of many lipid species that are poorly detected, if at all, with conventional MALDI[149–152] (*e.g.* some GSLs and sterols[152]). The increased ionization efficiency also allows for more complete coverage of lipids in a single experiment. For example, in the positive-ion mode, it enables detection and imaging of acidic phospholipids such as PS and PI that are not typically detected as protonated ions.[149,150]

A second benefit of MALDI-2 is the ability to improve the spatial resolution for lipid imaging experiments by virtue of the increased sensitivity. Recently MALDI-2 was combined with an oversampling approach to enable spatial resolutions of 6 μm from an initial laser diameter of ~12–15 μm.[150] From rat liver tissue, MALDI-2-MSI was able to detect and identify (at the sum-composition level) 66 lipid species across more than 15 lipid sub-classes. In contrast, conventional MALDI-MSI only detected 6 lipid species, all belonging to PC lipids. When applied to human brain tissue containing active multiple sclerosis lesions, MALDI-2 could detect and identify 147 lipid species at 6 μm pixel size and localized cholesterol esters to lesion-specific lipid droplets having individual sizes of ~5–20 μm in diameter. Using the transmission geometry approach (see Chapter 3), MALDI-2 enables high-quality lipid spectra to be generated from pixel sizes as small as 600 nm, enabling single-cell imaging (see below). The use of plasma as a source of post-ionization is another promising approach to improve the sensitivity of MALD-MSI experiments.[153] Plasma-based post-ionization results in similar benefits for lipid detection as observed with MALDI-2.

8.5.1.2 On-tissue Derivatization

On-tissue derivatization strategies[41] can also be employed to improve the sensitivity for targeted lipid classes. This method relies on chemical modification of targeted functional groups to introduce a more readily ionizable moiety (often a fixed positive charge) while preserving the location of analytes within the tissue.

Phos-tag, a zinc complex that binds to phosphate monoesters has been used to enable the MALDI-MSI of low abundance lipids such as LPA and sphingosine-1-phosphate (S1P).[154] This work provided amongst the first reports of S1P analysis from brain tissue with MALDI-MSI and was able to visualize region-specific accumulation of S1P lipid species in both *Sgpl1*$^{fl/fl/Nes}$ mice that are deficient in the lipid phosphate degradative enzyme S1P lyase, as well as *Ppap2b*$^{fl/fl/Nes}$ mice that are deficient in the LPP3/PAP2B enzyme that dephosphorylates lipids including LPA and S1P (Figure 8.4).[154]

Sterols are another class of lipids that typically ionize poorly with conventional MALDI. Using Girards Reagent T the carbonyl group of steroids such as testosterone, 5α-dihydrotestosterone, 11-dehydrocorticosterone, and corticosterone can be derivatized to enable their analysis using MALDI-MSI.[155–157] Interestingly, MS/MS of Girards T-derivatized sterols can also generate fragmentation patterns that allow the differentiation of some sterol isomers such as aldosterone from isomeric cortisone and cortisol from isomeric 18-hydroxycorticosterone.[158]

In addition to enabling differentiation of double bond positional isomers on tissue derivatisation,[79,80] Paternò–Büchi can also be used to improve the ionization efficiency of lipid species. Using acetylpyridine as the Paterno–Buchi reagent various non-polar lipids such as hydrocarbons, fatty acids, cholesterol esters, and TGs were derivatized and detected with significantly higher sensitivities due to the basicity of the pyridine moiety.[159] This approach has the added benefit of allowing double bond position assignment upon MS/MS.

8.5.2 Lipid Analysis and Imaging from Single Cells

As the sensitivity and spatial resolution of MALDI-MSI is constantly improving, it is increasingly being used to analyze and image single cells.[160] Lipids present a promising molecular class to target for single-cell MALDI due to the typically high signals observed for several abundant lipid classes. Single-cell MALDI analysis typically involves two main application areas: (i) imaging of lipids in single cells to reveal sub-cellular localizations;[161–163] or (ii) the profiling of multiple single cells for studying inter-cellular heterogeneity.[164,165]

One of the earliest examples of sub-cellular imaging of lipids was reported by Schober *et al.* and demonstrated the imaging of lipids from fixed HeLa cells with 7 μm pixel size.[162] Single-cell MALDI-MSI has also been used to study sphingolipid heterogeneity in human dermal fibroblasts, revealing both distinct sub-cellular distributions of sphingolipids.[163] More recently, transmission-mode MALDI-2 was deployed for the imaging of lipids in Vero B cells at a pixel size of 1 μm in both positive and negative-ion mode.[161] 3D-MALDI-MSI has also been applied to study lipidomics changes occurring during the embryonic development of zebrafish.[166]

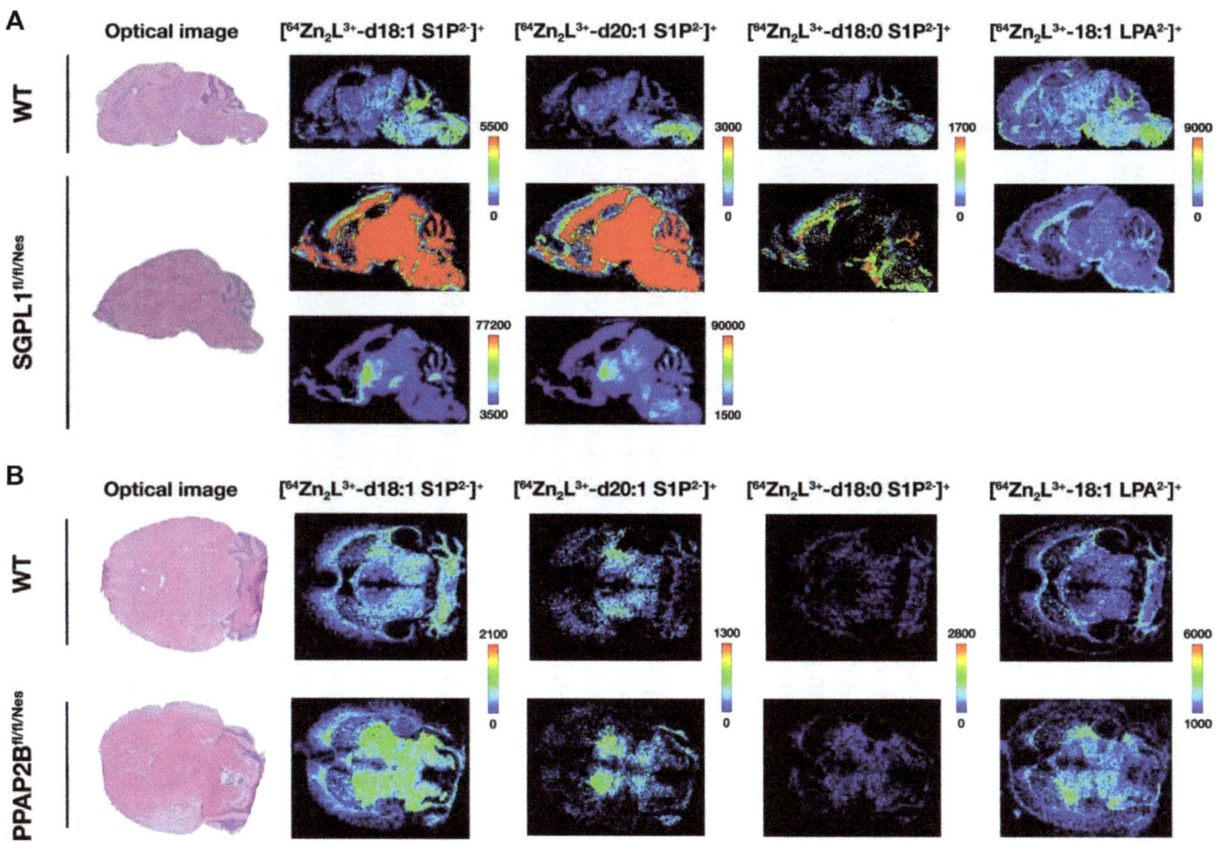

Figure 8.4 On-tissue derivatization and MALDI-MSI of sphingosine-1-phosphate (S1P) and lysophosphatidic acid (LPA) using Phos-tag (pixel size = 90 μm). (A) Distribution of several S1P and LPA lipids in *Sgpl1*$^{fl/fl/Nes}$ and wild-type (WT) mouse brain tissue. (B) Distribution of several S1P and LPA lipids in *Ppap2b*$^{fl/fl/Nes}$ and wild-type (WT) mouse brain tissue. Optical images of the H&E stained tissue sections are shown on the left.
Reproduced from ref. 154 with permission from American Chemical Society, Copyright 2021.

As an example of lipid profiling from multiple single cells, Neumann *et al.* reported a multimodal approach combining automated MALDI-MS profiling[167] of individual cells with subsequent immunocytochemistry.[165] This approach was used to profile 1800 cells consisting of both neurons and astrocytes that could be distinguished using immunocytochemistry. When combined with multivariate analysis of the MALDI-MS data using t-distributed stochastic neighbor embedding, the NF-L-positive cells (indicative of neurons) and the FFAP-positive cells (indicative of astrocytes) could be clustered into 5 and 4 sub-populations, respectively, based on the lipid profiles detected in positive-ion mode. A similar automated approach has also been used for the lipidomics profiling of 30 000 rat cerebellar cells, again revealing spectra signatures allowing for the differentiation of astrocyte-like, neuron-like, and other types of cells, including the identification of different sub-populations within each.[168]

8.5.3 Imaging Stable Isotope Labeled Lipids

The lipids detected by MSI are the result of a series of biochemical processes and represent a single intermediate time-point of dynamic lipid metabolism. The biosynthetic processes viewed by MSI are a static snapshot of the sample at the time of preservation (typically flash-freezing). Metabolic labeling using stable isotopes can allow these dynamic processes to be studied. By following the incorporation of labeled substrates over time the molecular specificities and time scales of lipid metabolic processes can be revealed. It is only recently that such methods have been combined with MALDI-MSI. The first study used deuterium labeling *via* heavy water to study lipid synthesis in mammary tumors from mice revealed different labeling efficiencies of lipids throughout the tissue.[169] A drawback of this approach is the non-specific deuterium labeling. The use of more specific labeled substrates can provide more detailed information of localized lipid biology. For example, d_6-palmitate has been used to study the synthesis and transport of palmitate-containing PC lipids in cultured neurons.[158] More recently, a combination of D_9-choline and universally ^{13}C labeled PC (16 : 0/16 : 0) ($U^{13}C$-DPPC) has been used to study PC synthesis and remodeling throughout mouse lung tissue using MALDI-MSI.[170] Visualizing the distribution of D_9-labeled PC lipids revealed the locations of PC lipids synthesized *via* the Kennedy pathway[171] over several hours. In contrast ($U^{13}C$-DPPC), administered as part of a therapeutic surfactant, provided a substrate by which acyl chain remodeling could occur. By detecting ions corresponding to $^{13}C_{24}$-PC(36 : 4) whereby one ^{13}C labeled palmitic acid is replaced with a 20 : 4 fatty acyl *via* the Lands cycle over 12–18 hours, the location of distinct acyl chain remodeling events was possible. Although the combination of metabolic labeling with MALDI-MSI is in its infancy, it will undoubtedly prove a powerful approach in the future to localize distinct metabolic events in time throughout heterogeneous biological tissues.

8.6 Conclusions and Future Directions

Lipids present a highly versatile and promising analyte class for analysis with MALDI-MSI. Not only are many key lipid classes readily detected by MALDI-MSI, but lipid distributions revealed by MSI have proven to provide many new insights regarding region-specific metabolic processes in healthy and diseased tissues. Given that abundant lipids such as phospholipids often give the most abundant signals during the MALDI-MSI analysis of biological material they are also widely used as the analyte class of choice to develop and evaluate new MSI methods. For example, most developments leading to MSI at low micron resolution and the analysis of single cells with MALDI-MSI have often first been demonstrated for lipid analysis. Although many challenges still exist, including the detection of low abundance or poorly ionizable lipids, the precise identification and isomer resolution of lipids, quantitation, and the ability for detailed biochemical interpretation of the observed lipid distributions, addressing these challenges is a very active area of research and much progress has been made in the past few years. With the ever-growing popularity of lipid MSI both the technological capabilities and breadth of possible biological applications are only expected to increase in the future.

Abbreviations

9-AA	9-aminoacridine
AD	Alzheimer's disease
ADP	adenosine diphosphate
AMP	adenosine monophosphate
ATP	adenosine triphosphate
Cer	ceramides
CerP	ceramide phosphates
CHCA	α-cyano-4-hydroxycinnamic acid
CMBT	5-chloro-2-mercaptobenzothiazole
DG	diacylglyceride
DAN	1,5-diaminonaphthalene
DDA	data-dependent acquisition
DESI	desorption electrospray ionization
DHA	2,5-dihydroxyacetophenone
DHB	2,5-dihydrobenzoic acid
ESI	electrospray ionization
GalCer	galactosylceramides
GluCer	glucosylceramides
GM	monosialogangliosides
HexCer	hexosylceramides
LDI	laser desorption ionization
LMW	low molecular weight
MALDI	matrix-assisted laser desorption/ionization

MALDI-2	laser post-ionization MALDI
mbar	millibar (metric unit of pressure)
MS	mass spectrometry
MS/MS	tandem mass spectrometry
MSI	mass spectrometry imaging
NAFLD	non-alcoholic fatty liver disease
NAPA	silicon nanospot arrays
NASH	non-alcoholic steatohepatitis
NIMS	nanostructure-initiator mass spectrometry
NPC	Niemann-Pick disease type C
PC	phosphatidylcholine
PE	phosphatidylethanolamine
PI	phosphatidylinositol
PS	phosphatidylserine
Q-TOF	quadrupole time-of-flight
SM	sphingomyelin
TBI	traumatic brain injury
TG	triacylglyceride
TIMS	trapped ion mobility
TOF	time-of-flight
SIMS	secondary ion mass spectrometry

Acknowledgements

S. R. E. acknowledges funding from the Australian Research Council Future Fellowship Scheme (grant number FT190100082).

References

1. E. Fahy, S. Subramaniam, R. C. Murphy, M. Nishijima, C. R. Raetz, T. Shimizu, F. Spener, G. van Meer, M. J. Wakelam and E. A. Dennis, *J. Lipid Res.*, 2009, **50**(Suppl), S9–S14.
2. G. Liebisch, E. Fahy, J. Aoki, E. A. Dennis, T. Durand, C. S. Ejsing, M. Fedorova, I. Feussner, W. J. Griffiths, H. Kofeler, A. H. Merrill, Jr, R. C. Murphy, V. B. O'Donnell, O. Oskolkova, S. Subramaniam, M. J. O. Wakelam and F. Spener, *J. Lipid Res.*, 2020, **61**, 1539–1555.
3. G. van Meer, D. R. Voelker and G. W. Feigenson, *Nat. Rev. Mol. Cell Biol.*, 2008, **9**, 112–124.
4. M. R. Wenk, *Nat. Rev. Drug Discovery*, 2005, **4**, 594–610.
5. R. W. Gross, *Biochim. Biophys. Acta, Mol. Cell Biol. Lipids*, 2017, **1862**, 731–739.
6. M. A. Aon, N. Bhatt and S. C. Cortassa, *Front. Physiol.*, 2014, **5**, 282.
7. J. Soppert, M. Lehrke, N. Marx, J. Jankowski and H. Noels, *Adv. Drug Delivery Rev.*, 2020, **159**, 4–33.
8. N. Schwartz, A. Verma, C. B. Bivens, Z. Schwartz and B. D. Boyan, *Biochim. Biophys. Acta*, 2016, **1863**, 2289–2298.

9. G. D'Angelo, S. Capasso, L. Sticco and D. Russo, *FEBS J.*, 2013, **280**, 6338–6353.
10. A. Shevchenko and K. Simons, *Nat. Rev. Mol. Cell Biol.*, 2010, **11**, 593–598.
11. J. Li and J. X. Cheng, *Sci. Rep.*, 2014, **4**, 6807.
12. G. D'Angelo, S. Capasso, L. Sticco and D. Russo, *FEBS J.*, 2013, **280**, 6338–6353.
13. L. Yetukuri, K. Ekroos, A. Vidal-Puig and M. Oresic, *Mol. BioSyst.*, 2008, **4**, 121–127.
14. S. J. Blanksby and T. W. Mitchell, *Annu. Rev. Anal. Chem.*, 2010, **3**, 433–465.
15. T. Porta Siegel, K. Ekroos and S. R. Ellis, *Angew. Chem., Int. Ed. Engl.*, 2019, **58**, 6492–6501.
16. M. R. Wenk, *Cell*, 2010, **143**, 888–895.
17. C. S. Ejsing, J. L. Sampaio, V. Surendranath, E. Duchoslav, K. Ekroos, R. W. Klemm, K. Simons and A. Shevchenko, *Proc. Natl. Acad. Sci. U. S. A.*, 2009, **106**, 2136–2141.
18. K. Scupakova, B. Balluff, C. Tressler, T. Adelaja, R. M. A. Heeren, K. Glunde and G. Ertaylan, *Clin. Chem. Lab. Med.*, 2020, **58**, 914–929.
19. F. F. Hsu, *Anal. Bioanal. Chem.*, 2018, **410**, 6387–6409.
20. T. Alexandrov, *Annu. Rev. Biomed. Data Sci.*, 2020, **3**, 61–87.
21. P. Inglese, G. Correia, P. Pruski, R. C. Glen and Z. Takats, *Anal. Chem.*, 2019, **91**, 6530–6540.
22. J. Zhang, J. Rector, J. Q. Lin, J. H. Young, M. Sans, N. Katta, N. Giese, W. Yu, C. Nagi, J. Suliburk, J. Liu, A. Bensussan, R. J. DeHoog, K. Y. Garza, B. Ludolph, A. G. Sorace, A. Syed, A. Zahedivash, T. E. Milner and L. S. Eberlin, *Sci. Transl. Med.*, 2017, **9**, eaan3968.
23. D. L. Phelps, J. Balog, L. F. Gildea, Z. Bodai, A. Savage, M. A. El-Bahrawy, A. V. M. Speller, F. Rosini, H. Kudo, J. S. McKenzie, R. Brown, Z. Takáts and S. Ghaem-Maghami, *Br. J. Cancer*, 2018, **118**, 1349–1358.
24. S. Vens-Cappell, I. U. Kouzel, H. Kettling, J. Soltwisch, A. Bauwens, S. Porubsky, J. Muthing and K. Dreisewerd, *Anal. Chem.*, 2016, **88**, 5595–5599.
25. H. Y. Wang, C. B. Liu and H. W. Wu, *J. Lipid Res.*, 2011, **52**, 840–849.
26. C. L. Carter, C. W. McLeod and J. Bunch, *J. Am. Soc. Mass Spectrom.*, 2011, **22**, 1991–1998.
27. D. R. N. Vos, A. P. Bowman, R. M. A. Heeren, B. Balluff and S. R. Ellis, *Int. J. Mass Spectrom.*, 2019, **446**, 116212.
28. A. Ly, A. Buck, B. Balluff, N. Sun, K. Gorzolka, A. Feuchtinger, K. P. Janssen, P. J. Kuppen, C. J. van de Velde, G. Weirich, F. Erlmeier, R. Langer, M. Aubele, H. Zitzelsberger, L. McDonnell, M. Aichler and A. Walch, *Nat. Protoc.*, 2016, **11**, 1428–1443.
29. J. X. M. Truong, X. Spotbeen, J. White, J. V. Swinnen, L. M. Butler, M. F. Snel and P. J. Trim, *Anal. Bioanal. Chem.*, 2021, **413**, 2695–2708.
30. P. M. Angel, J. M. Spraggins, H. S. Baldwin and R. Caprioli, *Anal. Chem.*, 2012, **84**, 1557–1564.
31. F. F. Hsu, A. Bohrer and J. Turk, *J. Am. Soc. Mass Spectrom.*, 1998, **9**, 516–526.

32. R. L. Griffiths and J. Bunch, *Rapid Commun. Mass Spectrom.*, 2012, **26**, 1557–1566.
33. R. L. Griffiths, J. Sarsby, E. J. Guggenheim, A. M. Race, R. T. Steven, J. Fear, P. F. Lalor and J. Bunch, *Anal. Chem.*, 2013, **85**, 7146–7153.
34. Z. Hall, Z. Ament, C. H. Wilson, D. L. Burkhart, T. Ashmore, A. Koulman, T. Littlewood, G. I. Evan and J. L. Griffin, *Cancer Res.*, 2016, **76**, 4608–4618.
35. X. Han and R. W. Gross, *J. Lipid Res.*, 2003, **44**, 1071–1079.
36. J. A. Bowden, C. J. Albert, O. S. Barnaby and D. A. Ford, *Anal. Biochem.*, 2011, **417**, 202–210.
37. D. Gode and D. A. Volmer, *Analyst*, 2013, **138**, 1289–1315.
38. J. Soltwisch, T. W. Jaskolla, F. Hillenkamp, M. Karas and K. Dreisewerd, *Anal. Chem.*, 2012, **84**, 6567–6576.
39. J. Zhou and T. D. Lee, *J. Am. Soc. Mass Spectrom.*, 1995, **6**, 1183–1189.
40. J. Leopold, Y. Popkova, K. M. Engel and J. Schiller, *Biomolecules*, 2018, **8**, 173.
41. Q. Zhou, A. Fulop and C. Hopf, *Anal. Bioanal. Chem.*, 2021, **413**, 2599–2617.
42. W. J. Perry, N. H. Patterson, B. M. Prentice, E. K. Neumann, R. M. Caprioli and J. M. Spraggins, *J. Mass Spectrom.*, 2020, **55**, e4491.
43. A. Thomas, J. L. Charbonneau, E. Fournaise and P. Chaurand, *Anal. Chem.*, 2012, **84**, 2048–2054.
44. S. R. Ellis, J. Cappell, N. O. Potocnik, B. Balluff, J. Hamaide, A. Van der Linden and R. M. Heeren, *Analyst*, 2016, **141**, 3832–3841.
45. I. Kaya, E. Jennische, S. Lange and P. Malmberg, *Anal. Methods*, 2018, **10**, 2428–2435.
46. J. Yang, J. L. Norris and R. Caprioli, *J. Mass Spectrom.*, 2018, **53**, 1005–1012.
47. J. M. Spraggins, D. G. Rizzo, J. L. Moore, M. J. Noto, E. P. Skaar and R. M. Caprioli, *Proteomics*, 2016, **16**, 1678–1689.
48. N. Ogrinc Potocnik, T. Porta, M. Becker, R. M. Heeren and S. R. Ellis, *Rapid Commun. Mass Spectrom.*, 2015, **29**, 2195–2203.
49. J. M. Spraggins and R. M. Caprioli, *J. Am. Soc. Mass Spectrom.*, 2011, **22**, 1022–1031.
50. R. C. Murphy, J. A. Hankin, R. M. Barkley and K. A. Zemski Berry, *Biochim. Biophys. Acta*, 2011, **1811**, 970–975.
51. S. Van Nuffel, N. Elie, E. Yang, J. Nouet, D. Touboul, P. Chaurand and A. Brunelle, *Anal. Chem.*, 2018, **90**, 1907–1914.
52. M. E. Dueñas, L. Carlucci and Y. J. Lee, *J. Am. Soc. Mass Spectrom.*, 2016, **27**, 1575–1578.
53. T. Morikawa-Ichinose, Y. Fujimura, F. Murayama, Y. Yamazaki, T. Yamamoto, H. Wariishi and D. Miura, *J. Am. Soc. Mass Spectrom.*, 2019, **30**, 1512–1520.
54. W. Bouschen, O. Schulz, D. Eikel and B. Spengler, *Rapid Commun. Mass Spectrom.*, 2010, **24**, 355–364.
55. M. Kompauer, S. Heiles and B. Spengler, *Nat. Methods*, 2017, **14**, 90–96.

56. S. Caughlin, D. H. Park, K. K. Yeung, D. F. Cechetto and S. N. Whitehead, *J. Vis. Exp.*, 2017, **121**, e55254.
57. J. Folch, M. Lees and G. H. Sloane Stanley, *J. Biol. Chem.*, 1957, **226**, 497–509.
58. M. Muthu, J. Gopal and S. Chun, *TrAC, Trends Anal. Chem.*, 2017, **97**, 96–103.
59. H. K. Woo, T. R. Northen, O. Yanes and G. Siuzdak, *Nat. Protoc.*, 2008, **3**, 1341–1349.
60. G. J. Patti, L. P. Shriver, C. A. Wassif, H. K. Woo, W. Uritboonthai, J. Apon, M. Manchester, F. D. Porter and G. Siuzdak, *Neuroscience*, 2010, **170**, 858–864.
61. K. B. Louie, B. P. Bowen, S. McAlhany, Y. Huang, J. C. Price, J. H. Mao, M. Hellerstein and T. R. Northen, *Sci. Rep.*, 2013, **3**, 1656.
62. A. R. Korte, N. J. Morris and A. Vertes, *Anal. Chem.*, 2019, **91**, 3951–3958.
63. J. A. Fincher, A. R. Korte, J. E. Dyer, S. Yadavilli, N. J. Morris, D. R. Jones, V. K. Shanmugam, R. K. Pirlo and A. Vertes, *J. Mass Spectrom.*, 2020, **55**, e4443.
64. J. K. Pauling, M. Hermansson, J. Hartler, K. Christiansen, S. F. Gallego, B. Peng, R. Ahrends and C. S. Ejsing, *PLoS One*, 2017, **12**, e0188394.
65. J. A. Hankin, S. E. Farias, R. M. Barkley, K. Heidenreich, L. C. Frey, K. Hamazaki, H. Y. Kim and R. C. Murphy, *J. Am. Soc. Mass Spectrom.*, 2011, **22**, 1014–1021.
66. I. A. Mulder, N. Ogrinc Potocnik, L. A. M. Broos, A. Prop, M. J. H. Wermer, R. M. A. Heeren and A. van den Maagdenberg, *Sci. Rep.*, 2019, **9**, 1090.
67. C. Hu, W. Luo, J. Xu and X. Han, *Mass Spectrom. Rev.*, 2020, **00**, 1–17.
68. E. Ryan and G. E. Reid, *Acc. Chem. Res.*, 2016, **49**, 1596–1604.
69. C. Bielow, G. Mastrobuoni, M. Orioli and S. Kempa, *Anal. Chem.*, 2017, **89**, 2986–2994.
70. A. P. Bowman, G. T. Blakney, C. L. Hendrickson, S. R. Ellis, R. M. A. Heeren and D. F. Smith, *Anal. Chem.*, 2020, **92**, 3133–3142.
71. R. L. Hansen and Y. J. Lee, *J. Am. Soc. Mass Spectrom.*, 2017, **28**, 1910–1918.
72. D. C. Perdian and Y. J. Lee, *Anal. Chem.*, 2010, **82**, 9393–9400.
73. S. R. Ellis, M. R. L. Paine, G. B. Eijkel, J. K. Pauling, P. Husen, M. W. Jervelund, M. Hermansson, C. S. Ejsing and R. M. A. Heeren, *Nat. Methods*, 2018, **15**, 515–518.
74. M. R. L. Paine, B. L. J. Poad, G. B. Eijkel, D. L. Marshall, S. J. Blanksby, R. M. A. Heeren and S. R. Ellis, *Angew. Chem., Int. Ed. Engl.*, 2018, **57**, 10530–10534.
75. R. Almeida, J. K. Pauling, E. Sokol, H. K. Hannibal-Bach and C. S. Ejsing, *J. Am. Soc. Mass Spectrom.*, 2015, **26**, 133–148.
76. B. M. Prentice, J. C. McMillen and R. M. Caprioli, *Int. J. Mass Spectrom.*, 2019, **437**, 30–37.
77. I. Lanekoff, K. Burnum-Johnson, M. Thomas, J. Short, J. P. Carson, J. Cha, S. K. Dey, P. Yang, M. C. Prieto Conaway and J. Laskin, *Anal. Chem.*, 2013, **85**, 9596–9603.

78. A. M. Moerman, M. Visscher, N. Slijkhuis, K. Van Gaalen, B. Heijs, T. Klein, P. C. Burgers, Y. B. De Rijke, H. M. M. Van Beusekom, T. M. Luider, H. J. M. Verhagen, A. F. W. Van der Steen, F. J. H. Gijsen, K. Van der Heiden and G. Van Soest, *J. Lipid Res.*, 2021, **62**, 100020.
79. A. Bednařík, S. Bölsker, J. Soltwisch and K. Dreisewerd, *Angew. Chem., Int. Ed.*, 2018, **57**, 12092–12096.
80. F. Waldchen, B. Spengler and S. Heiles, *J. Am. Chem. Soc.*, 2019, **141**, 11816–11820.
81. F. Waldchen, F. Mohr, A. H. Wagner and S. Heiles, *Anal. Chem.*, 2020, **92**, 14130–14138.
82. M. E. N. Born and B. M. Prentice, *Int. J. Mass Spectrom.*, 2020, **452**, 116338.
83. J. T. Specker, S. L. Van Orden, M. E. Ridgeway and B. M. Prentice, *Anal. Chem.*, 2020, **92**, 13192–13201.
84. D. R. Klein, C. L. Feider, K. Y. Garza, J. Q. Lin, L. S. Eberlin and J. S. Brodbelt, *Anal. Chem.*, 2018, **90**, 10100–10104.
85. T. H. Kuo, H. H. Chung, H. Y. Chang, C. W. Lin, M. Y. Wang, T. L. Shen and C. C. Hsu, *Anal. Chem.*, 2019, **91**, 11905–11915.
86. K. V. Djambazova, D. R. Klein, L. G. Migas, E. K. Neumann, E. S. Rivera, R. Van de Plas, R. M. Caprioli and J. M. Spraggins, *Anal. Chem.*, 2020, **92**, 13290–13297.
87. J. Soltwisch, B. Heijs, A. Koch, S. Vens-Cappell, J. Hohndorf and K. Dreisewerd, *Anal. Chem.*, 2020, **92**, 8697–8703.
88. J. Tu, Z. W. Zhou, T. Z. Li and Z. J. Zhu, *TrAC, Trends Anal. Chem.*, 2019, **116**, 332–339.
89. T. Fu, J. Oetjen, M. Chapelle, A. Verdu, M. Szesny, A. Chaumot, D. Degli-Esposti, O. Geffard, Y. Clément, A. Salvador and S. Ayciriex, *J. Mass Spectrom.*, 2020, **55**, e4531.
90. S. N. Jackson, D. Barbacci, T. Egan, E. K. Lewis, J. A. Schultz and A. S. Woods, *Anal. Methods*, 2014, **6**, 5001–5007.
91. P. O. Helmer, I. D. Nordhorn, A. Korf, A. Behrens, R. Buchholz, F. Zubeil, U. Karst and H. Hayen, *Anal. Chem.*, 2021, **93**, 2135–2143.
92. K. Skraskova, E. Claude, E. A. Jones, M. Towers, S. R. Ellis and R. M. Heeren, *Methods*, 2016, **104**, 69–78.
93. U. Daisy, Y. Ruichuan, S. Daniela Mesa, L. Yingju, S. Xiaofei, D. Sudhansu and L. Julia, *Nanospray Desorption Electrospray Ionization (nano-DESI) Mass Spectrometry Imaging of Drift Time-Separated Ions*, 2020.
94. C. L. Feider, N. Elizondo and L. S. Eberlin, *Anal. Chem.*, 2016, **88**, 11533–11541.
95. K. L. Leaptrot, J. C. May, J. N. Dodds and J. A. McLean, *Nat. Commun.*, 2019, **10**, 985.
96. R. A. Harris, K. L. Leaptrot, J. C. May and J. A. McLean, *Trends Anal. Chem.*, 2019, **116**, 316–323.
97. D. Hanahan and R. A. Weinberg, *Cell*, 2011, **144**, 646–674.
98. M. T. Snaebjornsson, S. Janaki-Raman and A. Schulze, *Cell Metab.*, 2020, **31**, 62–76.
99. T. Mashima, H. Seimiya and T. Tsuruo, *Br. J. Cancer*, 2009, **100**, 1369–1372.

100. L. M. Butler, Y. Perone, J. Dehairs, L. E. Lupien, V. de Laat, A. Talebi, M. Loda, W. B. Kinlaw and J. V. Swinnen, *Adv. Drug Delivery Rev.*, 2020, **159**, 245–293.
101. J. A. Menendez and R. Lupu, *Nat. Rev. Cancer*, 2007, **7**, 763–777.
102. S. Balaban, Z. D. Nassar, A. Y. Zhang, E. Hosseini-Beheshti, M. M. Centenera, M. Schreuder, H. M. Lin, A. Aishah, B. Varney, F. Liu-Fu, L. S. Lee, S. R. Nagarajan, R. F. Shearer, R. A. Hardie, N. L. Raftopulos, M. S. Kakani, D. N. Saunders, J. Holst, L. G. Horvath, L. M. Butler and A. J. Hoy, *Mol. Cancer Res.*, 2019, **17**, 949–962.
103. N. Koundouros and G. Poulogiannis, *Br. J. Cancer*, 2020, **122**, 4–22.
104. A. Hinsch, M. Buchholz, S. Odinga, C. Borkowski, C. Koop, J. R. Izbicki, M. Wurlitzer, T. Krech, W. Wilczak, S. Steurer, F. Jacobsen, E. C. Burandt, P. Stahl, R. Simon, G. Sauter and H. Schluter, *J. Mass Spectrom.*, 2017, **52**, 165–173.
105. R. Mirnezami, K. Spagou, P. A. Vorkas, M. R. Lewis, J. Kinross, E. Want, H. Shion, R. D. Goldin, A. Darzi, Z. Takats, E. Holmes, O. Cloarec and J. K. Nicholson, *Mol. Oncol.*, 2014, **8**, 39–49.
106. J. Garate, A. Maimo-Barcelo, J. Bestard-Escalas, R. Fernandez, K. Perez-Romero, M. A. Martinez, M. A. Payeras, D. H. Lopez, J. A. Fernandez and G. Barcelo-Coblijn, *Cancers*, 2021, **13**, 1350.
107. T. Bien, M. Perl, A. C. Machmüller, U. Nitsche, A. Conrad, L. Johannes, J. Müthing, J. Soltwisch, K.-P. Janssen and K. Dreisewerd, *Anal. Chem.*, 2020, **92**, 7096–7105.
108. M. Chen, J. Zhang, K. Sampieri, J. G. Clohessy, L. Mendez, E. Gonzalez-Billalabeitia, X. S. Liu, Y. R. Lee, J. Fung, J. M. Katon, A. V. Menon, K. A. Webster, C. Ng, M. D. Palumbieri, M. S. Diolombi, S. B. Breitkopf, J. Teruya-Feldstein, S. Signoretti, R. T. Bronson, J. M. Asara, M. Castillo-Martin, C. Cordon-Cardo and P. P. Pandolfi, *Nat. Genet.*, 2018, **50**, 206–218.
109. G. Deep and I. R. Schlaepfer, *Int. J. Mol. Sci.*, 2016, **17**, 1061.
110. L. Graham and M. T. Schweizer, *Med. Oncol.*, 2016, **33**, 44.
111. A. J. Armstrong, F. Saad, Phung, C. Dmuchowski, N. D. Shore, K. Fizazi, M. Hirmand, D. Forer, H. I. Scher and J. Bono, *Cancer*, 2017, **123**, 2303–2311.
112. D. F. Gleason, G. T. Mellinger and G. Veterans Administration Cooperative Urological Research, *J. Urol.*, 2017, **197**, S134–S139.
113. N. Mottet, J. Bellmunt, M. Bolla, E. Briers, M. G. Cumberbatch, M. De Santis, N. Fossati, T. Gross, A. M. Henry, S. Joniau, T. B. Lam, M. D. Mason, V. B. Matveev, P. C. Moldovan, R. C. N. van den Bergh, T. Van den Broeck, H. G. van der Poel, T. H. van der Kwast, O. Rouviere, I. G. Schoots, T. Wiegel and P. Cornford, *Eur. Urol.*, 2017, **71**, 618–629.
114. T. J. Wilt and I. M. Thompson, *BMJ*, 2006, **333**, 1102–1106.
115. T. Goto, N. Terada, T. Inoue, T. Kobayashi, K. Nakayama, Y. Okada, T. Yoshikawa, Y. Miyazaki, M. Uegaki, N. Utsunomiya, Y. Makino,

S. Sumiyoshi, T. Yamasaki, T. Kamba and O. Ogawa, *Prostate*, 2015, **75**, 1821–1830.

116. M. K. Andersen, T. S. Hoiem, B. S. R. Claes, B. Balluff, M. Martin-Lorenzo, E. Richardsen, S. Krossa, H. Bertilsson, R. M. A. Heeren, M. B. Rye, G. F. Giskeodegard, T. F. Bathen and M. B. Tessem, *Cancer Metab.*, 2021, **9**, 9.
117. X. Wang, J. Han, D. B. Hardie, J. Yang, J. Pan and C. H. Borchers, *Biochim. Biophys. Acta, Proteins Proteomics*, 2017, **1865**, 755–767.
118. E. C. Randall, G. Zadra, P. Chetta, B. G. C. Lopez, S. Syamala, S. S. Basu, J. N. Agar, M. Loda, C. M. Tempany, F. M. Fennessy and N. Y. R. Agar, *Mol. Cancer Res.*, 2019, **17**, 1155–1165.
119. L. Krasny, F. Hoffmann, G. Ernst, D. Trede, T. Alexandrov, V. Havlicek, O. Guntinas-Lichius, F. von Eggeling and A. C. Crecelius, *J. Am. Soc. Mass Spectrom.*, 2015, **26**, 36–43.
120. H. Hu, R. Yin, H. M. Brown and J. Laskin, *Anal. Chem.*, 2021, **93**, 3477–3485.
121. N. Verbeeck, R. M. Caprioli and R. Van de Plas, *Mass Spectrom. Rev.*, 2020, **39**, 245–291.
122. T. Alexandrov, *Annu. Rev. Biomed. Data Sci.*, 2020, **3**, 61–87.
123. E. Marien, M. Meister, T. Muley, T. G. del Pulgar, R. Derua, J. M. Spraggins, R. Van de Plas, F. Vanderhoydonc, J. Machiels, M. M. Binda, J. Dehairs, J. Willette-Brown, Y. L. Hu, H. Dienemann, M. Thomas, P. A. Schnabel, R. M. Caprioli, J. C. Lacal, E. Waelkens and J. V. Swinnen, *Oncotarget*, 2016, **7**, 12582–12597.
124. C. L. Carter, J. W. Jones, A. M. Farese, T. J. MacVittie and M. A. Kane, *Sci. Rep.*, 2017, **7**, 10343.
125. M. R. L. Paine, J. Liu, D. Huang, S. R. Ellis, D. Trede, J. H. Kobarg, R. M. A. Heeren, F. M. Fernandez and T. J. MacDonald, *Sci. Rep.*, 2019, **9**, 2205.
126. N. Plotegher, L. Bubacco, E. Greggio and L. Civiero, *Front. Neurosci.*, 2019, **13**, 330.
127. K. A. Keith and J. H. Huang, *Diagnostics (Basel)*, 2019, **10**, 4.
128. N. D. Osier and C. E. Dixon, *Front. Neurol.*, 2016, **7**, 134.
129. K. Mallah, J. Quanico, D. Trede, F. Kobeissy, K. Zibara, M. Salzet and I. Fournier, *Anal. Chem.*, 2018, **90**, 10568–10576.
130. K. Mallah, J. Quanico, A. Raffo-Romero, T. Cardon, S. Aboulouard, D. Devos, F. Kobeissy, K. Zibara, M. Salzet and I. Fournier, *Anal. Chem.*, 2019, **91**, 11879–11887.
131. A. Roux, L. Muller, S. N. Jackson, J. Post, K. Baldwin, B. Hoffer, C. D. Balaban, D. Barbacci, J. A. Schultz, S. Gouty, B. M. Cox and A. S. Woods, *J. Neurosci. Methods*, 2016, **272**, 19–32.
132. R. A. Nixon, A. M. Cataldo and P. M. Mathews, *Neurochem. Res.*, 2000, **25**, 1161–1172.
133. V. F. Vartabedian, P. B. Savage and L. Teyton, *Immunol. Rev.*, 2016, **272**, 109–119.
134. K. Belarbi, E. Cuvelier, M. A. Bonte, M. Desplanque, B. Gressier, D. Devos and M. C. Chartier-Harlin, *Mol. Neurodegener.*, 2020, **15**, 59.

135. A. Aqul, B. Liu, C. M. Ramirez, A. A. Pieper, S. J. Estill, D. K. Burns, B. Liu, J. J. Repa, S. D. Turley and J. M. Dietschy, *J. Neurosci.*, 2011, **31**, 9404–9413.
136. F. Tobias, M. T. Olson and S. M. Cologna, *J. Lipid Res.*, 2018, **59**, 2446–2455.
137. I. Kaya, W. Michno, D. Brinet, Y. Iacone, G. Zanni, K. Blennow, H. Zetterberg and J. Hanrieder, *Anal. Chem.*, 2017, **89**, 4685–4694.
138. I. Kaya, D. Brinet, W. Michno, M. Baskurt, H. Zetterberg, K. Blenow and J. Hanrieder, *ACS Chem. Neurosci.*, 2017, **8**, 2778–2790.
139. I. Kaya, H. Zetterberg, K. Blennow and J. Hanrieder, *ACS Chem. Neurosci.*, 2018, **9**, 1802–1817.
140. L. M. Munter, *J. Neurochem.*, 2020, **154**, 7–10.
141. I. Kaya, E. Jennische, J. Dunevall, S. Lange, A. G. Ewing, P. Malmberg, A. T. Baykal and J. S. Fletcher, *ACS Chem. Neurosci.*, 2020, **11**, 14–24.
142. I. Alecu and S. A. L. Bennett, *Front. Neurosci.*, 2019, **13**, 328.
143. E. E. Jones, W. Zhang, X. Zhao, C. Quiason, S. Dale, S. Shahidi-Latham, G. A. Grabowski, K. D. R. Setchell, R. R. Drake and Y. Sun, *SLAS Discovery*, 2017, **22**, 1218–1228.
144. Z. Hall, N. J. Bond, T. Ashmore, F. Sanders, Z. Ament, X. Wang, A. J. Murray, E. Bellafante, S. Virtue, A. Vidal-Puig, M. Allison, S. E. Davies, A. Koulman, M. Vacca and J. L. Griffin, *Hepatology*, 2017, **65**, 1165–1180.
145. K. Scupakova, Z. Soons, G. Ertaylan, K. A. Pierzchalski, G. B. Eijkel, S. R. Ellis, J. W. Greve, A. Driessen, J. Verheij, T. M. De Kok, S. W. M. Olde Damink, S. S. Rensen and R. M. A. Heeren, *Anal. Chem.*, 2018, **90**, 5130–5138.
146. H. Alamri, N. H. Patterson, E. Yang, P. Zoroquiain, A. Lazaris, P. Chaurand and P. Metrakos, *Anal. Bioanal. Chem.*, 2019, **411**, 885–894.
147. K. N. Robinson, R. T. Steven, A. M. Race and J. Bunch, *J. Am. Soc. Mass Spectrom.*, 2019, **30**, 1284–1293.
148. J. Soltwisch, H. Kettling, S. Vens-Cappell, M. Wiegelmann, J. Muthing and K. Dreisewerd, *Science*, 2015, **348**, 211–215.
149. J. C. McMillen, J. A. Fincher, D. R. Klein, J. M. Spraggins and R. M. Caprioli, *J. Mass Spectrom.*, 2020, **55**, e4663.
150. A. P. Bowman, J. F. J. Bogie, J. J. A. Hendriks, M. Haidar, M. Belov, R. M. A. Heeren and S. R. Ellis, *Anal. Bioanal. Chem.*, 2020, **412**, 2277–2289.
151. M. Niehaus, K. N. Robinson, T. Murta, E. A. Elia, A. M. Race, B. Yan, R. T. Steven and J. Bunch, *J. Am. Soc. Mass Spectrom.*, 2020, **31**, 2287–2295.
152. T. Bien, E. A. Hambleton, K. Dreisewerd and J. Soltwisch, *Anal. Bioanal. Chem.*, 2021, **413**, 2767–2777.
153. E. A. Elia, M. Niehaus, R. T. Steven, J. C. Wolf and J. Bunch, *Anal. Chem.*, 2020, **92**, 15285–15290.
154. T. Iwama, K. Kano, D. Saigusa, K. Ekroos, G. van Echten-Deckert, J. Vogt and J. Aoki, *Anal. Chem.*, 2021, **93**, 3867–3875.
155. D. F. Cobice, C. L. Mackay, R. J. Goodwin, A. McBride, P. R. Langridge-Smith, S. P. Webster, B. R. Walker and R. Andrew, *Anal. Chem.*, 2013, **85**, 11576–11584.

156. S. Shimma, H. O. Kumada, H. Taniguchi, A. Konno, I. Yao, K. Furuta, T. Matsuda and S. Ito, *Anal. Bioanal. Chem.*, 2016, **408**, 7607–7615.
157. D. F. Cobice, D. E. Livingstone, C. L. Mackay, R. J. Goodwin, L. B. Smith, B. R. Walker and R. Andrew, *Anal. Chem.*, 2016, **88**, 10362–10367.
158. E. Takeo, Y. Sugiura, T. Uemura, K. Nishimoto, M. Yasuda, E. Sugiyama, S. Ohtsuki, T. Higashi, T. Nishikawa, M. Suematsu, E. Fukusaki and S. Shimma, *Anal. Chem.*, 2019, **91**, 8918–8925.
159. P. Esch and S. Heiles, *J. Am. Soc. Mass Spectrom.*, 2018, **29**, 1971–1980.
160. I. S. Gilmore, S. Heiles and C. L. Pieterse, *Annu. Rev. Anal. Chem.*, 2019, **12**, 201–224.
161. M. Niehaus, J. Soltwisch, M. E. Belov and K. Dreisewerd, *Nat. Methods*, 2019, **16**, 925–931.
162. Y. Schober, S. Guenther, B. Spengler and A. Rompp, *Anal. Chem.*, 2012, **84**, 6293–6297.
163. L. Capolupo, I. Khven, L. Mazzeo, G. Glousker, F. Russo, J. P. Montoya, S. Ho, D. R. Bhandari, A. P. Bowman, S. R. Ellis, R. Guiet, J. Muthing, B. Spengler, R. M. A. Heeren, G. P. Dotto, G. La Manno and G. D'Angelo, *J. bioRxiv*, 2021, DOI: 10.1101/2021.02.23.432420.
164. B. Yang, N. H. Patterson, T. Tsui, R. M. Caprioli and J. L. Norris, *J. Am. Soc. Mass Spectrom.*, 2018, **29**, 1012–1020.
165. E. K. Neumann, T. J. Comi, S. S. Rubakhin and J. V. Sweedler, *Angew. Chem., Int. Ed. Engl.*, 2019, **58**, 5910–5914.
166. M. E. Duenas, J. J. Essner and Y. J. Lee, *Sci. Rep.*, 2017, **7**, 14946.
167. T. J. Comi, E. K. Neumann, T. D. Do and J. V. Sweedler, *J. Am. Soc. Mass Spectrom.*, 2017, **28**, 1919–1928.
168. E. K. Neumann, J. F. Ellis, A. E. Triplett, S. S. Rubakhin and J. V. Sweedler, *Anal. Chem.*, 2019, **91**, 7871–7878.
169. K. B. Louie, B. P. Bowen, S. McAlhany, Y. Huang, J. C. Price, J. H. Mao, M. Hellerstein and T. R. Northen, *Sci. Rep.*, 2013, **3**, 1656.
170. C. T. Ou, I. Matsumoto, J. Rozhin and T. T. Tchen, *Anal. Biochem.*, 1978, **88**, 357–366.
171. E. P. Kennedy and S. B. Weiss, *J. Biol. Chem.*, 1956, **222**, 193–214.

CHAPTER 9

MALDI Mass Spectrometry Imaging of Neuropeptides

JÖRG HANRIEDER[a,b]

[a] Department of Psychiatry and Neurochemistry, Sahlgrenska Academy at the University of Gothenburg, Mölndal Hospital, SE-431 80 Mölndal, Sweden; [b] Department of Neurodegenerative Disease, UCL Queen Square Institute of Neurology, University College London, Queen Square, London WC1N 3BG, United Kingdom
Email: jh@gu.se

9.1 Introduction

Neuropeptides, including peptide hormones, are key modulators in the brain that represent the largest class of chemical messengers for cell-to-cell communication.[1–5] Due to the complexity of the CNS, the spatial information of neuropeptide distribution patterns is of major interest in order to delineate basic cellular communication mechanisms in the brain as well as abnormal peptide signaling and interaction events in neurological disease pathology.

Consequently, some of the recent advances in the understanding of CNS-related molecular mechanisms directly benefit from the development of novel, advanced high-resolution imaging techniques. Biochemical imaging techniques are essential tools to gain in-depth understanding of molecular mechanisms and interactions at the subcellular scale. Here different molecular imaging modalities involving, *e.g.*, immunohistochemistry, *in situ* hybridization, chemical staining together with fluorescent microscopy, as

New Developments in Mass Spectrometry No. 12
MALDI Mass Spectrometry Imaging: From Fundamentals to Spatial Omics
Edited by Tiffany Siegel

Published by the Royal Society of Chemistry, www.rsc.org

well as spectroscopic methods are employed to retrieve topographical and temporal information of molecular abundance distributions.

Mass spectrometry imaging (MSI) has gained significant interest and generated broad impact in biomedical research and in particular in neuroscience research.[6–8]

MSI is an emerging, novel approach for spatial mapping of chemical species in biological tissue and single cells.[9,10] In contrast to common, antibody- or nucleotide-based detection approaches, MSI does not require any *a priori* knowledge of the potential target species and is furthermore not dependent on antibody or primer availability and specificity. The technology features superior molecular specificity, warranted by the mass spectrometric detection. This in turn allows comprehensive, multiplexed detection, identification, and localization of hundreds of different species including proteins, peptides, and lipids in complex biological tissue samples or single cells. The use of MSI to match histological features of a biological tissue sample enables comprehensive mapping of molecular localization patterns, also referred to as molecular histology (see Figure 9.1).[9,11]

During the last 20 years, particularly matrix-assisted laser desorption/ionization (MALDI)-based MSI has slowly evolved as a relevant, alternative approach in biochemical imaging for studying proteins, peptides, lipids, drugs, and metabolites in disease pathology as well as fundamental biological processes with particular focus on neurological disorders.[6,7]

MALDI-MS employs UV or IR laser irradiation for desorption and ionization of molecular species. This process is facilitated by precoating the sample with a crystalline, UV radiation absorbing matrix.[12] The technique is characterized by its expanded mass range allowing detection of polypeptides and small proteins. In the mid-1990s, Bernhard Spengler and Richard Caprioli pioneered the application of MALDI-MSI for spatial profiling of large molecular species in mammalian tissue samples.[13,14]

The present chapter focuses on MALDI-MSI and aims to provide a concise overview of its application to neuropeptide imaging in brain tissue and single cells. A number of key challenges with respect to sample retrieval, tissue preparation, data acquisition and peptide sequence validation are discussed. Finally, an overview of MSI-based neuropeptidomic applications is provided.

9.2 Methods

9.2.1 Sample Preparation for MALDI-MS Imaging of Neuropeptides

For MSI of proteins and peptides, various target preparation parameters are of critical relevance as they have significant impact on final data quality. This is particularly important with respect to signal intensity, reproducibility, as well as lateral (*i.e.*, spatial) resolution. The sample preparation workflow in MALDI-MSI comprises the tissue retrieval and tissue storage, tissue sectioning, tissue wash, and fixation followed by matrix application.

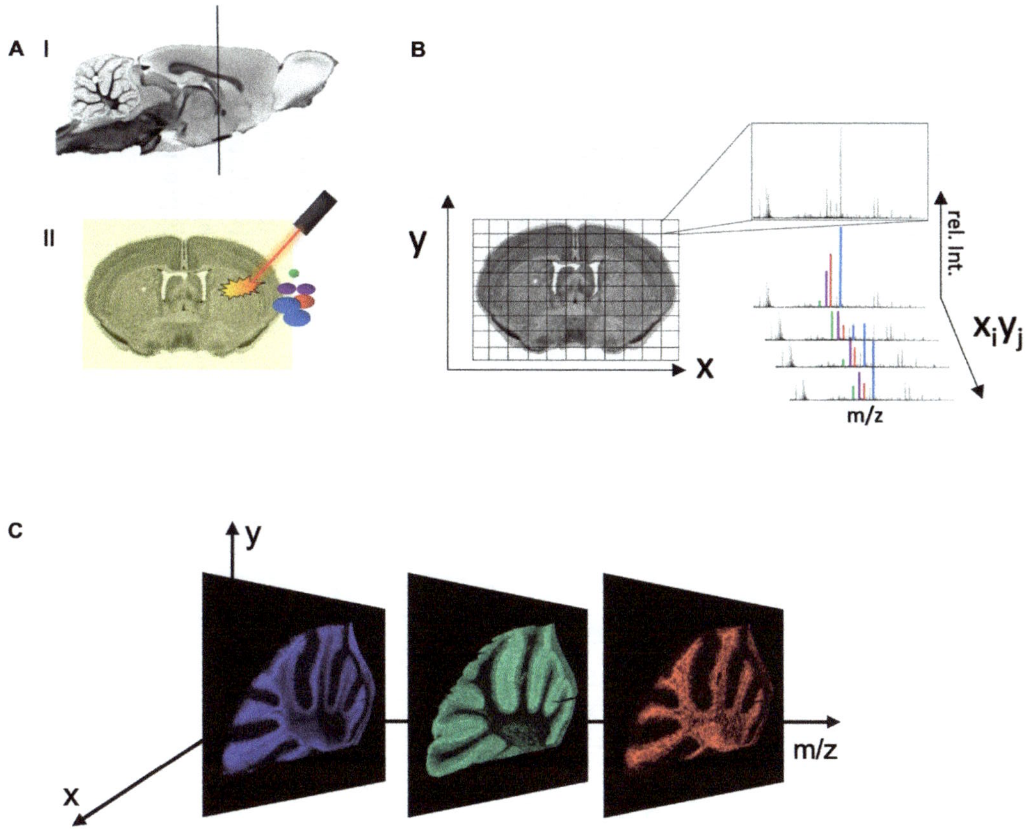

Figure 9.1 Principle of mass spectrometry imaging. (A) I, Tissue sections are collected and mounted on a sample plate for imaging MS. II, MALDI-MSI requires pre-coating with matrix (indicated in yellow) before systematic scanning with a laser probe. MALDI-based ionization allows generation of larger intact biomolecular species, including lipids, peptides, and proteins. (B) One mass spectrum is acquired for every x_i, y_j coordinate of the tissue section based on a pre-defined array corresponding to pixel size and spatial resolution. (C) Single ion images are generated by mapping the intensity of an individual ion signal (m/z; rel.Int) over a whole tissue slice. Reproduced from ref. 8 with permission from John Wiley & Sons, Copyright © 2018 International Society for Neurochemistry.

In particular, tissue treatment and matrix application are cornerstones for each MALDI-MSI experiment as they have to be adjusted for both the type of tissue and most importantly the molecular target of interest.

9.2.1.1 Tissue Collection

As for all tissue imaging experiments, tissue retrieval from animal or human sources is critical to data quality. This is of particular relevance, since commonly used perfusion and fixation strategies complicate applications for mass spectrometry due to interference of the polymeric fixation agents such as paraffin (*i.e.*, formalin fixed paraffin embedded, FFPE). However, FFPE tissues are the method of choice in biobanking and for generating tissue microarrays. This is superior due to the sample stability at room temperature and simplicity to process tissues for sectioning. FFPE tissues are therefore a still rather untapped source of large sample cohorts and hence of significant relevance for further introducing MSI to the clinical pathology community.

Solutions to overcome these obstacles include paraffin removal, antigen retrieval, and *in situ* trypsin digestion.[15,16] This approach, however, does not permit the analysis of intact neuropeptides due to extensive washing for polymer removal and enzymatic degradation. Paine *et al.* introduced an optimized protocol based on xylene and ethanol for deparaffination that facilitated detection of endogenous neuropeptides in a 30 year-old FFPE tissue sample of American cockroaches (*Periplaneta americana*).[17] Most recently, the Sweedler lab presented an elegant approach to rescue neuropeptide signals from *para*-formaldehyde fixed tissues by employing nucleophilic reagents to break the FA-induced methylene bridges.[18]

Given the complications introduced by fixation and embedding, fresh frozen tissue samples are, therefore, the most preferred type of preserved samples for MSI. In terms of sample dissection such as for intact brain, a quick procedure is essential, since post mortem delays by as little as 3 min result in severe degradation of neuropeptides.[19] One solution to overcoming this limitation is heat stabilization, where quick sample heating to 95 °C leads to efficient protease inactivation.[20] This approach, however, impacts the sample morphology, which in turn hampers its suitability in MSI. This is primarily due to difficulties in maintaining tissue integrity during sectioning although an elegant workaround using collection on conductive carbon tape has been presented recently.[21]

For fresh frozen tissue, cryosections are collected and thaw mounted onto conductive glass slides. Here, the collected sections need to be dried immediately before storage at −80 °C to prevent damage by water condensation during freezing.[22] A further complication in tissue collection is very small specimens that require some sort of mounting and embedding in MS compatible embedding media. Frozen sections are typically collected by mounting fresh frozen tissue within a cryostat using optimal cutting temperature media (OCT). However, OCT is a polymer interfering with MSI. Sosnowski *et al.* introduced an elegant workflow by using egg yolk for embedding rat pituitary glands to facilitate tissue sectioning and neuropeptide imaging.[23] Other embedding media used are gelatin and sucrose, which were demonstrated in the context of MSI.

9.2.1.2 Sample Cleanup and Tissue Derivatization Strategies

A further critical step in sample handling is the choice of appropriate washing protocols. Lipids typically do not require any advanced washing steps, whereas, drugs, neuropeptides, and proteins require optimized washing protocols for signal enhancement. These involve pH-sensitive cleanup as well as organic solvents for precipitation of peptides and proteins while washing off remaining lipids that potentially interfere with neuropeptide signals prior to matrix application.[22,24] Several washing protocols have been evaluated for enhancing protein signal in MALDI-MSI. Stepwise washing with gradient alcohol has been found to give the most significant improvement in signal quality.[25,26] Given the low concentration of neuropeptides and the diligent tissue preparation required, means to significantly enhance NP signals *in situ* would be of great benefit. Indeed, the Li Lab introduced a photochemical tissue derivatization strategy of primary amino groups using a photoactive compound, 2-nitrobenzaldehyde (NBA).[27] A further challenge in neuropeptide imaging includes analysis of single cells. The Sweedler lab has pioneered the *in situ* detection of neuropeptides in single nerve cells.[28] To overcome the inherent challenge with respect to spatial resolution, they introduced a stretched sample approach that facilitated mapping of the spatial localization patterns of individual neuropeptides within single nerve cells.[29]

9.2.1.3 Matrix Application

In terms of matrix application, there are currently three approaches available, including manual airbrush, automated pneumatic sprayers, and chemical inkjet printing. Each with its own strengths and limitations with respect to simplicity, cost, resulting signal intensity, spatial resolution, and signal reproducibility. In general, there is a tradeoff between signal quality and intensity and lateral resolution based on the correlation of vertical and lateral diffusion with increased droplet size. The most straightforward approach involves manual application of matrix solution with an airbrush sprayer. While this is a quite cost-effective solution it is severely hampered by a lack of reproducibility, susceptibility to sample diffusion, and limited extraction efficiency. A more controlled way to deposit matrix is to use automated, robotic spraying devices that combine pneumatic matrix deposition together with a robotic sprayer head that is automatically moved across the tissue in a well-defined, reproducible pattern. Such instruments include the *TM Sprayer* (HTX Technologies Carrboro, NC, USA) or *SunCollect* (Sunchrom, Napa, CA, USA), where an airbrush head is repeatedly moved over the tissue section thereby applying layer by layer of matrix. In the case of the TM Sprayer, the pneumatic spraying of the matrix is assisted by heating the spraying capillary.

Still, one major disadvantage of this solution remains in its rather low extraction efficiency, as a consequence of reduced lateral diffusion. The most controlled approach for sample application involves deposition by a chemical inkjet (*ChiP*, Shimadzu, Kyoto, Japan)[30] or a pneumatic vertical spotter (*Portrait*, Labcite, Sunnyvale, CA).[31] These methods use accurate deposition of distinct droplets in a geometrical pattern thereby avoiding lateral diffusion of the analyte.[32] The spotting approach is particularly beneficial for low abundance species, including neuropeptides.[33] Although this approach is limited in terms of spatial resolution, the best data quality in terms of signal intensity and reproducibility can be achieved.[30]

Over the last few years, several alternative approaches for matrix application have been reported to overcome the limitations in various matrix application protocols and to control crystal size, including dry coating, precoating and sublimation.[34,35] Although all these techniques provide major advantages in terms of crystal size, they generally suffer from their low extraction efficiency. Moreover, sublimation has been reported for protein imaging, despite limitations with respect to reproducibility.[34]

Several different matrices have been proposed for MALDI-MSI of various molecular species. Different matrices are well suited for different compounds and the choice depends largely on the targeted substance. Many advances in matrix developments have been aimed at overcoming various limitations of the standard compounds, including signal reproducibility, omission of interfering matrix cluster, extended mass range, and improved extraction and signal quality, respectively. For protein analysis, the most commonly used matrices are 2,5-dihydroxy-benzoic acid (DHB), sinapinic acid (SA), 2,5-dihydroxy-acetophenone (2,5-DHA), and 4-alpha-cyano-hydroxycinnamic acid (CHCA).

9.2.2 Data Acquisition and Data Analysis

9.2.2.1 MALDI-MSI Instrument Parameters for Acquisition

The MALDI-MSI experiment is detailed in a corresponding running sequence. Here, all the parameters of the acquisition and initial data processing are specified along with the acquisition region, acquisition pattern, and pixel-to-pixel distance, *i.e.* spatial resolution. The initial steps of setting up the imaging acquisition experiment involve image-registration and teaching for sequence setup, which is performed in the corresponding imaging software bundle provided by the vendor (*e.g.* FlexImaging, Bruker Daltonics; High Definition Imaging, Waters; Imagereveal, Shimadzu).

For the actual MALDI acquisition method the optimum number of good-quality spectra that can be obtained from each position needs to be specified including number of laser shots and laser energy. Too high laser fluence can influence the experiment dramatically leading to signal depletion due to oversampling effects. Here, the spatial resolution and laser focus settings have to be considered accordingly as too small spot-to-spot distances at too high laser fluence lead to the described oversampling phenomena. This is particularly challenging for protein and peptide imaging as typically higher laser fluences are needed for optimal desorption and ionization, thereby compromising the spatial resolution. For a typical DHA prepared tissue, a number of 50–200 laser shots need to be collected from each spot at minimum laser focus for ~25 pixel distance (*i.e.* lateral resolution).[36,37] Conversely, using DHB or DAN matrix for lipid analysis on an Ultraflextreme, a number of 5–10 shots at lower energies are sufficient allowing straightforward acquisition of imaging data at 5–10 μm-lateral resolution.[36,37] A further consideration, when specifying the pixel distance and spatial resolution, respectively, is the acquisition time. Acquisition of large areas at very high spatial resolution will take several hours. During this time, the matrix might sublimate away rendering the data acquired later during the experiment as low quality or worthless. Despite these challenges, significant improvements in obtaining higher spatial resolutions have been achieved by optimizing the laser optics.[38–41] Furthermore, using a transmission geometry instrument, the Caprioli Lab managed to perform subcellular imaging at 2 μm, which in turn allowed visualization of the peptide hormones, insulin and somatostatin in beta and delta cells in a human pancreatic islet.[42]

9.2.2.2 *Data Processing*

Imaging data acquisition is generally followed by subsequent data processing including baseline subtraction, peak picking, and mass re-calibration. On-the-fly MS processing is performed using a baseline removal algorithm that will not result in negative data (*e.g.*, ConvexHull 0.8 flatness; FlexAnalysis, Bruker Daltonics). Peak picking is performed using sensitive algorithms (Centroid, SNAP) considering peaks with a minimum signal to noise (S/N) >3.[43]

9.2.2.3 *Statistical Analysis*

The acquired images can involve enormous data size, as there is a complete spectrum for each pixel, hence these often comprise file sizes of multiple gigabytes. Therefore, spatial resolution and pixel number, respectively, need to be in a practical balance. The analysis of the entire imaging dataset can be approached in both supervised and unsupervised manners. Supervised, targeted analyses involve regions of interest (ROIs) annotation based on consecutively performed histology experiments.[10,31,43,44] This is followed by ROI spectral data extraction, data binning, and univariate statistics.[22,30,33,45] Furthermore, accurate quantification of single neuropeptides has been implemented in an elegant MALDI-MSI acquisition setup using isotopically labeled standards and matrix solution which is then processed and analyzed in a dedicated data analysis workflow available as msIQuant software.[46] In contrast to targeted analyses, complex MS imaging data can be interrogated in a less biased way using multiple multivariate statistical analysis (MVA) tools, such as principal component analysis (PCA), maximum autocorrelation factor analysis (MAF), as well as hierarchical cluster analysis (bisecting k-means).

These represent excellent approaches for unbiased interrogation of molecular intensity distributions and potential localization to anatomical features of interest. Multiple image analysis strategies have been reported for ToF-SIMS[47,48] and MALDI-MSI data.[49] This data handling approach allows segmentation of a biological sample in distinct regions of interest by detecting variances and correlations in the multivariate data that are encompassed in the MVA-factors. From the corresponding scores and loadings, the variables that are contributing the most (*i.e.*, mass peak values) to these variances can be deduced revealing histology-associated chemical changes. This approach also facilitates outlining potential ROI for subsequent statistical analysis of ROI spectral data. Similar to both approaches, ROIs are outlined or validated based on immunohistochemical staining subsequently performed on the same or consecutive tissue section. New bioinformatic tools have been introduced to combine and make use of the different imaging modality information in order to align and recognize tissue morphology.[50,51] These tools further allow unbiased, histology-based ROI annotation, segmentation, as well as ion distribution prediction and sharpening. The prediction and sharpening feature is further enhanced by segmentation of the histochemical staining and correlative analysis of the MSI data, also referred to as image fusion.[51–53]

9.2.3 Molecular Identification

While mass spectrometry imaging is a very powerful approach to delineate spatial distribution patterns of intact neuropeptide ions, there is a need for complementary validation strategies in order to fully exploit the potential of the technique. Initial mass peak annotation is based on accurate mass and reference values reported in the literature or in online depositories. This is complemented by subsequent immunostainings on the same or consecutive tissue sections. While this approach allows for verification of the spatial distribution data as well as indicates protein or peptide identity there are several limitations remaining with respect to chemical specificity and quantification.

9.2.3.1 Immunohistochemistry

The most common molecular histological approaches include immunohistochemistry (IHC) of the same tissue sections using fluorescence microscopy as well as chromogenic detection. Immunohistochemistry is, however, limited by the antibody specificity and is difficult to compare in the case of truncated protein isoforms leaving preliminary protein annotation questionable. Moreover, recent studies demonstrated that laser ablation induced distortion effects lead to antigen degradation and increased auto-fluorescence rendering fluorescent immunostaining on the same section challenging. Interestingly, chemical imaging with unspecific morphological staining techniques such as H&E (Hematoxylin and Eosin) and DAPI was not affected by laser ablation induced distortion. This is of great concern as most molecular distributions observed in imaging MS are aligned with histological stains.[36]

A particular caveat for neuropeptide IHC is the lack of specificity and sensitivity. Different neuropeptide species, such as opioid peptides including dynorphin and enkephalin species, possess the same 5 amino acid N-termini and differ in only a few amino acids at the C-termini. Consequently, there is substantial cross-reactivity for these kinds of target sequences, significantly hampering the application of IHC for *in situ* quantification and validation.

9.2.3.2 Mass Spectrometry-based Neuropeptide Identification

Due to the limitations of IHC based approaches with respect to selectivity and specificity as well as quantification, further efforts are needed for validating mass peak identities observed in MSI experiments.

In situ peak annotation using mass accuracy is particularly challenging for intact proteins as these experiments are performed in linear mode, where mass accuracy and mass resolution of the TOF instrumentation do not permit confident identification. Therefore, mass spectrometry-based MS/MS approaches either directly *in situ*, or on extracts of dissected tissues are the method of choice for unequivocal neuropeptide peak identification.[31,54,55]

The most convenient strategy would therefore involve direct *in situ* fragmentation and identification of neuropeptides. This top-down approach is, however, particularly challenging as only one or two MS/MS spectra can be retrieved from a single spot, since this MALDI-based MS/MS requires higher laser powers resulting in increased sample consumption.

A further common strategy to identify unknown peptide peaks is based on tissue extraction followed by pre-fractionation using a molecular weight cut-off (MWCO) filter to avoid convolution by degradation products of larger proteins.[3,56,57] For neuropeptide analysis the main focus has also been placed on limiting post-mortem degradation *in vivo*, *ex vivo*, and *in situ* using *e.g.*, heat stabilization and snap freezing[33,57,58] prior to sonication and mass cut-off filtration enabling large-scale neuropeptidomics.[3,56–59] Neuropeptide extracts are then directly analyzed by one- or two-dimensional capillary or nanoflow HPLC hyphenated to electrospray ionization tandem mass spectrometry.[3,56,57,60]

9.3 MALDI MS-based Neuropeptide Imaging in Biological Research

Over the last two decades, MALDI-MSI has gained significant popularity that enables the study of endogenous peptide regulation *in situ*. This comprises studies of invertebrate and vertebrate animals, mechanistic studies on neuropeptide processing *in vivo*, as well as peptide dynamics in neurodegenerative disorders.

9.3.1 Neuropeptide Imaging in Cells and Animal Models

MSI facilitates the visualization of neuropeptides at the single cell level. Here, the Sweedler lab has been at the forefront in developing MSI strategies in order to delineate spatial peptide patterns within single nerve cells from invertebrates such as *Aplysia califonica*.[61–64] The lab has further developed an array of elegant solutions to increase the spatial resolution of those measurements including stretched sample methods of fixed cells or laser oversampling.[29,65] The same group further pioneered the application of MSI for neuropeptide imaging in other invertebrates and insects.[66,67] Furthermore, research on neuropeptide imaging in honeybees (*Apis mellifera*) allowed the effects of ontogeny and aggressiveness to be evaluated.[68,69]

A wealth of very interesting research has been presented by the Li group for using MALDI-MSI to delineate neuropeptide profiles and regulatory mechanism in crustaceans.[70] This includes MSI studies on mapping the endogenous neuropeptidome in spiny lobster (*Panulirus interruptus*),[71] blue crab *Callinectes sapidus*, and European green crab *Carcinus* maenas,[72] as well as the Jonah crab, *Cancer borealis*.[73,74] This lab group also introduced distinct means for enhancing neuropeptide signals[27,70,75,76] and facilitated hyphenation to other peptide detection modalities such as capillary electrophoresis to expand the coverage of detected neuropeptide species.[77]

Neuropeptide imaging can further be employed for delineating neuronal circuits involving neuropeptide signaling[78] as well as to outline spatial mechanisms of peptide conversion.[79]

9.3.2 MALDI-MSI of Neuropeptides in Neurodegenerative Diseases

In neurodegenerative diseases, abnormal aggregation and deposition of proteins and peptides is a seminal pathological hallmark.[80,81] For instance, probing neuropeptide signaling (*i.e.* opioid peptides (*m/z* 600–2500)) is of essential relevance when investigating Parkinson's disease. Similarly, detection of amyloid beta (Aβ) peptide species (*m/z* 4000–4500) is important for studying AD pathology.

Commonly, antibody-based techniques are currently the dominant method used for *in situ* detection and quantification of brain peptides. This approach is, however, significantly limited by antibody specificity. This is particularly relevant for opioid peptides that differ in only a few C-terminal amino acids, which compromises the reliability of immunohistochemistry results significantly. Another limitation in antibody-based approaches is the inherent limited throughput, allowing only detection of a few species at a time. Furthermore, there are well-known quantification issues since the signal intensity in immunohistochemistry provides only semi-quantitative information. LC–MS-based peptide quantification in tissue extracts has been demonstrated to be a powerful approach for characterization of endogenous peptide species.[57] However, dissection and extraction of tissue samples result in loss of spatial information, which highlights the need for techniques that allow *in situ* analysis. MSI of proteins and peptides has significant advantages over immuno-based techniques particularly with respect to molecular specificity.

An important consideration for imaging of biological tissue samples is the inherent biological variation within the sample groups. This is particularly relevant for MSI of human material. Beyond the inherent patient-to-patient variation, one must account for technical issues that impact sample quality and increase variation, such as standardized sample collection, sample storage, and well-defined and short post-mortem delay. This highlights the need for appropriate study design, such as inclusion of sufficient individuals and matched controls.

An elegant way to avoid the effects of intra-sample group biological variation is to use hemispheric disease models such as a unilateral nigrostriatal lesion with 6-hydroxy-dopamine (6-OHDA) as a model for Parkinson's disease.[82] These animals develop PD pathology only on one side and the other side can serve as an internal control. This is particularly relevant for mass spectrometry-based studies to account for variation because of suppression effects. Using this model, MALDI-MSI has been employed to analyze spatial regulations of dynorphin neuropeptides in L-DOPA induced dyskinesia in experimental PD[22,33,45] (Figure 9.2A). Elevated levels of prodynorphin, the neuropeptide precursor protein, have been previously associated with L-DOPA induced dyskinesia (LID). The lack of specific antibodies permitting annotation of distinct prodynorphin processing products prevented precise mapping and accurate quantification of prodynorphin peptide products using conventional techniques. MALDI-MSI was used to demonstrate that dynorphin B and alpha neoendorphin were significantly elevated in the dorsal lateral striatum in the high dyskinetic group but not for low dyskinetic animals. In addition, both dynorphin species correlated positively with LID severity (Figure 9.2B).

MALDI-MSI has been further successfully applied for protein and peptide imaging in other studies on experimental PD. Using the same unilateral PD model, MALDI-MSI delineated differences in neuropeptide levels between different basal ganglia structures in LID. Here, striatal peptide changes were found to be primarily a consequence of dopamine depletion while neuropeptide levels in the globus pallidus were found to be associated with LID.[83]

In another study, MALDI-MSI was employed to study striatal protein localization in 1-methyl-4-phenyl-1,2,3,6-tetrahydropyridine (MPTP) injected rats.[44] MALDI-MSI revealed a striatal decrease of the neuronal calmodulin-binding protein Pep19, which suggests that altered calcium homeostasis might be associated with neuronal cell death in PD. In a 6-OHDA based model of experimental PD, ubiquitin, *trans*-elongation factor 1, hexokinase, and neurofilament M were found to be down-regulated in PD.[84]

Recently, the Andrén lab presented pioneering work on a MPTP injected macaque model of PD and LID.[85–87] They were able to employ MALDI-MSI to map the distribution of L-DOPA and monoaminergic pathways and showed that LID is associated with brain wide dysregulation of L-DOPA metabolism. More mechanistically they demonstrated that administration of a μ receptor agonist ameliorated LID.[86]

MALDI-MSI has further been demonstrated to be a very powerful technique to probe amyloid peptide pathology in Alzheimer's disease. Here, Aβ

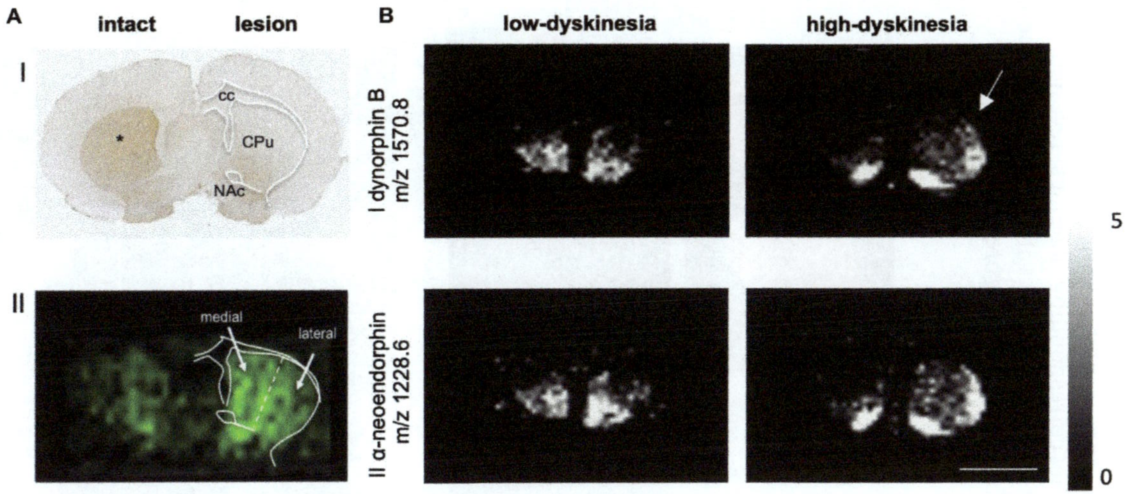

Figure 9.2 Mass spectrometry imaging of neuropeptides in an animal model of Parkinson's disease featuring L-DOPA-induced dyskinesia. (A.I) Unilateral 6-OHDA injection leads to dopamine depletion (as illustrated by tyrosine hydroxylase, TH-immunostaining *). L-DOPA therapy results in two distinct groups with low and high dyskinesia. (A.II) Striatal sections were analyzed by MALDI-MSI and regions of interest (ROI) were assigned for spectral data extraction. Scale bar: 2 mm; (b) Dynorphin peptides, I dynorphin B and II alpha neoendorphin, displayed a significant ($p < 0.05$) relative increase in abundance in the dorso-lateral striatum of high- (HD) compared to low-dyskinetic (LD) animals and lesion controls (LC).
Adapted from ref. 22 with permission from Journal of Visualized Experiments, Copyright 2012.

truncations in individual plaques have been characterized in APP23 transgenic mice.[88,89] The authors reported a significantly higher content of Aβ 1–40 than Aβ 1–42. Similarly, our group reported a comprehensive study profiling brain-wide Aβ profiles in transgenic animals carrying the Swedish and Arctic mutation in the *APP* gene (tgArcSwe and tgSwe).[37,90–92] To approach MSI data in tgAD mice in an unbiased way, multivariate image analysis was employed to outline pathological features that resemble Aβ plaques. Inspection of the corresponding loadings revealed distinct Aβ peptide species to be the most prominent variables causing the chemical differences obtained by cluster analysis (Figure 9.3A).[90,92]

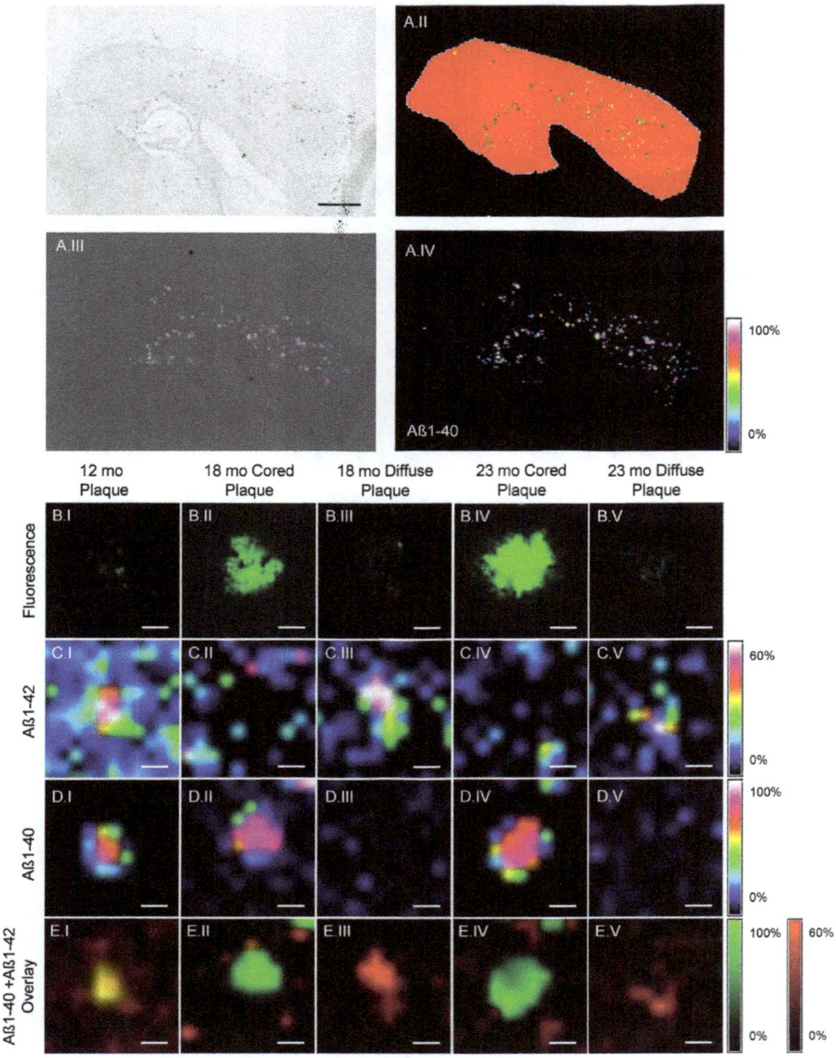

Recently, MALDI-MSI was employed for human AD tissue, where amyloid peptide profiles of individual plaques were delineated.[93] Here, Aβ1–42 and Aβ1–43 were found to be selectively deposited in senile plaques while other C-terminally truncated Aβ species including Aβ1–36, 1–37, 1–38, 1–39, 1–40, and Aβ1–41 were primarily detected in leptomeningeal blood vessels. In another study, MALDI-MSI identified amyloid polymorphism-associated peptide signatures. In this study, senile cored plaques that are specific to AD were found to be characterized by deposition of Aβ1–40 at the core and Aβ1–42 in the diffuse periphery. Similarly, diffuse plaques consisted solely of Aβ1–42 and N-terminal truncations of Aβ1–42, including Aβ4–42 and Aβ3pE–42. Most interestingly, the study identified deposition of the pyroglutamated, truncated version of full length Aβ1–42, Aβ3pE–42, to be specific to AD plaque pathology, while diffuse plaques that are also observed in cognitively unimpaired amyloid positive individuals did not show any Aβ3pE–42 content and consisted solely of Aβ1–42 and Aβ4–42. Time course experiments in mice further showed that Aβ1–40 deposition was associated with plaque maturation into cored deposits (Figure 9.3B–E).[92,94] Together these results suggest that initial non-toxic plaques consist of Aβ1–42, though in AD Aβ1–42 is pyroglutamated to Aβ3pE–42. This hydrophobic functionalization then facilitates subsequent Aβ1–40 deposition and plaque maturation into neurotoxicity associated cored amyloid plaques.[94]

Figure 9.3 MALDI-MS imaging of amyloid peptides in mouse models of Alzheimer's disease. (A.I) Brightfield image and (A.II) segmentation of sample MSI data from a 23 month-old tgSwe animal display localization of Aβ plaque corresponding clusters. (A.III) Partial overlay of tissue scan and the single ion signal of Aβ1–40 peptide (A.IV), which is the major variable underlying the plaque pseudo-objects in the segmentation analysis. (B–E) Correlative chemical imaging of amyloid plaque polymorphism. (B) Fluorescent amyloid staining of cortical plaques demonstrating (B.I) an intermediate plaque morphotype in a 12 month-old mouse, (B.II) a cored plaque in an 18 month-old mouse, (B.III) a diffuse plaque in an 18 month-old mouse, (B.IV) a cored plaque in a 23 month-old mouse, and (B.V) a diffuse plaque in a 23 month-old mouse. Plaques in the 12 month-old mouse appeared structurally similar to the diffuse plaques in older mice. Single ion images of (C.I–V) Aβ1–42 peptide and (D.I–V) Aβ1–40 for the respective plaque types and mice ages listed above. (E) Overlay images of the Aβ1–40 and Aβ1–42 peptide signal showed clear predominance of Aβ1–40 peptide in (D.II and D.IV) cored plaques and of Aβ1–42 in (C.III and C.V) diffuse plaques of the 18 and 23 month-old mice. (D.I and E.I) Plaques of a 12 month-old mouse displayed a much more mixed peptide profile, as seen in the composite yellow color. No. of animals per group at 18 and 23mo.: $n = 3$/timepoint. No. of plaques for each subtype and animal: 10–15 plaques. Univariate comparisons of the peak data between the groups were performed using paired, two-tailed t-tests. Error bars indicate SD. Scale bars B–E: 50 μm. Reproduced from ref. 92 with permission from John Wiley & Sons, Copyright © 2019 International Society for Neurochemistry.

9.4 Conclusions and Future Directions

The developments and applications of MS imaging presented here highlight the immense potential of the technique for studying spatial neuropeptide dynamics in complex biological tissues. Major challenges associated with sample preparation, sample throughput, and protein identification are continuously being addressed by recent developments. Moreover, standardized protocols and data formats significantly facilitate MSI-based applications in clinical settings. In summary, MALDI MS imaging has been emerging as a highly promising and valuable alternative for protein imaging and has been proven to show great potential for delineating molecular mechanisms underlying neurodegenerative disease pathology. This would be of great value for further investigating neuropathologies in both other animal models as well as in expanded post-mortem tissue cohorts. Given its here demonstrated potential, MSI will eventually become a key contender as a routine tool in molecular histology.

Acknowledgements

Funding was provided by The Swedish Research Council VR (2018-02181), Swedish Alzheimer Foundation (Alzheimerfonden), Alzheimer Research UK (ARUK-PPG2018B-025), Swedish Dementia Foundation (Demensfonden), Ahlén Stiftelsen, Stohnes Stiftelse, Stiftelsen Gamla Tjänarinnor.

References

1. R. Carr and S. Frings, *Cell Tissue Res.*, 2019, **375**, 217–225.
2. T. E. Thiele, *Int. Rev. Neurobiol.*, 2017, **136**, 1–3.
3. M. Svensson, K. Skold, P. Svenningsson and P. E. Andren, *J. Proteome Res.*, 2003, **2**, 213–219.
4. T. Hökfelt, C. Broberger, Z. Q. Xu, V. Sergeyev, R. Ubink and M. Diez, *Neuropharmacology*, 2000, **39**, 1337–1356.
5. M. R. Moore and P. M. Black, *Neurosurg. Rev.*, 1991, **14**, 97–110.
6. J. Hanrieder, P. Malmberg and A. G. Ewing, *Biochim. Biophys. Acta*, 2015, **1854**, 718–731.
7. J. Hanrieder, N. T. Phan, M. E. Kurczy and A. G. Ewing, *ACS Chem. Neurosci.*, 2013, **4**, 666–679.
8. W. Michno, P. M. Wehrli, K. Blennow, H. Zetterberg and J. Hanrieder, *J. Neurochem.*, 2019, **151**, 488–506.
9. L. A. McDonnell and R. M. A. Heeren, *Mass Spectrom. Rev.*, 2007, **26**, 606–643.
10. D. S. Cornett, M. L. Reyzer, P. Chaurand and R. M. Caprioli, *Nat. Methods*, 2007, **4**, 828–833.
11. E. H. Seeley and R. M. Caprioli, *Proc. Natl. Acad. Sci. U. S. A.*, 2008, **105**, 18126–18131.
12. M. Karas and F. Hillenkamp, *Anal. Chem.*, 1988, **60**, 2299–2301.

13. R. M. Caprioli, T. B. Farmer and J. Gile, *Anal. Chem.*, 1997, **69**, 4751–4760.
14. B. Spengler, M. Hubert and R. Kaufmann, in *Proceedings of the 42nd Annual Conference on Mass Spectrometry and Allied Topics (Chicago, Illinois)*, 1994.
15. J. O. Gustafsson, M. K. Oehler, S. R. McColl and P. Hoffmann, *J. Proteome Res.*, 2010, **9**, 4315–4328.
16. S. Meding, K. Martin, O. J. Gustafsson, J. S. Eddes, S. Hack, M. K. Oehler and P. Hoffmann, *J. Proteome Res.*, 2013, **12**, 308–315.
17. M. R. L. Paine, S. R. Ellis, D. Maloney, R. M. A. Heeren and P. Verhaert, *Anal. Chem.*, 2018, **90**, 9272–9280.
18. D. K. Lee, S. S. Rubakhin, I. Kusmartseva, C. Wasserfall, M. A. Atkinson and J. V. Sweedler, *Angew. Chem., Int. Ed. Engl.*, 2020, **59**, 22584–22590.
19. R. J. A. Goodwin, J. C. Dungworth, S. R. Cobb and A. R. Pitt, *Proteomics*, 2008, **8**, 3801–3808.
20. R. J. A. Goodwin, A. M. Lang, H. Allingham, M. Boren and A. R. Pitt, *Proteomics*, 2010, **10**, 1751–1761.
21. R. J. A. Goodwin, A. Nilsson, D. Borg, P. R. R. Langridge-Smith, D. J. Harrison, C. L. Mackay, S. L. Iverson and P. E. Andren, *J. Proteomics*, 2012, **75**, 4912–4920.
22. J. Hanrieder, A. Ljungdahl and M. Andersson, *J. Visualized Exp.*, 2012, **60**, e3445.
23. P. Sosnowski, T. Zera, B. Wilenska, E. Szczepanska-Sadowska and A. Misicka, *Rapid Commun. Mass Spectrom.*, 2015, **29**, 327–335.
24. M. Shariatgorji, P. Kallback, L. Gustavsson, N. Schintu, P. Svenningsson, R. J. A. Goodwin and P. E. Andren, *Anal. Chem.*, 2012, **84**, 4603–4607.
25. E. H. Seeley, S. R. Oppenheimer, D. Mi, P. Chaurand and R. M. Caprioli, *J. Am. Soc. Mass Spectrom.*, 2008, **19**, 1069–1077.
26. M. Martin-Lorenzo, B. Balluff, A. Sanz-Maroto, R. J. van Zeijl, F. Vivanco, G. Alvarez-Llamas and L. A. McDonnell, *J. Proteomics*, 2014, **108**, 465–468.
27. G. Li, F. Ma, Q. Cao, Z. Zheng, K. DeLaney, R. Liu and L. Li, *Nat. Commun.*, 2019, **10**, 4697.
28. R. W. Garden, S. A. Shippy, L. Li, T. P. Moroz and J. V. Sweedler, *Proc. Natl. Acad. Sci. U. S. A.*, 1998, **95**, 3972–3977.
29. T. A. Zimmerman, S. S. Rubakhin, E. V. Romanova, K. R. Tucker and J. V. Sweedler, *Anal. Chem.*, 2009, **81**, 9402–9409.
30. A. Ljungdahl, J. Hanrieder, J. Bergquist and M. Andersson, *Methods Mol. Biol.*, 2013, **1023**, 121–136.
31. M. Andersson, M. R. Groseclose, A. Y. Deutch and R. M. Caprioli, *Nat. Methods*, 2008, **5**, 101–108.
32. H. R. Aerni, D. S. Cornett and R. M. Caprioli, *Anal. Chem.*, 2006, **78**, 827–834.
33. J. Hanrieder, A. Ljungdahl, M. Falth, S. E. Mammo, J. Bergquist and M. Andersson, *Mol. Cell. Proteomics*, 2011, **10**, M111 009308.
34. J. Yang and R. M. Caprioli, *Anal. Chem.*, 2011, **83**, 5728–5734.
35. S. M. Puolitaival, K. E. Burnum, D. S. Cornett and R. M. Caprioli, *J. Am. Soc. Mass Spectrom.*, 2008, **19**, 882–886.

36. I. Kaya, W. Michno, D. Brinet, Y. Iacone, G. Zanni, K. Blennow, H. Zetterberg and J. Hanrieder, *Anal. Chem.*, 2017, **89**, 4685–4694.
37. I. Kaya, D. Brinet, W. Michno, M. Baskurt, H. Zetterberg, K. Blenow and J. Hanrieder, *ACS Chem. Neurosci.*, 2017, **8**, 2778–2790.
38. A. Römpp, S. Guenther, Y. Schober, O. Schulz, Z. Takats, W. Kummer and B. Spengler, *Angew. Chem., Int. Ed. Engl.*, 2010, **49**, 3834–3838.
39. A. Römpp and B. Spengler, *Histochem. Cell Biol.*, 2013, **139**, 759–783.
40. M. Kompauer, S. Heiles and B. Spengler, *Nat. Methods*, 2017, **14**, 1156–1158.
41. M. Kompauer, S. Heiles and B. Spengler, *Nat. Methods*, 2017, **14**, 90–96.
42. G. Thiery-Lavenant, A. I. Zavalin and R. M. Caprioli, *J. Am. Soc. Mass Spectrom.*, 2013, **24**, 609–614.
43. J. L. Norris, D. S. Cornett, J. A. Mobley, M. Andersson, E. H. Seeley, P. Chaurand and R. M. Caprioli, *Int. J. Mass Spectrom.*, 2007, **260**, 212–221.
44. K. Skold, M. Svensson, A. Nilsson, X. Q. Zhang, K. Nydahl, R. M. Caprioli, P. Svenningsson and P. E. Andren, *J. Proteome Res.*, 2006, **5**, 262–269.
45. A. Ljungdahl, J. Hanrieder, M. Faelth, J. Bergquist and M. Andersson, *PLoS One*, 2011, **6**, e25653.
46. P. Källback, M. Shariatgorji, A. Nilsson and P. E. Andrén, *J. Proteomics*, 2012, **75**, 4941–4951.
47. D. J. Graham and D. G. Castner, *Biointerphases*, 2012, **7**, 49.
48. A. Henderson, J. S. Fletcher and J. C. Vickerman, *Surf. Interface Anal.*, 2009, **41**, 666–674.
49. S.-O. Deininger, M. P. Ebert, A. Fuetterer, M. Gerhard and C. Roecken, *J. Proteome Res.*, 2008, **7**, 5230–5236.
50. P. M. Wehrli, W. Michno, K. Blennow, H. Zetterberg and J. Hanrieder, *J. Am. Soc. Mass Spectrom.*, 2019, **30**, 2278–2288.
51. R. Van de Plas, J. Yang, J. Spraggins and R. M. Caprioli, *Nat. Methods*, 2015, **12**, 366–372.
52. N. Verbeeck, R. M. Caprioli and R. Van de Plas, *Mass Spectrom. Rev.*, 2020, **39**, 245–291.
53. J. E. Cassat, J. L. Moore, K. J. Wilson, Z. Stark, B. M. Prentice, R. Van de Plas, W. J. Perry, Y. Zhang, J. Virostko, D. C. Colvin, K. L. Rose, A. M. Judd, M. L. Reyzer, J. M. Spraggins, C. M. Grunenwald, J. C. Gore, R. M. Caprioli and E. P. Skaar, *Sci. Transl. Med.*, 2018, **10**, eaan6361.
54. M. R. Groseclose, M. Andersson, W. M. Hardesty and R. M. Caprioli, *J. Mass Spectrom.*, 2007, **42**, 254–262.
55. J. Hanrieder, T. Ekegren, M. Andersson and J. Bergquist, *J. Neurochem.*, 2013, 695–707.
56. K. Skold, M. Svensson, A. Kaplan, L. Bjorkesten, J. Astrom and P. E. Andren, *Proteomics*, 2002, **2**, 447–454.
57. M. Svensson, K. Skold, A. Nilsson, M. Falth, P. Svenningsson and P. E. Andren, *Biochem. Soc. Trans.*, 2007, **35**, 588–593.
58. F. Y. Che, J. Lim, H. Pan, R. Biswas and L. D. Fricker, *Mol. Cell. Proteomics*, 2005, **4**, 1391–1405.
59. L. D. Fricker, *Endocrinology*, 2007, **148**, 4185–4190.

60. P. Yin, X. Hou, E. V. Romanova and J. V. Sweedler, *Methods Mol. Biol.*, 2011, **789**, 223–236.
61. T. A. Zimmerman, S. S. Rubakhin and J. V. Sweedler, *J. Am. Soc. Mass Spectrom.*, 2011, **22**, 828–836.
62. R. Kruse and J. V. Sweedler, *J. Am. Soc. Mass Spectrom.*, 2003, **14**, 752–759.
63. S. S. Rubakhin, W. T. Greenough and J. V. Sweedler, *Anal. Chem.*, 2003, **75**, 5374–5380.
64. M. Zhong, C. Y. Lee, C. A. Croushore and J. V. Sweedler, *Lab Chip*, 2012, **12**, 2037–2045.
65. J. C. Jurchen, S. S. Rubakhin and J. V. Sweedler, *J. Am. Soc. Mass Spectrom.*, 2005, **16**, 1654–1659.
66. A. B. Hummon, A. Amare and J. V. Sweedler, *Mass Spectrom. Rev.*, 2006, **25**, 77–98.
67. T. H. Ong, E. V. Romanova, R. H. Roberts-Galbraith, N. Yang, T. A. Zimmerman, J. J. Collins, 3rd, J. E. Lee, N. L. Kelleher, P. A. Newmark and J. V. Sweedler, *J. Biol. Chem.*, 2016, **291**, 8109–8120.
68. M. Pratavieira, A. R. da Silva Menegasso, A. M. Garcia, D. S. Dos Santos, P. C. Gomes, O. Malaspina and M. S. Palma, *J. Proteome Res.*, 2014, **13**, 3054–3064.
69. M. Pratavieira, A. Menegasso, F. G. Esteves, K. U. Sato, O. Malaspina and M. S. Palma, *J. Proteome Res.*, 2018, **17**, 2358–2369.
70. N. Q. Vu, A. R. Buchberger, J. Johnson and L. Li, *Anal. Bioanal. Chem.*, 2021, **413**, 2665–2673.
71. H. Ye, J. Wang, Z. Zhang, C. Jia, C. Schmerberg, A. D. Catherman, P. M. Thomas, N. L. Kelleher and L. Li, *J. Proteome Res.*, 2015, **14**, 4776–4791.
72. Y. Zhang, K. DeLaney, L. Hui, J. Wang, R. M. Sturm and L. Li, *J. Am. Soc. Mass Spectrom.*, 2018, **29**, 948–960.
73. R. Chen, C. Ouyang, M. Xiao and L. Li, *Rapid Commun. Mass Spectrom.*, 2014, **28**, 2437–2444.
74. H. Ye, L. Hui, K. Kellersberger and L. Li, *J. Am. Soc. Mass Spectrom.*, 2013, **24**, 134–147.
75. N. Q. Vu, K. DeLaney and L. Li, *Curr. Protein Pept. Sci.*, 2020, **22**, 158–169.
76. A. R. Buchberger, N. Q. Vu, J. Johnson, K. DeLaney and L. Li, *J. Am. Soc. Mass Spectrom.*, 2020, **31**, 1058–1065.
77. K. DeLaney and L. Li, *Analyst*, 2019, **145**, 61–69.
78. A. Hishimoto, H. Nomaru, K. Ye, A. Nishi, J. Lim, J. T. Aguilan, E. Nieves, G. Kang, R. H. Angeletti and N. Hiroi, *Biol. Psychiatry*, 2016, **79**, 415–420.
79. E. Bivehed, R. Strömvall, J. Bergquist, G. Bakalkin and M. Andersson, *Peptides*, 2017, **87**, 20–27.
80. K. Blennow, M. J. de Leon and H. Zetterberg, *Lancet*, 2006, **368**, 387–403.
81. M. Jucker and L. C. Walker, *Nature*, 2013, **501**, 45–51.
82. U. Ungerstedt, *Eur. J. Pharmacol.*, 1968, **5**, 107.
83. H. Hulme, E. Fridjonsdottir, H. Gunnarsdottir, T. Vallianatou, X. Zhang, H. Wadensten, R. Shariatgorji, A. Nilsson, E. Bezard, P. Svenningsson and P. E. Andrén, *Neurobiol. Dis.*, 2020, **137**, 104738.

84. J. Stauber, R. Lemaire, J. Franck, D. Bonnel, D. Croix, R. Day, M. Wisztorski, I. Fournier and M. Salzet, *J. Proteome Res.*, 2008, **7**, 969–978.
85. E. Fridjonsdottir, R. Shariatgorji, A. Nilsson, T. Vallianatou, L. R. Odell, L. S. Schembri, P. Svenningsson, P. O. Fernagut, A. R. Crossman, E. Bezard and P. E. Andrén, *Sci. Adv.*, 2021, **7**, eabe5948.
86. E. Bezard, Q. Li, H. Hulme, E. Fridjonsdottir, A. Nilsson, E. Pioli, P. E. Andren and A. R. Crossman, *J. Neurosci.*, 2020, **40**, 6812–6819.
87. M. Bourdenx, A. Nilsson, H. Wadensten, M. Fälth, Q. Li, A. R. Crossman, P. E. Andrén and E. Bezard, *Neurobiol. Dis.*, 2014, **62**, 307–312.
88. E. H. Seeley and R. M. Caprioli, *Proc. Natl. Acad. Sci. U. S. A.*, 2008, **105**, 18126–18131.
89. M. Stoeckli, D. Staab, M. Staufenbiel, K. H. Wiederhold and L. Signor, *Anal. Biochem.*, 2002, **311**, 33–39.
90. L. Carlred, W. Michno, I. Kaya, P. Sjovall, S. Syvanen and J. Hanrieder, *J. Neurochem.*, 2016, **138**, 469–478.
91. I. Kaya, W. Michno, D. Brinet, Y. Iacone, G. Zanni, K. Blennow, H. Zetterberg and J. Hanrieder, *Anal. Chem.*, 2017.
92. W. Michno, P. Wehrli, S. R. Meier, D. Sehlin, S. Syvänen, H. Zetterberg, K. Blennow and J. Hanrieder, *J. Neurochem.*, 2020, **152**, 602–616.
93. N. Kakuda, T. Miyasaka, N. Iwasaki, T. Nirasawa, S. Wada-Kakuda, J. Takahashi-Fujigasaki, S. Murayama, Y. Ihara and M. Ikegawa, *Acta Neuropathol. Commun.*, 2017, **5**, 73.
94. W. Michno, S. Nystrom, P. Wehrli, T. Lashley, G. Brinkmalm, L. Guerard, S. Syvanen, D. Sehlin, I. Kaya, D. Brinet, K. P. R. Nilsson, P. Hammarstrom, K. Blennow, H. Zetterberg and J. Hanrieder, *J. Biol. Chem.*, 2019, **294**, 6719–6732.

CHAPTER 10

MALDI Mass Spectrometry Imaging and Glycomics

CALVIN R. K. BLASCHKE AND RICHARD R. DRAKE*

Department of Cell and Molecular Pharmacology, Medical University of South Carolina, 173 Ashley Avenue, Charleston, 29425 South Carolina, United States
*Email: draker@musc.edu

10.1 Introduction

The metabolic utilization and interconversion of saccharides (also called glycans) are essential components of all living organisms. Glucose is used as an energy source and ribose and deoxyribose sugars are key components of the nucleic acid genetic code found in every organism ranging from the simplest prokaryote to the most complex eukaryotes. Complex biochemical arrangements of glycans as polysaccharides serve as structural components in plants, fungi, and animals. The attachment of a single saccharide or chain of saccharides to another class of molecule, together termed a glycoconjugate (or polysaccharide), is often relevant to the function of that molecule. The study of the biological structure and function of glycans in living organisms is thus termed glycobiology. The totality of an organism's glycan structure is the glycome, and the comprehensive study of the constituents that comprise the glycome is called glycomics.

Historically, the structural complexities and non-template directed synthesis of saccharides have made them more difficult to study than the other major classes of biomolecules.[1] Glycobiology has relied on technological advancements to discover and explore the different classes of glycan structures

New Developments in Mass Spectrometry No. 12
MALDI Mass Spectrometry Imaging: From Fundamentals to Spatial Omics
Edited by Tiffany Siegel

Published by the Royal Society of Chemistry, www.rsc.org

and their functions. Matrix-assisted laser desorption/ionization (MALDI) mass spectrometry imaging (MSI) is a recently developed technique applied to the field of glycobiology. This technique involves applying matrix to a sample and ionizing the analyte molecules with a laser pulse in a raster across the sample. A mass spectrum is generated for each coordinate, and the intensity of each ion can be visualized across the sample in a heat map. Because of the great biodiversity and ubiquitous utilization of glycans and glycoconjugates in living systems, specific MALDI-MSI workflows have been developed to characterize different glycan structural classes and glycoconjugates.

This chapter will cover the recent applications and methodological improvements of MALDI-MSI for the analysis of *N*-glycans, glycosphingolipids, glycosaminoglycans, glycosides, and non-conjugated saccharides. The glycan structures evaluated in the studies summarized in this chapter are displayed in Figure 10.1. An overview of sample preparation and MALDI-MSI analysis workflows are provided for each glycan class, as some require the use of an enzyme to generate the analyte for detection, while others require only matrix applications. Focus is given to studies published from the last four years, and the intent is to provide representative descriptions of published reports using MALDI-MSI for the biological characterization of these diverse glycan structural classes. For further reference, Table 10.1 provides representative reviews and book chapters covering the application of mass spectrometry techniques for each topic in broader biological detail.[2–13] Studies using MALDI-MSI to evaluate monosaccharides like glucose in cellular energy utilization are more appropriate to a metabolomics topic and are not discussed in this chapter. There is no doubt that many other glycan structures found in nature are also not discussed, but it is our expectation that the different MALDI-MSI methods and applications reviewed herein can be adapted to the study of other glycan classes.

10.2 *N*-glycans

The analysis of *N*-glycans has been one of the most targeted and robust applications of MALDI-MSI. *N*-glycans are non-linear oligosaccharides covalently bound to proteins at asparagine residues. Structurally, they are quite diverse and arise biosynthetically from a complex series of enzymatic modifications following transfer of a lipid-linked oligosaccharide donor to proteins co-translationally in the endoplasmic reticulum.[14] The N-linked glycans are modified by different glycosidases and glycosyltransferases to yield several structural classes (shown in Figure 10.1A) extending from a common core of two *N*-acetylglucosamine (GlcNAc) and three mannose (Man) residues. High-mannose *N*-glycans have only mannose residues. Complex *N*-glycans have two to four branches or "antennas" of residues with GlcNAc and galactose (Gal) residues that can be repeated or modified with other saccharides like fucose and sialic acid, or chemical groups like sulfate or phosphate. Hybrid *N*-glycans have both complex branches and mannose residues extending from the core. The wide variety of masses, charges, and

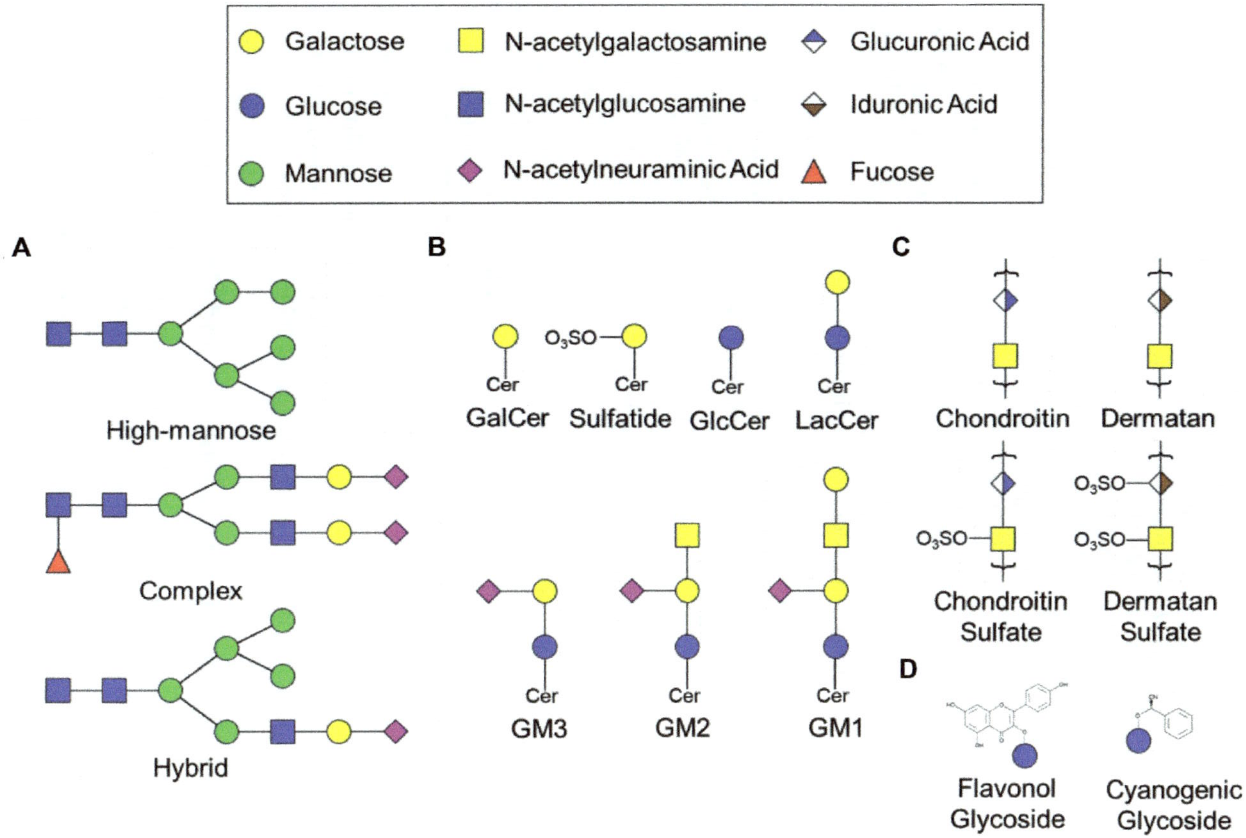

Figure 10.1 Representative structures of (A) *N*-glycans, (B) glycosphingolipids (Cer = ceramide), (C) glycosaminoglycans, and (D) glycosides that are covered in this chapter.

Table 10.1 Reviews and book chapters for further reference.

Topic	Title	Reference (#)
Glycomics	Mass spectrometry in clinical glycomics: The path from biomarker identification to clinical implementation	de Haan *et al.*, 2020 (2)
	Mass spectrometry approaches to glycomic and glycoproteomic analyses	Ruhaak *et al.*, 2018 (3)
N-glycans	MALDI mass spectrometry imaging of N-linked glycans in tissues	Drake *et al.*, 2018 (4)
	Sensitive and structure-informative N-glycosylation analysis by MALDI-MS; ionization, fragmentation, and derivatization	Nishikaze, 2017 (5)
Glycolipids	Visualization of brain gangliosides using MALDI imaging mass spectrometry	Sugiyama & Setou, 2018 (6)
	Detection and distribution of sphingolipids in tissue by FTICR MALDI-imaging mass spectrometry	Ellen Jones *et al.*, 2015 (7)
Saccharides	Oligosaccharide analysis by mass spectrometry: A review of recent developments	Kailemia *et al.*, 2014 (8)
	Sample preparation for mass spectrometry imaging of plant tissues: A review	Dong *et al.*, 2016 (9)
Glycosides	Mass spectrometry of natural products: Current, emerging and future technologies	Bouslimani *et al.*, 2014 (10)
	Imaging mass spectrometry for natural products discovery: A review of ionization methods	Spraker *et al.*, 2020 (11)
Glycosaminoglycans	Glycosaminoglycan detection methods: Applications of mass spectrometry	Kubaski *et al.*, 2017 (12)
	Developments in mass spectrometry for glycosaminoglycan analysis: A review	Pepi *et al.*, 2020 (13)

abundances of *N*-glycans can make analysis challenging, but MALDI-MSI has proven to be a valuable tool in this area. One key factor is the stability of the *N*-glycan structures and their ability to be easily detected following enzymatic release in formalin-fixed paraffin-embedded (FFPE) tissues.

10.2.1 *N*-glycan Method Development and Optimization

MALDI-MSI of *N*-glycans was first applied to fresh/frozen (FF) and FFPE tissues by Powers and colleagues in 2013 and 2014, respectively.[15,16] The main components of the workflows outlined in these studies have been maintained in most *N*-glycan MALDI-MSI studies, and further optimized over time.[17] For FFPE, tissue sections are dewaxed and rehydrated with a series of xylene/ethanol/water washes, undergo antigen retrieval in citraconic anhydride, sprayed and incubated with recombinant peptide *N*-glycosidase F (PNGase F), and sprayed with matrix. For FF tissue, a de-lipidation step using Carnoy's solution has been a key improvement from the original report. Crucial to the workflow, PNGase F is an enzyme that specifically cleaves *N*-glycans between the core GlcNAc and asparagine, releasing the complete oligosaccharide regardless of *N*-glycan class.[17] Several published protocols have described coupling MALDI-MSI with liquid chromatography (LC) tandem mass spectrometry (MS/MS) or on-tissue MS/MS to confirm *N*-glycan structures.[17,18] In the brief existence of *N*-glycan MALDI-MSI, improvements in instrumentation and sample preparation have increased the sensitivity, speed, and reproducibility, incorporated *N*-glycan structural analysis, and adapted the methodology to cell and biofluid analysis.

Using only MALDI-MSI, detected ions can be assigned *N*-glycan compositions by their *m/z* ratio but recent studies have developed methods for distinguishing isomeric species by structural components. Holst and colleagues developed an approach for detecting structural information by incorporating an on-tissue dimethylamidation–amidation derivatization into the workflow.[19] Sialic acids are attached to galactose residues on the terminus of *N*-glycan branches in either an α2,3 or α2,6 linkage. The labile nature of these bonds makes sialic acids susceptible to fragmentation during laser-induced ionization. The dimethylamidation–amidation derivatization modifies and stabilizes the sialic acids in a linkage-specific manner, which prevents fragmentation and creates distinct mass shifts that can be used to distinguish isomeric sialylated *N*-glycans. Fucose residues can be attached to *N*-glycans either at the core GlcNAc bound to the asparagine (core fucosylated) or to GlcNAc residues on the branches of the *N*-glycan (outer arm fucosylated). Recombinant endoglycosidase F3 (Endo F3) is an enzyme that preferentially cleaves core fucosylated *N*-glycans between the two GlcNAc residues, leaving a core GlcNAc-Fucose still attached to the protein. Detecting mass shifts in *N*-glycans treated with Endo F3 compared to PNGase F has been utilized to distinguish core and outer arm fucosylated *N*-glycans.[20] The recent advances in integrating ion mobility with MSI have yet to be applied to *N*-glycan imaging but present an additional promising method for identifying isomeric species.[21]

To increase ionization efficiency, Zhang and colleagues derivatized PNGase F released enzymes with Girard's P reagent, a positively-charged hydrazine reagent, prior to matrix application.[22] In *N*-glycans detected on human laryngeal cancer and ovarian cancer tissue sections, the derivatization increased the signal intensity and *m/z* range and improved the signal-to-noise ratio. MALDI-2 is a technique in which a second laser beam is pulsed through the MALDI-particle plume created by the initial primary laser pulse to increase the ionization efficacy of analytes.[23] After being optimized for the analysis of *N*-glycans in negative ion mode, MALDI-2 was utilized to image human cerebellum tissues and compared to traditional MALDI in positive ion mode.[24] While the overall spectra looked similar, MALDI-2 displayed a slight bias for smaller *N*-glycans. The investigators hypothesize this is due to the increased in-source decay of *N*-glycans in negative ion mode. However, MALDI-2 was able to produce high-quality MS/MS spectra that can facilitate structural analysis of *N*-glycans from minimal sample amounts.

For *N*-glycan analysis, α-cyano-4-hydroxycinnamic acid (CHCA) or 2,5 dihydroxybenzoic acid (DHB) are the typically used matrices. Trifluoroacetic acid (TFA) is usually added in the matrix solution, as well as NaCl for DHB. Several studies have investigated improving the matrices used in *N*-glycan analysis. A novel MALDI matrix hydrazinonicotinic acid (HYNIC) was found to have a higher sensitivity for *N*-glycans and increased salt tolerance than DHB.[25] When comparing a binary matrix of 2,5 DHB and 2,6 DHB or solely 2,5 DHB with no additives, TFA, or phosphoric acid, the binary matrix with phosphoric acid had the best detection of *N*-glycans released from ovalbumin.[26] MALDI-MSI has proven to be an adaptable platform that can be used to analyze the *N*-glycans of a variety of types of samples with a range of instruments. While the MALDI sources used for *N*-glycan analysis are typically under vacuum pressure, a sub-atmospheric pressure MALDI source was adapted for MSI of *N*-glycans on FFPE tissues.[27] One of the benefits of using a sub-atmospheric pressure MALDI source is that it is independent of the mass spectrometer and can be switched with an ESI source, providing greater instrument versatility. With MALDI-MSI becoming possible on a wider range of instruments, it is important to maintain quality controls for sample preparation and data acquisition. Hard-boiled egg white was shown to be an effective external control for monitoring PNGase F digestion, mass accuracy, and signal integrity, as well as being easy to source and prepare.[28] Eggs were used to verify the effectiveness of gelatin and chromium potassium sulfate dodecahydrate-coated ITO slides for *N*-glycan analysis.[29] The modified slides were developed as a cost-effective sample preparation for hard tissues, such as cartilage–bone, that can lose morphology during heat-induced epitope retrieval on ITO slides.

The MALDI-MSI workflow developed for *N*-glycans has been adapted for multiplexed analysis of glycosaminoglycan (GAG), *N*-glycans, and extracellular matrix (ECM) peptides from a single tissue.[30] To do so, the tissue is analyzed for each analyte individually with washing steps in between. The sequence of the enzymes applied to the tissue was Chondroitinase ABC, PNGase F, elastase, and collagenase type III. Digesting and removing the chondroitin sulfate GAGs was found to improve the signal of subsequently analyzed *N*-glycans, possibly due to increased enzyme access of PNGase F to N-glycosylation sites of proteoglycans. This information-dense approach is particularly useful when samples are limited.

Hydrophilic interaction liquid chromatography (HILIC) has commonly been used to separate released *N*-glycans from complex biological samples to avoid ion suppression from salts and other contaminants. In a novel technique developed by Chen and colleagues, *N*-glycans released from human serum by a filter-aided *N*-glycan separation (FANGS) were loaded into a HILIC column and the HILIC eluent was collected directly onto a stainless steel MALDI plate as the tip of the column was moved along the plate.[31] The flow was collected from 15 to 40 min on a 9.0 cm strip of the MALDI plate. Once dried, the plate was sprayed with matrix and imaged. The HILIC MALDI-MSI platform had higher peak intensities and detected more *N*-glycans than MALDI-MSI alone.

Recent adaptions of the tissue *N*-glycan workflow have utilized a slide-based format for the analysis of cells, glycoproteins, and biofluids. *N*-glycan MALDI-MSI was expanded to robust and reproducible analysis of cells by Angel and colleagues.[32] Cells were cultured overnight onto a chamber slide system and washed with PBS, neutral buffered formalin, and Carnoy's solution. The cells can then be processed *via* the typical PNGase and matrix application process and imaged. This method can analyze 8 chambers with as few as 3000 cells per chamber within 4 hours after the cells are cultured. *N*-glycans contributed from cell media were subtracted from the collected spectra to create reproducible cell *N*-glycan profiles. With no additional preparation time, stable isotope labeling can be used to determine the *N*-glycan turnover rates of the cultured cells.

One of the limitations of analyzing *N*-glycans released by PNGase F is that the detected *N*-glycans cannot be linked back to their carrier glycoprotein when investigating tissues or clinical biofluids. Black and colleagues addressed this issue with the development of antibody panel based (APB) *N*-glycan imaging.[33,34] In this method, antibodies are spotted in a grid onto nitrocellulose-coated microscope slides for the specific capture of glycoproteins from clinical biofluids, such as serum and plasma. The slide is then washed, and PNGase F and matrix are sprayed onto the slide. This method allows for the sensitive analysis of the *N*-glycan profiles of multiple glycoproteins from multiple samples in one imaging run.

N-glycan profiling of clinical biofluids typically requires derivatization or labeling followed by purification steps to isolate the released *N*-glycans from the other components that will cause ion suppression. A rapid slide-based MALDI-MSI method for the analysis of serum and plasma *N*-glycan profiles was developed to avoid these lengthy processing requirements.[35] In this workflow, serum or plasma is spotted onto a slide coated in an amine-binding hydrogel that immobilizes the proteins after an hour incubation period. The spots are washed with Carnoy's solution and water to remove salts and lipids and sprayed with PNGase F and matrix. Multiple slides can be prepared together, and up to 28 samples can be analyzed on one slide. This method was also modified to apply both PNGase and Endo F3 for identification of core fucosylated *N*-glycans.

These recent adaptions of the tissue MALDI-MSI workflow have capitalized on the minimal sample preparation steps to develop simple and rapid *N*-glycan profiling techniques. Additionally, the speed and sensitivity of MALDI-MSI analysis allow for high-throughput analysis with minimal sample usage. Together these slide-based methods represent the potential for MALDI-MSI of *N*-glycans to become a clinically relevant tool.

10.2.2 *N*-glycan Biological Applications

N-glycosylation is a common post-translational modification that impacts a diverse array of molecular processes, such as protein folding, cell signaling, and immune recognition.[36] The biological importance of N-glycosylation in human physiology is evidenced by the debilitating diseases caused by mutations in the glycosylation synthesis pathway, termed Congenital Disorders of Glycosylation.[37] Aberrant N-glycosylation has been observed in many diseases and has been investigated as a source of biomarkers. Most United States Food and Drug Administration (FDA)-approved cancer biomarkers are glycoproteins, demonstrating the potential for this approach.[38] By comparing the distribution of *N*-glycans in diseased tissue to adjacent normal tissue or tissue from a healthy patient, MALDI-MSI can identify disease-associated changes in the abundance of individual *N*-glycan species and/or structural classes of *N*-glycans. Because of the prevalence of cancer FFPE tissues obtained surgically and clinically annotated by pathologists, a majority of *N*-glycan MALDI-MSI analyses have focused on cancer. Several of these studies are highlighted in the following paragraphs.

When *N*-glycan profiles of specific regions of kidney tissues were examined by MALDI-MSI, multi-antennary fucosylated *N*-glycans with a bisecting GlcNAc were found to be abundant in the proximal tubules in the cortex region.[39] This set of *N*-glycans was notably not present in clear-cell renal cell carcinoma tissues. Previous studies of mouse kidney tissues with *N*-glycan MALDI-MSI reported similar findings.[40] Additional analysis of polycystic kidney disease tissues displayed a stark absence of fucosylated *N*-glycans in cyst fluid regions.[39] As shown in Figure 10.2, use of a new MALDI QTOF instrument for *N*-glycan analysis of a normal kidney tissue (reported in ref. 31) greatly improved the detection of glomerular and tubule *N*-glycans. Additionally, this tissue was processed for anti-Lewis Y antigen immunohistochemical staining that targets the extensive outer arm fucosylation reported for kidneys. Glomeruli notably lack these multi-fucosylated species and instead contain many complex tri-and tetra-antennary *N*-glycans that are sialylated.

Also common to these studies is the use of tissue microarrays (TMA) to increase the numbers of clinical samples that can be analyzed by *N*-glycan MALDI-MSI. A typical TMA slide can have 200 or more tissue core biopsies embedded in a single paraffin block. Thus, use of TMAs allows for the rapid analysis of a large number of samples in a single MALDI-MSI run. Although there is a possibility that the small tissue core is not representative of the whole diseased tissue or includes healthy tissue, especially with heterogeneous diseases such as cancer, the identical sample processing and data acquisition parameters of the tissue cores limit the experimental variability.

The abundance of hypercellular round cell areas and ECM can be used for grading and prognosis in myxoid liposarcoma (MLS) tumors. A TMA of morphologically graded MLS tumor tissues was analyzed by *N*-glycan MALDI-MSI to find *N*-glycans associated with MLS tumor progression.[41] The overall abundance and average number of mannoses on high-mannose *N*-glycans were increased in higher grade tissues. Increased abundance of tri-antennary *N*-glycans was associated with tumor progression and poor disease-specific survival.

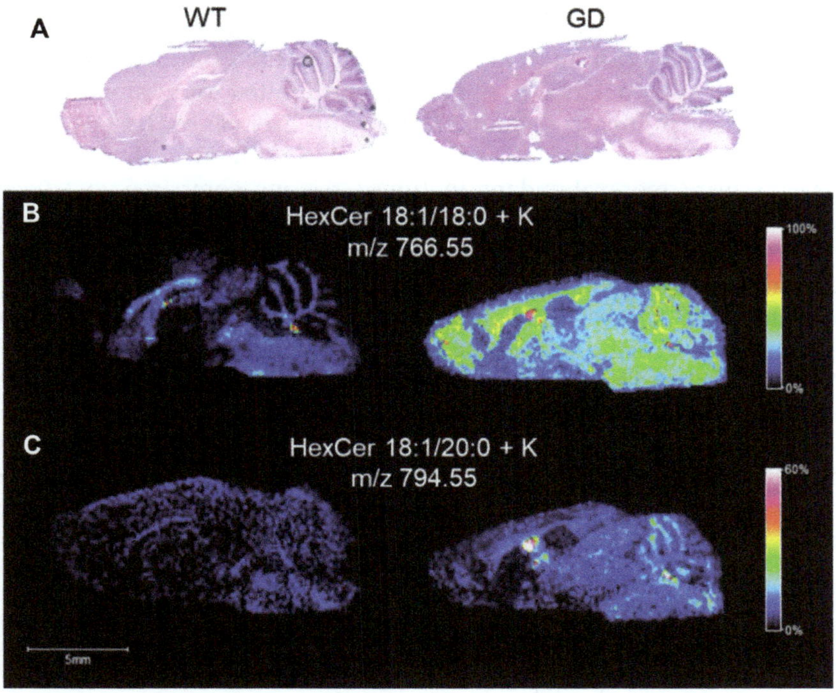

Figure 10.2 Example MALDI-MSI of a Gaucher disease sagittal brain slice. Shown are (A) the H&E stains of the imaged brain slices from a Gaucher disease (GD) mouse[70] and control mouse (WT). Representative potassium ion adducts of two hexosylceramide (HexCer) species elevated in the GD brain are shown in panels (B) and (C). Gaucher disease is a deficiency in the glucocerebrosidase responsible for degradation of glucosylceramides.

Most women diagnosed with ovarian cancer are in the metastatic late stages of the disease because ovarian cancer is asymptomatic. A high-mannose *N*-glycan ($Hex_6HexNAc_2$) and fucosylated biantennary *N*-glycan ($dHex_1Hex_5HexNAc_4$) were able to distinguish stage I ($n = 3$) and stage III ($n = 3$) ovarian cancer tissues.[42] In a TMA of 3 early-stage and 7 late-stage ovarian tissue cores, these *N*-glycans had 0.649 and 0.786 area under the curve (AUC) values after receiver operating characteristic (ROC) assessment, respectively. While this study had a relatively small sample size, previous MALDI-MSI studies of advanced ovarian cancer tissues also found high expression of high-mannose *N*-glycans.[43]

McDowell and colleagues correlated pancreatic ductal adenocarcinoma (PDAC) tissue *N*-glycan distributions to immunohistochemistry (IHC) staining of current and prospective pancreatic cancer biomarkers.[44] Combining *N*-glycan MSI data with biomarker characterization of tissues led to more accurate detection of PDAC than using the analyses individually. Additionally, high-resolution histopathological images were utilized for targeted analysis of pancreatic islets in normal pancreas tissue sections, and

5 sulfated N-glycans were found to be highly enriched in the islets. Bi-, tri-, and tetra-antennary structures without any fucose or sialic acid residues were found to be localized to necrotic regions in the PDAC tissues. Another study investigating necrotic tissue also found limited expression of fucosylated N-glycans in breast, thyroid, cervical, and liver cancer tissues.[45]

When analyzing 6 endometrial cancer (EC) tissues, high-mannose N-glycans were more abundant in tumor regions compared to adjacent normal regions.[46] A TMA with 8 EC tumor tissues from patients with lymph node metastasis and 20 EC tumor tissues from patients without lymph node metastasis was also analyzed by MALDI-MSI. The N-glycan profiling of the TMA identified a fucosylated biantennary N-glycan ($dHex_1Hex_5HexNAc_4$) that could distinguish tissues based on lymph node metastasis status.

Tetra-antennary N-glycans with repeating N-acetyllactosamine units (N-acetylglucosamine–galactose disaccharides) on one of the antennas have been associated with breast cancer tissues in separate studies.[47,48] In one study, the presence of a core fucosylated tetra-antennary N-glycan with a single N-acetyllactosamine unit in breast cancer tissues was found to predict increased risk of lymph node metastasis and disease recurrence and reduced survival.[47] Another study detected a series of N-glycans with repeating N-acetyllactosamine units in human epidermal growth factor receptor 2 (HER2) positive breast cancer and triple-negative breast cancer (TNBC).[48] Additionally, the abundance of these N-glycans was similar in primary and associated-metastatic breast cancer tissues.

Carter and colleagues analyzed non-human primate lung tissues 50 days and 180 days after radiation injury by MALDI-MSI and compared the N-glycan profiles to control animals that were not irradiated.[49] Histological analysis of the tissues identified a complex combination of pathologies in the tissues. A number of N-glycans were altered in areas of specific pathological presentations, such as areas of accumulated alveolar macrophages or fibrosis. Many of these N-glycan changes were independent of time post-exposure.

When comparing N-glycan abundances in tissues with hepatocellular carcinoma (HCC) to cirrhotic or adjacent normal tissue with MALDI-MSI, West and colleagues found an increase in tetra-antennary and fucosylated N-glycans in HCC tissues. This robust analysis included 5 HCC tissue sections, 3 normal tissue sections, 3 cirrhotic tissue sections, 138 HCC tissue cores, 95 normal tissue cores, and 22 cirrhotic tissue cores. For a subset of tissues, the N-glycan profiles were combined with clinical outcomes to determine associations of specific N-glycans with survival. Patients with high levels, defined as greater than the median expression in both HCC and normal tissues, of a bi-fucosylated tetra-antennary or mono-fucosylated tetra-antennary N-glycan were found to have shorter median survival times.

10.3 Glycosphingolipids

The majority of glycolipids found in vertebrates are glycosphingolipids, which consist of sphingosine lipid backbones bound to saccharides.[50]

Glycosphingolipids can be neutral (with no charged sugars or ionic groups) or charged *via* sialic acid or sulfate. Sialylated glycosphingolipids are called gangliosides, and sulfated glycosphingolipids are called sulfatides (sulfated galactosylceramide). Example structures of these glycosphingolipids are shown in Figure 10.1B. The nomenclature for gangliosides relies on abbreviations, such as GM1, where the G stands for ganglioside, the second letter refers to the number of sialic acids it contains (Mono, Di, Tri, *etc.*), and the number refers to the order of migration on thin-layer chromatography (ranging from 1 to 3).[50] As the nomenclature indicates, the complex pool of lipids found in tissues requires sensitive and robust methods to deconvolute glycolipid species with a wide array of attached saccharides, sphingosine chain lengths, and charges. MALDI-MSI has been successfully used to handle this analytical challenge, and recent technological advances have made improvements to the range of glycolipids that can be analyzed.

10.3.1 Glycosphingolipids Method Development and Optimization

The choice of matrix is very important for lipid analysis, and a wide variety of matrices have recently been developed or optimized for glycosphingolipid analysis.[51] Historically, dihydroxybenzoic acid (DHB) has been the most common matrix used for lipid and glycosphingolipid analyzes. More recently, the sublimation of DAN matrix has been used for analysis of glycosphingolipids in both positive and negative ion modes.[52] Sublimated 1,6-diphenyl-1,3,5-hexatriene (DPH) was demonstrated to be optimal for sulfatide analysis in mouse and rat brain tissue due to its minimal background signal in negative ion mode.[53] 2,6-Dihydroxyacetophenone (DHA) can also be used as a MALDI matrix for glycosphingolipid analysis in both positive and negative ion mode, but DHA sublimates under the vacuum and intermediate pressures required for several high-resolution mass spectrometers during long runs. Jackson and colleagues prevented this issue by coupling an atmospheric pressure MALDI source to an Orbitrap mass spectrometer and optimized the workflow for ganglioside analysis of tissues.[54] Sensitive and reproducible detection of neutral, acidic, and amphoteric glycosphingolipids on mouse brain sections was demonstrated with 3-aminophthalhydrazide as a matrix.[55]

Several innovative approaches have been developed to improve the ionization of cerebrosides, neutral glycosphingolipids with a single glucose or galactose head group. Silver nanoparticulate was implanted onto mouse brain tissues as a matrix for lipid analysis by Muller and colleagues.[56] In positive ion mode, 14 galactosylceramides were analyzed with minimal fragmentation and detected as primarily silver adducts, instead of the sodium and potassium adducts formed from endogenous salts. Another approach found significant increases in signal intensity of cerebrosides when mouse brain tissues were pre-treated with sodium carbonate prior to DHB sublimation and detected over 35 cerebrosides.[57] An infrared

MALD-mass spectrometer was utilized for analysis of thin-layer chromatography (TLC)-separated neutral glycosphingolipids directly on the TLC plate by Kouzel and colleagues.[58]

To improve the detection of gangliosides in tissue sections analyzed by MALDI-MSI, Yang, Dufresne, and Chaurand developed a novel sample preparation that increases the signal intensity of gangliosides on average 15-fold.[59] The tissues were washed and sprayed with an ammonium formate solution and DAN matrix was applied by sublimation. This sample preparation workflow was utilized by Harris and colleagues to analyze FFPE human brain tissue.[60] The fixation and deparaffinization processes involved with MALDI-MSI of FFPE tissues make lipid analysis challenging. By applying an ammonium formate wash to the tissues before matrix sublimation, the investigators were able to detect and visualize gangliosides in formalin-fixed clinical brain samples by MALDI-MSI for the first time. This workflow enhanced the ganglioside signal intensity in fresh frozen tissue as well.

As discussed previously, MALDI-2 is a technique to increase ionization of analytes with laser-induced post-ionization and has been used to account for the relatively low abundances and low ionization efficiencies of neutral glycosphingolipids in several tissue types.[61] MALDI-2 was coupled with transmission-mode geometry instrumentation in which a UV-laser transmitting microscope objective is used for fine adjustment of the primary ionizing laser beam by Niehaus and colleagues to analyze mouse brain tissues.[62] The increase in signal intensity provided by MALDI-2 compensates for the low amount of material ablated during transmission-mode MALDI and allows for visualization of the glycosphingolipid hexosylceramide at a subcellular resolution. MALDI-2 has also been coupled with an on-tissue Paterno-Buchi reaction for the analysis of carbon–carbon double bond isomeric glycosphingolipids.[63]

10.3.2 Biological Application of Glycosphingolipids

Other than adipose tissue, the brain contains more lipids than any other bodily organ.[64] The diverse cell functions and morphologies in the brain are reflected in the complexity and variety of its lipidome, and it is particularly enriched in cerebroside, ganglioside and sulfatide species. MALDI-MSI is a compatible method with brain glycosphingolipid analysis as it can resolve distributions in a variety of glycosphingolipids and localize them to specific brain regions and cell types. This detailed spatial analysis in brain tissue sections from rats or mice typically used in neurodevelopment and neurodegeneration studies is possible due to the increasing spatial resolution capabilities of mass spectrometers. Recent representative examples for analysis of glycosphingolipids in different neurodegenerative disease and stroke model tissues and non-brain tissues are as follows.

A study investigating the changes in monosialylated gangliosides over the lifespan of rats found unique neuroanatomically-specific changes.[65] An age-dependent increase in the ratio of d20:1/d18:1 GM3 gangliosides was observed in the peri-ventricular corpus callosum, and early development correlated with an increase in d20:1 species of GM1. There were significant regional differences in the abundances of the gangliosides between the superficial, intermediate, and deep layers of the cerebral cortex during aging. The possible differential regulation occurring in these regions during aging may be indicative of regional vulnerability to damage during aging.

Complex species of GM1 have been found to have protective effects against brain damaging event such as TBI and stroke, while in contrast, simple GM2 and GM3 gangliosides are linked to neurodegeneration.[66,67] The efficacy of chloroquine treatment prior to stroke injury in rats was examined by MALDI-MSI of brain gangliosides by Caughlin and colleagues.[68] Three days post-stroke the site of the stroke had a significant increase in d20:1 species of GM3, and other simple gangliosides increased across the brain tissues. Treatment with chloroquine prevented the depletion of complex GM1 at the site of injury that was seen in untreated rats, and partially restored complex GD2 species after 21 days.

The accumulation of lipids in the brain due to lysosomal storage disorders, such as Gaucher Disease (GD), Hunter's Disease (HD), and Niemann–Pick Disease, can lead to neurodegenerative conditions such as motor impairment and cognitive decline and can be fatal.[69] In one example, Jones and colleagues found a higher abundance of several hexosylceramides throughout the brain regions in GD mice, consistent with the defect of glucocerebrosidase in GD.[70] The cerebellum of the GD mice had elevated levels of lactosylceramides compared to control mice. By pairing the MALDI-MSI analysis with LC–ESI-MS/MS, H&E staining, and immunohistochemistry, this study correlated pathologies such as inflammation and autophagy with specific glycosphingolipid accumulation. As shown in Figure 10.3, an example of the accumulation of two hexosylceramide species in a sagittal section of GD brain relative to wild type is shown.

Another study using a Niemann-Pick disease mouse model found elevated levels of gangliosides GM2 and GM3 localized to the posterior lobules of the cerebellum that continued to increase over the course of the disease.[71] Hunter and colleagues found the myelin layer of the subventricular zone had reduced sulfatide abundance in brain sections from patients with Huntington's disease.[72] Sulfatides are normally highly enriched in myelin sheaths, so these findings may be indicative of the degradation of sulfatides during HD-associated myelination.

Alzheimer's disease (AD) is characterized by the aggregation of extracellular amyloid-β (Aβ) peptides that are thought to be neurotoxic.[73] Lipids have been implicated in AD pathogenesis and the lipid transporter protein, apolipoprotein E (APOE), ε4 allele is a major risk factor to develop sporadic AD.[74] MALDI-MSI of brain sections from the tgArcSwe AD mouse model found gangliosides GM2 and GM3 highly enriched and sulfatide levels decreased in Aβ deposits in the hippocampus and somatosensory cortex.[75] A trimodal MALDI-MSI approach was developed to analyze peptides and lipids in positive and negative ion mode on a single tissue for accurate

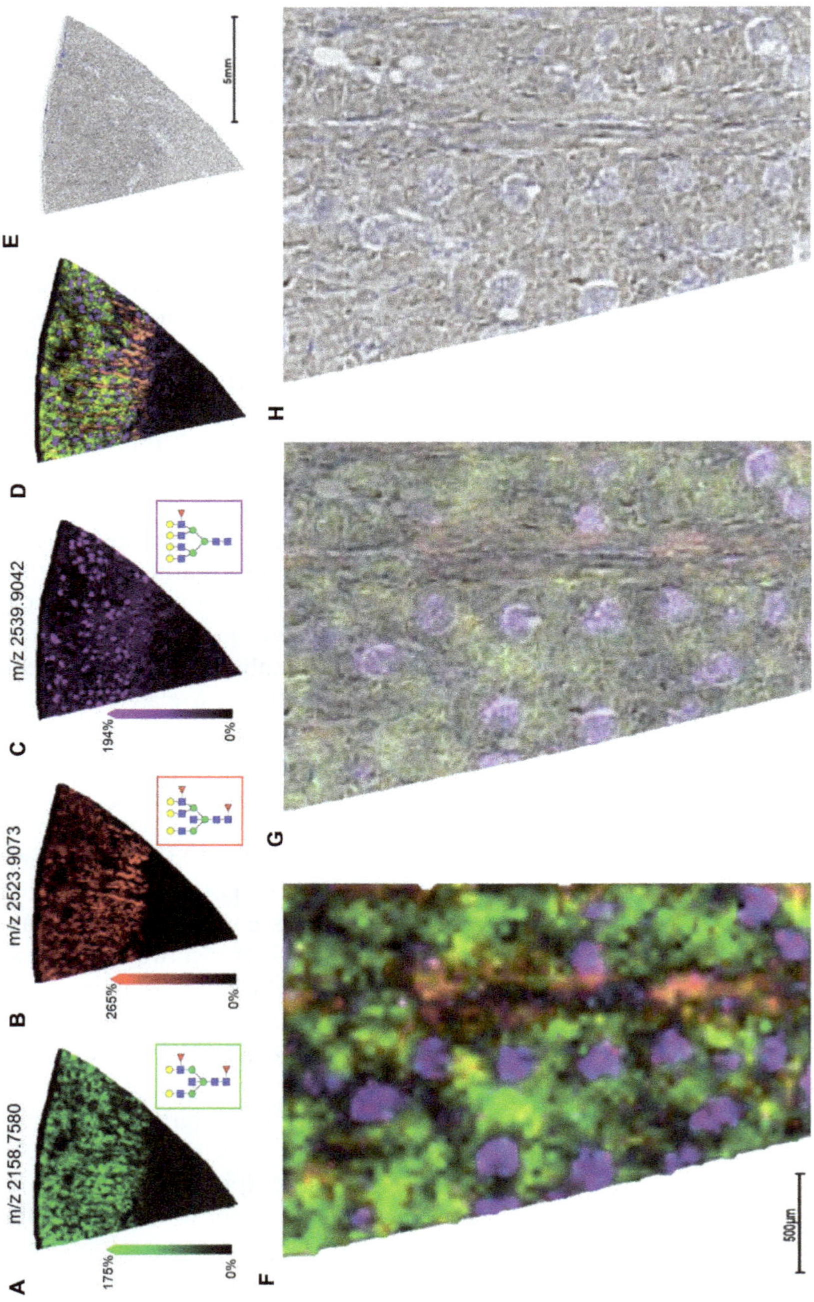
A
m/z 2158.7580
175%
0%
B
m/z 2523.9073
265%
0%
C
m/z 2539.9042
194%
0%
D
E
5mm
F
500µm
G
H

localization of neuronal lipids to Aβ isoforms using the same tgArcSwe AD mouse model.[76] In agreement with a previous study, the Aβ plaques and sulfatides did not colocalize and there was a substantial increase in ceramides. Optimization of this trimodal MALDI-MSI approach allowed for the detection of gangliosides in the same mouse model.[77] Several simple gangliosides (GM3 34:2, GM3 38:1, and GM2 36:1) were found to accumulate in the Aβ plaques.

The accumulation and exclusion of monosialylated-gangliosides and sulfatides in Aβ plaques, respectively, were also found in the tgAPPSwe and tgAPPArcSwe AD mouse model.[78,79] Another study compared the ganglioside profiles of wild-type and tgAPPSwe AD mice over the course of 20 months by MALDI-MSI.[80] The investigators found lower levels of GM1 and higher levels of GM3 and GM2 in the AD mice brain sections, in line with previous studies. The white matter and gray matter regions of the brain sections displayed unique ganglioside distributions during development. While control and AD mice had similar changes of GM2 during aging, the abundance of GM3 species did not show any consistent changes over time. Instead, GM3 species varied based on which region of the brain they were located. A decrease in GD1 d18:1 species in white matter regions during aging was uniquely observed in the AD mice.

The 5xFAD mouse is a widely used AD model that incorporates five of the common mutations used in other AD mouse models.[81] In a study by Kaya and colleagues, the distributions of gangliosides in the hippocampal and adjacent cortical regions of 5xFAD mice were analyzed by MALDI-MSI.[82] In concurrence with the other AD models, the monosialylated gangliosides were accumulated in Aβ plaques, but this study also found distributions of GMs specific to the sphingosine-base chain lengths. In the dentate gyrus, the subgranular zone only contains C18:1 GMs, while the molecular layer contains both C18:1 and C20:1 GMs. The coronal brain tissue sections of 5xFAD mice were also found to have Aβ deposit-associated depletion of sulfatides

Figure 10.3 Example MALDI-MSI of *N*-glycans in a normal human kidney also immunostained with anti-Lewis Y antigen. A FFPE tissue was initially immunostained with anti-Lewis Y antibody, optically imaged, then prepared for *N*-glycan imaging as described.[39] Data was processed and images were created in SCiLS software (Bruker Daltonics). The Lewis-Y stained image was used for co-localization of the *N*-glycans. In panels (A) and (B) are representative *N*-glycans with a bisecting GlcNAc and fucosylation indicative of tubule localization. In panel (C) is an *N*-glycan specific to glomeruli. Masses and glycan structure are included in each panel. (D) Overlay of the distribution of the three glycans shown in panels (A–C). (E) Lewis Y stained image (peroxidase stain). (F) An expanded area of the overlay image in panel (D). The colors and glycans correspond to those in panels (A–C). (G) Using SCiLS software, the overlay glycan image was visualized at 50% transparency with the Lewis Y stained image. (H) The Lewis Y stain at 100%, no MALDI image; tubules are stained brown from the Lewis Y antibody and glomeruli are unstained and appear blue.

and galactosylceramides.[83] The consistency in the findings from the AD mouse model study of glycolipid distributions attests to the robust and reproducible nature of MALDI-MSI.

Recent applications of MALDI-MSI for glycolipid analysis have not been limited to brain tissues. Mice and human pancreatic tissue were analyzed for lipid distributions within pancreatic islets by Prentice and colleagues.[84] Immunofluorescent-detected alpha and beta cells in serial tissue sections allowed investigators to localize MALDI-MSI signals to pancreatic islets. High-resolution (15 μm) imaging could be targeted to the pancreatic islets to reduce data acquisition times. The ganglioside GM3 (d34:1) was more abundant in the periphery of mice pancreatic islets, and several other GM3 lipids were detected in the islets. Immunostaining serial sections for GM3 synthase found the enzyme restricted to the islets as well. However, most of the GM3 lipids were ubiquitously distributed throughout the human pancreatic tissue. While GM3 (d34:1) was restricted to the islets, it was not localized to the periphery as seen in the mice tissues.

The distributions of glycolipids were visualized in damaged retinal tissue from rats to investigate the molecular changes of inflammation-induced optic nerve regeneration (IIR).[85] Several sulfatides associated with myelin membranes and inhibition of axon growth were found to decrease in abundance at the site of damage during the course of IIR. This could be indicative of the clearance of myelin debris. Additionally, GM3 gangliosides were enriched in the damaged tissue, possibly due to the lysosomal degradation of more complex gangliosides.

Bien and colleagues investigated glycosphingolipid distributions in tissue biopsies from patients with human colorectal cancer by MALDI-2-MSI.[61] Globotriaosylceramide (Gb3Cer) and globotetraosylceramide (Gb4Cer), cellular receptors of Shiga toxins, were enriched in the tumor-infiltrating blood vessels, dedifferentiated tumor cell areas, and tumor stroma. Fucosylated glycosphingolipids have been found to promote immune escape mechanisms in cancer. Notably in this study, the fucosylated glycosphingolipids were highly localized to dysplastic gland and cells infiltrating adipose tissue. The sensitivity of MALDI-2 was utilized to display various lipoforms of Gb3Cer and Gb4Cer, as well their precursors and fucosylated glycosphingolipids in a single run.

10.4 Glycosaminoglycans

Glycosaminoglycans (GAGs) are chains of repeating disaccharides that are attached to a core protein to form a proteoglycan. The glycan composition of the repeating units and the presence of sulfate groups differentiate the GAGs, the major types being hyaluronic acid, dermatan sulfate, keratan sulfate, chondroitin sulfate, heparin, and heparan sulfate. These GAGs are covalently linked to a core protein scaffold, except for hyaluronic acid, which is a non-sulfated linear polymer of repeating GlcNAc and glucuronic acid residues (and not attached to a protein). Proteoglycans can have multiple types of GAGs and differ in the number, length, and composition of their GAGs in different cell types.[86] In humans, proteoglycans are often secreted or incorporated into the cell membrane where they facilitate extracellular matrix assembly and cell signaling, requiring dynamic and regulated expression.[87]

Due to their roles in tissue repair and fibrosis, several GAGs have been found to be upregulated in liver, pulmonary, and cardiac fibrosis.[88–90] When Aichler and colleagues utilized MALDI-MSI to investigate TMAs from human patients with idiopathic pulmonary fibrosis and bleomycin-induced mice with fibrosis, a chondroitin/dermatan fragment and a chondroitin/dermatan sulfate fragment was commonly found in both sets of tissues.[91] The localization of the chondroitin sulfate fragments was confirmed by immunohistochemical staining and with a high degree of overlapping localization.

As discussed previously, Clift and colleague developed a method to target chondroitin sulfates for MALDI-MSI by applying chondroitinase ABC to tissues.[30] Chondroitin sulfate chains with 1–3 repeats in multiple sulfation states can be detected with chondroitinase ABC.[92,93] To confirm the mass values as chondroitin sulfate repeats, they performed on-tissue fragmentation *via* CID and ion mobility analysis. The chondroitinase ABC digestion was incorporated into a multiplexed workflow that can visualize and relatively quantify *N*-glycans, chondroitin sulfate GAGs, and ECM peptides and proteoglycans from a single tissue section utilizing serial enzyme treatments. This was the first enzymatically targeted study of chondroitin sulfate GAGs by MALDI-MSI.

The spatial and temporal regulation of GAGs suggests that MALDI-MSI studies could be immensely valuable. To the best of our knowledge, these are the only studies that have analyzed GAGs utilizing MALDI-MSI. There are several challenges inherent to GAG analysis that need to be addressed. In-source dissociation of sulfate groups historically has been an issue for GAG analysis by MALDI.[94] Strategies such as pairing GAG oligosaccharides with a basic peptide and utilizing ionic liquid matrices have been found to ameliorate sulfate group loss and improve signal.[95,96] The feasibility of applying these approaches to MALDI-MSI has not been evaluated.

10.5 Glycosides

Glycosides are a broad group of molecules composed of a saccharide (glycone) bound by a glycosidic linkage to another compound (aglycone), generally found in plants and microbes. Since the biological functions of the glycosides are heavily dependent on the aglycone, they are often classified by the chemical nature of the aglycone. The aglycones are often secondary metabolites that become bioactive when the glycosidic linkage is enzymatically hydrolyzed. Recent applications of MALDI-MSI to glycosides have focused on plant glycosides, namely flavonoid glycosides and cyanogenic glycosides.

The flavonoid glycosides found in plant extracts can be used medicinally and have been correlated with disease prevention.[97] Plants have been found to utilize flavonoid glycosides and flavonoids for a variety of functions including UV protection, plant defense, and hormone signaling.[98,99] *Ginkgo biloba* leaves are known to have a high concentration of a variety of flavonoids modified with glucose and rhamnose residues.[100] Utilizing MALDI-MSI, Beck and Stengel investigated the distribution of flavonoids and flavonoid glycosidases in *Ginkgo bilboa* leaves.[101] After determining 2,5-dihydroxy benzoic acid (DHB) as the most suitable matrix for positive ionization, thin leaf sections were imaged at a high spatial resolution (10 μm-raster). The flavonoid glycosides were found throughout the leaf, but a higher abundance was found near the vascular strands. The authors suggest that while UV protection is needed at the surface of the leaf, the chemical messaging and physiological regulation functions of flavonoid glycosidases would require rapid distribution and make the found distribution align with the current literature.

When analyzing a section of a ripe strawberry fruit by MALDI-MSI, Enomoto investigated the distribution of flavanol glycosides.[102] After identifying flavonols, ellagic acid, and the corresponding glycosides in a strawberry polyphenol extract by MALDI-MS/MS, the distribution of these molecules in frozen sections of strawberries were visualized. The flavonols and flavonol glycosides were primarily in the epidermis, while the ellagic acid and ellagic acid glycosidases were primarily distributed in the seeds and bottom of the fruit. Sun and colleagues conducted a similar study studying flavonoid glycosidases in *Scutellaria baicalensis* Georgi, a medicinal herb.[103] Using 9-AA as a matrix, MALDI-MSI of a section of *S. baicalensis* root identified 5 flavonol glycosides. The abundances of the glycosides were significantly higher in the phloem than in the xylem and pith. This finding is consistent with the need for the plant to rapidly transport these glycosides in the event of environmental stress.

As an effective defense against herbivores, some plants may enzymatically release toxic hydrogen cyanide that is stored in an inactive state as part of a glycoside.[104] These cyanogenic glycosides are spatially separated from the hydrolyzing enzymes either in a sub-cellular or tissue level.[105] Eight species of flowering plants from the *Lomatia* genus were analyzed by MALDI-MSI to investigate within-flower cyanogenic glycoside distributions.[106] The investigators found extremely high concentrations of cyanogenic glycosides within the pollen and the tissue directly surrounding the pollen. The relative abundance of the monoglycoside dhurrin and the diglycoside proteacin differed significantly throughout the flowers. Also utilizing MALDI-MSI, Hansen and colleagues identified a differential distribution of a prunasin, a monoglycoside, and amygdalin, a diglycoside, in the flowers of *Eucalyptus cladocalyx*.[107] Whereas the *Lomatia* flowers displayed a degree of colocalization of mono- and diglycosides, the *E. cladocalyx* flowers had highly distinct distributions of the detected mono- and diglycosides. The functional significance of the novel identifications and distributions of these important plant defense glycosides by MALDI-MSI merit further investigation.

10.6 Non-conjugated Saccharides

MALDI-MSI of non-conjugated mono-, di-, and oligosaccharides has been heavily applied and developed for plant tissue analysis. This class of low-molecular weight saccharides are utilized as energy sources and signals for adaptive and stress responses.[108,109] Investigation of the distribution of these molecules throughout plant tissue can provide a greater understanding of plant metabolism and defense. Recently, several groups have addressed MALDI-MSI-related issues of sensitivity and identification of isomer species.

Zhan and colleagues developed a workflow for examining the distribution of disaccharide isomers in plant tissues by coupling MS/MS with MSI.[110] First, the investigators performed MALDI-MS/MS on a series of disaccharide isomers dried with matrix. By comparing the fragment patterns and relative intensity of certain fragment ions, the isomers could be distinguished. Specifically, sucrose could be distinguished from other disaccharides by the relative intensities of the *m/z* 215 and 341 peaks. Initial MSI of purple onion bulb cross-sections indicated a higher abundance of saccharides in the inner epidermis than the outer epidermis. MS/MS analysis of the sections found that sucrose was localized to the inner epidermis while the other disaccharides were found mainly in the middle epidermis region. The MS/MS analysis of disaccharides was extended to sections of white onion and grape. While MSI of the sections indicates a homogeneous distribution of disaccharides throughout the plant sections, the MS/MS imaging indicates the uneven distribution of sucrose. This finding displays the potential for isomeric species to confound potentially biologically significant results. The innovation of this approach is accounting for this issue without requiring chemical modifications.

Several recent studies have aimed to improve the reliability and sensitivity of MALDI-MSI of saccharides, as well as integrate saccharide detection into developed workflows for broader metabolite analysis. The ionic liquid butylamine 2,5-dihydroxybenzoate (DHB-BuN) was optimized as a matrix to improve the reproducibility of disaccharide, trisaccharide, and tetrasaccharide analysis by MALDI MS, and applied to soybean and soybean leaf tissues for MALDI-MSI.[111] Multiple functional metabolites, including oligosaccharides, were detected in *Salvia miltiorrhiza Bge*, a medicinal herb, utilizing 1,5-diaminonaphthalene (DAN) and 1,1′-binaphthyl-2,2′diamine (BNDM) as MALDI matrices in positive and negative ion modes, respectively.[112] These matrices were selected for their low background interference and high sensitivity, even when analyzing a wide array of compounds. Additionally, the investigators found that an acetone washing of the root sections significantly improved the sensitivity of the workflow. Tetrasaccharides and pentasaccharides were found to be more abundant in the apex of the root of the plant rather than in the upper sections of the root, which is consistent with the known utilization of oligosaccharides in proliferating cells. Water-soluble gold nanoparticles were used as a matrix to image *Lepidium meyenii*, a herb found in Peru, fresh root tissue by Yang and colleagues.[113] The uniform size and minimal interference in the low molecular weight region make gold nanoparticles effective matrices for MALDI-MS studies. In the plant tissue, sucrose was found to have higher intensity and clearer spatial resolution when gold nanoparticles were used as a matrix instead of CHCA.

An emerging application of MALDI-MSI is the stacking of serial MS images into a 3D volume.[114] Fanuel and colleagues produced a 3D image of wheat grain cell walls utilizing this technique to identify structural variations of arabinoxylans and β-glucans, polysaccharides.[115] In this workflow, water-soluble fructo-oligosaccharides and starches are removed from sections of dried seeds by soaking the sections in distilled water and treating the sections with α-amylase, respectively. The large masses of these polysaccharides prohibit direct MALDI-MS analysis, so the investigators applied endoxylanase or endo-1,3(4)-β-D-glucanase to the section to degrade arabinoxylans and β-glucans, respectively. The enzymatic treatments were performed on two separate seeds but at the corresponding position of the seed along the *z*-axis. Using the 3D volume, formed from 30 2D images, the investigators found an uneven distribution of arabinoxylans and β-glucans throughout the seed that contribute to the physiochemical properties of the cell walls.

The polysaccharides in the cell wall of corn stalks were analyzed using a similar technique of enzymatic degradations.[116] Plant biorefineries are increasingly being developed as substitutes for carbon-based fossil fuel refineries. Corn stalks and other woody plants are a readily available non-food biomass that can be used in biorefineries, but the enzymatic conversion of corn stalks to fermentable sugars is physically inhibited by the amount of lignin, a heterogeneous and cross-linked phenyl-propane polymer, within the cell walls.[117] Arnaud and colleagues compared the lignin deposition to the rate and pattern of polysaccharide degradation of corn stalk cell walls at different stages of development.[116] By applying multiple enzymes to serial sections, the degradation of arabinoxylans, β-glucans, cellulose, and methylated glucuronoxylans was observed using MALDI-MSI. Regardless of the stage of development, all the classes of polysaccharides were preferentially degraded in tissue without lignin. While both arabinoxylans and methylated glucuronoxylans are heteroxylan products, they had distinct distributions in the center of the stem and more toward the periphery, respectively. MALDI-MSI is a particularly attractive technique for this area of study because it provides direct analysis of the local susceptibility of tissue to enzymatic degradation.

10.7 Conclusions and Future Directions

The structural complexity of the glycans described in this chapter has always presented unique challenges for mass spectrometry-based analyses. We attempted to highlight how MALDI-MSI applications to these different glycan classes offer particular advantages to other methods and provide immediate spatial and biological significance to the obtained results. The majority of the MALDI-MSI papers cited in this chapter are in the *N*-glycan and glycosphingolipid structural classes. They highlight two main methodological aspects central to successful MALDI-MSI analysis of these particular glycans – use of glycosidases for *N*-glycans and sample preparation/chemical matrix choices for glycosphingolipids. For *N*-glycans, the key component is the PNGase F enzyme that is needed to generate the glycan analyte from the tissue or spotted surface. The enzyme must be available in sufficient purity and amounts and have structural stability and robust enzymatic activity, such that it can be applied to slides using a solvent sprayer. Lack of any one of those parameters can be a hindrance to targeting a specific glycan structure. For example, a major glycan class in mammals is that of O-linked glycosylation whereby small glycan chains are linked to serine or threonine on protein carriers. This class of glycan was not discussed in the review, mainly due to lack of an enzyme with broad substrate properties like PNGase F to cleave the *O*-glycans from the protein. For MALDI-MSI of glycosphingolipids, biophysical parameters are important, *e.g.*, use of frozen tissues and how they are processed; choice of MALDI matrix used and ionization mode; sensitivity and resolution of the MALDI mass spectrometer.

Improvements in all aspects of these parameters continue to evolve. Notably, use of new generation MALDI-2 ion sources addresses the inherent poor ionization and relatively large mass of the glycosphingolipids. Advances in ion mobility mass spectrometry combined with MALDI-MSI capabilities will also greatly benefit all glycan structural classes. Lastly, there is a great opportunity to apply the different glycan MSI methods to non-mammalian organisms across the animal, plant, and protist kingdoms. The basic workflows and strategies described herein, combined with continued technological and instrumentation advancements, can readily be adapted to new glycan and glycoconjugate targets.

Acknowledgements

This work was partially supported by funding to RRD *via* grants from the National Institutes of Health (U01CA242096; U01 CA226052: U01 CA226158) and the State of South Carolina SmartState Endowed Chairs program.

References

1. A. Varki and S. Kornfeld, in *Essentials of Glycobiology*, ed. A. Varki, R. D. Cummings and J. D. Esko, *et al.*, Cold Spring Harbor Laboratory Press, Cold Spring Harbor, 2017, vol. 3, p. 1.
2. N. de Haan, M. Wuhrer and L. R. Ruhaak, *Clin. Mass Spectrom.*, 2020, **18**, 1.
3. L. R. Ruhaak, G. Xu, Q. Li, E. Goonatilleke and C. B. Lebrilla, *Chem. Rev.*, 2018, **118**, 7886.
4. R. R. Drake, C. A. West, A. S. Mehta and P. M. Angel, in *Glycobiophysics*, ed. Y. Yamaguchi and K. Kato, Springer, Singapore, 2018, p. 459.
5. T. Nishikaze, *Mass Spectrom.*, 2017, **6**, A0060.
6. E. Sugiyama and M. Setou, in *Gangliosides: Methods and Protocols*, ed. S. Sonnino and A. Prinetti, Humana Press, New York, 2018, vol. 10, p. 223.

7. E. E. Jones, S. Dworski, M. Kamani, D. Canals, M. Wada, L. Xiang, Y. A. Hannun, J. Norris, J. A. Medin and R. R. Drake, in *Bioactive Sphingolipids in Cancer Biology and Therapy*, ed. Y. A. Hannun, C. Luberto, C. Mao and L. Obeid, Springer, Switzerland, 2015, vol. 13, p. 339.
8. M. J. Kailemia, L. R. Ruhaak, C. B. Lebrilla and I. J. Amster, *Anal. Chem.*, 2014, **86**, 196.
9. Y. Dong, B. Li, S. Malitsky, I. Rogachev, A. Aharoni, F. Kaftan, A. Svatoš and P. Franceschi, *Front. Plant Sci.*, 2016, **7**, 60.
10. A. Bouslimani, L. M. Sanchez, N. Garg and P. C. Dorrestein, *Nat. Prod. Rep.*, 2014, **31**, 718.
11. J. E. Spraker, G. T. Luu and L. M. Sanchez, *Nat. Prod. Rep.*, 2020, **37**, 150.
12. F. Kubaski, H. Osago, R. W. Mason, S. Yamaguchi, H. Kobayashi, M. Tsuchiya, T. Orii and S. Tomatsu, *Mol. Genet. Metab.*, 2017, **120**, 67.
13. L. E. Pepi, P. Sanderson, M. Stickney and I. J. Amster, *Mol. Cell. Proteomics*, 2021, **20**, 100025.
14. P. Stanley and N. Taniguchi, in *Essentials of Glycobiology*, ed. A. Varki, R. D. Cummings and J. D. Esko, *et al.*, Cold Spring Harbor Laboratory Press, Cold Spring Harbor, 2017, vol. 3, p. 9.
15. T. W. Powers, E. E. Jones, L. R. Betesh, P. R. Romano, P. Gao, J. A. Copland, A. S. Mehta and R. R. Drake, *Anal. Chem.*, 2013, **85**, 9799.
16. T. W. Powers, B. A. Neely, Y. Shao, H. Tang, D. A. Troyer, A. S. Mehta, B. B. Haab and R. R. Drake, *PLoS One*, 2014, **9**, 1.
17. R. R. Drake, T. W. Powers, K. Norris-Caneda, A. S. Mehta and P. M. Angel, *Curr. Protoc. Protein Sci.*, 2018, **94**, 1.
18. M. T. Briggs, Y. Y. Ho, G. Kaur, M. K. Oehler, A. V. Everest-Dass, N. H. Packer and P. Hoffmann, *Rapid Commun. Mass Spectrom.*, 2017, **31**, 825.
19. S. Holst, B. Heijs, N. De Haan, R. J. M. Van Zeijl, I. H. Briaire-De Bruijn, G. W. Van Pelt, A. S. Mehta, P. M. Angel, W. E. Mesker, R. A. Tollenaar, R. R. Drake, J. V. M. G. Bovée, L. A. McDonnell and M. Wuhrer, *Anal. Chem.*, 2016, **88**, 5904.
20. C. A. West, H. Liang, R. R. Drake and A. S. Mehta, *J. Proteome Res.*, 2020, **19**, 2989.
21. E. S. Rivera, K. V. Djambazova, E. K. Neumann, R. M. Caprioli and J. M. Spraggins, *J. Mass Spectrom.*, 2020, **55**, e4614.
22. H. Zhang, X. Shi, N. Q. Vu, G. Li, Z. Li, Y. Shi, M. Li, B. Wang, N. V. Welham, M. S. Patankar, P. Weisman and L. Li, *Anal. Chem.*, 2020, **92**, 13361.
23. J. Soltwisch, H. Kettling, S. Vens-Cappell, M. Wiegelmann, J. Müthing and K. Dreisewerd, *Science*, 2015, **348**, 211.
24. B. Heijs, A. Potthoff, J. Soltwisch and K. Dreisewerd, *Anal. Chem.*, 2020, **92**, 13904.
25. J. Jiao, Y. Zhang, P. Yang and H. Lu, *Analyst*, 2015, **140**, 156.
26. I. Kim, S. Kim, D. Shin and J. Kim, *Bull. Korean Chem. Soc.*, 2016, **37**, 105.

27. Y. Shi, Z. Li, M. A. Felder, Q. Yu, X. Shi, Y. Peng, Q. Cao, B. Wang, L. Puglielli, M. S. Patankar and L. Li, *Anal. Chem.*, 2019, **91**, 12942.
28. M. R. Condina, P. Mittal, M. T. Briggs, M. K. Oehler, M. Klingler-Hoffmann and P. Hoffmann, *Anal. Chem.*, 2019, **91**, 14846.
29. Y. R. Lee, M. T. Briggs, J. S. Kuliwaba, P. H. Anderson, M. R. Condina and P. Hoffmann, *Anal. Bioanal. Chem.*, 2020, **413**, 1.
30. C. L. Clift, R. R. Drake, A. S. Mehta and P. M. Angel, *Anal. Bioanal. Chem.*, 2021, **413**, 2709.
31. Z. Chen, X. Zhong, C. Tie, B. Chen, X. Zhang and L. Li, *Anal. Bioanal. Chem.*, 2017, **409**, 4437.
32. P. M. Angel, J. Saunders, C. L. Clift, S. White-Gilbertson, C. Voelkel-Johnson, E. Yeh, A. Mehta and R. R. Drake, *J. Proteome Res.*, 2019, **18**, 3630.
33. A. P. Black, H. Liang, C. A. West, M. Wang, H. P. Herrera, B. B. Haab, P. M. Angel, R. R. Drake and A. S. Mehta, *Anal. Chem.*, 2019, **91**, 8429.
34. A. P. Black, P. M. Angel, R. R. Drake and A. S. Mehta, *Curr. Protoc. Protein Sci.*, 2019, **98**, e99.
35. C. R. K. Blaschke, A. Black, A. S. Mehta, P. M. Angel and R. R. Drake, *J. Am. Soc. Mass Spectrom*, 2020, **31**, 2511.
36. A. Varki and P. Gagneux, in *Essentials of Glycobiology*, ed. A. Varki, R. D. Cummings and J. D. Esko, *et al.*, Cold Spring Harbor Laboratory Press, Cold Spring Harbor, 2017, vol. 3, p. 7.
37. H. H. Freeze, *Nat. Rev. Genet.*, 2006, **7**, 537.
38. D. L. Meany and D. W. Chan, *Clin. Proteomics*, 2011, **8**, 1.
39. R. R. Drake, C. McDowell, C. West, F. David, T. W. Powers, T. Nowling, E. Bruner, A. S. Mehta, P. M. Angel, L. A. Marlow, H. W. Tun and J. A. Copland, *J. Mass Spectrom.*, 2020, **55**, e4490.
40. O. J. R. Gustafsson, M. T. Briggs, M. R. Condina, L. J. Winderbaum, M. Pelzing, S. R. McColl, A. V. Everest-Dass, N. H. Packer and P. Hoffmann, *Anal. Bioanal. Chem.*, 2015, **407**, 2127.
41. B. Heijs, S. Holst-Bernal, M. A. de Graaff, I. H. Briaire-de Bruijn, M. Rodriguez-Girondo, M. A. J. van de Sande, M. Wuhrer, L. A. McDonnell and J. V. M. G. Bovée, *Lab. Investig.*, 2020, **100**, 1252.
42. M. T. Briggs, M. R. Condina, Y. Y. Ho, A. V. Everest-Dass, P. Mittal, G. Kaur, M. K. Oehler, N. H. Packer and P. Hoffmann, *Proteomics*, 2019, **19**, 1800482.
43. A. V. Everest-Dass, M. T. Briggs, G. Kaur, M. K. Oehler, P. Hoffmann and N. H. Packer, *Mol. Cell. Proteomics*, 2016, **15**, 3003.
44. C. T. McDowell, Z. Klamer, J. Hall, C. A. West, L. Wisniewski, T. W. Powers, P. M. Angel, A. S. Mehta, B. B. Haab, R. R. Drake, D. N. Lewin, B. B. Haab and R. R. Drake, *Mol. Cell. Proteomics*, 2021, **20**, 100012.
45. D. A. Scott, K. Norris-Caneda, L. Spruill, E. Bruner, Y. Kono, P. M. Angel, A. S. Mehta and R. R. Drake, *Int. J. Mass Spectrom.*, 2019, **437**, 69.

46. P. Mittal, M. Briggs, M. Klingler-Hoffmann, G. Kaur, N. H. Packer, M. K. Oehler and P. Hoffmann, *Anal. Bioanal. Chem.*, 2020, **413**, 1.
47. H. Herrera, T. Dilday, A. Uber, D. Scott, J. N. Zambrano, M. Wang, P. M. Angel, A. S. Mehta, R. R. Drake, E. G. Hill and E. S. Yeh, *Int. J. Mol. Sci.*, 2019, **20**, 2528.
48. D. A. Scott, R. Casadonte, B. Cardinali, L. Spruill, A. S. Mehta, F. Carli, N. Simone, M. Kriegsmann, L. Del Mastro, J. Kriegsmann and R. R. Drake, *PROTEOMICS – Clin. Appl.*, 2019, **13**, 1800014.
49. C. L. Carter, G. A. Parker, K. G. Hankey, A. M. Farese, T. J. MacVittie and M. A. Kane, *Sci. Rep.*, 2020, **10**, 1.
50. R. L. Schnaar and T. Kinoshita, in *Essentials of Glycobiology*, ed. A. Varki, R. D. Cummings and J. D. Esko, *et al.*, Cold Spring Harbor Laboratory Press, Cold Spring Harbor, 2017, vol. 3, p. 11.
51. W. J. Perry, N. H. Patterson, B. M. Prentice, E. K. Neumann, R. M. Caprioli and J. M. Spraggins, *J. Mass Spectrom.*, 2020, **55**, e4491.
52. S. Caughlin, D. H. Park, K. K. C. Yeung, D. F. Cechetto and S. N. Whitehead, *J. Vis. Exp.*, 2017, **121**, e55254.
53. H. Ibrahim, K. Jurcic, J. S. H. Wang, S. N. Whitehead and K. K. C. Yeung, *Anal. Chem.*, 2017, **89**, 12828.
54. S. N. Jackson, L. Muller, A. Roux, B. Oktem, E. Moskovets, V. M. Doroshenko and A. S. Woods, *J. Am. Soc. Mass Spectrom.*, 2018, **29**, 1463.
55. B. Li, R. Sun, A. Gordon, J. Ge, Y. Zhang, P. Li and H. Yang, *Anal. Chem.*, 2019, **91**, 8221.
56. L. Muller, K. Baldwin, D. C. Barbacci, S. N. Jackson, A. Roux, C. D. Balaban, B. E. Brinson, M. I. McCully, E. K. Lewis, J. A. Schultz and A. S. Woods, *J. Am. Soc. Mass Spectrom.*, 2017, **28**, 1716.
57. M. Dufresne, N. H. Patterson, J. L. Norris and R. M. Caprioli, *Anal. Chem.*, 2019, **91**, 12928.
58. I. U. Kouzel, J. Soltwisch, G. Pohlentz, J. S. Schmitz, H. Karch, K. Dreisewerd and J. Müthing, *Int. J. Mass Spectrom.*, 2017, **416**, 53.
59. E. Yang, M. Dufresne and P. Chaurand, *Int. J. Mass Spectrom.*, 2019, **437**, 3.
60. A. Harris, A. Roseborough, R. Mor, K. K. C. Yeung and S. N. Whitehead, *J. Am. Soc. Mass Spectrom.*, 2020, **31**, 479.
61. T. Bien, M. Perl, A. C. MacHmüller, U. Nitsche, A. Conrad, L. Johannes, J. Müthing, J. Soltwisch, K. P. Janssen and K. Dreisewerd, *Anal. Chem.*, 2020, **92**, 7096.
62. M. Niehaus, J. Soltwisch, M. E. Belov and K. Dreisewerd, *Nat. Methods*, 2019, **16**, 925.
63. A. Bednařík, S. Bölsker, J. Soltwisch and K. Dreisewerd, *Angew. Chem.*, 2018, **130**, 12268.
64. G. Dawson, *Biochim. Biophys. Acta*, 2015, **1851**, 1026.
65. S. Caughlin, S. Maheshwari, N. Weishaupt, K. K. C. Yeung, D. F. Cechetto and S. N. Whitehead, *Sci. Rep.*, 2017, **7**, 1.

66. J. Zhang, X. Fang, Y. Zhou, X. Deng, Y. Lu, J. Li, S. Li, B. Wang and R. Xu, *J. Stroke Cerebrovasc. Dis.*, 2015, **24**, 1471.
67. H. Sohn, Y. S. Kim, H. T. Kim, C. H. Kim, E. W. Cho, H. Y. Kang, N. S. Kim, C. H. Kim, S. E. Ryu, J. H. Lee, J. H. Ko, H. Sohn, Y. S. Kim, H. T. Kim, C. H. Kim, E. W. Cho, H. Y. Kang, N. S. Kim, C. H. Kim, S. E. Ryu, J. H. Lee and J. H. Ko, *FASEB J.*, 2006, **20**, 1248.
68. S. Caughlin, J. Hepburn, Q. Liu, L. Wang, K. K. C. Yeung, D. F. Cechetto and S. N. Whitehead, *Mol. Neurobiol.*, 2019, **56**, 3552.
69. E. J. Parkinson-Lawrence, T. Shandala, M. Prodoehl, R. Plew, G. N. Borlace and D. A. Brooks, *Physiology*, 2010, **25**, 102.
70. E. E. Jones, W. Zhang, X. Zhao, C. Quiason, S. Dale, S. Shahidi-Latham, G. A. Grabowski, K. D. R. Setchell, R. R. Drake and Y. Sun, *SLAS Discovery*, 2017, **22**, 1218.
71. F. Tobias, K. C. Pathmasiri and S. M. Cologna, *Anal. Bioanal. Chem.*, 2019, **411**, 5659.
72. M. Hunter, N. J. Demarais, R. L. M. Faull, A. C. Grey and M. A. Curtis, *J. Neurochem.*, 2018, **146**, 613.
73. K. Blennow, M. J. de Leon and H. Zetterberg, *Lancet*, 2006, **368**, 387.
74. G. Di Paolo and T. W. Kim, *Nat. Rev. Neurosci.*, 2011, **12**, 284.
75. I. Kaya, D. Brinet, W. Michno, S. Syvänen, D. Sehlin, H. Zetterberg, K. Blennow and J. Hanrieder, *ACS Chem. Neurosci.*, 2017, **8**, 347.
76. I. Kaya, D. Brinet, W. Michno, M. Başkurt, H. Zetterberg, K. Blenow and J. Hanrieder, *ACS Chem. Neurosci.*, 2017, **8**, 2778.
77. I. Kaya, H. Zetterberg, K. Blennow and J. Hanrieder, *ACS Chem. Neurosci.*, 2018, **9**, 1802.
78. W. Michno, I. Kaya, S. Nyström, L. Guerard, K. P. R. Nilsson, P. Hammarström, K. Blennow, H. Zetterberg and J. Hanrieder, *Anal. Chem.*, 2018, **90**, 8130.
79. W. Michno, P. M. Wehrli, H. Zetterberg, K. Blennow and J. Hanrieder, *Biochim. Biophys. Acta, Proteins Proteomics*, 2019, **1867**, 458.
80. S. Caughlin, S. Maheshwari, Y. Agca, C. Agca, A. J. Harris, K. Jurcic, K. K. C. Yeung, D. F. Cechetto and S. N. Whitehead, *Biochim. Biophys. Acta, Gen. Subj.*, 2018, **1862**, 1327.
81. H. Oakley, S. L. Cole, S. Logan, E. Maus, P. Shao, J. Craft, A. Guillozet-Bongaarts, M. Ohno, J. Disterhoft, L. Van Eldik, R. Berry and R. Vassar, *J. Neurosci.*, 2006, **26**, 10129.
82. I. Kaya, E. Jennische, J. Dunevall, S. Lange, A. G. Ewing, P. Malmberg, A. T. Baykal and J. S. Fletcher, *ACS Chem. Neurosci.*, 2020, **11**, 14.
83. I. Kaya, E. Jennische, S. Lange, A. Tarik Baykal, P. Malmberg and J. S. Fletcher, *J. Neurochem.*, 2020, **154**, 84.
84. B. M. Prentice, N. J. Hart, N. Phillips, R. Haliyur, A. Judd, R. Armandala, J. M. Spraggins, C. L. Lowe, K. L. Boyd, R. W. Stein, C. V. Wright, J. L. Norris, A. C. Powers, M. Brissova and R. M. Caprioli, *Diabetologia*, 2019, **62**, 1036.

85. D. T. Stark, D. M. G. Anderson, J. M. K. Kwong, N. H. Patterson, K. L. Schey, R. M. Caprioli and J. Caprioli, *Invest. Ophthalmol. Visual Sci.*, 2018, **59**, 212.
86. U. Lindahl, J. Couchman, K. Kimata and J. D. Esko, in *Essentials of Glycobiology*, ed. A. Varki, R. D. Cummings and J. D. Esko, *et al.*, Cold Spring Harbor Laboratory Press, Cold Spring Harbor, 2017, vol. 3, p. 17.
87. L. Schaefer and R. M. Schaefer, *Cell Tissue Res.*, 2010, **339**, 237.
88. S. Guo, C. Xue, G. Li, X. Zhao, Y. Wang and J. Xu, *J. Carbohydr. Chem.*, 2015, **34**, 55.
89. R. R. Zhao, M. Ackers-Johnson, J. Stenzig, C. Chen, T. Ding, Y. Zhou, P. Wang, S. L. Ng, P. Y. Li, G. Teo, P. M. Rudd, J. W. Fawcett and R. S. Y. Foo, *Circulation*, 2018, **137**, 2497.
90. G. Westergren-Thorsson, U. Hedstrom, A. Nybom, E. Tykesson, M. Hornfelt, T. H. van Kuppevelt, G. Dellgren, L. Bjermer, O. Hallgren, X. H. Zhou, E. Åhrman, M. Maccarana, M. Wildt and L. Eriksson, *Int. J. Biochem. Cell Biol.*, 2017, **83**, 27.
91. M. Aichler, T. Kunzke, A. Buck, N. Sun, M. Ackermann, D. Jonigk, A. Gaumann and A. Walch, *Lab. Invest.*, 2018, **98**, 141.
92. R. Raghunathan, M. K. Sethi and J. Zaia, *MethodsX*, 2019, **6**, 2329.
93. A. Nimptsch, S. Schibur, M. Schnabelrauch, B. Fuchs, D. Huster and J. Schiller, *Anal. Chim. Acta*, 2009, **635**, 175.
94. P. Juhasz and K. Biemann, *Carbohydr. Res.*, 1995, **270**, 131.
95. Z. Shriver, R. Raman, G. Venkataraman, K. Drummond, J. Turnbull, T. Toida, R. Linhardt, K. Biemann and R. Sasisekharan, *Proc. Natl. Acad. Sci. U. S. A.*, 2000, **97**, 10359.
96. T. N. Laremore, F. Zhang and R. J. Linhardt, *Anal. Chem.*, 2007, **79**, 1604.
97. B. A. Graf, P. E. Milbury and J. B. Blumberg, *J. Med. Food*, 2005, **8**, 281.
98. G. Agati and M. Tattini, *New Phytol.*, 2010, **186**, 786.
99. L. P. Taylor and E. Grotewold, *Curr. Opin. Plant Biol.*, 2005, **8**, 317.
100. J. Kleijnen and P. Knipschild, *Lancet*, 1992, **340**, 1136.
101. S. Beck and J. Stengel, *Phytochemistry*, 2016, **130**, 201.
102. H. Enomoto, *Molecules*, 2020, **25**, 4600.
103. C. Sun, M. Zhang, H. Dong, W. Liu, L. Guo and X. Wang, *J. Pharm. Biomed. Anal.*, 2020, **179**, 113014.
104. R. M. Gleadow and B. L. Møller, *Annu. Rev. Plant Biol.*, 2014, **65**, 155.
105. J. Vetter, *Toxicon*, 2000, **38**, 11.
106. E. Ritmejerytė, B. A. Boughton, M. J. Bayly and R. E. Miller, *Ann. Bot.*, 2020, **126**, 387.
107. C. C. Hansen, M. Sørensen, T. A. M. Veiga, J. F. S. Zibrandtsen, A. M. Heskes, C. E. Olsen, B. A. Boughton, B. L. Møller and E. H. J. Neilson, *Plant Physiol.*, 2018, **178**, 1081.
108. S. Smeekens and H. A. Hellmann, *Front. Plant Sci.*, 2014, **5**, 113.
109. N. Shibuya and E. Minami, *Physiol. Mol. Plant Pathol.*, 2001, **59**, 223.
110. L. Zhan, X. Huang, J. Xue, H. Liu, C. Xiong, J. Wang and Z. Nie, *Food Chem.*, 2021, **338**, 127984.

111. X. Xu, P. Xing-Li, H. Yu-Yu and G. Can, *Chinese J. Anal. Chem.*, 2017, **45**, 1155.
112. C. Sun, W. Liu, S. Ma, M. Zhang, Y. Geng and X. Wang, *J. Chromatogr. A*, 2020, **1614**, 460704.
113. S. Yang, L. Zhan, C. Liu, L. Fu, R. Chen and Z. Nie, *Microchem. J.*, 2019, **150**, 104190.
114. E. H. Seeley and R. M. Caprioli, *Anal. Chem.*, 2012, **84**, 2105.
115. M. Fanuel, D. Ropartz, F. Guillon, L. Saulnier and H. Rogniaux, *Planta*, 2018, **248**, 1505.
116. B. Arnaud, S. Durand, M. Fanuel, F. Guillon, V. Méchin and H. Rogniaux, *J. Agric. Food Chem.*, 2020, **68**, 4042.
117. C. Huang, J. He, D. Min, C. Lai and Q. Yong, *Appl. Biochem. Biotechnol.*, 2016, **180**, 1508.

CHAPTER 11

MALDI Mass Spectrometry Imaging and Spatially-resolved Proteomics

EDUARDO CHICANO-GALVEZ,*[a] MICHEL SALZET[b,c] AND ISABELLE FOURNIER*[b,c]

[a] IMIBIC Mass Spectrometry and Molecular Imaging Unit, Maimonides Biomedical Research Institute of Cordoba (IMIBIC), Reina Sofia University Hospital, University of Cordoba (UCO), Córdoba, Spain; [b] Univ. Lille, Inserm, CHU Lille, U1192 - Protéomique Réponse Inflammatoire Spectrométrie de Masse - PRISM, F-59000 Lille, France; [c] Institut Universitaire de France (IUF), Paris, France
*Emails: eduardo.chicano@imibic.org; Isabelle.fournier@univ-lille.fr

11.1 From Protein MS Imaging to Spatially-resolved Proteomics: Combining Deep Proteome Identification with Spatial Information

Proteins are major players of biological regulation and participate in the phenotypes expressed by the organisms. Identifying the proteins, their regulation and spatial distribution is key to the understanding of physiological and pathophysiological pathways. Historically, proteomics was first defined in 1995 as the analysis of proteins present in a cell, tissue or organism under a defined condition.[1,2] Because of the highly dynamic and wide nature of the proteome (spanning over 7 orders of magnitude in terms of protein abundances or even presence of proteoforms), the development of proteomics represented a huge

New Developments in Mass Spectrometry No. 12
MALDI Mass Spectrometry Imaging: From Fundamentals to Spatial Omics
Edited by Tiffany Siegel

Published by the Royal Society of Chemistry, www.rsc.org

challenge to be overcome. It took two decades of technological development to allow the proteomic field to catch up with genomics. Initially, the field started with 2D-electrophoresis separation followed by spot-by-spot analysis in mass spectrometry (MS). Nowadays, it is largely based on LC–MS/MS due to the improvements made in hyphenated methods over the last decade. Many laboratories have now declared themselves as "gel-free" as the MS equipment and current sample preparation workflows are greatly simplified to avoid the use of gel technology which is time consuming and less reproducible.

In brief, all proteomic workflows are based on three main steps, namely: 1) separation of the complex protein/peptide mixture; 2) MS analysis, and 3) subsequent bioinformatics analysis of the acquired data.[3] One of the main challenges is the identification of proteins and their proteoforms; therefore, two main approaches (*i.e.* the bottom-up and top-down approaches) have been developed. The bottom-up approach, also referred to as the "shotgun" approach, is based on LC–MS/MS and is known as the "standard" massive proteomics analysis. Here, a peptide dilution from digested samples is analyzed by LC–MS/MS. Different MS instruments and approaches are employed ranging from triple quadrupole (QqQ) instruments for targeted approaches using the so-called single reaction monitoring (SRM) or multiple reaction monitoring (MRM); to hybrid instruments (*e.g.* Q-TOF, Q-Orbitrap) for non-targeted approaches using either data-dependent acquisition (DDA) or data-independent acquisition (DIA) modes. These are completed by a full bioinformatic analysis to detect new biomarkers, characterize post-translational modifications (PTMs) or screen diagnostic and prognostic markers. The bottom-up liquid-based proteomics has opened up a wide range of applications, such as oncoproteomics,[4] phosphoproteomics,[5,6] proteogenomics,[7,8] and metaproteomics.[9]

The second LC–MS/MS-based approach, the top-down proteomics, aims at characterizing intact proteins from samples using ideally Fourier Transform (FT)-MS as high spectral resolution and mass accuracy are needed for such high molecular weights (MW). All these approaches make use of electrospray ionization (ESI) to generate multiply charged ions, for on-line coupling to MS. The addition of ion mobility helped dig deeper into proteome identification from <10 ng sample amounts.[10,11]

It is noteworthy that the "conventional" proteomic workflow does not provide spatial information. Therefore, MS imaging (MSI) was developed to generate spatial information from biological samples. The first MS-based imaging experiments were carried out almost 60 years ago using secondary ion mass spectrometry (SIMS),[12] however, SIMS is not suited for large molecule analysis such as proteins due to extensive fragmentation. Richard Caprioli's group introduced the concept of MSI applied to proteins in the late 1990s with the introduction of matrix-assisted laser desorption/ionization (MALDI)-MSI.[13,14] This started a "race" in the protein MSI (pMSI, protein Mass Spectrometry Imaging) community to optimize and improve the analytical capabilities of the technique. Proteins are known to be more challenging to analyze by MS due to their higher MW (*i.e.* compared to low MS compounds such as lipids or drugs). Therefore, two strategies have naturally emerged in pMSI: one based on the analysis of intact proteins (*i.e.* top-down pMSI) and the second based on the analysis of digested peptides resulting from on-tissue digestion of proteins (*i.e.* bottom-up pMSI). Moreover, the need for the identification of the imaged proteins then promoted the development of spatial pMSI or spatially-resolved proteomics.[15] Since the initial attempts of pMSI, major advances have been made in terms of sample preparation workflows, which contributed to widespread integration by the community. These improvements were made toward the sample preparation protocols, including matrix application, or the MS analysis itself.[16] The pMSI workflow is strongly dependent on the preservation of the tissues. Indeed, pMSI can be applied to fresh-frozen, formalin-fixed (FF) or formalin-fixed and paraffin-embedded (FFPE)[17,18] samples, each of them with different approaches.[19] For fixed samples, on-tissue digestion, resulting in the analysis of generated digested peptides, remains the best option, while with fresh-frozen samples, proteins can be imaged either intact or after digestion. In all cases, the main steps of a pMSI experiment could be depicted as follows: 1) sectioning of the samples at the desired thickness (*i.e.* frozen, FF or FFPE); 2) section conditioning depending on the analysis to be carried out (*i.e.* intact protein or endogenous/tryptic peptides); 3) matrix application; 4) acquisition by MSI; and finally: 5) multivariate statistical analysis. Beyond imaging is the protein identification. Over the years, pMSI has expanded its application from basic research in plants[20,21] and animal tissues[22] to more applied fields like food,[23,24] or biomedical and clinical research.[25,26] Clinical pMSI is mainly focused on oncology and relies on the creation of molecular classification systems for diagnostic, prognostic, or patient stratification purposes in order to help make therapeutic decisions.[13,27–33] These classification systems greatly contribute to developing and integrating pMSI toward the next generation of molecular pathology.[34] One of the benefits of pMSI is the addition of molecular profiling to morphological and histological information, through co-registration of MSI data to histological staining with hematoxylin and eosin (H&E) with dedicated software. This is only possible because pMSI usually preserves spatial morphology.[35] Moreover, pMSI can be integrated to other -omics MSI analysis such as for lipids, metabolites, and/or elemental imaging to maximize the information collected from a sample through a multi-MSI integrative approach.[27–29,34] The ultimate goal of pMSI is to allow both the localization and the identification of the proteins over the analyzed tissue. Because protein identification from tissue has demonstrated to be a complex task, several strategies have developed which show their respective advantages and disadvantages,[32] namely: tissue homogenizations, *in-situ* enzymatic digestions, chemical derivatizations, laser capture microdissections, hydrogel protein extractions, or liquid extraction surface analysis. Figure 11.1 provides an overview of the MALDI-MSI workflow for pMSI and protein identification.

In this chapter, we will address all these aspects of pMSI ranging from MALDI-MSI to spatially-resolved proteomics and the respective sample preparation.

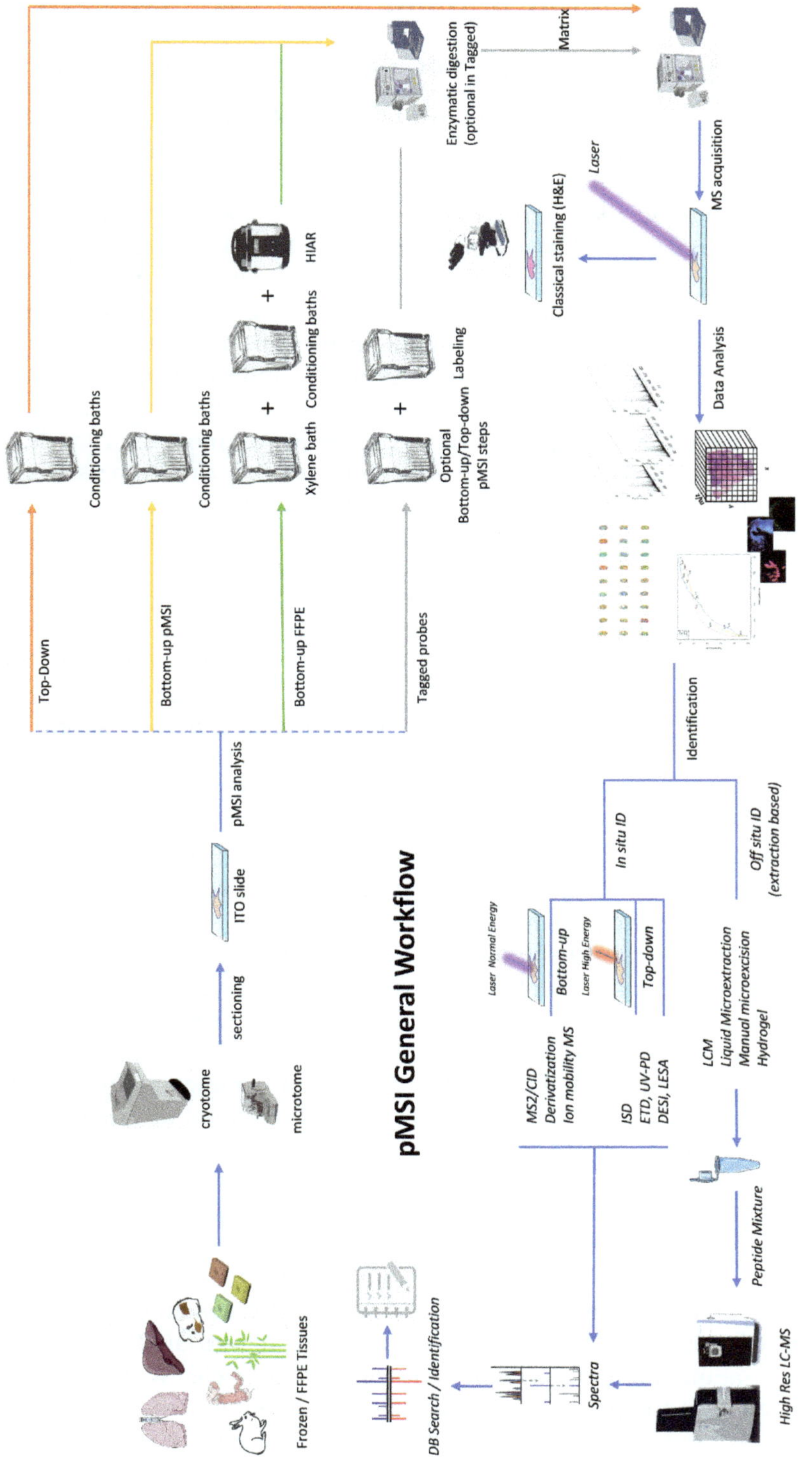

Figure 11.1 General workflow for pMSI from sample to generated results, including different strategies for the identification of proteins.

11.2 MALDI-MSI of Proteins

11.2.1 MALDI-MSI of Intact Proteins or Top-down pMSI

The main objective of intact pMSI is to obtain localization of low- or medium-size proteins for the rapid search of clinically relevant biomarkers.[31,36,37] This approach was initially demonstrated by Stoeckli *et al.*[13] in their work in 2001, where they showed the distribution of β-actin, S100A4, and Thymosin β4 proteins in human glioblastoma tissue sections. However, pMSI is by far the most challenging modality, in comparison to MSI analysis of smaller molecules (*e.g.* lipids). This is due to their physico-chemical properties (hydrophobicity and MW), their level of abundance in the cells, and the lack of sensitivity of the current detectors for higher MW. Indeed, protein MW range from 10 kDa to several hundreds of kDa and their abundances from 1 to 10 000 000 per cell with a dynamic range spanning over more than 7 orders of magnitude. This makes them more difficult to detect *in situ* from tissues in a highly complex environment without extraction and separation. Moreover, there are also some specificities with regards to the tissue preparation for proteins. Indeed, it was shown that several parameters of the MALDI process (*i.e.* initial ion velocities, effective temperature, internal energy) and most importantly ion yields are governed by the sample preparation, and the right parameters are trickier to adjust for proteins,[38,39] as opposed to other molecule classes (*i.e.* lipids or carbohydrates). The right compromise is to achieve sufficient extraction during the matrix deposition phase while avoiding their delocalization. This explains why the spatial resolution of images has always been a challenge with pMSI. Automatization of systems is a prerequisite for fast and robust matrix deposition. In the early days, matrix application was mainly based on spraying or micro-spotting of the matrix. Spraying devices definitively lead to higher spatial resolution as less solvent is used, and they provide a more homogeneous coverage of the tissue section with thinner matrix crystals. On the other hand, micro-spotting[40,41] devices were shown to provide better signals for protein detection than the dryer spraying systems; sublimation methods don't give good results for proteins unless a re-crystallization step is performed.[42] Modern spraying devices offer more control and tuning possibilities of the deposition parameters for pMSI like spray temperature, flow, and height (z-axis), although not all tissue preparations are suitable with these devices.

However, although different MALDI matrices were tested and optimized, intact pMSI gives a routine molecular mass coverage up to about 25–30 kDa from fresh-frozen samples[31] and closer to 50 kDa in specific-sample adapted protocols.[43] The main limitations of intact pMSI are due to the related problems of small amounts of analyte and detector sensitivity. The development of tissue washing procedures has largely contributed to advancing pMSI. The abundant lipids present in tissues have been clearly shown to limit the detection of the proteins by creating ion suppression, and this effect is even more pronounced if tissues are stored for a longer period of time.

Therefore, washing tissues prior to the matrix application with an adequate mixture of organic solvents[44] leads to great improvement in protein detection while maintaining their spatial localization as demonstrated by IHC.[44,45]

It is noteworthy that it is not possible to perform intact pMSI from FFPE samples. Indeed, formalin-fixation induces cross-linking of proteins by the formation of methylene bridges within the tissues;[46,47] and removing the methylene bridges is not trivial. However, FFPE samples are highly stable, and blocks can be stored at room temperature for a long period of time (*i.e.* over decades) even at room temperature (RT). FFPE offers high morphological preservation after sectioning, which renders FFPE the standard mode of preservation of precious patient samples in clinics. A protocol, based on an MSI-compatible nucleophile hydroxylamine hydrochloride (HA) reagent, was recently proposed to reverse cross-linking and perform intact protein imaging from FFPE, but further improvements are still needed for robust pMSI.[47] Alternatively, PAXgene fixation can be used,[48] and MSI of intact proteins became possible from the PFPE tissues,[49] though this fixation is not a standard preparation in routine clinical practice.

Finally, different groups have worked on improving intact protein mass coverage.[16,50–53] Using specific MALDI matrices in combination with more hydrophobic solvents or detergents showed an improvement in higher mass detection.[43,54] This is sometimes performed in combination with specific detectors which exhibit higher sensitivity for the detection of higher *m*/*z*. However, because of the very nature of the solvents, it is noteworthy that these protocols are more difficult to use for MSI as they are more suitable for the spotting method used for matrix deposition, and therefore do not provide the same spatial resolution.

Because of the typical feature of a MALDI source to provide a limited number of multiply charged ions, pMSI is limited to a certain type of instrumentation and more generally is performed using MALDI-TOF technology. This is limiting for experiments because for higher *m*/*z* the linear mode is used, and the spectral resolution, together with mass accuracy and the lack of sensitivity of the detectors penalizes the overall performance. Therefore, even though very attractive, a rather limited number of studies are using the top-down pMSI approach and improving this side of the field of pMSI not only in sample preparation but also in instrumentation.[55–57]

11.2.1.1 Sample Preparation Protocols for Intact Proteins pMSI

Sample preparation protocols for analysis of intact protein and its possible variants *via* pMSI have been available for a long time[58,59] but remain somewhat critical (and MS-Imaging in general). Therefore, the protocol should be carefully established in terms of the research question, the type of tissue (*i.e.* nature and fat content) and the targeted class of the molecules for analysis. The most important steps are the sectioning, tissue washing, and the matrix choice and its deposition method.

11.2.1.1.1 Tissue Sectioning. In the case of frozen samples, the sections to be obtained should have a thickness between 8 to 12 μm depending on the tissue and the difficulty in obtaining a sample without physical artifacts (folds, breaks, *etc.*).

11.2.1.1.2 Tissue Washing. The following combination is one of the most popular: ethanol/isopropanol in different proportions, followed by a step in a Carnoy solution (ethanol/chloroform/acetic acid), and an optional aqueous step to remove possible contaminants that might suppress the MS signal. Alternatively, methanol/chloroform also gives good results[45] and the washing needs to be optimized when starting with a different tissue.

A possible common protocol starting with the optimization could be as follows.[60]

After mounting the cryo-section (*i.e.* 10–12 μm-thickness) on the indium tin oxide (ITO) slide, the sample is placed in vacuum drying for 30 minutes. Then the following washing baths are recommended:

1. 70% fresh ethanol for 30 seconds
2. 100% ethanol for 30 seconds
3. Carnoy dehydrator (ethanol/chloroform/acetic acid: 6 : 3 : 1 v/v) for 2 min
4. 100% ethanol for 30 seconds
5. TFA 0.2% for 30 seconds
6. 100% ethanol for 30 seconds

Once the washing has been carried out, the slide is again placed in vacuum drying for 30 minutes and the sample is ready for the deposition of the matrix. At this point, the ITO slides can be stored at −80 °C if you have time limitations to carry out your analysis but taking into account that you have a limited time to analyze them.[61]

11.2.1.1.3 Choice of the MALDI Matrix. Sinapinic acid (SA) is the most commonly used matrix for imaging intact protein, although others have also demonstrated satisfying results, such as dihydroxyacetophenone (DHA),[32,59] dihydroxybenzoic acid (DHB), or even ferulic acid (FA) for higher mass proteins.[50] Solid ionic matrices based on SA also provide good protein signal intensity. This combination of SA to aniline (ANI) or *N,N*-dimethylaniline (DANI) ([SA/ANI] and [SA/DANI] in particular) enabled better signal detection than SA alone.[41] However, these matrices are not commercially available. Again, very importantly, the matrix must be optimized to the tissue and carefully tested to achieve the best performances (*i.e.* sensitivity, crystal size for lateral resolution). Some additives can also be used along with the matrix to improve protein detection, such as detergent (*e.g.* sodium dodecyl sulfate, (SDS)[54]).

As a guidance, SA concentrations are usually above 20 mg ml^{-1}, DHA concentrations above 100 − 150 mg ml^{-1}, DHB between 40 and 60 mg ml^{-1}

and FA at 12.5 mg ml^{-1}. The solvents of these matrices can also vary in their organic content from 60% acetonitrile (ACN) to 80% ACN and even optimized mixtures between organics and acids. For SA/ANI, the matrix is prepared just prior to use by adding 1 equivalent of ANI to a solution containing 20 mg mL^{-1} (1 equivalent) of SA in ACN/aqueous TFA 0.1% (6 : 4, v/v).

11.2.1.1.4 Matrix Deposition. In addition to the previous points raised concerning the matrix and use of additives and solution mixtures, one of the critical points is the deposition of this matrix. Manual methods with airbrush nebulizers can be used, but the use of commercially available robot nebulizers (*i.e.* TM sprayer from HTX Technologies, USA (https://bit.ly/3fqrtaR) or SunCollect from SunChrom, Germany (https://bit.ly/3oxJFDB)) is the ideal method for this critical step in sample preparation. This recommendation to the reader is made on the basis of three essential points at this stage of sample preparation: i) greater inter-sample reproducibility; ii) stability of the method to be used even if advanced experience and knowledge of the method of choice are required in this section of sample preparation; iii) obtaining smaller crystals as possible.

With regard to the latter, obtaining smaller crystals is desirable. First in MSI because it is mandatory to achieve the highest lateral resolution. Indeed, MSI at the cellular level down to 1 μm is only possible if the crystal is not larger than the laser spot size.[32] Second, because it was shown in fundamental studies on MALDI that the ion yield is increased in smaller crystals.[38,62]

11.2.2 MALDI-MSI of Digested Proteins or Bottom-up pMSI

This subcategory of pMSI is perhaps the most attractive from a clinical and advanced applications point of view due to the possibility of accessing not only fresh-frozen samples, but also formalin-fixed (FF) and formalin-fixed paraffin-embedded (FFPE) samples. The unquestionable advantages of FFPE samples in clinical studies lie mainly in three fundamental pillars: low maintenance costs, well-preserved morphology for the histology and/or immunohistochemistry (IHC) procedures; and the possibility to investigate large patient cohorts retrospectively.

Bottom-up pMSI requires additional steps in the sample preparation. As for standard bottom-up proteomics, enzymatic digestion is mandatory. Such strategies were already developed in 2007 for both fresh-frozen[63] and FFPE[64] samples. For FFPE the enzymatic digestion is a way to bypass the presence of methylene bridges induced by the paraformaldehyde (PFA) enabling the release of digested peptides. As for standard proteomics, the preferred enzyme is trypsin because of its high specificity (*i.e.* cleavage after lysine or arginine residues only) and because of the occurrence of these residues promoting the generation of peptides in a comfortable mass range (*i.e.* *m/z* 700–3500 Da) both for analytical and identification performances by direct MS. Trypsin can be replaced by a mixture of trypsin and LysC enzymes (Trypsin/Lys-C) which was shown in conventional proteomics to improve digestion yields from complex mixtures.[65] Optimized protocols have been developed to reach efficient on-tissue digestion, while preserving at best the localization of the digested peptides. Controlling the enzyme deposition with spotter or even better with sprayer is an important part but the incubation of the sample must also be well-controlled and could lead to the observation of delocalization if not. For fresh-frozen samples, tissue washing is used prior to the on-tissue digestion to improve the digestion rate. For FFPE tissues, additional tissue processing is to be included. The tissue is first deparaffinized and then rehydrated in a sequence of different baths with organic solvents. Then the tissue requires often to be summitted to antigen retrieval (AR). This is absolutely necessary for tissues which have been stored over longer periods (>1 year). The AR doesn't remove the methylene bridges, but it helps in "opening" the network of cross-linked proteins, providing in this way better access for the enzyme to its digestion sites. Only then is the tissue submitted to enzymatic digestion. Despite the more complex sample preparation, robust workflows have been developed both for fresh-frozen and FFPE tissues which are robust and reproducible and are applicable for clinical samples.[66–68] Regarding matrix deposition, the matrix is deposited using a sprayer though the choice of the matrix is different than for pMSI. Most often alpha-cyano-4-hydroxycinnamic acid (CHCA)[67,69,70] is preferred for tryptic peptides though in a few studies of fresh-frozen samples 2,5-dihydroxybenzoic acid (DHB)[70] was used. The solid ionic matrix CHCA/ANI[71,72] was found to improve the detection of digestion peptide for both sample types.

11.2.2.1 *Sample Preparation Protocols for Digested Proteins pMSI*

11.2.2.1.1 Tissue Sectioning. The first step in adequate sample preparation is the adaptation of a basic protocol, such as the one we are going to explain next, to the intrinsic characteristics of our samples. The first step is to cut the sample on cryotome or microtome for frozen or FFPE tissues, respectively.

As previously mentioned in intact pMSI, in the case of frozen samples, the sections should have a thickness between 8 and 12 μm always preventing physical artifacts (folds, breaks, *etc.*). If the sample is embedded in paraffin (FFPE), the thickness of the sections should be between 3 and 7 μm. As in the previous case, the characteristics of the tissue will largely determine this parameter though it was recently demonstrated that the finer the cut, the better the peak intensity in the MS spectrum.[73]

Generally, it is strongly recommended to coat the ITO slides with poly-L-lysine prior to mounting the sections, to ensure greater adherence of the tissue to the slides for the FFPE tissues. This is a cheap and cost-effective treatment that can save many headaches if the sample is somewhat "rebellious" *i.e.*, if some sample part detaches and is lost during the various washing processes of the tissue preparation. This is even more crucial for tissue microarray (TMA) samples, of which tiny tissue cores have the tendency to get lost during the different steps of the complex procedure (see below).

After cutting, the samples are deposited in a desiccator for 10 minutes at RT to remove any traces of moisture that may affect the tissues and hamper the spatial localization of analytes.

11.2.2.1.2 Tissue Washing. For frozen tissues, they go directly into an adjustment phase by washing with organic and aqueous solvents with increasing percentage of ethanol. A possible example of washing steps to be carried out is the following:

1. 30 s wash in 70% ethanol
2. 30 s wash in 90% ethanol
3. 30 s washing in 95% ethanol
4. Drying under ambient conditions.

An alternative protocol is:

1. 1 min in 70% ethanol
2. 1 min in 100% ethanol
3. 1 min in acetone
4. 30 s in chloroform (the 2 last steps being repeated)
5. Drying under ambient conditions.

For FFPE blocks, dewaxing is a mandatory first step. This is achieved using xylene (2 steps of 10 minutes each) or equivalent (*e.g.*, hexane, n-octane). Once the wax is removed, the tissues must be rehydrated using ethanol baths of increasing gradient, as follow:

1. 100% ethanol for 5 minutes
2. 96% ethanol for 5 minutes
3. 80% ethanol for 5 minutes
4. 70% ethanol for 5 minutes
5. Ammonium bicarbonate 10 mM or ultrapure water for 5 s (twice)

Then the FFPE still needs the additional step of AR which is the same step as the one applied to a sample when an IHC is performed in anatomic pathology. With the use of high temperature and specific buffers with adjusted pH, AR enables the proteins to unfold and gives access to digestion sites to the protease. AR procedures have to be optimized according to the nature of the tissue. Heat-induced antigen retrieval (HIAR) with 10 mM citrate at pH = 6.0 or with 10 mM Tris at pH = 9.0 buffer is generally used and left for 20–30 min at high temperature (95–100 °C).[48] Preferentially, this step is achieved in a dedicated system (*i.e.* decloaking chamber) for higher throughput and reproducibility. The AR time, temperature, as well as the buffer and its pH, must be optimized to the tissue.[48,74] Once the antigen retrieval process has been completed, the slides can be rinsed in 5 mM ammonium bicarbonate for 5 seconds (3 times).

11.2.2.1.3 On-tissue Digestion. The following step is common for both frozen and FFPE tissues. The most common protease for peptide generation is trypsin (*i.e.* MS, sequencing grade) at different concentrations that can range from 0.10 $\mu g\,mL^{-1}$ to 100 $\mu g\,mL^{-1}$ in ammonium acetate 10–50 mM or deionized water.[75] This concentration will depend, as indicated, on the intrinsic characteristics of the tissue under study. The use of trypsin can be combined with other proteases to optimize digestion and generate a multitude of new fragments obtained from the combination of these proteases for digestion. An example of this would be the use of trypsin combined with Lys-C, used in samples that generate little peptide diversity and that need a second protease to improve peptide richness.[76,77] However, other proteases or enzymes are also used to generate analytes of interest such as glycans using PNGase[75,78–80] and others or even a mix of them.[81]

The application of the enzyme needs to be well-controlled and reproducible and for this, the best by far is nebulizer robots since they allow us to optimize the deposition conditions in parameters such as flow, application speed, or temperature.[80] Although it can also be done manually, as in the previous case in the pMSI of intact protein, it is not recommended for optimal reproducibility. After enzyme deposition, the slides with the samples must be placed in a humid chamber under saturated conditions (*e.g.* with 50% methanol) and at the working temperature required by the protease. With trypsin, there are different approaches that can be optimized, from 2 h in a humid chamber to saturation containing K_2SO_4, to 18 h which means leaving the digestion overnight at RT, or incubating at higher temperature 37 °C.[71] Using a humid chamber and 37 °C incubation increases the digestion yield and reduces the digestion time, though the humidity of the chamber must be controlled to avoid delocalization issues.

11.2.2.1.4 Matrix Deposition. The final step corresponds to the matrix deposition. 4-alpha-cyano-hydroxycinnamic acid (CHCA) is the most popular matrix for digestion peptides but other matrices such as DHB or mixed matrices can also be used.[82] The final goal is the search of a superior performance when forming small crystals that can confer correct ionization as well as a lateral resolution adequate to your needs. Normally, 10 $mg\,ml^{-1}$ of CHCA is used, varying the organic content between 50 and 80% acetonitrile (ACN) and optimizing the application with a robotic nebulizer in each case. After the application of the matrix, a drying step is carried out and then the acquisition is made with the mass spectrometer.

11.2.3 Specific MALDI of Proteins Using Tagged Probes

Hunting of proteins in a complex mixture can also be performed *via* targeted approaches. For this purpose, a new concept using PC-linkers for the multiplex, specific detection and localization of biomolecules has been proposed.[83,84] This concept, called Tag-Mass, relies on affinity detection using a specifically designed probe that is later detected using mass spectrometry.[84] The Tag-Mass method is an affinity-based strategy wherein a probe is directed against a specific target. The probe bears a reporter group (Tag) that can be later imaged using MALDI-MSI to report the localization of the probe.[85] During Tag-Mass, the reporter group, which involves a photocleavable moiety that binds the reporter to the probe, is released *via* photodissociation as a result of the MALDI laser irradiation. Thus, the reporter is detached from the probe during the MALDI-MSI acquisition. Many different reporters can be used for this purpose, but peptides were often used. The photocleavable linker is selected to present a specific absorption band in the UV range at a wavelength (340 nm) that is very close to that of MALDI lasers (*i.e.*, 337–355 nm). Thus, after hybridization of the modified probe to its target, a classical MALDI-MSI sequence is performed. At a specific point during the acquisition, the presence of a probe will be indicated by the presence of the reporter that was released due to the MALDI laser irradiation, which is evident by the observation of a peak at the *m/z* value expected for the reporter.[37,84,86,87] This concept is compatible with all types of probes including mRNA probes, antibody probes, lectins, and aptamers, which can be used to selectively obtain images of mRNA, antigens, oligosaccharides (including glycosylated proteins), and drugs, respectively, as targets. Moreover, in this context, correlation of mRNA expression with the corresponding protein regulation, *i.e.*, the correlation of the transcriptome with the proteome is of special interest for increasing our understanding of biological mechanisms. In the Tag-Mass workflow, MALDI-MSI is combined with hybridization techniques including *in situ* hybridization (ISH) and IHC.[37,84,86,87] TAMSIM, another targeted technology for MALDI-MSI is based on an *N*-hydroxysuccinimide (NHS) linker coupled to 2,5-dioxopyrrolidin-1-yl-3-{[3-(6-(tertbutylamino)-6-oxohex-1-ynyl)-4-methoxyphenyl) bis (4-methoxyphenyl)] methylthio}propanoate, 2,5-dioxopyrrolidin-1-yl-3-{[3-(6-(benzylamino)-6-oxohex-1-ynyl)-4-methoxyphenyl) bis (4-methoxyphenyl)] methylthio}propanoateor2,5-dioxopyrrolidin-1-yl-3-{[3-(6-(phenylamino)-6-oxohex-1-ynyl)-4-methoxyphenyl) bis (4-methoxyphenyl)] methylthio}propanoate reporters.[88,89] Some improvements in the technology were proposed either based on recombinant single chain fragment variable (SCFV) antibodies[90] or with a dendrimer containing multiple mass tags.[91] These methods were then extended to other ion production sources, as laser ablation inductively coupled plasma mass spectrometry (LA-ICP-MS) probes bearing a monoatomic element (*e.g.*, gold) are easily detectable.[92,93] Finally, from this kind of technology has emerged multiplex imaging using a photocleavable link with metals (lanthanides)[94–98] and is now commercialized as the Hyperion imaging multiplex system.

11.3 Protein Identification in MALDI-MSI and Spatially-resolved Proteomics

The identification of the proteins imaged in MALDI-MSI is a major aspect since MSI reveals markers of physiological or pathophysiological processes. However, the identification of proteins remains much more complex than that of low molecular weight compounds, such as metabolites, lipids, and drugs, among others. For proteins, MALDI-MSI does not generate highly charged state ions (*i.e.* usually $z \leq 2$) and in most cases monocharged ions are observed. Therefore, ion activation methods for MS2 used in top-down approaches, *e.g.*, electron transfer dissociation (ETD), electron capture dissociation (ECD), and ultraviolet photodissociation (UVPD), are not suitable and MS/MS analysis by collision-induced dissociation (CID)/higher collision-induced dissociation (HCD) is severely limited.

These difficulties have led in the past to the development of databases such as the MSiMass List.[99] Today there are specific identification strategies which can be grouped into two main approaches, so-called *in situ* or off-tissue. These approaches can be selected according to the requirements of spatial resolution and depth of identification.

11.3.1 *In Situ* Identification

In situ protein identification approaches consist of detecting, locating, and identifying proteins directly on tissue sections. In the same way as large-scale proteomic analyses, bottom-up and top-down *in situ* approaches can also be considered. The choice of the strategy will depend on the information sought but also on the activation methods available. They are the most straightforward way to identification, but they unfortunately suffer from limitations, as detailed below.

11.3.1.1 In Situ *Bottom-up Approach*

If MS imaging of proteins is performed using the pMSI bottom-up approach, *i.e.*, based on the *in situ* enzymatic digestion of the proteins, then the digestion peptides can be submitted to conventional MS^2 analysis (most often CID) followed by database interrogation directly from the MS^2 spectrum. This strategy has the advantage that it can be applied to both fresh-frozen and FFPE tissues. As previously mentioned, this approach requires particular control over the digestion step in order to obtain an efficient digestion yield while limiting the delocalization of the generated peptides. However, the identification finds limitations in this approach mainly due to the limitation of the single charged ion of the MALDI source and the tissue context. It is well known that the quality of the peptide MS^2 spectra, namely the number of fragments and their intensity, is related to the intensity of the selected parent ions. In MALDI-MSI bottom-up approaches >500 digestion peptides are observed though only a limited number of them show high intensity because of the complexity of the tissue composition. Thus, weaker parent ions produce poor MS^2 spectra whose identification by database comparison will not lead to a possible identification. On the other hand, the formation of Schiff base and methylol adducts can arise as a result of formalin-fixation and can reduce protein IDs if they are not accounted for when performing the database search.[100] Moreover, it was observed that fragmentation yields from tissue appear to be lower than those from convectional dried droplet (DD) preparations presumably because the tissue limits the subjacent effect of the sample plate in the desorption/ionization process.[101] To improve peptide detection and subsequent identification, enzymatic digestion yield can be increased *via* the addition of MS compatible detergents such as octyl-α/β-glucoside.[102] Identification can also be improved by increasing fragmentation yields (monocharged ions) and orienting fragmentation. Strategies for *in situ* N-terminal derivatization of tryptic peptides have been proposed.[103] Different reagents can be used, such as TMPP-NHS. The addition of a derivatization step after digestion increases the number of identifications and limits the number of false positives.[103] Ion mobility (IMS) has also been shown to improve the analysis of complex mixtures by MALDI-MSI as it provides an additional dimension of separation based on the effective cross-sectional area of gas phase ions. It has been shown that IMS can differentiate isobaric ions and image them separately.[104] Such strategies have been applied, in particular to the study of pancreatic cancer, enabling the involvement of the GrP78 protein to be highlighted.[102]

11.3.1.2 In Situ *Top-down Approaches*

While *in situ* protein digestion remains the most commonly used strategy for protein identification, the approach is time consuming and requires careful automation of the various steps to ensure high reproducibility. Thus, top-down *in situ* approaches have progressively been developed for the identification of proteins with fewer preparation steps. This approach is limited to fresh-frozen samples, since in FFPE the proteins are cross-linked and currently can't be efficiently analyzed without on-tissue digestion. In Source Decay (ISD) is an activation method that consists of fragmenting the ions in the MALDI source very early during the MALDI process[105,106] and before the acceleration. ISD is achieved by increasing the laser energy and is independent of the molecular weight of the compound under study and therefore allows the sequencing of intact and high molecular weight proteins.[107] The peculiar fragmentation in ISD leads to strong orientation of the fragmentation toward a specific series of fragment ions. The yield of fragmentation and the type of ion fragment generated are dependent on the use, and certain MALDI matrixes were shown to be better suited for ISD than others. The 1,5-diaminonaphthalene matrix (1,5-DAN) in particular was found to be highly efficient in that respect[108,109] and was used for MALDI-MSI.[110] MALDI ISD imaging has been applied to pig eye sections and has allowed the identification and localization of S-crystalline β, B2-crystalline β, and α-crystalline.[108] Another study characterizing glioblastoma identified the involvement of thymosin β4 and calcycline (S100A6).[111] MALDI ISD imaging is an attractive method for the detection and identification of proteins in a single acquisition with little sample preparation. However, it is limited to the detection of the few most abundant proteins as it does not allow the selection of precursor ions, although it is possible to select ISD fragment ions for additional MS/MS analysis or pseudo MS^3, also known as "T3 sequencing".[107,112] In this context, an ionic matrix DAN/camphor has been developed and used in MALDI-MSI and prompt fragments were obtained from whole proteins in the section at each position where the matrix was added.[109]

As the proteins in FFPE tissues are fixed with bridging agents such as paraformaldehyde, top-down approaches are not applicable. In this context, top-down strategies were first developed using fresh-tissue sections. Recently, activation methods commonly used in top-down proteomics analyses including ETD and UV-PD have also been applied for *in situ* protein identification. These analyses could be performed with the use of ambient MS imaging sources such as DESI[113] and LESA.[114] Recently, a strategy to reverse the cross-linking of proteins in FFPE tissues has allowed the detection of intact proteins, though of limited molecular weight, by MSI.[47] These new developments may open the door to the different top-down proteomic strategies developed from fresh tissue in the future to FFPE.

11.3.2 Off-tissue Identification and Spatially-resolved Proteomics

In situ identification strategies, whether bottom-up or top-down, allow high spatial resolution to be maintained in the identification but are limited in the number of proteins that can be identified to the most predominant ones. In order to increase the depth of identification, one strategy is to correlate the *m/z* of the proteins or peptides detected by MS-MSI with those identified by conventional proteomic approaches after extracting the proteins from the whole tissue/tissue sections or from localized regions on the sections. This approach allows the generation of a library of *m/z* of identified compounds that can be back correlated with the *m/z* measured by MSI for their back identification. Over time the tissue approach has largely been developed and applied at different scales from whole sections to very limited regions. This is achieved using different sampling (or micro-sampling) methods that will be reviewed here. These developments have filled the gap between large-scale proteomic strategies that have no spatial dimension and MALDI-MSI for which *in situ* identification of proteins remains complex, giving rise to the field of so-called "Spatially-Resolved Proteomics".

11.3.2.1 Homogenization of the Tissue

This approach consists of extracting all proteins from a tissue or a tissue section and has long been used to compare the proteins identified by conventional large-scale proteomics with those detected by MS imaging from fresh-frozen tissue sections[18] or FFPE tissue.[64] The use of high spectral resolution mass spectrometers has greatly improved the strategy by bringing greater reliability to correlations between LC–MS/MS and MALDI-MSI data. One study compared the number of proteins identified by the bottom-up *in situ* approach and the re-identification of *m/z* detected by MALDI-Orbitrap imaging *via* the peptides identified by LC–MS/MS showing that only

4 proteins were identified *in situ* against 101 through the re-identification *via* the LC–MS/MS data considering a $\Delta m/z$ of 0.01 Th.[115] Some very high-resolution analyzers such as FT-ICR allow the detection of intact proteins with a MALDI source up to approximately *m/z* 20 000. Comparison of the *m/z* detected by the MALDI-MSI can then be performed directly from whole proteins without the need for *in situ* enzymatic digestion using two consecutive histological sections, one for MALDI-MSI and the second for extraction and top-down proteomics.[59,116] This strategy offers a new dimension for protein identification by MALDI-MSI but finds limitations for proteins specific to small, highly localized regions of tissue or minority due to the dilution effect at the scale of the total extract.

11.3.2.2 Laser Micro-dissection

Laser micro-dissection (LCM) is an approach that allows the analysis of a particular cell type or specific regions of tissue from histological sections by excising and then catapulting the cells out of the tissue using a UV or an IR laser.[15,117,118] The collected cells can then be used for any type of analysis and were mainly used initially for genomic applications; yet in proteomics it is now usually coupled to conventional bottom-up or top-down proteomics experiments. In particular, this strategy has been used with FFPE tissues to identify proteins that may be involved in different cancers such as liver, kidney, and lung cancer.[119] Following a bottom-up strategy, the authors identified more than 1000 proteins from regions containing less than 3000 cells. In another application the authors performed bottom-up identification from mouse glioma models micro-dissecting tumor and healthy regions of approximately 0.8 mm^2 to reassign the identifications to MALDI–FT-ICR imaging data.[31] Recently a nanodroplet processing platform (NanoPOTS)[120] which was developed to process very limited number of cells (<100) for proteomics was applied to LCM tissues and was shown to enable the identification 180, 695, and 1827 protein groups on average from tissue sections with respectively 50, 100, and 200 μm diameters.[121] Such approaches were also recently coupled with both DIA and isotope labeling with TMT for quantification purposes and was demonstrated to quantify the intra-tumor heterogeneity from FFPE tissues.[122]

11.3.2.3 Liquid Microextraction

As an alternative strategy, micro-extraction can be preferred to LCM. This strategy does not allow the analysis of a particular cell type but offers access to analyses at a very localized scale, limiting the damage to proteins and accessing a local micro-environment on the tissue. The micro-extraction is done through the creation of a liquid micro-junction (LMJ) as for example in the LESA technique. After micro-extraction the samples can be collected or injected directly into the MS instrument. If they are collected, they can then be subjected to bottom-up or top-down analysis. Initially this strategy was proposed by collection and bottom-up analysis on fresh tissue and then FFPE. In this case, micro-extraction is preceded by localized micro-digestion of the proteins over regions of about ≤1 mm^2 using a micro-spotter to deposit the enzyme (in general trypsin). The *in situ* generated digestion peptides are then extracted by LMJ and analyzed by LC–MS/MS. The *m/z* of the digested peptides identified in this way allows the re-identification of the peptides on the MALDI images. The first study carried out using this strategy showed the possibility of identifying more than 1200 protein groups from a region of about 600 μm corresponding to about 1900 cells,[123] with reassignment of 61 proteins on the MALDI-Orbitrap images thanks to the high spectral resolution and mass accuracy. This strategy has been applied in different clinical applications in oncology for ovarian cancer from FFPE tissue[124] or glioma.[71] The microextraction strategy was then applied to whole proteins *via* top-down approaches with or without prefractionation of the samples on a 1D gel.[125] A study in ovarian cancer identified proteins with post-translational modifications as well as proteins translated from non-canonical reading frames (Alternative Proteins) {Anonymous:2021gp} specific to ovarian cancer tumor regions.[126]

11.3.2.4 Manual Micro-excision

A very simple and inexpensive alternative to micro-extraction or micro-dissection techniques is the use of M-parafilm-coated histological slides on which tissue sections are mounted. This technique allows the manual micro-excision of pieces of M-parafilm (PAM) containing the specific regions of tissue to be subjected to bottom-up or top-down proteomic analysis. This strategy, which does not require any preparation equipment, has allowed the identification of more than 1000 proteins from regions of about 1 mm^2 [127,128] coupled to bottom-up large-scale analysis. Applied to the entire section, the PAM strategy allows image reconstruction based on the relative quantification of proteins identified by LC–MS/MS in LFQ. This concept was recently used in combination with the nanoPOTS technology to enable imaging based on protein quantification at a spatial resolution of 100 μm.[129] This strategy was also applied in top-down proteomics. In this context, 235 proteins were identified in a study on ovarian cancer,[130] 123 proteins from rat brain regions and back-correlated to the imaging data.[131]

11.3.2.5 Hydrogels

Hydrogels are super absorbent polymers capable of retaining large amounts of liquid.[132] Previously dehydrated hydrogels can be rehydrated with an enzyme solution and are then placed on specific regions of tissue sections allowing both protein digestion and concomitant extraction of digested peptides from fresh-frozen tissue[133] or FFPE.[134] The peptides are then extracted from the hydrogels for analysis by LC–MS/MS. For example, 3 mm diameter hydrogels have been prepared with acrylamide/bis-acrylamide containing trypsin solution which have been applied to regions of rat brain sections allowing the identification of membrane proteins.[16] A study on the impact of hydrogel diameter on protein identification yield led to the identification of 671 and 1051 proteins, respectively, for hydrogels of approximately 260 μm and 1.6 mm in diameter.[135]

11.3.3 MALDI-MSI Guided Spatially-resolved Proteomics

If the direct identification of proteins or digested peptides remains a challenge *in situ* from the sections used for the pMSI, MALDI-MSI can be advantageously used to guide the spatially-resolved proteomics. The MS spectra measured at different pixels of the image contain >500 signals, most of them corresponding to different analytes (except for various adducts of the same analyte). Individual images of each molecule can be retrieved alone to get the distribution of specific markers, but the global molecular profile can also be processed as a molecular barcoding reflecting the average phenotype of the cells present in the analyzed pixel. This information can be processed using multivariate statical analysis using different segmentation algorithms to rebuild images, highlighting the area with different molecular phenotypes inside the tissue. The molecular phenotyping might give a different result than the cell phenotyping as obtained after histology only looking at the morphology criteria of the cell. For example, it was shown in a study conducted in glioma that the molecular phenotyping was better at depicting the tumor heterogeneity than the histology annotation.[71,76] Thus, this segmentation information can be used to guide spatially-resolved proteomics and help discover markers associated with pathology. Various segmentation algorithms have been proposed for spatial segmentation[136–138] and despite the challenge of the important size of the data generated in MSI they are now included in software solutions (*e.g.* SCiLS (Bruker); Lipostar (Molecular Discovery); CardinalMSI R/Bioconductor package[139]) and can be routinely applied. They even enable the segmentation to be performed from several MALDI-MSI altogether enabling one to classify and stratify patients. This kind of approach was used in different applications for glioma[76] or breast cancer.[140]

11.4 Conclusions and Future Directions

Spatially-resolved proteomics guided by MALDI-MSI studies have greatly increased in number in recent years.[30,76,141–143] As such, well-standardized protocols are therefore necessary for translation of pMSI into clinics. Most of the efforts of the clinical pMSI community should therefore now be dedicated to this aspect. However, some clear lines have already been drawn by precursor groups.[35,87,125,130,144] Digital pathology will be the next step of this pMSI. The advent of whole-slide imaging, availability of faster networks, and cheaper storage solutions has made it easier for pathologists to manage digital slide images and share them for clinical use. Integration of pMSI into the pathology workflow, with deep learning, and computer-aided diagnostic techniques, will increase the possibility for a pathologist to give a better diagnosis. The major limitation is the number of samples that can be treated per day. Regarding the spatial resolution, one of the most outstanding applications and the next objective of pMSI is to reach the single cell scale for basic and clinical research as it is needed for pathologists to incorporate the molecular information of the analytes in their workflows for diagnosis and prognosis purposes. However, MSI-based omics will also be greatly increased by the development of high-resolution tissue expansion which will increase the limit of spatial resolution.[145] Moreover, MALDI-MSI has already reached single cells analyses for lipids and metabolites,[142,146–149] the next step will be pMSI. Taken together, pMSI is starting its translation to the clinic and the next improvement using atmospheric instruments based on DESI-MS[150] or SpiderMass technologies[142,146–149] will ease its complete deployment.

Abbreviations

ACN	Acetonitrile
ANI	Aniline
AR	Antigen retrieval
CHCA	Alpha-cyano-4-hydroxycinnamic acid
CID	Collision induced dissociation
DANI	*N,N*-dimethylaniline
DANI	1,5-diaminonaphthalene matrix
DDA	Data-dependent acquisition
DESI	Desorption electrospray ionization
DHA	Dihydroxyacetophenone
DHB	Dihydroxybenzoic acid
DIA	Data-independent acquisition
ECD	Electron capture dissociation
ESI	Electrospray
ETD	Electron transfer dissociation
FA	Ferulic acid
FF	Formalin-fixed
FFPE	Formalin-fixed paraffin-embedded
FT-MS	Fourier Transform mass spectrometer
HA	Hydroxylamine hydrochloride
HCD	Higher collision-induced dissociation
HE	Hematoxylin-Eosin
HIAR	Heat-induced antigen retrieval
IHC	Immunohistochemistry
IMS	Ion mobility mass spectrometry
ISD	In source decay
ISH	*In situ* hybridization
ITO	Indium tin oxide
LA-ICP-MS	Laser ablation inductively coupled plasma mass spectrometry
LC–MS/MS	Liquid Tandem Mass Spectrometry

LCM	Laser microdissection
LESA	Liquid Extraction Surface Analysis
LMJ	Liquid Micro-Junction
MALDI	Matrix-assisted laser desorption/ionization
MRM	Multiple Reaction Monitoring
MS	Mass Spectrometry
MW	Molecular Weight
NHS	*N*-hydroxysuccinimide
PAM	M-parafilm
PFA	Paraformaldehyde
pMSI	Protein Mass Spectrometry Imaging
PTM	Post-translational modifications
Q-TOF	Quadrupole Time of Flight
QqQ	Triple Quadrupole
SA	Sinapinic acid
SCFV	Single chain fragment variable
SDS	Sodium dodecyl sulfate
SIMS	Secondary ion mass spectrometry
SRM	Selected Reaction Monitoring
TFA	Trifluoroacetic acid
TMA	Tissue microarrays
TOF	Time of Flight
UVPD	Ultraviolet photodissociation

Acknowledgements

This research was supported by grants from the Ministère de L'Education Nationale, de L'Enseignement Supérieur et de la Recherche, SIRIC ONCOLille (IF), regional council Hauts de France, Université de lille, and INSERM. This work would never have been possible without funding from the Call for "Ayudas para la Vinculación de Técnicos de Apoyo a ECAIs"; Funding Agency: ANDALUSIAN HEALTH SERVICE; Reference: RF2-0003-2020. Eduardo Chicano would like to personally thank Tiffany Porta Siegel, Michel Salzet and Isabelle Fournier for the opportunity to work together on this chapter.

References

1. N. G. Anderson and N. L. Anderson, *Electrophoresis*, 1996, **17**, 443–453.
2. M. R. Wilkins, J. C. Sanchez, A. A. Gooley, R. D. Appel, I. Humpherysmith, D. F. Hochstrasser and K. L. Williams, *Biotechnol. Genet. Eng. Rev.*, 1996, **13**, 19–50.
3. R. Aebersold and M. Mann, *Nature*, 2016, **537**, 347–355.
4. Y. He, A. Mohamedali, C. Huang, M. S. Baker and E. C. Nice, *Clin. Chim. Acta*, 2019, **495**, 611–624.
5. Y. Liu and M. R. Chance, *Comput. Struct. Biotechnol. J.*, 2014, **10**, 90–97.
6. N. M. Riley and J. J. Coon, *Anal. Chem.*, 2016, **88**, 74–94.

7. T. Cardon, I. Fournier and M. Salzet, *Trends Biochem. Sci.*, 2021, **46**, 239–250.
8. A. I. Nesvizhskii, *Nat. Methods*, 2014, **11**, 1114–1125.
9. P. Wilmes, A. H. Buschart and P. L. Bond, *Proteomics*, 2015, **15**, 3409–3417.
10. D. B. Bekker-Jensen, A. Martínez-Val, S. Steigerwald, P. Rüther, K. L. Fort, T. N. Arrey, A. Harder, A. Makarov and J. V. Olsen, *Mol. Cell. Proteomics*, 2020, **19**, 716–729.
11. F. Meier, A.-D. Brunner, M. Frank, A. Ha, I. Bludau, E. Voytik, S. Kaspar-Schoenefeld, M. Lubeck, O. Raether, N. Bache, R. Aebersold, B. C. Collins, H. L. Röst and M. Mann, *Nat. Methods*, 2020, **17**, 1229–1236.
12. R. Castaing, G. S. S. De L, 1962, *Optique corpusculaire-premiers essais de microanalyse par emission ionique secondaire.*
13. M. Stoeckli, P. Chaurand, D. E. Hallahan and R. M. Caprioli, *Nat. Med.*, 2001, **7**, 493–496.
14. R. M. Caprioli, T. B. Farmer and J. Gile, *Anal. Chem.*, 1997, **69**, 4751–4760.
15. S. T. P. Mezger, A. M. A. Mingels, O. Bekers, R. M. A. Heeren and B. Cillero-Pastor, *Anal. Chem.*, 2021, **93**, 2527–2533.
16. J. J. Nicklay, G. A. Harris, K. L. Schey and R. M. Caprioli, *Anal. Chem.*, 2013, **85**, 7191–7196.
17. J. Stauber, R. Lemaire, J. Franck, D. Bonnel, D. Croix, R. Day, M. Wisztorski, I. Fournier and M. Salzet, *J. Proteome Res.*, 2008, **7**(3), 969–978.
18. R. Lemaire, S. A. Menguellet, J. Stauber, V. Marchaudon, J.-P. Lucot, P. Collinet, M.-O. Farine, D. Vinatier, R. Day, P. Ducoroy, M. Salzet and I. Fournier, *J. Proteome Res.*, 2007, **6**, 4127–4134.
19. M. Wisztorski, J. Franck, M. Salzet and I. Fournier, in *Mass Spectrometry Imaging*, Humana Press, Totowa, NJ, 2010, pp. 303–322.
20. C. Keller, E. Gemperline and L. Li, *Methods Mol. Biol.*, 2020, **2139**, 341–351.
21. E. Gemperline, C. Keller, D. Jayaraman, J. Maeda, M. R. Sussman, J.-M. Ane and L. Li, *J. Proteome Res.*, 2016, **15**, 4403–4411.
22. M. Winter, A. Tholey, A. Kristen and C. Röcken, *Proteomics*, 2017, **17**, 1700236.
23. M. Morisasa, K. Kimura, M. Sumida, S. Fukumoto, T. Tamura, R. Takeuchi, T. Mori and N. Goto-Inoue, *Foods*, 2020, **9**, 402.
24. Y. Yoshimura and N. Zaima, *Foods*, 2020 May, **9**(5), 575.
25. E. Berghmans, K. Boonen, E. Maes, I. Mertens, P. Pauwels and G. Baggerman, *J. Pers. Med.*, 2020, **10**, 54.
26. C. Meriaux, J. Franck, D. B. Park, J. Quanico, Y. H. Kim, C. K. Chung, Y. M. Park, H. Steinbusch, M. Salzet and I. Fournier, *Hippocampus*, 2014, **24**, 628–642.
27. M. Wisztorski, P. Saudemont, S. Aboulouard, T. Cardon, F. Narducci, Y.-M. Robin, A.-S. Lemaire, D. Bertin, F. Kobeissy, E. Leblanc, I. Fournier and M. Salzet, *Cell Reports Medicine*, 2021, **2**(6), 100318.
28. A. Walter, P. Paul-Gilloteaux, B. Plochberger, L. Sefc, P. Verkade, J. G. Mannheim, P. Slezak, A. Unterhuber, M. Marchetti-Deschmann,

M. Ogris, K. Bühler, D. Fixler, S. H. Geyer, W. J. Weninger, M. Glösmann, S. Handschuh and T. Wanek, *Front. Phys.*, 2020, **8**, 47.
29. A. Blutke, N. Sun, Z. Xu, A. Buck, L. Harrison, S. C. Schriever, P. T. Pfluger, D. Wiles, T. Kunzke, K. Huber, J. Schlegel, M. Aichler, A. Feuchtinger, K. Matiasek, S. M. Hauck and A. Walch, *Sci. Rep.*, 2020, **10**, 13.
30. D. J. Ryan, J. M. Spraggins and R. M. Caprioli, *Curr. Opin. Chem. Biol.*, 2019, **48**, 64–72.
31. M. Dilillo, R. Ait-Belkacem, C. Esteve, D. Pellegrini, S. Nicolardi, M. Costa, E. Vannini, E. L. de Graaf, M. Caleo and L. A. McDonnell, *Sci. Rep.*, 2017, **7**, 603–611.
32. A. Zavalin, J. Yang, K. Hayden, M. Vestal and R. M. Caprioli, *Anal. Bioanal. Chem.*, 2015, **407**, 2337–2342.
33. I. Kaya, S. Sämfors, M. Levin, J. Borén and J. S. Fletcher, *J. Am. Soc. Mass Spectrom.*, 2020, **31**, 2133–2142.
34. N. H. Patterson, M. Tuck, A. Lewis, A. Kaushansky, J. L. Norris, R. Van de Plas and R. M. Caprioli, *Anal. Chem.*, 2018, **90**, 12404–12413.
35. J. Quanico, J. Franck, M. Wisztorski, M. Salzet and I. Fournier, *Methods Mol. Biol.*, 2017, **1598**, 21–43.
36. S. Lahiri, N. Sun, V. Solis-Mezarino, A. Fedisch, J. Ninkovic, A. Feuchtinger, M. Goetz, A. Walch and A. Imhof, *Proteomics*, 2016, **16**, 437–447.
37. M. El Ayed, D. Bonnel, R. Longuespee, C. Castellier, J. Franck, D. Vergara, A. Desmons, A. Tasiemski, A. Kenani, D. Vinatier, R. Day, I. Fournier and M. Salzet, *Med. Sci. Monit.*, 2010, **16**, BR233–BR245.
38. T. W. Jaskolla, M. Karas, U. Roth, K. Steinert, C. Menzel and K. Reihs, *J. Am. Soc. Mass Spectrom.*, 2009, **20**, 1104–1114.
39. V. Horneffer, A. Forsmann, K. Strupat, F. Hillenkamp and U. Kubitscheck, *Anal. Chem.*, 2001, **73**, 1016–1022.
40. E. H. Seeley and R. M. Caprioli, *Anal. Chem.*, 2012, **84**, 2105–2110.
41. J. Franck, K. Arafah, A. Barnes, M. Wisztorski, M. Salzet and I. Fournier, *Anal. Chem.*, 2009, **81**, 8193–8202.
42. J. Yang and R. M. Caprioli, *Anal. Chem.*, 2011, **83**, 5728–5734.
43. B. D. Leinweber, G. Tsaprailis, T. J. Monks and S. S. Lau, *J. Am. Soc. Mass Spectrom.*, 2009, **20**, 89–95.
44. E. H. Seeley, S. R. Oppenheimer, D. Mi, P. Chaurand and R. M. Caprioli, *J. Am. Soc. Mass Spectrom.*, 2008, **19**, 1069–1077.
45. R. Lemaire, M. Wisztorski, A. Desmons, J. C. Tabet, R. Day, M. Salzet and I. Fournier, *Anal. Chem.*, 2006, **78**, 7145–7153.
46. E. A. Hoffman, B. L. Frey, L. M. Smith and D. T. Auble, *J. Biol. Chem.*, 2015, **290**, 26404–26411.
47. D.-K. Lee, S. S. Rubakhin, I. Kusmartseva, C. Wasserfall, M. A. Atkinson and J. V. Sweedler, *Angew. Chem., Int. Ed. Engl.*, 2020, **59**, 22584–22590.
48. B. Ergin, S. Meding, R. Langer, M. Kap, C. Viertler, C. Schott, U. Ferch, P. Riegman, K. Zatloukal, A. Walch and K.-F. Becker, *J. Proteome Res.*, 2010, **9**, 5188–5196.

49. C. Urban, A. Buck, J. T. Siveke, F. Lordick, B. Luber, A. Walch and M. Aichler, *Biochim. Biophys. Acta, Gen. Subj.*, 2018, **1862**, 51–60.
50. V. Mainini, G. Bovo, C. Chinello, E. Gianazza, M. Grasso, G. Cattoretti and F. Magni, *Mol. BioSyst.*, 2013, **9**, 1101–1107.
51. A. C. Grey, P. Chaurand, R. M. Caprioli and K. L. Schey, *J. Proteome Res.*, 2009, **8**, 3278–3283.
52. A. van Remoortere, R. J. M. van Zeijl, N. van den Oever, J. Franck, R. Longuespee, M. Wisztorski, M. Salzet, A. M. Deelder, I. Fournier and L. A. McDonnell, *J. Am. Soc. Mass Spectrom.*, 2010, **21**, 1922–1929.
53. J. Franck, R. Longuespee, M. Wisztorski, A. van Remoortere, R. Van Zeijl, A. Deelder, M. Salzet, L. McDonnell and I. Fournier, *Med. Sci. Monit.*, 2010, **16**, BR293–BR299.
54. V. Mainini, P. M. Angel, F. Magni and R. M. Caprioli, *Rapid Commun. Mass Spectrom.*, 2011, **25**, 199–204.
55. J. H. Jungmann, L. MacAleese, J. Visser, M. J. J. Vrakking and R. M. A. Heeren, *Anal. Chem.*, 2011, **83**, 7888–7894.
56. J. H. Jungmann, D. F. Smith, L. MacAleese, I. Klinkert, J. Visser and R. M. A. Heeren, *J. Am. Soc. Mass Spectrom.*, 2012, **23**, 1679–1688.
57. S. R. Ellis, J. Soltwisch and R. M. A. Heeren, *J. Am. Soc. Mass Spectrom.*, 2014, **25**, 809–819.
58. D. M. Anderson, M. G. Nye-Wood, K. L. Rose, P. J. Donaldson, A. C. Grey and K. L. Schey, *J. Mass Spectrom.*, 2020, **55**, e4473.
59. J. M. Spraggins, D. G. Rizzo, J. L. Moore, K. L. Rose, N. D. Hammer, E. P. Skaar and R. M. Caprioli, *J. Am. Soc. Mass Spectrom.*, 2015, **26**, 974–985.
60. L. Mourino-Alvarez, I. Iloro, F. de la Cuesta, M. Azkargorta, T. Sastre-Oliva, I. Escobes, L. F. Lopez-Almodovar, P. L. Sanchez, H. Urreta, F. Fernandez-Aviles, A. Pinto, L. R. Padial, F. Akerström, F. Elortza and M. G. Barderas, *Sci. Rep.*, 2016, **6**, 27106.
61. E. Patel, *Methods Mol. Biol.*, 2017, **1618**, 7–14.
62. M. Sadeghi and A. Vertes, *Appl. Surf. Sci.*, 1998, **127**, 226–234.
63. M. R. Groseclose, M. Andersson, W. M. Hardesty and R. M. Caprioli, *J. Mass Spectrom.*, 2007, **42**, 254–262.
64. R. Lemaire, A. Desmons, J. C. Tabet, R. Day, A. M. SalzetI Fournier, *J. Proteome Res.*, 2007, **6**(4), 1295–1305.
65. T. Glatter, C. Ludwig, E. Ahrne, R. Aebersold, A. J. R. Heck and A. Schmidt, *J. Proteome Res.*, 2012, **11**, 5145–5156.
66. A. Ly, R. Longuespee, R. Casadonte, P. Wandernoth, K. Schwamborn, C. Bollwein, C. Marsching, K. Kriegsmann, C. Hopf, W. Weichert, J. Kriegsmann, P. Schirmacher, M. Kriegsmann and S.-O. Deininger, *Special Issue: Proteomics in Pathology*, 2019, **13**(1).
67. A. M. Judd, D. B. Gutierrez, J. L. Moore, N. H. Patterson, J. Yang, C. E. Romer, J. L. Norris and R. M. Caprioli, *J. Mass Spectrom.*, 2019, **54**, 716–727.
68. N. Hajjaji, M. Abbouchi, L. A. Nguyen, S. Charles, S. Leclercq, D. Bertin, Y.-M. Robin, I. Fournier and M. Salzet, *J. Clin. Oncol.*, 2020, **38**, e13063.

69. A. Buck, B. Heijs, B. Beine, J. Schepers, A. Cassese, R. M. A. Heeren, L. A. McDonnell, C. Henkel, A. Walch and B. Balluff, *Anal. Bioanal. Chem.*, 2018, **410**, 5969–5980.
70. B. P. A. M. Heijs, W. M. Abdelmoula, S. Lou, I. H. Briaire-de Bruijn, J. Dijkstra, J. V. M. G. Bovée and L. A. McDonnell, *Anal. Chem.*, 2015, **87**(24), 11978–11983.
71. E. Le Rhun, M. Duhamel, M. Wisztorski, J.-P. Gimeno, F. Zairi, F. Escande, N. Reyns, F. Kobeissy, C.-A. Maurage, M. Salzet and I. Fournier, *Proteins Proteomics*, 2017, **1865**, 875–890.
72. R. Lemaire, J. C. Tabet, P. Ducoroy, J. B. Hendra, M. Salzet and I. Fournier, *Anal. Chem.*, 2006, **78**, 809–819.
73. R. Longuespee, K. Kriegsmann, M. Cremer, C. Zgorzelski, R. Casadonte, D. Kazdal, J. Kriegsmann, W. Weichert, K. Schwamborn, M. Fresnais, P. Schirmacher and M. Kriegsmann, *Special Issue: Proteomics in Pathology*, 2019, **13**(1).
74. T. Guo, W. Wang, P. A. Rudnick, T. Song, J. Li, Z. Zhuang, R. J. Weil, D. L. DeVoe, C. S. Lee and B. M. Balgley, *J. Histochem. Cytochem.*, 2007, **55**, 763–772.
75. B. Heijs, S. Holst, I. H. Briaire-de Bruijn, G. W. van Pelt, A. H. de Ru, P. A. van Veelen, R. R. Drake, A. S. Mehta, W. E. Mesker, R. A. Tollenaar, J. V. M. G. Bovée, M. Wuhrer and L. A. McDonnell, *Anal. Chem.*, 2016, **88**, 7745–7753.
76. L. Drelich, M. Duhamel, M. Wisztorski, S. Aboulouard, J.-P. Gimeno, P.-D. Caux, N. Ogrinc, P. Devos, T. Cardon, M. Weller, F. Escande, F. Zairi, C.-A. Maurage, I. Fournier, E. Le Rhun and M. Salzet, *Cell Reports Medicine*, 2021, DOI: 10.2139/ssrn.3770990.
77. B. Heijs, R. J. Carreira, E. A. Tolner, A. H. de Ru, A. M. J. M. van den Maagdenberg, P. A. van Veelen and L. A. McDonnell, *Anal. Chem.*, 2015, **87**, 1867–1875.
78. S. A. Malaker, J. Quanico, A. R. Romero, F. Kobeissy, S. Aboulouard, D. Tierny, C. R. Bertozzi, I. Fournier and M. Salzet, *Cell Chem. Biol.*, 2021, November 18, **28**, 1–13.
79. P. M. Angel, A. Mehta, K. Norris-Caneda and R. R. Drake, *Methods Mol. Biol.*, 2018, **1788**, 225–241.
80. R. R. Drake, T. W. Powers, K. Norris-Caneda, A. S. Mehta and P. M. Angel, *Curr. Protoc. Protein Sci.*, 2018, **94**, e68–21.
81. C. L. Clift, R. R. Drake, A. Mehta and P. M. Angel, *Anal. Bioanal. Chem.*, 2020, **123**, 1–11.
82. Q. Zhou, A. Fülöp and C. Hopf, *Anal. Bioanal. Chem.*, 2020, **48**, 1–19.
83. J. Stauber, M. E. Ayed, M. Wisztorski, M. Salzet and I. Fournier, *Methods Mol. Biol.*, 2010, **656**, 339–361.
84. R. Lemaire, J. Stauber, M. Wisztorski, C. Van Camp, A. Desmons, M. Deschamps, G. Proess, I. Rudlof, A. S. Woods, R. Day, M. Salzet and I. Fournier, *J. Proteome Res.*, 2007, **6**, 2057–2067.

85. J. Stauber, R. Lemaire, M. Wisztorski, S. Ait-Menguellet, J. P. Lucot, D. Vinatier, A. Desmond, M. Deschamps, G. Proess, I. Rudlof, M. Salzet and I. Fournier, *Mol. Cell. Proteomics*, 2006, **5**, S247.
86. J. Stauber, M. El Ayed, M. Wisztorski, R. Day, I. Fournier and M. Salzet, *Anal. Chem.*, 2009, **81**, 9512–9521.
87. J. Franck, K. Arafah, M. Elayed, D. Bonnel, D. Vergara, A. Jacquet, D. Vinatier, M. Wisztorski, R. Day, I. Fournier and M. Salzet, *Mol. Cell. Proteomics*, 2009, **8**, 2023–2033.
88. G. Thiery, M. S. Shchepinov, E. M. Southern, A. Audebourg, V. Audard, B. Terris and I. G. Gut, *Rapid Commun. Mass Spectrom.*, 2007, **21**, 823–829.
89. G. Thiery, E. Anselmi, A. Audebourg, E. Darii, M. Abarbri, B. Terris, J.-C. Tabet and I. G. Gut, *Proteomics*, 2008, **8**, 3725–3734.
90. G. Thiery, R. L. Mernaugh, H. Yan, J. M. Spraggins, J. Yang, F. F. Parl and R. M. Caprioli, *J. Am. Soc. Mass Spectrom.*, 2012, **23**, 1689–1696.
91. J. Yang, P. Chaurand, J. L. Norris, N. A. Porter and R. M. Caprioli, *Anal. Chem.*, 2012, **84**, 3689–3695.
92. J. S. Becker, M. Zoriy, A. Matusch, B. Wu, D. Salber, C. Palm and J. S. Becker, *Mass Spectrom. Rev.*, 2010, **29**, 156–175.
93. J. Seuma, J. Bunch, A. Cox, C. McLeod, J. Bell and C. Murray, *Proteomics*, 2008, **8**, 3775–3784.
94. T. C. de Bang and S. Husted, *Trends Anal. Chem.*, 2015, **72**, 45–52.
95. L. Mueller, A. J. Herrmann, S. Techritz, U. Panne and N. Jakubowski, *Anal. Bioanal. Chem.*, 2017, **409**, 3667–3676.
96. B. Neumann, S. Hoesl, K. Schwab, F. Theuring and N. Jakubowski, *J. Neurosci. Methods*, 2020, **334**, 108591.
97. B. Yang, Y. Zhang, B. Chen, M. He, X. Yin, H. Wang, X. Li and B. Hu, *Biosens. Bioelectron.*, 2017, **96**, 77–83.
98. C. Giesen, H. A. O. Wang, D. Schapiro, N. Zivanovic, A. Jacobs, B. Hattendorf, P. J. Schueffler, D. Grolimund, J. M. Buhmann, S. Brandt, Z. Varga, P. J. Wild, D. Guenther and B. Bodenmiller, *Nat. Methods*, 2014, **11**, 417.
99. L. A. McDonnell, A. Walch, M. Stoeckli and G. L. Corthals, *J. Proteome Res.*, 2014, **13**, 1138–1142.
100. K. Gorzolka and A. Walch, *Histol. Histopathol.*, 2014, **29**, 1365–1376.
101. B. Fatou, P. Saudemont, E. Leblanc, D. Vinatier, V. Mesdag, M. Wisztorski, C. Focsa, M. Salzet, M. Ziskind and I. Fournier, *Sci. Rep.*, 2016, **6**, 14.
102. M.-C. Djidja, S. Francese, P. M. Loadman, C. W. Sutton, P. Scriven, E. Claude, M. F. Snel, J. Franck, M. Salzet and M. R. Clench, *Proteomics*, 2009, **9**, 2750–2763.
103. J. Franck, M. El Ayed, M. Wisztorski, M. Salzet and I. Fournier, *Anal. Chem.*, 2009, **81**, 8305–8317.
104. J. Stauber, L. MacAleese, J. Franck, E. Claude, M. Snel, B. K. Kaletas, I. M. V. D. Wiel, M. Wisztorski, I. Fournier and R. M. A. Heeren, *J. Am. Soc. Mass Spectrom.*, 2010, **21**, 338–347.

105. A. R. Kelley, G. Perry, C. Bethea, R. J. Castellani and S. B. H. Bach, *Open Neurol. J.*, 2016, **10**, 88–98.
106. J. J. Lennon and K. A. Walsh, *Protein Sci.*, 1999, **8**, 2487–2493.
107. D. Suckau, A. Resemann, M. Schuerenberg, P. Hufnagel, J. Franzen and A. Holle, *Anal. Bioanal. Chem.*, 2003, **376**, 952–965.
108. D. Debois, V. Bertrand, L. Quinton, M.-C. De Pauw-Gillet and E. De Pauw, *Anal. Chem.*, 2010, **82**, 4036–4045.
109. D. Bonnel, R. Longuespee, J. Franck, M. Roudbaraki, P. Gosset, R. Day, M. Salzet and I. Fournier, *Anal. Bioanal. Chem.*, 2011, **401**, 149–165.
110. D. Calligaris, R. Longuespee, D. Debois, D. Asakawa, A. Turtoi, V. Castronovo, A. Noel, V. Bertrand, M.-C. De Pauw-Gillet and E. De Pauw, *Anal. Chem.*, 2013, **85**, 2117–2126.
111. R. Ait-Belkacem, D. Calligaris, L. Sellami, C. Villard, S. Granjeaud, T. Schembri, C. Berenguer, L. Ouafik, D. Figarella-Branger, O. Chinot and D. Lafitte, *J. Proteomics*, 2013, **79**, 172–179.
112. D. Asakawa, D. Calligaris, N. Smargiasso and E. De Pauw, *J. Phys. Chem. B*, 2013, **117**, 2321–2327.
113. D. J. Weston, *Analyst*, 2010, **135**, 661–668.
114. V. A. Mikhailov, R. L. Griffiths and H. J. Cooper, *Int. J. Mass Spectrom.*, 2017, **420**, 43–50.
115. Y. Schober, T. Schramm, B. Spengler and A. Roempp, *Rapid Commun. Mass Spectrom.*, 2011, **25**, 2475–2483.
116. J. M. Spraggins, D. G. Rizzo, J. L. Moore, M. J. Noto, E. P. Skaar and R. M. Caprioli, *Proteomics*, 2016, **16**, 1678–1689.
117. M. R. Emmert-Buck, R. F. Bonner, P. D. Smith, R. F. Chuaqui, Z. Zhuang, S. R. Goldstein, R. A. Weiss and L. A. Liotta, *Science*, 1996, **274**, 998–1001.
118. C. Stingl, F. G. I. van Vilsteren, C. Guzel, F. J. W. Ten Kate, M. Visser, K. K. Krishnadath, J. J. Bergman and T. M. Luider, *J. Proteome Res.*, 2011, **10**, 288–298.
119. R. Longuespee, D. Alberts, C. Pottier, N. Smargiasso, G. Mazzucchelli, D. Baiwir, M. Kriegsmann, M. Herfs, J. Kriegsmann, P. Delvenne and E. De Pauw, *Methods*, 2016, **104**, 154–162.
120. Y. Zhu, P. D. Piehowski, R. Zhao, J. Chen, Y. Shen, R. J. Moore, A. K. Shukla, V. A. Petyuk, M. Campbell-Thompson, C. E. Mathews, R. D. Smith, W.-J. Qian and R. T. Kelly, *Nat. Commun.*, 2018, **9**, 10.
121. Y. Zhu, M. Dou, P. D. Piehowski, Y. Liang, F. Wang, R. K. Chu, W. B. Chrisler, J. N. Smith, K. C. Schwarz, Y. Shen, A. K. Shukla, R. J. Moore, R. D. Smith, W.-J. Qian and R. T. Kelly, *Mol. Cell. Proteomics*, 2018, **17**, 1864–1874.
122. K. Buczak, J. M. Kirkpatrick, F. Truckenmueller, D. Santinha, L. Ferreira, S. Roessler, S. Singer, M. Beck and A. Ori, *Nat. Protoc.*, 2020, **15**, 2956–2979.
123. J. Quanico, J. Franck, C. Dauly, K. Strupat, J. Dupuy, R. Day, M. Salzet, I. Fournier and M. Wisztorski, *J. Proteomics*, 2013, **79**, 200–218.

124. M. Wisztorski, B. Fatou, J. Franck, A. Desmons, I. Farre, E. Leblanc, I. Fournier and M. Salzet, *Proteomics: Clin. Appl.*, 2013, **7**, 234–240.
125. M. Wisztorski, A. Desmons, J. Quanico, B. Fatou, J.-P. Gimeno, J. Franck, M. Salzet and I. Fournier, *Proteomics*, 2016, **16**, 1622–1632.
126. T. Cardon, F. Hervé, V. Delcourt, X. Roucou, M. Salzet, J. Franck and I. Fournier, *Anal. Chem.*, 2020, **92**, 1122–1129.
127. J. Franck, J. Quanico, M. Wisztorski, R. Day, M. Salzet and I. Fournier, *Anal. Chem.*, 2013, **85**, 130812115833000–130812115838134.
128. J. Quanico, J. Franck, M. Wisztorski, M. Salzet and I. Fournier, *Methods Mol. Biol.*, 2017, **1598**, 269–283.
129. P. D. Piehowski, Y. Zhu, L. M. Bramer, K. G. Stratton, R. Zhao, D. J. Orton, R. J. Moore, J. Yuan, H. D. Mitchell, Y. Gao, B.-J. M. Webb-Robertson, S. K. Dey, R. T. Kelly and K. E. Burnum-Johnson, *Nat. Commun.*, 2020, **11**, 8–12.
130. V. Delcourt, J. Franck, E. Leblanc, F. Narducci, Y.-M. Robin, J.-P. Gimeno, J. Quanico, M. Wisztorski, F. Kobeissy, J.-F. Jacques, X. Roucou, M. Salzet and I. Fournier, *EBioMedicine*, 2017, **21**, 55–64.
131. V. Delcourt, J. Franck, J. Quanico, J.-P. Gimeno, M. Wisztorski, A. Raffo-Romero, F. Kobeissy, X. Roucou, M. Salzet and I. Fournier, *Mol. Cell. Proteomics*, 2018, **17**, 357–372.
132. L. K. Fiddes, V. N. Luk, S. H. Au, A. H. C. Ng, V. Luk, E. Kumacheva and A. R. Wheeler, *Biomicrofluidics*, 2012, **6**, 14112–1411211.
133. D. Taverna, J. L. Norris and R. M. Caprioli, *Anal. Chem.*, 2015, **87**, 670–676.
134. D. Taverna, A. C. Pollins, L. B. Nanney, G. Sindona and R. M. Caprioli, *Exp. Dermatol.*, 2016, **25**, 143–146.
135. D. G. Rizzo, B. M. Prentice, J. L. Moore, J. L. Norris and R. M. Caprioli, *Anal. Chem.*, 2017, **89**, 2948–2955.
136. T. Alexandrov, *BMC Bioinf.*, 2012, **13 Suppl 16**, S11.
137. T. Alexandrov, S. Meding, D. Trede, J. H. Kobarg, B. Balluff, A. Walch, H. Thiele and P. Maass, *J. Proteomics*, 2011, **75**, 237–245.
138. L. Krasny, F. Hoffmann, G. Ernst, D. Trede, T. Alexandrov, V. Havlíček, O. Guntinas-Lichius, F. von Eggeling and A. C. Crecelius, *J. Am. Soc. Mass Spectrom.*, 2015, **26**, 36–43.
139. K. D. Bemis, A. Harry, L. S. Eberlin, C. Ferreira, S. M. van de Ven, P. Mallick, M. Stolowitz and O. Vitek, *Bioinformatics*, 2015, **31**, 2418–2420.
140. N. Hajjaji, S. Aboulouard, T. Cardon, D. Bertin, Y. M. Robin, I. Fournier and M. Salzet, *Cell Reports Medicine*, 2021, DOI: 10.2139/ssrn.3907620.
141. F. Dewez, J. Oejten, C. Henkel, R. Hebeler, H. Neuweger, E. De Pauw, R. M. A. Heeren and B. Balluff, *Proteomics*, 2020, **20**(23), 1900369.
142. M. J. Taylor, J. K. Lukowski and C. R. Anderton, *J. Am. Soc. Mass Spectrom.*, 2021, **32**, 872–894.
143. M. R. Eveque-Mourroux, B. Rocha, F. P. Y. Barré, R. M. A. Heeren and B. Cillero-Pastor, *J. Proteomics*, 2020, **215**, 103637.

144. M. Wisztorski, J. Quanico, J. Franck, B. Fatou, M. Salzet and I. Fournier, *Methods Mol. Biol.*, 2017, **1618**, 49–63.
145. L. Drelich, S. Aboulouard, J. Franck, M. Salzet, I. Fournier and M. Wisztorski, *Anal. Chem.*, 2021, DOI: 10.1021/acs.analchem.0c05372.
146. K. Scupakova, B. Balluff, C. Tressler, T. Adelaja, R. M. A. Heeren, K. Glunde and G. Ertaylan, *Clin. Chem. Lab. Med.*, 2020, **58**, 914–929.
147. I. S. Gilmore, S. Heiles and C. L. Pieterse, *Annu. Rev. Anal. Chem.*, 2019, **12**, 201–224.
148. D. R. Bhandari, G. Coliva, M. Fedorova and B. Spengler, *Methods Mol. Biol.*, 2020, **2064**, 103–111.
149. M. E. Dueñas, J. J. Essner and Y. J. Lee, *Sci. Rep.*, 2017, **7**, 10.
150. M. W. Towers, T. Karancsi, E. A. Jones, S. D. Pringle and E. Claude, *J. Am. Soc. Mass Spectrom.*, 2018, **29**, 1–11.

CHAPTER 12

The Role of Informatics and Data Analysis in MALDI Mass Spectrometry Imaging

BENJAMIN BALLUFF

Maastricht MultiModal Molecular Imaging Institute (M4i), Maastricht University, 6229 ER Maastricht, The Netherlands
Email: b.balluff@maastrichtuniversity.nl

12.1 Introduction – The Role of Informatics for the Success of Mass Spectrometry Imaging in Biomedical Research

Mass spectrometry imaging (MSI), and especially matrix-assisted laser desorption/ionization (MALDI), has become an integral part of biochemical sciences. Particularly biomedical research has embraced this technology to relate the cell-type specific molecular profiles obtained by MSI with the different disease states or exposures of the tissue to external conditions (chemo-, radiotherapy, environmental stress *etc.*).[1]

The success of MSI is also due to developments in hard- and software in that field, which made the technology accessible to life science.[2] This is mainly due to the hybrid character of MSI data, which is a mixture of mass spectral and imaging data. While many informatics approaches from the fields of mass spectrometry and image processing exist, in most cases it has not been possible to transfer these methods to MSI without adjustments.

New Developments in Mass Spectrometry No. 12
MALDI Mass Spectrometry Imaging: From Fundamentals to Spatial Omics
Edited by Tiffany Siegel

Published by the Royal Society of Chemistry, www.rsc.org

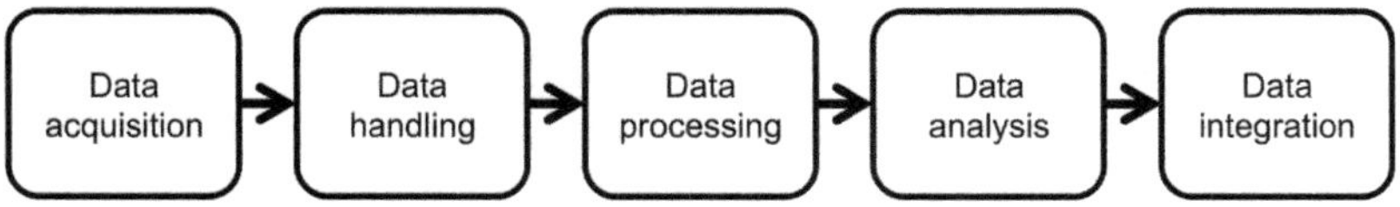

Figure 12.1 Sequential steps of a typical mass spectrometry imaging study from the viewpoint of the data.

In this chapter, I want to review the role of informatics in MSI in light of its main application, namely MALDI-MSI for biomedical research. This chapter will therefore follow the order of a typical MSI study, starting with software and informatics strategies for data acquisition and management, respectively. It will continue with data processing and analysis and will be followed by data integration, to place MSI results into a broader context of life-science-associated research areas (Figure 12.1).

The last aspect follows a current trend in life science to obtain a more complete and comprehensive view on the samples being investigated by integrating the different molecular levels of the central dogma of biology.[3] In that sense, MSI constitutes a powerful tool to complement other mature omics technologies by adding the spatial component to multi-omics studies. Likewise, it also complements other imaging technologies by adding a multi-molecular dimension to the images. New software solutions will further enable the integration into both directions where MSI can act as a bridge and therefore important pillar for future molecular investigations of biological tissue specimens.

12.2 Software for MSI Data Creation and Handling

12.2.1 Software for MSI Data Acquisition

The first moment where informatics comes into play during an MSI experiment is when an imaging run is set up at the mass spectrometer's control software. In that sense, a user-friendly graphical user interface increases the accessibility to a broader range of users including users from clinical and life-science-associated areas or students who will carry their experience with this technology to the research fields of the future. However, ease of use should not be traded in for functionality.

One central issue that acquisition software needs to address is the growing amount of data delivered by MSI, which is due to the continuous improvements in spatial resolution of MSI instrumentation.[4] Since tissues are rarely rectangular, imaging-run setup software should allow defining polygonal measurement areas. This reduces acquisition time, storage space, and computational time after acquisition significantly. Sophisticated tools to define tissue-fit measurement areas, similar to the Photoshop® magic wand tool, are powerful additions to polygon drawing tools.

In that line, it should also be possible to define several individual measurement areas within one dataset. This is relevant when there are several disjointed regions of interest within the tissue to be measured.[5] Heijs *et al.* have shown that histology-guided MSI experiments can result in a significant reduction of data loading and acquisition times.[5] For this reason, any imaging-run software should offer the opportunity to import high-resolution optical images and associated annotations into its environment. This would be beneficial for annotation of intense samples such as tissue microarrays (TMAs), where dozens to hundreds of several micrometer-sized tissue punches are assembled in one block. Sophisticated tissue annotation software such as QuPath allows easy annotation of complex tissue samples or TMAs.[6]

12.2.2 MSI Data Representation and Handling

With advances in MSI instrumentation in terms of higher spatial resolution and faster ionization rates, new data formats have been introduced, which try to reduce the redundancy of meta-information using *e.g.* relational database structures.

A good example of a redundancy saving data format in MSI is the open and common data format imzML, which was first introduced in 2012 and is now supported by most software in the field.[7] In this data format, which is based on the mzML mark-up language, the meta-data of all pixels belonging to one MSI dataset is stored only once in a separate XML file. Also, if the spectral data to store is not centroided, it is assumed that all spectra of one MSI dataset share the same abscissa, which is therefore stored only once, too. Other data formats have been proposed which use compression algorithms, "chunking of the data", or a controlled decrease of the numeric precision in the data to reduce the amount of data.[8,9] One thing to ensure when devising these data formats is the faster reading times when loading the data into the main memory of the computer. imzML, for instance, stores the spectral data in a binary data format, which is more efficient in terms of storing numeric values on disk and in terms of access speed, and OpenMSI makes use of the very efficient HDF5 format.[9]

If storage space is not an issue, creating data redundancy can also increase access speed. This has to do with the fact that table-shaped data, such as encountered in MSI (pixels in rows, *m*/*z* values in columns), is actually stored sequentially on hard disk. Arranging the data points in two different sequential ways makes the loading of spectra and the loading of images become equally fast, albeit for the price of twice the storage space consumption (Figure 12.2).[10]

However, hard disk memory is not the only constraint. The actual working memory of the PC is another important limitation, especially when several datasets have to be loaded simultaneously, as in the case of 3D MSI. Klinkert *et al.* proposed the mosaic datacube structure, which is used by DataCube Explorer and divides the dataset into smaller datacubes, which can be on demand recombined to volumes of interest which are usually smaller than the memory limit.[11] An important pillar to make MSI data more manageable is by processing it, which is explained in the next section.

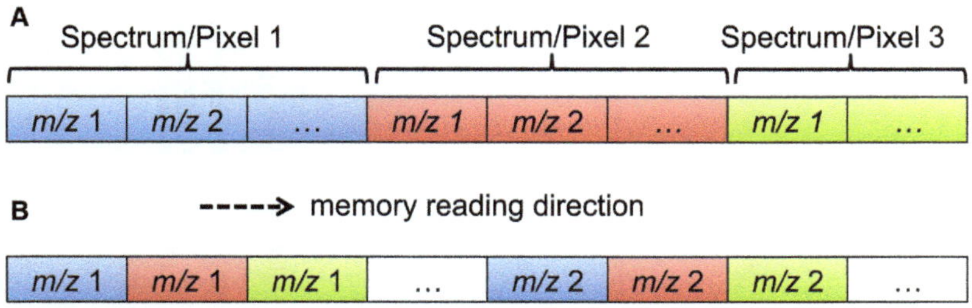

Figure 12.2 Creating redundancy of mass spectrometry imaging data accelerates data access times. (A) Storing spectra sequentially enables fast access of spectral information. (B) Storing *m/z* values sequentially by splitting the spectra accelerates image creation for single *m/z* values.

12.3 Data Processing

Data processing in MSI and in mass spectrometry in general is of central importance before moving into data analysis. Due to its hybrid character, MSI data can be processed both on a spectral level as well as on an image level. Most of the data processing steps have the common aim to extract the most relevant information out of the data while increasing the comparability between spectra by reducing technical variation. Chemical, physical, electronic, or human-related effects can cause this variation, which can, in the worst case, either mask the biological effect or can cause spurious correlations.[12]

Technical variation can manifest itself as noise in a spectrum, where one has to distinguish between chemical and electronic noise. The first usually introduces low-frequency components to the spectrum, whereas the electronic noise usually adds high-frequency components to the spectrum (Figure 12.3).

12.3.1 Baseline Subtraction

Chemical noise can appear as intensity offset from zero in a spectrum, termed *baseline*. The magnitude of this offset depends on the intensity of the desorption process and the detector's amplification settings.[13] Since this intensity offset can be dynamic across the mass spectrum, the estimation of the baseline is non-trivial. Moreover, in MSI, the baselines do not only differ between MSI dataset but also within pixels of the same MSI dataset,[12] which makes it necessary to apply time efficient baseline subtraction algorithm to process the spectra one-by-one. In general, there are many baseline subtraction methods[14] that are either based on sliding or segmenting windows, the fitting of mathematical functions to the baseline curve,[15] or a combination of both.[16] The most commonly used techniques in MSI are sliding-windows such as moving average, median or minimum,[17] Top-Hat filtering,[18] and Gaussian filtering.[19] In these filters, the most important parameter to set is the window size, which usually can be derived from the average peak width encountered in the spectra (Figure 12.4).

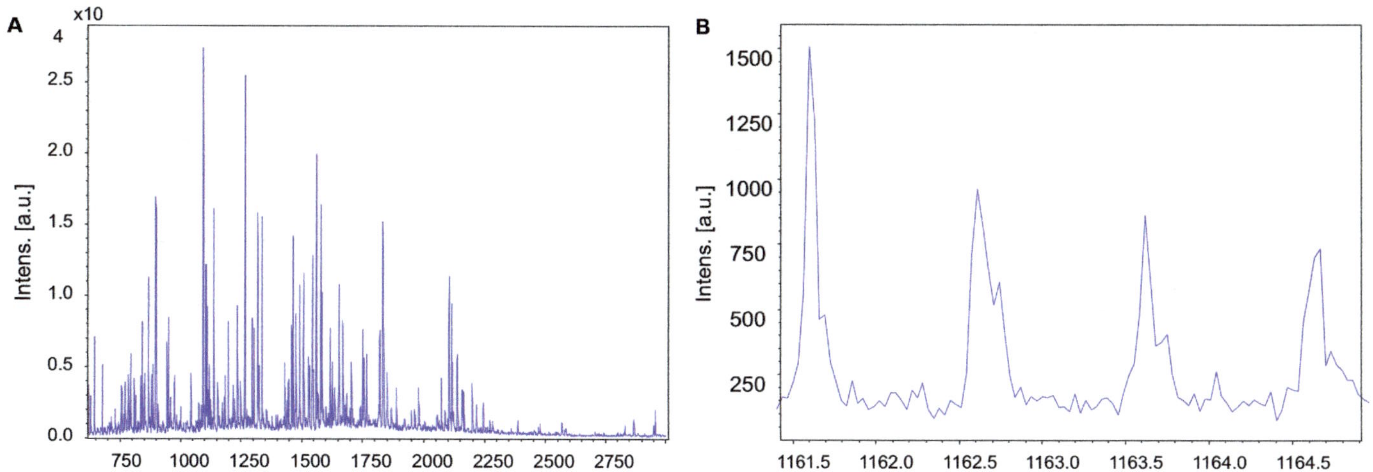

Figure 12.3 Different types of noise in a mass spectrometry spectrum. (A) Chemical noise is usually reflected as a low-frequency component in a mass spectrum, *e.g.* by elevating the baseline. (B) Electronic noise is usually reflected as a high-frequency component in spectra.

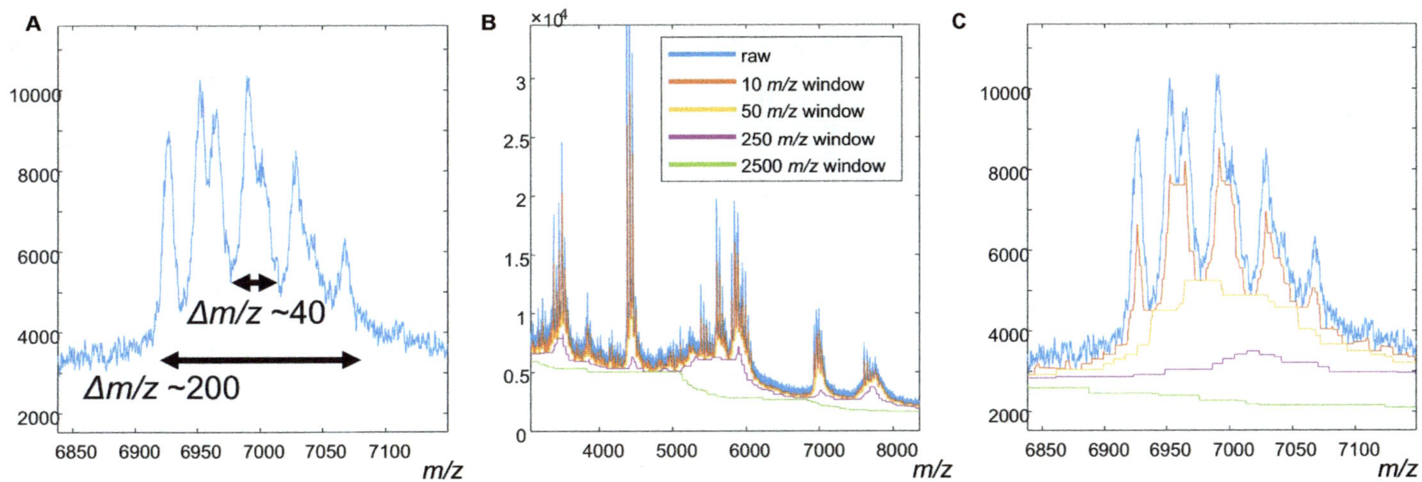

Figure 12.4 Window-based baseline subtraction in mass spectra. (A) The optimal window size can be deduced from the average peak width, in this example between 40 and 200 *m/z* units. (B) If the window selected is too wide (*e.g.* 2500 *m/z* units), the estimation of the baseline will be insufficient (green line). (C) If the window size chosen is too narrow (*e.g.* 10 *m/z* units) real peak signals will be affected negatively by the baseline (red line) subtraction.

12.3.2 Smoothing

While baseline subtraction removes low-frequency components of the spectrum, which is mainly chemical noise, smoothing tries to remove high-frequency noise components, as caused by electronics of the mass spectrometer. Like baseline subtraction, smoothing can be done with many algorithms.[14] The most used ones in the field of MSI are the moving average, Gaussian filtering, and Savitzky–Golay smoothing.[20] However, smoothing on a single pixel level is not always necessary, since: 1) it is computationally expensive, and 2) the subsequent peak picking is usually done on the average spectrum of an MSI dataset. Averaging spectra reduces random electronic noise by a factor, which is equivalent to the square root of the number of pixels/spectra in the MSI dataset.

12.3.3 Spectral Resampling

Averaging is only possible if all spectra share the same *m/z* range and digitization (sampling rate). If this is not the case, resampling (or binning) has to be performed, where the data points of a spectrum are mapped, usually by linear interpolation, onto a new common *m/z* axis (Figure 12.5A). The new *m/z* axis should orientate itself on the necessary mass resolution that one wants to achieve. As the digitizers sample the analog signal on a linear time scale, the resulting data point distribution of the *m/z* scale is usually non-linear and depends on the underlying physical separation principle of the mass analyzer (Figure 12.5B).

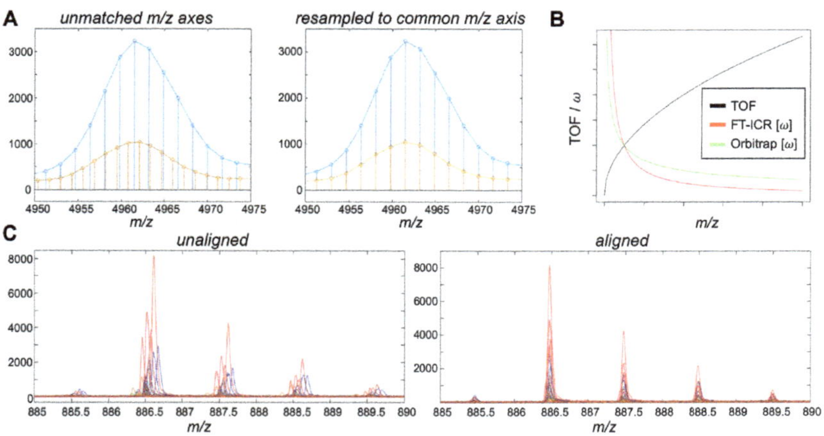

Figure 12.5 Spectral resampling and recalibration/alignment. (A) To enable a computational comparability between spectra (left), resampling maps the data points of a spectrum onto a new common *m/z* axis (right). (B) To exploit the full mass resolution while minimizing memory usage, the new *m/z* axis should account for the underlying physical separation principle of the mass analyzer by creating non-linear spaced bins. (C) During this step, unaligned spectra (left) can also be aligned, which enhances comparability (right). Abbreviations: TOF, time-of-flight; ω, angular frequency; FT-ICR, Fourier transform ion cyclotron resonance.

12.3.4 Spectral Alignment and Recalibration

However, even if spectra share the same *m/z* axis, it is not guaranteed that one can average spectra reliably. This is the case if spectra suffer from altered *m/z* calibrations, which becomes visible as unaligned peaks of the same molecule across different spectra (Figure 12.5C, left). The reasons are manifold and include outdated instrument calibrations, space-charge effects, temperature changes, charge accumulations, as well as topographic variations in the tissue.[21] To correct this spectral mismatch, alignment or recalibration of the spectra can be performed (Figure 12.5C, right). While the first approach concerns more the improvement of comparability between spectra, the second also aims at increasing the mass accuracy where it is important that the signals match at the chemically correct position. The latter is crucial for analyte annotation and therefore makes use of known molecule signals in the spectrum, known as lock-masses. In MALDI, the matrix and clusters of it constitute useful recalibration lock masses since they can be usually found in every spectrum, if the experiment has been done in the lower mass range ($m/z < 1000$).[22] Alternatively, any other known and detected endogenous compound can be used.[23]

Several spectral alignment tools have been developed for MSI, which enable to significantly reduce these mass shifts. Kulkarni *et al.* proposed the use of a minimum-spanning-tree based on peak list similarities to generate a consensus spectrum to which all peak lists are aligned using linear regression.[21] Eriksson *et al.* have recently addressed the correction of non-linear shifts in peak list data, as encountered in FT-based MS data, by developing the tool MSIWarp.[24] Both approaches require *a priori* peak picking on every spectrum since they work on peak lists. The approach by Ràfols *et al.*, in contrast, does not need peak picking beforehand since it determines linear mass shifts and scaling factors by maximizing the cross-correlation between Fourier-transformed spectra.[25] The method of Boskamp *et al.* uses the expected Kendrick mass defect of peptides to align the mass spectra.[26] The latter two techniques are available as R-package *rMSIproc*[27] and as plugin for the SCiLS Lab software, respectively. Other software packages such as Cardinal,[17] MALDIquant,[28] and LipostarMSI[29] offer similar options to align spectra to a reference peak list.

12.3.5 Normalization

Topographic height differences within or between tissue sections cannot only induce mass shifts but also intensity differences in MSI.[30] Moreover, inhomogeneous matrix application, analyte extraction, and crystal formation can add artificial variation in signal intensities within a sample too.[31] It is known that these effects are more pronounced in intense signals than in signals with lower intensity, known as heteroscedasticity.[32] As this happens on a pixel level, the single spectra are normalized individually by a factor that corrects for these detrimental effects. In TOF spectra, the chemical noise level in the baseline contains the information on the desorption intensity, which is best modeled by the total-ion-count.[31] In mass spectrometers with low chemical noise (as in FT-MS), desorption is mainly represented in the intensity of the peaks itself: root-mean-square or base peak normalization can then be used.[33] In targeted MSI, where MSI is performed to follow a specific compound's distribution, a structural analog can be applied to the sample for normalization.[34] Finally, heteroscedasticity is usually addressed by the application of variance stabilizing transformations during data analysis.

12.3.6 Peak Picking and Peak Filtering

In general, the detection of a peak in mass spectra can be based on one or more criteria including a minimum signal-to-noise ratio, relative or absolute intensity, slope of the peak, peak width, distance to next local minima *etc.*[14] It is important to mention that some recently proposed peak picking algorithms have extended these criteria by the incorporation of spatial information where neighboring pixels are more likely to share the same peak list than pixels far away from each other.[35,36]

Peak picking in MSI can be performed on the single spectra or on an abstracted spectrum. By averaging the spectra to form an abstracted spectrum, there is the danger that very localized signals in an MSI dataset can be diluted so much that they will not be detected.[37] One solution to increase the sensitivity for very localized signals is to perform a per pixel peak picking or on a random subset of pixels to reduce the computational load.[38] An alternative to per pixel peak picking is the use of quantile spectra. The most used quantile spectrum is the *skyline* or *basepeak spectrum* (100% quantile) (Figure 12.6).[39] Other quantile values, such as the median, can be taken to

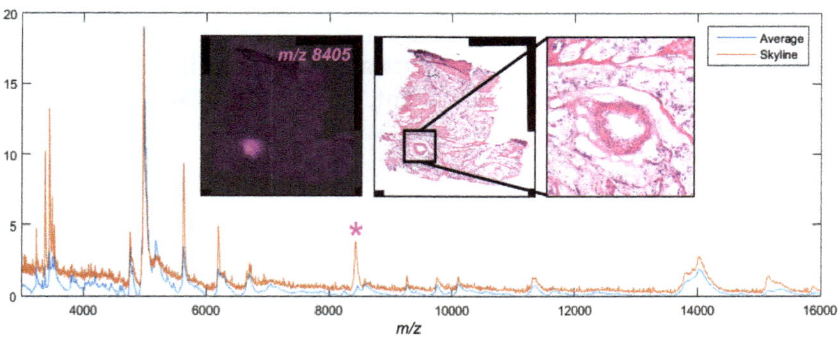

Figure 12.6 Use of quantile spectra to find localized peaks. The 100% quantile spectrum (a.k.a. basepeak, maximum, or skyline spectrum; orange spectrum) can be used to find very spatially confined molecular signals whose intensity would be diluted in the average spectrum (blue spectrum). Here a signal is shown that is exclusively located in the blood vessel of the tissue. As can be observed in the orange spectrum, not only the signal of localized peaks is amplified but also the noise.

balance the influence of outliers (with the consequence of noise amplification) against sensitivity to detect very localized signals.[40]

For many data analysis applications that follow, it can be beneficial to group or filter out non-informative or redundant peaks such as those corresponding to the same molecule but with different chemical adducts, isotopes, matrix peaks, *etc.* This can be done in several ways: accurate and high mass resolution MSI measurements can help in finding isotopes and chemical adducts with high certainty. This certainty can be increased by spatial congruence of the ion images, as exploited by METASPACE.[41] Correlated and anti-correlated peaks can also be assessed easily by a principal component analysis, which has been used to filter out matrix and off-tissue signals.[42] Furthermore, the mass defect in high mass resolution measurements can be used to find and filter chemically related compounds.[43] Finally, spatial chaos and pixel frequency filter have been proposed as additional peak filtering techniques to only retain peaks that exhibit a somehow structured visualization or a minimum presence in the sample.[44]

12.3.7 Image Processing and Visualization

Due to the dual nature of MSI data, image processing can also be applied to improve the quality of single *m/z* value images. One of the most common ways to improve contrast for the human eye in an image is by removing extreme values, which can occur and manifest themselves as visible hot spots in the image. The most common way to remove outliers in MSI is to use quantiles as thresholds, usually the 1% and 99% quantiles (Figure 12.7A).[37]

Since MSI data is purely digital, the correct interpretability of MSI images by a human observer can also depend on the selected color scheme (Figure 12.7B). Race and Bunch recommend abandoning rainbow-based color schemes and instead using perceptually linear color schemes whose linearity can be tested by the CIELAB distance metric.[45]

Additionally, the before-mentioned per-pixel normalization can increase contrast (Figure 12.7C). If after normalization technical variation remains between pixels, spatial smoothing, de-noising or interpolation can be applied.[46] These act locally in a defined window, since it can be assumed that, due to the spatial arrangement and interaction of cells in tissues, pixels that are close to each other are more similar than pixels further away from each other.[47] Smoothing, filtering, outlier removal, and contrast enhancement through the selection of appropriate color schemes or histogram equalization are implemented in most MSI software packages.[10,17,28]

Another aspect of image processing concerns the digital upscaling of the spatial resolution of MALDI-MSI data such as by super-resolution algorithms as initially proposed by Alexandrov *et al.*[48] This has now been complemented by image fusion algorithms, which can upscale the MALDI-MSI image resolutions by a factor of ten.[49,50] Unfortunately, these algorithms are not yet implemented in commercial or non-commercial MSI software and therefore require interaction with a programming environment.

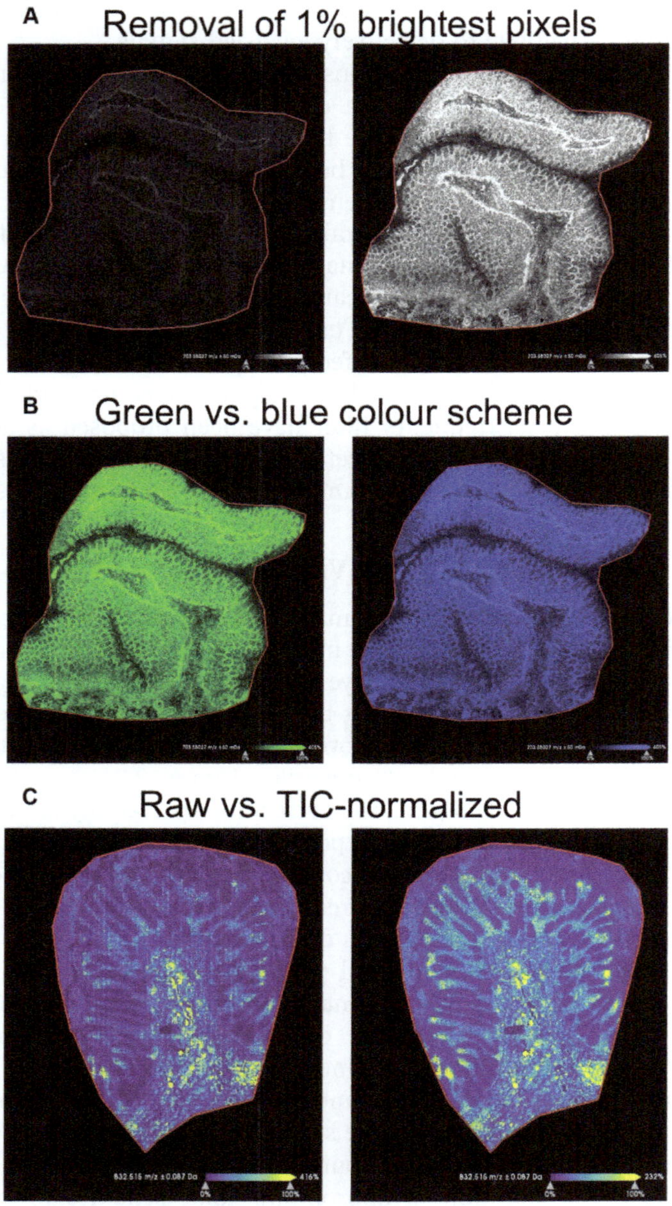

Figure 12.7 Enhancing contrast in mass spectrometry images. (A) Contrast can be enhanced by removing a defined percentage of the most intense or less intense pixels as shown here for a mouse colon based on the removal of pixels with total-ion-counts above the 99% quantile (right image). (B) The perceived contrast also depends on the selected color scheme with respect to the background colour. Since the black-to-blue distance in the CIE space is smaller than the black-to-green distance, the contrast is also lower for the black-blue color scheme (right image). (C) Pixel-based normalization can also increase contrast as shown in this mouse colon for the total-ion-count normalization (TIC, right image).

12.4 Data Analysis

Once data processing has been done appropriately, it can be assumed that technical variation has been minimized to extract reliable information from MSI data. This is also under the assumption that the experiments have been designed in such a way that disturbance variables cannot lead to false correlations.[12]

Similarly, to the processing step, the multivariate and image-forming nature of MSI data strongly influences the choice of the data analysis strategies. The strategy thereby depends on the aim of the study of which the most common are: the finding of discriminatory *m/z* values, the determination of their identity, dimensionality reduction, and the supervised or unsupervised classification of pixels or samples by multivariate algorithms. These aims and corresponding approaches are usually not stand-alone and intermingle to a non-linear workflow (Figure 12.8). It is noteworthy, however, that it is impossible to define the best modeling technique *a priori*. An initial benchmark study might be necessary, with a focus on providing interpretable models with adequate predictive ability.[51]

12.4.1 Finding Discriminatory *m/z* Values

12.4.1.1 Inferential Statistics

There are several approaches in MSI to assess the discriminatory power of individual *m/z* values of which inferential statistics constitutes the most common one.[46] Inferential statistics comprises a very broad group of procedures and the selection of the appropriate test depends mainly on the data type of the independent variable(s) which can be either of categorical nature (*e.g.* "disease group" *vs.* "control group"),[52] of ordinal nature (*e.g.* cancer stages I–IV, where a higher stage indicates a higher tumor burden),[53] or of

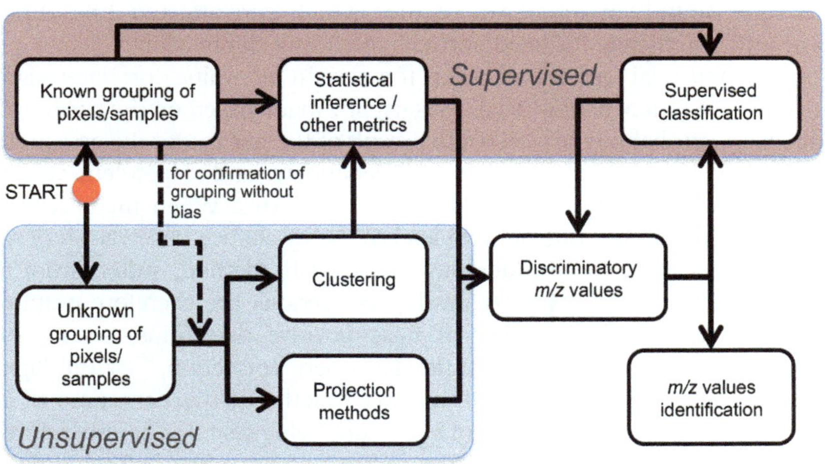

Figure 12.8 Common elements of supervised and unsupervised data analysis workflows.

metric nature (*e.g.*, time or blood levels of metabolites). In the latter case, regression or correlation analysis can be performed. Correlation analysis is especially used in MSI to find *m*/*z* values co-localized to specific regions[17] or to find *m*/*z* values with similar distributions.[41] Kunzke *et al.* used correlation to find spatially correlated and anti-correlated native glycan fragments and showed that some of them participate in the same biological processes.[54] Regression analysis can be useful to find molecules that change over time according to an expected mathematical function. Aljakna-Khan *et al.* used MSI and linear regression to detect metabolic changes in rat hearts over five different time durations of ischemia.[55]

An important criterion is the compliance of the MSI data with the assumption of the selected test. Most common parametric statistical tests (ANOVA or regression) assume normally distributed data and a certain independence of the observations from each other. In case the normality constraint is not met, non-parametric statistical tests can be employed or the data be transformed by taking the log or extracting the *n*th root.[31] If the independence assumption is not fulfilled – this is the case when single pixels or several regions from the same sample are submitted to the test – it has been shown that MSI data leads to the finding of false-positive results.[47] In that case, alternative statistical approaches must be applied to ensure reliable statistical results.[47,56]

MSI data can also lead to complex relations of the data in a statistical testing model when it comes to technical replicates, hierarchical relations, or paired designs. The first two were encountered in a round-robin study, where several technical replicates were measured across four different centers, which therefore constitutes a hierarchy.[57] Correct statistical modeling using mixed-effect models solved this scenario, where the tissue type was considered the fixed effect, and the replicate and center random effects. With respect to the usage of paired designs, Carreira *et al.* compared left *versus* right brain hemispheres of 32 mice, which were affected differently by a local and chemically-induced cortical spreading depression.[58]

For any statistical test, the decision if a certain *m*/*z* value correlates significantly with the independent variables is usually made based on the *p*-value. The *p*-value is the probability to observe the result of the test under the assumption that there is no correlation between independent and dependent variable. However, this assumption comes with a certain error. When many tests are done, this error can accumulate and lead to the by-chance observation of false positive results. MSI, as an inherently multivariate method, suffers from this statistical phenomenon. Multiple testing corrections are therefore employed routinely to reduce the number of false positive detections, such as the Bonferroni correction,[59] the Benjamini–Hochberg correction,[55] or the Significance Analysis of Microarrays.[60] In all these methods, the harshness of correction increases with the number of *m*/*z* values being tested. It can therefore be beneficial to exclude non-informative or redundant *m*/*z* values beforehand from testing, such as isotope and matrix peaks, to reduce the negative impact of the multiple-testing correction.

12.4.1.2 Other Discriminatory Metrics

Finally, if none of the statistical assumptions are met, the discriminatory power of single *m*/*z* values can also be assessed by other more agnostic approaches such as by receiver-operating-characteristics (ROC) analyses. Paine *et al.* used this method to compare lipid abundances between metastasizing and non-metastasizing brain tumor regions obtained by 3D-MSI of brains from six transgenic mice.[61] To fully exploit the 3D data, the comparisons were performed on a pixel level (441 152 *vs.* 1 010 805 individual mass spectra). Although this infringes the independence assumption for classical statistical testing, a ROC analysis still constitutes a valid method of choice.

Obtaining a list of discriminating *m*/*z* values can help in establishing patterns that can be used by supervised machine learning algorithms to create classifiers that can predict the label of unlabeled data.

12.4.2 Supervised Machine Learning

Supervised machine learning is quite a common approach in MALDI-MSI since obtaining directly multivariate patterns enables exploitation of the multi-dimensional space for separating different types of pixels. There are many machine-learning approaches such as discriminant analysis (DA), support vector machine (SVM), random forest (RF), and neuronal networks (NN). Comprehensive explanations on the mode of operation of these algorithms can be found elsewhere for SVM and RF,[62] and for DA and neuronal networks.[63] All of these methods distinguish themselves in their degree of interpretability, parameterization, and accuracy which depends very much on the task.[62] Interpretability is important in MSI due to the biomedical background of most MSI studies and the associated wish for an understanding of the underlying biological processes, which should be reflected in the classification model.

12.4.2.1 Supervised Machine Learning Algorithms in MSI

Linear DA (LDA), partial least squares DA (PLS-DA), and principal component analysis LDA (PCA–LDA) are the most common approaches in MSI, where the latter two reduce the dimensionality of the multivariate data before performing DA. In MSI, Mas *et al.* have used PLS-DA to discriminate highly similar but clinically different colorectal mouse xenografts on a pixel level.[64] Mascini *et al.* have used PCA–LDA to predict treatment response in patient-derived xenograft tissues of triple-negative breast cancers on a patient level[65] and Kriegsmann *et al.* have used LDA to distinguish subtypes of non-small cell lung carcinomas on a pixel level.[66] Recently, the same group has also directly compared the classification accuracy of different machine learning approaches (RF, LDA, SVM) for the same task, albeit on a larger cohort,[67] and for the separation of chromophobe renal cell carcinoma from renal oncocytomas.[68] In both studies, RF outperformed the other two algorithms. In another study, Klein *et al.* also compared LDA, SVM with different kernels, and different neural networks for the classification of epithelial ovarian cancer histotypes on a pixel and sample level.[69] In this study, neuronal networks slightly outperformed the other approaches. This is also the reason why more recently deep learning approaches (which are special forms of neuronal networks) have found their way into MSI. Again, since the nature of MSI data is unique, the architecture of deep neuronal networks needed to be adapted to the characteristics of MSI data, such as the vast existence of correlated *m*/*z* bins.[70]

12.4.2.2 Training and Testing

In this context, it is important to mention that very sophisticated machine learning algorithms need to be fed with lots of training data. This is on the one hand necessary to optimize the usually extensive parameterization, but also to overcome the curse of dimensionality where a non-linear separating plane will cut its way through the classes in a non-reproducible manner, which is termed over-fitting, as observed by Behrmann *et al.* for deep learning in MSI.[70] It is therefore needed to rigorously validate supervised classifiers on independent data. If external data is not available, internal cross-validation (CV) constitutes a valid approach to validate the model as long as the CV data split is done at the highest level of the sampling hierarchy, which is in MSI the biological replicate.[12] This guarantees the highest level of independence between the training and the validation datasets and therefore the most accurate estimate for the real classification error of the classifier. To solve the imbalance in MSI between the low number of biological replicates and the high number of pixels, Pérez-Guaita *et al.* recently proposed the use of constrained repeated random subsampling CV for MSI datasets.[71] In this pixel-based approach, the CV method ensures that the distance among train pixels and between test and train pixels is higher than a user-defined threshold, which leads to performance estimates that are closer to the real classification error.[71]

12.4.2.3 Feature Selection and Extraction

Another crucial aspect in multivariate classification is the feature selection, *i.e.*, the selection of *m*/*z* values that are used to train the classifier. As already described in the previous section, there are several statistical and non-statistical approaches to obtain a set of discriminating *m*/*z* values. Another approach is to use *m*/*z* values for classification that putatively belong to known biomarkers, as proposed by Cordero-Hernandez *et al.*[72] The rationale behind this idea is that the combination of known biomarkers might outperform the individual performances.

However, the steps of feature selection and machine learning are inherently connected to each other in the sense that *m*/*z* values might not exhibit discriminatory power at the univariate level, but in combination with each other.[46] In contrast to feature selection, feature extraction deals with the combination of features to generate less but more discriminatory features. Boskamp *et al.*, for instance, proposed the use of characteristic spectral patterns for feature extraction.[73] A non-negative matrix factorization algorithm thereby decomposes the MSI data, which leads to the reduction of the data to some defined number of multivariate components. These can then be used for further classification purposes. Likewise, Thomas *et al.* used deep learning methods to reduce the dimensionality of the original MSI data before feeding it into subsequent classifiers such as SVM, which improved the classification performance.[74]

12.4.3 Unsupervised Projection and Clustering Methods

High attention in MSI has been given to unsupervised approaches, which do not *a priori* require the data to carry a label. Based on the type of method used, MSI pixels or samples are either projected into a lower-dimensional space for a visual examination of similarity, termed *projection*, or directly grouped by their spectral similarity, termed *clustering*.[40] While all these unsupervised methods are mainly used to explore the inherent structure of the data, they can also be used to confirm the quality of the data by recreating the known grouping of the samples[75] or the underlying morphology of the tissue (Figure 12.8).[76]

12.4.3.1 Unsupervised Algorithms in MSI

The number of unsupervised methods is vast. The excellent review of Verbeeck *et al.* lists no less than 40 different algorithms that have been applied in the field of MSI.[40] Among these, the most commonly used projection methods are principal component analysis (PCA), non-negative matrix factorization (NMF), probabilistic latent semantic analysis (pLSA), self-organized map (SOM), and t-distributed stochastic neighborhood embedding (t-SNE), and uniform manifold approximation and projection (UMAP). The criteria for choosing the most suitable algorithm are manifold and pose a challenge for the researcher. Factors that play a role in the choice are the computational performance of the algorithms, the degree of parameterization (number of expected components, scaling, convergence criteria), the quality of the results, and the availability in user-friendly software packages.

One parameter that is important to all unsupervised methods is the selection of features (*m/z* values).[77] If no grouping is known initially, *m/z* values cannot be evaluated based on their discriminatory power. Ranking the features based on their variance after logarithmic transformation is commonly used in gene expression.[78] In MSI, Murta *et al.* have recently presented a hypothesis-driven peak selection based on previous tentative identification of the *m/z* values, which makes the resulting *a posteriori* easier to interpret.[77]

12.4.3.2 Linear Projection Techniques

In that line, most important to the researcher is the quality and the interpretability of the results. PCA has most likely been the earliest used multivariate technique in MALDI-MSI.[79] As it is based on variance and groups correlated *m/z* values into the same component, it can be used for finding region-specific *m/z* patterns[80] or cleaning the data by removing unwanted signals, such as off-tissue or matrix-related signals.[42] It has also been used on a patient level to find and confirm new and known clinical subgroups in pancreatic cancer.[81] An important parameter in PCA is the scaling of the data to ensure an equal contribution of the *m/z* values. Z-scoring and its lighter version "Pareto"-scaling are most commonly used, but many other methods come in, which might be more appropriate for every individual situation.[82] In contrast to PCA, NMF and pLSA require the user to enter an expected number of components before the analysis. While in a truly unsupervised scenario it is difficult to know the true number of components in advance, a way to determine this is by correlating the results with orthogonal information, such as clinical data of the patients.[83] Nonetheless, pLSA and NMF have found numerous applications in MALDI-MSI, especially to highlight histologically invisible, molecular changes in tissues such as intra-tumor heterogeneity[84] or molecular events in the rat and zebra finch brains associated with cortical spreading depression[85] or song learning behavior,[86] respectively.

12.4.3.3 Non-linear Projection Techniques

While PCA *et al.* are powerful methods that help explore MSI data, they can only capture linear relationships within the data. More modern techniques, such as UMAP, t-SNE, and SOM, can portray non-linear relationships.[87] In 2013 Fonville *et al.* have proven the superiority of t-SNE and SOM over linear techniques for MSI data.[88] Based on these results, t-SNE has increasingly been used in MSI. For instance, Abdelmoula *et al.* have used t-SNE for the reduction of datasets as a pre-processing step for automatic coregistration.[89] The same authors also used t-SNE to reveal molecular intra-tumor heterogeneity in several cancer types.[90] While t-SNE provides superior results, it puts high demands on runtime.[88] Alternatives, such as UMAP, have been proposed that provide comparable results to t-SNE, but allow the analysis of large MSI datasets (>100 000 pixels) with moderate computational requirements.[91]

12.4.3.4 Clustering Techniques

Clustering complements the projection approaches by grouping pixels into clusters, also called *image segmentation*. The most common methods are k-means, its hierarchical derivate bisecting k-means, and hierarchical clustering. Regular k-means is fast and has low memory demands but requires the predefinition of expected clusters. In advanced programming environments, the number of clusters can be optimized based on quality metrics such as the Silhouette criterion or Dunn's index.[92] Hierarchical clustering, in contrast, provides a more comprehensive overview in the form of a tree-like structure of the data. It puts however higher demands on memory and the results depend significantly on its parameterization. However, good starting parameters have been established for MSI data: correlation coefficient[92,93] or the cosine distance[94] as the distance metric, and Ward's linkage method.[95]

Bisecting k-means provides an interesting alternative since it combines the advantages of k-means and hierarchical clustering. It has been recently used by Lagarrigue *et al.* to segment a rat's epididymis into molecular clusters that are subdivided into known histological clusters.[96] Nevertheless, MSI data can also be transposed to group *m/z* values as recently shown by Hiratsuka *et al.*, who merged peptide distributions in human brain tissue based on spatial similarity and hierarchical clustering.[97]

A danger in the application of these distance-based methods is the curse of dimensionality, where distances in the high dimensional space become non-meaningful.[98] To overcome this, Alexandrov *et al.* proposed the use of High-Dimensional Space Clustering.[38] Due to the hybrid nature of MSI, spatially aware clustering algorithms have also been developed that take into account the spatial information within MSI data, which, in short, do not only look at individual pixels but also at their neighborhood when clustering them.[99,100]

While the number of available clustering techniques is growing, there is still a lack in the evaluation of the quality of these clustering algorithms.[101] As there is no ground truth, some algorithms have been tested on synthetic data still leaving the question open on how well these methods will work under real conditions.[93]

12.4.4 Identification of *m/z* Values

In MSI, the comprehensive identification of *m/z* values still poses a major challenge due to the lack of chromatographic separation before the MS analysis.[102] Because of this, most on-tissue MS/MS identifications of proteins/peptides have been limited to the most abundant ion signals in the tissue.[103] Approaches have been proposed throughout the years that aid in large-scale identification, which however either compromise the spatial resolution or the unambiguity, or lead to significant additional work.[102] The most common of these approaches is to extract material from the same or a consecutive section and submit it to liquid-based MS analyses with the aim to create a sample or cohort-specific molecular database which can be queried by using mass matching with high mass accuracy.[104,105] High accurate mass for querying databases is also the most common approach for MSI of metabolites. In METASPACE – an online platform dedicated to the exchange of MSI data – molecular annotations are based on several criteria, such as the accurate mass, the isotope pattern, the spatial congruence of isotope patterns and the quality of visualization.[106] This software queries many different metabolite databases, such as HMDB, ChEBI, or LipidMaps. That software and technological advances go hand in hand is also reflected in other software packages. LipostarMSI, for instance, provides MS/MS identification routines, which has been proven useful for lipid annotations.[29,107] MetaboScape utilizes libraries of molecule-specific ion mobility characteristics, such as collision cross-section values, to annotate MSI data of metabolites, lipids, or drugs.[104] Ion mobility not only holds great promise for identification, but also for classification due to the extra dimension for every *m/z* value (drift time).

A comprehensive identification of *m/z* values also increases the ability of integrating MSI data with other data or databases.

12.4.5 Data Integration

Data integration with databases from the field of systems biology or chemometrics has become the new standard in biomedical research to understand the molecular mechanisms behind the findings.[51,108] Queries against pathway databases usually require the large-scale identification of the involved molecules.[109] MSI is hence nowadays accompanied by other omics approaches. Mallah *et al.*, for instance, used lipid MSI to guide local liquid extractions for microproteomics to spatially map alterations in the proteome in brain regions after traumatic brain injury.[110] The authors used commercial software (Elsevier's Pathway Studio) to query the differential proteins and their involvement in pathways. However, there are also free tools which have been used substantially in MSI for mechanistic interpretations, such as Reactome (www.reactome.org),[105] MetaboAnalyst (www.metaboanalyst.ca),[111] or molecule–molecule interaction databases, such as STRING (https://string-db.org).[112]

An example that shows the potential of integrating non-MS omics data with MSI is the study by Ščupáková *et al.* who combined gene expression data from 15 human steatotic liver biopsies with the lipid MSI data from the same patients using GeneGo's MetaCore software.[113] This software identified several lipid–protein interaction networks related to inflammation processes and the grade of steatosis. However, the analysis was based on only four lipids. This also shows that a comprehensive list of identified molecules is needed to derive reliable mechanistic insights. This is the reason why MSI has so far been integrated more with other imaging data.

12.5 Image Registration

12.5.1 Integration with *Ex Vivo* Techniques

As a non-destructive *ex vivo* technique, MALDI-MSI has from the early days already been registered to imaging data from the very same tissue section using optical microscopy.[114] Since the latter is the gold standard in a clinical setting, correlating MALDI-MSI to histological information has become routine in clinically motivated MSI studies. For this, MSI and microscopic images are usually overlain semi-automatically using a manual control point co-registration followed by an automated geometric transformation of the images as offered by most MSI software packages with a graphical user interface.[115] The accuracy provided by this approach is usually sufficient to transfer information from the microscopic image to the MSI image.[89] Prade *et al.*, for instance, correlated in this way spatial metabolic patterns as obtained by MSI with multiple molecular statuses of the tissue as obtained by multiplex immunohistochemistry.[109]

Given further advances in the spatial resolution of MALDI-MSI, these co-registration techniques are currently being revised to achieve image alignment accuracies to enable single cell analyses.[116] Given also the increased throughput in MSI, automated co-registration approaches have been proposed.[89,117] The latter combined with digital pathology, *i.e.*, the automated annotation of histological images by machine learning algorithms, promises new means on how multi-modal datasets can be exploited by *e.g.*, correlating morphometric characteristics of cells with their molecular profile.[116]

Also, multiple MALDI-MSI datasets from the same sample can be overlayed since the choice of the matrix determines the molecular class that can be detected. In this way, Patterson *et al.* integrated five MALDI-MSI datasets based different sample preparation techniques and instrument settings to obtain a more complete molecular overview of single samples.[118] The authors developed a dedicated software package ("RegCombIMS") for this task.

A co-registration of the MSI data to the corresponding optical images can also enable microdissection of regions highlighted by MSI.[115] The dissected regions can be submitted to other *omics* techniques. This approach thereby combines the spatial information of MSI with the molecular profiling capabilities of genomics, transcriptomics, metabolomics, or proteomics.[119]

12.5.2 Integration with *In Vivo* Techniques

In vivo techniques, such as magnet resonance imaging (MRI), can offer several advantages over invasive techniques such as MSI. They allow longitudinal studies on the same subject and do not distort the original volumetric information of the tissue. The latter can be useful for 3D-MALDI-MSI since the original 3D volume is compromised by the sectioning. An *in vivo* technique can therefore act as a reference for arranging the MSI slices to a 3D volume.[120] If MSI images and optical images are already connected, the reconstruction can be done *via* the optical images. Sinha *et al.* followed this approach by first pre-aligning the optical images using an automatic intensity-based co-registration limited to translation and rotation since the optical images had been acquired during sectioning with a fixed camera.[121] The resulting volume was subsequently registered to the 3D-MRI data using a rigid registration. Using this approach, the same group investigated volumetric proteomic changes related to inflammation in kidneys of *S. aureus* infected mice.[122] Inflammation was reflected as visible abscess formation in MRI images. Likewise, Oetjen *et al.* first aligned the optical images of a mouse kidney using a rigid and elastic co-registration and then co-registered the resulting 3D MSI volume rigidly to the corresponding MRI dataset.[123] Recently, Abdelmoula *et al.* proposed an approach that is independent of the optical images and uses the t-SNE projected MSI images to create edge structures that guide the MSI–MRI registration.[120] The co-registration was performed on a section-to-section basis using affine transformation followed by a non-linear elastic transformation that allows correcting for local distortions. This allowed the volumetric distribution of erlotinib to be shown, as detected by MSI within the MRI volume. At last, *in vivo* techniques are continuously being used to create anatomical atlases from tissues and organs. Verbeeck *et al.* and Abdelmoula *et al.* have demonstrated the usefulness of registering 2D MSI data to these atlases, such as the Allen Brain Atlas or the Scalable Brain Atlas, to automatically extract molecular profiles from these distinct anatomical annotations.[124,125]

It is important to note that all co-registrations of MALDI-MSI with *in vivo* techniques have been done on tissues that exhibit very strong cellular and molecular structures (brain and kidney) and thereby simplify the co-registration problem for the algorithms. On the other hand, there are plenty of examples from DESI-MSI,[126] LA-ICP MSI,[127] and SIMS,[128] which demonstrate that accurate spatial co-registration might not be necessary. In that line, Aichler *et al.* used MALDI-MSI to describe the concentration and spatial distribution of the MRI-contrast agent gadofluorine M at a microscopic level in infarcted mouse hearts after MRI.[129] However, this study did not require the overlay of MRI and MSI images.

12.6 Outlook – MSI Software in the AI Age

As demonstrated throughout this chapter, software, data processing, and analysis algorithms are the pillars for the successful conduction of MALDI-MSI studies. As the field is further developing and gaining interest, more attention is given to MSI by researchers specialized in data analysis or image processing. More precisely, advanced machine learning and data science are now slowly coming to the forefront, as reflected in the growing number of reported deep learning-assisted data analysis. While deep learning can already be applied now, data science strategies are strongly dependent on the large-scale identification of the detected *m*/*z* values. MSI-guided spatial *omics* can thereby constitute a bridging technique until MSI is able to provide large-scale identifications directly from tissue.

While some of these algorithmic advances have already taken place, the average user has difficulty in exploiting these software tools since they are not integrated in most MSI software with a graphical user interface. This might also be due to the complicated parameterization of these solutions, which so far requires specialized expertise until general recommendations can be derived from the increased usage of these techniques.

Still the user is being offered nowadays, more than ever, a powerful collection of free and commercial software packages such as Cardinal, LipostarMSI, MALDIquant, Metabolite Explorer, MSIReader, rMSI, SCiLS, and SpectralAnalysis, just to name a few in alphabetical order. It is expected that all of these tools will become even more versatile and seamlessly integrate with other software such as those from the field of digital pathology.[117] This will further strengthen the position of MSI in the field of biomedical research.

References

1. P. M. Vaysse, R. M. A. Heeren, T. Porta and B. Balluff, *Analyst*, 2017, **142**, 2690.
2. M. Stoeckli, T. B. Farmer and R. M. Caprioli, *J. Am. Soc. Mass Spectrom.*, 1999, **10**, 67.
3. The Cancer Genome Atlas Research Network, *Cell*, 2017, **169**, 1327.
4. K. Scupakova, B. Balluff, C. Tressler, T. Adelaja, R. M. A. Heeren, K. Glunde and G. Ertaylan, *Clin. Chem. Lab. Med.*, 2020, **58**, 914.
5. B. Heijs, W. M. Abdelmoula, S. Lou, I. H. Briaire-de Bruijn, J. Dijkstra, J. V. Bovee and L. A. McDonnell, *Anal. Chem.*, 2015, **87**, 11978.
6. P. Bankhead, M. B. Loughrey, J. A. Fernandez, Y. Dombrowski, D. G. McArt, P. D. Dunne, S. McQuaid, R. T. Gray, L. J. Murray, H. G. Coleman, J. A. James, M. Salto-Tellez and P. W. Hamilton, *Sci. Rep.*, 2017, **7**, 16878.
7. T. Schramm, Z. Hester, I. Klinkert, J. P. Both, R. M. A. Heeren, A. Brunelle, O. Laprevote, N. Desbenoit, M. F. Robbe, M. Stoeckli, B. Spengler and A. Rompp, *J. Proteomics*, 2012, **75**, 5106.
8. P. Kallback, A. Nilsson, M. Shariatgorji and P. E. Andren, *Anal. Chem.*, 2018, **90**, 3676.
9. O. Rubel, A. Greiner, S. Cholia, K. Louie, E. W. Bethel, T. R. Northen and B. P. Bowen, *Anal. Chem.*, 2013, **85**, 10354.
10. P. Kallback, A. Nilsson, M. Shariatgorji and P. E. Andren, *Anal. Chem.*, 2016, **88**, 4346.
11. I. Klinkert, K. Chughtai, S. R. Ellis and R. M. A. Heeren, *Int. J. Mass Spectrom.*, 2014, **362**, 40.
12. B. Balluff, C. Hopf, T. Porta Siegel, H. I. Grabsch and R. M. A. Heeren, *J. Am. Soc. Mass Spectrom.*, 2021, **32**, 628.
13. K. Yang, X. Fang, R. W. Gross and X. Han, *J. Am. Soc. Mass Spectrom.*, 2011, **22**, 2090.
14. C. Yang, Z. He and W. Yu, *BMC Bioinf.*, 2009, **10**, 4.
15. Z. Wang, M. Zhang and B. Harrington Pde, *Anal. Chem.*, 2014, **86**, 9050.
16. B. Williams, S. Cornett, B. Dawant, A. Crecelius, B. Bodenheimer and R. Caprioli, in *Proceedings of the 43rd Annual Southeast Regional Conference*, Association for Computing Machinery, Kennesaw, Georgia, 2005, vol. 1, p. 137.
17. K. D. Bemis, A. Harry, L. S. Eberlin, C. Ferreira, S. M. van de Ven, P. Mallick, M. Stolowitz and O. Vitek, *Bioinformatics*, 2015, **31**, 2418.
18. T. E. Stanford, C. J. Bagley and P. J. Solomon, *Proteome Sci.*, 2016, **14**, 19.
19. D. Trede, S. Schiffler, M. Becker, S. Wirtz, K. Steinhorst, J. Strehlow, M. Aichler, J. H. Kobarg, J. Oetjen, A. Dyatlov, S. Heldmann, A. Walch, H. Thiele, P. Maass and T. Alexandrov, *Anal. Chem.*, 2012, **84**, 6079.

20. A. M. Race, A. D. Palmer, A. Dexter, R. T. Steven, I. B. Styles and J. Bunch, *Anal. Chem.*, 2016, **88**, 9451.
21. P. Kulkarni, F. Kaftan, P. Kynast, A. Svatos and S. Bocker, *Anal. Bioanal. Chem.*, 2015, **407**, 7603.
22. V. Garikapati, S. Karnati, D. R. Bhandari, E. Baumgart-Vogt and B. Spengler, *Sci. Rep.*, 2019, **9**, 3192.
23. J. O. R. Gustafsson, J. S. Eddes, S. Meding, T. Koudelka, M. K. Oehler, S. R. McColl and P. Hoffmann, *J. Proteomics*, 2012, **75**, 5093.
24. J. O. Eriksson, A. Sanchez Brotons, M. Rezeli, F. Suits, G. Marko-Varga and P. Horvatovich, *Anal. Chem.*, 2020, **92**, 16138.
25. P. Rafols, E. D. Castillo, O. Yanes, J. Brezmes and X. Correig, *Anal. Chim. Acta*, 2018, **1022**, 61.
26. T. Boskamp, D. Lachmund, R. Casadonte, L. Hauberg-Lotte, J. H. Kobarg, J. Kriegsmann and P. Maass, *Anal. Chem.*, 2020, **92**, 1301.
27. P. Rafols, B. Heijs, E. Del Castillo, O. Yanes, L. A. McDonnell, J. Brezmes, I. Perez-Taboada, M. Vallejo, M. Garcia-Altares and X. Correig, *Bioinformatics*, 2020, **36**, 3618.
28. S. Gibb and K. Strimmer, *Bioinformatics*, 2012, **28**, 2270.
29. S. Tortorella, P. Tiberi, A. P. Bowman, B. S. R. Claes, K. Scupakova, R. M. A. Heeren, S. R. Ellis and G. Cruciani, *J. Am. Soc. Mass Spectrom.*, 2020, **31**, 155.
30. R. Longuespee, K. Kriegsmann, M. Cremer, C. Zgorzelski, R. Casadonte, D. Kazdal, J. Kriegsmann, W. Weichert, K. Schwamborn, M. Fresnais, P. Schirmacher and M. Kriegsmann, *Proteomics: Clin. Appl.*, 2019, **13**, e1800074.
31. S. O. Deininger, D. S. Cornett, R. Paape, M. Becker, C. Pineau, S. Rauser, A. Walch and E. Wolski, *Anal. Bioanal. Chem.*, 2011, **401**, 167.
32. P. M. Wehrli, W. Michno, K. Blennow, H. Zetterberg and J. Hanrieder, *J. Am. Soc. Mass Spectrom.*, 2019, **30**, 2278.
33. J. M. Fonville, C. Carter, O. Cloarec, J. K. Nicholson, J. C. Lindon, J. Bunch and E. Holmes, *Anal. Chem.*, 2012, **84**, 1310.
34. F. Dewez, E. De Pauw, R. M. A. Heeren and B. Balluff, *Anal. Chem.*, 2021, **93**, 1393.
35. J. O. Eriksson, M. Rezeli, M. Hefner, G. Marko-Varga and P. Horvatovich, *Anal. Chem.*, 2019, **91**, 11888.
36. F. Lieb, T. Boskamp and H. G. Stark, *J. Proteomics*, 2020, **225**, 103852.
37. J. D. Watrous, T. Alexandrov and P. C. Dorrestein, *J. Mass Spectrom.*, 2011, **46**, 209.
38. T. Alexandrov, M. Becker, S. O. Deininger, G. Ernst, L. Wehder, M. Grasmair, F. von Eggeling, H. Thiele and P. Maass, *J. Proteome Res.*, 2010, **9**, 6535.
39. L. A. McDonnell, A. van Remoortere, N. de Velde, R. J. van Zeijl and A. M. Deelder, *J. Am. Soc. Mass Spectrom.*, 2010, **21**, 1969.
40. N. Verbeeck, R. M. Caprioli and R. Van de Plas, *Mass Spectrom. Rev.*, 2020, **39**, 245.

41. K. Ovchinnikova, L. Stuart, A. Rakhlin, S. Nikolenko and T. Alexandrov, *Bioinformatics*, 2020, **36**, 3215.
42. J. Cao, P. Goossens, M. Martin-Lorenzo, F. Dewez, B. S. R. Claes, E. A. L. Biessen, R. M. A. Heeren and B. Balluff, *J. Am. Soc. Mass Spectrom.*, 2020, **31**, 1825.
43. C. Kune, A. McCann, R. Raphael, A. A. Arias, M. Tiquet, D. Van Kruining, P. M. Martinez, M. Ongena, G. Eppe, L. Quinton, J. Far and E. De Pauw, *Anal. Chem.*, 2019, **91**, 13112.
44. T. Alexandrov and A. Bartels, *Bioinformatics*, 2013, **29**, 2335.
45. A. M. Race and J. Bunch, *Anal. Bioanal. Chem.*, 2015, **407**, 2047.
46. E. A. Jones, S. O. Deininger, P. C. W. Hogendoorn, A. M. Deelder and L. A. McDonnell, *J. Proteomics*, 2012, **75**, 4962.
47. A. Cassese, S. R. Ellis, N. Ogrinc Potocnik, E. Burgermeister, M. Ebert, A. Walch, A. M. van den Maagdenberg, L. A. McDonnell, R. M. Heeren and B. Balluff, *Anal. Chem.*, 2016, **88**, 5871.
48. T. Alexandrov, S. Meding, D. Trede, J. H. Kobarg, B. Balluff, A. Walch, H. Thiele and P. Maass, *J. Proteomics*, 2011, **75**, 237.
49. K. Scupakova, V. Terzopoulos, S. Jain, D. Smeets and R. M. A. Heeren, *Sci. Rep.*, 2019, **9**, 2915.
50. R. Van de Plas, J. Yang, J. Spraggins and R. M. Caprioli, *Nat. Methods*, 2015, **12**, 366.
51. S. Tortorella and S. Cinti, *Anal. Chem.*, 2021, **93**, 2713.
52. L. H. Cazares, D. Troyer, S. Mendrinos, R. A. Lance, J. O. Nyalwidhe, H. A. Beydoun, M. A. Clements, R. R. Drake and O. J. Semmes, *Clin. Cancer Res.*, 2009, **15**, 5541.
53. S. Steurer, C. Borkowski, S. Odinga, M. Buchholz, C. Koop, H. Huland, M. Becker, M. Witt, D. Trede, M. Omidi, O. Kraus, A. S. Bahar, A. S. Seddiqi, J. M. Singer, M. Kwiatkowski, M. Trusch, R. Simon, M. Wurlitzer, S. Minner, T. Schlomm, G. Sauter and H. Schluter, *Int. J. Cancer*, 2013, **133**, 920.
54. T. Kunzke, B. Balluff, A. Feuchtinger, A. Buck, R. Langer, B. Luber, F. Lordick, H. Zitzelsberger, M. Aichler and A. Walch, *Oncotarget*, 2017, **8**, 68012.
55. A. Aljakna Khan, N. Bararpour, M. Gorka, T. Joye, S. Morel, C. A. Montessuit, S. Grabherr, T. Fracasso, M. Augsburger, B. R. Kwak, A. Thomas and S. Sabatasso, *Sci. Rep.*, 2021, **11**, 5135.
56. K. A. Bemis, D. Guo, A. J. Harry, M. Thomas, I. Lanekoff, M. P. Stenzel-Poore, S. L. Stevens, J. Laskin and O. Vitek, *Int. J. Mass Spectrom.*, 2019, **437**, 49.
57. A. Buck, B. Heijs, B. Beine, J. Schepers, A. Cassese, R. M. A. Heeren, L. A. McDonnell, C. Henkel, A. Walch and B. Balluff, *Anal. Bioanal. Chem.*, 2018, **410**, 5969.
58. R. J. Carreira, R. Shyti, B. Balluff, W. M. Abdelmoula, S. H. van Heiningen, R. J. van Zeijl, J. Dijkstra, M. D. Ferrari, E. A. Tolner, L. A. McDonnell and A. M. van den Maagdenberg, *J. Am. Soc. Mass Spectrom.*, 2015, **26**, 853.

59. M. Marczyk, G. Drazek, M. Pietrowska, P. Widlak, J. Polanska and A. Polanski, *Procedia Comput. Sci.*, 2015, **51**, 693.
60. S. R. Oppenheimer, D. Mi, M. E. Sanders and R. M. Caprioli, *J. Proteome Res.*, 2010, **9**, 2182.
61. M. R. L. Paine, J. Liu, D. Huang, S. R. Ellis, D. Trede, J. H. Kobarg, R. M. A. Heeren, F. M. Fernandez and T. J. MacDonald, *Sci. Rep.*, 2019, **9**, 2205.
62. M. Sheykhmousa, M. Mahdianpari, H. Ghanbari, F. Mohammadimanesh, P. Ghamisi and S. Homayouni, *IEEE J. Sel. Top. Appl. Earth Obs. Remote Sens.*, 2020, **13**, 6308.
63. P. R. Harper, *Health Policy*, 2005, **71**, 315.
64. S. Mas, A. Torro, L. Fernandez, N. Bec, C. Gongora, C. Larroque, P. Martineau, A. de Juan and S. Marco, *Talanta*, 2020, **208**, 120455.
65. N. E. Mascini, G. B. Eijkel, P. ter Brugge, J. Jonkers, J. Wesseling and R. M. Heeren, *J. Proteome Res.*, 2015, **14**, 1069.
66. M. Kriegsmann, R. Casadonte, J. Kriegsmann, H. Dienemann, P. Schirmacher, J. Hendrik Kobarg, K. Schwamborn, A. Stenzinger, A. Warth and W. Weichert, *Mol. Cell. Proteomics*, 2016, **15**, 3081.
67. M. Kriegsmann, C. Zgorzelski, R. Casadonte, K. Schwamborn, T. Muley, H. Winter, M. Eichhorn, F. Eichhorn, A. Warth, S. O. Deininger, P. Christopoulos, M. Thomas, T. Longerich, A. Stenzinger, W. Weichert, C. Muller-Tidow, J. Kriegsmann, P. Schirmacher and K. Kriegsmann, *Cancers*, 2020, 12, 2704.
68. M. Kriegsmann, R. Casadonte, N. Maurer, C. Stoehr, F. Erlmeier, H. Moch, K. Junker, C. Zgorzelski, W. Weichert, K. Schwamborn, S. O. Deininger, M. Gaida, G. Mechtersheimer, A. Stenzinger, P. Schirmacher, A. Hartmann, J. Kriegsmann and K. Kriegsmann, *J. Cancer*, 2020, **11**, 6081.
69. O. Klein, F. Kanter, H. Kulbe, P. Jank, C. Denkert, G. Nebrich, W. D. Schmitt, Z. Wu, C. A. Kunze, J. Sehouli, S. Darb-Esfahani, I. Braicu, J. Lellmann, H. Thiele and E. T. Taube, *Proteomics: Clin. Appl.*, 2019, **13**, e1700181.
70. J. Behrmann, C. Etmann, T. Boskamp, R. Casadonte, J. Kriegsmann and P. Maass, *Bioinformatics*, 2018, **34**, 1215.
71. D. Perez-Guaita, G. Quintas and J. Kuligowski, *Anal. Chim. Acta*, 2020, **1097**, 30.
72. Y. Cordero Hernandez, T. Boskamp, R. Casadonte, L. Hauberg-Lotte, J. Oetjen, D. Lachmund, A. Peter, D. Trede, K. Kriegsmann, M. Kriegsmann, J. Kriegsmann and P. Maass, *Proteomics: Clin. Appl.*, 2019, **13**, e1700168.
73. T. Boskamp, D. Lachmund, J. Oetjen, Y. Cordero Hernandez, D. Trede, P. Maass, R. Casadonte, J. Kriegsmann, A. Warth, H. Dienemann, W. Weichert and M. Kriegsmann, *Biochim. Biophys. Acta, Proteins Proteomics*, 2017, **1865**, 916.
74. S. A. Thomas, Y. Jin, J. Bunch and I. S. Gilmore, *2017 IEEE Symposium Series on Computational Intelligence (SSCI)*, 2017.

75. A. Ly, R. Longuespee, R. Casadonte, P. Wandernoth, K. Schwamborn, C. Bollwein, C. Marsching, K. Kriegsmann, C. Hopf, W. Weichert, J. Kriegsmann, P. Schirmacher, M. Kriegsmann and S. O. Deininger, *Proteomics: Clin. Appl.*, 2019, **13**, e1800029.
76. H. C. Diehl, B. Beine, J. Elm, D. Trede, M. Ahrens, M. Eisenacher, K. Marcus, H. E. Meyer and C. Henkel, *Anal. Bioanal. Chem.*, 2015, **407**, 2223.
77. T. Murta, R. T. Steven, C. J. Nikula, S. A. Thomas, L. B. Zeiger, A. Dexter, E. A. Elia, B. Yan, A. D. Campbell, R. J. A. Goodwin, Z. Takats, O. J. Sansom and J. Bunch, *Anal. Chem.*, 2021, **93**, 2309.
78. C. H. Ding, *Bioinformatics*, 2003, **19**, 1259.
79. G. McCombie, D. Staab, M. Stoeckli and R. Knochenmuss, *Anal. Chem.*, 2005, **77**, 6118.
80. M. Nampei, M. Horikawa, K. Ishizu, F. Yamazaki, H. Yamada, T. Kahyo and M. Setou, *Sci. Rep.*, 2019, **9**, 13213.
81. M. C. Djidja, E. Claude, M. F. Snel, S. Francese, P. Scriven, V. Carolan and M. R. Clench, *Anal. Bioanal. Chem.*, 2010, **397**, 587.
82. R. A. van den Berg, H. C. Hoefsloot, J. A. Westerhuis, A. K. Smilde and M. J. van der Werf, *BMC Genomics*, 2006, **7**, 142.
83. B. Balluff, C. K. Frese, S. K. Maier, C. Schone, B. Kuster, M. Schmitt, M. Aubele, H. Hofler, A. M. Deelder, A. Heck, Jr, P. C. Hogendoorn, J. Morreau, A. F. Maarten Altelaar, A. Walch and L. A. McDonnell, *J. Pathol.*, 2015, **235**, 3.
84. E. A. Jones, A. van Remoortere, R. J. van Zeijl, P. C. Hogendoorn, J. V. Bovee, A. M. Deelder and L. A. McDonnell, *PLoS One*, 2011, **6**, e24913.
85. E. A. Jones, R. Shyti, R. J. M. van Zeijl, S. H. van Heiningen, M. D. Ferrari, A. M. Deelder, E. A. Tolner, A. van den Maagdenberg and L. A. McDonnell, *J. Proteomics*, 2012, **75**, 5027.
86. S. R. Ellis, J. Cappell, N. O. Potocnik, B. Balluff, J. Hamaide, A. Van, der Linden and R. M. Heeren, *Analyst*, 2016, **141**, 3832.
87. W. Gardner, R. Maliki, S. M. Cutts, B. W. Muir, D. Ballabio, D. A. Winkler and P. J. Pigram, *Anal. Chem.*, 2020, **92**, 10450.
88. J. M. Fonville, C. L. Carter, L. Pizarro, R. T. Steven, A. D. Palmer, R. L. Griffiths, P. F. Lalor, J. C. Lindon, J. K. Nicholson, E. Holmes and J. Bunch, *Anal. Chem.*, 2013, **85**, 1415.
89. W. M. Abdelmoula, K. Skraskova, B. Balluff, R. J. Carreira, E. A. Tolner, B. P. Lelieveldt, L. van der Maaten, H. Morreau, A. M. van den Maagdenberg, R. M. Heeren, L. A. McDonnell and J. Dijkstra, *Anal. Chem.*, 2014, **86**, 9204.
90. W. M. Abdelmoula, B. Balluff, S. Englert, J. Dijkstra, M. J. Reinders, A. Walch, L. A. McDonnell and B. P. Lelieveldt, *Proc. Natl. Acad. Sci. U. S. A.*, 2016, **113**, 12244.
91. T. Smets, N. Verbeeck, M. Claesen, A. Asperger, G. Griffioen, T. Tousseyn, W. Waelput, E. Waelkens and B. De Moor, *Anal. Chem.*, 2019, **91**, 5706.

92. P. Widlak, G. Mrukwa, M. Kalinowska, M. Pietrowska, M. Chekan, J. Wierzgon, M. Gawin, G. Drazek and J. Polanska, *Proteomics*, 2016, **16**, 1613.
93. A. Dexter, A. M. Race, I. B. Styles and J. Bunch, *Anal. Chem.*, 2016, **88**, 10893.
94. J. W. Lyron, K. Inge, J. R. G. Ove, M. Stephan and H. Peter, *Ann. Appl. Stat.*, 2015, **9**, 1973.
95. F. Lauer, S. Diehn, S. Seifert, J. Kneipp, V. Sauerland, C. Barahona and S. Weidner, *J. Am. Soc. Mass Spectrom.*, 2018, **29**, 2237.
96. M. Lagarrigue, R. Lavigne, B. Guevel, A. Palmer, K. Rondel, L. Guillot, J. H. Kobarg, D. Trede and C. Pineau, *J. Mass Spectrom.*, 2020, **55**, e4633.
97. T. Hiratsuka, Y. Arakawa, Y. Yajima, Y. Kakimoto, K. Shima, Y. Yamazaki, M. Ikegami, T. Yamamoto, H. Fujiwake, K. Fujimoto, N. Yamada and T. Tsuruyama, *Sci. Rep.*, 2020, **10**, 5757.
98. I. Assent, *Wiley Interdiscip. Rev. Data Min. Knowl. Discov.*, 2012, **2**, 340.
99. W. Zhang, M. Claesen, T. Moerman, M. R. Groseclose, E. Waelkens, B. De Moor and N. Verbeeck, *Anal. Bioanal. Chem.*, 2021, **413**, 2803.
100. T. Alexandrov and J. H. Kobarg, *Bioinformatics*, 2011, **27**, i230.
101. T. Porta Siegel, G. Hamm, J. Bunch, J. Cappell, J. S. Fletcher and K. Schwamborn, *Mol. Imaging Biol.*, 2018, **20**, 888.
102. D. J. Ryan, J. M. Spraggins and R. M. Caprioli, *Curr. Opin. Chem. Biol.*, 2019, **48**, 64.
103. S. K. Maier, H. Hahne, A. M. Gholami, B. Balluff, S. Meding, C. Schoene, A. K. Walch and B. Kuster, *Mol. Cell. Proteomics*, 2013, **12**, 2901.
104. C. Villette, L. Maurer, A. Wanko and D. Heintz, *Metabolomics*, 2019, **15**, 122.
105. B. Balluff, A. Buck, M. Martin-Lorenzo, F. Dewez, R. Langer, L. A. McDonnell, A. Walch and R. M. A. Heeren, *Proteomics: Clin. Appl.*, 2019, **13**, e1800137.
106. A. Palmer, P. Phapale, I. Chernyavsky, R. Lavigne, D. Fay, A. Tarasov, V. Kovalev, J. Fuchser, S. Nikolenko, C. Pineau, M. Becker and T. Alexandrov, *Nat. Methods*, 2017, **14**, 57.
107. S. R. Ellis, M. R. L. Paine, G. B. Eijkel, J. K. Pauling, P. Husen, M. W. Jervelund, M. Hermansson, C. S. Ejsing and R. M. A. Heeren, *Nat. Methods*, 2018, **15**, 515.
108. E. I. Vlachavas, J. Bohn, F. Uckert and S. Nurnberg, *Int. J. Mol. Sci.*, 2021, **22**, 2822.
109. V. M. Prade, T. Kunzke, A. Feuchtinger, M. Rohm, B. Luber, F. Lordick, A. Buck and A. Walch, *Mol. Metab.*, 2020, **36**, 100953.
110. K. Mallah, J. Quanico, A. Raffo-Romero, T. Cardon, S. Aboulouard, D. Devos, F. Kobeissy, K. Zibara, M. Salzet and I. Fournier, *Mol. Cell. Proteomics*, 2019, **18**, 1669.
111. M. Aichler, D. Borgmann, J. Krumsiek, A. Buck, P. E. MacDonald, J. E. M. Fox, J. Lyon, P. E. Light, S. Keipert, M. Jastroch, A. Feuchtinger, N. S. Mueller, N. Sun, A. Palmer, T. Alexandrov, M. Hrabe de Angelis, S. Neschen, M. H. Tschop and A. Walch, *Cell Metab.*, 2017, **25**, 1334.

112. U. Moginger, N. Marcussen and O. N. Jensen, *Oncotarget*, 2020, **11**, 3998.
113. K. Scupakova, Z. Soons, G. Ertaylan, K. A. Pierzchalski, G. B. Eijkel, S. R. Ellis, J. W. Greve, A. Driessen, J. Verheij, T. M. De Kok, S. W. M. Olde Damink, S. S. Rensen and R. M. A. Heeren, *Anal. Chem.*, 2018, **90**, 5130.
114. F. Deutskens, J. Yang and R. M. Caprioli, *J. Mass Spectrom.*, 2011, **46**, 568.
115. F. Dewez, M. Martin-Lorenzo, M. Herfs, D. Baiwir, G. Mazzucchelli, E. De Pauw, R. M. A. Heeren and B. Balluff, *Anal. Bioanal. Chem.*, 2019, **411**, 5647.
116. K. Scupakova, F. Dewez, A. K. Walch, R. M. A. Heeren and B. Balluff, *Angew. Chem., Int. Ed. Engl.*, 2020, **59**, 17447.
117. A. M. Race, D. Sutton, G. Hamm, G. Maglennon, J. P. Morton, N. Strittmatter, A. Campbell, O. J. Sansom, Y. Wang, S. T. Barry, Z. Takats, R. J. A. Goodwin and J. Bunch, *Anal. Chem.*, 2021, **93**, 3061.
118. N. H. Patterson, E. Yang, E. A. Kranjec and P. Chaurand, *Bioinformatics*, 2019, **35**, 1261.
119. F. Dewez, J. Oejten, C. Henkel, R. Hebeler, H. Neuweger, E. De Pauw, R. M. A. Heeren and B. Balluff, *Proteomics*, 2020, **20**, e1900369.
120. W. M. Abdelmoula, M. S. Regan, B. G. C. Lopez, E. C. Randall, S. Lawler, A. C. Mladek, M. O. Nowicki, B. M. Marin, J. N. Agar, K. R. Swanson, T. Kapur, J. N. Sarkaria, W. Wells and N. Y. R. Agar, *Anal. Chem.*, 2019, **91**, 6206.
121. T. K. Sinha, S. Khatib-Shahidi, T. E. Yankeelov, K. Mapara, M. Ehtesham, D. S. Cornett, B. M. Dawant, R. M. Caprioli and J. C. Gore, *Nat. Methods*, 2008, **5**, 57.
122. A. S. Attia, K. A. Schroeder, E. H. Seeley, K. J. Wilson, N. D. Hammer, D. C. Colvin, M. L. Manier, J. J. Nicklay, K. L. Rose, J. C. Gore, R. M. Caprioli and E. P. Skaar, *Cell Host Microbe*, 2012, **11**, 664.
123. J. Oetjen, M. Aichler, D. Trede, J. Strehlow, J. Berger, S. Heldmann, M. Becker, M. Gottschalk, J. H. Kobarg, S. Wirtz, S. Schiffler, H. Thiele, A. Walch, P. Maass and T. Alexandrov, *J. Proteomics*, 2013, **90**, 52.
124. W. M. Abdelmoula, R. J. Carreira, R. Shyti, B. Balluff, R. J. van Zeijl, E. A. Tolner, B. F. Lelieveldt, A. M. van den Maagdenberg, L. A. McDonnell and J. Dijkstra, *Anal. Chem.*, 2014, **86**, 3947.
125. N. Verbeeck, J. M. Spraggins, M. J. M. Murphy, H. D. Wang, A. Y. Deutch, R. M. Caprioli and R. Van de Plas, *Biochim. Biophys. Acta, Proteins Proteomics*, 2017, **1865**, 967.
126. L. S. Eberlin, I. Norton, D. Orringer, I. F. Dunn, X. Liu, J. L. Ide, A. K. Jarmusch, K. L. Ligon, F. A. Jolesz, A. J. Golby, S. Santagata, N. Y. Agar and R. G. Cooks, *Proc. Natl. Acad. Sci. U. S. A.*, 2013, **110**, 1611.
127. F. Lohofer, R. Buchholz, A. Glinzer, K. Huber, H. Haas, G. Kaissis, A. Feuchtinger, M. Aichler, P. B. Sporns, C. Holtke, M. Stolting,

F. Schilling, R. M. Botnar, M. A. Kimm, C. Faber, A. K. Walch, A. Zernecke, U. Karst and M. Wildgruber, *Sci. Rep.*, 2020, **10**, 79.

128. H. F. Wehrl, J. Schwab, K. Hasenbach, G. Reischl, G. Tabatabai, L. Quintanilla-Martinez, F. Jiru, K. Chughtai, A. Kiss, F. Cay, D. Bukala, R. M. Heeren, B. J. Pichler and A. W. Sauter, *Cancer Res.*, 2013, **73**, 1470.
129. M. Aichler, K. Huber, F. Schilling, F. Lohofer, K. Kosanke, R. Meier, E. J. Rummeny, A. Walch and M. Wildgruber, *Angew. Chem., Int. Ed. Engl.*, 2015, **54**, 4279.

Section 3

CHAPTER 13

MALDI Mass Spectrometry Imaging for Microbiology

JACOB E. PORTER AND LAURA M. SANCHEZ*

University of California Santa Cruz, 1156 High St, Santa Cruz, CA 95064, USA
*Email: lmsanche@ucsc.edu

13.1 Introduction to Microbiology as a Field of Study for MSI Development

Microbiology concerns the study of microorganisms. Within the broad field of microbiology, there are many subcategories and areas of study, which can be divided taxonomically (*e.g.* bacteriology, virology, mycology), or by topic (*e.g.* medical, food, and environmental microbiology).

There is a vast array of microorganisms, applications, and chemical classes for which MSI is useful in microbiology. Due to their small cell size (~0.5–5 microns in length), imaging microbes generally involves the study of microbial communities,[1] including bacterial colonies,[2,3] fungi, and algae.[4,5] There are a wide range of applications including environmental, food safety, and natural product chemistry. For instance, imaging toxins and natural products in unicellular species such as Cyanobacteria[6,7] can rapidly identify hazardous species in marine environments.[8–10] For the purposes of this chapter, the term "natural products" is used to remain consistent; many of the articles cited use the terms "secondary metabolites" or "specialized metabolites" to describe microbial natural products. In the field of microbial natural products, specific genera represent a substantial focus for discovering therapeutic compounds which, in an ecological niche, might be

New Developments in Mass Spectrometry No. 12
MALDI Mass Spectrometry Imaging: From Fundamentals to Spatial Omics
Edited by Tiffany Siegel

Published by the Royal Society of Chemistry, www.rsc.org

produced as defense mechanisms or feeding deterrents. Several high-profile antibiotics, including vancomycin,[11] were derived from the Actinobacteria phylum.[12] While natural products and their analogs can often be synthesized *de novo* once isolated and identified,[13,14] in many cases the complexity of these structures makes it more efficient and cost-effective to either extract the finished compound from cultured microbes, or to use semi-synthesis from the isolated natural product.[15] Additionally, microbial natural products can be advantageous over the environmentally destructive, laborious characterization of macroorganisms. With ever-increasing rates of antibiotic resistance, microbial species offer a promising avenue to identify much-needed new antimicrobial agents.[16,17]

The benefits of MSI in the field of microbiology are numerous. While classic techniques such as Gram staining can be used to identify genera, MSI provides information about the molecular composition of microbial colonies, including the influence of media composition on metabolite production, the distribution of metabolites across the colony, and the molecules implicated in host–microbe communication. MSI provides a snapshot of microbes at lifecycle stages (when appropriate); fungi and amoeba have been used to study the metabolomic and lipidomic changes that occur as microbes grow and decay.[18,19] Community-wide microbiome metabolomics shifts are also amenable to MSI: bacterial biofilms formed by foodborne or bloodborne pathogens present major risks in food production or to immunocompromised people in a hospital setting. Certain microbial species are especially prevalent in the literature; for instance, *Pseudomonas aeruginosa* is of high interest as a World Health Organization (WHO) critical priority pathogen owing to its ubiquity in clinical settings and its high rate of antibiotic resistance.[20] By imaging pathogenic bacteria cultured on media enriched with antimicrobial compounds, one can study the mechanisms of antibiotic resistance by identifying phenotypic changes associated with antimicrobial treatments.[21,22]

MSI is particularly useful to interrogate interactions between multiple species, either by co-culturing microbial species or by exploring host–microbe interactions. For example, bacteria can be co-cultured with other microbial species, such as fungi, and zones of inhibition interrogated to identify antimicrobial compounds.[23–26] Nematodes are useful model organisms for the study of host–microbe interactions, as a single nematode is large enough to be imaged individually.[27,28] Larger macroorganisms can also be studied for parasitic or symbiotic relationships with microorganisms; examples of symbiotic host–microbe relationships include the bioluminescent *Vibrio fischeri* in *Euprymna scolopes* (Hawaiian bobtail squid),[29,30] and antimicrobial bacteria on ants.[31,32] Host–parasite interactions include malarial parasites in mice[33] and *Schistosoma* in humans.[34] Opportunistic pathogens such as the aforementioned *P. aeruginosa* can be imaged on host tissue to potentially identify infection-specific biomarkers in clinical settings.[35]

Of course, microbial MSI presents unique challenges. For host tissues, many of the same preservation and preparation considerations described in Chapter 5 apply. In the following chapter, we will describe considerations for sample preparation and optimization in the cases of agar-based and host–microbe tissue imaging, followed by in-depth discussions of applications in samples of increasing complexity. We will outline experiments on individual microbial species, co-culturing experiments, and host–microbe interactions, followed by an overview of the utility of user-generated databases for data sharing and inter-lab communication. We will conclude with an overview of promising future technologies and applications for microbiological MSI, including multimodal imaging, advances in data processing and instrumentation such as matrix-free and atmospheric pressure imaging.

Microbial communities are plated and cultured on media, which contains compounds that can act as mass spectrometric interferents, detergents, or signal suppressants. MS-compatible media must be chosen, and factors such as media preparation, drying temperature, and humidity must be optimized and carefully controlled to ensure proper sample adhesion and homogeneity prior to analysis and to avoid issues with analyte migration.[36] Additionally, microbial metabolic profiles can change and bacterial cell culture viability decreases over time,[37,38] so experiments must be carried out expediently and care must be taken with sample storage to ensure reproducible results. Determining molecular species directly from signals at specific mass-to-charge ratio (m/z) values can also be challenging; common mass analyzers such as time-of-flight (ToF) instruments do not have high enough mass resolution to confidently assign formulae based on accurate mass. While MALDI-MS/MS can help with compound identification,[39,40] techniques such as LIFT (lifting the voltage of the collision cell above that of the source) require high signal intensity and can introduce mass error in the MS/MS stage without extensive internal calibration.[41] MSI is often integrated into a workflow with other techniques such as bioassay-guided fractionation[42] and LC–MS/MS[22,43,44] to identify compounds responsible for a physiological response from a fractionated mixture. Additionally, data sharing through publicly available user-generated databases such as Global Natural Product Social molecular networking (GNPS) can help pinpoint commonly-observed molecular species.[45]

13.2 Sample Preparation for Microbial MSI

Given the nature of culturing and fermenting microbes, these samples often require laborious preparation to be compatible with mass spectrometry-based analyses. For instance, LC–MS/MS based applications typically require cell lysis as a minimum, and whole cell lysates must often be fractionated to simplify analyses and ensure adequate signals from peptides or small molecules of interest. By comparison, sample preparation for MSI can be relatively simple, especially for atmospheric pressure techniques. Studying host–microbe interactions often requires tissue samples, which are subject to many of the same considerations covered in Chapter 5. Samples may be flash-frozen and directly sectioned or embedded in a matrix such as gelatin, carboxymethylcellulose (CMC), or optimal cutting temperature compound (OCT), which can interfere with analysis and suppress analyte signals. Gelatin, CMC, and ice are more compatible with MSI than other matrices such as paraffin.[46] Freshly harvested samples may need to be desiccated prior to introduction into an instrument, which is often recommended to ensure surface homogeneity, especially with specimens primarily composed of water such as the Hawaiian bobtail squid.[30] As is often the case with secondary ion mass spectrometry (SIMS) and laser desorption/ionization (LDI), tissue specimens may be gold-coated to improve the signals.[47–49]

Several factors must be considered when using MSI to analyze cultured microorganisms. Bacteria, fungi, and viruses are frequently grown as colonies (groups of cells rather than single cells) on a nutrient and moisture-rich solid substrate such as agar or agarose. For bacterial culturing, many labs use a basic protocol involving the growth of bacteria on agar in a petri dish, excision, and transfer to a target, prior to matrix application with a 53 μm stainless steel sieve and drying in an oven.[36,50] The most important part of this protocol is that microbes are cultured on 'thin' agar. This translates to approximately 10–11 mL of agar in a standard sized Petri dish (90 mm in diameter, 1 to 1.5 mm thickness). This height of wet agar can be dried in an oven to a uniform thinness compatible with commercial MALDI-TOF mass spectrometers. Using this procedure, samples are usually dried at 37 °C for 4–6 hours or overnight, and stored in a desiccator prior to analysis. Analyte migration through the wet agar captures diffusible small molecules and migration during the drying process has proven to be of minimal concern.[51] It is also possible to culture bacteria directly on agar poured onto the target, or to culture bacteria on an agar-embedded target, but this is not as commonly employed as this can alter bacterial growth due to trace contaminants on the target plates.[36] In contrast, culturing bacteria separately allows for the efficient study of multiple colonies and conditions on a single target in a single experiment. Of note, the sample topography is particularly important when ions are not collected into a funnel prior to release into the mass analyzer and sample topography heterogeneity is mitigated in other instruments that are FTICR based. However, it is still important to note that maintaining conductivity for MALDI is critical and agar can often be insulating, therefore using the thin agar culturing approach is still advantageous when the microbes can tolerate growth on thinner agar.

Whichever protocol is used, drying agar samples is of utmost importance for MALDI based applications, as placing wet samples into a source chamber under vacuum could result in rapid, explosive vaporization, which can damage the instrument. Care must be taken to avoid trapping air bubbles under the agar following transfer of agar slices onto the target. This can be corrected simply by gently lifting the agar up and carefully placing it back down onto the target plate with a spatula. Air bubbles that are allowed to dry under the agar slices can lead to improper adhesion and cause flaking inside the instrument; at best this will require source cleaning and at worst it may damage the optics. Detailed protocols and considerations, including using ground rather than polished steel and monitoring drying conditions, can be found in the supporting information of Yang *et al.*[36] A mixture of 1 : 1 α-cyano-4-hydroxycinnamic acid (CHCA) and 2,5-dihydroxybenzoic acid (DHB) is often used,[52] as this covers ionization for proteins, small peptides, carbohydrates, and glycopeptides. This exact combination of matrices was found to allow for strong adhesion of the agar pieces to steel target plates and it is evidenced by the resulting crystallization that matrix builds on the edges of agar pieces which likely results in strong adhesion. Some groups use CHCA or DHB alone.[6] Anderton *et al.* report both higher spatial resolution and sensitivity from DHB than CHCA using a sprayer.[53] Alternative compounds have been proposed to improve heterogeneity; for example, adding graphene oxide as an auxiliary matrix,[54] although these have not been tested for use on agar. Matrix can also be applied using an automated sprayer;[55] while this requires specialized equipment and careful optimization to ensure analytes of interest don't migrate in the liquid environment, it provides greater matrix homogeneity and higher signal-to-noise (*S*/*N*) ratio in the case of certain bacterial genera.[51] Matrix can be sprayed either prior to[56] or after the drying process, however, this will greatly depend on the type of small molecules specific microbes are known to produce.[51] For instance, *Pseudomonas aeruginosa* produces a large number of rhamnolipids which prevent this microbe from desiccation prior to matrix application as the colony and agar flake off the target slides during the drying process.[21,22] Another sample consideration for MALDI-based sample preparation is the agar itself. Some microbes require high salt content and buffers which may also impact the adhesion of the agarose as it dries to the target plates or slides, however most media can be adapted simply by adjusting salt and buffer percentages if the observed phenotypes and growth can be maintained. A comprehensive guide to spraying matrix, including explanations of analyte migration and preparation for microbial MSI can be found in the supporting material of Hoffman and Dorrestein.[51] While their supporting information includes in-depth consideration of analyte displacement and diffusion, many groups use modernized robotic sprayers[53,56] rather than the pneumatic sprayer described in Hoffman and Dorrestein's paper. Analyte migration is generally low even with the creation of droplets from matrix spraying, as analytes secreted by bacteria usually diffuse into the agar and are less affected by surface conditions. Li *et al.* observed analyte migration of 0.1–0.2 mm (~3–5%), within the raster size for MALDI-MSI.[56] Not all agar-based imaging requires matrix application; nanospray desorption electrospray ionization (nanoDESI) can obtain metabolite signals directly from agar in Petri dishes.[57] Laser ablation electrospray ionization (LAESI) can similarly detect metabolites directly from colonies on agar, and its ablative properties allow for depth profiling, detecting metabolites in 50-micron slices as a function of depth-wise diffusion through the agar medium.[58] Tian *et al.* described a matrix-free method for direct imaging of single bacterial cells deposited on Si wafers using C_{60}-SIMS.[59] Figure 13.1

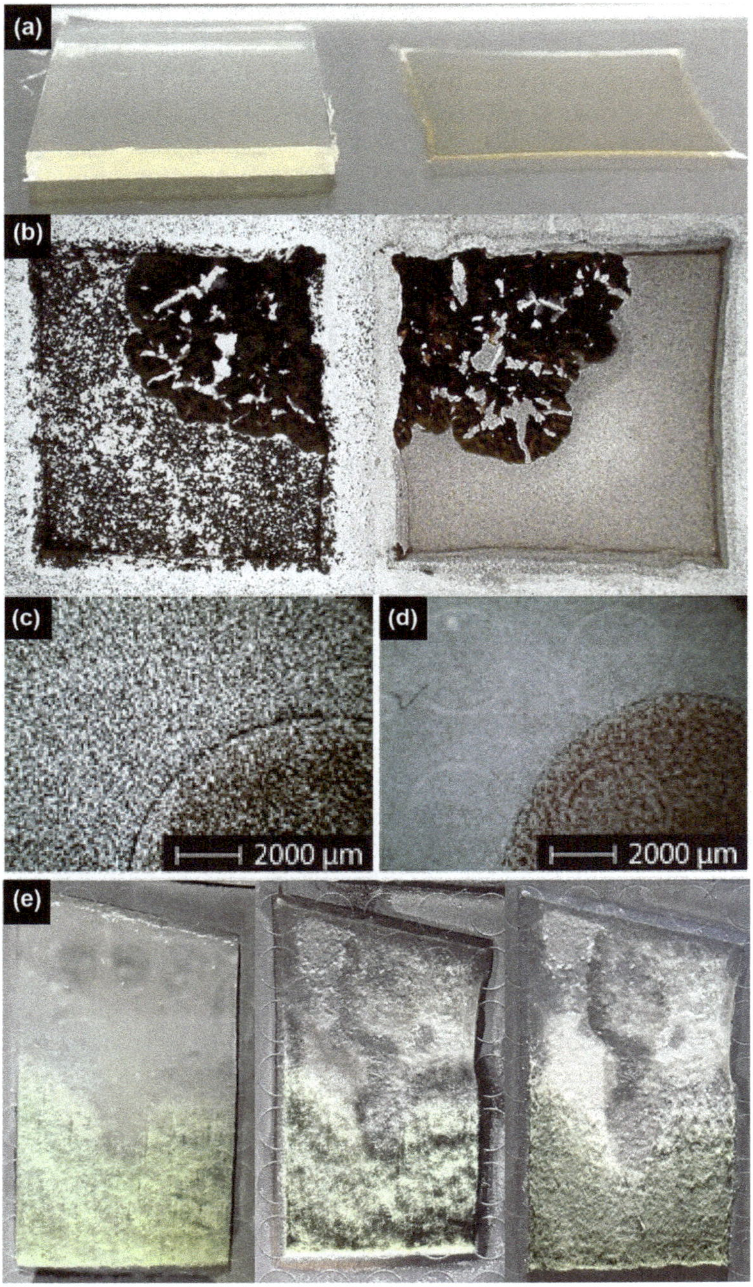

Figure 13.1 Sample preparation of agar slices: (a) shows an agar slice before (left) and after (right) drying at room temperature in a desiccator. (b) shows the application of matrix using a sieve (left) and a TM sprayer (right). Reproduced from ref. 60 with permission from HTX Imaging, Copyright 2015. (c) and (d) show matrix application using a sieve and a sprayer respectively on the µm scale. Reproduced from ref. 51 with permission from American Chemical Society, Copyright 2015. (e) shows a slice of agar with a bacterial-fungal co-culture before drying (left), after drying (middle), and after matrix sieving (right).

shows agar slices before and after drying, using a sieve and a TM sprayer. (a) and (b), reproduced with permission from Drs. Erin Gemperline and Lingjun Li,[60] show the reduction in volume and surface adherence of a drying agar slice. The magnified images (c) and (d), reproduced with permission from Hoffman and Dorrestein,[51] clearly show the much larger matrix crystals produced from sieving *vs* spraying. (e) shows an agar slice with a microbial co-culture, before drying (left), after drying (middle), and after matrix application (right).

Beyond agar-based imaging, there are protocols for multiple species using many different instruments. For example, bacteria on exotic substrates such as ant exoskeletons can be imaged by etching ITO-coated glass slides to provide a flat surface for matrix application.[32] Certain host tissues must be otherwise processed to ensure surface homogeneity for matrix application, such as the removal of ink sac, eyeballs, and mantle from *Euprymna scolopes* prior to dehydration.[30] Some smaller species can be coated in matrix directly prior to analysis.[27] MSI of bacterial biofilms may involve growth on agar or glass slides grown in liquid culture,[61,62] or alternative substrates such as mammalian tissue.[63,64] Several groups have described growing biofilms directly on ITO-coated glass slides in liquid media followed by dehumidification.[6,63] As with agar-based imaging, there are several papers that demonstrate analysis without matrix application; Urban *et al.* and Baumeister *et al.* describe methods for characterizing microalgae using matrix-free LDI-MS.[4,65] Garg *et al.* were also able to directly section lichens, which are communities of a photosymbiont with fungi and bacteria that grow in nature over extended periods of time.[66] Of note, for ambient ionization techniques the microbes can simply be grown on agar for analysis, but spatial resolution is typically sacrificed for simpler sample preparation. Some ambient ionization techniques involve applying high spray pressure to the sample which can damage malleable agar surfaces. Sica *et al.* describe extensive sample preparation optimization to use DESI-MS with fungus samples grown on agar; using solid substrates such as cheesecloth, balsa wood, and cardboard as inserts, they developed a method to culture fungi on agar with a solid cardboard insert to prevent sample deformation during analysis.[67]

13.3 Applications for Imaging Microbial Monocultures

While imaging individual single-celled organisms is limited to techniques with high (nm-scale) spatial resolution, microbial monocultures are typically several millimeters in diameter and can be imaged with a wide variety of instrumentation. These experiments often involve comparisons between different species or different conditions within a single species. Some of the simplest experiments analyze a microbial monoculture to explore the production of metabolites, proteins, or lipids as a function of cell growth or external stimuli. There are many considerations that must be considered prior to an imaging experiment. Metabolites can be excreted as a function of light, pH, salinity, or temperature. By analyzing the migration of these compounds across an agar-based substrate, colony-bound metabolites used as feeding deterrents or responsible for biological functions such as photosynthesis can be differentiated from secreted compounds that microbes use to interact with their environmental niches.[68]

Colony morphology is an important factor to consider when testing a microbial species for its compatibility with MSI. Bacteria and fungi often grow in homogeneous circular colonies with even distribution of metabolites, but both exhibit a wide range of colony morphologies. Filamentous bacteria, notably Actinobacteria,[69] and cyanobacteria in the orders Oscillatoriales, Synechococcales, Spirulinales, and Nostocales,[70,71] form large, elongated unicellular structures which can be imaged individually.[7] Certain microbial species have demonstrated specific migration patterns on agar which allows for the differentiation of compounds that are potentially responsible for behavior. Species such as *Bacillus subtilis*, *Salmonella enterica* and *Yersinia enterocolitica* exhibit swarming behavior in nutrient-rich conditions,[72] leading to filamentous colonies with dendrites showing the development and distribution of colonies over time. Debois *et al.* found small molecule-based surfactants localized in the edges and the center of these swarming patterns.[73] Many species of bacteria will attach to appropriate surfaces and excrete an extracellular polymeric substance (EPS), forming complex layered spherical 3D structures known as biofilms.[74] Biofilms irreversibly attach to surfaces, and their ubiquity on polymeric substrates presents microbial contamination risks in food safety, water processing, and medical contexts.[75] Santos *et al.* detailed a method for identifying proteins on the surface of *Listeria monocytogenes* biofilms.[76] They found proteomic differences between immersed and dehumidified proteins, under conditions representative of food workshop decontamination procedures. Fungal colonies also vary somewhat in morphology. Molds usually show a central circular area with hyphae or pseudohyphae extending outwards; yeasts can be similar to bacterial colonies, but may also show extended pseudohyphae patterns at the edges.[77] When a homogeneous sample is required, circular fungal plugs are often excised prior to analysis.[31]

Certain microbial species are considered model species with well-characterized genomes and behaviors that react to a range of stimuli and experimental conditions. *B. subtilis*, a ubiquitous Gram-positive species found in soil and human gastrointestinal tracts, is one notable example. It exhibits both cannibalistic and swarming behavior, excretes well-characterized natural products, and can form colony biofilms on solid agar.[73,78,79] Monoculture experiments are often carried out as a precursor to or in conjunction with experiments on co-cultures; for example, Araújo *et al.* collected data on endophytic pathogen-inhibiting bacteria, to confirm the presence of known cytotoxic metabolites (biosurfactant rhamnolipids with long-chain fatty acid tails), and several plant-pathogenic fungal/oomycetic species separately, to characterize the lipidomic profile in the absence of inhibition.[80] In other cases, MSI can identify bacterial species based on chemical signatures. Kim *et al.* used lipidomic profiles generated with laser desorption ionization (LDI)-MS to differentiate between nine bacterial species.[81] Bacterial colonies can also be studied as a function of time, to assess metabolic changes associated with cell growth. Watrous *et al.* used nano-DESI to image the intensity of characteristic signals at 6, 12, 24, and 48 hour timepoints.[82] By comparing spectral alignment and neutral losses between tandem MS spectra, they were able to generate molecular networks at each time point to identify the production of various molecular classes including polysaccharides, membrane lipids such as lysyl-phosphatidylglycerol (LPG) and digalactosyldiacylglycerol (DGDG), and surfactants (surfactin and linear/hydrolyzed derivatives). Identifying these compounds also in some cases differentiated between bacterially-derived natural products and background peaks; polysaccharides, for example, were unbranched polygalactose chains derived from the agar plugs rather than the bacteria.

Experiments on bacterial monocultures can involve differentiating between strains of a particular bacterial species. Phelan *et al.* subjected three *P. aeruginosa* strains to increasing concentrations of an antibiotic azithromycin and were able to image strain-specific colony survival as a function of natural product secretion.[22] In other cases, genetic manipulation can induce relevant changes. Biofilm formation and antibiotic production, for example, can be upregulated by inserting or deleting genes to alter phenotypes.[83,84] Behaviors such as motility, natural product excretion, and biofilm production can be moderated by genetic manipulation. Natural product gene clusters can be affected by upstream gene expression; the production of natural products can be moderated by deleting or overexpressing well-characterized, conserved genes coding for regulatory proteins, such as *laeA*,[85,86] *gcnE*,[87,88] and *kdmA*.[89,90] In some cases, antibiotic resistance is modulated by mutations in specific genes, as is the case with *rpoB*.[91,92] These genetic alterations can be exploited when designing MSI experiments. Si *et al.* imaged *B. subtilis* mutants *Δspo0A* and *ΔabrB*, with defective and enhanced biofilm formation respectively, to determine chemical signals responsible for cannibalistic behavior in biofilm formation. They observed increased levels of several metabolites in the *ΔabrB* mutant, including surfactants and plipostatins.[79]

Figure 13.2[21] shows the impact of genetic mutation and treatment with compounds on bacterial colony formation. Panel (a) shows the effect of a biofilm-inhibiting bile acid, taurolithocholic acid (TLCA), on colony morphology, with the wild type treated with TLCA exhibiting a similar wrinkled phenotype as the reduced biofilm phenazine-null *Δphz* mutant. Panels (b) and (c) show the change in intensity of two compounds, phenazine-1-carboxylic acid (PCA) and pyochelin respectively, upon treating the wild type and two phenazine-producing gene knockout *Δphz* and *ΔpqsH* mutants with TLCA. After TLCA treatment, both mutants showed a significant increase in pyochelin, a virulence-associated siderophore. Neither the wild type nor the mutants showed a significant change in the phenazine

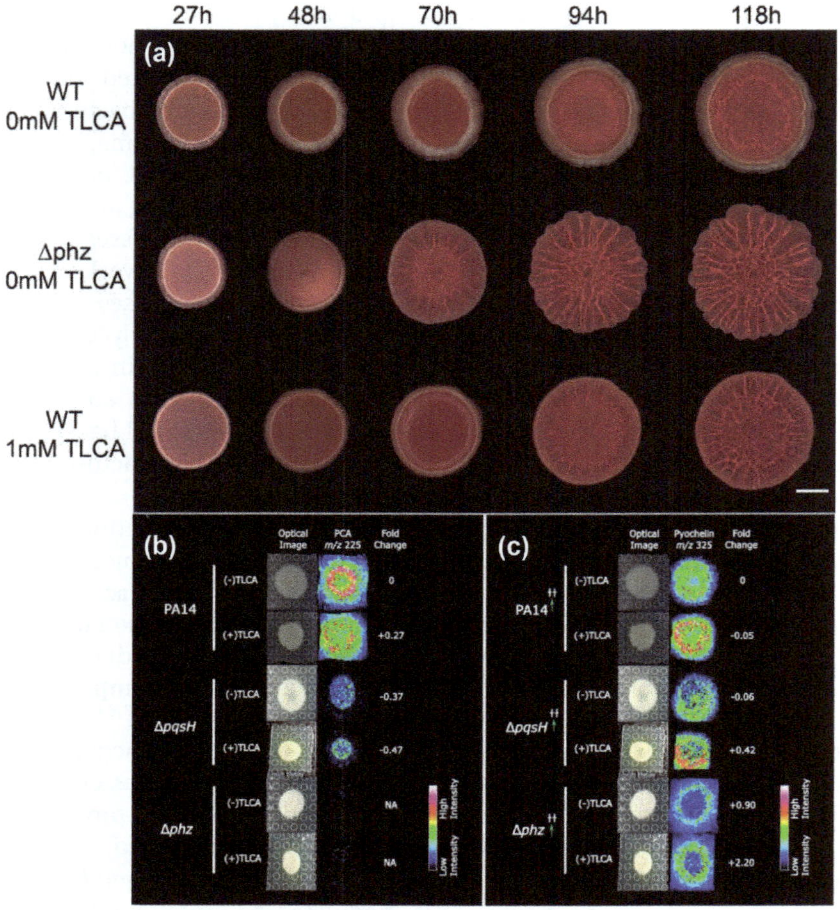

Figure 13.2 Microbial monocultures of *P. aeruginosa* PA14. Panel (a) shows the morphology of wild type (top) and *Δphz* (middle) strains, with the wild type exposed to TLCA on the bottom. Panels (b) and (c) show the fold change of the wild type and *Δphz* and *ΔpqsH* mutants, when monitoring PCA and pyochelin respectively.
Reproduced from ref. 21 with permission from American Chemical Society, Copyright 2019.

PCA, showing that TLCA can induce increased virulence in *P. aeruginosa* even without upregulated production of phenazines.

Higher spatial resolution instruments can image individual microbes to identify metabolic processes that occur within the cell. Imaging cyanobacteria to explore nitrogen fixation[93] and looking at interactions between silver nanoparticles and individual algal cells[94] are two applications of nanoscale SIMS at the cellular level. While colonies are generally imaged on the macro-scale (millimeters in diameter), high spatial resolution MALDI-MSI has also been applied with a 10 μM pixel size to image colonies on the micrometer scale, composed of ~50 cells.[95]

With the current prevalence of algal blooms posing environmental hazards, imaging microalgae can provide insight into their mechanism of action. Identifying changes in metabolic profiles associated with sporulation can help to generate hypotheses regarding the mechanisms of algal distribution.[96] Species of algae have also been used as model systems to study bio-nano interactions, by imaging silver nanoparticles at the single-cell level with nanoSIMS.[94] Schleyer *et al.* explored the influence of viral infection on algal growth by co-incubating *Emiliania huxleyi* cells with species-specific *Emiliania huxleyi* virus (EhV) virions and comparing the intensity of seven diagnostic lipid biomarkers across the radii of infected plaques.[97] They related the spatial distribution to the infection progress over time, with the plaque originating from a single infected cell in the center and radiating outwards. Several lipids, including phosphatidyl-*S*,*S*-dimethylpropanethiol (PDPT) 40:7 and diacylglyceryl-3-*O*-carboxy-(hydroxymethyl)-choline (DGCC) 40:7, increased in intensity over the course of infection, while betaine-like lipid (BLL) 38:6 decreased. Triacylglycerol (TAG) 48:1 showed a more complex modulating pattern, increasing and decreasing in concentric rings. The use of MSI in this case to document the process of a viral infection showcases the ability of spatial omics to explore the complex equilibria involved as lipid species in an algal host are broken down and reformed.

13.4 Applications for Imaging Microbial Co-cultures

Imaging multiple microbes at once presents many avenues for studying interactions between species, strains, and genera. Compounds responsible for a particular response can be identified by exploring the inhibition zone between species, as can compounds excreted under certain conditions. Many fungi and bacteria excrete antimicrobial agents as a defense mechanism, or as part of a symbiotic relationship with a host organism. Fungi and bacteria may produce antimicrobial compounds or other toxins to compete with other species for resources[98] or ward off predators,[99] but these historically classified antimicrobials may have a number of other proposed functions, including as signaling molecules.[100] The conditions and processes required to produce these compounds are complex and interlocking, and experiments must be carefully designed to consider these factors. As alluded to in Section 13.2, the lifecycle and optimal growth conditions of each microbial species differ; factors such as salt concentration, pH, and exposure to light must be controlled to ensure consistency. Culturing multiple species together poses additional factors: whether to grow species separately and introduce them at a later time point or inoculate multiple species together; when to start growing each species to ensure simultaneous maturation; the geometry of the two species in relation to each other. Each of these can impact the metabolic profile of species individually and as part of a co-culture. Wiener demonstrated the importance of choosing whether to co-inoculate species or introduce them at a later stage of growth; antibiotic-producing *Streptomyces griseus* struggled to survive when co-inoculated with

Bacillus subtilis, but showed a strong competitive advantage when *B. subtilis* was introduced to growing *S. griseus*.[101] Co-culturing can also have unintended consequences that should be taken into consideration, such as reducing phenotypic diversity.[102]

Co-culturing experiments often involve introducing a new microbial species or strain to a bacterial lawn. There are multiple approaches to do this, depending on the viability and optimal growth conditions of the microbes in question. Fungus can be grown directly on the same media as the bacteria, or grown separately, excised, and transferred to the bacterial lawn. In any scenario, growth conditions, timings, and quantities must all be considered and optimized to ensure reproducible results. Wakeman *et al.* spotted *Pseudomonas aeruginosa* onto paper disks directly on cultured *Staphylococcus aureus* lawns.[61] Gonzalez *et al.* demonstrated the importance of morphology and orientation on microbial interactions by comparing co-cultures of *B. subtilis* and *S. aureus*; T-shaped experiments showed a directional interaction between the two species, where they observed increased production of surfactin by *B. subtilis* and inhibition of *S. aureus*.

Figure 13.3[103] depicts a co-culture of *Penicillium* #12 with *Glutamicibacter arilaitensis*, along with each species separately, and an agar control. MSI identified several upregulated *m/z* values in the co-culture, one of which (*m/z* 655) was observed only in bacteria co-cultured with fungus, indicating a compound produced in response to inter-microbial interaction.

Co-culturing experiments can also be used to identify differences between particular strains or genetic mutants in the presence of inhibitory species.

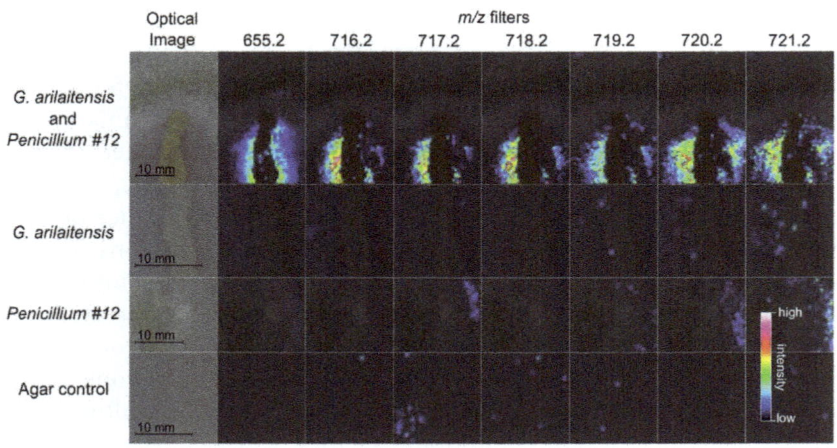

Figure 13.3 Bacterial–fungal co-cultures. A co-culture of *G. arilaitensis* and *penicillium* #12 is shown on the top, with the *G. arilaitensis* bacteria alone (second from top), *penicillium* #12 fungus alone (second from bottom) and agar control (bottom). The optical image is shown on the left, with MSI *m/z* filters applied in each column.

Chen *et al.* co-cultured multiple *Aspergillus* strains with a fungal pathogen, *Phellinus noxius*, to identify antifungal agents.[104] Yang *et al.* used mutants to explore the interaction between *B. subtilis* and *Streptomyces coelicolor*. The production of prodiginines by the latter species was inhibited in the presence of bacillaene-producing *B. subtilis*. While bacillaene itself was difficult to detect with MSI, *S. coelicolor* co-cultured with *B. subtilis* *Δpks* mutants incapable of bacillaene production showed no inhibition compared to the wild type. Compounds required for sporulation and hyphae formation in *S. coelicolor* were also inhibited by surfactin from the wild type *B. subtilis*, indicating the complexity and breadth of chemicals responsible for molecular communication between microbes. Liu *et al.* directly observed two compounds, sporulation delaying protein (SDP) and sporulation killing factor (SKF) in their experiments on *B. subtilis*.[78] Rather than co-culturing the *B. subtilis* with fungus, they cultured the wild type bacteria together with a *Δspo0A* non-sporulating mutant strain to explore molecular exchange. They hypothesized that SDP was responsible for the cannibalistic effect of the wild-type *B. subtilis*, as found in glassy areas of reduced growth at the interface between the two strains.

Microbial co-culturing experiments can give insight into symbiotic relationships between microbes and hosts. Boya *et al.* imaged a co-culture of *Streptomyces* bacteria and pathogenic *Escovopsis* fungi to identify compounds responsible for the antifungal activity of *Streptomyces* on the surface of fungus-growing ants.[31] Similarly, Martin *et al.* saw ions in a specific *m*/*z* range at the inhibition range between *Pseudomonas cichorii* and *Aspergillus fumigatus* and used MS/MS-based molecular networking to determine several lipopeptides responsible for antifungal activity produced by bacteria on the surface of certain frog skins.[105]

13.5 Applications for Imaging Host–Microbe Interactions

Beyond studying interactions between microbes, MSI has multiple applications for studying their interaction with host organisms, on the micro- and macro-scale. While microbial species of fungi have been discussed in previous sections, fungi can also act as a model for host–microbe interactions. Holzlechner *et al.* used *Trichoderma atroviride* as a model host organism to study the mechanism by which *Rhizoctonia solani* are inhibited.[106]

Host–microbe interactions are often studied in the context of bacterial infection of mammalian tissues. Many common diseases are caused by pathogenic microbes, and MSI offers a window into the spread of those diseases in infected tissue. Studies often look at specific microbes responsible for a disease, such as malaria[33] and leishmaniasis.[107] Neglected tropical diseases such as these are responsible for millions of infections per year, and over 1.7 billion people are vulnerable or at risk.[108] Developing methods to better understand the mechanism of action is imperative, as these diseases are often endemic to and rampant in low- and lower-middle-income countries. Perry *et al.* modeled host–pathogen iron competition, an important mechanism by which infectious agents obtain nutrients in a host, by using MALDI-MSI to explore the interaction of pathogenic *Staphylococcus aureus* cultures with host heart, liver, and kidney tissues.[64] In a targeted analysis, they found that chemically distinct siderophores were differentially expressed in each tissue, lending credence to the idea that these different chemical structures are not redundant, but each has a particular role in its biological niche. Hamerly *et al.* used a mouse model to explore the potential of MALDI-MSI as a diagnostic tool for monitoring infection in burn wounds.[35] They found specific ion signatures in thermally-injured mouse tissue post-inoculation, which were correlated with *Pseudomonas aeruginosa* control samples to propose biomarkers.

Beyond mammalian systems, there are a range of host–microbe interactions in creatures of other classes. The chemical relationships between insects and microbes are often studied as a model for host–parasite interactions. Mosquitoes are of particular interest due to their relevance as human disease vectors. Preliminary research has shown changes in phospholipid composition from infected mosquitoes.[33] Several other invertebrate organisms have been used as model systems for studying larger-scale host–microbe interactions, including the Hawaiian bobtail squid (*Euprymna scolopes*), which are colonized by bioluminescent *Vibrio fischeri* bacteria. While the challenges of imaging large multicellular organisms such as these have been previously described in the sample preparation section, they also provide a large surface area to analyze regions of bacteria for upregulated *m/z* signals.

Figure 13.4[29] shows the localization of a diketopiperazine, cyclo(D-histidyl-L-proline) at *m/z* 235 specifically in the light organ of bacteria-colonized *E. scolopes*. This compound is further upregulated in knockout *ΔbinK* organisms with enhanced biofilm production, showcasing the importance of biofilm formation in symbiotic host–microbe interactions.

Multimodal imaging approaches can increase the amount of available data from MSI experiments. By combining metabolomic information from MALDI-MSI experiments with fluorescent imaging, Geier *et al.* were able to visualize molecular species below the 3 μm spatial resolution limit of high-resolution AP-MALDI MSI.[109] They characterized the interaction between deep-sea *Bathymodiolus puteoserpentis* mussels and their oxidizing symbionts, correlating natural product secretion in symbiont-colonized mussels to phenotypic heterogeneity. Rath *et al.* combined MALDI-MSI experiments with high mass resolving power desorption electrospray ionization Fourier transform ion cyclotron resonance (DESI-FTICR) MS to directly identify metabolites from the surface of mouse gut microbiota.[110]

13.6 Compound Identification

Single-stage or linear MALDI-ToF experiments don't have the mass resolution or MS/MS capabilities to easily identify chemical species, so

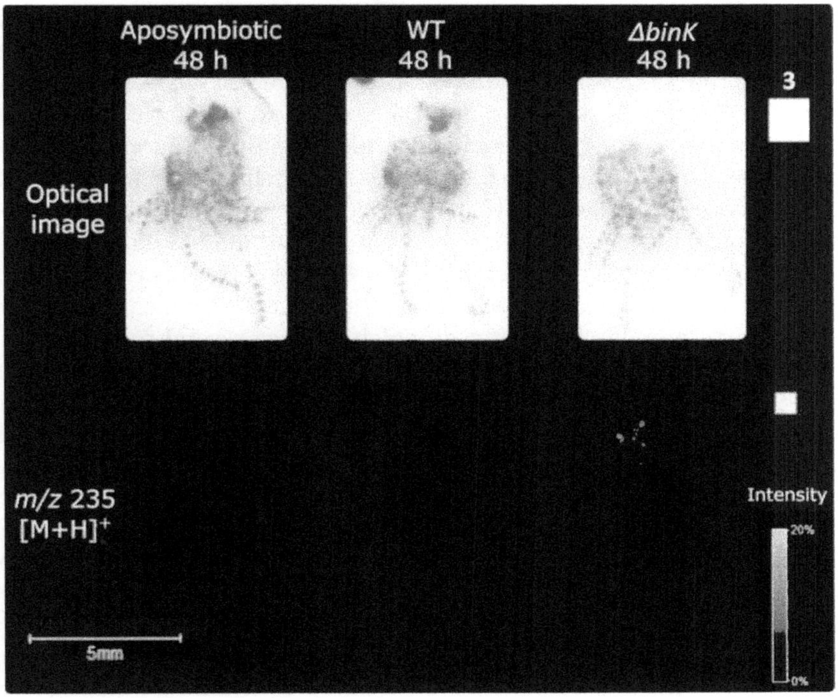

Figure 13.4 Juvenile aposymbiotic *E. scolopes* (left), colonized with wild type *V. fischeri* (middle), and colonized with biofilm-enhanced *V. fischeri* Δ*binK* (right) strains. Optical images are shown on the top, with MSI highlighting the species at *m/z* 235 on the bottom. A filter has been applied to enhance the contrast between *m/z* intensity in light organs and the rest of the organism.
Reproduced from ref. 29 with permission from Laura M. Sanchez, Copyright © The Authors.

compound identification often relies on multiple orthogonal experiments. Many MALDI-ToF instruments are capable of collecting MS/MS data to perform spatially-targeted tandem MS, but these experiments often involve splitting pixels into sub-pixels for individual analysis[111] and vaporizing the surface multiple times to acquire the necessary ion signal, causing issues with spatial resolution and sample ablation. Many approaches involve performing multiple experiments sequentially on a single sample; multimodal imaging techniques can pool the advantages of several analytical instruments. As discussed in the previous section, high resolving power in the MS domain is one strategy to identify species from selected *m/z* values. Similar to Rath *et al.* who used DESI-FTICR-MS,[110] Debois *et al.* described using the exact mass from MALDI-FTICR-MSI in conjunction with tandem MS experiments to determine the sequence of peptide-based antibiotics.[55]

These approaches are often time- and labor-intensive, necessitating compromises between the quality of MSI data and the number and accuracy of compound IDs. Another approach is to prepare separate samples for MSI and compound ID. Maier *et al.* prepared samples according to MALDI-MSI protocols and identified proteins by surface extraction followed by tryptic digestion.[112] Modern MS methods often use LC–MS/MS as a separate confirmatory technique, using in-solution digests or extracts for accurate mass and tandem MS data. When compound identifications cannot be directly extrapolated from data, shared databases can help to propose common species. These databases are often compiled from experiments like those described previously, using high-resolution MS and MS/MS data along with LC retention time matching to confirm IDs. Several databases cover different compound classes, including MSiMass[113] and the ProteomeXchange Consortium[114] for proteins (the latter of which collates several proteomic/peptidomic MS/MS databases), METLIN[115] and METASPACE[116] for metabolites.

GNPS,[117] collates user-generated untargeted raw MS/MS data, enabling information sharing between a broad coalition of scholars, institutions, instrumentation, and methodologies. This approach to sharing untargeted experimental data has multiple advantages for understanding the chemical composition of complex microbiological samples; while comparison to molecular databases allows for rapid confirmation of common species, most peaks in the spectra are still unknowns without conclusive database identifications. A key concept when considering compound identifications in natural products is dereplication, the process of identifying 'known unknowns'; since many metabolites are produced by multiple microbial species, distinguishing commonly found microbial products from novel natural products is a major focus of compound discovery.[118] Using databases such as the GNPS to mine MS/MS data for similarities to known spectra allows for comprehensive differentiation between unknown compounds and known natural products. Molecular features are categorized by spectral similarity to create clusters or "nodes" of peaks with similar MS properties. Species can then also be compared between sample types or conditions, identifying the chemical classes associated with biological processes. This approach, known as molecular networking, compares spectral data using computational algorithms to group ions based on *m/z* and fragmentation patterns. A number of papers have used this approach to characterize the molecular space of microbial species; Lybbert *et al.* found compound classes such as quinolones, rhamnolipids, and phenazines clustered together and grouped into a number of different families, pointing to their chemical diversity.[119] Pyochelin, a siderophore, was detected both unchanged and as a number of derivatives in several nodes. Comparison of these compounds to metabolomics databases found natural product matches, not only in other bacterial species but in diverse samples ranging from soil to ant fungus gardens.

For the purposes of MSI, compound identification often uses several of the techniques described here in conjunction. To confirm compound identifications from a microbial MSI experiment, Boya *et al.* used high resolution ToF-MS to obtain accurate masses, followed by tandem HPLC–MS/MS data and comparison with GNPS databases to dereplicate known compounds.[31] They assigned several compounds in the GNPS library to upregulated *m/z* values in bacterial–fungal co-cultures, identifying the antimicrobial compounds putatively responsible for the symbiotic ant-fungus relationship studied.

13.7 Conclusions and Future Directions

Many of the methods for compound identification described in the previous section involve cutting-edge uses of new technology, from new instrumentation with improved parameters to multimodal imaging techniques to large-scale data sharing. These are all areas of focus for improving the amount of information obtained from microbial MSI experiments. As molecular networking becomes more common as part of compound discovery workflows, authors such as Aron *et al.* have written detailed protocols to standardize the procedure.[120] The continual expansion of user-generated repositories such as GNPS *via* additional reference libraries and MS/MS spectra facilitates interrogating MSI data in greater depth, gleaning more chemical knowledge from existing data sets. Further tools to streamline data mining and data sharing are continually in development, including the Mass Spectrometry Search Tool (MASST) at the University of California, San Diego[121] and the Galaxy framework from Föll *et al.*[122] In addition to database tools, experimental data can be supplemented by computational *in silico* fragmentation tools. While theoretical MS/MS tools have been in use for over a decade,[123,124] coupling these tools to large-scale databases creates new avenues to improve the efficacy of annotating MS spectra and create molecular networks.[125] Advances in the fields of machine learning and artificial intelligence (AI)-assisted software over the past decade have opened up more opportunities to support experimental metabolomic data with algorithmically-generated candidate identifications.[126]

A large area of focus is in developing new instrumentation and coupled techniques to improve compound identification directly from experimental data. In addition to previously-discussed high mass resolving power instruments (MALDI–FTICR-MS,[127] DESI-Orbitrap MS[128]), sub-atmospheric pressure MALDI allows for greater spatial resolution,[129,130] which enables the study of bacterial microcolonies of ~50 cells,[95] and recent laser post-ionization (LPI) techniques such as MALDI-2 demonstrate improved sensitivity[131] and compound identification.[132] Many next-generation approaches to MSI experiments involve adding additional separational dimensions. Ion mobility spectrometry (IMS) separates ions by 3-dimensional collision cross section (CCS) in addition to m/z; multiple studies have shown the utility of ion mobility spectrometry for identifying bacterial genera[133,134] and for identifying natural products produced by microbes.[135] While Esquenazi *et al.*'s approach involved comparing data from separate mobility and imaging experiments,[135] coupled ion mobility-mass spectrometry (IM-MS) imaging allows for the separation of isobaric and isomeric species directly from a sample surface, confirming compound identifications by their 3-dimensional shape. Instrumentation and approach vary significantly between laboratories due to the relatively young age of high-resolution ion mobility spectrometry compared to long-established techniques such as LC, so IM-MS does not yet have access to the same level of user-generated confirmatory data. However, with increasingly accessible commercial high-resolution IM-MS instrumentation[136] and researchers continuously generating CCS databases,[137,138] the millisecond-scale mobility separations in IM-MS aim to provide a time-effective alternative to time and labor-intensive LC–MS/MS experiments. Many previously-discussed strategies for generating compound identifications, from tandem MS[139] to machine learning-facilitated theoretical prediction,[140] can also be integrated into an IM-MS workflow.

In addition to technological advances, biological questions are likely to continue to drive innovation in the MSI field. For instance, non-destructive techniques such as microscopy[141–143] and spectroscopy[144,145] can be integrated directly into a workflow to analyze the same microbial sample but add additional layers of data depending on which type of biomolecule is labeled. Incorporating different types of labels better facilitates multimodal imaging across the molecular scale. The use of multiple imaging techniques can also provide complementary data; Fincher *et al.* describe using nanopost array laser desorption ionization (NAPA-LDI) to observe neutral lipids such as hexosylceramides which are undetected by MALDI.[146] Many papers on multimodal techniques describe 3-dimensional tissue imaging,[147–150] which can be directly applied to study host–microbe interactions.

Looking to the future, it's clear that the breadth of emerging instrumental techniques, technology-facilitated cross-laboratory collaboration and algorithmically-enhanced data analysis will only improve the efficiency, speed, and comprehensiveness of natural products research, and mass spectrometry imaging is poised to continue providing researchers with a direct window into the molecular world of microbial communication.

Abbreviations

CCS	Collision Cross Section
CHCA	Cyano-hydroxycinnamic acid
CMC	Carboxymethylcellulose
DHB	Dihydroxybenzoic acid
EPS	Extracellular Polymeric Substrate
FTICR	Fourier Transform Ion Cyclotron Resonance
GNPS	Global Natural Product Social Molecular Networking
IMS	Ion Mobility Spectrometry
IM-MS	Ion Mobility-Mass Spectrometry
LAESI	Laser Ablation Electrospray Ionization
LDI	Laser Desorption Ionization
MS/MS	Tandem Mass Spectrometry
m/z	Mass to Charge ratio
nanoDESI	nanospray Desorption Electrospray Ionization
OCT	Optimal Cutting Temperature compound
PCA	Phenazine-1-Carboxylic Acid

SIMS	Secondary Ion Mass Spectrometry
TLCA	Taurolithocholic Acid
WHO	World Health Organization

Acknowledgements

This chapter was supported by the National Institute of General Medical Sciences of the National Institutes of Health (NIH) under Award Number R01GM125943, the National Cancer Institute of the NIH under Award Number R01CA240423, and by the National Science Foundation grant 2128044. We would also like to acknowledge Itzel Lizama Chamu and Jessica Cleary-Little for the research images utilized in Figure 13.1.

References

1. S. J. B. Dunham, J. F. Ellis, B. Li and J. V. Sweedler, *Acc. Chem. Res.*, 2017, **50**, 96–104.
2. T. Si, B. Li, T. J. Comi, Y. Wu, P. Hu, Y. Wu, Y. Min, D. A. Mitchell, H. Zhao and J. V. Sweedler, *J. Am. Chem. Soc.*, 2017, **139**, 12466–12473.
3. E. J. Lanni, R. N. Masyuko, C. M. Driscoll, J. T. Aerts, J. D. Shrout, P. W. Bohn and J. V. Sweedler, *Anal. Chem.*, 2014, **86**, 9139–9145.
4. T. U. H. Baumeister, M. Vallet, F. Kaftan, L. Guillou, A. Svatoš and G. Pohnert, *Metabolomics*, 2020, **16**, 28.
5. D. Barbano, R. Diaz, L. Zhang, T. Sandrin, H. Gerken and T. Dempster, *PLoS One*, 2015, **10**, e0135337.
6. B. B. Sandonato, V. G. Santos, M. F. Luizete, J. L. Bronzel Jr, M. N. Eberlin and H. M. S. Milagre, *J. Braz. Chem. Soc.*, 2017, **28**, 521–528.
7. E. Esquenazi, C. Coates, L. Simmons, D. Gonzalez, W. H. Gerwick and P. C. Dorrestein, *Mol. BioSyst.*, 2008, **4**, 562–570.
8. S. Y. Imanishi, T. Nakayama, H. Asukabe and K.-I. Harada, *Rapid Commun. Mass Spectrom.*, 2017, **31**, 325–332.
9. L.-W. Sun, W.-J. Jiang, H. Sato, M. Kawachi and X.-W. Lu, *PLoS One*, 2016, **11**, e0156275.
10. H. Asukabe, S. Akahori, E. Ueno, T. Nakayama, R. Yamashita, S. Arii, K.-I. Harada and S. Y. Imanishi, *J. Am. Soc. Mass Spectrom.*, 2020, **31**, 1572–1578.
11. J. E. Geraci, F. R. Heilman, D. R. Nichols, E. W. Wellman and G. T. Ross, *Antibiot. Annu.*, 1956, 90–106.
12. R. S. Griffith, *Rev. Infect. Dis.*, 1981, **3**, S200–S204.
13. D. A. Evans, M. R. Wood, B. W. Trotter, T. I. Richardson, J. C. Barrow and J. L. Katz, *Angew. Chem., Int.*, 1998, **37**, 2700–2704.
14. R. A. Holton, C. Somoza, H. B. Kim, F. Liang, R. J. Biediger, P. D. Boatman, M. Shindo, C. C. Smith and S. Kim, *J. Am. Chem. Soc.*, 1994, **116**, 1597–1598.

15. D. Giordano, D. Coppola, R. Russo, R. Denaro, L. Giuliano, F. M. Lauro, G. di Prisco and C. Verde, in *Advances in Microbial Physiology*, ed. R. K. Poole, Academic Press, 2015, vol. 66, ch. 4, pp. 357–428.
16. M. Igarashi, *J. Antibiot.*, 2019, **72**, 890–898.
17. S. E. Rossiter, M. H. Fletcher and W. M. Wuest, *Chem. Rev.*, 2017, **117**, 12415–12474.
18. J. D. DeBord, D. F. Smith, C. R. Anderton, R. M. A. Heeren, L. Paša-Tolić, R. H. Gomer and F. A. Fernandez-Lima, *PLoS One*, 2014, **9**, e99319.
19. W. Kim, J.-J. Park, F. M. Dugan, T. L. Peever, D. R. Gang, G. Vandemark and W. Chen, *Environ. Microbiol.*, 2017, **19**, 1822–1835.
20. P. D. Lister, D. J. Wolter and N. D. Hanson, *Clin. Microbiol. Rev.*, 2009, **22**, 582–610.
21. A. R. Condren, L. J. Kahl, G. Boelter, G. Kritikos, M. Banzhaf, L. E. P. Dietrich and L. M. Sanchez, *ACS Infect. Dis.*, 2020, **6**, 603–612.
22. V. V. Phelan, J. Fang and P. C. Dorrestein, *J. Am. Soc. Mass Spectrom.*, 2015, **26**, 873–877.
23. E. C. Pierce, M. Morin, J. C. Little, R. B. Liu, J. Tannous, N. P. Keller, B. E. Wolfe, K. Pogliano, L. M. Sanchez and R. J. Dutton, *Nat. Microbiol.*, 2021, **6**, 87–102.
24. J. E. Spraker, P. Wiemann, J. A. Baccile, N. Venkatesh, J. Schumacher, F. C. Schroeder, L. M. Sanchez and N. P. Keller, *mBio*, 2018, **9**, e00820-18.
25. J. E. Spraker, L. M. Sanchez, T. M. Lowe, P. C. Dorrestein and N. P. Keller, *ISME J.*, 2016, **10**, 2317–2330.
26. L. M. Sanchez, J. Cleary, *Large scale MALDI-TOF imaging of metabolites from filamentous fungi*, Bruker, 2018.
27. R. F. Menger, C. S. Clendinen, L. A. Searcy, A. S. Edison and R. A. Yost, *Curr. Metabolomics*, 2015, **3**, 130–137.
28. S. Bulgheresi, *FEMS Microbiol. Ecol.*, 2016, **92**, fiw007, 1–7.
29. K. E. Zink, D. A. Tarnowski, P. R. Lazzara, T. W. Moore, M. J. Mandel and L. M. Sanchez, 2020, DOI: 10.1101/2020.08.17.254854.
30. K. E. Zink, D. A. Tarnowski, M. J. Mandel and L. M. Sanchez, *J. Mass Spectrom.*, 2020, **55**, e4458.
31. C. A. Boya P, H. Fernández-Marín, L. C. Mejía, C. Spadafora, P. C. Dorrestein and M. Gutiérrez, *Sci. Rep.*, 2017, **7**, 5604.
32. E. Gemperline, H. A. Horn, K. DeLaney, C. R. Currie and L. Li, *ACS Chem. Biol.*, 2017, **12**, 1980–1985.
33. P. F. Scholl, D. Kongkasuriyachai, P. A. Demirev, A. B. Feldman, J. S. Lin, D. J. Sullivan Jr and N. Kumar, *Am. J. Trop. Med. Hyg.*, 2004, **71**, 546–551.
34. P. Kadesch, T. Quack, S. Gerbig, C. G. Grevelding and B. Spengler, *PLoS Neglected Trop. Dis.*, 2020, **14**, e0008145.
35. T. Hamerly, J. A. Everett, N. Paris, S. T. Fisher, A. Karunamurthy, G. A. James, K. P. Rumbaugh, D. D. Rhoads and B. Bothner, *Anal. Biochem.*, 2017, **539**, 144–148.

36. J. Y. Yang, V. V. Phelan, R. Simkovsky, J. D. Watrous, R. M. Trial, T. C. Fleming, R. Wenter, B. S. Moore, S. S. Golden, K. Pogliano and P. C. Dorrestein, *J. Bacteriol.*, 2012, **194**, 6023–6028.
37. S. Fukuda, Y. Nakanishi, E. Chikayama, H. Ohno, T. Hino and J. Kikuchi, *PLoS One*, 2009, **4**, e4893.
38. M. R. Barer, in *Molecular Medical Microbiology*, ed. Y.-W. Tang, M. Sussman, D. Liu, I. Poxton and J. Schwartzman, Academic Press, Boston, 2nd edn, 2015, vol. 1, ch. 10, pp. 181–199.
39. B. M. Prentice, C. W. Chumbley and R. M. Caprioli, *J. Mass Spectrom.*, 2015, **50**, 703–710.
40. D. Suckau, A. Resemann, M. Schuerenberg, P. Hufnagel, J. Franzen and A. Holle, *Anal. Bioanal. Chem.*, 2003, **376**, 952–965.
41. C. P. Santacruz, E. Ayala and C. Costa-Vera, *Braz. J. Phys.*, 2006, **36**, 789–794.
42. R. X. Poulin and G. Pohnert, *Anal. Bioanal. Chem.*, 2019, **411**, 13–19.
43. V. V. Phelan, W. J. Moree, J. Aguilar, D. S. Cornett, A. Koumoutsi, S. M. Noble, K. Pogliano, C. A. Guerrero and P. C. Dorrestein, *J. Bacteriol.*, 2014, **196**, 1683–1693.
44. A. Shobo, S. Baijnath, D. Bratkowska, S. Naiker, A. M. Somboro, L. A. Bester, S. D. Singh, T. Naicker, H. G. Kruger and T. Govender, *Drug Test. Anal.*, 2016, **8**, 832–838.
45. M. Wang, J. J. Carver, V. V. Phelan, L. M. Sanchez, N. Garg, Y. Peng, D. D. Nguyen, J. Watrous, C. A. Kapono, T. Luzzatto-Knaan, C. Porto, A. Bouslimani, A. V. Melnik, M. J. Meehan, W.-T. Liu, M. Crüsemann, P. D. Boudreau, E. Esquenazi, M. Sandoval-Calderón, R. D. Kersten, L. A. Pace, R. A. Quinn, K. R. Duncan, C.-C. Hsu, D. J. Floros, R. G. Gavilan, K. Kleigrewe, T. Northen, R. J. Dutton, D. Parrot, E. E. Carlson, B. Aigle, C. F. Michelsen, L. Jelsbak, C. Sohlenkamp, P. Pevzner, A. Edlund, J. McLean, J. Piel, B. T. Murphy, L. Gerwick, C.-C. Liaw, Y.-L. Yang, H.-U. Humpf, M. Maansson, R. A. Keyzers, A. C. Sims, A. R. Johnson, A. M. Sidebottom, B. E. Sedio, A. Klitgaard, C. B. Larson, C. A. Boya P, D. Torres-Mendoza, D. J. Gonzalez, D. B. Silva, L. M. Marques, D. P. Demarque, E. Pociute, E. C. O'Neill, E. Briand, E. J. N. Helfrich, E. A. Granatosky, E. Glukhov, F. Ryffel, H. Houson, H. Mohimani, J. J. Kharbush, Y. Zeng, J. A. Vorholt, K. L. Kurita, P. Charusanti, K. L. McPhail, K. F. Nielsen, L. Vuong, M. Elfeki, M. F. Traxler, N. Engene, N. Koyama, O. B. Vining, R. Baric, R. R. Silva, S. J. Mascuch, S. Tomasi, S. Jenkins, V. Macherla, T. Hoffman, V. Agarwal, P. G. Williams, J. Dai, R. Neupane, J. Gurr, A. M. C. Rodríguez, A. Lamsa, C. Zhang, K. Dorrestein, B. M. Duggan, J. Almaliti, P.-M. Allard, P. Phapale, L.-F. Nothias, T. Alexandrov, M. Litaudon, J.-L. Wolfender, J. E. Kyle, T. O. Metz, T. Peryea, D.-T. Nguyen, D. VanLeer, P. Shinn, A. Jadhav, R. Müller, K. M. Waters, W. Shi, X. Liu, L. Zhang, R. Knight, P. R. Jensen, B. O. Palsson, K. Pogliano, R. G. Linington, M. Gutiérrez, N. P. Lopes, W. H. Gerwick, B. S. Moore, P. C. Dorrestein and N. Bandeira, *Nat. Biotechnol.*, 2016, **34**, 828–837.

46. J. L. Norris and R. M. Caprioli, *Chem. Rev.*, 2013, **113**, 2309–2342.
47. G. L. Chadwick, F. Jiménez Otero, J. A. Gralnick, D. R. Bond and V. J. Orphan, *Proc. Natl. Acad. Sci. U. S. A.*, 2019, **116**, 20716–20724.
48. P. Chaurand, S. A. Schwartz and R. M. Caprioli, *Anal. Chem.*, 2004, **76**, 86 A–93 A.
49. S. N. Jackson, M. Ugarov, T. Egan, J. D. Post, D. Langlais, J. Albert Schultz and A. S. Woods, *J. Mass Spectrom.*, 2007, **42**, 1093–1098.
50. Y.-L. Yang, Y. Xu, P. Straight and P. C. Dorrestein, *Nat. Chem. Biol.*, 2009, **5**, 885–887.
51. T. Hoffmann and P. C. Dorrestein, *J. Am. Soc. Mass Spectrom.*, 2015, **26**, 1959–1962.
52. R. Liu, R. A. A. Khan, Q. Yue, Y. Jiao, Y. Yang, Y. Li and B. Xie, *Biochem. Biophys. Res. Commun.*, 2020, **527**, 689–695.
53. C. R. Anderton, R. K. Chu, N. Tolić, A. Creissen and L. Paša-Tolić, *J. Am. Soc. Mass Spectrom.*, 2016, **27**, 556–559.
54. Z. Zhu, J. Shen, Y. Xu, H. Guo, D. Kang, T. Yu, H. Wang, W. Xu, G. Wang and Y. Liang, *J. Mass Spectrom.*, 2019, **54**, 684–692.
55. D. Debois, M. Ongena, H. Cawoy and E. De Pauw, *J. Am. Soc. Mass Spectrom.*, 2013, **24**, 1202–1213.
56. B. Li, T. J. Comi, T. Si, S. J. B. Dunham and J. V. Sweedler, *J. Mass Spectrom.*, 2016, **51**, 1030–1035.
57. J. Watrous, P. Roach, B. Heath, T. Alexandrov, J. Laskin and P. C. Dorrestein, *Anal. Chem.*, 2013, **85**, 10385–10391.
58. H. Li, P. Balan and A. Vertes, *Angew. Chem., Int. Ed. Engl.*, 2016, **55**, 15035–15039.
59. H. Tian, D. A. Six, T. Krucker, J. A. Leeds and N. Winograd, *Anal. Chem.*, 2017, **89**, 5050–5057.
60. HTX Imaging, *Technical Note #36: Homogenous Spray Application of DHB on Dried Agar for MALDI Imaging of Microbial Cultures*, https://static1.squarespace.com/static/55eae94ae4b08810efb9cd87/t/560eb007e4b01ebc8cb5e8a6/1443803143077/TechNote36_d3%281%29+copy.pdf, (accessed 12 February 2021).
61. C. A. Wakeman, J. L. Moore, M. J. Noto, Y. Zhang, M. D. Singleton, B. M. Prentice, B. A. Gilston, R. S. Doster, J. A. Gaddy, W. J. Chazin, R. M. Caprioli and E. P. Skaar, *Nat. Commun.*, 2016, **7**, 11951.
62. K. A. Floyd, J. L. Moore, A. R. Eberly, J. A. D. Good, C. L. Shaffer, H. Zaver, F. Almqvist, E. P. Skaar, R. M. Caprioli and M. Hadjifrangiskou, *PLoS Pathog.*, 2015, **11**, e1004697.
63. S. K. Davies, S. Fearn, L. P. Allsopp, F. Harrison, E. Ware, S. P. Diggle, A. Filloux, D. S. McPhail and J. G. Bundy, *mSphere*, 2017, **2**, e00211–e00217.
64. W. J. Perry, J. M. Spraggins, J. R. Sheldon, C. M. Grunenwald, D. E. Heinrichs, J. E. Cassat, E. P. Skaar and R. M. Caprioli, *Proc. Natl. Acad. Sci. U. S. A.*, 2019, **116**, 21980–21982.

65. P. L. Urban, T. Schmid, A. Amantonico and R. Zenobi, *Anal. Chem.*, 2011, **83**, 1843–1849.
66. N. Garg, Y. Zeng, A. Edlund, A. V. Melnik, L. M. Sanchez, H. Mohimani, A. Gurevich, V. Miao, S. Schiffler, Y. W. Lim, T. Luzzatto-Knaan, S. Cai, F. Rohwer, P. A. Pevzner, R. H. Cichewicz, T. Alexandrov and P. C. Dorrestein, *mSystems*, 2016, **1**, e00139-16.
67. V. P. Sica, H. A. Raja, T. El-Elimat and N. H. Oberlies, *RSC Adv.*, 2014, **4**, 63221–63227.
68. D. J. Gonzalez, Y. Xu, Y.-L. Yang, E. Esquenazi, W.-T. Liu, A. Edlund, T. Duong, L. Du, I. Molnár, W. H. Gerwick, P. R. Jensen, M. Fischbach, C.-C. Liaw, P. Straight, V. Nizet and P. C. Dorrestein, *J. Proteomics*, 2012, **75**, 5069–5076.
69. E. Ortenberg and B. Telsch, in *Handbook of Water and Wastewater Microbiology*, ed. D. Mara and N. Horan, Academic Press, London, 2003, ch. 42, pp. 777–793.
70. K. Milferstedt, W. C. Kuo-Dahab, C. S. Butler, J. Hamelin, A. S. Abouhend, K. Stauch-White, A. McNair, C. Watt, B. I. Carbajal-González, S. Dolan and C. Park, *Sci. Rep.*, 2017, **7**, 17944.
71. J. Komárek and J. R. Johansen, in *Freshwater Algae of North America*, ed. J. D. Wehr, R. G. Sheath and J. P. Kociolek, Academic Press, Boston, 2nd edn, 2015, ch. 4, pp. 135–235.
72. D. B. Kearns, *Nat. Rev. Microbiol.*, 2010, **8**, 634–644.
73. D. Debois, K. Hamze, V. Guérineau, J.-P. Le Caër, I. B. Holland, P. Lopes, J. Ouazzani, S. J. Séror, A. Brunelle and O. Laprévote, *Proteomics*, 2008, **8**, 3682–3691.
74. R. M. Donlan, *Emerging Infect. Dis.*, 2002, **8**, 881–890.
75. C. Cattò and F. Cappitelli, *Int. J. Mol. Sci.*, 2019, **20**, 3794.
76. T. Santos, L. Théron, C. Chambon, D. Viala, D. Centeno, J. Esbelin and M. Hébraud, *J. Proteomics*, 2018, **187**, 152–160.
77. J. E. Sykes and S. C. Rankin, in *Canine and Feline Infectious Diseases*, ed. J. E. Sykes and W. B. Saunders, Saint Louis, 2014, ch. 4, pp. 29–36.
78. W.-T. Liu, Y.-L. Yang, Y. Xu, A. Lamsa, N. M. Haste, J. Y. Yang, J. Ng, D. Gonzalez, C. D. Ellermeier, P. D. Straight, P. A. Pevzner, J. Pogliano, V. Nizet, K. Pogliano and P. C. Dorrestein, *Proc. Natl. Acad. Sci. U. S. A.*, 2010, **107**, 16286–16290.
79. T. Si, B. Li, K. Zhang, Y. Xu, H. Zhao and J. V. Sweedler, *J. Proteome Res.*, 2016, **15**, 1955–1962.
80. F. D. S. Araújo, R. L. Vieira, E. P. L. Molano, H. J. Máximo, R. J. D. Dalio, P. H. Vendramini, W. L. Araújo and M. N. Eberlin, *RSC Adv.*, 2017, **7**, 29953–29958.
81. S. H. Kim, J. Kim, S.-H. Jo, J.-H. Kim, K. J. Kim and S. Yoon, *Biointerphases*, 2016, **11**, 041008.
82. J. Watrous, P. Roach, T. Alexandrov, B. S. Heath, J. Y. Yang, R. D. Kersten, M. van der Voort, K. Pogliano, H. Gross, J. M. Raaijmakers, B. S. Moore, J. Laskin, N. Bandeira and

P. C. Dorrestein, *Proc. Natl. Acad. Sci. U. S. A.*, 2012, **109**, E1743–E1752.
83. A. J. Sporer, C. Beierschmitt, A. Bendebury, K. E. Zink, A. Price-Whelan, M. C. Buzzeo, L. M. Sanchez and L. E. P. Dietrich, *Microbiology*, 2018, **164**, 790–800.
84. T. Horinouchi, T. Maeda, H. Kotani and C. Furusawa, *Sci. Rep.*, 2020, **10**, 4178.
85. J. W. Bok and N. P. Keller, *Eukaryotic Cell*, 2004, **3**, 527–535.
86. X. Zhao, J. E. Spraker, J. W. Bok, T. Velk, Z.-M. He and N. P. Keller, *Front. Microbiol.*, 2017, **8**, 1925.
87. C.-J. Lin, Y.-H. Hou and Y.-L. Chen, *Med. Mycol.*, 2020, **58**, 248–259.
88. H.-W. Nützmann, Y. Reyes-Dominguez, K. Scherlach, V. Schroeckh, F. Horn, A. Gacek, J. Schümann, C. Hertweck, J. Strauss and A. A. Brakhage, *Proc. Natl. Acad. Sci. U. S. A.*, 2011, **108**, 14282–14287.
89. A. Gacek-Matthews, L. M. Noble, C. Gruber, H. Berger, M. Sulyok, A. T. Marcos, J. Strauss and A. Andrianopoulos, *Mol. Microbiol.*, 2015, **96**, 839–860.
90. A. Gacek-Matthews, H. Berger, T. Sasaki, K. Wittstein, C. Gruber, Z. A. Lewis and J. Strauss, *PLoS Genet.*, 2016, **12**, e1006222.
91. Y. Tanaka, K. Kasahara, Y. Hirose, K. Murakami, R. Kugimiya and K. Ochi, *J. Bacteriol.*, 2013, **195**, 2959–2970.
92. T. Adékambi, M. Drancourt and D. Raoult, *Trends Microbiol.*, 2009, **17**, 37–45.
93. R. Popa, P. K. Weber, J. Pett-Ridge, J. A. Finzi, S. J. Fallon, I. D. Hutcheon, K. H. Nealson and D. G. Capone, *ISME J.*, 2007, **1**, 354–360.
94. R. Sekine, K. L. Moore, M. Matzke, P. Vallotton, H. Jiang, G. M. Hughes, J. K. Kirby, E. Donner, C. R. M. Grovenor, C. Svendsen and E. Lombi, *ACS Nano*, 2017, **11**, 10894–10902.
95. R. de Cassia Pessotti, B. L. Hansen, V. M. Zacharia, D. Polyakov and M. F. Traxler, *Anal. Chem.*, 2019, **91**, 14818–14823.
96. R. W. Kessler, A. C. Crecelius, U. S. Schubert and T. Wichard, *Anal. Bioanal. Chem.*, 2017, **409**, 4893–4903.
97. G. Schleyer, N. Shahaf, C. Ziv, Y. Dong, R. A. Meoded, E. J. N. Helfrich, D. Schatz, S. Rosenwasser, I. Rogachev, A. Aharoni, J. Piel and A. Vardi, *Nat. Microbiol.*, 2019, **4**, 527–538.
98. K. D. Hyde, J. Xu, S. Rapior, R. Jeewon, S. Lumyong, A. G. T. Niego, P. D. Abeywickrama, J. V. S. Aluthmuhandiram, R. S. Brahamanage, S. Brooks, A. Chaiyasen, K. W. T. Chethana, P. Chomnunti, C. Chepkirui, B. Chuankid, N. I. de Silva, M. Doilom, C. Faulds, E. Gentekaki, V. Gopalan, P. Kakumyan, D. Harishchandra, H. Hemachandran, S. Hongsanan, A. Karunarathna, S. C. Karunarathna, S. Khan, J. Kumla, R. S. Jayawardena, J.-K. Liu, N. Liu, T. Luangharn, A. P. G. Macabeo, D. S. Marasinghe, D. Meeks, P. E. Mortimer, P. Mueller, S. Nadir, K. N. Nataraja, S. Nontachaiyapoom, M. O'Brien, W. Penkhrue, C. Phukhamsakda,

U. S. Ramanan, A. R. Rathnayaka, R. B. Sadaba, B. Sandargo, B. C. Samarakoon, D. S. Tennakoon, R. Siva, W. Sriprom, T. S. Suryanarayanan, K. Sujarit, N. Suwannarach, T. Suwunwong, B. Thongbai, N. Thongklang, D. Wei, S. N. Wijesinghe, J. Winiski, J. Yan, E. Yasanthika and M. Stadler, *Fungal Divers.*, 2019, **97**, 1–136.

99. E. Calla-Quispe, H. L. Fuentes-Rivera, P. Ramírez, C. Martel and A. J. Ibañez, *Life*, 2020, **10**, 18.
100. J. Clardy, M. A. Fischbach and C. R. Currie, *Curr. Biol.*, 2009, **19**, R437–R441.
101. P. Wiener, *Evol. Ecol.*, 1996, **10**, 405–421.
102. J. Heyse, B. Buysschaert, R. Props, P. Rubbens, A. G. Skirtach, W. Waegeman and N. Boon, *Appl. Environ. Microbiol.*, 2019, **8**, e02814–e02818.
103. J. L. Cleary, S. Kolachina, B. E. Wolfe and L. M. Sanchez, *mSystems*, 2018, **3**, e00036-18.
104. P.-Y. Chen, C.-Y. Hsieh, C.-J. Shih, Y.-J. Lin, C.-W. Tsao and Y.-L. Yang, *J. Nat. Prod.*, 2018, **81**, 1527–1533.
105. H. C. Martin, R. Ibáñez, L. F. Nothias, P. C. A. Boya, L. K. Reinert, L. A. Rollins-Smith, P. C. Dorrestein and M. Gutiérrez, *Sci. Rep.*, 2019, **9**, 3019.
106. M. Holzlechner, S. Reitschmidt, S. Gruber, S. Zeilinger and M. Marchetti-Deschmann, *Proteomics*, 2016, **16**, 1742–1746.
107. C. F. Jaegger, F. Negrão, D. M. Assis, K. R. A. Belaz, C. F. F. Angolini, A. M. A. P. Fernandes, V. G. Santos, A. Pimentel, D. R. Abánades, S. Giorgio, M. N. Eberlin and D. F. O. Rocha, *Mol. BioSyst.*, 2017, **13**, 1946–1953.
108. World Health Organization, *Neglected Tropical Diseases progress dashboard 2011–2020*, https://www.who.int/teams/control-of-neglected-tropical-diseases/overview/progress-dashboard-2011-2020 (accessed 10 September 2021).
109. B. Geier, E. M. Sogin, D. Michellod, M. Janda, M. Kompauer, B. Spengler, N. Dubilier and M. Liebeke, *Nat. Microbiol.*, 2020, **5**, 498–510.
110. C. M. Rath, T. Alexandrov, S. K. Higginbottom, J. Song, M. E. Milla, M. A. Fischbach, J. L. Sonnenburg and P. C. Dorrestein, *Anal. Chem.*, 2012, **84**, 9259–9267.
111. A. D. Feenstra, R. L. Hansen and Y. J. Lee, *Analyst*, 2015, **140**, 7293–7304.
112. S. K. Maier, H. Hahne, A. M. Gholami, B. Balluff, S. Meding, C. Schoene, A. K. Walch and B. Kuster, *Mol. Cell. Proteomics*, 2013, **12**, 2901–2910.
113. L. A. McDonnell, A. Walch, M. Stoeckli and G. L. Corthals, *J. Proteome Res.*, 2014, **13**, 1138–1142.
114. J. A. Vizcaíno, E. W. Deutsch, R. Wang, A. Csordas, F. Reisinger, D. Ríos, J. A. Dianes, Z. Sun, T. Farrah, N. Bandeira, P.-A. Binz, I. Xenarios, M. Eisenacher, G. Mayer, L. Gatto, A. Campos, R. J. Chalkley, H.-J. Kraus, J. P. Albar, S. Martinez-Bartolomé, R. Apweiler,

G. S. Omenn, L. Martens, A. R. Jones and H. Hermjakob, *Nat. Biotechnol.*, 2014, **32**, 223–226.
115. C. A. Smith, G. O'Maille, E. J. Want, C. Qin, S. A. Trauger, T. R. Brandon, D. E. Custodio, R. Abagyan and G. Siuzdak, *Ther. Drug Monit.*, 2005, **27**, 747–751.
116. T. Alexandrov, *Annu. Rev. Biomed. Sci.*, 2020, **3**, 61–87.
117. R. A. Quinn, L.-F. Nothias, O. Vining, M. Meehan, E. Esquenazi and P. C. Dorrestein, *Trends Pharmacol. Sci.*, 2017, **38**, 143–154.
118. S. P. Gaudêncio and F. Pereira, *Nat. Prod. Rep.*, 2015, **32**, 779–810.
119. A. C. Lybbert, J. L. Williams, R. Raghuvanshi, A. D. Jones and R. A. Quinn, *Metabolites*, 2020, **10**, 445.
120. A. T. Aron, E. C. Gentry, K. L. McPhail, L.-F. Nothias, M. Nothias-Esposito, A. Bouslimani, D. Petras, J. M. Gauglitz, N. Sikora, F. Vargas, J. J. J. van der Hooft, M. Ernst, K. B. Kang, C. M. Aceves, A. M. Caraballo-Rodríguez, I. Koester, K. C. Weldon, S. Bertrand, C. Roullier, K. Sun, R. M. Tehan, C. A. Boya P, M. H. Christian, M. Gutiérrez, A. M. Ulloa, J. A. Tejeda Mora, R. Mojica-Flores, J. Lakey-Beitia, V. Vásquez-Chaves, Y. Zhang, A. I. Calderón, N. Tayler, R. A. Keyzers, F. Tugizimana, N. Ndlovu, A. A. Aksenov, A. K. Jarmusch, R. Schmid, A. W. Truman, N. Bandeira, M. Wang and P. C. Dorrestein, *Nat. Protoc.*, 2020, **15**, 1954–1991.
121. M. Wang, A. K. Jarmusch, F. Vargas, A. A. Aksenov, J. M. Gauglitz, K. Weldon, D. Petras, R. da Silva, R. Quinn, A. V. Melnik, J. J. J. van der Hooft, A. M. Caraballo-Rodríguez, L. F. Nothias, C. M. Aceves, M. Panitchpakdi, E. Brown, F. Di Ottavio, N. Sikora, E. O. Elijah, L. Labarta-Bajo, E. C. Gentry, S. Shalapour, K. E. Kyle, S. P. Puckett, J. D. Watrous, C. S. Carpenter, A. Bouslimani, M. Ernst, A. D. Swafford, E. I. Zúñiga, M. J. Balunas, J. L. Klassen, R. Loomba, R. Knight, N. Bandeira and P. C. Dorrestein, *Nat. Biotechnol.*, 2020, **38**, 23–26.
122. M. C. Föll, L. Moritz, T. Wollmann, M. N. Stillger, N. Vockert, M. Werner, P. Bronsert, K. Rohr, B. A. Grüning and O. Schilling, *Gigascience*, 2019, **8**, 1–12.
123. S. Wolf, S. Schmidt, M. Müller-Hannemann and S. Neumann, *BMC Bioinf.*, 2010, **11**, 148.
124. S. Clerens, W. Van den Ende, P. Verhaert, L. Geenen and L. Arckens, *Proteomics*, 2004, **4**, 629–632.
125. R. R. da Silva, M. Wang, L.-F. Nothias, J. J. J. van der Hooft, A. M. Caraballo-Rodríguez, E. Fox, M. J. Balunas, J. L. Klassen, N. P. Lopes and P. C. Dorrestein, *PLoS Comput. Biol.*, 2018, **14**, e1006089.
126. K. Qu, F. Guo, X. Liu, Y. Lin and Q. Zou, *Front. Microbiol.*, 2019, **10**, 827.
127. J. M. Spraggins, D. G. Rizzo, J. L. Moore, K. L. Rose, N. D. Hammer, E. P. Skaar and R. M. Caprioli, *J. Am. Soc. Mass Spectrom.*, 2015, **26**, 974–985.

128. N. E. Manicke, A. L. Dill, D. R. Ifa and R. G. Cooks, *J. Mass Spectrom.*, 2010, **45**, 223–226.
129. G. Li, Q. Cao, Y. Liu, K. DeLaney, Z. Tian, E. Moskovets and L. Li, *Rapid Commun. Mass Spectrom.*, 2019, **33**, 327–335.
130. Y. Shi, Z. Li, M. A. Felder, Q. Yu, X. Shi, Y. Peng, Q. Cao, B. Wang, L. Puglielli, M. S. Patankar and L. Li, *Anal. Chem.*, 2019, **91**, 12942–12947.
131. M. Niehaus, J. Soltwisch, M. E. Belov and K. Dreisewerd, *Nat. Methods*, 2019, **16**, 925–931.
132. J. C. McMillen, J. A. Fincher, D. R. Klein, J. M. Spraggins and R. M. Caprioli, *J. Mass Spectrom.*, 2020, **55**, e4663.
133. L. Hokkinen, A. Kesti, J. Lepomäki, O. Anttalainen, A. Kontunen, M. Karjalainen, J. Aittoniemi, R. Vuento, T. Lehtimäki, N. Oksala and A. Roine, *Future Microbiol.*, 2020, **15**, 233–240.
134. V. Bocos-Bintintan, C. L. P. Thomas and I. A. Ratiu, *Talanta*, 2020, **206**, 120233.
135. E. Esquenazi, M. Daly, T. Bahrainwala, W. H. Gerwick and P. C. Dorrestein, *Bioorg. Med. Chem.*, 2011, **19**, 6639–6644.
136. C. G. Vasilopoulou, K. Sulek, A.-D. Brunner, N. S. Meitei, U. Schweiger-Hufnagel, S. W. Meyer, A. Barsch, M. Mann and F. Meier, *Nat. Commun.*, 2020, **11**, 331.
137. L. Righetti, A. Bergmann, G. Galaverna, O. Rolfsson, G. Paglia and C. Dall'Asta, *Anal. Chim. Acta*, 2018, **1014**, 50–57.
138. W. B. Struwe, K. Pagel, J. L. P. Benesch, D. J. Harvey and M. P. Campbell, *Glycoconj. J.*, 2016, **33**, 399–404.
139. F. Meier, A.-D. Brunner, S. Koch, H. Koch, M. Lubeck, M. Krause, N. Goedecke, J. Decker, T. Kosinski, M. A. Park, N. Bache, O. Hoerning, J. Cox, O. Räther and M. Mann, *Mol. Cell. Proteomics*, 2018, **17**, i-2545.
140. Z. Zhou, X. Shen, J. Tu and Z.-J. Zhu, *Anal. Chem.*, 2016, **88**, 11084–11091.
141. N. H. Patterson, M. Tuck, R. Van de Plas and R. M. Caprioli, *Anal. Chem.*, 2018, **90**, 12395–12403.
142. A. Blutke, N. Sun, Z. Xu, A. Buck, L. Harrison, S. C. Schriever, P. T. Pfluger, D. Wiles, T. Kunzke, K. Huber, J. Schlegel, M. Aichler, A. Feuchtinger, K. Matiasek, S. M. Hauck and A. Walch, *Sci. Rep.*, 2020, **10**, 14461.
143. S. Shimma and T. Sagawa, *J. Agric. Food Chem.*, 2019, **67**, 9652–9657.
144. P. Lasch and I. Noda, *Anal. Chem.*, 2017, **89**, 5008–5016.
145. E. K. Neumann, T. J. Comi, N. Spegazzini, J. W. Mitchell, S. S. Rubakhin, M. U. Gillette, R. Bhargava and J. V. Sweedler, *Anal. Chem.*, 2018, **90**, 11572–11580.
146. J. A. Fincher, A. R. Korte, S. Yadavilli, N. J. Morris and A. Vertes, *Analyst*, 2020, **145**, 6910–6918.
147. S. Van Nuffel, M. Quatredeniers, A. Pirkl, J. Zakel, J.-P. Le Caer, N. Elie, Q. P. Vanbellingen, S. J. Dumas, M. K. Nakhleh, M.-R. Ghigna, E. Fadel,

M. Humbert, P. Chaurand, D. Touboul, S. Cohen-Kaminsky and A. Brunelle, *Anal. Chem.*, 2020, **92**, 12079–12087.
148. S. Chughtai, K. Chughtai, B. Cillero-Pastor, A. Kiss, P. Agrawal, L. MacAleese and R. M. A. Heeren, *Int. J. Mass Spectrom.*, 2012, **325–327**, 150–160.
149. B. Heijs, S. Holst, I. H. Briaire-de Bruijn, G. W. van Pelt, A. H. de Ru, P. A. van Veelen, R. R. Drake, A. S. Mehta, W. E. Mesker, R. A. Tollenaar, J. V. M. G. Bovée, M. Wuhrer and L. A. McDonnell, *Anal. Chem.*, 2016, **88**, 7745–7753.
150. D. Veličković, B. J. Agtuca, S. A. Stopka, A. Vertes, D. W. Koppenaal, L. Paša-Tolić, G. Stacey and C. R. Anderton, *ISME J.*, 2018, **12**, 2335–2338.

CHAPTER 14

MALDI Mass Spectrometry Imaging in Drug Discovery and Development

JEREMY A. BARRY,[a] PETER S. MARSHALL[b] AND M. REID GROSECLOSE*[a]

[a] GlaxoSmithKline, Collegeville, PA, USA; [b] GlaxoSmithKline, Stevenage, UK
*Email: reid.x.groseclose@gsk.com

14.1 Introduction

Preceding decades have revealed a general decrease in the efficiency of pharmaceutical research and development (R&D) as evidenced by a decline in productivity (measured in new drug approvals) concordant with rising R&D costs.[1,2] This trend is often attributed to the financial burden of late-stage failures where the most common cause of program termination is due to a lack of human efficacy.[3] As a result, the pharmaceutical industry has been striving to improve the success rate for new medicines. Several high-level strategies have been proposed to address this issue through improved predictivity, translation, and validation.[4–7] A common theme among these approaches is the reduction of late-stage attrition by identifying risk earlier in the drug development process therefore allowing for a more focused investment in projects with a higher likelihood of success. Also embedded in these frameworks is the incorporation of new scientific innovations which are imperative in providing the necessary evidence to make these early informed decisions. Mass spectrometry imaging (MSI) represents one of the many novel technologies

New Developments in Mass Spectrometry No. 12
MALDI Mass Spectrometry Imaging: From Fundamentals to Spatial Omics
Edited by Tiffany Siegel

Published by the Royal Society of Chemistry, www.rsc.org

that has demonstrated significant potential and is being integrated across the drug discovery and development pipeline to support critical decisions on the progression (or termination) of programs.[8–11]

14.1.1 Mass Spectrometry Imaging

MSI is a label-free technique capable of simultaneously visualizing the molecular distribution of a variety of analytes through the combination of spatial interrogation with the sensitivity and selectivity of mass spectrometry (MS) detection. There are several common approaches to the MSI experiment, distinguished primarily by the ionization source, including matrix-assisted laser desorption/ionization (MALDI), desorption electrospray ionization (DESI), and secondary ion mass spectrometry (SIMS). There are pros and cons for each method and the decision on which to use is primarily dependent upon instrument availability and the question at hand. While this chapter focuses primarily on MALDI-MSI, it should be noted that MSI data obtained from other ionization sources are often complementary despite some potential overlap.[12]

Biological applications of MSI involve collecting cryosections (usually 5–10 μm-thick) of fresh-frozen tissue onto a suitable sample support, most often a glass microscope slide. Indium tin oxide (ITO) coated microscope slides are a preferred option when a conductive sample support is needed. It should be noted that while fresh-frozen tissue is more commonly used, significant effort has been made to develop methods to make use of the widely available tissue banks of formalin-fixed paraffin-embedded (FFPE) tissues. One of the prerequisites specific to MALDI is the application of a laser-absorbing "matrix" which is often a small organic compound applied to facilitate desorption and ionization of tissue-related analytes. Matrix application is often conducted using a commercial automated deposition system (*e.g.* HTX TM Sprayer, SunChrom SunCollect, Bruker ImagePrep) to achieve a homogenous coating of matrix across the sample surface. In its most common form, MALDI-MSI utilizes an ultraviolet laser to analyze an array of discrete positions across the surface of a sample. The laser is pulsed to generate ions from each position or 'pixel' and a mass spectrum is collected. This process is then repeated to collect a two-dimensional array of mass spectra over the sample surface with a pre-defined raster width. The choice of raster width is limited by physical constraints such as the laser focus diameter and the movement resolution of the sample stage as well as practical constraints including sensitivity and the homogeneity of the sample preparation. By correlating the abundance of any ion in the mass spectra with the positional array of the pixels, a heatmap can be generated, which is representative of that ion's distribution in the sample.

When applied to sections of biological tissue, MSI can yield the molecular distribution of exogenous (*i.e.* drugs, drug-related material, delivery vehicles) and endogenous components (*i.e.* metabolites, lipids, peptides, and proteins). Further correlation of MSI data with histology is crucial to understand molecular processes occurring in particular cell types within a tissue[13,14] and has become an integral part of any MSI workflow.[15] Applications within pharma are commonly divided into two workflows including: i) targeted experiments, which primarily focus on the distribution of one or several distinct analytes, or ii) non-targeted workflow, where changes in the abundance or distribution of endogenous components are correlated with disease onset or pharmacology. A multimodal approach combining both the targeted and non-targeted workflows along with correlation to other imaging modalities provides a wealth of information and a more holistic view of the underlying biology.[16]

The targeted MSI workflow is often applied to the distribution assessment of drug-related material (*i.e.* parent drug, drug metabolites, formulation components, *etc.*) but can also be valuable in assessing specific biomarkers or pathways. While the distribution information alone can be immensely useful, it is difficult to draw specific conclusions without an indication of how much material is present. Advances in quantitative MSI have made it possible to provide distributions for key analytes of interest on a quantitative scale.[17–19] MSI quantification, discussed in greater detail in Chapter 6 of this book, is achievable by several methods where calibration curves are prepared either on-target,[20] on-tissue,[21,22] or in-tissue.[23–25] The spotting approaches on-target plate or on-tissue are easier to prepare but may fall short of accounting for the tissue matrix effects covered by the in-tissue methods. In addition, the benefits of reproducibility gained from the in-tissue methods make it better suited for application to large sample sets. As is true with most MS-based methods, the inclusion of stable isotopically labeled (SIL) standard improves the accuracy and reproducibility of the quantification. For MSI quantification, this SIL standard is homogeneously applied to the sample often using an automated deposition system as mentioned above. The ratio of analyte-to-SIL standard are used to account for local variations in matrix effects including ion suppression, which often varies among different cell types even within a tissue.[26,27] Successful implementation of a SIL into the quantification workflow requires careful consideration of the most appropriate application conditions[27] (*e.g.* pre-matrix *vs.* matrix mixed) and some prior knowledge of the drug levels in the target tissue is needed in order to apply the SIL at an appropriate concentration.[28] While the value of including an SIL standard for normalization is well understood, SIL standards are rarely available for compounds in the early stages of drug discovery.

14.1.2 The Drug Discovery and Development Pipeline

The development of a new drug involves a series of key stages and milestones that provide the framework along the path from target discovery through to the approval of a safe and efficacious medicine. A schematic of this process is outlined in Figure 14.1 and, while depicted as clearly defined stages, there is often some degree of overlap between these phases of development. The process typically starts with the identification of a novel druggable target that is hypothesized to be relevant for a particular disease indication. Target selection and validation are now largely informed by genomic evidence of the target's association to a specific

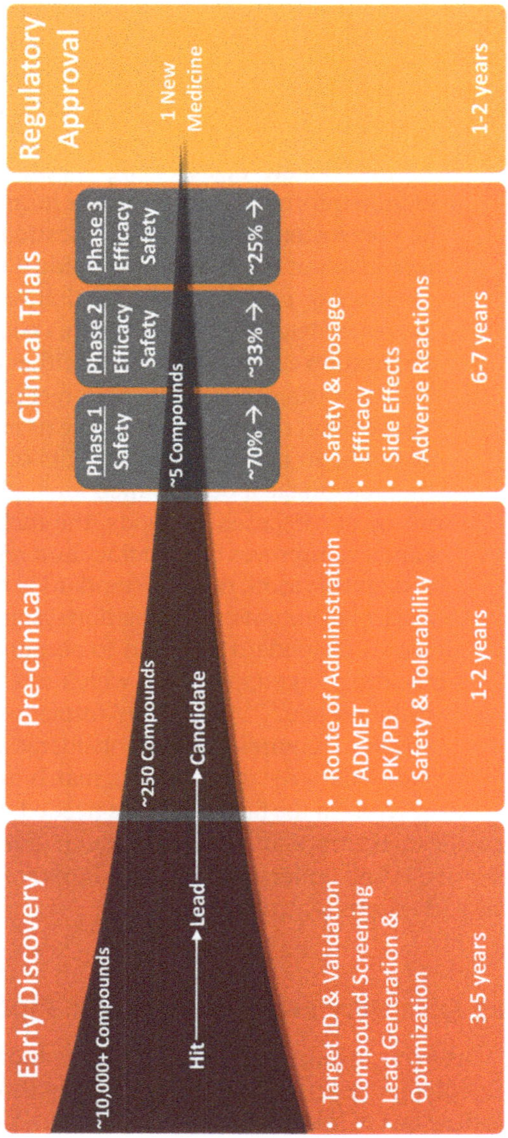

Figure 14.1 Overview of the drug discovery and development process.

disease. Compound libraries are then screened against the validated target to find molecules or classes of molecules which bind to the target. These 'hits' are validated to eliminate non-specific or irrelevant binding thus producing a refined series of lead molecules. These lead molecules are further optimized to improve upon their physiochemical properties yielding a single candidate compound to proceed with preclinical evaluation. In the preclinical phase *in vitro* and *in vivo* models are used to evaluate key pharmacokinetic (PK) parameters such as the absorption, distribution, metabolism, excretion, and toxicity (ADMET) as well as validating the pharmacodynamic (PD) parameters surrounding the mechanism of action and efficacy. This preclinical phase is largely used to identify potential risks and attempt to model behavior of the compound prior to the first dose in humans (FDIH), which is the first phase of clinical trials aimed at evaluating the dosage and the corresponding safety and tolerability of the drug in a small group of healthy volunteers. Phase 2 and 3 trials are designed to provide evidence of efficacy while continuing to monitor for potential adverse reactions in respectively larger cohorts of volunteers with the disease or condition. According to current data from the US Food and Drug Administration (FDA), roughly 70% of drugs move on from Phase 1, whereas roughly only 33% and 25% of drugs progress beyond Phase 2 and Phase 3, respectively.[29] After spending about 10 years collecting data on the safety and efficacy of the drug candidate, regulatory review hopefully results in the approval of a new medication.

Over the years, MSI has been strategically implemented at key points along this drug discovery process to generate the critical information needed to make informed decisions about project progression or termination. Throughout the rest of this chapter, we will highlight key examples where MALDI-MSI has demonstrated significant potential in the discovery and development of new medicines.

14.2 Early Discovery

A drug discovery program begins by identifying an unmet clinical need for a disease or clinical condition, which is the driving force behind the lengthy (12–15 year) and expensive (>$1 billion) endeavor to bring a new medicine to patients.[30] Initial research in early discovery focuses on the identification of a target protein or pathway that is hypothesized to play a critical role in the disease and thus can either be inhibited or activated to achieve a therapeutic effect on the disease state. Once an appropriate target is identified and validated, an intensive search begins to discover a chemical or biological molecule that achieves the desired target modulation. Discovering a "hit" molecule that has the desired activity is achieved by running vast libraries of compounds (10 000–1 000 000s) through an appropriate screening assay. Drug discovery screening approaches can be classified into two general categories: target-based screens, where target modulation serves as the readout, or phenotypic screens, in which the recovery of a phenotype in a disease model is the readout.[31,32] A number of different analytical technologies are used to generate the readouts from these assays. Recent advances in MALDI-MSI instrumentation and sample preparation have extended applications into screening which has shown promise as an alternative to more traditional fluorescence-based approaches.

High-throughput screening (HTS) involves screening of the entire compound library directly against the drug target or in a more complex assay system, such as a cell-based assay, whose activity is dependent upon the target but which would then also require secondary assays to confirm the site of action of compounds.[33]

14.2.1 High-throughput Screening

High-throughput screening (HTS) is a critical method for the rapid identification of active compounds that modulate a particular enzyme or biochemical pathway.[34–36] The compounds identified provide a platform from which to commence drug design. Traditionally, fluorescence-based technologies have been used for HTS against enzymatic or biochemical targets with sample readouts (usually in 1536 well-plates) of <1 second per sample and sample volumes of 10 μL or less. With the introduction of commercially available MALDI-MSI instruments with laser repetition rates of 10 kHz or greater, sampling times comparable to fluorescence-based methods can now be achieved. While some may argue that MALDI-based HTS does not constitute imaging, the principles of acquisition are conserved and in HTS a MALDI target plate consisting of enzymatic or biochemical assay spots in arrays of 96,[37] 384,[38–40] or 1536[41–43] spots is imaged. MALDI HTS of enzymatic reactions enables the rapid identification and kinetics of potential inhibitors. The major benefit of this approach is the ability to screen with direct label-free detection, thus reducing false positives (due to interference from colored or fluorescent compounds), saving time and costs, and eliminating the need for the extensive deconvolution assessments required for fluorescence assays.[44]

14.2.2 Complex *In Vitro* Models

HTS biochemical assays are optimized for speed and throughput, but as a result of their simplicity it is recognized that these screens do not accurately reflect the complexity of living systems.[45] To address this limitation, cell-based assays, traditionally cultured as 2-dimensional (2D) adherent mammalian cells, have been increasingly used in HTS campaigns with uses including quantitative assessments of biological activity, drug uptake, cytotoxicity, and off-target interactions. The relative simplicity of 2D monolayer cultures and amenability to HTS applications makes these systems attractive; however, the non-native architecture and rigidity of these systems can result in cellular behavior and responses that are not reflective of cells in their native tissue environment.[46] This recognition has prompted development of increasingly complex cell-based models over the past two decades including spheroids, organoids, three-dimensional (3D) bio-printed tissue, and organs-on-a-chip.[47–49] These complex *in vitro* models (CIVMs) aim to create a test system that more closely recapitulates the intricate microenvironment of an *in vivo* system and therefore are more physiologically relevant and clinically translatable.[50] Human disease-relevant *in vitro* models have a number of attractive advantages including reduced animal usage, which could in turn reduce study timelines and cost. While CIVMs have potentially broad applications in drug discovery and development, there is a need for advanced analytical tools to effectively monitor these systems and better understand the complex biology and elucidate mechanistic insight. MALDI-MSI is well positioned to be an integral part of the analytical strategy for investigation of CIVMs due to its ability to simultaneously image exogenous and endogenous compounds while maintaining the spatial integrity of the sample. Thus, with MALDI-MSI there is the opportunity to analyze these systems for both drug distribution and endogenous molecular changes associated with treatment or disease phenotype.

In the first report of using MALDI-MSI to analyze 3D cell cultures, Li and Hummon imaged HCT116 human carcinoma spheroids and demonstrated that at particular sizes they mimicked tumors by developing a region of proliferating cells on the outside and a layer of quiescent cells surrounding a central hypoxic region.[51] Liu *et al.*[52] used MALDI-MSI to investigate the time-dependent penetration and distribution of cetuximab, a monoclonal antibody epidermal growth factor receptor (EFGR) inhibitor, in spheroids derived from two different colon-cancer cell lines (HT-29 and DLD-1) and patient-derived colorectal-tumor organoids. Additionally, MALDI-MSI analysis revealed significantly elevated levels of ATP in the spheroids treated with cetuximab, potentially reflecting apoptotic changes in cells responding to treatment.

Despite the enormous potential of CIVM analysis by MALDI-MSI, there are considerable sample preparation challenges that must be resolved to facilitate efficient and robust analysis. For example, organoid and spheroid cultures grown in 96 or 384-well plates must be individually transferred to an embedding mold and frozen in a suitable embedding matrix to facilitate cryosectioning for MALDI-MSI analysis. This is currently a manual and labor-intensive process that negates the throughput benefits of these systems. Additionally, the cell cultures transferred to the embedding mold will typically have a variable position on the *z*-axis, therefore one must rely on visual inspection to determine when the appropriate plane has been reached to collect a section for analysis. To resolve this issue, Johnson *et al.*[53] developed a sample preparation strategy that incorporated a gelatin microarray microwell mold that could effectively align multiple organoids to the same *z*-axis, enabling tens to hundreds of organoids to be sampled in a single section. This type of sample preparation in combination with technological innovations of liquid/sample handling robotics and automation, or even MALDI-MSI compatible substrates will be critical to improve the efficiency of sample preparation for CIVMs and leverage the high-throughput capabilities of MALDI-MSI.

14.3 Preclinical Phase of Development

The successful progression of a drug candidate relies on a thorough understanding of exposure at the site of action (cell/tissue/organ containing the target), binding to the intended target (often a protein/enzyme/receptor), and evidence that target modulation elicits the desired pharmacological effect. These concepts are more commonly referred to as the 'three pillars of survival'.[4] The drug candidates remaining after lead optimization are further evaluated in the preclinical phase, where experiments are designed to advance the understanding of drug disposition with the aim of predicting efficacy and safety in human patients. Dosage, formulation, and route of administration can be key factors in balancing improved delivery to the intended target while limiting potential toxicity, which may arise from on-/off-target pharmacology. MALDI-MSI in the preclinical setting provides meaningful insight into drug distribution, pharmacology, and toxicity which is imperative in supporting program progression or termination. The latter being less glamorous but arguably of equal if not greater importance.

14.3.1 Drug Disposition and Pharmacokinetics

Developing a thorough understanding of drug disposition, specifically the availability of free drug at the site of action, is critical to the success of a drug candidate. As it is readily measurable, plasma drug concentration is often used as a surrogate for tissue concentration; however, several factors may result in a disequilibrium between tissue and plasma concentrations leading to potential misinterpretation of plasma-based estimates.[54] This highlights the importance of directly measuring drug concentration at the tissue level where applicable. The concentration of drug in tissue may also be presented as a ratio of tissue/plasma to help account for the plasma contribution in highly perfused tissues. Historically, quantitative whole-body radiography (QWBA) has been recommended by a number of regulatory agencies to provide supporting evidence of drug distribution.[55,56] While QWBA plays an important role in understanding drug disposition, the reliance on a radiolabel imposes cost restraints and a lack of selectivity between the parent drug and metabolites which may limit its use early in drug development. Bioanalysis of tissue homogenates by liquid chromatography coupled to mass spectrometry (LC–MS) provides almost unparalleled sensitivity with the selectivity to simultaneously quantify parent drug and potential drug metabolites. However, homogenization prior to LC–MS disregards the drug distribution within the tissue and if localized, homogenization can have a dilutive effect on the concentration. MSI bridges this gap between QWBA and LC–MS by providing selective and quantitative distribution of drug-related material within the context of histological tissue features. The availability of drug, metabolite, and endogenous distributions provided by MSI drives further mechanistic understanding of drug ADMET.

14.3.1.1 Administration and Absorption

Drug administration refers to the method and route by which a pharmaceutical compound is delivered to achieve a therapeutic effect.[57] Most drugs are administered orally as a tablet or capsule because this is the most convenient and economical route of administration; however, a number of other delivery routes are commonly used including intravenous, intramuscular, intranasal, and intradermal. Each route of administration faces unique challenges and barriers that must be considered during development to maximize drug efficacy while minimizing side effects.

For oral dosing, MALDI-MSI has been used to visualize the absorption and uptake along the intestinal track, which can be critical for selecting the appropriate dosage form. One study looked at the influence of drug permeability at the site of absorption,[58] where two highly permeable drugs (propranolol and metoprolol) and one less permeable drug (atenolol) were co-administered and their absorption profiles probed by MALDI-MSI. The more permeable drugs showed greater uptake in the proximal sections of the small intestine compared to atenolol, which was absorbed further along the jejunum. In addition to determining the site of absorption along the GI tract, this study took a closer look at the cross-sectional absorption along the intestinal villi. Propranolol and metoprolol showed gradient absorption with high concentration at the tip of the villi decreasing towards the crypt, in contrast to atenolol, which was fairly evenly distributed throughout. While LC–MS tissue homogenate concentrations may be valuable in assessing axial absorption, evidence from a recent MALDI-MSI study indicates that these may be misleading due to contributions from the drug remaining in the intestinal lumen – even after perfusion.[59]

For some chronic diseases, such as HIV, adherence to high frequency dosing regimens can be problematic. The development of long-acting parenteral formulations eases some of this burden by creating an intramuscular or subcutaneous drug depot for sustained drug release over longer periods.[60] The kinetics of a long-acting parenteral formulation for cabotegravir, an integrase inhibitor for the treatment of HIV-1, was evaluated using a multimodal approach combining MRI, immunohistochemistry, and MALDI-MSI.[61] The combined approach helped identify the role of macrophage infiltration at the periphery of the depot in the absorption and prolonged systemic exposure of cabotegravir. A follow-up study using higher resolution MALDI-MSI highlighted the differential distribution of the cabotegravir salt adducts as well as provided evidence of drug localization in multi-nucleated giant cells supporting the previously proposed hypothesis of the role of macrophages in the dissolution and sustained release of cabotegravir.[62] On the topic of long-term controlled drug release, drug-eluting beads were used to deliver the anti-angiogenic drug sunitinib to implanted tumors in rabbit liver.[63] A combination of classical histology, fluorescence, and MALDI-MSI were used to map the release of sunitinib over time, with MALDI-MSI providing additional detail on the localization of sunitinib metabolites.

14.3.1.2 Distribution

Not long after it was demonstrated that MALDI-MSI could be used to detect drug compound directly from sections of biological tissue,[64] the technology was further applied to the visualization of drug distribution in tissue.[65] Direct evidence of drug localization to the intended site of action is an area where MALDI-MSI has been invaluable and a number of recent reviews nicely summarize these contributions.[8–11,18]

Likely due to the aforementioned historical influence of QWBA, some of the earliest examples of MSI applied to drug distribution studies were performed as whole-body analyses[66,67] often using QWBA as a comparator.[67–71] Most examples utilize a full mass range with accurate mass measurements enabling the simultaneous analysis of drugs and their metabolites[66,68,70,72–76] though there are several instances where a more targeted tandem MS approach was used.[77,78] While whole-body MSI can be an excellent method to obtain important information on the distribution of novel drug molecules,[79–83] there are several limitations that have restricted its applicability. Whole-body sectioning is notoriously tedious despite some advances in tape collection.[72] While giving an overview of distribution across a number of tissues, the orientation of the sectioning plane through each individual organ is typically less than desirable. Variation in tissue matrix effects, such as ion suppression, can also complicate the interpretation of drug intensity from different organs[84,85] often requiring normalization to deconvolute the influence of these effects.[68,86,87] Despite these challenges, whole-body MSI combined with other sampling modalities can be a useful technique to evaluate the distribution of drug-related material in preclinical species. A recent example used whole-body MSI to evaluate the distribution of the cyclic peptide melanotan II and several of its metabolites.[82] This analysis was combined with liquid microjunction surface sampling for structural confirmation of the drug-related material and validation of the observed relative intensities across several organs (Figure 14.2).

Most often, select tissues are collected for a more targeted understanding of drug distribution in key organs. For example, MALDI-MSI was used to provide evidence of ibrexafungerp, a compound under investigation for the treatment of multiple fungal infections, localization at the site of infection in an intra-abdominal candidiasis mouse model.[88] By evaluating the tissue distribution of multiple drug candidates or even comparison of candidates to current standards of care, MALDI-MSI can provide evidence to aid in candidate selection and program progression by revealing localization to the target site of action.[89–92] The distribution of multiple antibiotics commonly used in the treatment of tuberculosis was evaluated by high spatial resolution MALDI-MSI, which was capable of resolving drug localization to small lipid deposits around airways.[93] In an animal model of tuberculosis, evidence of the limited penetration of rifapentine in cavitary lung lesions helped to explain the reduced clinical benefit observed in patients with similar prognosis.[90] Barring the risk of potential drug–drug interactions, cassette dosing (simultaneous dosing of multiple drugs in a single formulation) is one approach to evaluate the disposition properties of several compounds while significantly reducing the number of animals used, which is an industry-wide initiative. This approach was used to visualize the distribution of eight compounds (4 per dose group) in multiple tissues.[94]

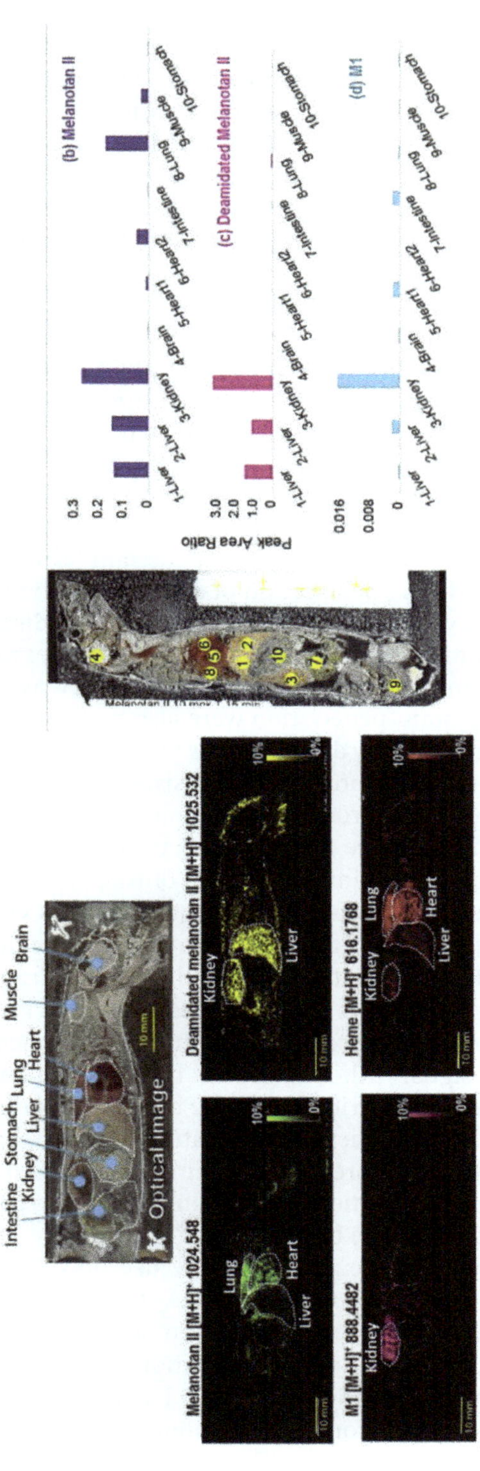

Figure 14.2 MALDI-MSI of whole-body tissue sections depicting the distribution of the cyclic peptide Melanotan II and two of its primary metabolites (left panels). On the right, a liquid microjunction surface sampling probe was used for structural confirmation and validation of the relative abundance of the target analyte across several organs. Reproduced from ref. 82 with permission from Elsevier, Copyright 2020.

This study also highlighted the benefit of using complementary MSI technologies (MALDI, DESI, and LESA) to allow for detection of all compounds across the panel. A cassette dosing strategy was used to evaluate penetration of four topically applied drugs and two formulations in *ex vivo* human skin.[95] MALDI-MSI not only provided insight into the tissue penetration for each of the drugs but also helped to elucidate the route of penetration. Interestingly, the physiochemical properties of each of the drugs alone were not sufficient to predict the penetration without accounting for the influence of the formulation.

Once absorbed, drug distribution is dependent upon several factors including vascular permeability, blood flow, and perfusion rate into the surrounding tissue. However, selective membranes, such as the blood–brain barrier (BBB), restrict tissue permeability to protect from potential toxins and pathogens. From a drug delivery perspective, this can be beneficial to limit the potential for central nervous system (CNS) toxicity for targets elsewhere in the body, but it can also be an impediment for drugs targeting specific areas of the brain. For example, Swales *et al.* used an array of MSI technologies to visualize the distribution of several penetrant and "nonpenetrant" drugs in brain tissues to better understand drug penetration of the blood–brain barrier.[94] Liu *et al.*[96] demonstrated the value of MALDI-MSI to evaluate the distribution and metabolism of oncology drugs in the brain. Some key outcomes relating to the validation of heme as a marker for brain vasculature as well as the impact of vasculature on verifying brain penetration were also presented. Despite the ability of some species to passively diffuse across the BBB, efflux transporters, such as P-glycoprotein (P-gp) and breast cancer resistance protein (Bcrp1), can actively pump substrates back across the membrane. One example by Choo *et al.*[97] highlighted the importance of these efflux transporters by comparing the brain penetration of cobimetinib in a series of wild type and knock-out animals (P-gp knock-out and/or Bcrp1 knock-out) as well as wild-type mice treated with elacridar (an efflux transporter inhibitor). They demonstrated that cobimetinib was a P-gp substrate and showed increased drug penetration in the P-gp knock-out compared to the wild-type animals as evidenced by a higher brain to plasma ratio and distribution to the brain parenchyma. Again, correlation of drug ion signal with endogenous markers like heme aided in the interpretation of the drug distribution in the brain to support evidence of brain penetration *versus* confinement in vasculature. The permeability of the BBB can also be impacted by the presence of brain tumors which may impose localized effects on the tight junctions thereby increasing the penetration of components that would otherwise be restricted.[98]

Oncology is one area in particular where the use of MSI for assessing distribution has been expanding to help gauge drug accessibility to the tumor microenvironment.[99] A recent publication suggests an approach to correlate tumor heterogeneity with drug distribution to aid in interpretation of drug penetration.[91] Another study collected and analyzed a series of tissue sections to generate a 3-dimensional representation of the heterogeneous paclitaxel distribution within the tumor.[100] A quantitative MSI approach was used to evaluate the tumor penetration of the PARP inhibitors niraparib and olaparib.[89] The distribution of niraparib was found to be heterogeneous but demonstrated much greater tumor penetration compared to olaparib. These represent some recent examples but there are a number of reports on the application of MALDI-MSI in oncology drug development.[20,101–113] Many of those published involve the use of platinum-containing drugs, *e.g.* cisplatin or oxaliplatin.[114–116]

14.3.1.3 Metabolism

Traditionally, LC–MS is used for preliminary metabolite profiling and identification but with the role of MALDI-MSI pushing earlier in drug discovery, potential drug metabolites are often queried as a part of drug distribution studies.[117] The dilutive effect of tissue homogenization required for LC–MS analysis has the potential to limit the detection or even mask the significance of species that are highly localized. Through direct analysis of the tissue, MALDI-MSI was able to highlight the distribution of a previously unidentified metabolite which had accumulated primarily in the bile duct of the liver.[118] Additionally, aside from differences in abundance, which can be determined by LC–MS, parent drug and metabolites may also demonstrate different distributions in tissue and a number of examples indicate that this is often the case.[98,119] The ability to determine and compare the distributions of the parent drug and related metabolites can be imperative in understanding mechanisms of efficacy or toxicity. This information is particularly useful in the case of prodrugs or instances where the drug metabolites retain activity.[106] The presence of isomeric drug metabolites, such as multiple unique sites of oxidation, can be a challenge as without further deconvolution (*i.e.* MS/MS or ion mobility), their contributions may be merged into a single ion image. A recent review nicely summarizes MALDI-MSI applications specifically relating to the evaluation of drug metabolism.[83]

14.3.1.4 Pharmacokinetics

Understanding the PK parameters for a drug candidate typically involves measuring drug concentration in plasma or tissue across a series of time points to assess the temporal drug exposure. Compared to plasma PK where samples can be collected at multiple time points from the same animal, the ex-vivo nature of tissue PK implies multiple animals are needed for data at each time point which can add a high level of inter-individual variability. Additionally, the reduced throughput of MALDI-MSI compared to LC–MS further limits the practicality of exclusively using MALDI-MSI to measure tissue concentration across all time points. Therefore, LC–MS is often used to highlight relevant time points to sample by MALDI-MSI. There is, however, at least one example in the literature where MALDI-MSI tissue concentrations are used to model PK parameters.[119] The distribution of pirfenidone, a treatment for idiopathic pulmonary fibrosis, in mice lungs was evaluated by MALDI-MSI over six time points from 10 to 240 minutes post-dose. The distribution and concentrations of the parent drug as well as two drug-related metabolites were used to model the drug PK and estimate the maximum concentration (C_{max}) and corresponding time point (t_{max}).

14.3.2 Targeted Drug Delivery

Traditionally, drug delivery strategies have focused on administering the pharmaceutical compound in such a way that it reaches systemic circulation and is transported *via* the blood to the target tissue(s). One issue with this approach is that all tissues are exposed to the drug and therefore off-target toxicity or other undesired side effects are more likely to occur. An alternative approach is targeted drug delivery, where a drug is delivered specifically to a target site or tissue, thereby reducing off-target effects and potentially enhancing therapeutic efficacy.[120] Advancements in nanotechnology over the past two decades have driven development of innovative nanoparticle-based drug delivery systems with the ability to deliver pharmaceutical agents selectively to target tissues.[121] The most common and clinically promising nanotherapeutic delivery vehicles are liposomes, which are nanometer-sized spherical vesicles consisting of a lipid bilayer encapsulating an aqueous core. Liposome formulations are non-toxic and non-immunogenic and can be optimized to accumulate drugs in diseased tissues through the enhanced permeability and retention (EPR) effect. The EPR effect is based on the abnormally wide fenestrations in the blood vessel endothelial cells in diseased tissues such as tumors, which allows for extravasation and accumulation of particles up to several hundred nanometers including liposomes.[122] Though promising, the clinical success of liposomal formulations remains quite small and therefore innovative analytical technologies are needed to attain a better mechanistic understanding of liposomal delivery.[123] MALDI-MSI is uniquely suited to provide molecular images of the liposome components in tissues including the phospholipid bilayer and the encapsulated drug. For example, Fülöp *et al.*[124] used MALDI-MSI to investigate the distribution of a liposomal formulation consisting of 1,2-dipalmitoyl-*sn-glycero*-3-phosphoglycerol (DPPG) and 1,2-distearoyl-*sn-glycero*-3-phosphoethanol-amine conjugated with PEG (PEG_{36}-DSPE) and indocyanine green as the payload. Mouse liver, kidneys, and brain were imaged in both positive- and negative-ion modes by MALDI-MSI to elucidate the tissue distribution of each liposome component. The authors performed a pixel-to-pixel colocalization analysis of the liposomal lipids and heme to distinguish localization between the vasculature and tissue parenchyma. They also investigated the potential for using colocalization of the two lipid markers and drug in a given pixel as a surrogate read-out for monitoring the liposomal integrity. In a similar application, Ryu *et al.*[113] reported using MALDI-MSI to investigate the intra-tumoral tissue distribution of nanoparticles encapsulating an antineoplastic drug in three preclinical xenograft models. The analysis revealed heterogeneous distribution within the tumors, but the MALDI-MSI intensity correlated well with the total drug concentration measured by LC–MS.

Zhao *et al.*[125] used MALDI-MSI as part of a multimodal imaging strategy to trace the distribution of an azoreductase-responsive prodrug that has the potential to serve as a colon-targeted delivery platform. In this delivery system, cleavage of the prodrug activated a NIR fluorophore enabling *in vivo* fluorescence tracking of the drug delivery to the colon. MALDI-MSI was then used *ex vivo* to map the tissue distribution of the payload and fluorophore to enable a comprehensive assessment of the drug release.

14.3.3 Pharmacodynamics

Even among the more targeted MALDI-MSI experiments, there is still a wealth of information about endogenous species that is often underutilized. Changes in the abundance or even the distribution of endogenous components can be visualized and linked with drug pharmacology. Specific metabolic pathways can be targeted and mapped to gain a deeper understanding of the changing biology.[126–128] More challenging non-targeted or 'omics'-based approaches involve the evaluation of global changes in endogenous species between control, dosed, and diseased groups or some combination thereof.[129] Additionally, endogenous species that correlate with the histological architecture of the tissue provide landmarks within the MSI data set, which aid in the interpretation of drug disposition or pharmacology.[96,130–132]

Sample preparation plays an important role in experiments mapping endogenous species. It was demonstrated that the selection of matrix can have a tremendous impact on the classes of endogenous species that are observed and, in some cases, multiple matrix preparations were used to increase the breadth of coverage.[133] While multiplexing with different matrices is one option, the ideal scenario is a single matrix that works well in both polarities.[134] A recent review highlights some of the key matrices and their applications to the analysis of different molecular classes.[135] It is important to note that the choice of matrix can impose effects on the analysis[136] including potential errors in assignment.[137] Differences in ionization efficiency inevitably impact which molecular classes are detectable under 'normal' preparation conditions. On-tissue derivatization strategies have been proposed to aid in the visualization of these otherwise elusive molecules.[138–141] Another often overlooked but important aspect of sample preparation is to ensure that tissues are oriented similarly such that sections are collected from similar planes of the tissue when comparing between two or more groups to avoid artifactual changes in abundance or distribution that are associated with tissue architecture rather than true differences among the groups. Lastly, storage and preparation conditions should be considered given the potential impact on the stability of endogenous molecules.[142–147] The inclusion of quality control samples, replicates, and randomizing acquisition order can help to mitigate measurement variability.

The observance of endogenous changes associated with drug administration can be advantageous by providing indication of pharmacodynamic effects including efficacy and toxicity. Evidence of the favorable pharmacology of sunitinib in tumor-bearing mice was provided through the correlation of MALDI-MSI derived drug localization with a higher density of VEGFR-2 expression determined by immunofluorescence from a serial tissue section.[148] Despite being unable to directly detect drug penetration, one recent study looked at the impact of the immunomodulatory treatment teriflunomide on several metabolic pathways in the brain and they observed decreased levels of several nucleotides as well as two carbohydrate intermediates.[149] A derivatization approach was used to map the catecholaminergic metabolic pathway and provide evidence of dysregulated L-DOPA metabolism linked to L-DOPA-induced dyskinesia in primates.[150]

The pharmacometabolomic effects of pirfenidone treatment were investigated where a correlative change in abundance of several ions was observed when comparing the treated and control groups.[119] However, as these ions were not identified it is difficult to interpret their significance. A follow-up report provided a more extensive interpretation of these endogenous changes citing specific alterations in metabolite pathways in lungs of mice subjected to bleomycin-induced fibrosis and correlation to human patients with idiopathic pulmonary fibrosis (IPF).[151] Importantly, pirfenidone treatment was found to restore the redox imbalance and glycolysis in IPF tissue as well as down regulate metabolites involved in fibrotic pathways. The PD effects of epacadostat, an inhibitor of indoleamine-2,3-dioxygenase which converts tryptophan to kynurenine, was evaluated in the CT26 mouse tumor model using a combination of MALDI-MSI and LC–MS/MS.[152] Epacadostat treatment was shown to correlate with a reduction in the kynurenine/tryptophan ratio as determined by both methods. In addition, regions of localized epacadostat related to decreased kynurenine abundance as well as reduced lactate levels indicating a potential downstream impact on glycolysis (Figure 14.3).

14.3.4 Disease Pathogenesis

More broadly, another application of non-targeted MALDI-MSI with significant potential in the pharmaceutical industry is to evaluate endogenous changes associated with the onset or progression of disease to either improve mechanistic understanding or in some instances identify potential disease biomarkers that can be used to track pharmacodynamic efficacy. In an ischemic stroke rat brain model, MRI followed by positron emission tomography (PET) imaging were performed 48 hours after ischemic stroke and 3 months post-stroke respectively, to image neuroinflammation and neurodegeneration in the infarcted regions of the rat brain.[153] Following PET imaging, the rat brain was analyzed by MALDI-MSI and clear differences in the lipid profiles between the infarct region and the remainder of the brain were observed. In a separate study metabolic changes associated with neonatal stroke were evaluated by MSI including the recovery potential of several

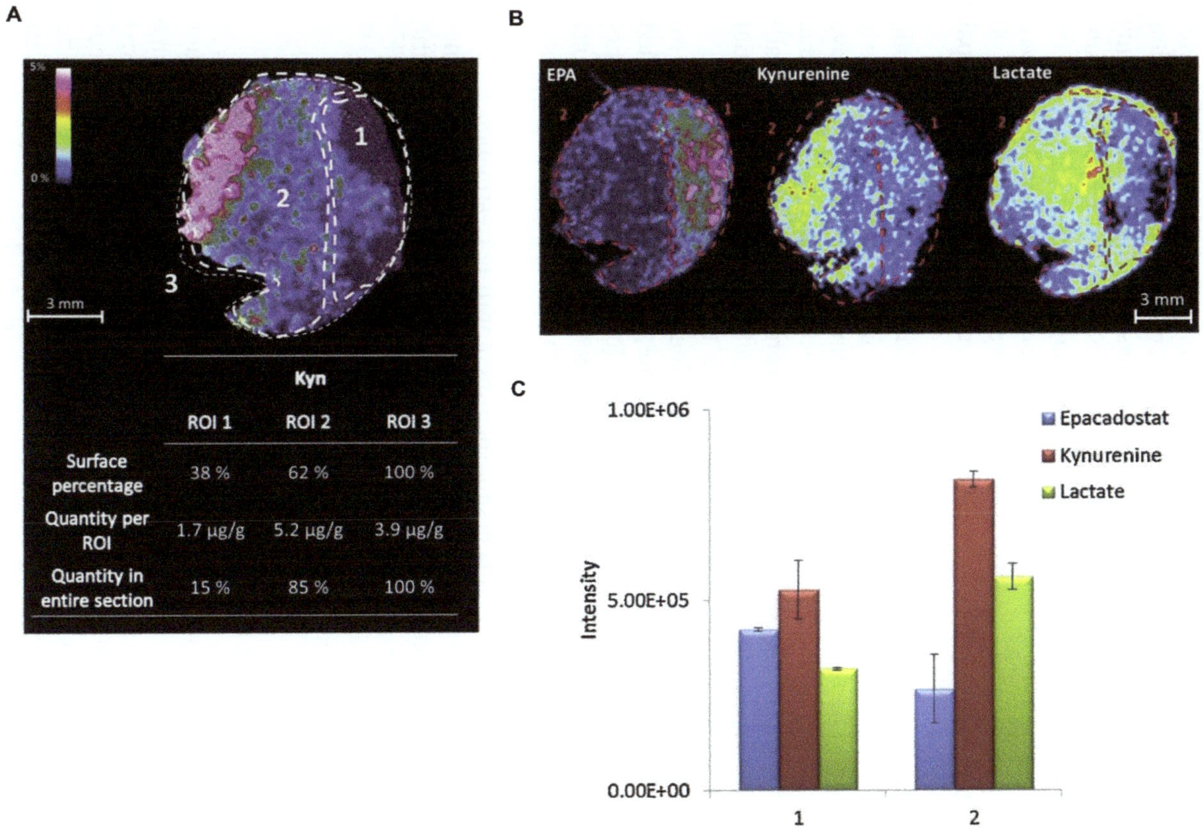

Figure 14.3 The localization of epacadostat, an inhibitor of kynurenine metabolism, is shown to correlate with regions of lower kynurenine abundance.
Reproduced from ref. 152 with permission from Elsevier, Copyright 2019.

cell therapies.[154] The injury resulted in the upregulation of several amino acids and the downregulation of intermediates in the TCA cycle. While the treatments appeared to reverse the impacts on the TCA cycle, the other injury-induced metabolic changes did not recover. Utilizing an on-tissue derivatization strategy, MALDI-MSI was used to extensively map the metabolic pathways of key neurotransmitters and demonstrated the modulation of these species incited by Parkinson's disease.[140] A multimodal study comparing control and diseased tissue was able to identify and validate diacylglycerophosphoinositol lipid as a potential biomarker for pulmonary arterial hypertension.[155] In transgenic Alzheimer's disease mice (tgArcSwe), a combination of MALDI-MSI and immunofluorescence highlighted several sphingolipid species which correlated with cortical and hippocampal amyloid plaques.[156]

The non-targeted MSI approach has been useful in mapping the metabolic signatures of a tumor microenvironment.[157,158] In correlation with the early observations of Warburg[159] describing altered tumor metabolism, there have been numerous publications mapping metabolic differences in endogenous species between tumor, stroma, and non-cancerous tissue.[160–166] These advances in understanding tumor vitality are useful in identifying potential druggable approaches to slow or ideally reverse their progression.

MALDI-MSI analysis of endogenous metabolites in cells has also been reported,[128,167] with ATP, fructose-1,6-bisphosphate, and citrate being identified from human-derived HeLa cells. Metabolites from human acute lymphoblastic leukemia Jurkat cells treated with three different anticancer drugs (epigallocatechin-3-*O*-gallate, 5-fluorouracil, and methotrexate) have also been analyzed by MALDI-MSI and principal component analysis was used to identify clear independent clusters for the different drug treatments.[128]

In addition to those mentioned previously, MALDI-MSI of endogenous metabolite markers has been evaluated across a number of different therapeutic areas including oncology,[112,168] atherosclerosis,[169] Alzheimer's disease,[170] dermatology,[171,172] brain ischemia/stroke,[173–175] muscular dystrophy,[176] osteoarthritis,[177] and retinal degeneration.[178–180]

14.3.5 Toxicology and Safety Investigations

Adverse events that arise during preclinical or clinical evaluation often trigger investigations to develop a thorough understanding of the underlying mechanism of toxicity and to assess future risk. One classification of toxicological findings are those related to sequestered agents whereby accumulation of the drug or drug-related material is ultimately the cause of the adverse event.[181,182] In these areas, MALDI-MSI is being increasingly used to provide molecular context to histopathological findings related to safety events.[183]

Being a major site of drug metabolism and a primary route for drug clearance, the liver is often a target organ for safety-related findings. An investigation into the mechanism of amitriptyline-induced liver toxicity highlighted the upregulation of several lipids localized to the periportal region which, combined with histological findings, helped indicate drug-induced macrovesicular steatosis as the primary driver of liver injury.[184] A clinical finding of hepatotoxicity in a small percentage of patients with prior liver inflammation being treated with the antiretroviral therapy nevirapine sparked a multimodal investigation into the potential causes.[185] A combination of gene expression, immunohistochemistry, histopathology, and MALDI-MSI were used to evaluate the progression of fibrosis in a preclinical model simulating liver damage associated with viral hepatitis. The distribution of nevirapine was found to be lobular with higher intensity regions surrounding sub-lobular veins, and regions of fibrosis were highlighted by mapping the distribution of a sphingomyelin species.

As the kidney is often a final pathway for the excretion of drug-related material, it is another organ commonly implicated in toxicology investigations. A recent review highlighted the application of imaging technologies to investigate nephrotoxicity relating specifically to crystalline deposits in the kidney.[186] During a rat renal toxicology study for two potential drug candidates, crystalline deposits were observed in the kidney linked to tubular degeneration.[187] MALDI-MSI was used to provide evidence that a drug metabolite common to both drug candidates was a primary component of the crystals. In a separate study, several drug metabolites were found to be associated with the crystalline deposits in FFPE tissue sections of rabbit kidneys.[187–189] A similar toxicological finding of tubular crystalline deposits was observed in a nephrotoxicity study for dabrafenib, however the drug-related material, while in the lumen of damaged collecting ducts, was not detected directly within the deposits.[190] Further non-targeted investigation with MALDI and laser desorption/ionization (LDI), the latter of which involves direct laser interrogation of the sample without the addition of an exogenous matrix, indicated that the deposits were primarily composed of calcium phosphate clusters. A potential mechanism for the age-related nephrotoxicity was proposed where the localization of a carboxylic acid drug metabolite may have disrupted calcium homeostasis resulting in the downstream formation of calcium phosphate calculi. A series of polymyxin antibiotics analyzed by MALDI-MSI highlighted analogs which were more prone to dose-limited nephrotoxicity localized to the cortex of the kidney whereas the less nephrotoxic analog was more homogeneously distributed.[189] To further understand the metabolomic dynamics of cisplatin-induced acute kidney injury, a study combining LC–MS and MALDI-MSI correlated abundance and spatial distribution changes across several key metabolic pathways including glycolysis, TCA cycle, nucleotide metabolism, and amino acid metabolism.[191]

Unexpected exposure across the blood brain barrier can potentially result in neurological findings. CNS exposure of fosdevirine was evaluated in several preclinical species after several patients experienced seizures in a Phase IIb study.[192] Similar neurobehavioral signs were observed in rabbit and minipig after dosing but no events were found in monkeys. In both rabbit and minipig, a glutathione metabolite of fosdevirine was highly localized to

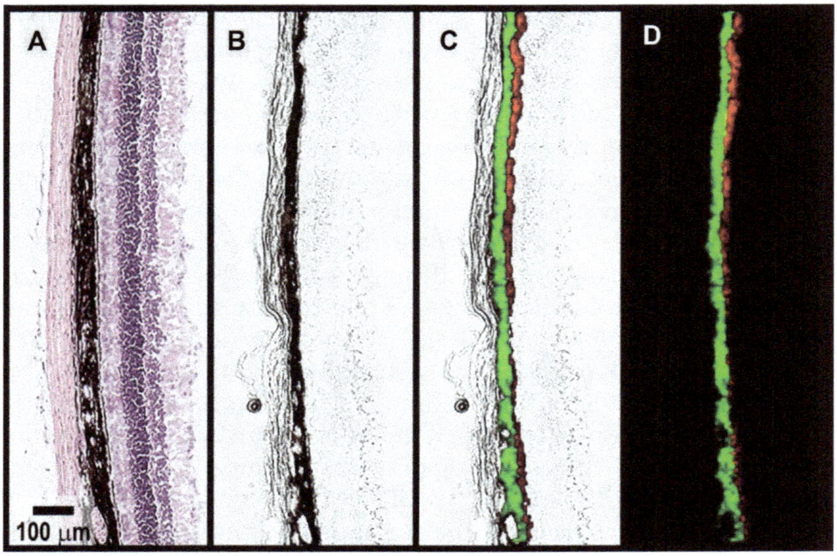

Figure 14.4 The distribution of retigabine dimer (green) and A2E (red), a marker of the retinal pigment epithelium layer, indicating the restriction of the retigabine metabolite from crossing the blood–retinal barrier. Reproduced from ref. 194 with permission from American Chemical Society, Copyright 2019.

the white matter of the brain, whereas in the monkey the parent drug was the most prominent component and was localized to the gray matter. A hypothesis indicating the role of this glutathione metabolite in the observation of seizures was proposed.

Ocular findings, though rare, are especially concerning given their potential for phototoxicity and associated vision loss. In one study, MALDI-MSI highlighted the potential for benzalkonium chloride, a preservative excipient common in eyedrops used for glaucoma treatment, to penetrate beyond just the ocular surface to areas related to glaucoma pathophysiology.[193] Retinal dyspigmentation associated with long-term administration of retigabine, an adjunctive treatment for refractory partial-onset seizures, initiated a large preclinical effort to understand the origins and severity of these findings.[194] The localization of several retigabine dimers to high melanin-containing regions of ocular tissues aided in the proposal of a mechanism of their formation. Perhaps most importantly, MALDI-MSI helped demonstrate that while these species were likely responsible for the observed dyspigmentation, they did not appear to cross the blood–retinal barrier mitigating the risk of phototoxicity (Figure 14.4).

14.4 Clinical and Beyond

The invasive nature of *ex vivo* imaging technologies is a major limitation to routine clinical application and there are limited examples of MALDI-MSI being utilized specifically for clinical applications in pharmaceutical development.[195,196] Oncology and dermatology are two areas where the opportunity for obtaining clinical samples (*e.g.* biopsies) is relatively high compared to other therapeutic areas. For oncology applications, the role of MALDI-MSI in tumor classification and diagnosis has received significant attention. Protein expression data generated by MALDI-MSI of various cancerous tissue types has been shown to successfully achieve diagnostic tumor classification.[197–205] For example, Balluff *et al.*[206] used MALDI-MSI in the *de novo* discovery of phenotypic intratumor heterogeneity in human gastric and breast cancer tissues. The approach taken enabled the discovery of clinically detrimental and molecularly distinct tumor sub-populations, which can be indistinguishable histologically. These unique molecular phenotypes were shown to be associated with tumor progression and ultimately disease outcome. Another notable publication was the utilization of MALDI-MSI in the field of gastric cancer to distinguish early-stage and late-stage cancer patients into different prognostic groups.[207] A seven-protein signature relating to three survival-related protein markers, namely HNP-1, CRIP1, and S100-A6 was identified. The over-expression of the proteins HNP-1 and S100-A6 was characterized as an indicator of poor prognosis. Interestingly, the involvement of S100 family proteins (A6 and A10) has been observed in several other cancer types, *e.g.* S100-A10 (along with the proteins COX7 A2 and TAGLN2) were identified as prognostic markers in Barrett's adenocarcinoma.[208] High expression levels of S100-A6 have also been correlated with poor prognosis for glioma patients[209] and also in metastatic papillary thyroid carcinoma[210] using MALDI-MSI. More recently, a novel truncated form of S100 protein has been identified that predicts disease-free survival in patients with lymph node-positive breast cancer.[115] Whilst not fully understood, the role of S100 proteins in cancer has been highlighted by Bresnick *et al.*[211] and is known to be involved in many other cancer types including breast, colorectal, renal cell carcinoma, skin, and thyroid cancer. Other clinical studies focusing on different cancer types investigated by MALDI-MSI include brain,[212] breast,[213] ovarian,[214] prostate,[206,215] and gastrointestinal cancers.[207,208,216,217] These examples highlight the potential for MALDI-MSI to be implemented in clinical trials of new therapeutics to help stratify patients for treatment and better understand treatment response.

In addition to fresh-frozen clinical specimens, efforts to incorporate MALDI-MSI analysis to elucidate proteomic signatures on FFPE tissue, including FFPE tissue microarrays (TMAs), have also been reported.[218–223] TMAs were originally developed to enable high throughput (1000 cylindrical tissue core biopsies at a time) molecular and pathological analysis of tissue specimens.[224–226] Groseclose *et al.*[227] developed a MALDI-MSI method which incorporated antigen retrieval and *in-situ* enzymatic digestion to analyze the protein content of FFPE TMAs containing human non-small cell lung-cancer biopsies. This strategy has now been successfully applied for TMAs of breast cancer,[228] colon cancer,[229,230] endometrial cancer,[231] gastric cancer,[220,232,233] head and neck cancer,[234] lung cancer,[235,236] esophageal cancer,[237] ovarian cancer,[214,224–226,238,239] pancreatic cancer,[240,241] prostate cancer,[242,243] renal cell cancer,[244–246] skin cancer,[247] and thyroid cancer.[210,240,241]

Though powerful, several limitations have emerged from the clinical studies reported utilizing MALDI-MSI. For example, the majority of specimens collected from patients in clinical trials are FFPE tissues. Concerns that the tissue processing steps involved in the production of FFPE tissue may alter the distribution of exogenous and endogenous small molecules in tissues have generally limited the efforts to image small molecule drugs in these samples. In contrast to this assumption, Ly *et al.*[248] developed a protocol for the *in-situ* analysis of endogenous metabolites from FFPE tissue. Importantly, the authors showed that 72% of the *m/z* species in the range *m/z* 50–1000 detected in the FFPE tissues coincided with those detected in the fresh-frozen tissue samples, suggesting that most of the metabolites were conserved in FFPE tissue samples. These results indicate that the low molecular weight analytes in FFPE tissue, including drugs, may be accessible for MALDI-MSI analysis. Another key consideration for future clinical studies involving biomarker detection is the importance of proper study design, particularly, the need for training and validation sets. The clinical samples obtained for the study should be well defined, present in large numbers, and divided into training and validation sets (and ideally analyzed blind).

14.5 Conclusion and Future Directions

As evident from the numerous examples above, the multiplexed and label-free nature of MALDI-MSI allows for both targeted and non-targeted analyses of xenobiotics and endogenous components, which holds tremendous potential for a broad range of applications across the pharmaceutical industry. The development and implementation of standardized workflows from preparation through reporting, as well as validation and quality control measures, will aid in broadening the acceptance of MALDI-MSI data in both regulated and non-regulated environments.[249–253] Several multi-site cross validation studies have been performed highlighting encouraging results on the reproducibility of MALDI-MSI that is achievable when following strict protocols.[28,254–258] Along the same vein of improving the power of MALDI-MSI data, would be the adoption of stereological approaches to sampling aimed at increasing confidence in conclusions derived about observed molecular distributions. An often overlooked obstacle of handling clinical, as well as certain preclinical samples, is the safety precautions necessary to protect the analyst from the presence of virulent pathogens or other harmful infectious components of the samples.[259,260]

It is encouraging to see that a majority of studies are now designed to integrate and correlate MALDI-MSI data with other *in vivo* and *ex vivo* imaging modalities as well as orthogonal non-imaging approaches to provide greater depth of biological understanding while in turn helping to validate MALDI-MSI.[16,261–266] Aiding in this goal, a recent report suggested a universal sample preparation workflow to allow for multiple imaging analyses on the same tissue.[143] The co-registration of multimodal data is another area that remains a considerable challenge but there has been some recent headway.[267,268] In addition, the field of MSI continues to make advancements in instrumentation such as the incorporation of ion mobility[269] or secondary ionization,[270] which has the potential to uncover elusive molecular species while providing an additional layer of confidence to identification.

A noticeable void in all the examples above relating to the use of MALDI-MSI to provide invaluable information about drug disposition is the absence of applications towards large molecule biologics. While small molecule drugs (<1 kDa) serve a significant place in pharma,[271] the development of biologic therapies (*e.g.* monoclonal antibodies (mAbs), therapeutic proteins, oligonucleotides) has resulted in significant improvements in the treatments of high-burden diseases such as autoimmune disorders and cancer.[272] In the case of mAbs, their size, often exceeding 150 kDa, imposes several limitations. One being the practical mass range on most MALDI mass spectrometers. In addition, mAbs often exhibit limited ability to penetrate and localize within tissues, implying their presence at fairly low concentrations which are often below MALDI-MSI detection limits.[273] To the best of our knowledge, there is only one example of MALDI-MSI analysis of a therapeutic protein in tissue[274] and no reports of analyzing a therapeutic oligonucleotide by MALDI-MSI. One potentially promising method was reported by Liu *et al.*[52] where an on-tissue reduction approach was used to break the disulfide bonds of the therapeutic mAb cetuximab in spheroids derived from colon cancer cell lines. This on-tissue reduction generates two heavy chains (50 kDa) and two light chains (25 kDa), the latter of which could be directly detected by MALDI-MSI. The cells in this report were incubated with relatively high concentrations of cetuximab, therefore it will be important to establish the feasibility of this method for MALDI-MSI analysis of tissues collected from *in vivo* studies, where mAb concentrations might be considerably lower. Lastly, other *ex vivo* imaging methodologies such as immunofluorescence are currently utilized to map mAbs, albeit after labeling, with subcellular resolution that will be difficult to match by MALDI-MSI without significant breakthroughs in methodologies.

References

1. J. W. Scannell, A. Blanckley, H. Boldon and B. Warrington, *Nat. Rev. Drug Discovery*, 2012, **11**, 191.
2. S. M. Paul, D. S. Mytelka, C. T. Dunwiddie, C. C. Persinger, B. H. Munos, S. R. Lindborg and A. L. Schacht, *Nat. Rev. Drug Discovery*, 2010, **9**, 203.
3. M. Hay, D. W. Thomas, J. L. Craighead, C. Economides and J. Rosenthal, *Nat. Biotechnol.*, 2014, **32**, 40–51.
4. P. Morgan, P. H. Van Der Graaf, J. Arrowsmith, D. E. Feltner, K. S. Drummond, C. D. Wegner and S. D. A. Street, *Drug Discovery Today*, 2012, **17**, 419–424.

5. D. Cook, D. Brown, R. Alexander, R. March, P. Morgan, G. Satterthwaite and M. N. Pangalos, *Nat. Rev. Drug Discovery*, 2014, **13**, 419–431.
6. M. R. Nelson, H. Tipney, J. L. Painter, J. Shen, P. Nicoletti, Y. Shen, A. Floratos, P. C. Sham, M. J. Li, J. Wang, L. R. Cardon, J. C. Whittaker and P. Sanseau, *Nat. Genet.*, 2015, **47**, 856–860.
7. H. Dolgos, M. Trusheim, D. Gross, J.-P. Halle, J. Ogden, B. Osterwalder, E. Sedman and L. Rossetti, *Drug Discovery Today*, 2016, **21**, 517–526.
8. R. J. A. Goodwin, Z. Takats and J. Bunch, *SLAS Discovery*, 2020, **25**, 963–976.
9. E. Davoli, M. Zucchetti, C. Matteo, P. Ubezio, M. D'Incalci and L. Morosi, *Mass Spectrom. Rev.*, 2021, **40**, 201–214.
10. J. G. Swales, G. Hamm, M. R. Clench and R. J. A. Goodwin, *Int. J. Mass Spectrom.*, 2019, **437**, 99–112.
11. S. Schulz, M. Becker, M. R. Groseclose, S. Schadt and C. Hopf, *Curr. Opin. Biotechnol*, 2019, **55**, 51–59.
12. A. Taylor, A. Dexter and J. Bunch, *Anal. Chem.*, 2018, **90**, 5637–5645.
13. K. Schwamborn and R. M. Caprioli, *Nat. Rev. Cancer*, 2010, **10**, 639–646.
14. M. Aichler and A. Walch, *Lab. Invest.*, 2015, **95**, 422–431.
15. A. Walch, S. Rauser, S. O. Deininger and H. Hofler, *Histochem. Cell Biol.*, 2008, **130**, 421–434.
16. T. Porta Siegel, G. Hamm, J. Bunch, J. Cappell, J. S. Fletcher and K. Schwamborn, *Mol. Imaging Biol.*, 2018, **20**, 888–901.
17. F. Tobias and A. B. Hummon, *J. Proteome Res.*, 2020, **19**, 3620–3630.
18. B. Prideaux, A. Lenaerts and V. Dartois, *Curr. Opin. Chem. Biol.*, 2018, **44**, 93–100.
19. I. Rzagalinski and D. A. Volmer, *Biochim. Biophys. Acta*, 2017, **1865**, 726–739.
20. L. Signor, E. Varesio, R. F. Staack, V. Starke, W. F. Richter and G. Hopfgartner, *J. Mass Spectrom.*, 2007, **42**, 900–909.
21. A. Nilsson, T. E. Fehniger, L. Gustavsson, M. Andersson, K. Kenne, G. Marko-Varga and P. E. Andrén, *PLoS One*, 2010, **5**, e11411.
22. R. J. A. Goodwin, C. L. Mackay, A. Nilsson, D. J. Harrison, L. Farde, P. E. Andren and S. L. Iverson, *Anal. Chem.*, 2011, **83**, 9694–9701.
23. M. R. Groseclose and S. Castellino, *Anal. Chem.*, 2013, **85**, 10099–10106.
24. J. A. Barry, M. R. Groseclose, D. D. Fraser and S. Castellino, in *Protocol Exchange*, Nature Publishing Group, 2018.
25. J. A. Barry, M. R. Groseclose and S. Castellino, *Bioanalysis*, 2019, **11**, 1099–1116.
26. R. F. Reich, K. Cudzilo, J. A. Levisky and R. A. Yost, *J. Am. Soc. Mass Spectrom.*, 2010, **21**, 564–571.
27. C. W. Chumbley, M. L. Reyzer, J. L. Allen, G. A. Marriner, L. E. Via, C. E. Barry 3rd and R. M. Caprioli, *Anal. Chem.*, 2016, **88**, 2392–2398.
28. J. A. Barry, R. Ait-Belkacem, W. M. Hardesty, L. Benakli, C. Andonian, H. Licea-Perez, J. Stauber and S. Castellino, *Anal. Chem.*, 2019, **91**, 6266–6274.

29. Clinical Research, https://www.fda.gov/node/366029, (accessed 10-FEB-2021).
30. J. P. Hughes, S. Rees, S. B. Kalindjian and K. L. Philpott, *Br. J. Pharmacol.*, 2011, **162**, 1239–1249.
31. J. G. Moffat, F. Vincent, J. A. Lee, J. Eder and M. Prunotto, *Nat. Rev. Drug Discovery*, 2017, **16**, 531–543.
32. I. Boussaad, G. Cruciani, S. Bolognin, P. Antony, C. M. Dording, Y. J. Kwon, P. Heutink, E. Fava, J. C. Schwamborn and R. Kruger, *Sci. Rep.*, 2021, **11**, 1439.
33. S. Fox, S. Farr-Jones, L. Sopchak, A. Boggs, H. W. Nicely, R. Khoury and M. Biros, *J. Biomol. Screen.*, 2006 Oct, **11**(7), 864–869.
34. L. Kategaya, P. Di Lello, L. Rouge, R. Pastor, K. R. Clark, J. Drummond, T. Kleinheinz, E. Lin, J. P. Upton, S. Prakash, J. Heideker, M. McCleland, M. S. Ritorto, D. R. Alessi, M. Trost, T. W. Bainbridge, M. C. M. Kwok, T. P. Ma, Z. Stiffler, B. Brasher, Y. Tang, P. Jaishankar, B. R. Hearn, A. R. Renslo, M. R. Arkin, F. Cohen, K. Yu, F. Peale, F. Gnad, M. T. Chang, C. Klijn, E. Blackwood, S. E. Martin, W. F. Forrest, J. A. Ernst, C. Ndubaku, X. Wang, M. H. Beresini, V. Tsui, C. Schwerdtfeger, R. A. Blake, J. Murray, T. Maurer and I. E. Wertz, *Nature*, 2017, **550**, 534–538.
35. M. Machalkova, J. Schejbal, Z. Glatz and J. Preisler, *Anal. Bioanal. Chem.*, 2018, **410**, 7441–7448.
36. D. Weigt, C. A. Parrish, J. A. Krueger, C. A. Oleykowski, A. R. Rendina and C. Hopf, *Cell Chem. Biol.*, 2019, **26**, 1322–1331, e1324.
37. K. Guitot, T. Drujon, F. Burlina, S. Sagan, S. Beaupierre, O. Pamlard, R. H. Dodd, C. Guillou, G. Bolbach, E. Sachon and D. Guianvarc'h, *Anal. Bioanal. Chem.*, 2017, **409**, 3767–3777.
38. R. E. Heap, A. G. Hope, L. A. Pearson, K. Reyskens, S. P. McElroy, C. J. Hastie, D. W. Porter, J. S. C. Arthur, D. W. Gray and M. Trost, *SLAS Discovery*, 2017, **22**, 1193–1202.
39. M. Winter, T. Bretschneider, C. Kleiner, R. Ries, J. P. Hehn, N. Redemann, A. H. Luippold, D. Bischoff and F. H. Büttner, *SLAS Discovery: Adv. Life Sci. R*, 2018, **23**, 561–573.
40. M. Winter, T. Bretschneider, S. Thamm, C. Kleiner, D. Grabowski, S. Chandler, R. Ries, J. T. Kley, D. Fowler and C. Bartlett, *SLAS Discovery: Adv. Life Sci. R*, 2019, **24**, 766–777.
41. K. Beeman, J. Baumgärtner, M. Laubenheimer, K. Hergesell, M. Hoffmann, U. Pehl, F. Fischer and J.-C. Pieck, *SLAS Discovery: Adv. Life Sci. R*, 2017, **22**, 1203–1210.
42. V. De Cesare, C. Johnson, V. Barlow, J. Hastie, A. Knebel and M. Trost, *Cell Chem. Biol.*, 2018, **25**, 1117–1127, e1114.
43. M. Winter, R. Ries, C. Kleiner, D. Bischoff, A. H. Luippold, T. Bretschneider and F. H. Büttner, *SLAS Technology: Transl. Life Sci. Innov.*, 2019, **24**, 209–221.
44. C. Haslam, J. Hellicar, A. Dunn, A. Fuetterer, N. Hardy, P. Marshall, R. Paape, M. Pemberton, A. Resemannand and M. Leveridge, *J. Biomol. Screening*, 2016, **21**, 176–186.

45. G. Nierode, P. S. Kwon, J. S. Dordick and S. J. Kwon, *J. Microbiol. Biotechnol.*, 2016, **26**, 213–225.
46. S. A. Langhans, *Front. Pharmacol.*, 2018, **9**, 6.
47. P. E. Feist, S. Sidoli, X. Liu, M. M. Schroll, S. Rahmy, R. Fujiwara, B. A. Garcia and A. B. Hummon, *Anal. Chem.*, 2017, **89**, 2773–2781.
48. X. Liu, E. M. Weaver and A. B. Hummon, *Anal. Chem.*, 2013, **85**, 6295–6302.
49. X. Liu and A. B. Hummon, *Anal. Chem.*, 2015, **87**, 9508–9519.
50. J. E. Ekert, J. Deakyne, P. Pribul-Allen, R. Terry, C. Schofield, C. G. Jeong, J. Storey, L. Mohamet, J. Francis, A. Naidoo, A. Amador, J. L. Klein and W. Rowan, *SLAS Discovery*, 2020, **25**, 1174–1190.
51. H. Li and A. B. Hummon, *Anal. Chem.*, 2011, **83**, 8794–8801.
52. X. Liu, J. K. Lukowski, C. Flinders, S. Kim, R. A. Georgiadis, S. M. Mumenthaler and A. B. Hummon, *Anal. Chem.*, 2018, **90**, 14156–14164.
53. J. Johnson, J. T. Sharick, M. C. Skala and L. Li, *J. Mass Spectrom.*, 2020, **55**, e4452.
54. D. Zhang, C. E. C. A. Hop, G. Patilea-Vrana, G. Gampa, H. K. Seneviratne, J. D. Unadkat, J. R. Kenny, K. Nagapudi, L. Di, L. Zhou, M. Zak, M. R. Wright, N. N. Bumpus, R. Zang, X. Liu, Y. Lai and S. C. Khojasteh, *Drug Metab. Dispos.*, 2019, **47**, 1122.
55. E. G. Solon, A. Schweitzer, M. Stoeckli and B. Prideaux, *AAPS J.*, 2010, **12**, 11–26.
56. E. G. Solon, *Cell Tissue Res.*, 2015, **360**, 87–107.
57. Y. H. Bae and K. Park, *Adv. Drug Delivery Rev.*, 2020, **158**, 4–16.
58. A. Nilsson, A. Peric, M. Strimfors, R. J. A. Goodwin, M. A. Hayes, P. E. Andrén and C. Hilgendorf, *Sci. Rep.*, 2017, **7**, 6352.
59. L. R. S. Huizing, J. McDuffie, F. Cuyckens, M. van Heerden, T. Koudriakova, R. M. A. Heeren and R. J. Vreeken, *Anal. Chem.*, 2021, **93**, 2144–2151.
60. W. R. Spreen, D. A. Margolis and J. C. Pottage, Jr, *Curr. Opin. HIV AIDS*, 2013, **8**, 565–571.
61. B. M. Jucker, H. Alsaid, M. Rambo, S. C. Lenhard, B. Hoang, F. Xie, M. R. Groseclose, S. Castellino, V. Damian, G. Bowers and M. Gupta, *J. Controlled Release*, 2017, **268**, 102–112.
62. M. R. Groseclose and S. Castellino, *Int. J. Mass Spectrom.*, 2019, **437**, 92–98.
63. K. Fuchs, A. Kiss, P. E. Bize, R. Duran, A. Denys, G. Hopfgartner, G. Borchard and O. Jordan, *J. Controlled Release*, 2018, **269**, 128–135.
64. F. J. Troendle, C. D. Reddick and R. A. Yost, *J. Am. Soc. Mass Spectrom.*, 1999, **10**, 1315–1321.
65. M. L. Reyzer, Y. Hsieh, K. Ng, W. A. Korfmacher and R. M. Caprioli, *J. Mass Spectrom.*, 2003, **38**, 1081–1092.
66. T. C. Rohner, D. Staab and M. Stoeckli, *Mech. Ageing Dev.*, 2005, **126**, 177–185.

67. S. Khatib-Shahidi, M. Andersson, J. L. Herman, T. A. Gillespie and R. M. Caprioli, *Anal. Chem.*, 2006, **78**, 6448–6456.
68. M. Stoeckli, D. Staab and A. Schweitzer, *Int. J. Mass Spectrom.*, 2007, **260**, 195–202.
69. Y. Hsieh, J. Chen, I. Knemeyer, L. Crossman and W. A. Korfmacher, *Drug Metab. Lett.*, 2008, **2**, 1–4.
70. P. J. Trim, C. M. Henson, J. L. Avery, A. McEwen, M. F. Snel, E. Claude, P. S. Marshall, A. West, A. P. Princivalle and M. R. Clench, *Anal. Chem.*, 2008, **80**, 8628–8634.
71. D. M. Drexler, S. H. Tannehill-Gregg, L. Wang and B. J. Brock, *J. Pharmacol. Toxicol. Methods*, 2011, **63**, 205–208.
72. R. J. A. Goodwin, A. Nilsson, D. Borg, P. R. R. Langridge-Smith, D. J. Harrison, C. L. Mackay, S. L. Iverson and P. E. Andren, *J. Proteomics*, 2012, **75**, 4912–4920.
73. G. Hamm, D. Bonnel, R. Legouffe, F. Pamelard, J.-M. Delbos, F. Bouzom and J. Stauber, *J. Proteomics*, 2012, **75**, 4952–4961.
74. M. L. Reyzer, P. Chaurand, P. M. Angel and R. M. Caprioli, in *Mass Spectrometry Imaging: Principles and Protocols*, ed. S. S. Rubakhin and J. V. Sweedler, Humana Press, Totowa, NJ, 2010, pp. 285–301.
75. D. Saigusa, R. Saito, K. Kawamoto, A. Uruno, K. Kano, J. Aoki, M. Yamamoto and T. Kawamoto, *Anal. Chem.*, 2019, **91**, 8979–8986.
76. S. K. Shahidi-Latham, S. M. Dutta, M. C. Prieto Conaway and P. J. Rudewicz, *Anal. Chem.*, 2012, **84**, 7158–7165.
77. N. Takai, Y. Tanaka, K. Inazawa and H. Saji, *Rapid Commun. Mass Spectrom.*, 2012, **26**, 1549–1556.
78. P. J. Trim, in *Imaging Mass Spectrometry: Methods and Protocols*, ed. L. M. Cole, Springer New York, New York, NY, 2017, pp. 175–189.
79. P. M. Angel and R. M. Caprioli, *Biochemistry*, 2013, **52**, 3818–3828.
80. D. Bonnel, R. Legouffe, N. Willand, A. Baulard, G. Hamm, B. Deprez and J. Stauber, *Bioanalysis*, 2011, **3**, 1399–1406.
81. P. Chaurand, D. S. Cornett, P. M. Angel and R. M. Caprioli, *Mol. Cell. Proteomics*, 2011, **10**, O110 004259.
82. B. Chen, M. Vavrek, R. Gundersdorf, W. Zhong and M. T. Cancilla, *Anal. Chim. Acta*, 2020, **1125**, 279–287.
83. Y. Sugihara, K.-I. Watanabe and Á. Végvári, *Bioanalysis*, 2016, **8**, 575–588.
84. D. A. Pirman, A. Kiss, R. M. A. Heeren and R. A. Yost, *Anal. Chem.*, 2013, **85**, 1090–1096.
85. L. Tomlinson, J. Fuchser, A. Futterer, M. Baumert, D. G. Hassall, A. West and P. S. Marshall, *Rapid Commun. Mass Spectrom.*, 2014, **28**, 995–1003.
86. P. Källback, M. Shariatgorji, A. Nilsson and P. E. Andrén, *J. Proteomics*, 2012, **75**, 4941–4951.
87. D. A. Pirman, R. F. Reich, A. Kiss, R. M. A. Heeren and R. A. Yost, *Anal. Chem.*, 2013, **85**, 1081–1089.
88. A. Lee, B. Prideaux, M. Zimmerman, C. Carter, S. Barat, D. Angulo, V. Dartois, D. S. Perlin and Y. Zhao, *Antimicrob. Agents Chemother.*, 2020, **64**, e02268–02219.

89. L. Morosi, C. Matteo, T. Ceruti, S. Giordano, M. Ponzo, R. Frapolli, M. Zucchetti, E. Davoli, M. D'Incalci and P. Ubezio, *Int. J. Biol. Sci.*, 2020, **16**, 1363–1375.
90. D. Rifat, B. Prideaux, R. M. Savic, M. E. Urbanowski, T. L. Parsons, B. Luna, M. A. Marzinke, A. A. Ordonez, V. P. DeMarco, S. K. Jain, V. Dartois, W. R. Bishai and K. E. Dooley, *Sci. Transl. Med.*, 2018, **10**, eaai7786.
91. M. Prasad, G. Postma, P. Franceschi, L. Morosi, S. Giordano, F. Falcetta, R. Giavazzi, E. Davoli, L. M. C. Buydens and J. Jansen, *GigaScience*, 2020, **9**, 1–11.
92. H. K. Seneviratne, A. N. Hamlin, C. J. S. Heck and N. N. Bumpus, *ACS Pharmacol. Transl. Sci.*, 2020, **3**, 655–665.
93. A. Treu, J. Kokesch-Himmelreich, K. Walter, C. Holscher and A. Rompp, *J. Am. Soc. Mass Spectrom.*, 2020, **31**, 2277–2286.
94. J. G. Swales, J. W. Tucker, N. Strittmatter, A. Nilsson, D. Cobice, M. R. Clench, C. L. Mackay, P. E. Andren, Z. Takáts and P. J. Webborn, *J. Anal. Chem.*, 2014, **86**, 8473–8480.
95. I. S. Sørensen, C. Janfelt, M. M. B. Nielsen, R. W. Mortensen, N. Ø. Knudsen, A. H. Eriksson, A. J. Pedersen and K. T. Nielsen, *Anal. Bioanal. Chem.*, 2017, **409**, 4993–5005.
96. X. Liu, J. L. Ide, I. Norton, M. A. Marchionni, M. C. Ebling, L. Y. Wang, E. Davis, C. M. Sauvageot, S. Kesari, K. A. Kellersberger, M. L. Easterling, S. Santagata, D. D. Stuart, J. Alberta, J. N. Agar, C. D. Stiles and N. Y. R. Agar, *Sci. Rep.*, 2013, **3**, 2859.
97. E. F. Choo, J. Ly, J. Chan, S. K. Shahidi-Latham, K. Messick, E. Plise, C. M. Quiason and L. Yang, *Mol. Pharmaceutics*, 2014, **11**, 4199–4207.
98. E. C. Randall, K. B. Emdal, J. K. Laramy, M. Kim, A. Roos, D. Calligaris, M. S. Regan, S. K. Gupta, A. C. Mladek, B. L. Carlson, A. J. Johnson, F.-K. Lu, X. S. Xie, B. A. Joughin, R. J. Reddy, S. Peng, W. M. Abdelmoula, P. R. Jackson, A. Kolluri, K. A. Kellersberger, J. N. Agar, D. A. Lauffenburger, K. R. Swanson, N. L. Tran, W. F. Elmquist, F. M. White, J. N. Sarkaria and N. Y. R. Agar, *Nat. Commun.*, 2018, **9**, 4904.
99. R. J. A. Goodwin, J. Bunch and D. F. McGinnity, in *Advances in Cancer Research*, ed. R. R. Drake and L. A. McDonnell, Academic Press, 2017, vol. 134, pp. 133–171.
100. S. Giordano, L. Morosi, P. Veglianese, S. A. Licandro, R. Frapolli, M. Zucchetti, G. Cappelletti, L. Falciola, V. Pifferi, S. Visentin, M. D'Incalci and E. Davoli, *Sci. Rep.*, 2016, **6**, 37027.
101. S. J. Atkinson, P. M. Loadman, C. Sutton, L. H. Patterson and M. R. Clench, *Rapid Commun. Mass Spectrom.*, 2007, **21**, 1271–1276.
102. A. Rompp, S. Guenther, Z. Takats and B. Spengler, *Anal. Bioanal. Chem.*, 2011, **401**, 65–73.
103. G. Marko-Varga, T. E. Fehniger, M. Rezeli, B. Dome, T. Laurell and A. Vegvari, *J. Proteomics*, 2011, **74**, 982–992.

104. S. M. Maira, S. Pecchi, A. Huang, M. Burger, M. Knapp, D. Sterker, C. Schnell, D. Guthy, T. Nagel, M. Wiesmann, S. Brachmann, C. Fritsch, M. Dorsch, P. Chene, K. Shoemaker, A. De Pover, D. Menezes, G. Martiny-Baron, D. Fabbro, C. J. Wilson, R. Schlegel, F. Hofmann, C. Garcia-Echeverria, W. R. Sellers and C. F. Voliva, *Mol. Cancer Ther.*, 2012, **11**, 317–328.
105. M. Yasunaga, M. Furuta, K. Ogata, Y. Koga, Y. Yamamoto, M. Takigahira and Y. Matsumura, *Sci. Rep.*, 2013, **3**, 3050.
106. A. Buck, S. Halbritter, C. Spath, A. Feuchtinger, M. Aichler, H. Zitzelsberger, K. P. Janssen and A. Walch, *Anal. Bioanal. Chem.*, 2015, **407**, 2107–2116.
107. L. Salphati, S. Shahidi-Latham, C. Quiason, K. Barck, M. Nishimura, B. Alicke, J. Pang, R. A. Carano, A. G. Olivero and H. S. Phillips, *Drug Metab. Dispos.*, 2014, **42**, 1110–1116.
108. J. J. Connell, Y. Sugihara, S. Torok, B. Dome, J. Tovari, T. E. Fehniger, G. Marko-Varga and A. Vegvari, *Anal. Bioanal. Chem.*, 2015, **407**, 2245–2253.
109. A. Vegvari, A. S. Shavkunov, T. E. Fehniger, D. Grabau, E. Nimeus and G. Marko-Varga, *Clin. Transl. Med.*, 2016, **5**, 10.
110. I. H. Bartelink, B. Prideaux, G. Krings, L. Wilmes, P. R. E. Lee, P. Bo, B. Hann, J.-P. Coppé, D. Heditsian, L. Swigart-Brown, E. F. Jones, S. Magnitsky, R. J. Keizer, N. de Vries, H. Rosing, N. Pawlowska, S. Thomas, M. Dhawan, R. Aggarwal, P. N. Munster, L. J. Esserman, W. Ruan, A. H. B. Wu, D. Yee, V. Dartois, R. M. Savic, D. M. Wolf and L. van 't Veer, *Breast Cancer Res.*, 2017, **19**, 107.
111. M. Nishidate, K. Yamamoto, C. Masuda, H. Aikawa, M. Hayashi, T. Kawanishi and A. Hamada, *Sci. Rep.*, 2017, **7**, 16763.
112. Y. Sugihara, D. Rivas, J. Malm, M. Szasz, H. Kwon, B. Baldetorp, H. Olsson, C. Ingvar, M. Rezeli, T. E. Fehniger and G. Marko-Varga, *Clin. Transl. Med.*, 2018, **7**, 22.
113. S. Ryu, M. Ohuchi, S. Yagishita, T. Shimoi, K. Yonemori, K. Tamura, Y. Fujiwara and A. Hamada, *Sci. Rep.*, 2020, **10**, 15535.
114. A. Bouslimani, N. Bec, M. Glueckmann, C. Hirtz and C. Larroque, *Rapid Commun. Mass Spectrom.*, 2010, **24**, 415–421.
115. A. C. K. Chung, X. Li, W.-C. Li, T. Wang, H.-K. Lee, L. Jin, Z. Cai and K. C.-F. Leung, *Nanoscale Adv.*, 2020, **2**, 5857–5865.
116. X. Liu and A. B. Hummon, *Sci. Rep.*, 2016, **6**, 38507.
117. D. S. Wagner, J. L. Pirhalla and G. D. Bowers, *Bioanalysis*, 2013, **5**, 463–479.
118. S. Castellino, M. R. Groseclose and D. Wagner, *Bioanalysis*, 2011, **3**, 2427–2441.
119. N. Sun, I. E. Fernandez, M. Wei, Y. Wu, M. Aichler, O. Eickelberg and A. Walch, *Histochem. Cell Biol.*, 2016, **145**, 201–211.
120. Z. Zhao, A. Ukidve, J. Kim and S. Mitragotri, *Cell*, 2020, **181**, 151–167.
121. C. Zylberberg and S. Matosevic, *Drug Delivery*, 2016, **23**, 3319–3329.

122. S. K. Golombek, J. N. May, B. Theek, L. Appold, N. Drude, F. Kiessling and T. Lammers, *Adv. Drug Delivery Rev.*, 2018, **130**, 17–38.
123. D. Rosenblum, N. Joshi, W. Tao, J. M. Karp and D. Peer, *Nat. Commun.*, 2018, **9**, 1410.
124. A. Fulop, D. A. Sammour, K. Erich, J. von Gerichten, P. van Hoogevest, R. Sandhoff and C. Hopf, *Sci. Rep.*, 2016, **6**, 33791.
125. X. B. Zhao, W. Ha, K. Gao and Y. P. Shi, *Anal. Chem.*, 2020, **92**, 9039–9047.
126. T. J. A. Dekker, E. A. Jones, W. E. Corver, R. J. M. van Zeijl, A. M. Deelder, R. A. E. M. Tollenaar, W. E. Mesker, H. Morreau and L. A. McDonnell, *Anal. Bioanal. Chem.*, 2015, **407**, 2167–2176.
127. T. Murta, R. T. Steven, C. J. Nikula, S. A. Thomas, L. B. Zeiger, A. Dexter, E. A. Elia, B. Yan, A. D. Campbell, R. J. A. Goodwin, Z. Takáts, O. J. Sansom and J. Bunch, *Anal. Chem.*, 2021, **93**, 2309–2316.
128. D. Miura, Y. Fujimura, M. Yamato, F. Hyodo, H. Utsumi, H. Tachibana and H. Wariishi, *Anal. Chem.*, 2010, **82**, 9789–9796.
129. D. Miura, Y. Fujimura, H. Tachibana and H. Wariishi, *Anal. Chem.*, 2010, **82**, 498–504.
130. B. Flinders, L. Huizing, M. Van Heerden, F. Cuyckens, U. Neumann, L. van der Laan, S. W. M. Olde Damink, R. M. A. Heeren, F. G. Schaap and R. J. Vreeken, *Anal. Chem.*, 2018, **90**, 11835–11846.
131. Z. Ablonczy, N. Smith, D. M. Anderson, A. C. Grey, J. Spraggins, Y. Koutalos, K. L. Schey and R. K. Crouch, *Proteomics*, 2014, **14**, 936–944.
132. N. E. Mascini, M. Cheng, L. Jiang, A. Rizwan, H. Podmore, D. R. Bhandari, A. Römpp, K. Glunde and R. M. A. Heeren, *Anal. Chem.*, 2016, **88**, 3107–3114.
133. A. S. Woods and S. N. Jackson, *AAPS J.*, 2006, **8**, E391–E395.
134. A. Thomas, J. L. Charbonneau, E. Fournaise and P. Chaurand, *Anal. Chem.*, 2012, **84**, 2048–2054.
135. C. D. Calvano, A. Monopoli, T. R. I. Cataldi and F. Palmisano, *Anal. Bioanal. Chem.*, 2018, **410**, 4015–4038.
136. W. J. Perry, N. H. Patterson, B. M. Prentice, E. K. Neumann, R. M. Caprioli and J. M. Spraggins, *J. Mass Spectrom.*, 2020, **55**, e4491.
137. B. Fuchs, A. Bischoff, R. Suss, K. Teuber, M. Schurenberg, D. Suckau and J. Schiller, *Anal. Bioanal. Chem.*, 2009, **395**, 2479–2487.
138. Q. Zhou, A. Fülöp and C. Hopf, *Anal. Bioanal. Chem.*, 2021, **413**, 2599–2617.
139. C. Harkin, K. W. Smith, F. L. Cruickshank, C. Logan Mackay, B. Flinders, R. M. A. Heeren, T. Moore, S. Brockbank and D. F. Cobice, *Mass Spectrom. Rev.*, 2021, 1–33.
140. M. Shariatgorji, A. Nilsson, E. Fridjonsdottir, T. Vallianatou, P. Källback, L. Katan, J. Sävmarker, I. Mantas, X. Zhang, E. Bezard, P. Svenningsson, L. R. Odell and P. E. Andrén, *Nat. Methods*, 2019, **16**, 1021–1028.

141. M. Shariatgorji, A. Nilsson, R. J. Goodwin, P. Källback, N. Schintu, X. Zhang, A. R. Crossman, E. Bezard, P. Svenningsson and P. E. Andren, *Neuron*, 2014, **84**, 697–707.
142. J. G. Swales, A. Dexter, G. Hamm, A. Nilsson, N. Strittmatter, F. Michopoulos, C. Hardy, P. Morentin-Gutierrez, M. Mellor, P. E. Andren, M. R. Clench, J. Bunch, S. E. Critchlow and R. J. A. Goodwin, *Anal. Chem.*, 2018, **90**, 6051–6058.
143. A. Dannhorn, E. Kazanc, S. Ling, C. Nikula, E. Karali, M. P. Serra, J. L. Vorng, P. Inglese, G. Maglennon, G. Hamm, J. Swales, N. Strittmatter, S. T. Barry, O. J. Sansom, G. Poulogiannis, J. Bunch, R. J. Goodwin and Z. Takats, *Anal. Chem.*, 2020, **92**, 11080–11088.
144. R. J. A. Goodwin, J. C. Dungworth, S. R. Cobb and A. R. Pitt, *Proteomics*, 2008, **8**, 3801–3808.
145. R. J. A. Goodwin, S. L. Iverson and P. E. Andren, *Rapid Commun. Mass Spectrom.*, 2012, **26**, 494–498.
146. J. K. Lukowski, A. Pamreddy, D. Velickovic, G. Zhang, L. Pasa-Tolic, T. Alexandrov, K. Sharma, C. R. Anderton and P. Kidney Precision Medicine, *J. Am. Soc. Mass Spectrom.*, 2020, **31**, 2538–2546.
147. B. Flinders, J. Morrell, P. S. Marshall, L. E. Ranshaw, R. M. A. Heeren and M. R. Clench, *Rapid Commun. Mass Spectrom.*, 2021, **35**, e8957.
148. S. Torok, A. Vegvari, M. Rezeli, T. E. Fehniger, J. Tovari, S. Paku, V. Laszlo, B. Hegedus, A. Rozsas, B. Dome and G. Marko-Varga, *Br. J. Pharmacol.*, 2015, **172**, 1148–1163.
149. I. Rzagalinski, N. Hainz, C. Meier, T. Tschernig and D. A. Volmer, *Anal. Bioanal. Chem.*, 2019, **411**, 353–365.
150. E. Fridjonsdottir, R. Shariatgorji, A. Nilsson, T. Vallianatou, L. R. Odell, L. S. Schembri, P. Svenningsson, P.-O. Fernagut, A. R. Crossman, E. Bezard and P. E. Andrén, *Sci. Adv.*, 2021, **7**, eabe5948.
151. N. Sun, I. E. Fernandez, M. Wei, M. Witting, M. Aichler, A. Feuchtinger, G. Burgstaller, S. E. Verleden, P. Schmitt-Kopplin, O. Eickelberg and A. Walch, *Eur. Respir. J.*, 2018, **52**, 1702314.
152. L. Poncelet, R. Ait-Belkacem, R. Marillier, B. Gomes and J. Stauber, *J. Pharm. Biomed. Anal.*, 2019, **170**, 220–227.
153. F. Henderson, P. J. Hart, J. M. Pradillo, M. Kassiou, L. Christie, K. J. Williams, H. Boutin and A. McMahon, *Rapid Commun. Mass Spectrom.*, 2018, **32**, 721–729.
154. E. Tanaka, Y. Ogawa, R. Fujii, T. Shimonaka, Y. Sato, T. Hamazaki, T. Nagamura-Inoue, H. Shintaku and M. Tsuji, *Sci. Rep.*, 2020, **10**, 21881.
155. S. Van Nuffel, M. Quatredeniers, A. Pirkl, J. Zakel, J. P. Le Caer, N. Elie, Q. P. Vanbellingen, S. J. Dumas, M. K. Nakhleh, M. R. Ghigna, E. Fadel, M. Humbert, P. Chaurand, D. Touboul, S. Cohen-Kaminsky and A. Brunelle, *Anal. Chem.*, 2020, **92**, 12079–12087.
156. I. Kaya, W. Michno, D. Brinet, Y. Iacone, G. Zanni, K. Blennow, H. Zetterberg and J. Hanrieder, *Anal. Chem.*, 2017, **89**, 4685–4694.
157. D. Tarin, *Cancer Microenviron.*, 2012, **5**, 95–112.

158. National Cancer Institute Dictionary of Cancer Terms, https://www.cancer.gov/publications/dictionaries/cancer-terms/def/tumor-micro environment, (accessed 10-FEB-2021).
159. O. Warburg, K. Posener and E. Negelein, *Biochem. Z.*, 1924, **152**, 319–344.
160. M. K. Andersen, T. S. Hoiem, B. S. R. Claes, B. Balluff, M. Martin-Lorenzo, E. Richardsen, S. Krossa, H. Bertilsson, R. M. A. Heeren, M. B. Rye, G. F. Giskeodegard, T. F. Bathen and M. B. Tessem, *Cancer Metab.*, 2021, **9**, 9.
161. M. Dilillo, R. Ait-Belkacem, C. Esteve, D. Pellegrini, S. Nicolardi, M. Costa, E. Vannini, E. L. Graaf, M. Caleo and L. A. McDonnell, *Sci. Rep.*, 2017, **7**, 603.
162. K. Chughtai, L. Jiang, T. R. Greenwood, K. Glunde and R. M. Heeren, *J. Lipid Res.*, 2013, **54**, 333–344.
163. M. Kawashima, N. Iwamoto, N. Kawaguchi-Sakita, M. Sugimoto, T. Ueno, Y. Mikami, K. Terasawa, T. A. Sato, K. Tanaka, K. Shimizu and M. Toi, *Cancer Sci.*, 2013, **104**, 1372–1379.
164. R. Mirnezami, K. Spagou, P. A. Vorkas, M. R. Lewis, J. Kinross, E. Want, H. Shion, R. D. Goldin, A. Darzi, Z. Takats, E. Holmes, O. Cloarec and J. K. Nicholson, *Mol. Oncol.*, 2014, **8**, 39–49.
165. S. Guo, Y. Wang, D. Zhou and Z. Li, *Sci. Rep.*, 2014, **4**, 5959.
166. Y. Muranishi, T. Sato, S. Ito, J. Satoh, A. Yoshizawa, S. Tamari, Y. Ueda, Y. Yutaka, T. Menju, T. Nakamura and H. Date, *Sci. Rep.*, 2019, **9**, 8916.
167. G. Sun, K. Yang, Z. Zhao, S. Guan, X. Han and R. W. Gross, *Anal. Chem.*, 2008, **80**, 7576–7585.
168. B. M. Grüner, I. Winkelmann, A. Feuchtinger, N. Sun, B. Balluff, N. Teichmann, A. Herner, E. Kalideris, K. Steiger, R. Braren and M. Aichler, *Mol. Cancer Ther.*, 2016, **15**, 1145–1152.
169. T. Matsumoto, T. Kobayashi and K. Kamata, *Curr. Med. Chem.*, 2007, **14**, 3209–3220.
170. E. Gonzalez de San Roman, I. Manuel, M. T. Giralt, I. Ferrer and R. Rodriguez-Puertas, *Biochim. Biophys. Acta, Biomembr.*, 2017, **1859**, 1604–1614.
171. B. Enthaler, J. K. Pruns, S. Wessel, C. Rapp, M. Fischer and K.-P. Wittern, *Anal. Bioanal. Chem.*, 2012, **402**, 1159–1167.
172. A. Harvey, L. M. Cole, R. Day, M. Bartlett, J. Warwick, R. Bojar, D. Smith, N. Cross and M. R. Clench, *Proteomics*, 2016, **16**, 1718–1725.
173. K. Hattori, M. Kajimura, T. Hishiki, T. Nakanishi, A. Kubo, Y. Nagahata, M. Ohmura, A. Yachie-Kinoshita, T. Matsuura, T. Morikawa, T. Nakamura, M. Setou and M. Suematsu, *Antioxid. Redox Signaling*, 2010, **13**, 1157–1167.
174. S. Koizumi, S. Yamamoto, T. Hayasaka, Y. Konishi, M. Yamaguchi-Okada, N. Goto-Inoue, Y. Sugiura, M. Setou and H. Namba, *Neuroscience*, 2010, **168**, 219–225.
175. H. Liu, R. Chen, J. Wang, S. Chen, C. Xiong, J. Wang, J. Hou, Q. He, N. Zhang, Z. Nie and L. Mao, *Anal. Chem.*, 2014, **86**, 10114–10121.

176. D. Touboul, H. Piednoel, V. Voisin, S. De La Porte, A. Brunelle, F. Halgand and O. Laprevote, *Eur. J. Mass Spectrom.*, 2004, **10**, 657–664.
177. M. R. Eveque-Mourroux, P. J. Emans, R. R. M. Zautsen, A. Boonen, R. M. A. Heeren and B. Cillero-Pastor, *Analyst*, 2019, **144**, 5953–5958.
178. H. E. Bowrey, D. M. Anderson, P. Pallitto, D. B. Gutierrez, J. Fan, R. K. Crouch, K. L. Schey and Z. Ablonczy, *Proteomics: Clin. Appl.*, 2016, **10**, 391–402.
179. M. G. Nye-Wood, J. M. Spraggins, R. M. Caprioli, K. L. Schey, P. J. Donaldson and A. C. Grey, *Exp. Eye Res.*, 2017, **154**, 70–78.
180. M. Rujoi, R. Estrada and M. C. Yappert, *Anal. Chem.*, 2004, **76**, 1657–1663.
181. A. M. Monro, *Regul. Toxicol. Pharmacol.*, 1990, **12**, 137–160.
182. M. Pellegatti and S. Pagliarusco, *Expert Opin. Drug Metab. Toxicol.*, 2011, **7**, 137–146.
183. O. Karlsson and J. Hanrieder, *Arch. Toxicol.*, 2016, **91**, 2283–2294.
184. J. M. Kampa, M. Sahin, M. Slopianka, M. Giampa, H. Bednarz, R. Ernst, B. Riefke, K. Niehaus and A. Fatangare, *Toxicol. Lett.*, 2020, **325**, 43–50.
185. H. R. Brown, S. Castellino, M. R. Groseclose, C. S. Elangbam, K. Mellon-Kusibab, L. W. Yoon, L. D. Gates, D. L. Krull, N. F. Cariello, L. Arrington-Brown, T. Tillman, S. Fowler, V. Shah, D. Bailey and R. T. Miller, *Toxicol. Pathol.*, 2016, **44**, 112–131.
186. B. Lenz, A. Brink, M. Siam, A. De Paepe, S. Bassett, A. Eichinger-Chapelon, P. Maliver, R. Neff, U. Niederhauser, B. Steinhuber, R. Zurbach, T. Singer, C. Funk, F. Schuler, M. Albassam and S. Schadt, *Toxicol. Sci*, 2018, **163**, 409–419.
187. A. Nilsson, B. Forngren, S. Bjurström, R. J. A. Goodwin, E. Basmaci, I. Gustafsson, A. Annas, D. Hellgren, A. Svanhagen, P. E. Andrén and J. Lindberg, *PLoS One*, 2012, **7**, e47353.
188. A. L. Bruinen, C. van Oevelen, G. B. Eijkel, M. Van Heerden, F. Cuyckens and R. M. A. Heeren, *J. Am. Soc. Mass Spectrom.*, 2016, **27**, 117–123.
189. A. Nilsson, R. J. A. Goodwin, J. G. Swales, R. Gallagher, H. Shankaran, A. Sathe, S. Pradeepan, A. Xue, N. Keirstead, J. C. Sasaki, P. E. Andren and A. Gupta, *Chem. Res. Toxicol.*, 2015, **28**, 1823–1830.
190. M. R. Groseclose, S. B. Laffan, K. S. Frazier, A. Hughes-Earle and S. Castellino, *J. Am. Soc. Mass Spectrom.*, 2015, **26**, 887–898.
191. M. Irie, E. Hayakawa, Y. Fujimura, Y. Honda, D. Setoyama, H. Wariishi, F. Hyodo and D. Miura, *Biochem. Biophys. Res. Commun.*, 2018, **496**, 140–146.
192. S. Castellino, M. R. Groseclose, J. Sigafoos, D. Wagner, M. de Serres, J. W. Polli, E. Romach, J. Myer and B. Hamilton, *Chem. Res. Toxicol.*, 2013, **26**, 241–251.
193. D. Bonnel and J. Stauber, in *Drug Safety Evaluation: Methods and Protocols*, ed. J.-C. Gautier, Springer New York, New York, NY, 2017, pp. 129–140.

194. M. R. Groseclose and S. Castellino, *Chem. Res. Toxicol.*, 2019, **32**, 294–303.
195. M. Aichler, M. Elsner, N. Ludyga, A. Feuchtinger, V. Zangen, S. K. Maier, B. Balluff, C. Schone, L. Hierber, H. Braselmann, S. Meding, S. Rauser, H. Zischka, M. Aubele, M. Schmitt, M. Feith, S. M. Hauck, M. Ueffing, R. Langer, B. Kuster, H. Zitzelsberger, H. Hofler and A. K. Walch, *J. Pathol.*, 2013, **230**, 410–419.
196. T. E. Fehniger, Á. Végvári, M. Rezeli, K. Prikk, P. Ross, M. Dahlbäck, G. Edula, R. Sepper and G. Marko-Varga, *Anal. Chem.*, 2011, **83**, 8329–8336.
197. S. O. Deininger, M. P. Ebert, A. Futterer, M. Gerhard and C. Rocken, *J. Proteome Res.*, 2008, **7**, 5230–5236.
198. L. A. McDonnell, G. L. Corthals, S. M. Willems, A. van Remoortere, R. J. van Zeijl and A. M. Deelder, *J. Proteomics*, 2010, **73**, 1921–1944.
199. S. Meding, U. Nitsche, B. Balluff, M. Elsner, S. Rauser, C. Schone, M. Nipp, M. Maak, M. Feith, M. P. Ebert, H. Friess, R. Langer, H. Hofler, H. Zitzelsberger, R. Rosenberg and A. Walch, *J. Proteome Res.*, 2012, **11**, 1996–2003.
200. C. Schone, H. Hofler and A. Walch, *Clin. Biochem.*, 2013, **46**, 539–545.
201. M. A. Rodrigo, O. Zitka, S. Krizkova, A. Moulick, V. Adam and R. Kizek, *J. Pharm. Biomed. Anal.*, 2014, **95**, 245–255.
202. S. Rauser, S. O. Deininger, D. Suckau, H. Hofler and A. Walch, *Expert Rev. Proteomics*, 2010, **7**, 927–941.
203. T. Gemoll, U. J. Roblick and J. K. Habermann, *Mol. Med. Rep.*, 2011, **4**, 1045–1051.
204. R. D. Addie, B. Balluff, J. V. Bovee, H. Morreau and L. A. McDonnell, *Anal. Chem.*, 2015, **87**, 6426–6433.
205. Y. Ucal, Z. A. Durer, H. Atak, E. Kadioglu, B. Sahin, A. Coskun, A. T. Baykal and A. Ozpinar, *Biochim. Biophys. Acta*, 2017, **1865**, 795–816.
206. B. Balluff, C. K. Frese, S. K. Maier, C. Schone, B. Kuster, M. Schmitt, M. Aubele, H. Hofler, A. M. Deelder, A. Heck, Jr, P. C. Hogendoorn, J. Morreau, A. F. Maarten Altelaar, A. Walch and L. A. McDonnell, *J. Pathol.*, 2015, **235**, 3–13.
207. B. Balluff, S. Rauser, S. Meding, M. Elsner, C. Schone, A. Feuchtinger, C. Schuhmacher, A. Novotny, U. Jutting, G. Maccarrone, H. Sarioglu, M. Ueffing, H. Braselmann, H. Zitzelsberger, R. M. Schmid, H. Hofler, M. P. Ebert and A. Walch, *Am. J. Pathol.*, 2011, **179**, 2720–2729.
208. M. Elsner, S. Rauser, S. Maier, C. Schone, B. Balluff, S. Meding, G. Jung, M. Nipp, H. Sarioglu, G. Maccarrone, M. Aichler, A. Feuchtinger, R. Langer, U. Jutting, M. Feith, B. Kuster, M. Ueffing, H. Zitzelsberger, H. Hofler and A. Walch, *J. Proteomics*, 2012, **75**, 4693–4704.
209. S. A. Schwartz, R. J. Weil, R. C. Thompson, Y. Shyr, J. H. Moore, S. A. Toms, M. D. Johnson and R. M. Caprioli, *Cancer Res.*, 2005, **65**, 7674–7681.

210. M. Nipp, M. Elsner, B. Balluff, S. Meding, H. Sarioglu, M. Ueffing, S. Rauser, K. Unger, H. Hofler, A. Walch and H. Zitzelsberger, *J. Mol. Med.*, 2012, **90**, 163–174.
211. A. R. Bresnick, D. J. Weber and D. B. Zimmer, *Nat. Rev. Cancer*, 2015, **15**, 96–109.
212. S. A. Schwartz, R. J. Weil, M. D. Johnson, S. A. Toms and R. M. Caprioli, *Clin. Cancer Res.*, 2004, **10**, 981–987.
213. D. S. Cornett, J. A. Mobley, E. C. Dias, M. Andersson, C. L. Arteaga, M. E. Sanders and R. M. Caprioli, *Mol. Cell. Proteomics*, 2006, **5**, 1975–1983.
214. R. Lemaire, S. A. Menguellet, J. Stauber, V. Marchaudon, J. P. Lucot, P. Collinet, M. O. Farine, D. Vinatier, R. Day, P. Ducoroy, M. Salzet and I. Fournier, *J. Proteome Res.*, 2007, **6**, 4127–4134.
215. L. H. Cazares, D. Troyer, S. Mendrinos, R. A. Lance, J. O. Nyalwidhe, H. A. Beydoun, M. A. Clements, R. R. Drake and O. J. Semmes, *Clin. Cancer Res.*, 2009, **15**, 5541–5551.
216. P. Chaurand, B. B. DaGue, R. S. Pearsall, D. W. Threadgill and R. M. Caprioli, *Proteomics*, 2001, **1**, 1320–1326.
217. B. Balluff, M. Elsner, A. Kowarsch, S. Rauser, S. Meding, C. Schuhmacher, M. Feith, K. Herrmann, C. Rocken, R. M. Schmid, H. Hofler, A. Walch and M. P. Ebert, *J. Proteome Res.*, 2010, **9**, 6317–6322.
218. R. Lemaire, A. Desmons, J. C. Tabet, R. Day, M. Salzet and I. Fournier, *J. Proteome Res.*, 2007, **6**, 1295–1305.
219. M.-C. Djidja, E. Claude, M. F. Snel, P. Scriven, S. Francese, V. Carolan and M. R. Clench, *J. Proteome Res.*, 2009, **8**, 4876–4884.
220. Y. Morita, K. Ikegami, N. Goto-Inoue, T. Hayasaka, N. Zaima, H. Tanaka, T. Uehara, T. Setoguchi, T. Sakaguchi, H. Igarashi, H. Sugimura, M. Setou and H. Konno, *Cancer Sci.*, 2010, **101**, 267–273.
221. S. Meding, B. Balluff, M. Elsner, C. Schone, S. Rauser, U. Nitsche, M. Maak, A. Schafer, S. M. Hauck, M. Ueffing, R. Langer, H. Hofler, H. Friess, R. Rosenberg and A. Walch, *J. Pathol.*, 2012, **228**, 459–470.
222. R. Casadonte and R. M. Caprioli, *Nat. Protoc.*, 2011, **6**, 1695–1709.
223. R. Casadonte, R. Longuespée, J. Kriegsmann and M. Kriegsmann, in *Advances in Cancer Research*, ed. R. R. Drake and L. A. McDonnell, Academic Press, 2017, vol. 134, pp. 173–200.
224. J. Kononen, L. Bubendorf, A. Kallioniemi, M. Barlund, P. Schraml, S. Leighton, J. Torhorst, M. J. Mihatsch, G. Sauter and O. P. Kallioniemi, *Nat. Med.*, 1998, **4**, 844–847.
225. H. Battifora, *Lab. Invest.*, 1986, **55**, 244–248.
226. H. Battifora and P. Mehta, *Lab. Invest.*, 1990, **63**, 722–724.
227. M. R. Groseclose, P. P. Massion, P. Chaurand and R. M. Caprioli, *Proteomics*, 2008, **8**, 3715–3724.
228. M. E. Sanders, E. C. Dias, B. J. Xu, J. A. Mobley, D. Billheimer, H. Roder, J. Grigorieva, M. Dowsett, C. L. Arteaga and R. M. Caprioli, *J. Proteome Res.*, 2008, **7**, 1500–1507.

229. P. H. Pevsner, J. Melamed, T. Remsen, A. Kogos, F. Francois, P. Kessler, A. Stern and S. Anand, *Biomarkers Med.*, 2009, **3**, 55–69.
230. A. Hinsch, M. Buchholz, S. Odinga, C. Borkowski, C. Koop, J. R. Izbicki, M. Wurlitzer, T. Krech, W. Wilczak, S. Steurer, F. Jacobsen, E. C. Burandt, P. Stahl, R. Simon, G. Sauter and H. Schlüter, *J. Mass Spectrom.*, 2017, **52**, 165–173.
231. P. Mittal, M. Klingler-Hoffmann, G. Arentz, L. Winderbaum, N. A. Lokman, C. Zhang, L. Anderson, J. Scurry, Y. Leung, C. J. Stewart, J. Carter, G. Kaur, M. K. Oehler and P. Hoffmann, *Proteomics*, 2016, **16**, 1793–1801.
232. H. K. Kim, M. L. Reyzer, I. J. Choi, C. G. Kim, H. S. Kim, A. Oshima, O. Chertov, S. Colantonio, R. J. Fisher, J. L. Allen, R. M. Caprioli and J. E. Green, *J. Proteome Res.*, 2010, **9**, 4123–4130.
233. K. Erich, K. Reinle, T. Muller, B. Munteanu, D. A. Sammour, I. Hinsenkamp, T. Gutting, E. Burgermeister, P. Findeisen, M. P. Ebert, J. Krijgsveld and C. Hopf, *Mol. Cell. Proteomics*, 2019, **18**, 151–161.
234. N. E. Mascini, J. Teunissen, R. Noorlag, S. M. Willems and R. M. A. Heeren, *Methods*, 2018, **151**, 21–27.
235. M. Kriegsmann, R. Casadonte, J. Kriegsmann, H. Dienemann, P. Schirmacher, J. Hendrik Kobarg, K. Schwamborn, A. Stenzinger, A. Warth and W. Weichert, *Mol. Cell. Proteomics*, 2016, **15**, 3081–3089.
236. M. Kriegsmann, C. Zgorzelski, R. Casadonte, K. Schwamborn, T. Muley, H. Winter, M. Eichhorn, F. Eichhorn, A. Warth, S. O. Deininger, P. Christopoulos, M. Thomas, T. Longerich, A. Stenzinger, W. Weichert, C. Muller-Tidow, J. Kriegsmann, P. Schirmacher and K. Kriegsmann, *Cancers*, 2020, **12**, 2704.
237. A. Quaas, A. S. Bahar, K. von Loga, A. S. Seddiqi, J. M. Singer, M. Omidi, O. Kraus, M. Kwiatkowski, M. Trusch, S. Minner, E. Burandt, P. Stahl, W. Wilczak, M. Wurlitzer, R. Simon, G. Sauter, A. Marx and H. Schluter, *Histopathology*, 2013, **63**, 455–462.
238. M. El Ayed, D. Bonnel, R. Longuespee, C. Castelier, J. Franck, D. Vergara, A. Desmons, A. Tasiemski, A. Kenani, D. Vinatier, R. Day, I. Fournier and M. Salzet, *Med. Sci. Monit.*, 2010, **16**, BR233–BR245.
239. O. Klein, F. Kanter, H. Kulbe, P. Jank, C. Denkert, G. Nebrich, W. D. Schmitt, Z. Wu, C. A. Kunze, J. Sehouli, S. Darb-Esfahani, I. Braicu, J. Lellmann, H. Thiele and E. T. Taube, *Proteomics: Clin. Appl.*, 2019, **13**, e1700181.
240. M.-C. Djidja, E. Claude, M. F. Snel, S. Francese, P. Scriven, V. Carolan and M. R. Clench, *Anal. Bioanal. Chem.*, 2010, **397**, 587–601.
241. R. Casadonte, M. Kriegsmann, F. Zweynert, K. Friedrich, G. Baretton, M. Otto, S. O. Deininger, R. Paape, E. Belau, D. Suckau, D. Aust, C. Pilarsky and J. Kriegsmann, *Proteomics*, 2014, **14**, 956–964.
242. K. Schwamborn, R. C. Krieg, M. Reska, G. Jakse, R. Knuechel and A. Wellmann, *Int. J. Mol. Med.*, 2007, **20**, 155–159.
243. S. Steurer, C. Borkowski, S. Odinga, M. Buchholz, C. Koop, H. Huland, M. Becker, M. Witt, D. Trede, M. Omidi, O. Kraus, A. S. Bahar,

A. S. Seddiqi, J. M. Singer, M. Kwiatkowski, M. Trusch, R. Simon, M. Wurlitzer, S. Minner, T. Schlomm, G. Sauter and H. Schluter, *Int. J. Cancer*, 2013, **133**, 920–928.
244. T. M. Morgan, E. H. Seeley, O. Fadare, R. M. Caprioli and P. E. Clark, *J. Urol.*, 2013, **189**, 1097–1103.
245. S. Steurer, A. S. Seddiqi, J. M. Singer, A. S. Bahar, C. Eichelberg, M. Rink, R. Dahlem, H. Huland, G. Sauter, R. Simon, S. Minner, E. Burandt, P. R. Stahl, T. Schlomm, M. Wurlitzer and H. Schluter, *Anticancer Res.*, 2014, **34**, 2255–2261.
246. M. Kriegsmann, R. Casadonte, N. Maurer, C. Stoehr, F. Erlmeier, H. Moch, K. Junker, C. Zgorzelski, W. Weichert, K. Schwamborn, S. O. Deininger, M. Gaida, G. Mechtersheimer, A. Stenzinger, P. Schirmacher, A. Hartmann, J. Kriegsmann and K. Kriegsmann, *J. Cancer*, 2020, **11**, 6081–6089.
247. W. M. Hardesty, M. C. Kelley, D. Mi, R. L. Low and R. M. Caprioli, *J. Proteomics*, 2011, **74**, 1002–1014.
248. A. Ly, A. Buck, B. Balluff, N. Sun, K. Gorzolka, A. Feuchtinger, K. P. Janssen, P. J. Kuppen, C. J. van de Velde, G. Weirich, F. Erlmeier, R. Langer, M. Aubele, H. Zitzelsberger, L. McDonnell, M. Aichler and A. Walch, *Nat. Protoc.*, 2016, **11**, 1428–1443.
249. O. J. R. Gustafsson, L. J. Winderbaum, M. R. Condina, B. A. Boughton, B. R. Hamilton, E. A. B. Undheim, M. Becker and P. Hoffmann, *GigaScience*, 2018, **7**, 1–13.
250. L. A. McDonnell, A. Römpp, B. Balluff, R. M. A. Heeren, J. P. Albar, P. E. Andrén, G. L. Corthals, A. Walch and M. Stoeckli, *Anal. Bioanal. Chem.*, 2015, **407**, 2035–2045.
251. B. Balluff, C. Hopf, T. Porta Siegel, H. I. Grabsch and R. M. A. Heeren, *J. Am. Soc. Mass Spectrom.*, 2021, **32**, 628–635.
252. Y. Ucal, A. Coskun and A. Ozpinar, *Expert Rev. Proteomics*, 2019, **16**, 521–532.
253. A. Bespalov, M. C. Michel and T. Steckler, *Good Research Practice in Non-Clinical Pharmacology and Biomedicine*, Springer Nature, 2020.
254. T. J. Dekker, B. D. Balluff, E. A. Jones, C. D. Schone, M. Schmitt, M. Aubele, J. R. Kroep, V. T. Smit, R. A. Tollenaar, W. E. Mesker, A. Walch and L. A. McDonnell, *J. Proteome Res.*, 2014, **13**, 4730–4738.
255. A. Römpp, J.-P. Both, A. Brunelle, R. M. A. Heeren, O. Laprévote, B. Prideaux, A. Seyer, B. Spengler, M. Stoeckli and D. F. Smith, *Anal. Bioanal. Chem.*, 2015, **407**, 2329–2335.
256. A. Buck, B. Heijs, B. Beine, J. Schepers, A. Cassese, R. M. A. Heeren, L. A. McDonnell, C. Henkel, A. Walch and B. Balluff, *Anal. Bioanal. Chem.*, 2018, **410**, 5969–5980.
257. A. Ly, R. Longuespée, R. Casadonte, P. Wandernoth, K. Schwamborn, C. Bollwein, C. Marsching, K. Kriegsmann, C. Hopf, W. Weichert, J. Kriegsmann, P. Schirmacher, M. Kriegsmann and S.-O. Deininger, *Proteomics: Clin. Appl.*, 2019, **13**, e1800029.

258. P. Kallback, T. Vallianatou, A. Nilsson, R. Shariatgorji, N. Schintu, M. Pereira, F. Barre, H. Wadensten, P. Svenningsson and P. E. Andren, *Anal. Chem.*, 2020, **92**, 14676–14684.
259. L. H. Cazares, S. A. Van Tongeren, J. Costantino, T. Kenny, N. L. Garza, G. Donnelly, D. Lane, R. G. Panchal and S. Bavari, *BMC Microbiol.*, 2015, **15**, 101.
260. A. Dannhorn, S. Ling, S. Powell, E. McCall, G. Maglennon, G. N. Jones, A. J. Pierce, N. Strittmatter, G. Hamm, S. T. Barry, J. Bunch, R. J. A. Goodwin and Z. Takats, *Anal. Chem.*, 2021, **93**, 2767–2775.
261. M. Tuck, L. Blanc, R. Touti, N. H. Patterson, S. Van Nuffel, S. Villette, J.-C. Taveau, A. Römpp, A. Brunelle, S. Lecomte and N. Desbenoit, *Anal. Chem.*, 2021, **93**, 445–477.
262. N. Verbeeck, J. M. Spraggins, M. J. M. Murphy, H. D. Wang, A. Y. Deutch, R. M. Caprioli and R. Van de Plas, *Biochim. Biophys. Acta, Proteins Proteomics*, 2017, **1865**, 967–977.
263. T. W. Bocklitz, A. C. Crecelius, C. Matthaus, N. Tarcea, F. von Eggeling, M. Schmitt, U. S. Schubert and J. Popp, *Anal. Chem.*, 2013, **85**, 10829–10834.
264. J. Kriegsmann, R. Casadonte, K. Kriegsmann, R. Longuespée and M. Kriegsmann, *Pathol., Res. Pract.*, 2018, **214**, 1057–1063.
265. K. Kriegsmann, R. Longuespée, M. Hundemer, C. Zgorzelski, R. Casadonte, K. Schwamborn, W. Weichert, P. Schirmacher, A. Harms, D. Kazdal, J. Leichsenring, A. Stenzinger, A. Warth, M. Fresnais, J. Kriegsmann and M. Kriegsmann, *Proteomics: Clin. Appl.*, 2019, **13**, 1800035.
266. S. T. P. Mezger, A. M. A. Mingels, O. Bekers, R. M. A. Heeren and B. Cillero-Pastor, *Anal. Chem.*, 2021, **93**, 2527–2533.
267. A. M. Race, D. Sutton, G. Hamm, G. Maglennon, J. P. Morton, N. Strittmatter, A. Campbell, O. J. Sansom, Y. Wang, S. T. Barry, Z. Takats, R. J. A. Goodwin and J. Bunch, *Anal. Chem.*, 2021, **93**, 3061–3071.
268. T. Smets, T. De Keyser, T. Tousseyn, E. Waelkens and B. De Moor, *Anal. Chem.*, 2021, **93**, 3452–3460.
269. G. Paglia, A. J. Smith and G. Astarita, *Mass Spectrom. Rev.*, 2021, 1–44.
270. J. Soltwisch, H. Kettling, S. Vens-Cappell, M. Wiegelmann, J. Muthing and K. Dreisewerd, *Science*, 2015, **348**, 211–215.
271. H. X. Ngo and S. Garneau-Tsodikova, *MedChemComm*, 2018, **9**, 757–758.
272. Y. Ingrasciotta, P. M. Cutroneo, I. Marciano, T. Giezen, F. Atzeni and G. Trifiro, *Drug Saf.*, 2018, **41**, 1013–1022.
273. S. Awwad and U. Angkawinitwong, *Pharmaceutics*, 2018, **10**, 83.
274. R. Ait-Belkacem, C. Berenguer, C. Villard, L. Ouafik, D. Figarella-Branger, A. Beck, O. Chinot and D. Lafitte, *MAbs*, 2014, **6**, 1385–1393.

CHAPTER 15

MALDI Mass Spectrometry Imaging in the Forensic Arena

ROBERT BRADSHAW AND SIMONA FRANCESE*

Biomolecular Sciences Research Centre (BMRC), Sheffield Hallam University, Sheffield, UK
*Email: s.francese@shu.ac.uk

15.1 Introduction

While still niche, the application of Matrix Assisted Laser Desorption Ionization – Mass Spectrometry Imaging (MALDI-MSI) in the forensic arena is exciting due to interesting developments which have led or may lead to implementation into casework and new forensic provision. The first example of application of this MSI technique in the forensic domain was shown by Francese's group in 2009 with the visualization of lipids and exogenous contaminants in latent fingermarks.[1] Since then, a number of areas have been explored by MALDI-MSI, with hair analysis being the largest field of forensic application, after fingermarks. This chapter will review such applications by comparing, where appropriate, advantages and disadvantages against gold standard techniques for particular types of forensic evidence. The authors offer their views on potential uptake of MALDI-MSI in additional forensic science applications as well as barriers preventing implementation.

For practical use, to quickly convey the developmental stage or maturity of each of the MALDI-MSI applications described in this chapter, a technology readiness level (TRL) has been assigned. Different TRL scales have been evaluated included from the UK Ministry of Defense, the UK Research Council EPSRC, the chart from the UK House of Commons, Technology and

New Developments in Mass Spectrometry No. 12
MALDI Mass Spectrometry Imaging: From Fundamentals to Spatial Omics
Edited by Tiffany Siegel

Published by the Royal Society of Chemistry, www.rsc.org

Table 15.1 Technology Readiness Level (TRL) scale used to grade the different MALDI-MSI applications to forensics and toxicology covered in this chapter.[2] Italicized text signifies technology/application in its infancy; bold text relates to developments toward deployment in an operational context; the standard font represents the different levels of technology/application viability in an operational context.

TRL	Description
1	*Published papers reporting technology or data potentially relevant to the area being considered*
2	*Published papers reporting technology or data relevant to the area of application being considered*
3	*Initial application of the technology being assessed to samples of forensic relevance in experimental settings*
4	**Focused studies showing feasibility of the technology for the proposed application**
5	**Focused studies, assessing factors such as sensitivity and repeatability for the proposed application**
6	**Critical mass studies showing clear potential for operational use**
7	Pseudo-operational trials, analysis of evidence under pseudo-operational conditions
8	Operational trials on evidence obtained from casework
9	Publication of data from live casework
10	Publication of data showing the application of the technology in casework over several years

Innovation Center's, Science and Technology (NASA), and from the UK Home Office edited Fingermark Visualization Manual (FVM). The authors have selected an adaptation of the FVM TRL scale[2] which is shown in Table 15.1.

While some types of evidence and applications generally fall under the umbrella of "forensic science", the authors have subdivided this chapter into MALDI-MSI applications in forensic science and in forensic toxicology and have indicated the use of the derived knowledge in both fields.

15.2 MALDI-MSI Application in Forensics

15.2.1 Fingermarks

For over a century, fingermarks have been regarded solely as a source of biometric information, at least until the advent of DNA. Several fingermark enhancement techniques (FET) are available to crime scene investigators (CSI) and crime labs to visualize the ridge flow and *minutiae* making a fingermark unique and thus enabling personal identification. However, since 2008, the analytical community has invested significant efforts to support FET with additional imaging techniques, above all mass spectrometry. In the field of fingermarks, still to date, the most published and operationally feasible MSI technique is MALDI-MSI.[3,4] MALDI-MSI works on fingermarks present on any thin (up to 200 microns) and flat surfaces; for any other surfaces, from curved to those large in size, the fingermark can be recovered with a common CSI lifting tape. As a result, MALDI-MSI has been proven to

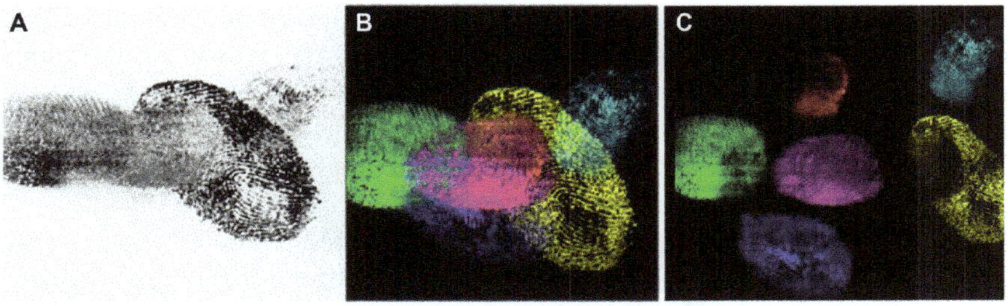

Figure 15.1 Separation of overlapping fingermarks by MALDI-MSI. (A) Optical image of overlapping inked marks. (B) Overlapping MALDI-MS images of *m/z* signals unique to each mark. (C) Separate recalling of the MALDI-MS image showing six different fingermarks.

enable fingermark imaging from glass, varnished wood, leather,[5] plastic,[5–7] paper,[7] polymer banknotes,[8] metallic surfaces,[5] parcel and clear tape,[9] and marks obtained directly from crime scenes (textured plug sockets and the PVC interior side of a window frame).[7] In terms of retrievable physical information, it is possible to improve the ridge pattern continuity by superimposing complementary molecular images[10] as well as separating overlapping fingermarks[11] (Figure 15.1).

While the former capability increases the chance of suspect identification using the National database (if the suspect's fingerprint is on record), the latter may be very useful in the process of "fingermark elimination" and/or to distinguish an assailant's fingermark from the victim's fingermark. To be operationally exploited, these capabilities need to be compatible with the prior application of commonly used FET. For this reason, work has been undertaken over the last decade showing compatibility with a number of FET, namely powders,[1,6,12,13] Vacuum Metal Deposition,[6,14] Cyanoacrylate Fuming,[6,14–16] powder suspension,[7] acid dyes,[3,17,18] ninhydrin,[3,6] as well as with sequential processes'[9,14,16,19] which are advocated to maximize, when appropriate, the amount of biometric information recovered from a fingermark.

As MALDI MS images are generated from the detected ions, much work has also been undertaken to prove the concept of "*criminal chemical profiling*",[20] whereby information concerning the identity of a suspect,[13,21] their lifestyle,[22,23] state of mind[7] and possible "activities" prior to and during committing the crime can be inferred/suggested.[19] This opportunity has been largely demonstrated using MALDI MS in profiling mode, for example for the determination of the sex of an individual through endogenous peptide/protein content in a fingermark, useful in triaging marks at crime scenes,[13] and in the detection of drugs and metabolites[14,24] (the latter achieved also in operational casework[7]).

In forensic toxicology investigations, blood is the predominant routine analytical specimen and the gold standard techniques are gas chromatography (GC) and liquid chromatography (LC) MS/MS. However, since the work of Bailey *et al.* and Groeneveld *et al.* in 2015,[14,23] it is well reported that toxicological information derived from the consumption of drugs can also be obtained from fingermarks by MALDI-MS enabling the provision of evidence on the lifestyle and/or state of mind of the individual. For this reason, the MALDI-MS detection and imaging of excreted drugs and metabolites within fingermarks is covered under the forensic science section of the chapter rather than in the forensic toxicology section.

To date, MALDI-MS has been successfully tested to detect and/or map seven common drugs of abuse and their metabolites in enhanced and not enhanced fingermarks.[14,23,25] Table 15.2 offers a breakdown list of drugs and metabolites which have been profiled and/or imaged, as gathered from the literature.

Although there is no quantitative study reporting detection limits in relation to drug and metabolite excretion, in the work by Bailey *et al.*,[23] benzoylecgonine was detected in a fingermark of a drug rehabilitating patient at a level above the 16 $ngmL^{-1}$ cut-off in the corresponding oral fluid, as determined by complementary GC–MS analysis. Adding the imaging dimension (*i.e.* detecting a drug on a fingermark ridge pattern) provides investigators with associative evidence that may further help informing a case. However, while all the investigated drugs and metabolites could be profiled, not all provided a suitable quality fingermark image.[14] Cannabinoids are such examples due to the poor ionization efficiency in MALDI, as shown by Groeneveld *et al.*[14] A possible solution to combat this issue would be through *in situ* derivatization and this strategy has proven successful for imaging these species in hair, as shown by Beasley *et al.*[26]

Table 15.2 Detection and mapping of drugs and metabolites in fingermarks using MALDI mass spectrometry.

Drug/metabolite	MALDI-MS Profiling	Reference(s)	MALDI-MS Imaging	Reference(s)
• **amphetamine**	x	14	x	14
methamphetamine	x	25	x	14, 25
4-methylmethamphetamine	x			
• **cocaine**	x	7, 14, 24, 25	x	7, 14, 25
ecgonine methyl ester	x	7, 14	x	14
benzoylecgonine	x	14, 24	x	14
norcocaine	x	7		
• **5-dimethoxy-4-bromophenethylamine**	x	14		
• **meta-chloropheny-lpiperazine**	x	14		
• **ecstasy**	x	14		14
methylenedioxyamphetamine	x	14		14
• **heroin**	x	14, 25	x	14, 25
6-acetylmorphine	x	14	x	14
morphine	x	14	x	14
• **Δ 9-tetrahydrocannabinol**	x	14	x	14
11-nor-9-carboxy-THC	x	14	x	14
• **methadone**	x	14	x	Unpublished

Less explored is the detection of pharmaceuticals and metabolites;[15,16] unpublished MALDI-MS profiling data by Francese's group, acquired through MALDI ion mobility separation (IMS)-MS/MS show the possibility to detect propranolol (a prescribed beta blocker) and its main metabolite (4-hydroxypropranolol) in fingermarks, 2 hours following administration. There is no reason why other medications could not be profiled (and imaged) by MALDI-MS and the authors speculate that this is simply an unexplored area.

Of interest to *criminal chemical profiling* is also the proposition by Gorka *et al.*[27] (echoed by Lauzon *et al.*,[28] albeit using Ag-LDI) according to which profiling the chemistry of an unidentifiable fingermark could help to establish the correspondence with an identifiable mark of the same donor at the crime scene. This knowledge would primarily aid in the reconstruction of the dynamics of the crime. However, the additional concept proposed by Gorka *et al.*,[27] that the molecular makeup of a fingermark can act as a biological passport to identify an individual, needs much research to prove that the intra-donor fingermark molecular composition variability allows it.

While MALDI-MS profiling data is important intelligence, establishing a link between biometric information (ridge pattern through the MALDI-MS image) and the *chemical criminal profiling* information would provide additional associative evidence, that can become corroborative through parallel standard investigations. This concept has been experimented particularly with the detection and mapping of exogenous substances such as condom lubricants,[19,29] drugs,[14] and blood.[12,17,18,30–32]

In the case of sexual assaults or rapes, where the perpetrator wears a condom to avoid leaving biological traces, condom lubricant polymers are particularly helpful macromolecules that can be detected in fingermarks, given their generally high ionization efficiency. Francese's group reported on the visualization of these polymers in condom contaminated fingermarks and the extracted profiles showed marked differences in the mixtures of polymers for each condom brand/type investigated.[19] This capability has been adapted to MALDI-MS profiling casework commissioned by a European forensic provider (*Simona Francese, personal communication*). Furthermore, as polymers persist through time, their detection and mapping are possible in older marks too, thus potentially aiding with the late reporting of this kind of crime.

Blood is frequently encountered at the scene of violent crimes and its detection in fingermarks is important to establish not only the dynamics but also the nature of the crime. Francese's group has developed MALDI-MS protocols enabling significant contributions in this area,[17,18,30–32] followed by an Australian research group adding to the demonstration of these capabilities.[12] Bloodstains and blood marks can be analyzed as extracts following swabbing or *in situ*. Surfaces of deposition have included so far aluminum slides,[17,18,30–32] glass slides,[31] fabric,[3] and ceramic tiles;[30] work is in progress to expand the analysis to tape-lifted blood evidence and to additional surfaces of deposition. Intact analysis of heme and hemoglobin is straightforward.[17] However, to increase specificity of the method, a proteomic approach in combination with MALDI-MSP or MSI[3,30–32] has been devised bearing the obvious sensitivity challenge of detecting and mapping blood-specific proteins within one of the protein-richest biological matrices and without protein-depletion or pre-purification steps. In Profiling mode, it was possible to confirm the possibility to detect blood from fresh and very old specimens (9 year-old palm print developed using Acid Black 1[30] and a 37 year-old ninhydrin developed fingermark[3]). Kamanna *et al.*[12] devised a method to differentiate human and non-human hemoglobin; the method developed by Francese's group included additional blood proteins and was pre-validated on a set of 56 mixed blind samples through a combined bottom-up proteomics approach. Such approach enabled the determination of blood provenance, down to animal species.[32] These capabilities have been confirmed through MALDI-MSI in both fresh[17,31] and 4 year-old, developed marks.[18]

Other forensically interesting exogenous substances are explosives. Kaplan *et al.*[16] published a report showing a spot MALDI-MS image in negative-mode of both trinitrotoluene (TNT) at *m/z* 226.0 and of the RDX ion fragment at *m/z* 102.1 using α-cyano-4-hydroxycinnamic acid (CHCA) as a matrix. In 2015, the authors of this chapter had invested some efforts in both the detection and mapping of explosives particularly on the components of Semtex, namely TNT, pentaerythritol tetranitrate (PETN) and the nitramide RDX. Detection of PETN has never been possible whereas low intensity negative-ion mode MS and MS/MS spectra were obtained for TNT and RDX, using 9-aminoacridine (9AA) as a matrix. When imaging experiments were conducted, explosives were spotted at a concentration between 5 and 60 $\mu g\,mL^{-1}$; the lowest detectable concentration was 20 $\mu g\,mL^{-1}$ for TNT (*m/z* 265.0) and 10 $\mu g\,mL^{-1}$ for RDX (*m/z* 337.3), thus making the detection limits in the imaging mode 20 ng and 5 ng, respectively (unpublished data). Very recently, Musah's group published a further report on the application of MALDI-MSI to the touch chemistry of explosives[33] covering TNT, tetryl and picric acid. Here, both profiling and imaging data were shown with these three explosives mapped onto the fingermark ridge pattern using 9AA, in what appears to be, from the mass images, high spatial resolution (and with a pixel size for imaging experiments reported to be between 70 and 80 μm). While very interesting, this application is in its infancy and whether the volatility of explosives will impinge on the ability to detect and map them in non-freshly deposited marks remains to be seen.

In contrast to other MSI techniques, MALDI-MSI has been used in police casework (major crimes).[7] Together with secondary ion mass spectrometry (SIMS) imaging, MALDI-MSI has also been included in the Fingermark Visualization Manual (FVM), edited by the Home Office UK[2] under Category C processes.[†] Currently MALDI-MSI is in the process of being promoted to Category B[b]. Based on instances of operational implementation and on the recognition by the Home Office UK, the application of MALDI-MSI to fingermarks is considered by the authors of this chapter as having the highest

[†]A Category C process indicates *"a process at a developmental stage exhibiting potential [...] for occasional operational use [...] when Category A processes have been exhausted."*

Table 15.3 TLR grading for different areas of MALDI-MSI and MS profiling application to fingermarks based on the TRL scale shown in Table 15.1.

Application of MALDI-MSI to fingermarks	TRL	Application of MALDI-MS profiling to fingermarks	TRL
Visualization of the ridge detail	9	—	
Mapping of drugs and metabolites	8	Detection and identification of drugs and metabolites	8
Mapping of condom lubricants	5	Detection and identification of condom lubricants	8
Mapping of blood	7	Detection of blood	8
Mapping of pharmaceuticals and metabolites	2	Detection and identification of pharmaceuticals and metabolites	2
Mapping of explosives	2	Detection and identification of explosives	2

technology readiness level (TRL). The TRL breakdown into the different applications covered is shown in Table 15.3.

Despite the maturity, wider implementation is slowed down by a combination of a few factors namely: (i) poor knowledge among the end users of the benefit and operational feasibility of MALDI-MSI in fingerprinting; (ii) general lack of funding in forensic science which would enable further investigation of the wider effectiveness and robustness of the technique in a pseudo-operational context; and (iii) potentially damaging (or rejection of) expert witness testimony in court due to lack of accreditation. Breaking these barriers would facilitate and speed up implementation of MALDI-MSI in fingermark analysis.

15.2.2 Biological Tissues (*Post-mortem* Analysis)

The determination of *post-mortem* intervals (PMIs) has always been challenging and is the subject of ongoing technological developments. At least for short PMI, the measurement of the rectal body temperature remains the most used method of PMI determination. However, other methods have been investigated, proposed or employed; they are based on forensic entomology,[34] RNA/DNA degradation[35,36] and thanatochemistry.[37] These methods are largely based on the principle that, upon death, tissues undergo degradation/changes. However, these changes are impacted by the mode of death and by the different environmental conditions to which the body was exposed; environmental factors might accelerate or decrease tissue degradation, thus confounding the determination of the real PMI and rendering these methods inaccurate. The work of Li *et al.*[38,39] on determination of PMIs by a combination of MALDI-MSI and multivariate analysis, applied to human and mouse liver tissue[38] and rat muscle tissue,[39] is very interesting and it represents the first of its kind. In these reports, Li *et al.* demonstrate the opportunity to correctly discriminate between PMIs of 0, 2, 4, and 6 days, exploiting the detection of peptides resulting from protein and peptide degradation. The decreasing ion intensity of markers with increasing PMI was clearly observable in the molecular images of the liver tissues (Figure 15.2) and the rat muscle tissue examined.

Though the data presented are exciting, the authors themselves acknowledge the present limitations due to the challenges of the sample preparation potentially affecting the accuracy of the PMI estimation as well as of a lack of a systematic study including all the necessary variables.

Being the first study in this area, this is considered a demonstrated proof of principle and therefore the determination of PMI through the application of MALDI-MSI to tissues has been assigned a TRL of 4.

15.3 MALDI-MSI Application in Forensic Toxicology

In toxicology, the primary aim is to understand the effects of drug consumption, in particular its biotransformation and distribution to both the target and other tissues. Information is provided at large by Whole Body Autoradiography (WBA) and LC–MS analysis.[40] As an imaging technique, MALDI-MSI was first applied to toxicology studies in 2011[41] and increasingly since then. In addition to the target-free character of the technique, MALDI-MSI also presents the advantage of detecting and discriminating between drugs and their metabolites as opposed to whole body autoradiography (WBA), which only monitors the radioactive probe. Also, differently from LC–MS, there is no requirement for tissue homogenization and preparation, thus reducing sample preparation time, in addition to preserving spatial information. While not quantitatively rivalling its use in biomedicine and biology research, the application of MALDI-MSI to toxicology is considered emerging since it has contributed in many cases to implementable knowledge, from mechanistic understanding to drug toxicity.[41–44] However, the translation of the MALDI-MSI "know how" to *forensic toxicology* is slow, with no apparent justification. The main application of MALDI-MSI in this area is in drug detection, distribution, and quantification in hair which is, therefore, the focus of this section. While in its infancy, this section will also cover the potential for MALDI-MSI application to tissues (including bones), in the context of *post-mortem* analysis from a forensic toxicology perspective. This is expected to be a future expanding area for the application of MALDI-MSI. While blood, urine, and oral fluids are the target specimen for forensic toxicology investigations, hair, tissues, and fingermarks present significant opportunities to provide quicker and spatially resolved intelligence through the application of MALDI-MSI.

15.3.1 MALDI-MSI of Drugs and Metabolites in Hair

Although hair can be considered as forensic evidence, this specific MALDI-MSI application will be discussed in a forensic toxicology context as this sample type has predominantly been used as a matrix to determine drug abuse or pharmaceutical drug use.

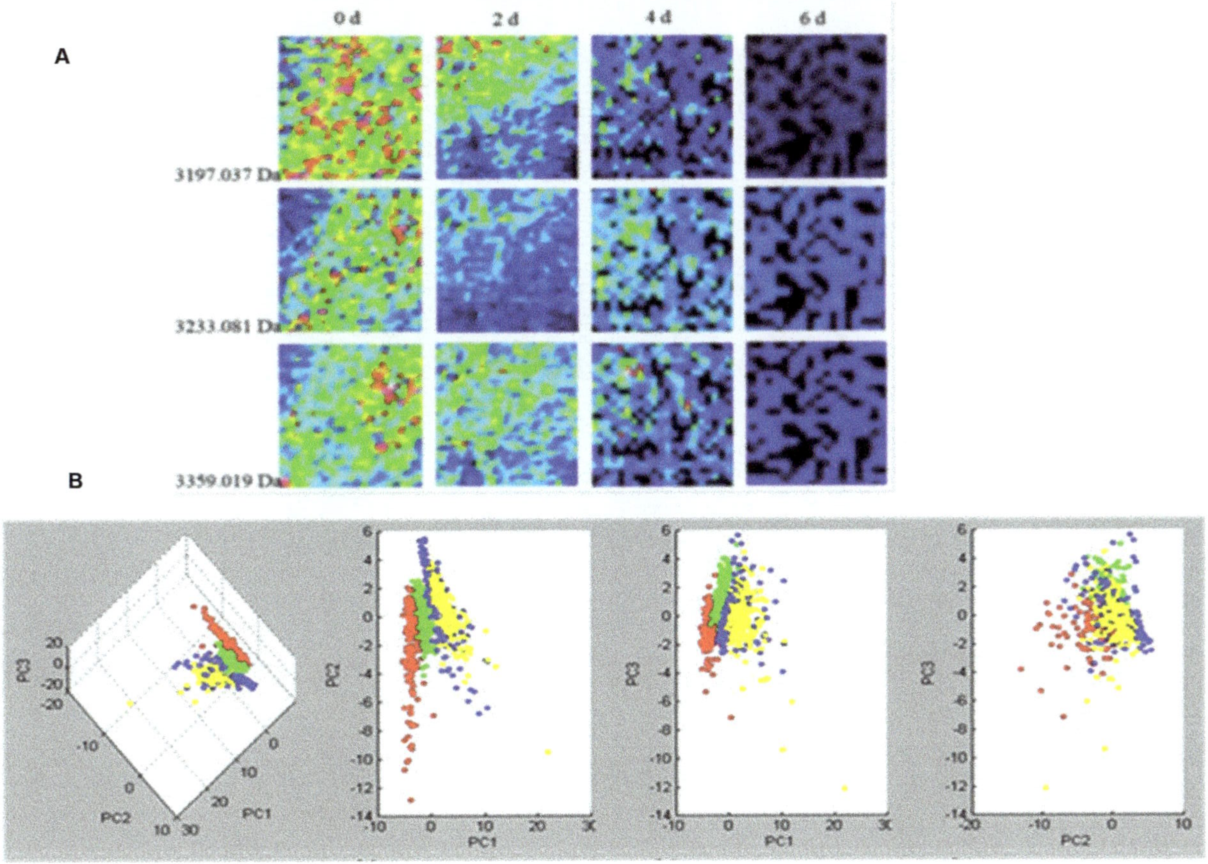

Figure 15.2 Determination of PMIs through a combined approach of MALDI-MSI and PCA. (A) MALDI-MSI of rat liver at 0, 2, 4, and 6-days post-mortem. (B) Principal component analysis (PCA) of MALDI-IMS data acquired in the liver samples at 0 (red), 2 (green), 4 h (blue), and 6 days (yellow) post-mortem in human liver samples.
Reproduced from ref. 38, https://doi.org/10.1038/s41598-017-05216-0, under the terms of the CC BY 4.0 licence, http://creativecommons.org/licenses/by/4.0/.

Hair is particularly useful for the analysis of drugs of abuse (DOA)/pharmaceuticals and their metabolites. These molecules can be incorporated into the hair matrix *via* the bloodstream, where they have increased stability in comparison to other biological matrices (such as blood, urine, *etc.*) and have been shown to allow for detection of these compounds from weeks to years after use.[45] As well as the increased longevity of this forensic evidence, knowledge of the average growth rate of hair can allow for a timeframe at which the molecules have been incorporated into this biological matrix to be determined.[46] Furthermore, from a sample collection perspective, hair collection is much less invasive and can be stored more easily than other sample types which make it ideal as a forensic specimen.

Traditionally, the analysis of drugs in hair has been performed by LC–MS and GC–MS. These approaches are well established, considered to be reliable, and allow quantification of a range of chemical species with high sensitivity, though they do also have some pitfalls:

(*a*) *Spatial resolution:* although segmentation of hair samples can provide a timeline of when a DOA/pharmaceutical had been used, the conventional practice is to cut ~2 cm sections of hair and extract the analytes prior to analysis. Due to the rate at which hair grows (~0.4 mm per day), this could result in large errors (up to 40 days).[47] To obtain higher spatial resolution, smaller sections of hair are required, though inevitably this results in a reduction of analyte concentration and loss of sensitivity;

(*b*) *Sample preparation:* this is complex, requires relatively large amounts of hair (10–100 mg),[48] and laborious steps such as sample homogenization, digestion, and analyte extraction.[45] It may also include sample clean-up (*e.g.*, liquid–liquid or solid phase extraction) and derivatization to make the samples more amenable for analysis.

The first application of MALDI-MS to hair samples was encountered over a decade ago.[49,50] In this initial work, profiling analyses allowed for the rapid detection of cocaine and its metabolites from user hairs. Much like the conventional LC–MS and GC–MS applications, hair segmentation and analyte extraction steps were required prior to analysis, reducing spatial resolution capabilities. Although MSI was attempted, sampling directly from the hair surface was unsuccessful and was attributed to low analyte concentrations and strong binding to the hair matrix.

The first MALDI-MSI applications for the mapping of drugs in hair samples were established in 2011.[5,47] Porta *et al.*[45] visualized the distribution of cocaine and its metabolites (*i.e.* benzoylecgonine, ethylcocaine, and norcocaine) in single hairs derived from chronic users. The use of a high repetition rate laser provided rapid image acquisition times and a MALDI-selected reaction monitoring MS (MALDI-SRM/MS) methodology enabled a greater amount of sensitivity to be obtained in comparison to the work of Vogliardi and co-workers,[49,50] showing a limit of detection (LOD) of 5 ng mg^{-1}. This is within the typical concentration range (0.5–216 ng mg^{-1}) expected from an individual following cocaine abuse.[50] Analyte identification was achieved through the application of MS/MS and MS^3 analyses.

In the same year, Miki *et al.*[47] focused on the detection and visualization of methamphetamine (MA), a powerful central nervous system stimulant. MALDI-SRM/MS allowed for unequivocal detection of MA, with irregular ion intensities observed across the lengths of the hairs, reflecting differing amounts of MA incorporated following repeated administration over time. Typical concentrations of MA within the hairs, determined by complementary LC–MS/MS analyses, were between 15–75 ng mg^{-1}. Although quantification was not achieved *via* MALDI-MSI, it was noted that the signal intensity decreased from the root to the tip of the hair; this was attributed to the loss of analyte over time due to physical processes such as hair washing. Importantly, these novel applications allowed for visualization directly within intact hair samples and removed the analyte extraction step characterizing sample preparation within the traditional so-called hyphenated approaches.

Following these initial studies, MALDI-MSI has since been applied for the detection of numerous analytes contained within hair samples, these include; DOA (cocaine,[45,48,57] methamphetamine,[47] ketamine,[52] and cannabinoids[26,53,54]), pharmaceuticals (zolpidem[55,56] and tilidine[57]), and recreational drugs (nictotine[58]). Analytes which exhibit a characteristically 'poor' MALDI ionization can also be investigated, as shown through the application of relatively simple derivatization procedures to enable the analysis of cannabinoids in hair samples.[26] However inevitably, this protocol increased the number of sample preparation steps required for the analyses of these compounds directly within intact hair samples.

The extensive research conducted in this niche area of science over the past decade has also allowed for important questions/potential problems to be investigated. Firstly, in the context of MALDI-MSI, a potential sensitivity issue due to the morphology of hair was addressed in initial work of Miki *et al.*[47] As molecules of interest are incorporated into hair *via* the bloodstream, they will be embedded into the keratin matrix[51] and bind to melanin in the inner core of the hair,[48] causing potential issues with analyte extraction. Therefore, they performed longitudinal sectioning of hairs by manual blade cutting to allow for the molecules of interest to be exposed, with the aim to improve analyte extraction and sensitivity. The manual cutting approach was selected over freezing/cryosectioning or resin embedding/sectioning to avoid unwanted denaturing, loss of the target analyte (MA), external contamination, and fracturing of hairs (specifically in frozen samples). Different research groups have since performed their own variations on either hair sectioning[48,51,53,56,58,59] or hair 'scraping',[52,54] to expose the analytes of interest. The work of Flinders *et al.*[51] focused on the development of a hair cutting device which provided reproducibility in the longitudinal sectioning of hair and minimal damage to the hair structure. This was confirmed by the application of scanning electron microscopy (SEM).

With regards to analyte quantification, the application of MALDI-MSI has long remained inferior in this capacity in comparison to LC–MS and GC–MS. This is due to MSI being affected by numerous factors, including: (a) sample heterogeneity, creating differences in ionization efficiency across a sample, and (b) the necessity to generate suitable "matrix matched standards" to create a calibration model which relates ion signal intensities to a quantitative value.[60] Different groups have employed the use of a homogenous coating of isotopically labeled standards dissolved within the MALDI matrix to allow for signal normalization.[45,58] Although this methodology does not allow for absolute quantitation to be achieved, relative quantitation is possible based on signal intensities observed across intact hair samples. Further developments by Wang and Wang[59] showed the potential to create a MALDI MS calibration curve for methamphetamine using olanzapine as an internal standard. However, from user hair samples, only semi (or relative) quantitation could be achieved due to an inability to completely extract the target analyte from the hair samples. More recently, Flinders *et al.*[48] developed the closest example toward an absolute quantitative methodology whereby control (blank) hairs were sprayed with a cocaine dilution series (0.02–2.35 ng mm^{-2}) to provide homogenous and reproducible calibration standards within the hair tissue (Figure 15.3).

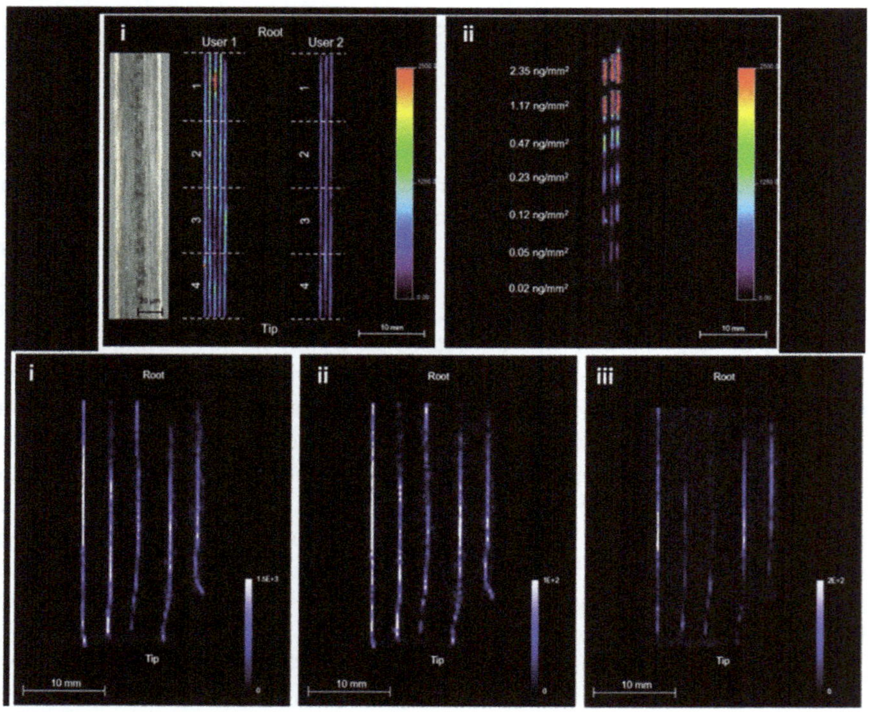

Figure 15.3 MALDI-MSI of hair from cocaine users. Top panel: MALDI MS/MS images of a cocaine product ion (*m/z* 182) in (i) the longitudinally sectioned hair of drug users and (ii) control hair samples sprayed with a cocaine dilution series. Bottom panel: MALDI MS/MS images of cocaine users' longitudinally sectioned hair samples showing (i) cocaine, (ii) benzoylecgonine, and (iii) cocaethylene. (2017).
Reproduced from ref. 48, https://doi.org/10.1007/s13361-017-1766-0, with permission from American Chemical Society, Copyright © The Authors 2017.

Nonetheless, it was noted that in practice, the hair color, ethnicity, and gender of the donor of the control hairs should match the "real" samples to ensure that the endogenous composition and morphology of the hairs are consistent.

Although these examples show a clear increase in knowledge toward more reliable quantification of compounds in hair using MALDI-MSI, there are additional considerations which cannot be overlooked. One issue with quantification from this sample type is the potential for external contamination (such as from sweat, sebum, or handling of drugs) resulting in falsely elevated signals and the generation of false positives. To investigate this issue, Poetsch *et al.*[55] designed an experiment whereby LC–MS segmental hair analysis was performed on hair sampled at 24 hours and 30 days post-administration of a single dose of tilidine (an opioid painkiller). Interestingly, tilidine and its major metabolite were detected in all hair segments analyzed from both timepoints, suggesting that external tilidine contamination had occurred. This is of stark importance in a forensic context as it could be incorrectly indicative of chronic use. Indeed, to reduce contamination-derived false positives, hair washing is routinely incorporated into sample preparation protocols prior to analysis. However, in this study both LC–MS/MS and MALDI-MSI were used to show that washing protocols did not completely remove external contaminants which had been incorporated over time. It was also noted that the extent of hair contamination is affected by hair damage, often seen to be accentuated toward the tips. Indeed, further research into the effect of cosmetic treatments on hair samples by Cuypers *et al.*[46] supports this statement. They showed that peroxide treatment of cocaine users' hairs removed the drug and metabolites from the hair matrix. In fact, a complete removal of these analytes was observed in contaminated hairs following 15 minutes of treatment. This was attributed to the breaking of melanin–cocaine bonds and the degradation of cocaine into its reaction products. Not only do these examples show that quantification from hair samples should be approached with caution, but it also provides overwhelming evidence that forensic toxicological hair evidence should be carefully scrutinized before reaching any final conclusions.

Finally, an important point to note is that many of the previously described studies have focused on single hair analysis. This would be beneficial in terms of speed and reducing sample amount, although it must be noted there are different growth phases of hair. These include the anagen phase (active hair growth with incorporation of substances from the bloodstream), the catagen phase (a transition period lasting 2–3 weeks) and the telogen phase (a period of 2–6 months before hairs fall out).[55] These growth phases affect the distribution of compounds within hair samples and cannot be differentiated in cut hair due to the absence of the root.[45] Due to this circumstance, numerous hairs from the same donor should be analyzed to confirm results,[48] thus extending analysis time. Flinders *et al.*,[48] therefore, optimized a continuous laser firing "raster imaging" methodology, focusing specifically on rastering speed and spatial resolution (minimizing over sampling and maximizing sensitivity). They also investigated the orientation of analysis (horizontal or vertical) and determined that resolution would need to be changed accordingly due to the elliptical shape of the laser spot which affected the number of pixels generated from longitudinally sectioned hairs, though both could be applied successfully.

In conclusion, forensic/toxicological hair analysis by MALDI-MSI is an area that has gained significant research focus over the past 10 years. This novel approach does provide many advantages over traditional hyphenated techniques (LC–MS and GC–MS) such as higher spatial resolution for drug localization to infer on a drug use timeframe, much simpler sample preparation, and rapid analysis times. However, there are still essential aspects which need to be addressed prior to this technique being applied in forensic casework, some of which have been discussed here. Other overarching points that have not been considered within this chapter are (a) differences in hair growth between individuals and (b) differences in the incorporation of specific drugs and metabolites into hairs based on the chemical/physical properties of specific molecules.[45] Indeed, if this technology was to be employed in forensic investigations, we assume that it would need to be used as a complementary technique to the traditionally used approaches to offer additional information in cases. An extensive review for this niche but emerging application of MALDI-MSI has been provided by Flinders *et al.*[61] Hair analysis services using MS-based methodologies (namely, MALDI-MSI and inductively coupled plasma MS (ICP-MS)) are now being offered at companies such as *Hair Diagnostix* (https://www.hairdiagnostix.com), part of the Dutch Screening Group. Here they offer both a forensic screening service but also 'personalized health analytics', providing information on nutrition and the presence of toxins (among other applications). Clearly the technology has now advanced beyond "research-only" and has numerous applications which can benefit both the forensic and the healthcare arenas. For this reason, the authors of this chapter consider this application relatively advanced and assign a TRL of 6.

15.3.2 Potential use of MALDI-MSI of Drugs and Metabolites in Tissues

Acetaminophen (paracetamol) is a common antipyretic and anti-inflammatory drug. However, acetaminophen overdose is also commonly observed in suicide cases. Segzin *et al.*[62] reported the application of MALDI-MSI to determine the distribution of paracetamol and its metabolites in mouse livers, contributing to the understanding of the "aggravated loss of clearance capacity".[63] The study assists the common toxicology practice to screen for adverse effects through histopathological examination. Although to this day, GC or LC MS/MS remain the gold standard for drug detection and quantification in forensic toxicology, MALDI-MSI has the advantage of retaining and conveying spatial information which may inform on the dynamics of drug consumption. While the scope of the work by Segzin *et al.*[62] clearly fits the aims of clinical toxicology, the application of MALDI-MSI to determine drug-related causes of death in *post-mortem* examination of livers (or other organs/tissues) would not be an unlikely leap.

Another interesting application of MALDI-MSI to tissues in the domain of forensic toxicology has been recently reported by Vandenbosch *et al.*[64] The skeletal tissue has been scarcely accessible due to a lack of appropriate protocols for sectioning undecalcified bones. Taking stock from the very few reports on the application of MALDI-MSI to skeletal tissue,[65–68] these authors report the detection and distribution of methadone and its metabolite 2-ethylidene-1,5-dimethyl-3,3-diphenylpyrrolidine (EDDP) in a human clavicle from an individual who died from methadone overdose. The distribution of both methadone and EDDP enabled differentiation of the bone marrow from the surrounding bone tissue that could also clearly be distinguished from the surrounding matrix (Figure 15.4).

This study rides the wave of a renewed interest in forensic toxicology toward bones since it has been shown that they act as storage for certain drugs and metabolites.[69,70] Given this study and the very few reports in the area, mainly on protocol optimization, MALDI-MSI for drug and medication detection in bones is currently rated at a TRL 4. However, this application is still interesting as sample preparation is much less time consuming than for the gold standard techniques and hence, it could provide spatially resolved and more rapid answers on the analytes of interest. Furthermore, this specimen may be the only evidence available to examine bodies in advanced states of decomposition.

15.4 Prospective Applications of MALDI-MSI in the Forensic Arena

In this section, the authors present an overview of potentially interesting/developing forensic areas of application for MALDI-MSI. In some cases, knowledge obtained through techniques other than MALDI-MS is reported as it is relevant or potentially relevant to the area being investigated. In other cases, this section covers applications of MALDI-MS profiling and/or imaging that are currently not considered in the forensic domain but have the potential to be translated and adapted to address forensic queries.

15.4.1 Forensic Entomology

Forensic entomology is the exploitation of insect biology and their development stages to inform criminal investigations.[71] There is a range of information that this discipline can provide, such as time of tissue infestation (in elderly/child and animal neglect/abuse cases), how death occurred in murders/suspect suicides, possibly where, and also if the victim has ingested/been given drugs or medications prior to death, through detecting their incorporation into the insects (forensic entomotoxicology[72]). However, the determination of the PMI remains the most common application of this discipline and it is based on the principle that different insect species colonize a cadaver at different body decomposition stages, thus revealing the time since death. Despite the arrival

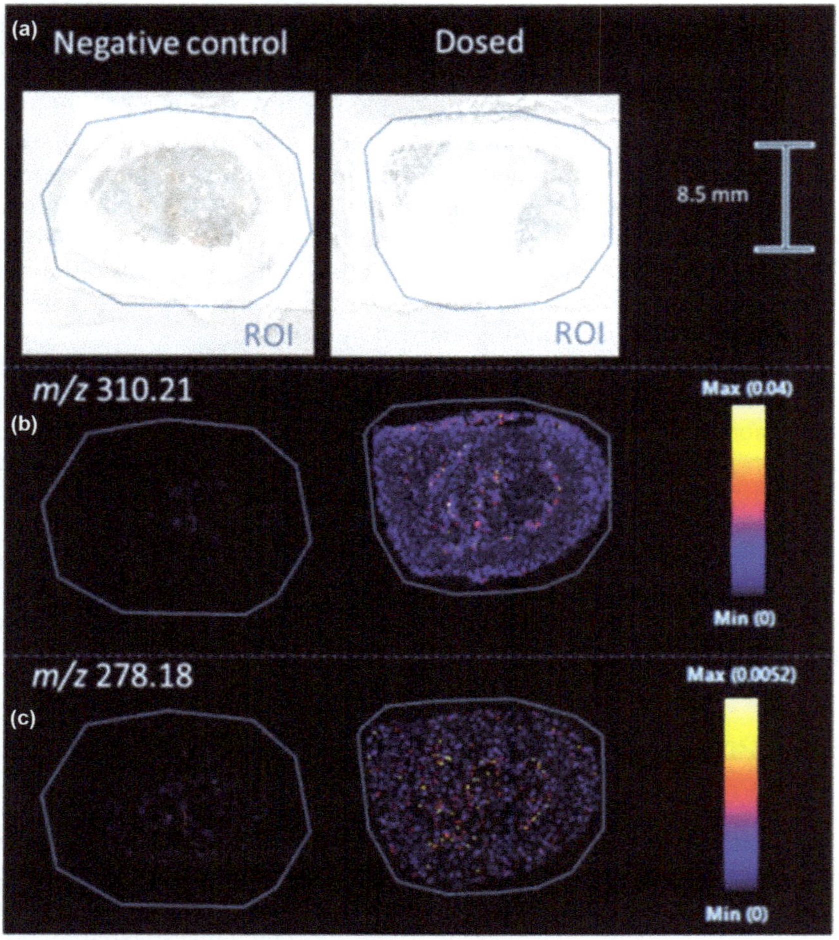

Figure 15.4 MALDI-MS images of distribution images of methadone and EDDP in a human clavicle from a methadone overdose individual. (a) Cross-section of the clavicle with the region of interest (ROI) in blue. (b) Molecular image of methadone (*m*/*z* 310.21). (c) Molecular image of EDDP (*m*/*z* 278.18). The distribution images are total ion current (TIC)-normalized. Reproduced from ref. 64, https://doi.org/10.1007/s00216-020-02920-1, under the terms of the CC BY 4.0 licence, http://creativecommons.org/licenses/by/4.0/.

of the first colonizing insects in a matter of minutes,[71] these estimations are considered more reliable for longer PMI, whereas classic pathology techniques are instead used in the early hours following death.[73] In addition to insect identity, the development stage of a certain recovered population of insects crucially increases the accuracy of PMI estimation.[74] The development stage is determined by observing the physical characteristics of the specimen by Scanning Electron Microscopy (SEM) which is a standard morphological examination technique.

The work of Bhandari *et al.*[75] offers an example of how MALDI-MSI could contribute to forensic entomology. In their report, the detection and visualization of lipids were exploited to identify morphological structures with a higher specificity than histology or immunohistology; while the authors intended to study the biochemistry of the beetle *Paederus riparius,* their work indicates the potential to answer forensic entomology questions. In 2020, Yang *et al.*[76] reported a comprehensive analysis of the application of MSI to map small molecules (endogenous, exogenous, venom and toxins) and larger biomolecules in insects, with 22/31 papers employing MALDI-MS as the imaging technique. Although the number of such papers is "respectable", these reports focus on better understanding of either the physiology and biology of the insects, their social interaction, or of the interaction with plants. More "forensic entomology-specific" MS-based research has been undertaken by Musah's group;[77,78] in the context of determining the PMI based on insect identity, they showed the opportunity to distinguish and identify maggots of different species by the lipid compositions of their eggs[77] and amino acid content of *larvae*.[78] This application is important as it avoids the time-consuming rearing of the maggots to an adult state, which is necessary to enable a reliable distinction of the key morphological characteristics. However, this research did not employ MALDI-MS or imaging but direct analysis in real time–high resolution mass spectrometry (DART–HRMS), either in combination with multivariate statistical analysis or machine learning algorithms, with the latter permitting species discrimination and identification even in mixtures.

Considering the reports of Bhandari *et al.*,[75] Musah's group[77,78] and Yang *et al.*,[76] it is these authors' opinion that, although from a technical point of view, adequate knowledge of MALDI-MSI appears to be available to be exploited in forensic entomology, there is not a real drive (or possibly a need) to expand its application in this area. This is probably due to the much more rapid and less laborious forensic response obtained using either MS profiling methods or routinely employed techniques (SEM, pathology) depending on the forensic query. Based on the above discussion, the authors have assigned TRL 1 to the application of MALDI-MSI in forensic entomology.

15.4.2 Microbial Forensics

Microbial forensics exploits the identification of microorganisms and/or their products to provide intelligence in bioterrorism,[79] biocrimes or in an accidental release, by associating the source and/or perpetrator or a group of individuals.[79,80] The detection of the *botulinum* neurotoxin produced by the bacterium *Clostridium botulinum* and of anthrax (an infection caused by the bacterium *Bacillus anthracis*) represent known instances in which microbial forensics has contributed to aid bioterrorism investigations. Schmedes *et al.*[81] published a comprehensive review of all the areas of investigation in which microbial forensics has contributed including on PMI determination.[82]

Microbial forensic investigations have been classically conducted using genetic techniques which have received a boost with the advent of massive parallel sequencing.[83] However, a number of non-genetic techniques have also emerged and, within this group, MS has certainly been playing its part in microbial characterization. In one remarkable application, Kapono *et al.*[84] combined 3D molecular cartography with MS and microbial libraries to obtain individual, specific chemical and microbial signatures, as well as providing information on individuals' interactions in an office space. Both the chemical and the microbial signatures directly indicated who had touched the various office objects such as computers, cell phones, desk phone, and other. This work would very clearly aid investigators in the reconstruction of crime scene dynamics by characterizing human–environment interaction. Among the MS techniques, MALDI-MS has been prominent. However, bacterial typing by MALDI-MS has been used at large in clinical environments since the landmark papers from Mellmann *et al.*[85] and Seng *et al.*,[86] in 2008 and 2009, respectively. In an application better fitting microbial forensics, MALDI-MS was used to detect and quantify protein toxins from *C. botulinum* and *B. anthracis* and their enzymatic activity.[87] While MALDI-MS profiling of microorganisms and of their toxins has been reported widely in the literature, the imaging of cultures/biofilms proved to be more technically challenging from a sample preparation point of view[88] and explains the lower number of reports in the literature.[89–93] Biofilms can form on different surfaces from hospital environments to a patient's body, and they have been used to understand the chemical interactions between intra- and inter-microorganism communities,[94] as well as with animal and plant tissues.[95,96] Recently, Geier *et al.* combined MALDI-MSI with fluorescence *in situ* hybridization (FISH) microscopy to visualize host–microbe interactions at a micrometric scale.[97] Although these reports provide the *know how* to employ MALDI-MSI in this context, microbial forensics does remain, to date, an unexplored field of application. But is there a need and a rational for the adoption of MALDI-MSI within this discipline? Inspired by a recent review,[98] the authors of this chapter hypothesize that, although clearly challenging and very constrained, MALDI-MSI could be used to image *post-mortem* tissue sections, such as from skin (where the epinecrotic microbial consortia proliferate) and internal organs (the so-called thanatomicrobiome) to inform on the manner of death and even determine PMI. However, the current state of the art of MALDI-MSI applied to microbial forensics only warrants TRL 2.

15.4.3 Forensic Botany

Forensic botany (or plant forensics) is the use of plants, their chemistry, biology, ecology, and environment in criminal investigations. Several parts of a plant can be investigated including the flowers, leaves, seeds, spores, and even the plant microbiome. The most popular example of the use of forensic botany is in placing a suspect at a crime scene through the analysis of pollen or spores (forensic palynology) found on their clothing or footwear.[99] Although less frequent than microbial toxins, plant toxins can also be used in a bioterrorism act and the identification of such threats also fits within the area of forensic botany. The most potent plant toxin in this context is ricin, which is extracted from castor beans (in *Ricinus communis*)[100] and has been listed as a category B agent of potential bioterrorism risk by the Center for Disease Control and Prevention (CDC) [100]. This toxin has already been used in political assassination attempts (through contaminated letters) as well as by the military during World War 2 and beyond.[101] Structural characterization represents the entirety of the work on this toxin[101–104] but MALDI-MS, although providing either a central or assistive role, was only employed in profiling mode. Despite the technical *knowhow*, there is no application of MALDI-MSI in a forensic context for this or for any other plant-derived toxins. While the imaging of a plant's natural chemical compounds could in principle provide the dynamics of the crime and associative evidence (for example if found in an individual's fingermark) or could be exploited in *post-mortem* analysis of tissues, the application of MALDI-MSI in forensic botany remains unexplored; as such, a TRL grading of 1 is assigned.

15.4.4 Food Forensics

Forensic food analysis covers the use of various methods for the determination of food authenticity and provenance.[105] Due to the ability of MALDI-MSI to provide chemical information simultaneously with showing analyte distribution, it is unsurprising that it has been used in the analysis of food within a wide range of scenarios.[106] However, to the authors knowledge, the only example of MALDI-MSI used in food forensics (specifically for traceability) assessed the authenticity of beef using lipid metabolomics.[107] However, it was remarked that it should only be applied as a simple method of authenticity assessment prior to conventional isotope ratio analysis. Although other papers have not focused on food analysis within a forensic domain, there are specific examples which suggest the potential application of MALDI-MSI for this purpose. For example, Goto-Inoue *et al.*[108] have recently used MALDI-MSI for discrimination of wild and farmed red sea bream based on differences in molecular markers. This included molecular classes such as triacylglycerols which were more abundant in farmed fish in comparison to the wild counterparts. We speculate that this application could be particularly useful within food fraud cases, whereby farmed fish may be marketed as the more expensive wild variety. A similar observation can be made about the recent work of Morisasa *et al.*[109] who used MALDI-MSI to assess fish quality based on chemical changes during fish aging, even following freezing. Specifically, they focused on the peptide composition of fish fillets treated with different concentrations of formaldehyde (FA), a powerful protein denaturant which causes fish hardening. They also described the increased commercial use of protease treatment on fish fillets to improve texture (softening) prior to consumption and were able to detect molecular differences following this treatment. We speculate that by using the identified molecular species, forensic information could be gathered on the freshness of fish and even the ability to identify when proteases have been applied to improve fish texture.

Of particular interest are developments in food forensics using other forms of (non-imaging) MS. As an example, rapid evaporative ionization mass spectrometry (REIMS) was combined with multivariate statistical analysis to discriminate white fish species.[110] This approach requires no sample preparation and can provide increased discriminatory power and more rapid screening over genomic profiling. We speculate that similar statistical analysis could be applied to data collected by MALDI-MSI in food forensics to allow for increased sample-type discrimination. Considering this potential, but also accounting for the limited work that has been conducted in this area, we assign a TRL of 3.

15.4.5 Forensic Analysis of Fibers

The analysis of fibers is an important discipline within forensic science, providing associative evidence between an individual and another person/object, based on Locard's exchange principle that "every contact leaves a trace". Therefore, it is important that the approaches used for fiber identification (i) are accurate enough to ensure viability in court, (ii) have a quick turnaround, (iii) are widely available to the forensic community,[111] and (iv) allow for the integrity of the sample to be maintained.

Traditionally, fiber analysis is separated into non-destructive and destructive methodologies. Non-destructive approaches include microscopy (used to examine morphological characteristics and damage to the fibers) and spectroscopy (such as Raman or Fourier-transform infrared (FT-IR)) which can be used to determine the fiber polymer and dye composition.[111] However, spectroscopy is not as specific as MS approaches. Therefore, destructive methodologies such as LC–MS have gained increasing popularity, allowing for improved chemical characterization, though necessitating time-consuming extraction steps prior to analysis.[111,112] Furthermore, extraction protocols can only recover low concentrations of dye from the fibers,[113] requiring sensitivity of the applied analytical techniques.

The first use of a matrix-assisted laser desorption-based approach directly from a fiber surface was performed by Cochran *et al.* (2013) using infrared-matrix assisted laser desorption electrospray ionization (IR-MALD-ESI).[111] Here, the conventional MALDI "matrix" was substituted for water. The specific wavelength of the IR laser irradiation corresponded to the absorption maximum of stretching modes for the hydroxyl group contained within the water (causing resonant excitation), enabling extraction and desorption of dye molecules (such as Acid Black 58) prior to ionization using ESI. Using water ensured that generated spectra were devoid of interfering signals and allowed for sample reuse, which is important in a forensic context. Furthermore, sample preparation/analysis time was reduced significantly (seconds for analysis time), rather than the hour(s) that may be required for LC–MS. Cochran and co-workers transferred this methodology to IR-MALDESI imaging, enabling visualization of dye distribution within single fibers (as small as ~10 µm in diameter and 100 µm in length) which had been recovered by tape lifting (mimicking a real forensic sample recovery scenario).[112] An interesting aspect of this work was the switching of MS polarity modes to allow the analysis of dyes with different ionization characteristics (acidic/basic dyes). In this instance, they used ice as the desorption matrix, which reduced damage to the fibers caused by laser irradiation.

MALDI-MS forensic fiber analysis is clearly a subject area which has had minimal research interest. From a sample preparation perspective, there may be restrictions due to the necessary sample pre-treatment steps (such as matrix deposition) characterizing MALDI-MS based approaches.[114] Certainly, due to the wide variety of fiber compositions and morphologies a consistent matrix deposition approach on "real-world" forensic fibers would prove to be problematic. This would necessitate matching an appropriate matrix to the suspected chemical composition of the fibers, while also choosing a matrix deposition technique which allows maximum extraction of analytes. Furthermore, deposition of a matrix onto the sample may render the approach (semi)-destructive depending on whether the matrix can be removed following analysis. From a forensic perspective, the ability to reuse the sample, or at the very least only destroy a small amount, would be more advantageous. Until these issues are resolved, we assume that MS application to fiber analysis will be limited to approaches which are minimally invasive and require limited sample preparation. Presently, the application of MALDI-MSI to fibers is assigned a TRL of 2.

15.4.6 Ink/Documents

In forensic analysis of ink/documents, there are many pieces of information which may be useful. This can include, but is not limited to, the age of documents/ink, forgery, text manipulation, and the order of ink deposition. Much like the analysis of dyes from fibers, traditional methodologies include microscopy and non-invasive spectroscopic examination, although MS applications have been increasingly used to determine specific molecular mass information. A detailed overview of both traditional and MS-based technologies employed in this area, including different types of MSI, can be found elsewhere.[115]

There are several applications showing the use of MALDI-MS for profiling analysis of inks, a selection of which will be discussed here. In 2005, Chen was able to identify and characterize different ball point pen inks which had been extracted from paper, though it was found that interfering signals were obtained from the MALDI matrix, paper substrate, and also from the TLC plate extracts which were used to separate ink components prior to analysis.[116] Weyermann *et al.*[117] used laser desorption ionization (LDI) and MALDI-MS to attempt estimation of the age of ball point inks through monitoring of dye degradation products over time. However, degradation was affected by environmental factors (such as humidity, heat, and light) and therefore the storage conditions and the initial composition of the ink are required to allow for forensically relevant conclusions to be drawn. Interestingly, it was observed that the best results were obtained by analyzing inks directly from the paper surface rather than from extracts, thus reducing sample preparation and document destructiveness. Dunn and Allison[118] highlighted the suitability of MALDI-MS profiling for polyionic dyes (containing multiple charge groups) which can be particularly problematic to analyze by LDI. They also highlighted the ability to analyze inks directly from the paper surface. Most recently, a minimally-destructive approach was developed by Huynh *et al.*[119] who used direct analyte-probed nano-extraction (DAPNe) to extract analytes of interest directly from paper prior to MALDI MS ink characterization. Much like other solvent-based extraction protocols, the composition of solvent influenced the extraction of different molecules.

Beyond these MALDI-MS publications and despite the abundance of different MSI approaches applied to documentation/ink analysis, to the authors knowledge, there are currently no additional reports of MALDI-MSI application to this type of evidence and it is important to understand the reasons. A pan European round robin study employing MALDI-MSI as one of the investigative techniques for detection of forgery was recently conducted as part of the COST Action CA16101 (https://multiforesee.com) (Fischer *et al.* 2021 in preparation). From the results obtained, it was concluded that other imaging techniques, including LDI as well as spectroscopic methodologies, and also non-imaging techniques, perform much more effectively and quickly in detecting forgery through ink analysis. This observation may mean that the application of MALDI-MSI will never be developed. In the current scenario, a TRL of 1 for MALDI-MSI analysis of documents/ink has been assigned.

15.5 Conclusions

While the application of MALDI-MSI in the forensic arena is not as wide as in other life science fields, there are clearly interesting and emerging areas. This chapter highlights the application to fingermarks and hair as being the most advanced. Indeed, in the former case, there are examples of operational implementation to provide both biometric and criminal profiling intelligence supported by a solid body of knowledge on feasibility, versatility, and robustness. In the latter case, drug imaging in hair by MALDI-MSI, while still challenging in some instances, has evolved into an additional provision which can well address forensic, forensic toxicology, and general health/lifestyle queries. While in its infancy, the application of MALDI-MSI in forensic pathology and forensic toxicology using tissues and bones is gaining interest. Beyond these three areas of application, there are additional disciplines, that could clearly benefit from the application of MALDI-MSI, but they remain unexplored. In some cases, the authors speculate that a lack of

MALDI-MSI research in these areas reflects a low drive due to spatial information not being required, the relative destructiveness of the technique in a forensic context, and faster answers to the forensic query being offered by other techniques.

References

1. R. Wolstenholme, R. Bradshaw, M. R. Clench and S. Francese, *Rapid Commun. Mass Spectrom.*, 2009, **23**, 3031.
2. H. Bandey, V. Bowman, S. Bleay, R. Downham and V. H. Sears *Fingermark Visualisation Manual*, ed. H. Bandey, CAST, Home Office, Sandridge, UK, 2014.
3. S. Francese, *Aust. J. Forensic Sci.*, 2019, **51**, 623.
4. S. M. Bleay, M. J. Bailey, R. S. Croxton and S. Francese, *WIREs Forensic Sci.*, 2020, e1403.
5. L. Ferguson, R. Bradshaw, R. Wolstenholme, M. Clench and S. Francese, *Anal. Chem.*, 2011, **83**, 5585.
6. R. Bradshaw, S. Bleay, R. Wolstenholme, M. R. Clench and S. Francese, *Forensic Sci. Int.*, 2013, **232**, 111–124.
7. R. Bradshaw, N. Denison and S. Francese, *Analyst*, 2017, **142**, 1581.
8. K. Scotcher and R. Bradshaw, *Sci. Rep.*, 2018, **8**, 8765.
9. R. Bradshaw, G. Wilson, N. Denison and S. Francese, *Forensic Sci. Int.*, 2021, **319**, 110643.
10. S. Francese, R. Bradshaw, L. S. Ferguson, R. Wolstenholme, S. Bleay and M. R. Clench, *Analyst*, 2013, **138**, 4215.
11. R. Bradshaw, W. Rao, R. Wolstenholme, M. R. Clench, S. Bleay and S. Francese, *Forensic Sci. Int.*, 2012, **222**, 318.
12. S. Kamanna, J. Henry, N. Voelcker, A. Linacre and K. P. Kirkbride, *Rapid Commun. Mass Spectrom.*, 2017, **31**, 1927.
13. C. Heaton, C. Bury, E. Patel, R. Bradshaw, F. Wulfert, R. M. Heeren, L. Marchant, N. Denison, R. McColm and S. Francese, *Forensic Chem.*, 2020, **20**, 1.
14. G. Groeneveld, M. dePuit, S. Bleay, R. Bradshaw and S. Francese, *Sci. Rep.*, 2015, **5**, 11716.
15. L. Sundar and F. Rowell, *Analyst*, 2014, **139**, 633.
16. K. Kaplan-Sandquist, M. A. LeBeau and M. L. Miller, *Forensic Sci. Int.*, 2014, **235**, 68.
17. R. Bradshaw, S. Bleay, M. R. Clench and S. Francese, *Sci. Justice*, 2014, **54**, 110.
18. M. Witt, K. Kennedy, C. Heaton, G. Langenburg and S. Francese, *Bruker Daltonik*, 2021, application note, MSI-22, https://www.bruker.com/en/products- and-solutions/mass-spectrometry/ms-software/scils-lab/_jcr_content/root/sections/more_information/sectionpar/search_copy_copy_cop.download-asset.pdf/ c74307d7-facb-48a8-b702-76d89d408c2a/1885564-msi-22-imaging-fingerprints-ebook.pdf.

19. R. Bradshaw, R. Wolstenholme, L. S. Ferguson, C. Sammon, K. Mader, E. Claude, R. D. Blackledge, M. R. Clench and S. Francese, *Analyst*, 2013, **138**, 2546.
20. S. Francese, R. Bradshaw and N. Denison, *Analyst*, 2017, **142**, 2518.
21. L. S. Ferguson, F. Wulfert, R. Wolstenholme, J. M. Fonville, M. R. Clench, V. A. Carolan and S. Francese, *Analyst*, 2012, **137**, 4686.
22. P. Hinners, K. C. O'Neill and Y. J. Lee, *Sci. Rep.*, 2018, **8**, 5149.
23. K. C. O'Neill, P. Hinners and Y. J. Lee, *Anal. Methods*, 2020, **12**, 792.
24. M. J. Bailey, R. Bradshaw, S. Francese, T. L. Salter, C. Costa, M. Ismail, R. Webb, I. Bosman, K. Wolff and M. de Puit, *Analyst*, 2015, **140**, 6254.
25. A. Skriba and V. Havlicek, *Eur. J. Mass Spectrom.*, 2018, **24**, 124.
26. E. Beasley, S. Francese and T. Bassindale, *Anal. Chem.*, 2016, **88**, 10328.
27. M. Gorka, M. Augsburger, A. Thomas and A. Bécue, *Forensic Chem.*, 2019, **12**, 99.
28. N. Lauzon and P. Chaurand, *Analyst*, 2018, **143**, 3586.
29. R. Bradshaw, R. Wolstenholme, R. D. Blackledge, M. R. Clench, L. S. Ferguson and S. Francese, *Rapid Commun. Mass Spectrom.*, 2011, **25**, 415.
30. E. Patel, P. Cicatiello, L. Deininger, M. R. Clench, G. Marino, P. Giardina, G. Langenburg, A. West, P. Marshall, V. Sears and S. Francese, *Analyst*, 2015, **141**, 191–198.
31. L. Deininger, E. Patel, M. R. Clench, V. Sears, C. Sammon and S. Francese, *Proteomics*, 2016, **16**, 1707.
32. K. Kennedy, C. Heaton, G. Langenburg, L. Cole, T. Clark, M. R. Clench, V. Sears, M. Sealey, R. McColm and S. Francese, *Sci. Rep.*, 2020, **10**, 17087.
33. C. M. Longo and R. A. Musah, *Forensic Chem.*, 2020, **20**, 100269.
34. K. Brown, A. Thorne and M. Harvey, *Int. J. Legal Med.*, 2015, **129**, 835.
35. J. Hansen, I. Lesnikova, A. M. Funder and J. Banner, *Forensic Sci., Med., Pathol.*, 2014, **10**, 322.
36. J. D. Wells, M. C. Lecheta, M. O. Moura and L. R. LaMotte, *Int. J. Legal Med.*, 2015, **129**, 405.
37. A. Salam, E. A. Shaat, M. Hassan, A. Aziz, A. A. M. Sheta and H. A. S. M. Hussein, *Alex. J. Med.*, 2012, **48**, 335.
38. C. Li, Z. Li, Y. Tuo, D. Ma, Y. Shi, Q. Zhang, X. Zhuo, K. Deng, Y. Chen, Z. Wang and P. Huang, *Sci. Rep.*, 2017, **7**, 4887.
39. C. Li, D. Ma, K. Deng, Y. Chen, P. Huang and Z. Wang, *J. Forensic Sci.*, 2017, **62**, 1345.
40. S. Castellino, M. R. Groseclose and D. Wagner, *Bioanalysis*, 2011, **3**, 2427.
41. O. Karlsson and J. Hanrieder, *Arch. Toxicol.*, 2017, **91**, 2283.
42. A. L. Bruinen, C. van Oevelen, G. B. Eijkel, M. Van Heerden, F. Cuyckens and R. M. Heeren, *J. Am. Soc. Mass Spectrom.*, 2016, **27**, 117.
43. M. R. Groseclose, S. B. Laffan, K. S. Frazier, A. Hughes-Earle and S. Castellino, *J. Am. Soc. Mass Spectrom.*, 2015, **26**, 887.

44. A. Nilsson, B. Forngren, S. Bjurström, R. J. A. Goodwin, E. Basmaci, I. Gustafsson, A. Annas, D. Hellgren, A. Svanhagen, P. E. Andrén and J. Lindberg, *PLoS One*, 2012, **7**, e47353.
45. T. Porta, C. Grivet, T. Kraemer, E. Varesio and G. Hopfgartner, *Anal. Chem.*, 2011, **83**, 4266.
46. E. Cuypers, B. Flinders, I. Bosmanc, K. Lusthof, A. Van Asten, J. Tytgat and R. Heeren, *Forensic Sci. Int.*, 2014, **242**, 103.
47. A. Miki, M. Katagi, N. Shima, H. Kamata, M. Tatsuno, T. Nakanishi, H. Tsuchihashi, T. Takubo and K. Suzuki, *Forensic Toxicol.*, 2011, **29**, 111.
48. B. Flinders, E. Beasley, R. Verlaan, E. Cuypers, S. Francese, T. Bassindale, M. Clench and R. Heeren, *J. Am. Soc. Mass Spectrom.*, 2017, **28**, 2462.
49. S. Vogliardi, D. Favretto, G. Frison, S. Ferrara, R. Seraglia and P. Traldi, *J. Mass Spectrom.*, 2009, **44**(1), 18.
50. S. Vogliardi, D. Favretto, G. Frison, S. Maietti, G. Viel, R. Seraglia, P. Traldi and S. Ferrara, *Anal. Bioanal. Chem.*, 2010, **396**(7), 2435.
51. B. Flinders, E. Cuypers, H. Zeijlemaker, J. Tytgat and R. Heeren, *Drug Test. Anal.*, 2015, **7**, 859.
52. M. Shen, P. Xiang, Y. Shi, H. Pu, H. Yan and B. Shen, *Anal. Bioanal. Chem.*, 2014, **406**, 4611.
53. A. Kernalléguen, C. Enjalbal, J. Alvarez, O. Belgacem, G. Léonetti, D. Lafitte and A. Pélissier-Alicot, *Anal. Chim. Acta*, 2018, **24**(1041), 87.
54. H. Lin, X. Zeng, Q. Wang, Y. Li, B. Sun, Y. Wang and H. Wang, *Forensic Toxicol.*, 2020, **38**, 216.
55. M. Poetzsch, A. Steuer, A. Roemmelt, M. Baumgartner and T. Kraemer, *Anal. Chem.*, 2014, **86**, 11758.
56. N. Shima, K. Sasaki, T. Kamata, S. Matsuta, M. Wada, H. Kakehashi, S. Nakano, H. Kamata, H. Nishioka, T. Sato, H. Tsuchihashi, A. Miki and M. Katagi, *Drug Metab. Dispos.*, 2017, **45**(3), 286.
57. M. Poetzsch, M. Baumgartner, A. Steuer and T. Kraemer, *Drug Test. Anal.*, 2015, **7**, 143.
58. T. Nakanishi, T. Nirasawa and T. Takubo, *J. Anal. Toxicol.*, 2014, **38**, 349.
59. H. Wang and Y. Wang, *Anal. Chim. Acta*, 2017, **975**, 42–51.
60. C. Russo, N. Brickebank, C. Duckett, S. Mellor, S. Rumbelow and M. Clench, *Anal. Chem.*, 2018, **90**, 10031.
61. B. Flinders, T. Bassindale and R. M. A. Heeren, Emerging Technologies for the Analysis of Forensic Traces. in *Advanced Sciences and Technologies for Security Applications*, ed. S. Francese Springer, Cham, 2019.
62. S. Segzin, R. Hassan, S. Zühlke, L. Kuepfer, J. G. Hengstler, M. Spiteller and A. Ghallab, *Arch. Toxicol.*, 2018, **92**, 2963.
63. A. Schenk, A. Ghallab, U. Hofmann, R. Hassan, M. Schwarz, A. Schuppert, L. O. Schwen, A. Braeuning, D. Teutonico, J. G. Hengstler and L. Kuepfer, *Sci. Rep.*, 2017, **7**, 6224.
64. M. Vandenbosch, S. P. Nauta, A. Svirkova, M. Poeze, R. M. A. Heeren, T. Porta Siege, E. Cuypers and M. Marchetti-Deschmann, *Anal. Bioanal. Chem.*, 2020, DOI: 10.1007/s00216-020-02920-1.

65. K. Schaepe, D. R. Bhandari, J. Werner, A. Henss, A. Pirkl, M. Kleine-Boymann, M. Rohnke, S. Wenisch, E. Neumann, J. Janek and B. Spengler, *Anal. Chem.*, 2018, **90**, 8856.
66. E. H. Seeley, K. J. Wilson, T. E. Yankeelov, R. W. Johnson, J. C. Gore, R. M. Caprioli, L. M. Matrisian and J. A. Sterling, *Bone*, 2014, **61**, 208.
67. A. Svirkova, A. Turyanskaya, L. Perneczky, C. Streli and M. Marchetti-Deschmann, *Analyst*, 2018, **143**, 2587.
68. D. C. Muddiman, S. Khodjaniyazova, N. J. Hanne and J. H. Cole, *Anal. Methods*, 2019, **11**, 5929.
69. I. M. McIntyre, C. V. King, M. Boratto and O. H. Drummer, *Drug Monit.*, 2000, **22**, 79.
70. M. Vandenbosch, L. Rooseleers, W. Van Den Bogaert, J. Wuestenbergs, W. Van de Voorde and E. Cuypers, *Forensic Sci. Int.*, 2020, **309**, 110225.
71. I. Joseph, D. G. Mathew, P. Sathyan and G. Vargheese, *J. Forensic Dent. Sci.*, 2011, **3**, 89.
72. V. Kapil and P. Reject, *Entomol. Ornithol. Herpetol.*, 2013, **2**, 1000104.
73. R. Shrestha, T. Kanchan and K. Krishan, in *StatPearls [Internet]*, Treasure Island (FL): StatPearls Publishing, 2020.
74. P. M. Mendonça, J. R. dos Santos-Mallet, R. P. de Mello, L. Gomes and M. M. de Carvalho Queiroz, *Micron*, 2008, **39**, 802.
75. D. R. Bhandari, M. Schott, A. Römpp, A. Vilcinskas and B. Spengler, *Anal. Bioanal. Chem.*, 2015, **407**, 2189.
76. F. Y Yang, W. Y He and M. S You, Current Advances in Mass Spectrometry Imaging for Insect Physiology and Metabolism, in *Pests, Weeds and Diseases in Agricultural Crop and Animal Husbandry Production*, 2020, DOI: 10.5772/intechopen.92584.
77. J. E. Giffen, J. Y. Rosati, C. M. Longo and R. A. Musah, *Anal. Chem.*, 2017, **89**, 7719.
78. S. Beyramysoltan, M. I. Ventura, J. Y. Rosati, J. E. Giffen-Lemieux and R. A. Musah, *Anal. Chem.*, 2020, **92**, 5439.
79. B. Budowle, S. E. Schutzer, A. Einseln, L. C. Kelley, A. C. Walsh, J. A. L. Smith, B. L. Marrone, J. Robertson and J. Campos, *Science*, 2003, **301**, 1852.
80. N. Fierer, C. L. Lauber, N. Zhou, D. McDonald, E. K. Costello and R. Knight, *Proc. Natl. Acad. Sci. U. S. A.*, 2010, **107**, 6477.
81. S. E. Schmedes, A. Sajantila and B. Budowle, *J. Clin. Microbiol.*, 2016, **54**, 1964.
82. J. L. Pechal, T. L. Crippen, M. E. Benbow, A. M. Tarone, S. Dowd and J. Tomberlin, *Int. J. Legal Med.*, 2014, **128**, 193.
83. T. Tucker, M. Marra and J. M. Friedman, *Am. J. Hum. Genet.*, 2009, **85**, 142.
84. C. A. Kapono, J. T. Morton, A. Bouslimani, A. V. Melnik, K. Orlinsky, T. Luzzatto Knaan, N. Garg, Y. Vázquez-Baeza, I. Protsyuk, S. Janssen, Q. Zhu, T. Alexandrov, L. Smarr, R. Knight and P. C. Dorrestein, *Sci. Rep.*, 2018, **8**, 3669.

85. A. Mellmann, J. Cloud, T. Maier, U. Keckevoet, I. Ramminger, P. Iwen, J. Dunn, G. Hall, D. Wilson, P. Lasala, M. Kostrzewa and D. Harmsen, *J. Clin. Microbiol.*, 2008, **46**, 1946.
86. P. Seng, M. Drancourt, F. Gouriet, B. La Scola, P. E. Fournier, J. M. Rolain and D. Raoult, *Clin. Infect. Dis.*, 2009, **15**, 543.
87. R. Kalb, A. E. Boyer and J. R. Barr, *Toxins*, 2015, **7**, 3497.
88. B. Li, T. J. Comi, T. Si, S. J. B. Dunham and J. V. Sweedler, *J. Mass Spectrom.*, 2016, **51**, 1030.
89. E. Esquenazi, Y. L. Yang, J. Watrous, W. H. Gerwick and P. C. Dorrestein, *Nat. Prod. Rep.*, 2009, **26**, 1521.
90. S. Vergeiner, L. Schafferer, H. Haas and T. Müller, *J. Am. Soc. Mass Spectrom.*, 2014, **25**, 1498.
91. T. Hoffmann and P. C. Dorrestein, *J. Am. Soc. Mass Spectrom.*, 2015, **26**, 1959.
92. C. R. Anderton, R. K. Chu, N. Tolić, A. Creissen and L. Paša-Tolić, *J. Am. Soc. Mass Spectrom.*, 2016, **27**, 556.
93. E. U. Brockmann, D. Steil, A. Bauwens, J. Soltwisch and K. Dreisewerd, *Anal. Chem.*, 2019, **91**, 15081.
94. P. C. A. Boya, H. Fernández-Marín, L. C. Mejía, C. Spadafora, P. C. Dorrestein and M. Gutiérrez, *Sci. Rep.*, 2017, **7**, 5604.
95. S. J. B. Dunham, J. F. Ellis, B. Li and J. V. Sweedler, *Acc. Chem. Res.*, 2017, **50**, 96.
96. J. D. Watrous and P. C. Dorrestein, *Nat. Rev. Microbiol.*, 2011, **9**, 683.
97. B. Geier, E. M. Sogin, D. Michellod, M. Janda, M. Kompauer, B. Spengler, N. Dubilier and M. Liebeke, *Nat. Microbiol.*, 2020, **5**, 498.
98. E. V. Spagnolo, C. Stassi, C. Mondello, S. Zerbo, L. Milone and A. Argo, *Leg. Med.*, 2019, **36**, 73.
99. H. Miller Coyle, C. L. Lee, W. Y. Lin, H. C. Lee and T. M. Palmbach, *Croat. Med. J.*, 2005, **46**, 606.
100. G. J. Moran, *Emerg. Med. Clin. N. Am.*, 2002, **20**, 311.
101. S. Worbs, M. Skiba, M. Söderström, M. L. Rapinoja, R. Zeleny, H. Russmann, H. Schimmel, P. Vanninen, S. A. Fredriksson and B. G. Dorne, *Toxins*, 2015, **7**, 4906.
102. C. S. Brinkworth, *Anal. Chem.*, 2010, **82**, 5246.
103. D. Chen, W. A. Bryden and C. Fenselau, *J. Mass Spectrom.*, 2018, **53**, 1013.
104. R. B. Sousa, K. S. C. Lima, C. G. M. Santos, T. C. C. França, E. Nepovimova, K. Kuca, M. R. Dornelas and A. L. S. Lima, *Toxins*, 2019, **11**, 201.
105. M. Aceto, Food Forensics, *Compr. Anal. Chem.*, 2015, **68**, 441.
106. M. Morisasa, T. Sato, K. Kimura, T. Mori and N. Goto-Inoue, *Foods*, 2019, **8**, 633.
107. N. Zaima, N. Goto-Inoue, T. Hayasaka, H. Enomoto and M. Setou, *Anal. Bioanal. Chem.*, 2011, **400**, 1865.
108. N. Goto-Inoue, T. Sato, M. Morisasa, Y. Igarashi and T. Mori, *J. Agric. Food Chem.*, 2019, **67**, 7197.

109. M. Morisasa, K. Kimura, M. Sumida, S. Fukumoto, T. Tamura, R. Takeuchi, T. Mori and N. Goto-Inoue, *Foods*, 2020, **9**, 402.
110. C. Black, O. Chevallier, S. Haughey, J. Balog, S. Stead, S. Pringle, M. Riina, F. Martucci, P. Acutis, M. Morris, D. Nikolopoulos, Z. Takats and C. Elliott, *Metabolomics*, 2017, **13**, 153.
111. K. Cochran, J. Barry, D. Muddiman and D. Hinks, *Anal. Chem.*, 2013, **85**, 2.
112. K. Cochran, J. Barry, G. Robichaud and D. Muddiman, *Anal. Bioanal. Chem.*, 2015, **407**(3), 813.
113. R. Macrae and K. W. Smalldon, *J. Forensic Sci.*, 1979, **24**, 109.
114. F. Van Geenen, M. Franssen, A. Schotman, H. Zuilhof and M. Nielen, *Anal. Chem.*, 2017, **89**, 4031.
115. C. Almeida, D. Sales, F. Tosato, N. Santosa, J. Filho, C. Macrino, F. Pinto, P. Filgueiras and W. Romão, *Microchem. J.*, 2019, **148**, 220.
116. H. Chen, *J. Forensic Sci.*, 2005, **4**, 15.
117. C. Weyermann, D. Kirsch, C. Costa-Vera and B. Spengler, *J. Am. Soc. Mass Spectrom.*, 2006, **17**, 297.
118. J. Dunn and J. Allison, *J. Forensic Sci.*, 2007, **52**(5), 1205.
119. V. Huynh, M. Phelps, T. Golden and G. Verbeck, *Forensic Chem.*, 2017, **2**, 86.

CHAPTER 16

MALDI Mass Spectrometry Imaging for Toxicological Evaluation of Environmental Pollutants

MÉLANIE LAGARRIGUE,*[a,b] MIRA MERDAS[a,b,c] AND CHARLES PINEAU[a,b]

[a] Univ Rennes, Inserm, EHESP, Irset (Institut de recherche en santé, environnement et travail), UMR_S 1085, Rennes Cedex F-35042, France; [b] Protim, Univ Rennes, Rennes Cedex F-35042, France; [c] Technologie Servier, 25/27 rue Eugène Vignat – CS 11749, 45007 Orléans Cedex 1, France
*Email: melanie.lagarrigue@univ-rennes1.fr

16.1 Introduction

Chemical environmental pollutants are substances released into our environment (water, air, soil) as a result of human activities and can have harmful effects on humans and other living organisms.[1] They can come from industry, agriculture, or transports, and are also omnipresent in everyday products (food, cosmetics, cleaning products, drugs, packaging, *etc.*). We are therefore continually exposed through inhalation, ingestion, or through the skin to a wide variety of chemicals that can be classified as pesticides, other organics (persistent organic pollutants (POP), hydrocarbons, and plasticizers), inorganics (metals, carbon-containing inorganics, nutrients) and fibers, and nanomaterials according to the United States Environmental Protection Agency.[2]

New Developments in Mass Spectrometry No. 12
MALDI Mass Spectrometry Imaging: From Fundamentals to Spatial Omics
Edited by Tiffany Siegel

Published by the Royal Society of Chemistry, www.rsc.org

Risk assessment associated with this exposure to chemical pollutants is a major concern for authorities and citizens. The Organization for Economic Co-operation and Development guidelines describe methods that have to be used by governments, academic laboratories, and industry to test the potential effects of chemicals on human health (toxicity) and the environment (eco-toxicity). These internationally validated and regularly updated tests allow the *in vitro* or *in vivo* evaluation of various adverse effects (irritation, neurotoxicity, mutagenicity, carcinogenicity, developmental toxicity, reprotoxicity over one or two generations, *etc.*) after oral, respiratory or dermal administration of the studied molecule.[3] However, the development of new approaches is essential to enrich the panoply of tools available for risk assessment and for a better understanding of the mechanisms of toxicity of environmental pollutants. In this field, matrix-assisted laser desorption ionization-mass spectrometry imaging (MALDI-MSI) is very complementary to standard techniques such as autoradiography, immunohistochemistry, or analysis of biological fluids or tissue homogenates. MALDI-MSI can thus provide answers to essential questions: What are the target organs or tissues of the environmental pollutant and its potential metabolites or degradation products? Which molecular responses are induced by the exposure?

The interest in MALDI-MSI for the study of low molecular weight molecule (*i.e.* $<1000\ g\ mol^{-1}$) distribution has been demonstrated in numerous publications.[4–8] Most of them concern the localization of drugs and their potential metabolites in a pharmaceutical context, however, the same approach can be applied to chemical pollutants.[4] Contrary to autoradiography, MALDI-MSI makes it possible to simultaneously determine the identity and distribution of xenobiotics and their metabolites as well as endogenous substances in biological samples without any labeling. This not only reduces the risks and costs associated with radioactivity, but also allows the simultaneous monitoring of the parent molecule and its metabolites. The characterization of metabolites or degradation products of environmental pollutants is particularly important because they may have different targets and toxicity compared with the parent molecule. Moreover, MALDI-MSI is one of the few imaging techniques that enables the localization of all types of compounds (*e.g.*, small molecules, proteins, peptides, lipids, sugars). This represents a remarkably interesting opportunity to study molecular responses associated with the exposure to environmental pollutants while preserving spatial information, thus perfectly complementing standard techniques based on the analysis of tissue homogenates. However, as with all techniques, MALDI-MSI has some limitations. First, its main limitation is sensitivity.[9–11] Indeed, due to the absence of extraction and separation steps, the sensitivity of MALDI-MSI is often low compared to approaches based on the analysis of tissue homogenates. As a result, doses administered to study the distribution of small molecules by MALDI-MSI are often very high. Although they are often close to doses used in acute toxicity studies (between 5 and 2000 $mg\ kg^{-1}$), they are far from the environmental exposure estimated for daily intakes in the ng–μg kg^{-1} range.[5,12,13] Moreover, the absence of extraction and separation steps limits the identification capabilities of MALDI-MSI to abundant endogenous compounds. This is particularly the case for proteins that are very important for a better understanding of molecular responses associated with the exposure to chemical pollutants.[14]

In this chapter, the sample preparation steps to consider for the study of environmental pollutant distribution will be presented. Several examples of the application of MALDI-MSI for the localization of environmental pollutants in different types of tissue and for the identification of molecular responses associated with the exposure will be described and illustrate the great potential of MALDI-MSI for a better understanding of the mechanisms of toxicity (see Table 16.1). Finally, recent developments that offer interesting possibilities of overcoming the limits of this approach will be discussed.

16.2 Sample Preparation for the Study of Environmental Pollutant Distribution

Sample preparation for MALDI-MSI of environmental pollutants is very similar to that used for any small molecule. Only the steps that are particularly critical and related to pollutants will therefore be discussed here.

16.2.1 Tissue Collection

The harmful effects of chemical pollutants must be studied in humans and other living organisms present in the environment. The distribution of environmental pollutants can therefore be carried out on extremely varied samples (*e.g.* mammalian organs, spheroids, plants, fish, mollusks), each of them requiring appropriate preparation methods.[15–18] For animal samples, standard methods based on snap-freezing in liquid nitrogen and cryo-section can be used. For fragile or very small tissues, inclusion in a medium compatible with MALDI-MSI is generally required. Carboxymethylcellulose (CMC) and gelatin are the most used. Liquid nitrogen can also be replaced by cold-isopentane or dry ice. The preparation of plant tissue sections is more difficult and often requires a significant optimization of the protocol.[18] Indeed, the plant cell walls and the cuticles can strongly hinder the extraction and ionization processes.[19] Moreover, the high water content in plant tissues (leaves, fruits) can make cryosectioning particularly difficult. Therefore, embedding can thus be required to preserve tissue integrity during freezing and cryosectioning. The thickness of the sections can also be increased (*i.e.* up to 20–50 μm) to facilitate the obtention of intact tissue sections. Dry and hard tissues such as seeds can be sectioned at room temperature using a microtome or a vibratome. Washing of tissue sections should be avoided for MALDI-MSI of environmental pollutants. Indeed, this step, which is usually performed before MALDI-MSI of proteins to remove salts and lipids, could lead to the loss of small molecules of interest.

Table 16.1 Application of MALDI-MSI to environmental toxicology.

Ref.	First author	Year	Tissue	Molecule(s) detected by MALDI-MSI	Dose of exposure	Administration method	Matrix	Ion mode	Matrix deposition	Instrument	Lateral resolution (μm)
20	Lagarrigue	2014	Mouse liver	Chlordecone	5 or 8 mg kg^{-1} per day for 1 to 25 days	Gavage	DCTB	Negative	Spraying	MALDI–TOF/TOF	50 or 100
53	Yang	2019	Mouse kidney	PFOS	10 mg kg^{-1} per day for 14 days	Gavage	CHCA	Negative	Spraying	MALDI–TOF	80
54	Mullen	2005	Soya plant	Mesotrione	20 ppm	Spiked hydroponic systems	SA	Negative	Electrospray deposition	MALDI–QqTOF	500
54	Mullen	2005	Soya plant	Azoxystrobin	20 ppm	Spiked hydroponic systems	CHCA	Positive	Electrospray deposition	MALDI–QqTOF	500
55	Annangudi	2015	Wheat leaf	Epoxiconazole, azoxystrobin, and pyraclostrobin	1 μg to 1 mg mL^{-1}	Track sprayer calibrated at 100 gai per ha in 200 L	DHB	Positive	Airbrush	MALDI–TOF/TOF	300
56	Anderson	2009	Sunflower	Nicosulfuron	40 ppm (root uptake) 1.25 mg mL^{-1} (foliar uptake)	Root or foliar uptakes	CHCA	Positive	Airbrush	MALDI-QTOF	200
57	Anderson	2010	Sunflower	Sulfonylurea	1 mg mL^{-1}	Manual spotting on leaves	CHCA	Positive	Airbrush	MALDI–QTOF	500
59	Nayek	2020	lung, brain, kidney, liver, heart, and stomach of rats	Silver nanoparticles	4 h per day (=200 ppb per day) for 5 days	Inhalation	DHB	Positive	Spraying	MALDI-LTQ-Orbitrap	60

17	Pedersen	2020	Quagga mussels	HDPE microplastics	0.1 g L^{-1} for 24 h	Water in the aquarium	DHA or CHCA	Positive	Sublimation + recrystallization	Not precised	Not precised
60	Rivas	2017	Encapsuled polymer	Polycaprolactone diol and transformation products	3–6 g per capsule for 7 days	Polymer capsule in the reactor	DCTB	Positive	Sublimation	MALDI-TOF/TOF	200
16	Xie	2020	Spheroids of breast cancer cells	Bisphenol S	1 pM to 10 μM for 18 days	Culture medium	DAN	Negative	Spraying	MALDI-TOF/TOF	50
43	Zhao	2018	Mouse kidney	Lipids in animals exposed to BPS	10 or 100 μg of PBS kg^{-1} per day for 56 days	Gastric infusion	NEDC	Negative	Spraying	MALDI-TOF	50
66	Fernandes	2018	Green beans	Lipids in plants exposed to chlorpyrifos (CPF)	2 to 8 mg of CPF every 3 days for 45 days	Watering	DHB	Positive	Sublimation	MALDI-TOF/TOF	Not precised
68	Villette	2019	Leaves of Salix alba	Xenobiotics and predicted metabolites	Environmental exposure	Environmental exposure	CHCA	Positive	Spraying	MALDI FT-ICR	50
84	Lin	2018	Mouse kidney	Bisphenols	50 mg kg^{-1}	Gastric infusion	SALDI	Negative	NA	SALDI-TOF	50

16.2.2 Quantification (Optional)

Different quantitative MALDI-MSI methods have been developed and mainly applied to drugs but they can also be transferred for environmental pollutant quantification.[20] In order to obtain a relatively reliable quantification, it is highly recommended to normalize the signal to an internal standard which is typically an isotope-labeled form of the targeted molecule.[8,21,22] This method is the most used to correct signal variations due to irregularities in the matrix layer and to ion suppression effects which can vary greatly, especially in heterogeneous tissues. The internal standard can either be directly introduced into the matrix solution or applied on or under the section before the matrix. The calibration curve can be obtained by simply depositing increasing amounts of the molecule on the slide or on a blank tissue. Another approach (*i.e.* mimetic tissue[23]) is based on different layers of tissue homogenate spiked with increasing amounts of the quantified molecule. A section of the frozen mimetic tissue block is then placed near the tissue to be analyzed. The reader can refer to Chapter 6 of this book for additional information related to quantification.

16.2.3 On-tissue Chemical Derivatization (Optional)

Chemical derivatization is widely used in liquid chromatography coupled with electrospray ionization mass spectrometry (LC–ESI-MS) analysis to enable or improve the detection of compounds that are poorly ionized.[24,25] This technique is thus increasingly used in MALDI-MSI to improve the detection sensitivity of endogenous or exogenous small molecules.[26–28] It could be included in the sample preparation protocol for MALDI-MSI of environmental pollutants to detect the compounds at more relevant doses. On-tissue chemical derivatization (OTCD) enhances the detection sensitivity by adding a positively- or negatively-charged or easily protonated or deprotonated moiety to the molecule of interest. Another advantage of OTCD is the significant increase in molecular mass, which results in a shift out of the mass range where ions from conventional organic matrices are often detected. Finally, as each derivative reagent specifically modifies a particular chemical function, OTCD is particularly interesting to target MALDI-MSI species bearing a given chemical function. Many OTCD protocols are now available to target different chemical functions (primary amine,[29,30] ketone,[31,32] carboxylic acid,[33,34] phenol,[35] thiol,[36] *etc.*) and could thus be applied to a wide range of chemical pollutants.

However, the implementation of chemical derivatization on biological tissue sections presents several constraints compared to derivatization in solution. To preserve the tissue section, the derivatization reaction should preferably be carried out at room temperature with short reaction times and high derivatization efficiency. The concentration of derivatization reagent must also be optimized to find a good compromise between derivatization efficiency and ion suppression effects that could be due to an excess of derivatization reagent. The solution containing the derivatizing reagent is usually sprayed directly on tissue sections before matrix deposition by automated sprayers,[30,37] airbrushes,[38,39] or electrospray.[40,41] Deposition conditions must be controlled to limit the delocalization during the derivatization. Slides or targets can also be pre-coated with the derivatization reagent, which simplifies the sample preparation procedure by allowing the tissues to be directly thaw-mounted and derivatized.[42] Some derivatization reagents are also able to assist the desorption and ionization phenomena during the MALDI process and can thus play the role of reactive matrices.[26,27]

16.2.4 Matrix Deposition

The matrix deposition step must ensure crucial tasks: (i) promote ionization, (ii) provide a good extraction, and (iii) prevent the delocalization of the compounds of interest. The choice of the matrix is therefore crucial and has to be optimized for each molecule. As the chemical structures of environmental pollutants are quite diverse, it may be necessary to test several matrices and to consider unusual matrices such as DCTB or NEDC.[20,27,43] If the environmental pollutant remains undetectable or detected with a very low sensitivity, derivatization may be the solution. Spraying is currently the most common application method and provides a good compromise between extraction and crystal size. However, sublimation can also be used with or without recrystallization (see Table 16.1).

16.2.5 Acquisition

TOF-type analyzers have been widely used in MALDI-MSI for various applications.[44–46] However, high-resolution mass spectrometers are particularly adapted to MALDI-MSI of small molecules because they limit signal overlapping with matrix peaks and other interfering ions that are abundant in the low-mass-to-charge range (*i.e.* $m/z < 1000$). A high mass accuracy (*i.e.* <3 ppm) is also advantageous for the annotation of molecules and facilitates the correlation with eventual LC–MS/MS data acquired with a high mass resolution. Quadrupole-type analyzers benefit from the possibility of performing acquisitions in selected or multiple reaction monitoring modes, which contribute to the improvement of detection sensitivity and specificity.[22,47]

16.3 Application of MALDI-MSI for the Study of Environmental Pollutant Distribution

16.3.1 Quantitative MALDI-MSI of Environmental Pollutants

16.3.1.1 Chlordecone

Extensively used in the French West Indies as an organochlorine insecticide in banana plantations from 1973 to 1993, chlordecone ($C_{10}Cl_{10}O$) was the cause of persistent pollution and contamination of the local population for several decades with effects demonstrated on human health.[48–50] Chlordecone was therefore listed as a POP at the Stockholm convention. It accumulates mainly in the liver and is known to potentiate the action of hepatotoxic agents in rodents.[20] However, the localization of chlordecone in the liver was unknown until 2014. A method of quantitative MALDI-MSI based on the isotopic labeled form of chlordecone was thus developed to study its accumulation in the liver of exposed mice.[20] Chlordecone is a good example of environmental pollutants with a very peculiar chemical structure that requires specific detection conditions. Indeed, due to the high number of chlorine atoms, the ketone group of chlordecone is rapidly hydrated in the presence of water, resulting in the formation of chlordecone hydrate, which is probably the dominant form in tissues. Among the large number of matrices tested, chlordecone hydrate was only detected with DCTB in negative-ion mode. The internal standard was directly added into the DCTB matrix solution and the calibration curve was calculated from increasing amounts of chlordecone manually deposited on a blank tissue. The quantities of chlordecone determined by MALDI-MSI from liver sections appeared to be different from the absolute quantities determined by gas chromatography (GC), but were perfectly correlated ($R^2 = 0.995$) (see Figure 16.1A). This correlation was therefore used to convert the quantities measured by MALDI-MSI into absolute quantities. This example shows that quantitative MALDI-MSI has to be compared to an orthogonal technique. Quantitative MALDI-MSI was then used to determine the accumulation profile of chlordecone in the mouse liver at different exposure times. MALDI-MSI then revealed a particular distribution of chlordecone in the pathological liver of a mouse co-treated with chlordecone and CCl_4 and provided local quantifications in necrotic and healthy areas of the tissue (see Figure 16.1B and C). The sensitivity of this method was sufficient to detect chlordecone at a dose traditionally used in rodent toxicology studies, with the advantage of being able to extract quantitative information at the pixel level.

16.3.1.2 *Perfluorooctanesulfonic Acid (PFOS)*

PFOS is a fluorinated surfactant that has been largely used in coatings, electric and electronic parts, firefighting foam, and packing materials. Due to its extremely high persistence and its substantial bioaccumulating properties, PFOS is listed as a POP. Its production and use are thus forbidden but are still authorized for photo-imaging, semi-conductors and aviation hydraulic fluids for which technically feasible alternatives to PFOS are not available. Different epidemiological studies found that accumulation of PFOS in humans was associated with renal diseases, where the glomerular filtration rate was decreased.[51,52] However, the distribution of PFOS in the kidney was unknown until 2019. Yang *et al.* thus developed a protocol for quantitative MALDI-MSI of PFOS in a mouse kidney.[53] PFOS was only

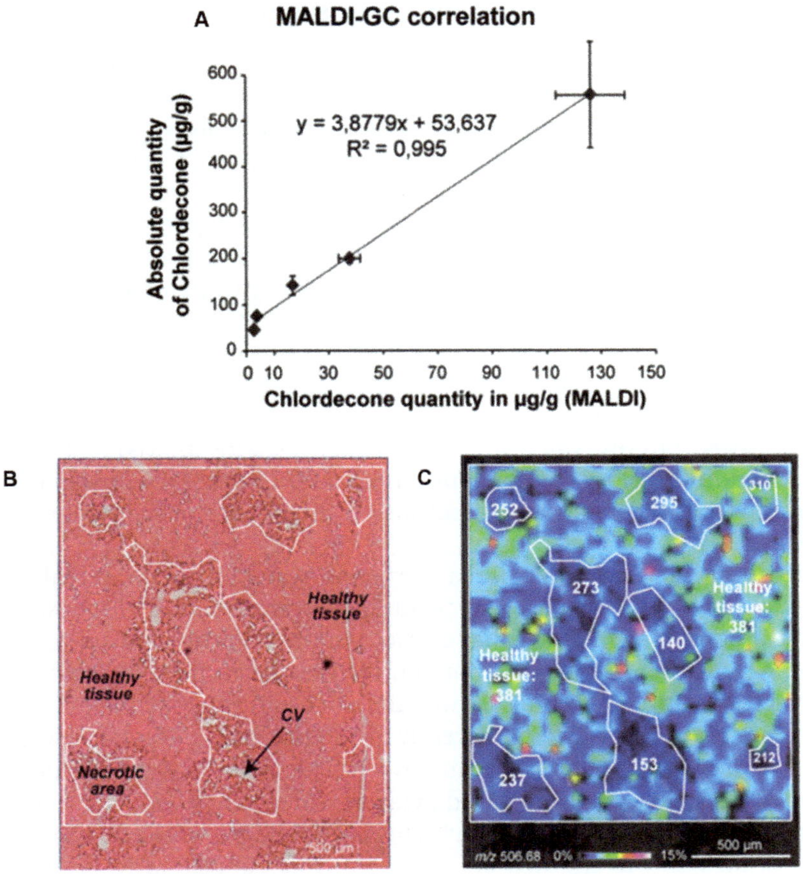

Figure 16.1 *In situ* absolute quantification of chlordecone in a mouse liver. (A) Correlation between quantities of chlordecone determined by MALDI-MSI from tissue sections and absolute quantities determined by GC from tissue homogenates. (B) Microscopic image of a liver section from a mouse co-exposed to chlordecone and CCl4 and stained with H&E. (C) Local absolute quantities measured in necrotic and healthy areas measured by MALDI-MSI on a serial section.

detected in negative-ion mode and the best sensitivity was obtained with the CHCA matrix. Calibration solutions containing increasing concentrations of PFOS and a constant amount of internal standard ($^{13}C_8$-PFOS) were manually deposited on a blank kidney section. A solution composed of PFOS and $^{13}C_8$-PFOS was used as a quality control. Quantitative MALDI-MSI revealed that PFOS was not uniformly distributed in the kidney but mainly located in the cortex region. This information, that would not have been obtained by the quantification of PFOS from tissue homogenates, could explain the glomerular atrophy observed in PFOS-exposed kidneys.

16.3.2 MALDI-MSI of Pesticides in Plants

MALDI-MSI can also be used to study the absorption, distribution, and metabolism of agrochemicals in plants. This information is essential to better understand and control their action. The feasibility of MALDI-MSI to localize herbicides and fungicides in plants was first demonstrated by Mullen *et al.*[54] They were able to detect mesotrione herbicide and azoxystrobin fungicide deposited on the surface of soybean leaves. Other MALDI-MSI experiments were carried out by adding azoxystrobin in the nutrient solution of a soybean plant growing in a hydroponic system in order to localize this fungicide in the soybean stem.[54] This study revealed that azoxystrobin was absorbed from the roots to the stem of the soybean plants. Annangudi *et al.* used MALDI-MSI to study the distribution of three fungicides (*i.e.* epoxiconazole, azoxystrobin, and pyraclostrobin) on the surface of wheat leaves.[55] In this study, all three substances were sprayed on the leaves under conditions close to field application.[55] MALDI-MSI revealed a uniform distribution of the three agrochemicals on the surface of the treated wheat leaves.

Nicosulfuron is a sulfonylurea herbicide widely used in cereal crops. Anderson *et al.* used MALDI-MSI to study the translocation of nicosulfuron in the sunflower stem after foliar or root uptake.[56] For this purpose, MALDI image acquisitions were carried out on sections taken from the root bundle to the growing tip. This study showed that nicosulfuron and one of its phase I metabolites were localized differently depending on the type of absorption. The distribution of four other sulfonylurea herbicides (chlorimuron-ethyl, chlorsulfuron, imazosulfuron, and pyrazosulfuron-ethyl) was also studied in sunflowers after foliar absorption.[57] The amplitude of translocation from the application point to the tips and/or to the roots appeared to be different depending on the herbicide.

16.3.3 MALDI-MSI of Nanoparticles and Plastics

16.3.3.1 Nanoparticles

Nanomaterials are defined by the European commission as a natural, incidental, or manufactured material containing at least 50% of particles with dimensions between 1 and 100 nm (recommendation 2011/696/EU). They are increasingly used in everyday products (cosmetics, food, drugs, construction materials, *etc.*) and raise many questions about their potential risks to human health and the environment, especially because the regulatory framework governing their use is not clearly established. Among them, silver nanoparticles (AgNPs) are used for their antimicrobial properties and for the treatment of cancers.[58] Nayek *et al.* studied the accumulation and the mechanisms of toxicity induced by AgNPs in different organs (lung, brain, kidney, liver, heart, and stomach) of rats exposed to subacute inhalation of AgNPs.[59] Quantitative inductively coupled plasma-mass spectrometry

(ICP-MS) revealed that AgNPs accumulate primarily in the lung, followed by the liver and kidneys. MALDI-MSI experiments carried out on these organs showed the presence of distinct hot spots of silver isotopes. An increase in circulating concentrations of lactate and glutathione disulfide was evidenced by LC–MS in rats exposed to AgNPs, indicating an increase in anaerobic respiration and reactive oxygen species production. These results, combined with the demonstration of bone marrow cytotoxicity, confirm the systemic toxicity of subacute inhalation of AgNPs and illustrate the complementarity of different techniques.

16.3.3.2 Plastics

The impact of the accumulation of microplastics in marine systems on aquatic organisms is widely studied. Although significant concentrations of microplastics are also found in lakes and rivers, little is known about the effects of this exposure on freshwater organisms. Quagga mussels have been used as model freshwater organisms to study the inclusion of high-density polyethylene (HDPE) microplastics (*i.e.* particle size 10–45 μm).[17] MALDI-MSI experiments were conducted on quagga mussels exposed to fluorescent HDPE particles at concentrations ranging from 0 to 0.8 g L^{-1}. HDPE microplastics were not detected directly, but three unassigned ions (*m/z* 475, 1130, and 1396) were co-localized with the HDPE microspheres observed on a fluorescence microscopy image. The *m/z* 475 ion was particularly interesting because it was specifically correlated with tissue lesions near the plastic inclusion. Although these characteristic ions were not identified, they demonstrated that quagga mussels accumulate and retain microplastics after acute exposure and revealed the potential of MALDI-MSI for the localization of microplastic inclusions.

Biodegradation of plastics in aquatic systems is another major concern. Rivas *et al.* studied the biodegradation of encapsulated polycaprolactone diol (PCLD) by bacteria in aerobic and denitrifying reactors at a wastewater treatment plant (WWTP).[60] The biodegradation products of PCLD were first identified by LC–MS/MS. Most of them were then detected by MALDI-MSI to analyze their distribution inside the capsule. Three different degradation processes were identified: two hydrolytic cleavages and one cyclization. MALDI-MSI highlighted the spatial heterogeneity of the degradation at the polymer surface with clear differences between aerobic and denitrifying reservoirs.

16.4 Combination of MALDI-MSI and LC–MS/MS for the Identification of Responses Induced by the Exposure to Environmental Pollutants

The characterization of molecular responses associated with the exposure to environmental pollutants is essential for understanding the biological processes involved in the mechanisms of toxicity and for the identification of toxicity markers.[61] Approaches based on mass spectrometry are particularly effective in determining responses induced at the proteome and metabolome levels. LC–MS/MS analysis of tissue homogenates is the standard method because of its high sensitivity and identification capabilities. Nevertheless, MALDI-MSI offers the possibility of detecting discreet events localized in very specific tissue areas that can be totally missed/dissipated/diluted during the analysis of tissue homogenates. Due to the high complementarity of both techniques, the most promising approach seems to be the combination of LC–MS/MS with MALDI-MSI to provide in-depth analysis on the one hand, and spatial information on the other hand.

16.4.1 Alteration of Energy and Lipid Metabolisms after Exposure to Bisphenol S

Bisphenol A (BPA) is a well-known plasticizer that has been extensively used in daily-use consumer products (*e.g.* plastic bottles, polycarbonates, baby bottles, epoxy resins). BPA has been banned in several countries because of its endocrine disrupting activity. Bisphenol S (BPS) is used as a substitute for BPA and can be found in products labeled "BPA-free". Like several other bisphenols, BPS is unfortunately suspected to have a similar or superior estrogenic activity than BPA.[62] Although it has greater heat stability than BPA, BPS is widely released in human body fluids, including breast milk.[63] Different toxicological studies suggest that BPS could alter the development of mammary glands[63] and promote the proliferation of estrogen receptor α (ERα)-negative normal breast cells (MCF-10A) and ERα-positive breast cancer cells (MCF-7).[64] Xie *et al.* studied the effects of BPS on spheroids of MDA-MB-231 ERα-negative breast cancer cells.[16] Three-dimensional *in vitro* models more accurately reproduce the spatial organization and interactions of tumor tissues with their microenvironment than conventional two-dimensional cell cultures. They are thus increasingly used for anticancer drug screening, but could also be very useful to study the mechanisms of action of environmental pollutants. After showing that BPS exposure led actually to an increase in cell proliferation, Xie *et al.* performed MALDI-MSI experiments to localize BPS in breast cancer cell spheroids (Figure 16.2).[16] They demonstrated that BPS rapidly penetrated in the whole spheroids after 5 min but was then gradually distributed towards the core of the spheroids. This phenomenon was attributed to differential metabolic activities in the different regions of the spheroids. LC–MS/MS-based metabolomics analyses revealed that ATP, ADP, GMP, and AMP were more abundant in spheroids exposed to BPS. MALDI-MSI complemented this information by revealing that these energy-related metabolites were mainly located in the proliferative zone. The authors suggested that BPS exposure significantly increased energy supply in the outer layer of spheroids, which could explain the observed proliferation.

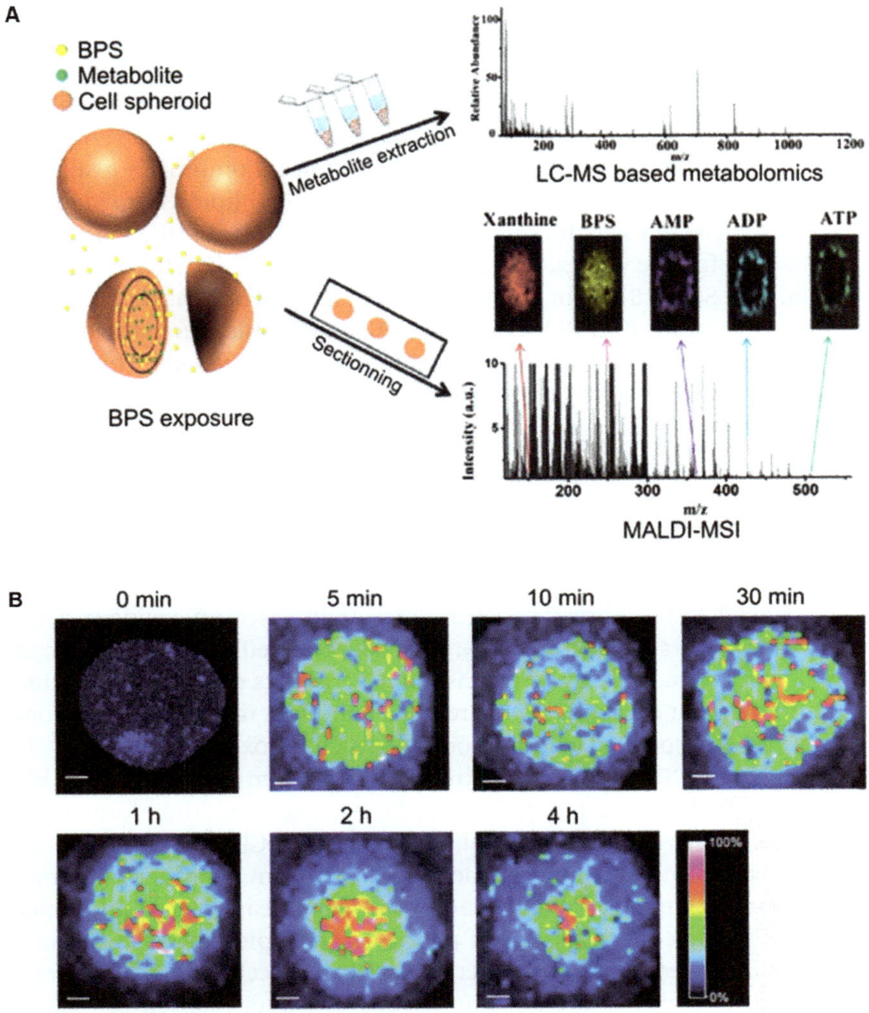

Figure 16.2 Effects of BPS exposure on spheroids of MDA-MB-231 ERα-negative breast cancer cells. (A) Combination of LC–MS/MS analysis and MALDI imaging of energy related metabolites, and (B) MALDI-MSI ion images of BPS at different exposure time in MDA-MB-231 spheroids treated with 10 μM BPS for 0 min, 5 min, 10 min, 30 min, 1 h, 2 h, and 4 h. Scale bar: 200 μm.
Reproduced from ref. 16 with permission from American Chemical Society, Copyright 2020.

Another study was conducted on BPS to better understand its nephrotoxicity. Lipid analyses were performed by LC–MS/MS and MALDI-MSI on kidneys of mice exposed or not to BPS.[62] LC–MS/MS data revealed significant lipidome perturbations in the kidney of mice exposed to BPS. MALDI-MSI experiments highlighted changes in the spatial distribution and the relative abundances of lipids. Spatial segmentation and probabilistic latent semantic analysis revealed three different lipid profiles corresponding to the renal cortex, medulla, and pelvis regions. Further data analysis of MALDI-MSI showed that the variation and accumulation of lipids were stronger in the renal cortex, suggesting that the cortex was more sensitive to BPS exposure than the medulla or pelvis regions. Combining results obtained by LC–MS/MS and MALDI-MSI, five differential lipids were identified as nephrotoxicity markers for the different regions: DAG (18 : 0/22 : 6) and SM (d22 : 0/20 : 4) for the cortex, Cer (d18 : 2/24 : 1) and PC (20 : 0/22 : 6) for the medulla, and So (d18 : 0) for the pelvis. To more precisely assess the tissue heterogeneity induced by BPS exposure, a 3D-model was reconstructed from MALDI images acquired on consecutive mouse kidney sections. This allowed the identification of new substructures in the renal cortex and pelvis presenting differential lipid profiles.

16.4.2 Metabolic Responses in Plants

16.4.2.1 Chlorpyrifos

Chlorpyrifos (CPF) is a broad-spectrum insecticide used on a wide variety of crops (fruits, vegetables, cotton, *etc.*). Because of its neurotoxicity to humans, its use is limited to uninhabited areas, but CPF and its degradation products contaminate water and soil. Exposure to CPF induces toxic effects in aquatic systems and is associated with neurodevelopmental toxicity in infants.[65] The phytotoxicity of CPF is also well studied, but the impact of CPF on lipid metabolism in plants is poorly known. Fernandes *et al.* therefore studied lipidomic changes in green bean plants exposed to CPF.[66] LC–MS/MS analyses were performed on lipid extracts prepared from six different parts of the plant (roots, stem base, stem, leaves, pods, and bean seeds) and revealed a decrease in the concentrations of pigments and lipids involved in photosynthesis in the leaves and a reduction in triacylglycerols and unsaturated phosphatidylcholines in the pods and seeds of plants exposed to CPF (see Figure 16.3). These results were complemented with MALDI-MSI experiments performed on pods and seeds treated with different doses of CPF. The distribution of lipids detected by LC–MS/MS could thus be precisely determined and confirmed the results obtained. The overall results show that treatment with CPF can reduce crop yield and the nutritional value of green beans.

16.4.2.2 Micropollutants

Micropollutants are defined by the European Environment Agency as pollutants which exist in very small traces in water.[67] Different living organisms including plants can thus be exposed to a multitude of micropollutants. However, plants have remarkable capacities of adaptation and defense against various threats that often involve a highly active metabolism.

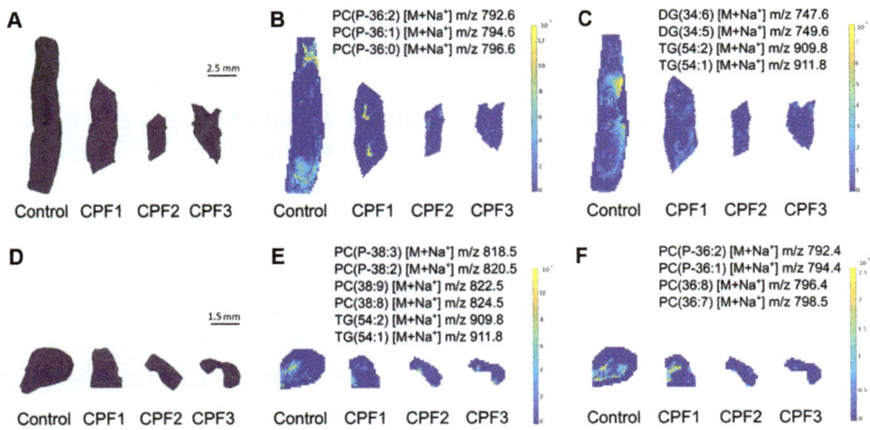

Figure 16.3 Distribution of lipids detected by MALDI-MSI in pods and seeds of green bean plants. (A and D) Optical images of beans analyzed by MALDI-MSI. Spatial distribution of two different components of the MALDI images and the identified compounds of the respective components in (B and C) pods, and in (E and F) seeds. Plants watered with water containing 0 (control), 0.02%, (CPF1) 0.06% (CPF2), and 0.08% (CPF3) of CPF.
Reproduced from ref. 66 with permission from Elsevier, Copyright 2018.

The study of metabolic processes induced by the exposure to a cocktail of micropollutants is a crucial issue to better understand the accumulation and detoxification capabilities of plants, which play an important role in the elimination of xenobiotics from polluted environments. Villette *et al.* combined LC–MS/MS and MALDI-MSI to study the metabolism of the leaves of a white willow (*S. alba*) growing at the exit of a WWTP.[68] Metabolite analysis was performed by LC–MS/MS in surface water, sludge, branches, and leaves. Surprisingly, LC–MS/MS revealed that exogenous metabolites represented more than a third of the identified compounds and corresponded to drugs and their metabolites (21%), pesticides (5%), cosmetics (1%), and other toxic substances from industrial activities (7%). MALDI-MSI experiments were then carried out on leaf sections embedded in CMC and revealed a wide variety of distribution patterns of xenobiotics, ranging from diffuse to very specifically localized in the leaf. The authors found that the identifications obtained in LC–MS/MS and MALDI-MSI were very complementary. Among xenobiotics detected by MALDI-MSI, telmisartan (anti-hypertensive) and betamethasone dipropionate (corticosteroid) were studied in more detail as the parent molecules and several of their first- and second-generation metabolites were observed. Depending on the metabolization pathway, the metabolites were observed in different areas. This study further illustrates the complementarity of MALDI-MSI and LC–MS/MS and demonstrates the feasibility of detecting a cocktail of micropollutants accumulated in plants under natural conditions.

16.5 Conclusions and Future Directions

MALDI-MSI has great potential in improving our understanding of the mechanisms of toxicity of environmental pollutants. Several examples show that the integration of MALDI-MSI and LC–MS/MS data can be particularly powerful by combining spatial information and in-depth analysis.[69–73] The absence of separation prior to ion detection limits the sensitivity and identification capabilities of MALDI-MSI. However, several recent developments offer remarkably interesting possibilities. On-tissue chemical derivatization of small molecules is increasingly used and could be very interesting to significantly improve the detection sensitivity of environmental pollutants and their metabolites or degradation products. It also offers the advantage of targeting the analysis towards a given chemical function. Laser-induced post-ionization (MALDI-2) significantly improves the detection sensitivity of MALDI-MSI and represents a considerable advance in many fields, including environmental toxicology.[10,74,75] Indeed, the combination of OTCD and MALDI-2 could potentially allow the localization of chemical pollutants at doses closer to environmental exposure. It could also offer the possibility of investigating the famous "cocktail effect" defined as the potentiation of the toxicity of a mixture of pollutants in trace quantities. This effect is a major concern in environmental toxicology because it better reflects the reality of our exposure to chemical pollutants but remains extremely challenging to study. Moreover, MALDI-2 in transmission-mode geometry (t-MALDI) can provide images with a spatial resolution lower than 1 μm, giving access to information at the sub-cellular level.[74] The implementation of ion mobility in MALDI-MSI represents another interesting possibility of overcoming the absence of separation often cited to explain the limited sensitivity and identification capabilities of direct analysis by MALDI-MSI. Thus, several examples show that ion mobility improves the sensitivity, specificity, and identification capabilities of MALDI-MSI and allows the differentiation of isobaric or isomeric species.[76,77] To our knowledge, MALDI-MSI has been rarely used to identify protein responses associated with the exposure to environmental pollutants resulting from human activities. This is probably due to the difficulty in identifying proteins detected by MALDI-MSI. Fortunately, several strategies can be used and are still the subject of important developments such as the integration of MALDI-MSI and laser capture micro-dissection or liquid micro-extractions, and the extraction of proteins with hydrogels placed onto the surface of the tissue.[14,72,78–82] Interferences due to the important signals detected in the low-mass range is another problem often mentioned for MALDI-MSI of small molecules. Their impact can be reduced by high-resolution mass spectrometers, but the use of surface-assisted LDI (SALDI) based on nanostructured surfaces composed of porous silicon, carbon, or metal nanoparticles could be another interesting solution.[83] Lin *et al.* thus demonstrated the feasibility of SALDI imaging for the localization of BPS and four other bisphenols (BPA, BPB, BPC, and

BPAF) that were undetectable by MALDI in mouse kidneys.[84] The application of MALDI-MSI in environmental toxicology is still in its infancy but we foresee a bright future with very interesting perspectives.

Abbreviations

AgNPs	Silver nanoparticles
BPA	Bisphenol A (or 4-[2-(4-hydroxyphenyl)propan-2-yl]phenol)
BPB	Bisphenol B (or 4-[2-(4-hydroxyphenyl)butan-2-yl]phenol)
BPC	Bisphenol C (or 4-[2,2-dichloro-1-(4-hydroxyphenyl)ethenyl]-phenol)
BPS	Bisphenol S (or 4-(4-hydroxyphenyl)sulfonylphenol)
BPAF	Bisphenol AF (or 4-[1,1,1,3,3,3-hexafluoro-2-(4-hydroxyphenyl)-propan-2-yl]phenol
Cer	Ceramide
CHCA	α-cyano-4-hydroxycinnamic acid
CMC	Carboxymethylcellulose
CPF	Chlorpyrifos
DAG	Diacylglycerol
DCTB	Trans-2-[3-(4-*tert*-butylphenyl)-2-methyl-2-propenylidene]malononitrile
HDPE	High-density polyethylene
DHB	2,5-dihydroxybenzoic acid
ER	Estrogen receptor
ICP-MS	Inductively coupled plasma-mass spectrometry
NEDC	*N*-(1-naphthyl)-ethylenediamine dihydrochloride
OTCD	On-tissue chemical derivatization
PC	Phosphatidylcholine
PCLD	Polycaprolactone diol
POP	Persistent organic pollutant
PFOS	Perfluorooctanesulfonic acid
SA	Sinapinic acid
SALDI	Surface-assisted laser desorption/ionization
SM	Sphingolipid
So	Sphingosine
WWTP	Wastewater treatment plant

References

1. I. Manisalidis, E. Stavropoulou, A. Stavropoulos and E. Bezirtzoglou, *Front. Public Health*, 2020, **8**, 14.
2. U.S. EPA, *Exposure Factors Handbook 2011 Edition (Final Report)*, U.S. Environmental Protection Agency, Washington, DC, 2011.
3. S. Parasuraman, *J. Pharmacol. Pharmacother.*, 2011, **2**, 74–79.
4. O. Karlsson and J. Hanrieder, *Arch. Toxicol.*, 2017, **91**, 2283–2294.

5. M. Nishidate, M. Hayashi, H. Aikawa, K. Tanaka, N. Nakada, S.-I. Miura, S. Ryu, T. Higashi, Y. Ikarashi, Y. Fujiwara and A. Hamada, *Drug Metab. Pharmacokinet.*, 2019, **34**, 209–216.
6. P. J. Trim and M. F. Snel, *Methods*, 2016, **104**, 127–141.
7. M. Morisasa, T. Sato, K. Kimura, T. Mori and N. Goto-Inoue, *Foods*, 2019, **8**, 633–649.
8. I. Rzagalinski and D. A. Volmer, *Biochim. Biophys. Acta*, 2017, **1865**, 726–739.
9. P. Chaurand, *Bioanalysis*, 2015, **7**, 2279–2281.
10. F. P. Y. Barré, M. R. L. Paine, B. Flinders, A. J. Trevitt, P. D. Kelly, R. Ait-Belkacem, J. P. Garcia, L. B. Creemers, J. Stauber, R. J. Vreeken, B. Cillero-Pastor, S. R. Ellis and R. M. A. Heeren, *Anal. Chem.*, 2019, **91**, 10840–10848.
11. I. Rzagalinski, B. Kovačević, N. Hainz, C. Meier, T. Tschernig and D. A. Volmer, *Anal. Chem.*, 2018, **90**, 12592–12600.
12. OECD, *Test No. 423: Acute Oral toxicity - Acute Toxic Class Method*, OECD, 2002.
13. M. Lu, G. Li, Y. Yang and Y. Yu, *Sci. Total Environ*, 2021, **752**, 142001.
14. D. J. Ryan, J. M. Spraggins and R. M. Caprioli, *Curr. Opin. Chem. Biol.*, 2019, **48**, 64–72.
15. K. A. Nelson, G. J. Daniels, J. W. Fournie and M. J. Hemmer, *J. Biomol. Tech.*, 2013, **24**, 119–127.
16. P. Xie, X. Liang, Y. Song and Z. Cai, *Anal. Chem.*, 2020, **92**, 11341–11348.
17. A. F. Pedersen, K. Gopalakrishnan, A. G. Boegehold, N. J. Peraino, J. A. Westrick and D. R. Kashian, *Environ. Pollut.*, 2020, **260**, 113964.
18. Y. Dong, B. Li, S. Malitsky, I. Rogachev, A. Aharoni, F. Kaftan, A. Svatoš and P. Franceschi, *Front. Plant Sci.*, 2016, **7**, 60.
19. K. Takahashi, T. Kozuka, A. Anegawa, A. Nagatani and T. Mimura, *Plant Cell Physiol.*, 2015, **56**, 1329–1338.
20. M. Lagarrigue, R. Lavigne, E. Tabet, V. Genet, J.-P. Thomé, K. Rondel, B. Guével, L. Multigner, M. Samson and C. Pineau, *Anal. Chem.*, 2014, **86**, 5775–5783.
21. F. Tobias and A. B. Hummon, *J. Proteome Res.*, 2020, **19**, 3620–3630.
22. T. Porta, A. Lesur, E. Varesio and G. Hopfgartner, *Anal. Bioanal. Chem.*, 2015, **407**, 2177–2187.
23. M. R. Groseclose and S. Castellino, *Anal. Chem.*, 2013, **85**, 10099–10106.
24. Y. Zhu, P. Deng and D. Zhong, *Bioanalysis*, 2015, **7**, 2557–2581.
25. T. Santa, *Biomed. Chromatogr.*, 2011, **25**, 1–10.
26. C. Harkin, K. W. Smith, F. L. Cruickshank, C. Logan Mackay, B. Flinders, R. M. A. Heeren, T. Moore, S. Brockbank and D. F. Cobice, *Mass Spectrom. Rev.*, 2021, 1–33.
27. Q. Zhou, A. Fülöp and C. Hopf, *Anal. Bioanal. Chem.*, 2020, 1.
28. M. Merdas, M. Lagarrigue, Q. Vanbellingen, T. Umbdenstock, G. Da Violante and C. Pineau, *J. Mass Spectrom.*, 2021, e4731.
29. S. Toue, Y. Sugiura, A. Kubo, M. Ohmura, S. Karakawa, T. Mizukoshi, J. Yoneda, H. Miyano, Y. Noguchi, T. Kobayashi, Y. Kabe and M. Suematsu, *Proteomics*, 2014, **14**, 810–819.

30. M. Shariatgorji, A. Nilsson, P. Kallback, O. Karlsson, X. Zhang, P. Svenningsson and P. E. Andren, *J. Am. Soc. Mass Spectrom.*, 2015, **26**, 934–939.
31. D. F. Cobice, C. L. Mackay, R. J. Goodwin, A. McBride, P. R. Langridge-Smith, S. P. Webster, B. R. Walker and R. Andrew, *Anal. Chem.*, 2013, **85**, 11576–11584.
32. D. F. Cobice, D. E. Livingstone, C. L. Mackay, R. J. Goodwin, L. B. Smith, B. R. Walker and R. Andrew, *Anal. Chem.*, 2016, **88**, 10362–10367.
33. S.-S. Wang, Y.-J. Wang, J. Zhang, T.-Q. Sun and Y.-L. Guo, *Anal. Chem.*, 2019, **91**, 4070–4076.
34. C. Sun, W. Liu, Y. Geng and X. Wang, *Anal. Chem.*, 2020, **92**, 12126–12131.
35. E. Beasley, S. Francese and T. Bassindale, *Anal. Chem.*, 2016, **88**, 10328–10334.
36. A. Fülöp, T. Bausbacher, S. Rizzo, Q. Zhou, H. Gillandt, C. Hopf and M. Rittner, *Anal. Chem.*, 2020, **92**, 6224–6228.
37. A. Chacon, I. Zagol-Ikapitte, V. Amarnath, M. L. Reyzer, J. A. Oates, R. M. Caprioli and O. Boutaud, *J. Mass Spectrom.*, 2011, **46**, 840–846.
38. Y. Sugiura, E. Takeo, S. Shimma, M. Yokota, T. Higashi, T. Seki, Y. Mizuno, M. Oya, T. Kosaka, M. Omura, T. Nishikawa, M. Suematsu and K. Nishimoto, *Hypertension*, 2018, **72**, 1345–1354.
39. H. Enomoto, T. Sensu, E. Yumoto, T. Yokota and H. Yamane, *Rapid Commun. Mass Spectrom.*, 2018, **32**, 1565–1572.
40. Q. Wu, T. J. Comi, B. Li, S. S. Rubakhin and J. V. Sweedler, *Anal. Chem.*, 2016, **88**, 5988–5995.
41. M. E. Dueñas, E. A. Larson and Y. J. Lee, *Front. Plant Sci.*, 2019, **10**, 860.
42. M. L. Manier, M. L. Reyzer, A. Goh, V. Dartois, L. E. Via, C. E. Barry and R. M. Caprioli, *J. Am. Soc. Mass Spectrom.*, 2011, **22**, 1409–1419.
43. C. Zhao, P. Xie, T. Yong, H. Wang, A. C. K. Chung and Z. Cai, *Anal. Chem.*, 2018, **90**, 3196–3204.
44. R. R. Drake, C. A. West, A. S. Mehta and P. M. Angel, in *Glycobiophysics*, ed. Y. Yamaguchi and K. Kato, Springer Singapore, Singapore, 2018, vol. 1104, pp. 59–76.
45. J. Kriegsmann, M. Kriegsmann and R. Casadonte, *Int. J. Oncol.*, 2015, **46**, 893–906.
46. M. A. Merlos Rodrigo, O. Zitka, S. Krizkova, A. Moulick, V. Adam and R. Kizek, *J. Pharm. Biomed. Anal.*, 2014, **95**, 245–255.
47. S. Khatib-Shahidi, M. Andersson, J. L. Herman, T. A. Gillespie and R. M. Caprioli, *Anal. Chem.*, 2006, **78**, 6448–6456.
48. O. Boucher, M.-N. Simard, G. Muckle, F. Rouget, P. Kadhel, H. Bataille, V. Chajès, R. Dallaire, C. Monfort, J.-P. Thomé, L. Multigner and S. Cordier, *Neurotoxicology*, 2013, **35**, 162–168.
49. R. Dallaire, G. Muckle, F. Rouget, P. Kadhel, H. Bataille, L. Guldner, S. Seurin, V. Chajès, C. Monfort, O. Boucher, J. Pierre Thomé,

S. W. Jacobson, L. Multigner and S. Cordier, *Environ. Res.*, 2012, **118**, 79–85.
50. L. Multigner, J. R. Ndong, A. Giusti, M. Romana, H. Delacroix-Maillard, S. Cordier, B. Jégou, J. P. Thome and P. Blanchet, *J. Clin. Oncol.*, 2010, **28**, 3457–3462.
51. A. Kataria, H. Trachtman, L. Malaga-Dieguez and L. Trasande, *Environ. Health*, 2015, **14**, 89.
52. D. J. Watkins, J. Josson, B. Elston, S. M. Bartell, H.-M. Shin, V. M. Vieira, D. A. Savitz, T. Fletcher and G. A. Wellenius, *Environ. Health Perspect.*, 2013, **121**, 625–630.
53. C. Yang, H. K. Lee, Y. Zhang, L.-L. Jiang, Z.-F. Chen, A. C. K. Chung and Z. Cai, *Anal. Chem.*, 2019, **91**, 8783–8788.
54. A. K. Mullen, M. R. Clench, S. Crosland and K. R. Sharples, *Rapid Commun. Mass Spectrom.*, 2005, **19**, 2507–2516.
55. S. P. Annangudi, K. Myung, C. Avila Adame and J. R. Gilbert, *Environ. Sci. Technol.*, 2015, **49**, 5579–5583.
56. D. M. Anderson, V. A. Carolan, S. Crosland, K. R. Sharples and M. R. Clench, *Rapid Commun. Mass Spectrom.*, 2009, **23**, 1321–1327.
57. D. M. G. Anderson, V. A. Carolan, S. Crosland, K. R. Sharples and M. R. Clench, *Rapid Commun. Mass Spectrom.*, 2010, **24**, 3309–3319.
58. P. Malik and T. K. Mukherjee, *Int. J. Pharm.*, 2018, **553**, 483–509.
59. S. Nayek, I. W. De Silva, R. Aguilar, A. K. Lund and G. F. Verbeck, *J. Appl. Toxicol.*, 2020, 1–15.
60. D. Rivas, B. Zonja, P. Eichhorn, A. Ginebreda, S. Pérez and D. Barceló, *Anal. Bioanal. Chem.*, 2017, **409**, 5401–5411.
61. M. G. Lionetto, R. Caricato and M. E. Giordano, *TOBIOMJ*, 2019, **9**, 1–9.
62. D. Chen, K. Kannan, H. Tan, Z. Zheng, Y.-L. Feng, Y. Wu and M. Widelka, *Environ. Sci. Technol.*, 2016, **50**, 5438–5453.
63. Y. Deceuninck, E. Bichon, P. Marchand, C.-Y. Boquien, A. Legrand, C. Boscher, J. P. Antignac and B. Le Bizec, *Anal. Bioanal. Chem.*, 2015, **407**, 2485–2497.
64. D. K. Tucker, S. Hayes Bouknight, S. S. Brar, G. E. Kissling and S. E. Fenton, *Environ. Health Perspect.*, 2018, **126**, 087003.
65. R. D. Burke, S. W. Todd, E. Lumsden, R. J. Mullins, J. Mamczarz, W. P. Fawcett, R. P. Gullapalli, W. R. Randall, E. F. R. Pereira and E. X. Albuquerque, *J. Neurochem.*, 2017, **142**, 162–177.
66. C. Fernandes, E. Figueira, R. Tauler and C. Bedia, *Ecotoxicol. Environ. Saf.*, 2018, **156**, 25–33.
67. P. H. Collin, *Dictionary of Ecology and The Environment*, P. Collin, Middlesex, Teddington, 3rd edn, 1995.
68. C. Villette, L. Maurer, A. Wanko and D. Heintz, *Metabolomics*, 2019, **15**, 122.
69. F. Dewez, J. Oejten, C. Henkel, R. Hebeler, H. Neuweger, E. De Pauw, R. M. A. Heeren and B. Balluff, *Proteomics*, 2020, **20**, 1900369.
70. F. Hoffmann, C. Umbreit, T. Krüger, D. Pelzel, G. Ernst, O. Kniemeyer, O. Guntinas-Lichius, A. Berndt and F. Eggeling, *Proteomics: Clin. Appl.*, 2019, **13**, 1700173.

71. Š. Strnad, V. Vrkoslav, Z. Klimšová, J. Zemenová, J. Cvačka, L. Maletínská and D. Sýkora, *Bioanalysis*, 2018, **10**, 71–81.
72. J. Quanico, J. Franck, M. Wisztorski, M. Salzet and I. Fournier, *Methods Mol. Biol.*, 2017, **1598**, 269–283.
73. M. Irie, Y. Fujimura, M. Yamato, D. Miura and H. Wariishi, *Metabolomics*, 2014, **10**, 473–483.
74. M. Niehaus, J. Soltwisch, M. E. Belov and K. Dreisewerd, *Nat. Methods*, 2019, **16**, 925–931.
75. E. C. Spivey, J. C. McMillen, D. J. Ryan, J. M. Spraggins and R. M. Caprioli, *J. Mass Spectrom.*, 2019, **54**, 366–370.
76. E. S. Rivera, K. V. Djambazova, E. K. Neumann, R. M. Caprioli and J. M. Spraggins, *J. Mass Spectrom.*, 2020, **55**, e4614.
77. P. O. Helmer, I. D. Nordhorn, A. Korf, A. Behrens, R. Buchholz, F. Zubeil, U. Karst and H. Hayen, *Anal. Chem.*, 2021, **93**, 2135–2143.
78. D. G. Rizzo, B. M. Prentice, J. L. Moore, J. L. Norris and R. M. Caprioli, *Anal. Chem.*, 2017, **89**, 2948–2955.
79. M. Dilillo, R. Ait-Belkacem, C. Esteve, D. Pellegrini, S. Nicolardi, M. Costa, E. Vannini, E. L. de Graaf, M. Caleo and L. A. McDonnell, *Sci. Rep.*, 2017, **7**, 603.
80. M. Wisztorski, A. Desmons, J. Quanico, B. Fatou, J.-P. Gimeno, J. Franck, M. Salzet and I. Fournier, *Proteomics*, 2016, **16**, 1622–1632.
81. D. J. Ryan, D. Nei, B. M. Prentice, K. L. Rose, R. M. Caprioli and J. M. Spraggins, *Rapid Commun. Mass Spectrom.*, 2018, **32**, 442–450.
82. L. Lamont, M. Baumert, N. Ogrinc Potočnik, M. Allen, R. Vreeken, R. M. A. Heeren and T. Porta, *Anal. Chem.*, 2017, **89**, 11143–11150.
83. H. He, Z. Guo, Y. Wen, S. Xu and Z. Liu, *Anal. Chim. Acta*, 2019, **1090**, 1–22.
84. Z. Lin, J. Wu, Y. Dong, P. Xie, Y. Zhang and Z. Cai, *Anal. Chem.*, 2018, **90**, 10872–10880.

CHAPTER 17

MALDI Mass Spectrometry Imaging in the Clinical Landscape

KRISTINA SCHWAMBORN

Institute of Pathology, Technical University of Munich (TUM), Munich, Germany
Email: kschwamborn@tum.de

17.1 Introduction

Pathology is a cornerstone in modern medicine and has always been central in the fight against diseases.[1] According to the Oxford Dictionary, pathology is the science of effects and causes of diseases.[2] It also represents a subspecialty of medicine that deals with the laboratory examination of samples of body tissues for diagnostic purposes.[2] Therefore, it underpins every aspect of patient care, from diagnostic testing, quality control, and treatment advice to the use of cutting-edge genetic technologies and the prevention of disease.[3] Since pathologists are positioned at the hub of laboratory testing they intersect and interact with all major medical specialties.[1] Especially in cancer diagnostics, the analysis of cytology, biopsy, or resection specimens represent the final step in the diagnostic process and has therefore remained the linchpin for over 100 years.[4]

Historically, gross (macroscopic) examination of tissues as well as light microscopic evaluation and interpretation of changes in hematoxylin and eosin (H&E) stained tissue sections has substantially contributed to the understanding of disease, thus replacing the myth and superstition of traditional medicine.[5] Since this approach, in particular the morphologic

New Developments in Mass Spectrometry No. 12
MALDI Mass Spectrometry Imaging: From Fundamentals to Spatial Omics
Edited by Tiffany Siegel

Published by the Royal Society of Chemistry, www.rsc.org

analysis of H&E stained tissue sections by a skilled pathologist possesses so much power and represents a windfall in terms of data quality, quantity, and cost compared to other techniques, it will remain invaluable and irreplaceable.[6] The old saying that "a picture is worth a thousand words" will doubtlessly remain true in pathology.

The growing understanding of the molecular basis of diseases has led to the transformation of health care and heralded the area of personalized medicine. Also, the field of pathology has seen monumental changes in recent years.[1] Ancillary testing beyond the aforementioned evaluation of H&E stained sections has become part of the daily routine. Immunohistochemistry is expanding and enables more precise diagnoses as well as information on treatment and prognosis.[7] Many additional techniques such as microsatellite instability testing, clonality analysis, fluorescence/chromogen *in situ* hybridization, and conventional and next-generation sequencing (NGS) have been incorporated in the pathologist's toolbox, leading toward molecular histology.[8,9] The underlying biochemistry of tissues and organs is supposed to be revealed, while additional information on how therapeutics or toxins influence the function or misfunction of an organ is provided.[3,10] The pathology report can be seen as a knowledge service intended to guide the physician in finding the optimal treatment for a given patient.

These factors have led to a continuous increase in complexity and extensiveness of diagnostic pathology work-up. The changes in tissue preparation guidelines for lung cancer surgical specimens within the last decades illustrate this.[11] Whereas simple histochemistry stains such as H&E and periodic acid-Schiff stain were sufficient in the 1960s, the modern pathology work-up requires a plethora of additional testing in order to correctly subtype lung cancer. Therefore, a number of immunohistochemical stains and often additional molecular analyses have to be performed. This increase is reflected by a more than 60% increase in the number of tissue blocks per case and slides per tissue block over the last decade (*e.g.*, immunohistochemistry (IHC), and PCR).[11,12] Depending on the type of cancer, different actionable mutations have been identified and are routinely tested for in pathology sample work-ups. For example, for breast cancer, estrogen, progesterone, and human epidermal growth factor receptor 2 (HER2) are routinely tested for by IHC. For lung cancer, programmed cell death ligand 1 (PD-L1), EGFR, BRAF, ALK, ROS1, KRAS, NTRK, MET exon 14, RET, and HER2 are tested for by IHC and NGS.[1,13]

Although extraction-based molecular diagnostics such as NGS are molecular tests, the first step is selecting the appropriate tissue sample for testing. Thus, this step relies on the evaluation of the sample by a trained pathologist to ensure sample adequacy and avoid necrotic, degraded, or poorly preserved tissue areas.[7] The correct annotation of tumor areas for tumor cell enrichment together with estimation of the tumor cell percentage is a crucial step, and inadequate sampling can result in a false negative test result due to insufficient tumor DNA yield and ultimately in the wrong therapy.[14] This is also true for other tissue-based analysis, including proteomics and matrix-assisted laser desorption/ionization mass spectrometry imaging (MALDI-MSI).[3]

Although molecular testing has become essential for the classification of central nervous tumors,[15] most other tumors are still classified based on morphology even though the National Cancer Institute sought to develop new classifications based on molecular parameters: the Cancer Genome Atlas Project.[16] However, the genetic complexity of cancer stunned the researchers involved and made it difficult to identify the genetic changes that actually cause disease.[17,18] On the one hand, many cancer-driver genes could be identified. On the other hand, tumor complexity became more evident, since mutations varied tremendously in their allelic frequencies within the same tumor.[19]

Thus, morphology and with it, pathology, will remain at the center and continue to play a crucial role in modern medicine. However, multiple studies have shown that growing demand due to population growth in general and increase in disease incidence due to an aging population as well as more complex diagnostics stands in sharp contrast to the declining numbers of practicing board certified pathologists in the US, Canada, and Europe.[20–22] A key toward solving this dilemma is a more focused integration of novel techniques to counteract upcoming challenges and avoid negative implications for patient care. Apart from the aforementioned armamentarium the advent of digital pathology opened the door for artificial intelligence as well as machine learning tools to mine subvisual morphometric phenotypes, aid in histopathological diagnosis, and ultimately, improve patient management.[23–25] Thus far, an under-recognized field in pathology practice is proteomics in general, apart from its most simple technique immunohistochemistry. However, a broader knowledge about disease development, progression, and treatment could be gained by combining genetic, histomorphological, and proteomic information.[26]

17.2 Possibilities and Advantages of MALDI-MSI in Pathology Research and Practice

Many possible proteomic and metabolomic applications are possible when it comes to their integration into a clinical/pathological workflow. However, most proteomic approaches require the sample to be investigated in solution. In terms of tissue samples in pathology this means disrupting the tissue sample's integrity in order to dissolve the analytes of interest. This leads to a complete loss of tissue morphology and all the information that could be gained from it. For example, differences between tumor cells at the tumor margin and tumor cells located in the center of the tumor as well as differences in the adjacent normal tissue close to the tumor margin and further away from the tumor,[27] in particular gradient changes decreasing or increasing with distance, are nearly impossible to capture with solution-based approaches. Also, in order to analyze pure cell populations, tissue samples have to be laser-micro-dissected to isolate the desired cells. This is a cumbersome and time-consuming preparation approach.[28] Herein lies one major advantage and strength of MALDI-MSI which allows the measurement of molecules with direct correlation to morphological features at the cellular level. Utilizing an intact tissue section, mass spectra are acquired at pre-defined x and y coordinates within a tissue section. Thus, each spectrum can be mapped to a discrete tissue location providing a molecular map across the tissue section.[29] Latest improvements in sample preparation as well as technologies enable the investigation of tissue sections at or almost at cellular level. This is of particular interest for the analysis of small histological features, *e.g.* precursor lesions. Classic in-solution based approaches would require many sections to be micro-dissected to acquire the necessary number of cells for subsequent analysis. Also, MALDI-MSI is ideally suited to study the spatial organization of different cell types within tissue sections of samples from healthy donors as well patients with various diseases.[30] For example, the spatial organization of immune cells within cancer samples has become of particular interest, wherein differences in inter-cell distances are of importance.[31] Another ideal application is the study of intra-tumor heterogeneity.[32,33] In particular in routine pathology, possible applications of MALDI-MSI are detection, subtyping, prognosis, and treatment prediction of diseases, particularly tumors.

17.3 Tumor and Disease Detection

Accurate disease detection is the bread and butter in every pathology lab. This concerns the presence of inflammatory, immunological, genetic, and idiopathic disorders as well as malignant and benign tumors. In particular, early studies utilizing MALDI-MSI in a clinical setting tried to aid in this task and aimed to outperform traditional methods in pathology. Mainly differences between tumor and normal tissue have been studied in a number of different organs including prostate cancer,[34–38] head and neck squamous cell carcinoma,[39,40] lung cancer,[41–43] gastric cancer,[44] kidney cancer,[27,45–47] hepatocellular carcinoma,[48–50] ovarian cancer,[51] cervical cancer,[52] lymphoma,[53] and sarcoma.[33] Also, MALDI-MSI has been utilized to characterize different types of tissue-resident immune cells[30] and to study traumatic brain and spinal cord injuries.[54,55]

A recently published study ventured a new and exciting application of MALDI-MSI, the investigation of changes within the extracellular matrix in lung cancer.[41] Tissue samples from patients with early stage, low-grade adenocarcinoma of the lung ($n = 20$) and samples from normal lung ($n = 10$, 2 samples per patient) from a tissue microarray (TMA) were subjected to on-tissue digestion with collagenase type III to establish signature patterns in normal and tumor samples. These were afterwards compared to a sample from a patient with advanced pulmonary adenocarcinoma. Unsupervised analysis by hierarchical clustering was able to distinguish between cores harboring normal lung tissue and adenocarcinoma. Statistical analysis based on the area under the receiver-operating curve (AUC) could identify 25 ions (*i.e.* mass-to-charge (*m*/*z*) ratios/values) that could discriminate between normal lung tissue and adenocarcinoma. Even though adenocarcinoma induces a change in microenvironment with an increase in stromal compartment (so-called desmoplastic stromal reaction), at least 10 ions were of higher intensity in normal lung tissue. Also, significant fold changes could be observed between adenocarcinoma and normal lung samples. The analysis of a whole slide rendering a resection specimen from an intermediate grade lung adenocarcinoma yielded similar results. Conventional LC–MS/MS was utilized to identify post-translational modifications of collagens present in the extracellular matrix as described previously.[56] Five ions could be identified with hydroxylated lysine residues, among them a peptide from collagen 3A1 modified by hydroxylation of proline (*m*/*z* 919.4884). This showed a 3.5-fold increase in intensity in samples with adenocarcinoma compared to normal lung (AUC 0.914, p-value <0.001).

To the author's knowledge, almost all studies on tumor detection have been focused on differentiating normal/healthy tissue from tumors. It is noteworthy that the difference between normal/healthy tissues has not yet been clarified at the molecular level, which renders the clinical investigation challenging. However, some of the most difficult tasks in pathology are the distinctions between reactive changes in tissue samples (*e.g.* due to inflammation) and neoplastic/dysplastic changes. Especially in small biopsy samples, that are sometimes even obscured by crush or cauterization artifacts, a clear distinction can be impossible. For example, in Barrett's esophagus (BE) there are gradual changes: BE negative for dysplasia, BE indefinite for dysplasia, BE with low-grade dysplasia, BE with high-grade dysplasia, and BE with carcinoma.[57] All these different stages are usually obscured by various degrees of inflammation, making the distinction between BE with reactive changes (thus negative for dysplasia) and BE with low-grade dysplasia particularly challenging. Therefore, inter-observer and intra-observer variation is reported to be high.[58] This is just one of many examples wherein MALDI-MSI could be of help. Other techniques such as liquid chromatography electrospray ionization tandem mass spectrometry (LC–ESI-MS/MS) would not be suitable for this, especially with regards to a possible routine application since the lesions are typically small (<1 mm).

17.4 Tumor and Disease Subtyping

A very important and challenging application is tumor subtyping. Nowadays, "just" diagnosing a tumor is only the first step in completing the pathology report. As mentioned above, complexity in almost all areas of pathology has increased over recent decades. More morphologically and/or molecularly defined subtypes of solid cancers as well as lymphomas and leukemia are being discovered. The accurate distinction between these subtypes is of utmost importance as different subtypes warrant different treatment regimes.

One perfect example is the changing landscape of renal tumor entities. They comprise a heterogeneous group of tumors with distinct clinical, pathological, and molecular features.[59] Historically, four different subtypes were known. The latest World Health Organization (WHO) classification on renal tumors in 2016 already incorporated more than ten different subtypes, and the upcoming new WHO classification is expected to include even more emerging subtypes.[60,61] Therefore, the challenge of diagnosing the correct subtype has increased, especially in small pre-therapeutic biopsies.[59] Of particular difficulty is the distinction between tumors that exhibit oncocytic and eosinophilic cells, notably the differentiation between renal oncocytoma (RO), a benign renal tumor, and other malignant renal tumors that can display a similar histomorphology.[62] The three most common malignant renal tumor subtypes are clear cell renal cell carcinoma (ccRCC), papillary renal cell carcinoma (pRCC), and chromophobe renal cell carcinoma (chRCC).[63]

Many research studies have investigated proteomic changes in renal cell carcinoma (RCC) however, most have either analyzed differences between (adjacent) normal kidney tissue and a single subtype of RCC, or differences between two different subtypes.[27,46,64–66] One previous study analyzed a large TMA-based sample set with 789 samples utilizing MALDI-MSI.[67] Samples mainly belonged to two of the main subtypes, ccRCC ($n = 528$) and pRCC ($n = 112$), whereas only 41 chRCC, 42 other subtypes, and no oncocytomas were included. Several *m/z* values could be identified which were associated with different features of RCC, such as histological subtype, stage, grade, and patient prognosis. However, none of these *m/z* values were identified by MS/MS in order to link them to proteins or peptides. In a recent study, Moginger *et al.* analyzed whole slides from formalin-fixed and paraffin-embedded (FFPE) ccRCC ($n = 12$), chRCC ($n = 5$), and RO ($n = 11$) by MALDI-MSI combined with rapid LC–MS/MS based micro-proteomics technology to characterize and classify renal tumor subtypes.[68] Generating a classifier based on partial least square discriminant analysis, MALDI-MSI data could be classified on a pixel-based approach with an overall accuracy of 87% (93% for ccRCC, and 88% for chRCC and RO). For each tumor type, the two most important *m/z* values were listed: *m/z* 704.5 and *m/z* 723.5 for ccRCC, *m/z* 1039.5 and *m/z* 1169.5 for chRCC, and *m/z* 806.5 and *m/z* 1640.0 for RO, respectively. Unfortunately, the authors were not able to identify these tryptic peptides due to a lack of sensitivity. However, 100% of all samples could be correctly classified utilizing the rapid LC–MS/MS-based micro-proteomics approach.

Another challenging diagnostic area is the distinction between different melanocytic lesions, in particular the distinction between Spitz nevi (SN) and Spitzoid malignant melanoma (SMM).[69,70] Atypical Spitzoid neoplasms represent the most challenging cases with features of both SN and SMM. Inter-observer variability in these lesions is typically high, and it can be difficult to predict clinical behavior from assessment of morphological features.[71] Multiple studies utilizing MALDI-MSI to differentiate between different melanocytic lesions have been published and emphasized the strength of this technique to improve diagnosis, predict outcome and enable risk stratification.[72–76] In the earliest study, FFPE tissue sections from 56 SN and 58 SMM were analyzed by MALDI-MSI for tumor and tumor microenvironment.[73] For statistical analysis and classification, the sample set was divided into a training set (26 SN and 25 SMM) and a validation set (30 SN and 33 SMM). Four cases of SMM had to be excluded from the validation set since they did not contain enough tumor, also, 5 cases of SN and 2 cases of SMM had to be excluded since they did not contain sufficient tumor microenvironment. Utilizing the data from the training set, a molecular signature for both tumor and tumor microenvironment was determined and used to classify the validation set. Therein, 29 out of 30 cases (97%) with SN could be correctly classified utilizing the tumor signature. The one case with SN that was incorrectly classified only had limited amount of tumor tissue, thus was represented by significantly fewer spectra compared to the other cases. Out of the 29 cases with SMM harboring sufficient tumor material 26 (90%) were correctly classified. When using the signature of the tumor microenvironment, sensitivity and specificity were 64% and 90%, respectively. Two of the tryptic peptides differentially expressed between SN and SMM that were used for classification were identified as actin (*m*/*z* 976.49) and vimentin (*m*/*z* 1428.61). In the latest study, FFPE samples from 357 cases with malignant melanoma and 399 benign nevi (including samples from biopsies and TMAs) were analyzed. For classification purposes, the sample set was divided into a training set (111 nevi and 100 melanomas) and a validation set (288 nevi and 257 melanomas). Utilizing the training set, a linear discriminant analysis (LDA) based classification algorithm was built and applied to all samples of the validation set. Classification output was converted to a numerical score ranging from -10 to $+10$, in order to classify each sample as likely benign (range from -10 to -0.371), intermediate (range from -0.370 to $+0.370$), and malignant (range from $+0.371$ to $+10$). In the validation set, 239 out of 259 (93%) of all melanoma samples were correctly classified as malignant, and 282 out of 288 (98%) of all nevi samples were correctly classified as benign.

Other studies benefiting from the strength of MALDI-MSI to differentiate between tumor subtypes include studies on ovarian cancer,[77] breast cancer,[78] lung cancer,[12,79,80] pancreatic tumors,[81] gastric cancer,[82] adenocarcinomas of different primary origin,[83–86] sarcomas,[87] and pituitary adenomas.[88] Also, grading in cancer samples has been achieved by MALDI-MSI.[89–91]

Apart from tumor subtyping, the distinction of different inflammatory diseases or benign tumors can also pose a challenge in everyday pathology practice. Thus, MALDI-MSI has also been utilized to delineate different non-tumor diseases such as chronic inflammatory bowel disease,[92,93] kidney diseases,[94] osteoarthritis,[95,96] non-alcoholic fatty liver disease,[97–99] autoimmune liver diseases,[100] weight-related changes in thyroid tissue,[101] age-related changes in the human lens,[102] and neurodegenerative diseases.[103–107]

17.5 Outcome and Response Prediction

In general, localized cancer has a better prognosis and requires different treatment compared to metastatic cancer.[108–112] The latter often requires extensive surgery coupled with neo-adjuvant and/or adjuvant therapy (chemotherapy, radiation, immunotherapy, or various combinations thereof).[109,113] Therefore, adequate patient management relies on the correct assignment of disease extent and spread.[114] In the pre-therapeutic setting, this is performed by radiological imaging. However, different radiological imaging modalities suffer from low sensitivity.[115,116] Being able to accurately predict the presence or absence of lymph nodes metastasis based on the proteomic profile of the primary tumor would therefore improve patient care. Mittal *et al.* identified potential discriminatory markers to delineate primary endometrial cancers with ($n = 16$) and without ($n = 27$) lymph node metastasis by MALDI-MSI.[117] Utilizing a LDA-based model an overall classification accuracy of 88.4% could be achieved. Out of the 20 highly ranked *m/z* values, 2 could be verified as peptides from alpha-actin-2 (*m/z* 976.42 and *m/z* 1198.67) by direct off-tissue MS/MS analysis. The *m/z* 967.42 could be identified as plectin by nano-LC–ESI-MS/MS. Results from MALDI-MSI for both alpha-actin 2 and plectin could be validated by data-independent acquisition nano-LC–ESI-MS/MS on laser-micro-dissected normal and tumor tissue from patients with and without lymph node metastasis (4 patients each). Normalized data from tumor samples showed an increased expression in primary tumors without lymph node metastases for both proteins. Additionally, quantitative IHC analysis was performed on the same sample set. Whereas no difference in plectin staining between primary tumors with and without lymph node metastases could be detected, alpha-actin-2 showed a significant lower staining intensity in tumors with lymph node metastases.

Other studies have also shown the power of MALDI-MSI to identify prognostic biomarkers or predict outcomes in patients with colon cancer,[118,119] lung cancer,[120] thyroid cancer,[121] endometrial cancer,[122,123] ovarian cancer,[124,125] breast cancer,[126–128] prostate cancer,[129,130] bladder cancer,[131] renal cancer,[67,132] esophageal cancer,[133,134] gastric cancer,[135,136] head and neck squamous cell carcinoma,[137] hepatocellular carcinoma,[138] adrenocortical carcinoma,[139] malignant melanoma,[140] brain tumors,[141] sarcomas,[33,142–145] and also non-cancer diseases.[146–148]

Within the last two decades, it has become evident that patients suffering from the very same tumor, even down to the same subtype, tumor stage, grade, and mutation spectrum still show divergent responses to treatment. Thus, the holy grail in cancer treatment has become personalized medicine, meaning a treatment tailored to the individual patient and their tumor. This task calls for proteomic techniques to fully understand cancer mechanisms, particularly with regards to progression and possible treatment options.[149]

A very promising approach in cancer therapy is immune checkpoint inhibition, a therapy wherein the interaction between programmed death-1

(PD-1) and its ligand PD-L1 is blocked.[150,151] While many patients show a good and even complete response to this treatment a high number of patients do not respond, and some might even progress rapidly under treatment.[152] In a recent study, pre-treatment FFPE biopsy specimens from 25 patients with non-small cell lung cancer (NSCLC) were subjected to MALDI-MSI analysis.[153] All patients had a high PD-L1 expression (>50%) and sufficient tumor-infiltrating lymphocytes. During treatment, 9 (36%) patients exhibited tumor regression (=responders) and 16 (64%) showed no tumor regression (=non-responders). Average mass spectra from the 9 responders and 16 non-responders revealed differences in *m/z* values corresponding to neutrophil defensins 1–3 (*m/z* 3369.5, *m/z* 3440.6, and *m/z* 3484.6). These *m/z* values were present in 7 responders and only 2 non-responders, thus suggesting a correlation between neutrophil defensin expression and positive response to immunotherapy. In order to verify these findings, immunohistochemistry was performed on serial sections from the same sample collective as well as sections previously analyzed by MALDI-MSI utilizing a defensin 1/3 polyclonal antibody. Presence of defensin staining was scored (percentage of positive cells) for tumor cells and immune cells individually. Performing a Mann–Whitney U test revealed a significant association between percentage of tumor ($p = 0.027$) as well as immune cell ($p = 0.043$) defensin expression and response to immunotherapy.

Further studies have investigated the possibility to predict treatment response utilizing MALDI-MSI in breast cancer,[154,155] esophageal cancer,[156] and colorectal cancer liver metastases.[157]

17.6 Challenges and Considerations

As highlighted by the previous paragraphs, MALDI-MSI has the potential to be incorporated into the pathologist's toolbox and revolutionize day-to-day pathology practice. The first hurdle toward the implementation of MALDI-MSI in pathology practice was taken when protocols to analyze not only fresh-frozen but also FFPE tissue samples were published.[80,158,159] Fresh-frozen tissue samples are available in pathology labs worldwide but they account for only a small amount of samples compared to the vast amount of available FFPE tissue samples. These represent treasures in pathology archives all around the world. Especially when investigating rare diseases or tumors, FFPE samples can yield the necessary sample number for significant analysis. Also, FFPE samples allow for the analysis of TMAs that enable the analysis of many patient samples under the same experimental conditions. The second hurdle was taken when performance of commercially available mass spectrometers improved. Nowadays, mass spectrometers operate with high repetition-rate lasers, thus enabling the acquisition of spectra even from whole tissue sections in a matter of hours.[160] Also, enhanced lateral/spatial resolution allows for analysis almost at the single cell level.[161–163] The third hurdle was sample preparation. While in the early beginnings matrix was applied manually, thus not yielding optimal and most importantly reproducible results, nowadays commercially available instruments for matrix as well as enzyme application allow for a fast and reproducible sample preparation.[164–167] Compared to standard IHC approaches that usually need a single section for each antibody examined, MALDI-MSI only needs a single thin tissue section to evaluate hundreds of analytes in parallel.[84,149,168,169] This approach can yield the same or even better results in differentiating tumor subtypes from one another when compared to standard immunohistochemistry approaches.[84] Also, sections previously analyzed by MALDI-MSI can afterward even be used for H&E staining, IHC, immunofluorescence, or DNA-analysis.[170–172] Thus, valuable and sometimes scarce patient tissue material can be saved and therefore be made available for additional diagnostic testing. Since MALDI-MSI can provide classification algorithms based on multiple *m/z* values rather than a single (bio-)marker molecule it can be more accurate for predicting disease, prognosis and treatment response.[173]

However, there are still several challenges that need to be addressed before MALDI-MSI can be implemented in pathology practice. A summary of these challenges and possible solutions is available in Table 17.1.

The first issue is with regards to the number of samples in published studies. Most of the studies presented above have only analyzed a limited number of samples; most are below 100 per group, some even below 10. MALDI-MSI-based patterns are used to differentiate between healthy and disease, disease or tumor types, predict outcome or treatment response present clinical biomarkers. These need to be tested in large groups of patients, ideally with different ethnicity and geographical distribution, and validated in independent sample sets.[26,174] The number of patients should be inversely proportional to the number of *m/z* values measured to prevent overfitting of classification algorithms.[26,175] This represents a major problem in biomarker discovery studies wherein the number of samples is relatively small in comparison to a relatively large number of *m/z* values.[176] Thus, during classification analysis a "chance" pattern is identified that can perfectly discriminate samples from one another in the given sample set but is not reproducible in a second sample set.[177] In other proteomic approaches, the propagated solution is the following biomarker development pipeline: a discovery phase is followed by a verification step and a final validation is performed utilizing a different methodology (*e.g.* immunoassay).[178] For MALDI-MSI this pipeline can also be adopted (Figure 17.1), as previously proposed.[175] The initial discovery phase is performed to identify discriminatory *m/z* values between different well-defined disease states or healthy and disease states. These discriminatory *m/z* values should be verified using an independent second sample set (=test set) followed by identification of as many significant *m/z* values as possible that were used to delineate the different sample groups. As a final step, a third sample set (=validation set) should be used to confirm the diagnostic power of the potential biomarkers utilizing different techniques such as immunohistochemistry. Additionally, identification of *m/z* values enables

Table 17.1 Summary of challenges and possible solutions.

Challenges	Possible solutions
Study design	– Randomized sample measurements; with regards to TMAs, samples from all groups should be randomized across TMAs – Include control samples to monitor technical variation in sample preparation and measurements – Appropriate sample size to enable split in training and test set for statistical analysis – Utilize TMAs to measure more samples with less technical variation
Sample size	– Use FFPE instead of fresh-frozen material in order to have a larger sample cohort – Consider cooperation between multiple labs to gain access to more samples – Utilize TMAs
Sample preparation	– Create standard operating procedures to ensure reproducible sample preparation
Spatial resolution	– Optimize mass spectrometers to achieve more precise laser focus and sample stage step – Optimize detectors – Optimize sample preparation to achieve smaller matrix crystals and better analyte extraction
Co-registration	– Utilize multiple fiducials and image warping to account for differences in tissue position and dimensions due to staining – Create readily available software solutions
Statistical analysis	– Create readily available software solutions that offer different statistical methods – Create algorithms that can account for and correct differences in spectra (*e.g.* total ion count, peak alignment) from different measurements within a study cohort
Reproducibility	– Round robin test between different labs

further systematic analysis of the potential biomarkers.[163] This can include pathway analysis and visualization as well as a literature search.

Another potential systematic source of technical variation, the so-called batch effect, can be a reason for failure of implementation of MALDI-MSI into the clinic and has recently been discussed in detail.[179] Batch effects can act at many levels and time points during the MALDI-MSI work-flow and even before, *e.g.* during sample collection. Cooperative approaches among laboratories will be necessary to address batch effects and should include samples from different laboratories, first to avoid a bias stemming from certain FFPE processing protocols only used in a particular laboratory, and second to enlarge the sample collective. Identifying, avoiding, and counteracting these are key to circumvent detrimental consequences. Briefly, an optimized study design is the first step, and randomization of the sample measurements should not be omitted. Additionally, to the points regarding sample size discussed above, samples from different groups should be

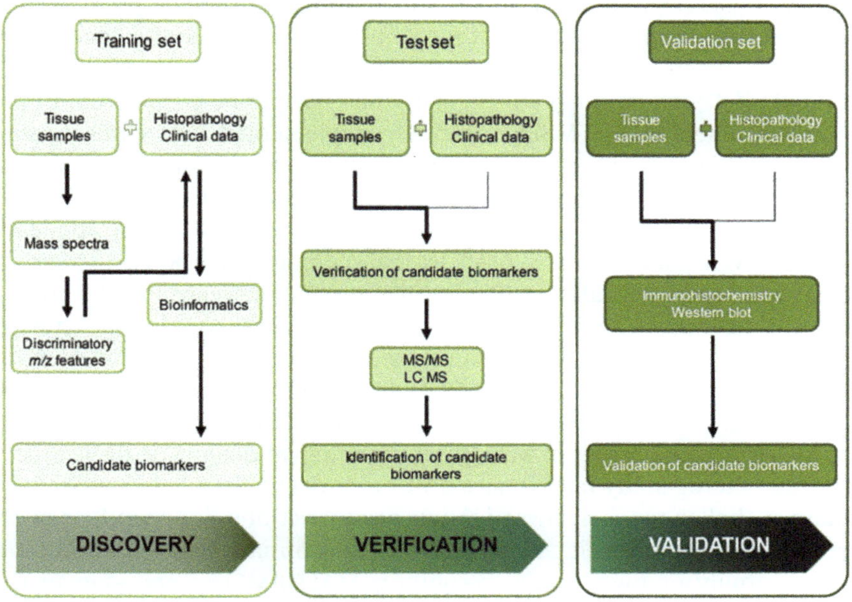

Figure 17.1 Proposed workflow for MALDI-MSI-based biomarker discovery and validation. Discovery of potential discriminatory *m/z* features should only be the first step. It should be followed by verification utilizing a second sample set (test set) and MS/MS-based identification of as many candidate biomarkers as possible. This enables validation by an independent third sample set (validation set) employing different techniques such as IHC to confirm the biomarker's potential.
Reproduced from ref. 175 with permission from Elsevier, Copyright 2012.

randomly distributed on the slides and technical replicates should also be included.[180] A second step is outlier detection on a data as well as sample level to exclude extreme data and thus increase sensitivity.

Reproducibility is another crucial factor that needs to be addressed.[178] This includes reproducibility within a single lab as well as reproducibility between different labs. One step toward achieving this goal is a standardized and robust experimental workflow.[166,181,182] Additionally, on-slide quality controls to ascertain and monitor instrument performance, effectiveness of on-tissue digestion, quality of matrix application, and standardized reporting in publications should be mandatory. Different tools to achieve this have been proposed and instrument vendors have started to implement tools to check for and ascertain continuous instrument performance.[181,183] Inter-laboratory performance and reproducibility should also be evaluated in multicenter/round robin studies following examples from routine quality performance studies that have to be performed regularly in a pathology lab to ascertain diagnostic standards.[184] A few round robin studies have already been performed,[181,185–187] but more should follow.

Another challenge leads on from the increasing capability to produce high-spatially-resolved images in less time. Thus, data size will also increase exponentially when cellular and eventually sub-cellular image resolution is achieved.[163] Therefore, data storage solutions have to be considered, in particular for when MALDI-MSI is routinely implemented into clinical/pathological practice. Governmental regulations in many countries require pathology labs to store glass slides for a minimal period of ten years.[188] Thus, it is to be expected, that a similar minimal storage time will also be applicable to MALDI-MSI data if this is used to make clinical decisions. Increase in spatial resolution will also require software solutions capable of handling larger file sizes as well as optimized solutions for image correlation. Thus far, only custom-built solutions to achieve accurate image correlation to study differences at a cellular level exist.[163] However, widely available software solutions are urgently needed in order to streamline this process and allow more labs to harness the power of MALDI-MSI for clinical diagnostic decisions.

Tissue heterogeneity (as discussed in more detail elsewhere[3]) will always remain a challenge when investigating tissue sections, independent of the method used. However, it can be met with appropriate strategies and tools. The possibility of high spatial resolution MALDI-MSI is one step in this direction. However, its full potential can only be achieved when accurate co-registration (as discussed in the previous paragraph) is possible. Additionally, correct interpretation and annotation of stained histology images by trained pathologists is mandatory for precise supervised classification analysis since classification results are influenced by the scrupulousness of the histological annotation. For example, Mascini *et al.* noticed a dramatic decrease in classification accuracy of TMA data when the tumor cell percentage was decreased from 80% down to 50% within each tissue core sample.[137] As pathologists are often overloaded with cases to examine and are declining in numbers (as discussed in the beginning), digital pathology solutions offer new and exciting possibilities.[163]

17.7 Conclusions and Future Directions

The outstanding molecular recognition, high spatial resolution, and high-throughput capabilities of MALDI-MSI have the potential to substantially benefit pathology research and practice. As highlighted above, many studies have proven its feasibility and versatility to aid in detection, subtyping, prognosis, and treatment prediction of diseases, particularly tumors. In order to achieve a true transition from bench to bedside more comprehensive and multicenter studies with hundreds of samples from different labs are necessary to validate the robustness and applicability of MALDI-MSI in a clinical setting and in support of patient-tailored treatment. The technique not only holds great promise to establish different analyte patterns by themselves, but the combination of different analyte patterns from metabolomic, glycomic, lipidomic, peptidomic, and proteomic signatures can even improve classification accuracy.[133] In looking forward, the next steps toward the implementation for MALDI-MSI need a collaborative effort of many fields: pathologists to define areas of diagnostic need and provide adequate samples, engineers and physicists to build even better instrumentation, chemists and biologists to improve sample preparation, as well as mathematicians, statisticians, and computer scientists to advance data analysis, storage, and statistical tools. This can truly bring the art and practice of modern molecular pathology to a more effective level.

References

1. M. J. Misialek, *Am. J. Clin. Pathol.*, 2014, **142**, 584–585.
2. A. Stevenson, *Oxford Dictionary of English*, 3rd edn, 2010, Oxford University Press, New York, NY.
3. K. Schwamborn, *Adv. Cancer Res.*, 2017, **134**, 1–26.
4. S. R. Lakhani and A. Ashworth, *Nat. Rev. Cancer*, 2001, **1**, 151–157.
5. P. A. Humphrey, *Mo. Med.*, 2010, **107**, 76–77.
6. J. Rosai, *Lab. Invest.*, 2007, **87**, 403–408.
7. L. Sun and J. D. Pfeifer, *Semin. Diagn. Pathol.*, 2019, **36**, 342–354.
8. V. Angerilli, F. Galuppini, F. Pagni, N. Fusco, U. Malapelle and M. Fassan, *Diagnostics*, 2021, **11**, 339.
9. S. Masood, *Breast J.*, 2020, **26**, 27–34.
10. R. W. Vachet, *Nat. Nanotechnol.*, 2015, **10**, 103–104.
11. A. Warth, A. Stenzinger, M. Andrulis, W. Schlake, G. Kempny, P. Schirmacher and W. Weichert, *Virchows Arch.*, 2016, **468**, 101–108.
12. M. Kriegsmann, C. Zgorzelski, R. Casadonte, K. Schwamborn, T. Muley, H. Winter, M. Eichhorn, F. Eichhorn, A. Warth, S. O. Deininger, P. Christopoulos, M. Thomas, T. Longerich, A. Stenzinger, W. Weichert, C. Muller-Tidow, J. Kriegsmann, P. Schirmacher and K. Kriegsmann, *Cancers*, 2020, **12**, 2704.
13. N. A. Pennell, M. E. Arcila, D. R. Gandara and H. West, *Am. Soc. Clin. Oncol. Educ. Book*, 2019, **39**, 531–542.
14. E. Thunnissen, K. M. Kerr, F. J. Herth, S. Lantuejoul, M. Papotti, R. C. Rintoul, G. Rossi, B. G. Skov, B. Weynand, L. Bubendorf, G. Katrien, L. Johansson, F. Lopez-Rios, V. Ninane, W. Olszewski, H. Popper, S. Jaume, P. Schnabel, L. Thiberville and F. Laenger, *Lung Cancer*, 2012, **76**, 1–18.
15. A. Z. Feldman, L. J. Jennings, N. R. Wadhwani, D. J. Brat and C. M. Horbinski, *Curr. Neurol. Neurosci. Rep.*, 2020, **20**, 23.
16. N. Cancer Genome Atlas Research, J. N. Weinstein, E. A. Collisson, G. B. Mills, K. R. Shaw, B. A. Ozenberger, K. Ellrott, I. Shmulevich, C. Sander and J. M. Stuart, *Nat. Genet.*, 2013, **45**, 1113–1120.
17. K. Hede, *J. Natl. Cancer Inst.*, 2008, **100**, 1566–1569.
18. M. Beckman, *J. Natl. Cancer Inst.*, 2006, **98**, 1758–1759.
19. M. Janiszewska, *Oncogene*, 2020, **39**, 2031–2039.
20. B. Markl, L. Fuzesi, R. Huss, S. Bauer and T. Schaller, *Virchows Arch.*, 2021, **478**, 335–341.

21. S. J. Robboy, S. Weintraub, A. E. Horvath, B. W. Jensen, C. B. Alexander, E. P. Fody, J. M. Crawford, J. R. Clark, J. Cantor-Weinberg, M. G. Joshi, M. B. Cohen, M. B. Prystowsky, S. M. Bean, S. Gupta, S. Z. Powell, V. O. Speights, Jr., D. J. Gross and W. S. Black-Schaffer, *Arch. Pathol. Lab. Med.*, 2013, **137**, 1723–1732.
22. S. J. Robboy, S. Gupta, J. M. Crawford, M. B. Cohen, D. S. Karcher, D. G. Leonard, B. Magnani, D. A. Novis, M. B. Prystowsky, S. Z. Powell, D. J. Gross and W. S. Black-Schaffer, *Arch. Pathol. Lab. Med.*, 2015, **139**, 1413–1430.
23. K. Bera, K. A. Schalper, D. L. Rimm, V. Velcheti and A. Madabhushi, *Nat. Rev. Clin. Oncol.*, 2019, **16**, 703–715.
24. S. W. Jahn, M. Plass and F. Moinfar, *J. Clin. Med.*, 2020, **9**, 3697.
25. M. K. K. Niazi, A. V. Parwani and M. N. Gurcan, *Lancet Oncol.*, 2019, **20**, e253–e261.
26. T. Kowalczyk, M. Ciborowski, J. Kisluk, A. Kretowski and C. Barbas, *Biochim. Biophys. Acta, Mol. Basis Dis.*, 2020, **1866**, 165690.
27. S. R. Oppenheimer, D. Mi, M. E. Sanders and R. M. Caprioli, *J. Proteome Res.*, 2010, **9**, 2182–2190.
28. M. R. Emmert-Buck, R. F. Bonner, P. D. Smith, R. F. Chuaqui, Z. Zhuang, S. R. Goldstein, R. A. Weiss and L. A. Liotta, *Science*, 1996, **274**, 998–1001.
29. K. Schwamborn and R. M. Caprioli, *Nat. Rev. Cancer*, 2010, **10**, 639–646.
30. M. Holzlechner, K. Strasser, E. Zareva, L. Steinhauser, H. Birnleitner, A. Beer, M. Bergmann, R. Oehler and M. Marchetti-Deschmann, *J. Proteome Res.*, 2017, **16**, 65–76.
31. N. Zwing, H. Failmezger, C. H. Ooi, D. P. Hibar, M. Canamero, B. Gomes, F. Gaire and K. Korski, *Front. Immunol.*, 2020, **11**, 550250.
32. A. Marusyk, V. Almendro and K. Polyak, *Nat. Rev. Cancer*, 2012, **12**, 323–334.
33. S. Lou, B. Balluff, M. A. de Graaff, A. H. Cleven, I. Briaire-de Bruijn, J. V. Bovee and L. A. McDonnell, *Proteomics*, 2016, **16**, 1802–1813.
34. M. K. Andersen, T. S. Hoiem, B. S. R. Claes, B. Balluff, M. Martin-Lorenzo, E. Richardsen, S. Krossa, H. Bertilsson, R. M. A. Heeren, M. B. Rye, G. F. Giskeodegard, T. F. Bathen and M. B. Tessem, *Cancer Metab.*, 2021, **9**, 9.
35. J. D. Pallua, G. Schaefer, C. Seifarth, M. Becker, S. Meding, S. Rauser, A. Walch, M. Handler, M. Netzer, M. Popovscaia, M. Osl, C. Baumgartner, H. Lindner, L. Kremser, B. Sarg, G. Bartsch, C. W. Huck, G. K. Bonn and H. Klocker, *J. Proteomics*, 2013, **91**, 500–514.
36. P. M. Angel, L. Spruill, M. Jefferson, J. R. Bethard, L. E. Ball, C. Hughes-Halbert and R. R. Drake, *Prostate*, 2020, **80**, 1071–1086.
37. L. H. Cazares, D. Troyer, S. Mendrinos, R. A. Lance, J. O. Nyalwidhe, H. A. Beydoun, M. A. Clements, R. R. Drake and O. J. Semmes, *Clin. Cancer Res.*, 2009, **15**, 5541–5551.
38. K. Schwamborn, R. C. Krieg, M. Reska, G. Jakse, R. Knuechel and A. Wellmann, *Int. J. Mol. Med.*, 2007, **20**, 155–159.

39. Y. Yuan, X. Xie, Y. Jiang, Z. Wei, P. Wang, F. Chen, X. Li, C. Sun, H. Zhao, X. Zeng, L. Jiang, Y. Zhou, H. Dan, M. Feng, R. Liu, Z. Wang and Q. Chen, *Cell Death Dis.*, 2017, **8**, e3035.
40. X. Xie, Y. Jiang, Y. Yuan, P. Wang, X. Li, F. Chen, C. Sun, H. Zhao, X. Zeng, L. Jiang, Y. Zhou, H. Dan, M. Feng, R. Liu and Q. Chen, *Oncotarget*, 2016, **7**, 59987–60004.
41. P. M. Angel, E. Bruner, J. Bethard, C. L. Clift, L. Ball, R. R. Drake and C. Feghali-Bostwick, *J. Mass Spectrom.*, 2020, **55**, e4450.
42. S. M. Rahman, A. L. Gonzalez, M. Li, E. H. Seeley, L. J. Zimmerman, X. J. Zhang, M. L. Manier, S. J. Olson, R. N. Shah, A. N. Miller, J. B. Putnam, Y. E. Miller, W. A. Franklin, W. J. Blot, D. P. Carbone, Y. Shyr, R. M. Caprioli and P. P. Massion, *Cancer Res.*, 2011, **71**, 3009–3017.
43. E. Marien, M. Meister, T. Muley, S. Fieuws, S. Bordel, R. Derua, J. Spraggins, R. Van de Plas, J. Dehairs, J. Wouters, M. Bagadi, H. Dienemann, M. Thomas, P. A. Schnabel, R. M. Caprioli, E. Waelkens and J. V. Swinnen, *Int. J. Cancer*, 2015, **137**, 1539–1548.
44. H. K. Kim, M. L. Reyzer, I. J. Choi, C. G. Kim, H. S. Kim, A. Oshima, O. Chertov, S. Colantonio, R. J. Fisher, J. L. Allen, R. M. Caprioli and J. E. Green, *J. Proteome Res.*, 2010, **9**, 4123–4130.
45. R. R. Drake, C. McDowell, C. West, F. David, T. W. Powers, T. Nowling, E. Bruner, A. S. Mehta, P. M. Angel, L. A. Marlow, H. W. Tun and J. A. Copland, *J. Mass Spectrom.*, 2020, **55**, e4490.
46. E. E. Jones, T. W. Powers, B. A. Neely, L. H. Cazares, D. A. Troyer, A. S. Parker and R. R. Drake, *Proteomics*, 2014, **14**, 924–935.
47. T. M. Morgan, E. H. Seeley, O. Fadare, R. M. Caprioli and P. E. Clark, *J. Urol.*, 2013, **189**, 1097–1103.
48. C. A. West, M. Wang, H. Herrera, H. Liang, A. Black, P. M. Angel, R. R. Drake and A. S. Mehta, *J. Proteome Res.*, 2018, **17**, 3454–3462.
49. J. Le Faouder, S. Laouirem, M. Chapelle, M. Albuquerque, J. Belghiti, F. Degos, V. Paradis, J. M. Camadro and P. Bedossa, *J. Proteome Res.*, 2011, **10**, 3755–3765.
50. C. Marquardt, T. Tolstik, C. Bielecki, R. Kaufmann, A. C. Crecelius, U. S. Schubert, U. Settmacher, A. Stallmach and O. Dirsch, *Z. Gastroenterol.*, 2015, **53**, 33–39.
51. R. Lemaire, S. A. Menguellet, J. Stauber, V. Marchaudon, J. P. Lucot, P. Collinet, M. O. Farine, D. Vinatier, R. Day, P. Ducoroy, M. Salzet and I. Fournier, *J. Proteome Res.*, 2007, **6**, 4127–4134.
52. K. Schwamborn, R. C. Krieg, S. Uhlig, H. Ikenberg and A. Wellmann, *Int. J. Mol. Med.*, 2011, **27**, 417–421.
53. K. Schwamborn, R. C. Krieg, P. Jirak, G. Ott, R. Knuchel, A. Rosenwald and A. Wellmann, *J. Cancer Res. Clin. Oncol.*, 2010, **136**, 1651–1655.
54. K. Mallah, J. Quanico, A. Raffo-Romero, T. Cardon, S. Aboulouard, D. Devos, F. Kobeissy, K. Zibara, M. Salzet and I. Fournier, *Mol. Cell. Proteomics*, 2019, **18**, 1669–1682.
55. J. Quanico, L. Hauberg-Lotte, S. Devaux, Z. Laouby, C. Meriaux, A. Raffo-Romero, M. Rose, L. Westerheide, J. Vehmeyer, F. Rodet,

P. Maass, D. Cizkova, N. Zilka, V. Cubinkova, I. Fournier and M. Salzet, *Sci. Rep.*, 2018, **8**, 16083.
56. P. M. Angel, S. Comte-Walters, L. E. Ball, K. Talbot, A. Mehta, K. G. M. Brockbank and R. R. Drake, *J. Proteome Res.*, 2018, **17**, 635–646.
57. P. Kinra, G. P. S. Gahlot, R. Yadav, V. Baloda, G. K. Makharia, S. D. Gupta and P. Das, *Pathol., Res. Pract.*, 2018, **214**, 993–999.
58. P. V. Kaye, S. A. Haider, M. Ilyas, P. D. James, I. Soomro, W. Faisal, J. Catton, S. L. Parsons and K. Ragunath, *Histopathology*, 2009, **54**, 699–712.
59. S. S. Salami, A. K. George and A. M. Udager, *Curr. Opin. Neurol.*, 2018, **28**, 383–391.
60. N. J. Rupp and H. Moch, *Urologe A*, 2020, **59**, 135–141.
61. K. Trpkov and O. Hes, *Histopathology*, 2019, **74**, 31–59.
62. C. Luders and G. Kristiansen, *Pathologe*, 2016, **37**, 153–158.
63. H. Moch, International Agency for Research on Cancer and World Health Organization, *WHO Classification of Tumours of the Urinary System and Male Genital Organs*, 4th edn, 2016, International Agency for Research on Cancer, Lyon.
64. C. Chinello, V. L'Imperio, M. Stella, A. J. Smith, G. Bovo, A. Grasso, M. Grasso, F. Raimondo, M. Pitto, F. Pagni and F. Magni, *Expert Rev. Proteomics*, 2016, **13**, 1103–1120.
65. M. Kriegsmann, R. Casadonte, N. Maurer, C. Stoehr, F. Erlmeier, H. Moch, K. Junker, C. Zgorzelski, W. Weichert, K. Schwamborn, S. O. Deininger, M. Gaida, G. Mechtersheimer, A. Stenzinger, P. Schirmacher, A. Hartmann, J. Kriegsmann and K. Kriegsmann, *J. Cancer*, 2020, **11**, 6081–6089.
66. C. H. Na, J. H. Hong, W. S. Kim, S. R. Shanta, J. Y. Bang, D. Park, H. K. Kim and K. P. Kim, *Mol. Cells*, 2015, **38**, 624–629.
67. S. Steurer, A. S. Seddiqi, J. M. Singer, A. S. Bahar, C. Eichelberg, M. Rink, R. Dahlem, H. Huland, G. Sauter, R. Simon, S. Minner, E. Burandt, P. R. Stahl, T. Schlomm, M. Wurlitzer and H. Schluter, *Anticancer Res.*, 2014, **34**, 2255–2261.
68. U. Moginger, N. Marcussen and O. N. Jensen, *Oncotarget*, 2020, **11**, 3998–4015.
69. R. L. Barnhill, Z. B. Argenyi, L. From, L. F. Glass, J. C. Maize, M. C. Mihm, Jr., M. S. Rabkin, S. G. Ronan, W. L. White and M. Piepkorn, *Hum. Pathol.*, 1999, **30**, 513–520.
70. M. Miteva and R. Lazova, *Semin. Cutaneous Med. Surg.*, 2010, **29**, 165–173.
71. L. Cerroni, R. Barnhill, D. Elder, G. Gottlieb, P. Heenan, H. Kutzner, P. E. LeBoit, M. Mihm, Jr., J. Rosai and H. Kerl, *Am. J. Surg. Pathol.*, 2010, **34**, 314–326.
72. R. Lazova, K. Smoot, H. Anderson, M. J. Powell, A. S. Rosenberg, F. Rongioletti, L. Pilloni, S. D'Hallewin, R. Gueorguieva, I. Tantcheva-Poor, O. Obadofin, C. Camacho, A. Hsi, H. H. Kluger, O. Fadare and E. H. Seeley, *J. Cutaneous Pathol.*, 2020, **47**, 226–240.
73. R. Lazova, E. H. Seeley, M. Keenan, R. Gueorguieva and R. M. Caprioli, *Am. J. Dermatopathol.*, 2012, **34**, 82–90.

74. A. K. Alomari, E. J. Glusac, J. Choi, P. Hui, E. H. Seeley, R. M. Caprioli, K. L. Watsky, J. Urban and R. Lazova, *J. Cutaneous Pathol.*, 2015, **42**, 757–764.
75. R. Lazova, E. H. Seeley, H. Kutzner, R. A. Scolyer, G. Scott, L. Cerroni, I. Fried, M. E. Kozovska, A. S. Rosenberg, V. G. Prieto, B. M. Shehata, M. M. Durham, G. Henry, J. L. Rodriguez-Peralto, E. Riveiro-Falkenbach, J. T. Schaefer, R. Danialan, S. Fraitag, S. Vollenweider-Roten, A. Sepehr, M. Sangueza, N. Hijazi, Y. Corredoira, R. Kowal, O. M. Harris, F. Bravo, A. S. Boyd, R. Gueorguieva and R. M. Caprioli, *J. Am. Acad. Dermatol.*, 2016, **75**, 1176–1186, e1174.
76. R. Lazova, Z. Yang, C. El Habr, Y. Lim, K. A. Choate, E. H. Seeley, R. M. Caprioli and L. Yangqun, *Am. J. Dermatopathol.*, 2017, **39**, 689–695.
77. O. Klein, F. Kanter, H. Kulbe, P. Jank, C. Denkert, G. Nebrich, W. D. Schmitt, Z. Wu, C. A. Kunze, J. Sehouli, S. Darb-Esfahani, I. Braicu, J. Lellmann, H. Thiele and E. T. Taube, *Proteomics: Clin. Appl.*, 2019, **13**, e1700181.
78. S. Rauser, C. Marquardt, B. Balluff, S. O. Deininger, C. Albers, E. Belau, R. Hartmer, D. Suckau, K. Specht, M. P. Ebert, M. Schmitt, M. Aubele, H. Hofler and A. Walch, *J. Proteome Res.*, 2010, **9**, 1854–1863.
79. M. Kriegsmann, R. Casadonte, J. Kriegsmann, H. Dienemann, P. Schirmacher, J. H. Kobarg, K. Schwamborn, A. Stenzinger, A. Warth and W. Weichert, *Mol. Cell. Proteomics*, 2016, **15**, 3081–3089.
80. M. R. Groseclose, P. P. Massion, P. Chaurand and R. M. Caprioli, *Proteomics*, 2008, **8**, 3715–3724.
81. R. Casadonte, M. Kriegsmann, A. Perren, G. Baretton, S. O. Deininger, K. Kriegsmann, T. Welsch, C. Pilarsky and J. Kriegsmann, *Proteomics: Clin. Appl.*, 2019, **13**, e1800046.
82. B. Balluff, M. Elsner, A. Kowarsch, S. Rauser, S. Meding, C. Schuhmacher, M. Feith, K. Herrmann, C. Rocken, R. M. Schmid, H. Hofler, A. Walch and M. P. Ebert, *J. Proteome Res.*, 2010, **9**, 6317–6322.
83. M. Kriegsmann, R. Longuespee, P. Wandernoth, C. Mohanu, K. Lisenko, W. Weichert, A. Warth, H. Dienemann, E. De Pauw, T. Katzenberger, D. Aust, G. Baretton, J. Kriegsmann and R. Casadonte, *Biochim. Biophys. Acta, Proteins Proteomics*, 2017, **1865**, 858–864.
84. K. Schwamborn, G. Weirich, K. Steiger, G. Zimmermann, M. Schmidmayr, W. Weichert and R. M. Caprioli, *Proteomics: Clin. Appl.*, 2019, **13**, e1800064.
85. R. Casadonte, M. Kriegsmann, F. Zweynert, K. Friedrich, G. Baretton, M. Otto, S. O. Deininger, R. Paape, E. Belau, D. Suckau, D. Aust, C. Pilarsky and J. Kriegsmann, *Proteomics*, 2014, **14**, 956–964.
86. S. Meding, U. Nitsche, B. Balluff, M. Elsner, S. Rauser, C. Schone, M. Nipp, M. Maak, M. Feith, M. P. Ebert, H. Friess, R. Langer, H. Hofler, H. Zitzelsberger, R. Rosenberg and A. Walch, *J. Proteome Res.*, 2012, **11**, 1996–2003.
87. S. M. Willems, A. van Remoortere, R. van Zeijl, A. M. Deelder, L. A. McDonnell and P. C. Hogendoorn, *J. Pathol.*, 2010, **222**, 400–409.

88. D. Calligaris, D. R. Feldman, I. Norton, O. Olubiyi, A. N. Changelian, R. Machaidze, M. L. Vestal, E. R. Laws, I. F. Dunn, S. Santagata and N. Y. Agar, *Proc. Natl. Acad. Sci. U. S. A.*, 2015, **112**, 9978–9983.
89. E. C. Randall, G. Zadra, P. Chetta, B. G. C. Lopez, S. Syamala, S. S. Basu, J. N. Agar, M. Loda, C. M. Tempany, F. M. Fennessy and N. Y. R. Agar, *Mol. Cancer Res.*, 2019, **17**, 1155–1165.
90. E. Le Rhun, M. Duhamel, M. Wisztorski, J. P. Gimeno, F. Zairi, F. Escande, N. Reyns, F. Kobeissy, C. A. Maurage, M. Salzet and I. Fournier, *Biochim. Biophys. Acta, Proteins Proteomics*, 2017, **1865**, 875–890.
91. R. F. Oezdemir, N. T. Gaisa, K. Lindemann-Docter, S. Gostek, R. Weiskirchen, M. Ahrens, K. Schwamborn, C. Stephan, D. Pfister, A. Heidenreich, R. Knuechel and C. Henkel, *Clin. Biochem.*, 2012, **45**, 7–11.
92. A. E. M'Koma, E. H. Seeley, M. K. Washington, D. A. Schwartz, R. L. Muldoon, A. J. Herline, P. E. Wise and R. M. Caprioli, *Inflammatory Bowel Dis.*, 2011, **17**, 875–883.
93. E. H. Seeley, M. K. Washington, R. M. Caprioli and A. E. M'Koma, *Proteomics: Clin. Appl.*, 2013, **7**, 541–549.
94. A. Smith, V. Iablokov, M. Mazza, S. Guarnerio, V. Denti, M. Ivanova, M. Stella, I. Piga, C. Chinello, B. Heijs, P. A. van Veelen, H. Benediktsson, D. A. Muruve and F. Magni, *Kidney Blood Pressure Res.*, 2020, **45**, 233–248.
95. M. R. Eveque-Mourroux, P. J. Emans, A. Boonen, B. S. R. Claes, F. G. Bouwman, R. M. A. Heeren and B. Cillero-Pastor, *J. Proteome Res.*, 2021, **20**, 2973–2982.
96. M. J. Peffers, B. Cillero-Pastor, G. B. Eijkel, P. D. Clegg and R. M. Heeren, *Arthritis Res. Ther.*, 2014, **16**, R110.
97. K. Scupakova, Z. Soons, G. Ertaylan, K. A. Pierzchalski, G. B. Eijkel, S. R. Ellis, J. W. Greve, A. Driessen, J. Verheij, T. M. De Kok, S. W. M. Olde Damink, S. S. Rensen and R. M. A. Heeren, *Anal. Chem.*, 2018, **90**, 5130–5138.
98. J. Wattacheril, E. H. Seeley, P. Angel, H. Chen, B. P. Bowen, C. Lanciault, R. M. Caprioli, N. Abumrad and C. R. Flynn, *PLoS One*, 2013, **8**, e57165.
99. H. Alamri, N. H. Patterson, E. Yang, P. Zoroquiain, A. Lazaris, P. Chaurand and P. Metrakos, *Anal. Bioanal. Chem.*, 2019, **411**, 885–894.
100. C. L. Bowlus, E. H. Seeley, J. Roder, J. Grigorieva, H. Roder, R. M. Caprioli and M. Gershwin, *Cell. Mol. Immunol.*, 2011, **8**, 237–242.
101. A. A. Alfadda, H. Benabdelkamel, A. A. Fathaddin, I. O. Alanazi, N. Lauzon, P. Chaurand and A. Masood, *J. Mass Spectrom.*, 2021, **56**, e4671.
102. A. C. Grey and K. L. Schey, *Invest. Ophthalmol. Visual Sci.*, 2009, **50**, 4319–4329.
103. I. Kaya, D. Brinet, W. Michno, M. Baskurt, H. Zetterberg, K. Blenow and J. Hanrieder, *ACS Chem. Neurosci.*, 2017, **8**, 2778–2790.
104. I. Kaya, H. Zetterberg, K. Blennow and J. Hanrieder, *ACS Chem. Neurosci.*, 2018, **9**, 1802–1817.

105. W. Michno, S. Nystrom, P. Wehrli, T. Lashley, G. Brinkmalm, L. Guerard, S. Syvanen, D. Sehlin, I. Kaya, D. Brinet, K. P. R. Nilsson, P. Hammarstrom, K. Blennow, H. Zetterberg and J. Hanrieder, *J. Biol. Chem.*, 2019, **294**, 6719–6732.
106. M. Shariatgorji, A. Nilsson, E. Fridjonsdottir, T. Vallianatou, P. Kallback, L. Katan, J. Savmarker, I. Mantas, X. Zhang, E. Bezard, P. Svenningsson, L. R. Odell and P. E. Andren, *Nat. Methods*, 2019, **16**, 1021–1028.
107. H. Hulme, E. Fridjonsdottir, H. Gunnarsdottir, T. Vallianatou, X. Zhang, H. Wadensten, R. Shariatgorji, A. Nilsson, E. Bezard, P. Svenningsson and P. E. Andren, *Neurobiol. Dis.*, 2020, **137**, 104738.
108. C. H. Zhang, Y. Y. Li, Q. W. Zhang, A. Biondi, V. Fico, R. Persiani, X. C. Ni and M. Luo, *Front. Oncol.*, 2018, **8**, 628.
109. A. P. Brown, D. K. Gaffney, M. K. Dodson, A. P. Soisson, T. W. Belnap, K. Alleman and W. T. Sause, *Int. J. Gynecol. Cancer*, 2013, **23**, 861–868.
110. B. L. Seagle, D. Gilchrist-Scott, S. Graves, A. E. Strohl, W. Nieves-Neira and S. Shahabi, *JCO Clin. Cancer Inform.*, 2017, **1**, 1–15.
111. Y. K. Zhang, Z. D. Chai, L. L. Tan, Z. Y. Wang, Z. J. Chen, H. B. Le and W. Y. Zhu, *World J. Surg. Oncol.*, 2017, **15**, 64.
112. D. Hawes, A. M. Neville and R. J. Cote, *Biomed. Pharmacother.*, 2001, **55**, 229–242.
113. H. Pan, H. Wang, M. Qian, X. Mao, G. Shi, G. Ma, M. Yu, H. Xie, L. Ling, Q. Ding, K. Zhang, S. Wang and W. Zhou, *JAMA Netw. Open*, 2021, **4**, e211809.
114. N. Peyravian, P. Larki, E. Gharib, E. Nazemalhosseini-Mojarad, F. Anaraki, C. Young, J. McClellan, M. Ashrafian Bonab, H. Asadzadeh-Aghdaei and M. R. Zali, *Biomedicines*, 2018, **6**, 27.
115. A. Girard, H. Vila Reyes, H. Shaish, J. F. Grellier, L. Dercle, P. Y. Salaun, O. Delcroix and M. Rouanne, *Front. Oncol.*, 2020, **10**, 565086.
116. S. Woo, C. H. Suh, S. Y. Kim, J. Y. Cho and S. H. Kim, *AJR, Am. J. Roentgenol.*, 2018, **210**, W95–W109.
117. P. Mittal, M. Klingler-Hoffmann, G. Arentz, L. Winderbaum, N. A. Lokman, C. Zhang, L. Anderson, J. Scurry, Y. Leung, C. J. Stewart, J. Carter, G. Kaur, M. K. Oehler and P. Hoffmann, *Proteomics*, 2016, **16**, 1793–1801.
118. F. Boyaval, R. Van Zeijl, H. Dalebout, S. Holst, G. van Pelt, A. Farina-Sarasqueta, W. Mesker, R. Tollenaar, H. Morreau, M. Wuhrer and B. Heijs, *Mol. Cell. Proteomics*, 2021, 100057.
119. A. Hinsch, M. Buchholz, S. Odinga, C. Borkowski, C. Koop, J. R. Izbicki, M. Wurlitzer, T. Krech, W. Wilczak, S. Steurer, F. Jacobsen, E. C. Burandt, P. Stahl, R. Simon, G. Sauter and H. Schluter, *J. Mass Spectrom.*, 2017, **52**, 165–173.
120. K. Yanagisawa, S. Tomida, Y. Shimada, Y. Yatabe, T. Mitsudomi and T. Takahashi, *J. Natl. Cancer Inst.*, 2007, **99**, 858–867.
121. M. Nipp, M. Elsner, B. Balluff, S. Meding, H. Sarioglu, M. Ueffing, S. Rauser, K. Unger, H. Hofler, A. Walch and H. Zitzelsberger, *J. Mol. Med.*, 2012, **90**, 163–174.

122. P. Mittal, M. Klingler-Hoffmann, G. Arentz, L. Winderbaum, G. Kaur, L. Anderson, J. Scurry, Y. Leung, C. J. Stewart, J. Carter, P. Hoffmann and M. K. Oehler, *Biochim. Biophys. Acta, Proteins Proteomics*, 2017, **1865**, 846–857.
123. P. Mittal, M. Briggs, M. Klingler-Hoffmann, G. Kaur, N. H. Packer, M. K. Oehler and P. Hoffmann, *Anal. Bioanal. Chem.*, 2021, **413**, 2721–2733.
124. M. T. Briggs, M. R. Condina, Y. Y. Ho, A. V. Everest-Dass, P. Mittal, G. Kaur, M. K. Oehler, N. H. Packer and P. Hoffmann, *Proteomics*, 2019, **19**, e1800482.
125. R. Longuespee, C. Boyon, C. Castellier, A. Jacquet, A. Desmons, O. Kerdraon, D. Vinatier, I. Fournier, R. Day and M. Salzet, *Histochem. Cell Biol.*, 2012, **138**, 141–154.
126. D. A. Scott, R. Casadonte, B. Cardinali, L. Spruill, A. S. Mehta, F. Carli, N. Simone, M. Kriegsmann, L. Del Mastro, J. Kriegsmann and R. R. Drake, *Proteomics: Clin. Appl.*, 2019, **13**, e1800014.
127. L. Phillips, A. J. Gill and R. C. Baxter, *Front. Oncol.*, 2019, **9**, 379.
128. H. Herrera, T. Dilday, A. Uber, D. Scott, J. N. Zambrano, M. Wang, P. M. Angel, A. S. Mehta, R. R. Drake, E. G. Hill and E. S. Yeh, *Int. J. Mol. Sci.*, 2019, **20**, 2528.
129. T. Goto, N. Terada, T. Inoue, T. Kobayashi, K. Nakayama, Y. Okada, T. Yoshikawa, Y. Miyazaki, M. Uegaki, N. Utsunomiya, Y. Makino, S. Sumiyoshi, T. Yamasaki, T. Kamba and O. Ogawa, *Prostate*, 2015, **75**, 1821–1830.
130. S. Steurer, C. Borkowski, S. Odinga, M. Buchholz, C. Koop, H. Huland, M. Becker, M. Witt, D. Trede, M. Omidi, O. Kraus, A. S. Bahar, A. S. Seddiqi, J. M. Singer, M. Kwiatkowski, M. Trusch, R. Simon, M. Wurlitzer, S. Minner, T. Schlomm, G. Sauter and H. Schluter, *Int. J. Cancer*, 2013, **133**, 920–928.
131. S. Steurer, J. M. Singer, M. Rink, F. Chun, R. Dahlem, R. Simon, E. Burandt, P. Stahl, L. Terracciano, T. Schlomm, W. Wagner, W. Hoppner, M. Omidi, O. Kraus, M. Kwiatkowski, O. Doh, M. Fisch, A. Soave, G. Sauter, M. Wurlitzer, H. Schluter and S. Minner, *Urol. Oncol.*, 2014, **32**, 1225–1233.
132. A. J. Murphy, J. Pierce, E. H. Seeley, L. M. Sullivan, E. D. Ruchelli, M. L. Nance, R. M. Caprioli and H. N. Lovvorn, 3rd, *J. Surg. Res.*, 2015, **196**, 332–338.
133. B. Balluff, A. Buck, M. Martin-Lorenzo, F. Dewez, R. Langer, L. A. McDonnell, A. Walch and R. M. A. Heeren, *Proteomics: Clin. Appl.*, 2019, **13**, e1800137.
134. M. Elsner, S. Rauser, S. Maier, C. Schone, B. Balluff, S. Meding, G. Jung, M. Nipp, H. Sarioglu, G. Maccarrone, M. Aichler, A. Feuchtinger, R. Langer, U. Jutting, M. Feith, B. Kuster, M. Ueffing, H. Zitzelsberger, H. Hofler and A. Walch, *J. Proteomics*, 2012, **75**, 4693–4704.
135. T. Kunzke, B. Balluff, A. Feuchtinger, A. Buck, R. Langer, B. Luber, F. Lordick, H. Zitzelsberger, M. Aichler and A. Walch, *Oncotarget*, 2017, **8**, 68012–68025.

136. B. Balluff, S. Rauser, S. Meding, M. Elsner, C. Schone, A. Feuchtinger, C. Schuhmacher, A. Novotny, U. Jutting, G. Maccarrone, H. Sarioglu, M. Ueffing, H. Braselmann, H. Zitzelsberger, R. M. Schmid, H. Hofler, M. P. Ebert and A. Walch, *Am. J. Pathol.*, 2011, **179**, 2720–2729.
137. N. E. Mascini, J. Teunissen, R. Noorlag, S. M. Willems and R. M. A. Heeren, *Methods*, 2018, **151**, 21–27.
138. N. Pote, T. Alexandrov, J. Le Faouder, S. Laouirem, T. Leger, M. Mebarki, J. Belghiti, J. M. Camadro, P. Bedossa and V. Paradis, *Hepatology*, 2013, **58**, 983–994.
139. N. Sun, T. Kunzke, S. Sbiera, S. Kircher, A. Feuchtinger, M. Aichler, S. Herterich, C. L. Ronchi, I. Weigand, N. Schlegel, J. Waldmann, M. Candida Villares Fragoso, T. G. Whitsett, A. J. Gill, M. Fassnacht, A. Walch and M. Kroiss, *Clin. Chem.*, 2019, **65**, 1276–1286.
140. W. M. Hardesty, M. C. Kelley, D. Mi, R. L. Low and R. M. Caprioli, *J. Proteomics*, 2011, **74**, 1002–1014.
141. S. A. Schwartz, R. J. Weil, R. C. Thompson, Y. Shyr, J. H. Moore, S. A. Toms, M. D. Johnson and R. M. Caprioli, *Cancer Res.*, 2005, **65**, 7674–7681.
142. B. Heijs, S. Holst-Bernal, M. A. de Graaff, I. H. Briaire-de Bruijn, M. Rodriguez-Girondo, M. A. J. van de Sande, M. Wuhrer, L. A. McDonnell and J. Bovee, *Lab. Invest.*, 2020, **100**, 1252–1261.
143. S. Lou, B. Balluff, A. H. G. Cleven, J. Bovee and L. A. McDonnell, *J. Am. Soc. Mass Spectrom*, 2017, **28**, 376–383.
144. S. Lou, A. H. Cleven, B. Balluff, M. de Graaff, M. Kostine, I. Briaire-de Bruijn, L. A. McDonnell and J. V. Bovee, *Clin. Sarcoma Res.*, 2016, **6**, 17.
145. D. Baumhoer, M. Elsner, J. Smida, S. Zillmer, S. Rauser, C. Schoene, B. Balluff, S. Bielack, G. Jundt, A. Walch and M. Nathrath, *Oncotarget*, 2011, **2**, 970–975.
146. I. Alves, B. Santos-Pereira, H. Dalebout, S. Santos, M. M. Vicente, A. Campar, M. Thepaut, F. Fieschi, S. Strahl, F. Boyaval, R. Vizcaino, R. Silva, S. Holst-Bernal, C. Vasconcelos, L. Santos, M. Wuhrer, A. Marinho, B. Heijs and S. S. Pinho, *Arthritis Rheumatol.*, 2021, DOI: 10.1002/art.41768.
147. M. Ivanova, O. Dyadyk, D. Ivanov, F. Clerici, A. Smith and F. Magni, *Virchows Arch.*, 2020, **476**, 903–914.
148. M. Murakami, Y. Rhayem, T. Kunzke, N. Sun, A. Feuchtinger, P. Ludwig, T. M. Strom, C. Gomez-Sanchez, T. Knosel, T. Kirchner, T. A. Williams, M. Reincke, A. K. Walch and F. Beuschlein, *JCI Insight*, 2019, **4**, e130356.
149. E. Berghmans, K. Boonen, E. Maes, I. Mertens, P. Pauwels and G. Baggerman, *J. Pers. Med.*, 2020, **10**, 54.
150. D. M. Pardoll, *Nat. Rev. Cancer*, 2012, **12**, 252–264.
151. F. K. Dermani, P. Samadi, G. Rahmani, A. K. Kohlan and R. Najafi, *J. Cell. Physiol.*, 2019, **234**, 1313–1325.
152. X. Wang, F. Wang, M. Zhong, Y. Yarden and L. Fu, *Mol. Cancer*, 2020, **19**, 81.

153. E. Berghmans, J. Jacobs, C. Deben, C. Hermans, G. Broeckx, E. Smits, E. Maes, J. Raskin, P. Pauwels and G. Baggerman, *Cancers*, 2020, **12**, 863.
154. N. E. Mascini, G. B. Eijkel, P. ter Brugge, J. Jonkers, J. Wesseling and R. M. Heeren, *J. Proteome Res.*, 2015, **14**, 1069–1075.
155. J. A. Bauer, A. B. Chakravarthy, J. M. Rosenbluth, D. Mi, E. H. Seeley, N. De Matos Granja-Ingram, M. G. Olivares, M. C. Kelley, I. A. Mayer, I. M. Meszoely, J. A. Means-Powell, K. N. Johnson, C. J. Tsai, G. D. Ayers, M. E. Sanders, R. J. Schneider, S. C. Formenti, R. M. Caprioli and J. A. Pietenpol, *Clin. Cancer Res.*, 2010, **16**, 681–690.
156. M. Aichler, M. Elsner, N. Ludyga, A. Feuchtinger, V. Zangen, S. K. Maier, B. Balluff, C. Schone, L. Hierber, H. Braselmann, S. Meding, S. Rauser, H. Zischka, M. Aubele, M. Schmitt, M. Feith, S. M. Hauck, M. Ueffing, R. Langer, B. Kuster, H. Zitzelsberger, H. Hofler and A. K. Walch, *J. Pathol.*, 2013, **230**, 410–419.
157. N. H. Patterson, B. Alabdulkarim, A. Lazaris, A. Thomas, M. M. Marcinkiewicz, Z. H. Gao, P. B. Vermeulen, P. Chaurand and P. Metrakos, *Sci. Rep.*, 2016, **6**, 36814.
158. R. Lemaire, A. Desmons, J. C. Tabet, R. Day, M. Salzet and I. Fournier, *J. Proteome Res.*, 2007, **6**, 1295–1305.
159. R. Casadonte and R. M. Caprioli, *Nat. Protoc.*, 2011, **6**, 1695–1709.
160. J. M. Spraggins, D. G. Rizzo, J. L. Moore, M. J. Noto, E. P. Skaar and R. M. Caprioli, *Proteomics*, 2016, **16**, 1678–1689.
161. D. M. Anderson, J. M. Spraggins, K. L. Rose and K. L. Schey, *J. Am. Soc. Mass Spectrom*, 2015, **26**, 940–947.
162. K. Huber, P. Khamehgir-Silz, T. Schramm, V. Gorshkov, B. Spengler and A. Rompp, *Anal. Bioanal. Chem.*, 2018, **410**, 5825–5837.
163. K. Scupakova, F. Dewez, A. K. Walch, R. M. A. Heeren and B. Balluff, *Angew. Chem., Int. Ed. Engl.*, 2020, **59**, 17447–17450.
164. M. Stoeckli and D. Staab, *J. Am. Soc. Mass Spectrom*, 2015, **26**, 911–914.
165. P. M. Angel, K. Norris-Caneda and R. R. Drake, *Curr. Protoc. Protein Sci.*, 2018, **94**, e65.
166. A. M. Judd, D. B. Gutierrez, J. L. Moore, N. H. Patterson, J. Yang, C. E. Romer, J. L. Norris and R. M. Caprioli, *J. Mass Spectrom.*, 2019, **54**, 716–727.
167. B. Beine, H. C. Diehl, H. E. Meyer and C. Henkel, *Methods Mol. Biol.*, 2016, **1394**, 129–150.
168. T. Porta Siegel, G. Hamm, J. Bunch, J. Cappell, J. S. Fletcher and K. Schwamborn, *Mol. Imaging Biol.*, 2018, **20**, 888–901.
169. K. Schwamborn, M. Kriegsmann and W. Weichert, *Biochim. Biophys. Acta, Proteins Proteomics*, 2017, **1865**, 776–783.
170. I. Kaya, W. Michno, D. Brinet, Y. Iacone, G. Zanni, K. Blennow, H. Zetterberg and J. Hanrieder, *Anal. Chem.*, 2017, **89**, 4685–4694.
171. K. Kriegsmann, R. Longuespee, M. Hundemer, C. Zgorzelski, R. Casadonte, K. Schwamborn, W. Weichert, P. Schirmacher, A. Harms, D. Kazdal, J. Leichsenring, A. Stenzinger, A. Warth, M. Fresnais, J. Kriegsmann and M. Kriegsmann, *Proteomics: Clin. Appl.*, 2019, **13**, e1800035.

172. D. Kazdal, R. Longuespee, S. Dietz, R. Casadonte, K. Schwamborn, A. L. Volckmar, J. Kriegsmann, K. Kriegsmann, M. Fresnais, A. Stenzinger, H. Sultmann, A. Warth and M. Kriegsmann, *Proteomics: Clin. Appl.*, 2019, **13**, e1800034.
173. A. Srivastava and D. J. Creek, *Proteomics*, 2019, **19**, e1700448.
174. J. M. Koomen, E. B. Haura, G. Bepler, R. Sutphen, E. R. Remily-Wood, K. Benson, M. Hussein, L. A. Hazlehurst, T. J. Yeatman, L. T. Hildreth, T. A. Sellers, P. B. Jacobsen, D. A. Fenstermacher and W. S. Dalton, *Mol. Cell. Proteomics*, 2008, **7**, 1780–1794.
175. K. Schwamborn, *J. Proteomics*, 2012, **75**, 4990–4998.
176. D. F. Ransohoff, *Nat. Rev. Cancer*, 2004, **4**, 309–314.
177. S. G. Baker, B. S. Kramer, M. McIntosh, B. H. Patterson, Y. Shyr and S. Skates, *Clin. Trials*, 2006, **3**, 43–56.
178. E. S. Boja, T. E. Fehniger, M. S. Baker, G. Marko-Varga and H. Rodriguez, *J. Proteome Res.*, 2014, **13**, 5325–5332.
179. B. Balluff, C. Hopf, T. Porta Siegel, H. I. Grabsch and R. M. A. Heeren, *J. Am. Soc. Mass Spectrom*, 2021, **32**, 628–635.
180. A. L. Oberg and O. Vitek, *J. Proteome Res.*, 2009, **8**, 2144–2156.
181. A. Ly, R. Longuespee, R. Casadonte, P. Wandernoth, K. Schwamborn, C. Bollwein, C. Marsching, K. Kriegsmann, C. Hopf, W. Weichert, J. Kriegsmann, P. Schirmacher, M. Kriegsmann and S. O. Deininger, *Proteomics: Clin. Appl.*, 2019, **13**, e1800029.
182. K. Erich, D. A. Sammour, A. Marx and C. Hopf, *Biochim. Biophys. Acta, Proteins Proteomics*, 2017, **1865**, 907–915.
183. D. R. N. Vos, I. Jansen, M. Lucas, M. R. L. Paine, O. J. de Boer, S. L. Meijer, C. D. Savci-Heijink, H. A. Marquering, D. M. de Bruin, R. M. A. Heeren, S. R. Ellis and B. Balluff, *J. Proteomics*, 2019, **193**, 184–191.
184. E. E. Torlakovic, C. C. Cheung, C. D'Arrigo, M. Dietel, G. D. Francis, C. B. Gilks, J. A. Hall, J. L. Hornick, M. Ibrahim, A. Marchetti, K. Miller, J. H. van Krieken, S. Nielsen, P. E. Swanson, M. Vyberg, X. Zhou, C. R. Taylor, From the International Society for Immunohistochemistry and Molecular Morphology (ISIMM) and International Quality Network for Pathology (IQN Path), *Appl. Immunohistochem. Mol. Morphol.*, 2017, **25**, 151–159.
185. A. Buck, B. Heijs, B. Beine, J. Schepers, A. Cassese, R. M. A. Heeren, L. A. McDonnell, C. Henkel, A. Walch and B. Balluff, *Anal. Bioanal. Chem.*, 2018, **410**, 5969–5980.
186. T. J. Dekker, B. D. Balluff, E. A. Jones, C. D. Schone, M. Schmitt, M. Aubele, J. R. Kroep, V. T. Smit, R. A. Tollenaar, W. E. Mesker, A. Walch and L. A. McDonnell, *J. Proteome Res.*, 2014, **13**, 4730–4738.
187. A. Rompp, J. P. Both, A. Brunelle, R. M. Heeren, O. Laprevote, B. Prideaux, A. Seyer, B. Spengler, M. Stoeckli and D. F. Smith, *Anal. Bioanal. Chem.*, 2015, **407**, 2329–2335.
188. S. N. Kapila, K. Boaz and S. Natarajan, *Int. J. Appl. Basic Med. Res.*, 2016, **6**, 3–7.

CHAPTER 18

MALDI Mass Spectometry Imaging for Eye Diseases

ANGUS C. GREY*[a] AND KEVIN L. SCHEY[b]

[a] School of Medical Sciences, University of Auckland, New Zealand;
[b] Department of Biochemistry, Vanderbilt University, USA
*Email: ac.grey@auckland.ac.nz

18.1 Ocular Structure, Function, and Common Diseases

18.1.1 General Introduction to the Eye

The eye is a highly specialized organ that contains a variety of tissues and fluids that facilitate vision. In the anterior eye, major structures include the transparent light focusing elements (cornea and lens), and the pigmented iris which controls the amount of light entering the eye. In the posterior eye, the photoreceptor cells which transduce light into electrical signals are supported by a range of neuronal cells for local visual processing (retina) and connected by the optic nerve to the brain's primary visual cortex for higher visual processing. The ocular fluids (humors) bathe the ocular tissues and constitute the majority of the ocular globe volume. The vitreous humor (VH) provides structural properties to the ocular globe and interfaces posteriorly with the retina, and anteriorly with the posterior surface of the lens. The aqueous humor (AH) bathes the anterior surface of the lens and nourishes the anterior eye to maintain intraocular pressure and the refractive and optical properties of the eye (see Figure 18.1A). When light enters this healthy, integrated organ, it is focused by the cornea and lens.

New Developments in Mass Spectrometry No. 12
MALDI Mass Spectrometry Imaging: From Fundamentals to Spatial Omics
Edited by Tiffany Siegel

Published by the Royal Society of Chemistry, www.rsc.org

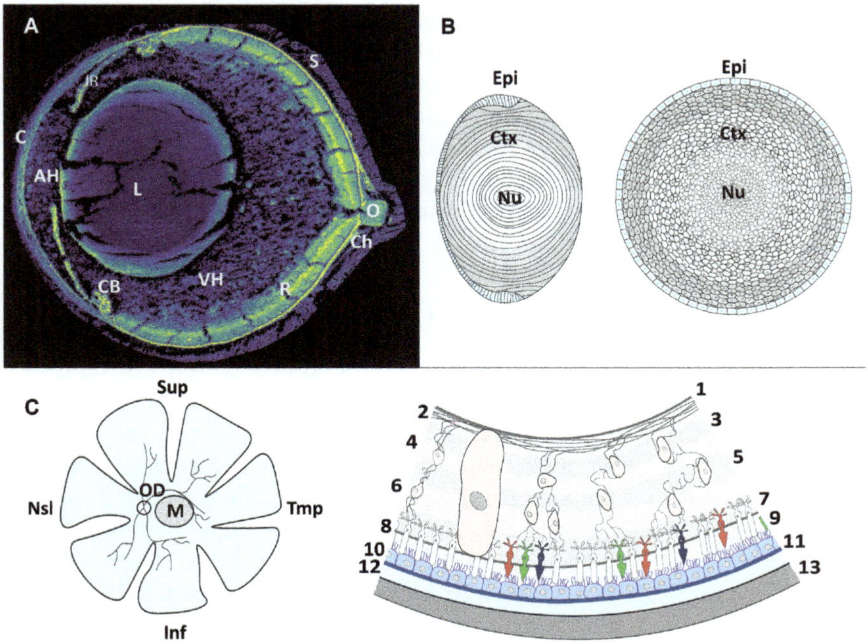

Figure 18.1 Schematic diagram of ocular tissues analyzed my MALDI-MSI. (A) A MALDI image of lipid (*m/z* 327.2309), docosahexaenoic acid depicts the macro-anatomy of a mouse eye. Major structures are labeled; cornea (C), iris (IR), ciliary body (CB), lens (L), sclera (S), choroid (Ch), retina (R), optic nerve (O), and the ocular fluids aqueous humor (AH), and vitreous humor (VH). Reproduced from ref. 16 with permission from John Wiley & Sons, Copyright © 2019 John Wiley & Sons, Ltd. (B) Schematic diagrams of the ocular lens sectioned in either axial (left) or equatorial (right) orientation. The lens is divided into specific regions; epithelium (Epi), cortex (Ctx), and nucleus (Nu). (C) Schematic diagram of a retinal flat mount preparation (left) shows the anatomical regions of the mammalian retina. Sup = superior, Inf = inferior, Nsl = nasal, Tmp = temporal, OD = optic disc, M = macula. Schematic diagram of the posterior eye, which when cut in a cross-section (right) has a characteristic layered structure, consisting of (1) inner limiting membrane, (2) nerve fiber layer, (3) ganglion cell layer, (4) inner plexiform layer, (5) inner nuclear layer, (6) outer plexiform layer, (7) cell bodies of rods (white) and cones (red, green, blue), (8) outer limiting membrane, (9) photoreceptors, (10) retinal pigmented epithelium, (11) Bruch's membrane, (12) choroid, and (13) sclera.

Light strikes the retina, activating photoreceptor cells which communicate with our brain's primary visual cortex to form our sense of sight. There are many diseases that affect visual acuity, which has significant implications for our quality of life and independence. While some major eye diseases are treatable with drugs (*e.g.* uveitis, glaucoma), others remain intractable, despite significant global efforts to develop novel pharmacological treatments, and require surgical solutions (*e.g.* cataracts). Moreover, the implications of drug treatment of one specific tissue of the eye on other ocular tissues are unclear. Matrix-assisted laser desorption/ionization mass spectrometry imaging (MALDI-MSI) has been used extensively to characterize the bio-molecular structure of ocular tissues, investigate aging and disease-related changes, and to trace delivery and interaction of ocular pharmaceuticals with specific tissues of the eye. Before focusing on ocular diseases and treatments, a brief discussion of normal ocular tissue structure and function is required.

18.1.2 Anterior Eye – Cornea, Lens, and Cataract

The cornea consists of three main layers; a multi-layered epithelium, a thicker, collagenous stromal layer that contains keratocytes, and the single cell layered endothelium.[1] To maintain tissue transparency, the cornea is avascular and nourished by the tear film on its epithelial surface, and from the AH on its endothelial surface. Its convex shape is maintained not only by its bio-molecular constituents, but also by intraocular pressure (IOP).[2] IOP is determined by the volume of AH, which is produced by the ciliary body and flows across the lens anterior surface, forward through the iris and drains through the trabecular meshwork into Schlemm's canal. Some corneal dystrophies threaten a patient's eyesight. For example, keratoconus results when the lower part of the cornea thins, and over time bulges outward.[3] This can cause optical distortions such as blur and astigmatism and, in severe cases, replacement of the affected cornea is required.

The lens (see Figure 18.1B) is an avascular, biconvex transparent tissue suspended in the anterior eye by ligaments and is surrounded by the ciliary body.[4] The lens cellular material is contained within a thin, collagenous capsule. Within the capsule, the anterior surface of the lens is covered by an epithelial cell monolayer. Throughout life, epithelial cells migrate to the lens equator where they differentiate and elongate into lens fibers, which make up the bulk of the lens. Continued addition of cells at the lens surface creates a cell age gradient where fiber cells that were formed *in utero* are localized to the lens center (nucleus), while young, elongating fiber cells are found at the lens edge. Dramatic changes in the expression of both soluble and membrane proteins, and lipid components of lens fibers,[5] occur during lens cell differentiation. Remarkably, the ocular lens can remain transparent over many decades of life.[6] However, lens cataract, which is a clouding of the lens, is common and can affect specific regions.[7] It currently has no pharmaceutical therapy, with the only treatment option being surgical removal of the diseased lens and implantation of a synthetic intraocular lens (IOL) within the retained lens capsule. The major risk factors for cataracts are diabetes, which can cause opacities in the lens cortex, and age, which generally causes opacities in the lens nucleus. Age-related nuclear (ARN) cataract is the most prevalent form and is strongly correlated to a loss of the antioxidant glutathione (GSH), which leads to the irreversible cross-linking of highly modified proteins which then form insoluble aggregates.[8]

18.1.3 Posterior Eye – Retina, Macular Degeneration and Optic Nerve

The large, viscous VH separates the lens from the retina.[9] The retina consists of ten distinct layers and is rich in phospholipids (see Figure 18.1C). The photoreceptor cells, called rods and cones, lie on the posterior surface of the retina on top of the retinal pigmented epithelium (RPE). Cones are abundant in the central region (macula/fovea) and facilitate high visual acuity, while rods, which are more light-sensitive and play a role in night vision, are more abundant in the periphery. Degeneration of photoreceptors in the macular region is the cause of multiple blinding diseases.[10] The outer segments of photoreceptors contain high levels of the visual pigment molecules, such as rhodopsin, and are supported and replenished by the RPE. Interestingly, axons from the photoreceptor cells project anteriorly, where the visual signal is modulated by bipolar and ganglion cells before passing along axons of the nerve fiber layer, exiting the eye at the optic nerve head and traveling *via* the optic nerve to the primary visual cortex.

The retinoid all-*trans*-retinal and its related compounds play a central role in visual cycle signaling, interacting with opsin proteins within the photoreceptor cells and RPE. This results in the conversion of a photon of light into electrical signals in the photoreceptors.[11] In addition to its role in the visual cycle, all-*trans*-retinal can react non-enzymatically with phosphatidylethanolamine to form the bis-retinoid A2E.[12] A2E is a major component of lipofuscin, the autofluorescent aging pigment that accumulates in all tissues with age.[13] While A2E is formed in the photoreceptor outer segments, it undergoes transfer to the RPE as a result of the normal outer segment phagocytic process.[12] Its accumulation[12] and toxicity to RPE cells[14] has suggested that it plays a role in macular degeneration.[15] Clinically, retinal autofluorescence is imaged ophthalmoscopically, and aids in the diagnosis and follow-up of many retinal disorders. Another clinical marker of retinal disease is the appearance of drusen and other types of retinal deposits such as subretinal drusenoid deposits and basal laminar deposits, that accumulate extracellular material above and below the RPE. The presence of such deposits represents the highest risk for macular degeneration.

From the brief descriptions above, it is clear that the major ocular diseases involve a wide range of bio-molecular classes. Moreover, these disease-related changes occur in specific tissue regions that can be difficult to isolate and enrich for, making molecular mapping approaches essential. Therefore, MALDI-MSI has been applied to study major ocular diseases and has been instrumental in biomedical studies that have furthered our understanding of the molecular mechanisms behind retinal degeneration, lens cataracts, and clinically in reviewing the appropriate use and effects of drugs on different ocular tissues.

18.2 MALDI-MSI in the Anterior Eye

18.2.1 Corneal Structure, Corneal Dystrophies and Keratoconus

In addition to its light focusing function, the cornea provides a robust barrier to the passage of molecules into the eye.[1] There are a number of MALDI-MSI studies that have analyzed whole eyes,[16–18] including corneal tissue. Tissue preparation for whole eye MALDI-MSI is difficult due to the heterogeneity of ocular tissue and large fluid-filled compartments, necessitating tissue pre-treatment[17] or tape-transfer methods[16,19] to generate high quality tissue sections. These methods allow spatial sampling to be tailored to different ocular regions. For example, low resolution sampling can be performed in the ocular fluids, while higher spatial sampling is required for tissues such as the cornea. Studies providing systematic characterization of normal corneal structure and bio-molecular distributions by MALDI-MSI are limited to mouse and human tissue. In mice, both wild type (WT) corneas and those with a corneal crystallin (ALDH3A1) knocked out (KO), which develop corneal haze and lens cataracts, have been studied.[20] Lipid MALDI-MSI described the three distinct layers of the mouse cornea (*i.e.* epithelium, stroma, and endothelium), with no significant differences observed between the WT and KO. In contrast, protein mapping suggested that differences in putative histone H4.4, acyl-CoA binding protein, and one other unidentified protein, were present in the ALDH3A1 KO mice. It is not clear whether the proteins mapped by MALDI-MSI play a role in the observed pathologies. However, this study showed for the first time that MALDI-MSI had the technological capability to map spatial distributions in corneal tissue. In humans, key proteins from patients with corneal dystrophy, a protein aggregation pathology that causes corneal opacity and decreased visual acuity, have been mapped *via* an on-tissue digestion approach.[21] The differential expression of seven proteins in two patients with different mutations compared to healthy control tissue was identified. Interestingly, these protein distributions co-localized with the amyloid protein deposits that are characteristic of corneal dystrophy. While the number of samples included in this study was low, these results suggested that MALDI-MSI could be performed to further understand the causes of corneal dystrophy.

18.2.2 The Lens in Aging and Cataracts

Due to its inherent cell age gradient and lack of cell turnover, the lens is a unique tissue for studying the aging process and its effect on small molecules, lipids, peptides, and proteins. There are many MALDI-MSI studies that have mapped these molecules in the normal ocular lens, and how these biomolecules change within a single lens section and with increasing age and disease. One challenge is that human tissue is difficult to source. Commonly performed surgery for the lens removes the entire cell mass once it has become opaque (cataract), and replaces it with a synthetic IOL inserted into the remaining lens capsular bag. The routinely used method of phacoemulsification breaks up the structure of the lens, yielding waste tissue in which spatial information is lost. However, as part of this surgery, the anterior portion of the lens capsule is removed as a circular disk. The anterior surface of this tissue has been analyzed by tryptic peptide MALDI-MSI to map proteins in this transparent basement membrane. Interestingly, concentric distributions of apolipoprotein E (ApoE) and collagen IV alpha 1 were detected.[22] In equivalent tissue from patients with pseudoexfoliation syndrome, a systemic extracellular matrix disease, post-translational modifications to ApoE and lysyl oxidase-like 1, found in capsular fibrillary deposits that are characteristic of pseudoexfoliation syndrome, were detected.[23] While MALDI-MSI is slowly pushing into the clinical space, particularly in its application to cancer diagnosis and prognosis,[24] this study is one of the only examples of the use of MALDI-MSI in an ocular clinical setting.

To analyze human cataract formation, old lenses are available through donor programs, while there are far fewer young human lenses donated and available for spatially-resolved analysis. However, this makes study of the normal aging process, and its distinction from ARN cataract formation more challenging. Therefore, both animal and human tissue have been analyzed. For example, the antioxidant GSH plays a central role in the protection of the lens against oxidative stress.[25] While it is abundant in the lens, its concentration is known to decrease with increasing age, and this decrease is augmented in ARN cataracts.[26,27] GSH is a tripeptide of cysteine, glycine, and glutamate. Through the free sulfhydryl group on cysteine, it is able to reversibly bind protein sulfhydryl groups and prevent protein mis-folding. Its distribution has been mapped by MALDI–FT-ICR MSI in bovine[28] and human[29] lenses. In the normal bovine lens, free GSH was most abundant in the lens cortex and found at much lower levels in the lens nucleus. In the bovine lens model of aging, a decrease in cortical GSH levels was evident, while oxidized GSH (GSSG) and other GSH conjugates increased in specific lens cortex regions.[28] While this lens aging model appears accurate in some biochemical and optical parameters,[30] the results for GSH are different from what is observed in human lens aging. An absolute quantitation MALDI-MSI approach has been used to confirm previous non-imaging approaches[25] that show that cortical GSH remains relatively constant as we age, while nuclear GSH decreases dramatically (see Figure 18.2A).[29] This decrease in nuclear GSH may be due to other bio-molecular changes that occur in the aging lens which separates the lens into two distinct regions (cortex and nucleus). This results in an inability of the lens to transport free GSH or its component amino acids into the lens nucleus.[31] In addition, conjugation of GSH to other biomolecules takes place with age. This conjugation has been shown by MALDI-MSI with respect to lens UV filters, which are tryptophan derivatives found in the lenses of primates and humans.[32] In young lenses, free UV filters protect other ocular tissues from UV light and are most abundant in

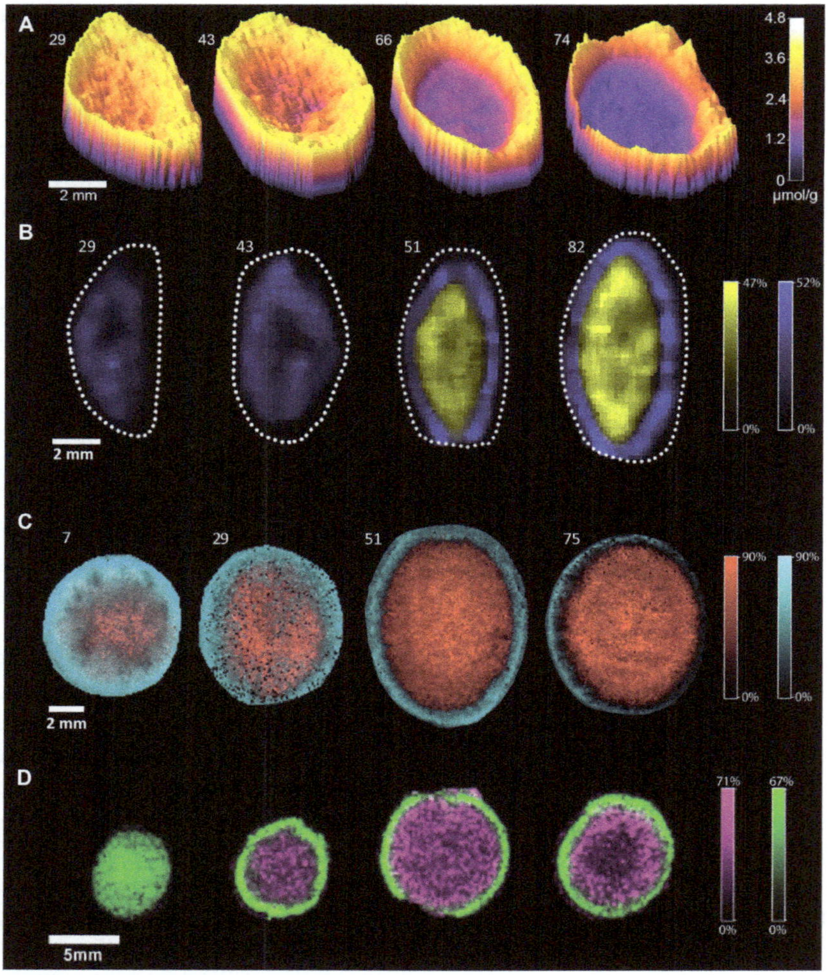

Figure 18.2 Bio-molecular changes in the aging human lens. (A) Absolute quantitative MALDI-MSI of GSH in axial sections of the aging human lens. Reproduced from ref. 29 with permission from Elsevier, Copyright 2017. (B) Ceramide lipid class (*yellow*) and ceramide phosphate lipid class (*blue*) show a marked increase in the human lens nucleus and perinuclear region, respectively, with increasing age, as demonstrated in axial tissue sections. (C) The A subunit of α-crystallin is full-length (*cyan*) in equatorial sections from young lenses and most abundant in the lens cortex. In aging human lenses, the full-length protein is truncated extensively. The αA-crystallin truncation product containing residues 1–40 (*red*) increases in the lens nucleus. (D) Peptide MALDI-MSI of equatorial lens sections show AQP0 modification by lipid addition (*green*) or C-terminal truncation (*purple*) is low in young human lenses, but increases in the old human lens. Reproduced from ref. 61 with permission from The Association for Research in Vision and Ophthalmology, Inc., Copyright 2015.

the lens nucleus. With advancing age, these molecules become linked to proteins[33] and GSH[34] and may promote cataract formation.[35]

One of the changes associated with the barrier to GSH transport is to lens lipids. It is well established that lens lipids differ between species,[5] and across fiber cell ages in a single lens.[5,36] MSI has spatially mapped these changes and allowed a new level of spatial correlation with other lens biomolecules to be performed. While the specific lipids found in animal lenses differ, the characteristic concentric ring patterns of lipid expression are a common feature in all lenses analyzed by MSI. Specific lipid classes have been spatially mapped by MALDI-MSI in bovine,[37] porcine,[17,38,39] and fish[16] lenses, while normal human lenses have been analyzed by MALDI[39–41] and desorption electrospray ionization (DESI).[42,43] DESI-MSI was first used to show dramatic age-related changes to sphingolipids, such as ceramides and dehydroceramides, which increase significantly in the nucleus of old human lenses.[43] This has also been observed by MALDI, and spatially correlated with GSH changes, showing that ceramides increased in the same lens nuclear region that GSH decreased, while ceramide phosphates appeared in a perinuclear ring with increasing age (see Figure 18.2B).[29] It has been proposed that this change in lipids may alter the interaction of lens crystallin proteins with the fiber cell membranes, and contribute to the barrier to GSH transport into the aging lens nucleus.[31]

Initial protein studies focused on mapping the distribution of α-crystallin in normal lenses to establish a point of comparison for cataract model studies. The lens contains three major classes of crystallin proteins, α, β, and γ. Crystallins are particularly amenable to MALDI-MSI analysis since they are relatively small (~20 kDa) and highly abundant in lenses (>300 mg mL^{-1}). Techniques to map each major class have been developed,[44–46] with a particular focus on α-crystallin since its extensive post-translation modifications, which are difficult to detect *via* antibodies, are thought to play a role in cataract formation. α-Crystallin is a heterodimer of A and B chains, which in the lens are structural proteins that also serve as a small heat shock protein to protect against protein mis-folding, aggregation, and subsequent cataract formation. Early studies characterized these two subunits, and their extensive C-terminal truncation and phosphorylation, in bovine[45] and rabbit lenses[47] using MALDI-TOF. Intact αA- and αB-crystallin were most abundant in the cortex, while C-terminal truncation increased in the nucleus. Phosphorylated, full-length αA- and αB-crystallin was most abundant in a specific region in the cortex, although the functional significance of this is not clear. Using similar tissue preparation approaches and instrumentation, a subsequent study of a broad range of ages of human lenses showed similar modifications to α-crystallin in the aging human lens, where full-length α-crystallin is detected in the lens cortex, and truncated α-crystallin is detected in the lens nucleus (see Figure 18.2C).[48] The relative amount and number of different truncation products in human lenses were significantly higher than that found in animal lenses, possibly due to both protein sequence differences between the species, and the considerably older human lens tissue. While truncation is thought to alter the function of α-crystallin, and its molecular chaperone activity in particular,[49] the role that these peptides play in the maintenance of tissue transparency and cataract formation in the aging lens is unclear. In a rodent model of ARN cataract, many α-crystallin protein fragments were mapped in both pre- and post-cataractous lenses.[50] While these fragments were localized to specific lens regions, there were no unique α-crystallin forms that localized to the opaque nuclear region. The same result was found in human lenses which compared low molecular weight (<3 kDa) crystallin peptides in normal aging and an age-matched mildly cataractous human lens.[51] In contrast, a study of human nuclear cataract showed localization of multiple γS-, βA3-, and γD-crystallin peptides in the opaque nucleus.[52] An additional study of cataractous human lens tissue showed elevated α-crystallin protein phosphorylation in the cataractous lens nucleus.[53] These studies utilized very few cataractous lenses, and more extensive studies are required to map protein modifications in human lens cataracts.

While ARN cataract is the most prevalent form, other forms exist. Numerous genetic mutations in the α-crystallin genes have been linked to cataract formation.[54] MALDI–TOF MSI revealed a novel amino acid substitution in the αA-crystallin protein that may be linked to cataract formation.[48] While no details were available regarding the type or severity of the cataract, the L52F substitution was localized to a region of the protein important for its molecular chaperone function. Spatial mapping of the mutant form showed that it co-localized with the wild type, and was not associated with an opaque lens region.

In addition to the critical importance of α-crystallin proteins in the lens, the integral membrane protein AQP0 also plays essential roles in lens tissue transparency.[55] It is a small (~26 kDa), abundant protein that functions as a water channel and cell adhesion molecule. AQP0 was the first integral membrane protein to be spatially mapped in any tissue by MALDI-MSI. Delicate, water-based tissue-washing was used to deplete the soluble crystallin proteins, and acoustic micro-droplet ejection matrix application used to prepare the samples for analysis.[56] MALDI-TOF was able to detect intact AQP0, predominantly in cortical regions, and C-terminally truncated forms, localized to the nucleus, in normal bovine and human tissue. Interestingly, a novel lipid modification (myristoylation) to AQP0 has also been mapped in bovine and human lenses.[57] This modification, which was found in both N- and C-terminal domains, may play a role in targeting AQP0 to specific membrane domains in the normal lens, but whether it is involved in cataract formation is unknown.

The C-terminal domain of AQP0 is the main regulatory domain of the protein, and it interacts with a range of other lens proteins. The AQP0 C-terminus is known to be highly modified, which is thought to be a mechanism used in the lens to alter protein function in specific lens regions. Specifically, C-terminal truncation may alter its water permeability through loss of interaction with calmodulin,[58] and with cytoskeletal proteins, such as filensin.[59,60] In some lenses, such as in rodents, AQP0 C-terminal truncation is likely enzymatic, while in the long-lived human, this process is thought to result primarily from deamidation of asparagine residues and subsequent spontaneous protein backbone cleavage. To spatially map deamidation, which results in a mass shift that is too small (~1 Da) to be resolved at the whole protein level, an on-tissue digestion MALDI-MSI strategy has been used. Following tissue washing and trypsin treatment, unmodified and modified peptides of AQP0 have been spatially mapped in the aging human lens using MALDI–FT-ICR (see Figure 18.2D).[61] This analysis confirmed the previous localization of truncated and lipid-modified AQP0 detected at the intact protein level, and also found differential localization of un-, singly-, and doubly-deamidated AQP0 C-terminal peptides. This study also indicated that AQP0 fatty acid acylation and deamidation occur during early post-partum development, suggesting that AQP0 functions may change throughout lens development and aging.

Recently, a new lens tissue preparation method was developed to detect lower abundance, low solubility cytoskeletal proteins. Using a detergent washing solution and a multiple enzyme strategy, peptides of vimentin, CP49, and filensin were spatially mapped in bovine lenses.[62] Interestingly, truncation of filensin was localized to a similar lens region in which AQP0 and filensin are known to associate. Myristoylation of filensin, which is believed to be important in mediating its interaction with AQP0,[63,64] was also detected. Since these changes were accurately localized to the lens cortex, application of this approach to a cataract model may yield further insight into whether this class of lens proteins, and their interaction with AQP0, is part of a cortical cataract formation process.

18.3 MALDI-MSI in the Posterior Eye

18.3.1 Retina and Macular Degeneration

The layered structure of the retina is exquisite, with multiple distinct yet interconnected cell layers integrating to support transduction of light into electrical signals for transmission *via* the optic nerve to the primary visual cortex. MALDI-MSI studies have defined small molecule, lipid, and protein distributions in retinal cross-sections, and also in flat-mounted retinal tissue to correlate bio-molecular distributions with clinically measurable markers of normal and abnormal retinal function, such as fundus autofluorescence and presence of drusen. These approaches have been used in both animal and human tissue, revealing both structural similarities and important differences, particularly with respect to a major disease of the retina, macular degeneration.

Since the primary function of the retina is to transduce light into electrical signals, understanding changes that take place in light- and dark-adapted retinas is fundamental to understanding retinal function. The spatial metabolome of this tissue has been investigated by MALDI–FT-ICR MSI in porcine tissue.[65] Negative-ion mode analysis of retinal cross-sections allowed a range of metabolites to be detected and localized to specific retinal layers. Interestingly, the spatial distribution and intensity of several central metabolism compounds were different between light- and dark-treated retinas. For example, under the light-treated samples, glucose-6-phosphate was detected at high intensities in both the outer and inner retina, but fructose-1,6-bisphosphate was completely shifted to the inner retina, demonstrating a change in metabolic state of specific retinal layers in the different treated tissue.

The spatial distributions of lipids in normal retinal cross-sections have been defined in fish,[16] salamander,[66] mouse, (see Figure 18.3A)[67–70] and human[71,72] tissue at high spatial resolution. These studies demonstrated the ability of MALDI-MSI to resolve lipid signals in distinct retinal layers, within sub-populations of the same cell type, and in the process generate atlases of the complexity of lipids present in the normal retina. These atlases can be used to compare the lipidomes of specific retinal layers,[73] and also to tissue in pathological states. The spatial lipidome of the macula, a circular-shaped region found in the central retina, and more peripheral retinal areas, has been determined in retinal flat mounts.[74,75] Comparisons of the amounts of the lipid-like bis-retinoids between central and peripheral retina in the single RPE cell layer in normal aged human tissue has also been possible (see Figure 18.3B).[76,77] Bis-retinoids have been implicated in age-related macular degeneration (AMD), where the macula degenerates with age resulting in photoreceptor and vision loss. While all aging retinas contain some level of A2E, other forms of retinal diseases have been studied to further understand its role in retinal pathology. For example, Stargardt disease is a juvenile form of macular degeneration and is caused by mutations in the *ABCA4* gene. Abca4 is an integral membrane protein predominantly localized to photoreceptor outer segments. It functions to transport intermediates in the visual cycle across the photoreceptor cell membrane. The retinas of patients with Stargardt disease accumulate the aging pigment lipofuscin in the RPE cell layer. A genetic knockout mouse model of the *Abcr4* gene has been used to study changes in lipids and lipid-like bis-retinoids such as A2E.

Since A2E is abundant in lipofuscin and is regarded as toxic to RPE cells *in vitro*,[14] a key question in the study of macular degeneration has been whether A2E is responsible *in vivo* for the loss of photoreceptor cells which characterizes both Stargardt disease, and AMD. To address this, autofluorescence measured by confocal microscopy was used to map lipofuscin distribution, and compared to A2E distribution detected in the same retinal flat mounts by subsequent MALDI-MSI analysis. In comparison to WT mice, retinas from the *Abcr4−/−* mouse were found to have elevated levels of A2E in both young and old animals,[78] and also showed high correlation between lipofuscin and A2E distributions (see Figure 18.3C).[79] This approach was also used to identify additional components of lipofuscin and demonstrated that oxidized forms of A2E, which may lead to the formation of toxic A2E

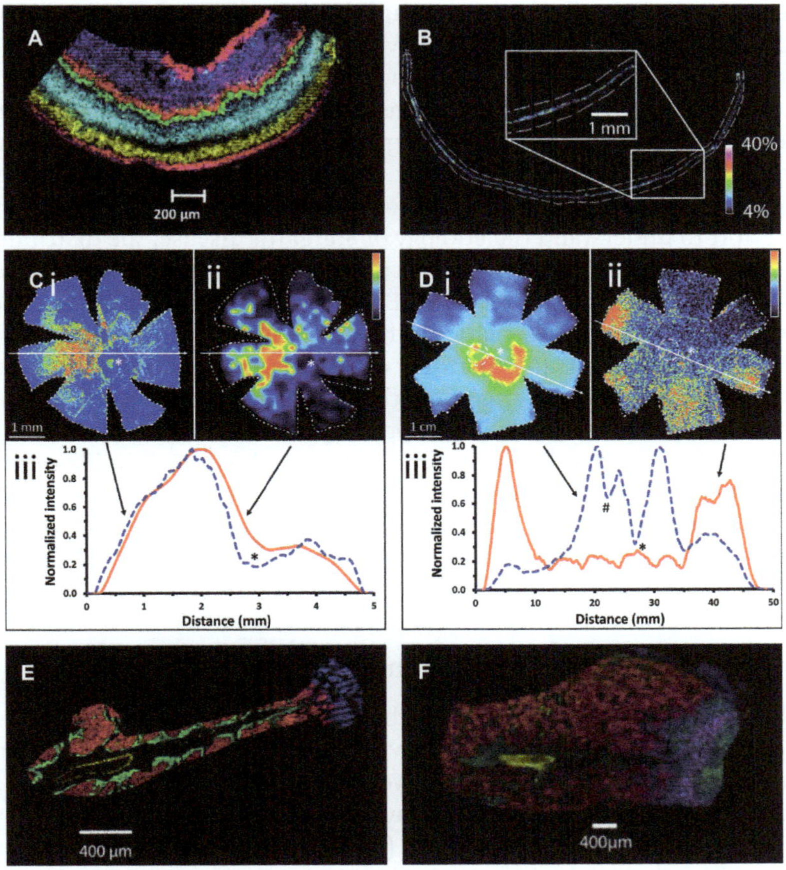

Figure 18.3 Retinal structure and disease-related changes in the posterior eye. (A) High-resolution lipid imaging reveals the elegant layered structure of the murine retina. Reproduced from ref. 70 with permission from American Chemical Society, Copyright 2014. (B) A2E accumulates in the RPE of aging human retina. Reproduced from ref. 76, https://doi.org/10.1038/s41598-017-17402-1, under the terms of the CC BY 4.0 license, http://creativecommons.org/licenses/by/4.0/. (C–D) A2E (i) correlates with (ii) fluorescence-based imaging of the aging pigment lipofuscin in a mouse model of Stargardt disease (C) as shown by line plots (iii, A2E = blue dash, lipofuscin = red), but not in aging human tissue (D). Reproduced from ref. 79 with permission from Elsevier, Copyright 2013. (E) Human optic nerve anatomy is also defined by phospholipid, and (F) protein MALDI-MSI. Reproduced from ref. 85 with permission from American Chemical Society, Copyright 2015.

epoxides,[80] were also present in these retinae.[81] Importantly, when this correlative imaging approach was applied to macaque[82] or human tissue, there was a distinct lack of correlation between lipofuscin fluorescence and A2E distribution (see Figure 18.3D),[83] likely due to anatomical features specific to primates, such as the presence of the macula. Moreover, using

HPLC and UV/VIS spectroscopy, a sampling of 10 eyes showed that the levels of A2E in the central area were on average 3–6 times[84] lower than in peripheral areas of the same eye. Confirmed by MALDI-MSI, this result suggests that the levels of A2E cannot by themselves provide an explanation for the higher lipofuscin fluorescence and degeneration found in the central area of the human RPE.[77] Taken further, the data lead to the question of whether A2E plays a role in the degeneration of macular photoreceptors. Therefore, in an effort to understand the etiology of Stargardt disease, the distributions of a different class of small molecules, bis(monoacylglycero)phosphate (BMP) lipids, were mapped to the RPE in retinas from the *Abcr* −/− mouse model.[76] Due to their association with the endosomal/lysosomal storage diseases Niemann–Pick and neuronal ceroid lipofuscinosis, the presence of BMP lipids in these tissues suggested that perturbations in lipid metabolism associated with late endosomal/lysosomal dysfunction may play a role in the pathogenesis of Stargardt disease. Interestingly, these BMP lipids were also present in aged, healthy human retinas.[76] Although significant progress has been made in understanding the potential roles bis-retinoids and BMPs play in human macular degeneration diseases with MALDI-MSI, their precise roles remain unclear.

18.3.2 Optic Nerve, Trauma and Glioma

In addition to retinal ganglion cells, the optic nerve contains glial cells that produce the myelin that wraps nerve fibers, and together they are ensheathed by the three meningeal layers dura, arachnoid, and pia mater. Spatial maps of lipids and proteins have been defined in rodent and human optic nerve tissue by high-resolution (up to 10 μm) MALDI-MSI. In rodent tissues, phosphatidylcholine species were observed throughout the nerve tissue while sulfatide species were observed in high abundance in the meninges surrounding the optic nerve.[69] In human posterior eye cross-sections, common phospholipids that are abundant in most tissues were not highly localized and were observed throughout the optic nerve and retina. However, sulfatide and plasmalogen glycerophosphoethanolamine (PE) lipids with a monounsaturated fatty acid were found enriched in the optic nerve.[72] A more detailed picture of the distribution of both lipids and proteins was revealed by cutting the human optic nerve in both sagittal and coronal planes. As with the rodent optic nerve, specific lipids were localized to glial cells, nerve fiber bundles, central artery, and dura mater (see Figure 18.3E).[85] In addition, intact protein MALDI-MSI was conducted on sagittal optic nerve sections, and successfully localized proteins such as myelin basic protein and hemoglobin alpha to specific regions (see Figure 18.3F). These proteins were identified by complementary ETD fragmentation-based LC–MS/MS analysis, which has proved to be an efficient way of assigning protein identity in MALDI-MSI datasets.[46] Interestingly, an N-terminal fragment of vimentin was mapped to the optic nerve head region. Vimentin is associated with astrocytes, which are the main glial cell type in the non-myelinated region of the optic nerve head. Therefore, while the localization of this protein in the optic nerve head is consistent with known biology, the significance of the truncation is unknown.[85]

With these lipid and protein maps of the optic nerve defined in rodents and humans, a spatially-resolved assessment of the bio-molecular changes associated with optic nerve pathologies is possible. In the first example, the ability of inflammation-induced optic nerve regeneration (IIR) to limit molecular changes associated with optic nerve crush injury was assessed.[86] Using intravitreal injection to induce IIR, and monitoring changes in lipids by MALDI-MSI, the area of the crush site that was occupied by mass spectral features consistent with lysosome-specific lipids increased, suggesting the presence of a robust phagocytic response that improves clearance of myelin and axon debris. These results extended the current understanding of the mechanism of IIR in optic nerve crush injury.

A second example used a 3-D MALD-MSI approach to map protein distribution changes in optic glioma.[87] Neurofibromatosis type 1 (NF1) is a common neurogenetic disorder, in which affected individuals develop tumors of the nervous system. Children with NF1 are particularly prone to brain tumors (gliomas) involving the optic pathway that can result in impaired vision. This study identified molecular similarities and differences between glioma-bearing and WT tissues. For example, while some myelin basic protein isoforms were similar between WT and optic glioma samples, signals for myelin basic protein (1–43), and other abundant proteins such as cytochrome oxidase 6B1 and dynein light chain 1, showed differences in different anatomical sub-regions, providing potential new clues to tissue changes induced by cancerous glioma cells.

18.4 MALDI-MSI for Ocular Therapeutics Detection

18.4.1 Advantages and Applications in Pre-clinical Settings

While a broad range of endogenous biomolecules has been studied extensively in the eye with MALDI-MSI, the number of studies tracing exogenous biomolecules and pharmaceuticals is more limited. Previous techniques to map the tissue distribution of pharmaceuticals, such as quantitative whole-body autoradiography (QWBA), rely on the synthesis of radioactive forms of the compound and are unable to distinguish the parent compound from its *in vivo* metabolites. In contrast, MALDI-MSI requires no radioactive compounds, and can readily trace the dosed pharmaceutical and its metabolites due to the resulting *m/z* differences generated by metabolism. The advantages of MALDI-MSI were evident in a combined imaging study that showed that neither a dosed, proprietary drug candidate nor its metabolites accumulated in the anterior eye, mitigating the concern over ocular toxicity.[88]

In *ex vivo* studies, penetration of glucose[89,90] and an array of common ocular drugs into the isolated lens[91] have been studied to understand normal lens function and its role in ocular drug pharmacokinetics. In porcine lenses, 5 drugs with a range of lipophilicities were detected in lenses incubated in a drug cocktail for 4 hours (see Figure 18.4A). Interestingly, despite these differences, all detected drugs were excluded from the nucleus, suggesting a barrier to the movement of these drugs into the lens region most commonly affected by lens cataracts. However, whether this potential barrier

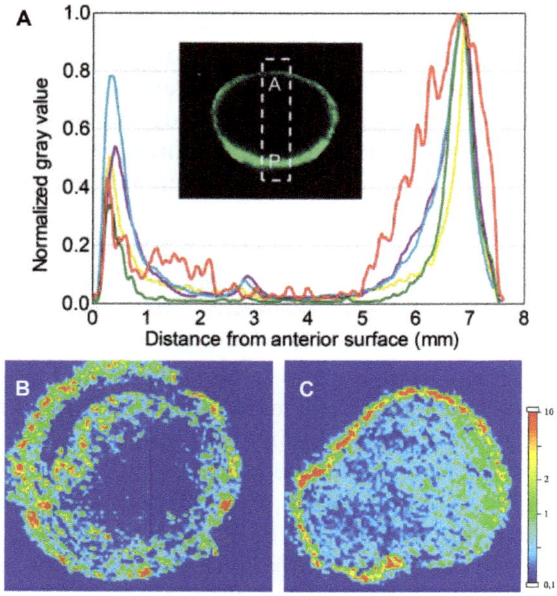

Figure 18.4 MALDI-MSI of pharmaceutical compounds in ocular tissue. (A) Penetration of ocular drugs into isolated pig lenses to investigate the role of the lens in ocular drug pharmacokinetics. Line plots of each drug detected by MALDI-MSI (*red = pilocarpine, yellow = atropine, purple = tizanidine, blue = propranolol, green = pindolol*) show exclusion from the lens nucleus. Reproduced from ref. 91, https://doi.org/10.1021/acs.molpharmaceut.9b00585, with permission from American Chemical Society, Copyright 2019. (B) Distribution of riboflavin in human cornea sections after standard corneal collagen cross-linking therapy, and (C) iontophoresis-enhanced corneal collagen cross-linking. Reproduced from ref. 95, https://doi.org/10.1155/2014/404587, under the terms of the CC BY 3.0 license, https://creativecommons.org/licenses/by/3.0/. (D) Retigabine and *N*-acetyl metabolite of retigabine (E) are localized to the pigmented tissues of the eye. Reproduced from ref. 96 with permission from American Chemical Society, Copyright 2019. (F) *In vivo* experiments assessing the uptake and routes of delivery of brimonidine for glaucoma treatment at 15 min, and (G) 60 min post-dosing. Reproduced from ref. 97 with permission from SAGE Publishing, Copyright 2017. (H) Atropine distributions in the whole mouse eye in an intensity-manipulated image to highlight the signal at the back of the eye, and (I) at optimized intensity to show relative amounts of atropine across the whole eye. Reproduced from ref. 18, https://doi.org/10.1177%2F2472555217724780, under the terms of the CC BY 4.0 license, https://creativecommons.org/licenses/by/4.0/.

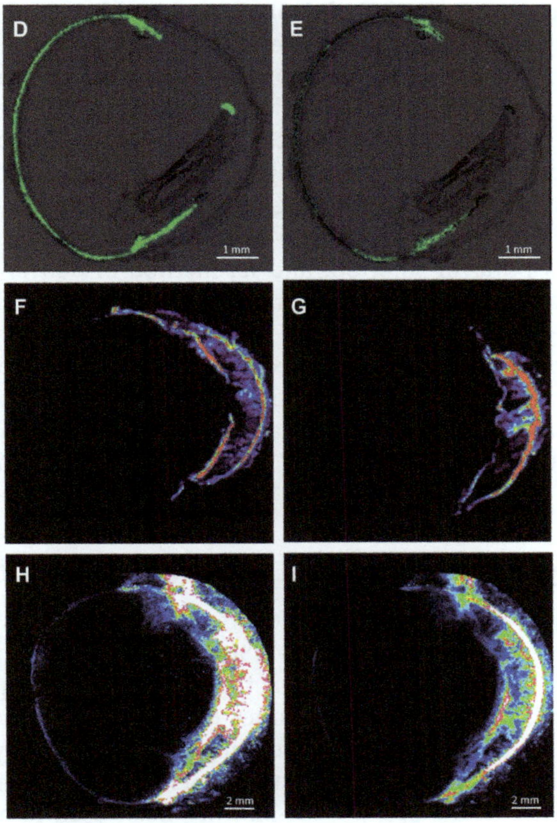

Figure 18.4 Continued.

to drug movement is the same as the barrier to GSH movement observed in the human lens is unknown.[31] In addition, the anterior surface appeared to be more resistant to drug uptake than the posterior lens surface, possibly due to the anterior epithelial cell layer. While there is currently no effective pharmaceutical treatment for cataracts in humans, these studies offer additional functional information to support the development of future, novel anti-cataract therapies.

The preventative effect that a healthy diet can have on the development of major diseases, including cataract and AMD, is well known. In particular, the high dietary intake of omega-3 ($n-3$) long chain polyunsaturated fatty acids (C18–22) is associated with a lower risk of developing AMD.[92] In addition, a higher intake of eicosapentaenoic acid (EPA) (20 : 5, $n-3$) and docosahexaenoic acid (DHA) (22 : 6, $n-3$) is protective against the progression of AMD.[93] A recent MALDI-MSI study defined the efficacy of different dietary formulations of triglycerides and/or phospholipids with EPA and DHA for their incorporation into the RPE cells, and retina in the rat.[94] Interestingly, both EPA and DHA enhanced DHA increase in the retina, with elevated DHA-containing phospholipids localized around the optic nerve. This provided the first spatial quantitative data on very long chain polyunsaturated fatty acid changes in the retina following different $n-3$ rich diets.

In an example of the application of MALDI-MSI in clinical and translational research, a combination of animal model and human donor tissue has been used to assess the effectiveness of different therapeutic methods to address keratoconus.[95] Corneal collagen cross-linking, which is the only effective treatment to halt disease progression by increasing corneal rigidity, is catalyzed by riboflavin. Therefore, MALDI-MSI was used to determine the most effective riboflavin solution, and its method of delivery, by mapping riboflavin penetration into both rabbit and human cadaveric corneas. While the combination of MALDI-MSI and stress–strain measurements showed that the standard protocol, which required the removal of the corneal epithelium, remained the best method to arrest keratoconus *via* collagen cross-linking, MALDI-MSI showed that the iontophoresis-based riboflavin penetration, which retains the corneal epithelium, showed promise as a new approach (see Figure 18.4B–C).

18.4.2 Drug Detection in the Eye for Off-target Effects

The spatial distribution of pharmaceuticals and their metabolites have been studied in the whole eye with MALDI-MSI using *in vivo* approaches, dosing the animals either systemically[96] or topically[18,97,98] as appropriate, to understand both off-target and on-target effects. In the case of off-target effects, some drugs are known to have a high binding affinity for melanin. Therefore, these drugs can accumulate in melanin-containing tissues and cause pathologies. In the eye, MALDI-MSI has been used to map the tissue distribution of two drugs that show ocular pathologies. First, chloroquine, an anti-malarial drug associated with corneal deposits and pigmentary retinopathy, was spatially mapped using a Q-TOF instrument running in MS/MS mode.[99] This approach confirmed that chloroquine could be detected in the retina of rats, which was consistent with its known binding to melanin in the RPE cell layer[100] and the pigmentary retinopathy that is associated with long-term chloroquine use.

More recently, the distribution of retigabine, an anti-epileptic drug, and its metabolites was investigated in whole eyes of rats dosed systemically.[96] In humans, long-term use of retigabine has been associated with potential vision loss due to pigment changes in the retina. The rat eye study formed part of a long-term observational study by the drug manufacturer, GlaxoSmithKline, to understand the mechanism of retinal dyspigmentation associated with retigabine treatment. Ions representing retigabine and an *N*-acetyl metabolite of retigabine were detected, predominantly in melanin-containing layers of the uveal tract including the choroid, ciliary body, and iris (see Figure 18.4D–E). Higher spatial resolution analysis (5–10 μm) of the retinal layers showed that retigabine and its metabolites were restricted to the choroid, and were not present in the RPE layer. Importantly, this demonstrated that the blood–retina barrier excluded retigabine and its metabolites, and showed that their observed accumulation was not phototoxic. While this study was not directly translatable to observations in humans, it contributed to a reassessment of retigabine use, which led to permanent discontinuation of the drug. This study is an excellent example of how MALDI-MSI is informing critical decisions that impact the clinical space.

18.4.3 On-target Ocular Drug Detection

To treat eye diseases, particularly those that affect the anterior eye, topical administration is the most popular method of drug delivery due to its relative simplicity, targeted nature, and reduced potential for off-target effects in comparison to systemic delivery. Treating tissues in the posterior eye by this delivery method is more challenging, and MALDI-MSI has been used to assess the efficacy of this approach. One study has traced the delivery of brimonidine, a selective α2-adrenergic receptor agonist that is used in the treatment of glaucoma.[97] *In vivo* dosing of rabbit eyes with brimonidine provided high quality samples that allowed localization of the drug in whole eye sections at two time-points following dosing (see Figure 18.4F–G). Brimonidine lowers intraocular pressure by decreasing AH outflow and increasing AH clearance, likely due to an effect on the ciliary body in the anterior eye. This study showed that brimonidine localized to this area in the same time frame as its therapeutic effect (~60 min after dosing). Interestingly, brimonidine was also detected at low levels in the posterior eye, although the precise tissue localization was unclear. The uveal-scleral route was proposed as the primary route of transit of brimonidine to the posterior eye. A more recent study, also in rabbits dosed *in vivo*, investigated the distribution of beta blockers across the ocular surface, and their uptake into the anterior eye over time.[101] Interestingly, these drugs were rapidly taken up in the cornea and AH (at 10 min), and were found at a lower concentration 60 min after dosing. Furthermore, the three beta blockers were distributed unevenly across the ocular surface. A comparison of drug levels in different corneal layers, which was validated by LC–MS methods, showed that their concentrations were more similar in corneal stroma than epithelium. Together, these studies confirm that MALDI-MSI can be used to study the behavior of topically-administered ocular drugs.

Another common topically-administered ocular drug, atropine, is used to treat several eye conditions, such as uveitis, early amblyopia, and to slow the progress of myopia in children. Myopia is rising in prevalence and is predicted to affect over one billion people worldwide by 2050. While atropine is a known muscarinic acetylcholine receptor antagonist, surprisingly its sites and mechanism of action in slowing myopia progression are unknown, despite its regular prescription. To address this, two recent MALDI-MSI studies have used topical administration of atropine (1%) into *in vivo* rabbit and mouse eyes. Using either MALDI–TOF/TOF[98] or MALDI–FT-ICR[18] approaches, atropine was detected primarily in the tissues of the anterior eye (*i.e.* cornea, ciliary body, lens, conjunctiva, and sclera) and the AH (see Figure 18.4H). Interestingly, both studies found evidence of atropine also reaching regions of the posterior eye, such as the VH and retina, either 30 min or 5–72 h following topical application (see Figure 18.4I). While the study by Wang *et al.* was consistent with the absorption of atropine *via* the transcorneal and transconjunctival-scleral routes, the route for transit to the posterior eye was not discussed in this study. However, the results presented by Mori *et al.* supported the assertion that atropine reaches the posterior eye *via* a periocular-scleral route. Together these studies are consistent with the possibility that atropine acts in multiple regions of the eye to slow the progression of myopia in children.

In addition to the active compound, eye-drop formulations contain additives such as physiological salts and preservatives. Benzalkonium chloride (BAK) has been used as a preservative in anti-glaucoma eye-drops and is thought to enhance penetration of the active pharmaceutical.[102] However, BAK can lead to eye irritation, inflammation, and cell death processes, especially in long-term treatment.[103] MALDI-MSI has detected BAK in New Zealand white rabbits dosed with two different concentrations *via* topical eye-drops.[104] MALDI-MSI of rabbit eye cross-sections showed presence of BAK on the cornea, and on the retina in the vicinity of the optic nerve. Immunohistochemistry of the same tissue suggested inflammatory cell infiltration in the BAK-treated eyes. Together these results suggested that BAK may be having deleterious tissue effects. The authors suggested that preservative-free eye-drops should be considered for glaucoma patients and other diseases requiring long-term treatment with eye-drops.

For more severe ocular diseases, alternative routes of therapeutic administration are required. Retinoblastoma is a rare form of cancer that develops in the retina and predominantly affects young children. Melphalan is a chemotherapeutic drug used to treat retinoblastoma. The tissue distribution of melphalan was measured by MALDI-MSI after intra-arterial injection into the ophthalmic artery of rabbits.[105] Drug distribution was highest in the retina compared to VH, with little drug detected in supporting scleral tissues. This approach gave excellent vitreous and retinal tissue penetration of the drug, and also achieved drug levels in these regions that are sufficient to kill human retinoblastoma cells. Furthermore, this approach could be used to test the efficacy and targeting of novel drugs to treat retinoblastoma.

18.5 Conclusions and Future Directions

It is clear that MALDI-MSI of ocular tissue has revealed novel biological and biochemical information. Spatial maps of a range of biomolecules have been generated to define normal tissue structure, and identify molecular changes that take place in aging and age-related diseases of the retina and lens. With these established protocols, further information to spatially resolve biomarkers of ocular pathologies should be forthcoming. For example, the presence of retinal deposits such as drusen, subretinal drusenoid deposits, and basal laminar deposits are major risk factors for AMD. Although such features can be detected by routine clinical imaging, the molecular composition of these deposits and how they differ remains limited. MALDI-MSI provides an ideal tool to determine such molecular compositions and differences. In the lens, MALDI-MSI offers the opportunity to determine cataract specific signals (proteins, lipids, and metabolites) in different types of lens cataract. Finally, the use of MALDI-MSI technologies to determine drug and metabolite distributions in ocular tissues is ongoing and will increase our understanding of ocular drug delivery and efficacy. Continued development of instrumentation to enhance spatial resolution, increase sensitivity, broaden the classes of detected biomolecules, and software tools to interpret and integrate the resulting multi-omic data will undoubtedly enhance our efforts to develop therapies to prevent the onset of debilitating eye diseases.

Acknowledgements

The authors thank Ms Melody Tang for creating Figure 18.1 schematics and adapting Figures 18.2–18.4.

References

1. A. O. Eghrari, S. A. Riazuddin and J. D. Gottsch, *Prog. Mol. Biol. Transl. Sci.*, 2015, **134**, 7–23.
2. J. W. Kiel, *The Ocular Circulation*, Morgan & Claypool Life Sciences, San Rafael CA, 2010.
3. Y. S. Rabinowitz, *Surv. Ophthalmol.*, 1998, **42**, 297–319.
4. J. F. Hejtmancik and A. Shiels, *Prog. Mol. Biol. Transl. Sci.*, 2015, **134**, 119–127.
5. D. Borchman and M. C. Yappert, *J. Lipid Res.*, 2010, **51**, 2473–2488.
6. P. J. Donaldson, A. C. Grey, B. Maceo Heilman, J. C. Lim and E. Vaghefi, *Prog. Retinal Eye Res.*, 2017, **56**, e1–e24.
7. C. Schmitt and O. Hockwin, *J. Inherited Metab. Dis.*, 1990, **13**, 501–508.
8. M. F. Lou, *Prog. Retinal Eye Res.*, 2003, **22**, 657–682.
9. H. E. Grossniklaus, E. E. Geisert and J. M. Nickerson, *Prog. Mol. Biol. Transl. Sci.*, 2015, **134**, 383–396.
10. M. E. Boulton, S. K. Mitter, H. V. Rao and W. A. Dunn Jr, in *Retina*, ed. A. Schachat, Elsevier, 5th edn, 2013, pp. 537–552.
11. A. Tsin, B. Betts-Obregon and J. Grigsby, *J. Biol. Chem.*, 2018, **293**, 13016–13021.
12. J. R. Sparrow, N. Fishkin, J. Zhou, B. Cai, Y. P. Jang, S. Krane, Y. Itagaki and K. Nakanishi, *Vision Res.*, 2003, **43**, 2983–2990.
13. A. Terman and U. T. Brunk, *Int. J. Biochem. Cell Biol.*, 2004, **36**, 1400–1404.
14. L. E. Lamb and J. D. Simon, *Photochem. Photobiol.*, 2004, **79**, 127–136.

15. J. R. Sparrow, E. Gregory-Roberts, K. Yamamoto, A. Blonska, S. K. Ghosh, K. Ueda and J. Zhou, *Prog. Retinal Eye Res.*, 2012, **31**, 121–135.
16. B. A. Boughton, O. R. B. Thomas, N. J. Demarais, D. Trede, S. E. Swearer and A. C. Grey, *J. Mass Spectrom.*, 2020, **55**, e4460.
17. A. D. Palmer, R. Griffiths, I. Styles, E. Claridge, A. Calcagni and J. Bunch, *J. Mass Spectrom.*, 2012, **47**, 237–241.
18. N. Mori, T. Mochizuki, F. Yamazaki, S. Takei, H. Mano, T. Matsugi and M. Setou, *PLoS One*, 2019, **14**, e0211376.
19. D. M. G. Anderson, K. A. Floyd, S. Barnes, J. M. Clark, J. I. Clark, H. McHaourab and K. L. Schey, *Anal. Bioanal. Chem.*, 2015, **407**, 2311–2320.
20. Y. Chen, J. V. Jester, D. M. Anderson, S. A. Marchitti, K. L. Schey, D. C. Thompson and V. Vasiliou, *Chem. - Biol. Interact.*, 2017, **276**, 9–14.
21. A. Venkatraman, G. Hochart, D. Bonnel, J. Stauber, S. Shimmura, L. Rajamani, K. Pervushin and J. S. Mehta, *Proteomics Clin Appl*, 2019, **13**, e1800053.
22. M. Ronci, S. Sharma, T. Chataway, K. P. Burdon, S. Martin, J. E. Craig and N. H. Voelcker, *J. Proteome Res.*, 2011, **10**, 3522–3529.
23. M. Ronci, S. Sharma, S. Martin, J. E. Craig and N. H. Voelcker, *J. Proteomics*, 2013, **82**, 27–34.
24. A. Kurreck, L. A. Vandergrift, T. L. Fuss, P. Habbel, N. Y. R. Agar and L. L. Cheng, *Prostate Cancer Prostatic Dis.*, 2018, **21**, 297–305.
25. F. J. Giblin, *J. Ocul. Pharmacol. Ther.*, 2000, **16**, 121–135.
26. J. J. Harding, *Biochem. J.*, 1970, **117**, 957–960.
27. H. Pau, P. Graf and H. Sies, *Exp. Eye Res.*, 1990, **50**, 17–20.
28. M. G. Nye-Wood, J. M. Spraggins, R. M. Caprioli, K. L. Schey, P. J. Donaldson and A. C. Grey, *Exp. Eye Res.*, 2016, **154**, 70–78.
29. A. C. Grey, N. J. Demarais, B. J. West and P. J. Donaldson, *Int. J. Mass Spectrom.*, 2019, **437**, 58–68.
30. J. C. Lim, E. Vaghefi, B. Li, M. G. Nye-Wood and P. J. Donaldson, *Invest. Ophthalmol. Visual Sci.*, 2016, **57**, 1961–1973.
31. R. J. W. Truscott, *Ophthalmic Res.*, 2000, **32**, 185–194.
32. N. J. Demarais, P. J. Donaldson and A. C. Grey, *Exp. Eye Res.*, 2019, **184**, 146–151.
33. A. Korlimbinis, J. A. Aquilina and R. J. Truscott, *Invest. Ophthalmol. Visual Sci.*, 2007, **48**, 1718–1723.
34. B. Garner, S. Vazquez, R. Griffith, R. A. Lindner, J. A. Carver and R. J. Truscott, *J. Biol. Chem.*, 1999, **274**, 20847–20854.
35. J. Mizdrak, P. G. Hains, R. J. Truscott, J. F. Jamie and M. J. Davies, *Free Radical Biol. Med.*, 2008, **44**, 1108–1119.
36. M. Rujoi, R. Estrada and M. C. Yappert, *Anal. Chem.*, 2004, **76**, 1657–1663.
37. C. H. Le, J. Han and C. H. Borchers, *Anal. Chem.*, 2012, **84**, 8391–8398.
38. V. Vidová, J. Pól, M. Volný, P. Novák, V. Havlícek, S. K. Wiedmer and J. M. Holopainen, *J. Lipid Res.*, 2010, **51**, 2295–2302.

39. J. Pól, V. Vidová, T. Hyötyläinen, M. Volný, P. Novák, M. Strohalm, R. Kostiainen, V. Havlíček, S. K. Wiedmer and J. M. Holopainen, *PLoS One*, 2011, **6**, e19441.
40. J. Pol, H. Faltyskova, L. Krasny, M. Volný, O. Vlacil, M. Hajduch, K. Lemr and V. Havlicek, *Eur. J. Mass Spectrom.*, 2015, **21**, 297–303.
41. J. A. Seng, J. R. Nealon, S. J. Blanksby and T. W. Mitchell, *Biomolecules*, 2018, **8**, 156.
42. S. R. Ellis, C. Wu, J. M. Deeley, X. Zhu, R. J. Truscott, M. in het Panhuis, R. G. Cooks, T. W. Mitchell and S. J. Blanksby, *J. Am. Soc. Mass Spectrom*, 2010, **21**, 2095–2104.
43. J. M. Deeley, J. A. Hankin, M. G. Friedrich, R. C. Murphy, R. J. Truscott, T. W. Mitchell and S. J. Blanksby, *J. Lipid Res.*, 2010, **51**, 2753–2760.
44. D. M. Anderson, M. G. Nye-Wood, K. L. Rose, P. J. Donaldson, A. C. Grey and K. L. Schey, *J. Mass Spectrom.*, 2019, **55**, e4473.
45. J. Han and K. L. Schey, *Invest. Ophthalmol. Visual Sci.*, 2006, **47**, 2990–2996.
46. K. L. Schey, D. M. Anderson and K. L. Rose, *Anal. Chem.*, 2013, **85**, 6767–6774.
47. A. C. Grey and K. L. Schey, *Mol. Vision*, 2008, **14**, 171–179.
48. A. C. Grey and K. L. Schey, *Invest. Ophthalmol. Visual Sci.*, 2009, **50**, 4319–4329.
49. P. J. Groenen, K. B. Merck, W. W. de Jong and H. Bloemendal, *Eur. J. Biochem.*, 1994, **225**, 1–19.
50. D. R. Stella, K. A. Floyd, A. C. Grey, M. B. Renfrow, K. L. Schey and S. Barnes, *Invest. Ophthalmol. Visual Sci.*, 2010, **51**, 5153–5161.
51. S. P. Su, J. D. McArthur and J. Andrew Aquilina, *Exp. Eye Res.*, 2010, **91**, 97–103.
52. K. L. Schey, K. Rose and D. M. G. Anderson, *Association for Research in Vision and Ophthalmology Annual Meeting*, Seattle, USA, 2013.
53. J. Jiao, A. Miao, Y. Zhang, Q. Fan, Y. Lu and H. Lu, *Analyst*, 2015, **140**, 4284–4290.
54. J. Horwitz, *Exp. Eye Res.*, 2003, **76**, 145–153.
55. A. B. Chepelinsky, *Handb. Exp. Pharmacol.*, 2009, 265–297.
56. A. C. Grey, P. Chaurand, R. M. Caprioli and K. L. Schey, *J. Proteome Res.*, 2009, **8**, 3278–3283.
57. K. L. Schey, D. B. Gutierrez, Z. Wang, J. Wei and A. C. Grey, *Biochemistry*, 2010, **49**, 9858–9865.
58. J. B. Fields, K. L. Németh-Cahalan, J. A. Freites, I. Vorontsova, J. E. Hall and D. J. Tobias, *J. Biol. Chem.*, 2017, **292**, 185–195.
59. Y. Nakazawa, M. Oka, K. Furuki, A. Mitsuishi, E. Nakashima and M. Takehana, *Mol. Vision*, 2011, **17**, 3191–3199.
60. K. M. Lindsey Rose, R. G. Gourdie, A. R. Prescott, R. A. Quinlan, R. K. Crouch and K. L. Schey, *Invest. Ophthalmol. Visual Sci.*, 2006, **47**, 1562–1570.
61. J. L. Wenke, K. L. Rose, J. M. Spraggins and K. L. Schey, *Invest Ophthalmol. Visual Sci.*, 2015, **56**, 7398–7405.

62. Z. Wang, D. J. Ryan and K. L. Schey, *Exp. Eye Res.*, 2020, **198**, 108134.
63. A. Sandilands, A. R. Prescott, A. M. Hutcheson, R. A. Quinlan, J. T. Casselman and P. G. FitzGerald, *Eur. J. Cell Biol.*, 1995, **67**, 238–253.
64. Z. Wang and K. L. Schey, *Exp. Eye Res.*, 2017, **159**, 23–29.
65. N. Sun, A. Ly, S. Meding, M. Witting, S. M. Hauck, M. Ueffing, P. Schmitt-Kopplin, M. Aichler and A. Walch, *Proteomics*, 2014, **14**, 913–923.
66. M. C. Roy, H. Nakanishi, K. Takahashi, S. Nakanishi, S. Kajihara, T. Hayasaka, M. Setou, K. Ogawa, R. Taguchi and T. Naito, *J. Lipid Res.*, 2011, **52**, 463–470.
67. T. Hayasaka, N. Goto-Inoue, Y. Sugiura, N. Zaima, H. Nakanishi, K. Ohishi, S. Nakanishi, T. Naito, R. Taguchi and M. Setou, *Rapid Commun. Mass Spectrom.*, 2008, **22**, 3415–3426.
68. T. Hayasaka, N. Goto-Inoue, N. Zaima, K. Shrivas, Y. Kashiwagi, M. Yamamoto, M. Nakamoto and M. Setou, *J. Am. Soc. Mass Spectrom*, 2010, **21**, 1446–1454.
69. D. M. G. Anderson, D. Mills, J. Spraggins, W. S. Lambert, D. J. Calkins and K. L. Schey, *Mol. Vision*, 2013, **19**, 581–592.
70. D. M. G. Anderson, Z. Ablonczy, Y. Koutalos, J. Spraggins, R. K. Crouch, R. M. Caprioli and K. L. Schey, *J. Am. Soc. Mass Spectrom*, 2014, **25**, 1394–1403.
71. D. M. G. Anderson, J. D. Messinger, N. H. Patterson, E. S. Rivera, A. Kotnala, J. M. Spraggins, R. M. Caprioli, C. A. Curcio and K. L. Schey, *J. Am. Soc. Mass Spectrom*, 2020, **31**, 2426–2436.
72. K. A. Zemski Berry, W. C. Gordon, R. C. Murphy and N. G. Bazan, *J. Lipid Res.*, 2014, **55**, 504–515.
73. X. Pereiro, R. Fernández, G. Barreda-Gómez, N. Ruzafa, A. Acera, J. Araiz, E. Astigarraga and E. Vecino, *Sci. Rep.*, 2020, **10**, 20053.
74. T. J. Garrett and W. W. Dawson, *Methods Mol. Biol.*, 2009, **579**, 247–260.
75. T. J. Garrett, R. F. Menger, W. W. Dawson and R. A. Yost, *Anal. Bioanal. Chem.*, 2011, **401**, 103–113.
76. D. M. G. Anderson, Z. Ablonczy, Y. Koutalos, A. M. Hanneken, J. M. Spraggins, M. W. Calcutt, R. K. Crouch, R. M. Caprioli and K. L. Schey, *Sci. Rep.*, 2017, **7**, 17352.
77. L. Adler, N. P. Boyer, D. M. Anderson, J. M. Spraggins, K. L. Schey, A. Hanneken, Z. Ablonczy, R. K. Crouch and Y. Koutalos, *Photochem. Photobiol. Sci.*, 2015, **14**, 1983–1990.
78. A. C. Grey, R. K. Crouch, Y. Koutalos, K. L. Schey and Z. Ablonczy, *Invest. Ophthalmol. Visual Sci.*, 2011, **52**, 3926–3933.
79. Z. Ablonczy, D. Higbee, A. C. Grey, Y. Koutalos, K. L. Schey and R. K. Crouch, *Arch. Biochem. Biophys.*, 2013, **539**, 196–202.
80. J. R. Sparrow, H. R. Vollmer-Snarr, J. Zhou, Y. P. Jang, S. Jockusch, Y. Itagaki and K. Nakanishi, *J. Biol. Chem.*, 2003, **278**, 18207–18213.

81. Z. Ablonczy, N. Smith, D. M. Anderson, A. C. Grey, J. Spraggins, Y. Koutalos, K. L. Schey and R. K. Crouch, *Proteomics*, 2014, **14**, 936–944.
82. P. Pallitto, Z. Ablonczy, E. E. Jones, R. R. Drake, Y. Koutalos, R. K. Crouch, J. Donello and J. Herrmann, *Photochem. Photobiol. Sci.*, 2015, **14**, 1888–1895.
83. Z. Ablonczy, D. Higbee, D. M. Anderson, M. Dahrouj, A. C. Grey, Y. Koutalos, K. L. Schey, D. B. Gutierrez, A. Hanneken and R. K. Crouch, *Invest Ophthalmol. Visual Sci.*, 2013, **54**, 5535–5542.
84. P. Bhosale, B. Serban and P. S. Bernstein, *Arch. Biochem. Biophys.*, 2009, **483**, 175–181.
85. D. M. G. Anderson, J. M. Spraggins, K. L. Rose and K. L. Schey, *J. Am. Soc. Mass Spectrom*, 2015, **26**, 940–947.
86. D. T. Stark, D. M. G. Anderson, J. M. K. Kwong, N. H. Patterson, K. L. Schey, R. M. Caprioli and J. Caprioli, *Invest. Ophthalmol. Visual Sci.*, 2018, **59**, 212–222.
87. D. M. G. Anderson, R. Van de Plas, K. L. Rose, S. Hill, K. L. Schey, A. C. Solga, D. H. Gutmann and R. M. Caprioli, *J. Proteomics*, 2016, **149**, 77–84.
88. D. M. Drexler, S. H. Tannehill-Gregg, L. Wang and B. J. Brock, *J. Pharmacol. Toxicol. Methods*, 2011, **63**, 205–208.
89. A. Zahraei, G. Guo, R. D. Perwick, P. J. Donaldson, N. J. Demarais and A. C. Grey, *J. Mass Spectrom.*, 2020, **56**, e4666.
90. A. C. Grey, M. Tang, A. Zahraei, G. Guo and N. J. Demarais, *Anal. Bioanal. Chem.*, 2021, **413**, 2637–2653.
91. E. M. Heikkinen, S. Auriola, V.-P. Ranta, N. J. Demarais, A. C. Grey, E. M. del Amo, E. Toropainen, K.-S. Vellonen, A. Urtti and M. Ruponen, *Mol. Pharmaceutics*, 2019, **16**, 3968–3976.
92. G. Querques, R. Forte and E. H. Souied, *J. Nutr. Metab.*, 2011, **2011**, 748361.
93. A. U. Kishan, B. S. Modjtahedi, E. N. Martins, S. P. Modjtahedi and L. S. Morse, *Surv. Ophthalmol.*, 2011, **56**, 195–213.
94. E. Vidal, B. Jun, W. C. Gordon, M. A. Maire, L. Martine, S. Grégoire, S. Khoury, S. Cabaret, O. Berdeaux, N. Acar, L. Bretillon and N. G. Bazan, *J. Lipid Res.*, 2020, **61**, 1733–1746.
95. P. Vinciguerra, R. Mencucci, V. Romano, E. Spoerl, F. I. Camesasca, E. Favuzza, C. Azzolini, R. Mastropasqua and R. Vinciguerra, *BioMed Res. Int.*, 2014, **2014**, 404587.
96. M. R. Groseclose and S. Castellino, *Chem. Res. Toxicol.*, 2019, **32**, 294–303.
97. K. J. Grove, V. Kansara, M. Prentiss, D. Long, M. Mogi, S. Kim and P. J. Rudewicz, *SLAS Discovery*, 2017, **22**, 1239–1245.
98. L. Z. Wang, N. Syn, S. Li, V. A. Barathi, L. Tong, J. Neo, R. W. Beuerman and L. Zhou, *Acta Ophthalmol.*, 2019, **97**, e238–e247.
99. Y. Yamada, K. Hidefumi, H. Shion, M. Oshikata and Y. Haramaki, *Rapid Commun. Mass Spectrom.*, 2011, **25**, 1600–1608.

100. P. Eves, L. Smith-Thomas, S. Hedley, M. Wagner, C. Balafa and S. Mac Neil, *Pigm. Cell Res.*, 1999, **12**, 22–35.
101. A. Balla, S. Auriola, A. C. Grey, N. J. Demarais, A. Valtari, E. M. Heikkinen, E. Toropainen, A. Urtti, K.-S. Vellonen and M. Ruponen, *Pharmaceutics*, 2021, **13**, 658.
102. K. Okabe, H. Kimura, J. Okabe, A. Kato, H. Shimizu, T. Ueda, S. Shimada and Y. Ogura, *Invest. Ophthalmol. Visual Sci.*, 2005, **46**, 703–708.
103. C. Baudouin, A. Labbé, H. Liang, A. Pauly and F. Brignole-Baudouin, *Prog. Retinal Eye Res.*, 2010, **29**, 312–334.
104. F. Brignole-Baudouin, N. Desbenoit, G. Hamm, H. Liang, J.-P. Both, A. Brunelle, I. Fournier, V. Guerineau, R. Legouffe, J. Stauber, D. Touboul, M. Wisztorski, M. Salzet, O. Laprevote and C. Baudouin, *PLoS One*, 2012, **7**, e50180.
105. A. B. Daniels, M. T. Froehler, J. M. Pierce, A. H. Nunnally, M. W. Calcutt, T. M. Bridges, D. C. LaNeve, P. E. Williams, K. L. Boyd, M. L. Reyzer, C. W. Lindsley, D. L. Friedman and A. Richmond, *Invest. Ophthalmol. Visual Sci.*, 2018, **59**, 446–454.

CHAPTER 19

Correlative Multimodal Mass Spectometry Imaging – Imaging Across the Scales

MARTINA MARCHETTI-DESCHMANN

Institute of Chemical Technologies and Analytics, TU Wien, Getreidemarkt 9/164, 1060 Vienna, Austria
Email: martina.marchetti-deschmann@tuwien.ac.at

19.1 Introduction

Multimodal imaging combines two or more imaging techniques. Its palette contains a variety of specialized visualization methods for imaging different structures, microenvironments, biomolecules, and probes present in the sample under investigation. Multimodal imaging is gaining importance in more and more imaging centers worldwide, as it allows deep insight into spatially-resolved processes occurring on various surfaces and in different objects. For biosciences this means *in vivo* and *in vitro* function and structure are of interest for a better understanding of biological processes and events. Beyond life sciences correlative imaging also plays a crucial role in material sciences.[1] Its capability to take a specific look at selected areas provides information on surface reactions or alterations occurring at material interfaces, giving access to a deep understanding of structure–function or structure–activity relations in novel materials. Although this chapter will have a strong focus on biological and medical applications, some exceptional examples will be given in the context of material sciences.

New Developments in Mass Spectrometry No. 12
MALDI Mass Spectrometry Imaging: From Fundamentals to Spatial Omics
Edited by Tiffany Siegel

Published by the Royal Society of Chemistry, www.rsc.org

Multimodal imaging is well known and has been performed for more than a decade in medical applications.[2] Imaging technologies based on MS or other concepts can provide sophisticated insights with unique contributions on the chemical and structural level. The fact that each technology has its distinct advantages and its own potential in terms of lateral and/or spatial resolution, respectively the 2- and 3-dimensional arrangements, leads to synergistic combinations of the highest value.

The aim of this chapter is to give the reader an understanding of the different terminologies used in the field of imaging and to provide a short summary of state-of-the-art multimodal imaging concepts. This chapter lays emphasis on MSI and its use for correlative approaches to show its outstanding impact on the field. The aim is to give an overview of the potential if different ionization techniques are combined to collect comprehensive sample data, to mention the benefits of using more than one mass analyzer, and to highlight aspects of MS for improved, or sometimes intentionally reduced, lateral resolutions. Fusing two highly complementary imaging methods, one being MS-based and the other coming from a field not related to MS, will highlight future trends and potentials for fusing multiple modalities in one instrument to produce functional hybrid instruments.

This chapter summarizes a personal choice of topics and may have missed some of the next transformative multimodal imaging approaches. But in 5 to 10 years time, we will see how this chapter meets actual trends and developments in the field of MSI.

19.2 Correlative, Hybrid, Multisensor, Multimodal Imaging – Is There a Difference?

The concept of correlative imaging has been around for over 40 years. The first paper introducing "correlative" in the context of "imaging" appeared in 1969[3] in the field of microscopy, and in 1974[4] the challenges for a hypothetical "correlative microscope" were summarized for the first time. Correlative imaging refers to the concept of measuring a sample with two different instruments and subsequently correlating the generated data in a subsequent data analysis process. Correlative microscopy can give structural, functional, dynamical, and specific chemical information of a single sample for a well-defined time point or even time lapse series across all relevant length scales and levels of biological organization. Modern imaging technologies cover functional and structural features of a sample across the scales, some can be applied label-free measuring physico-chemical properties of a sample to visualize distributions, others demand *a priori* labeling of the sample or the incorporation of probes with specific interaction (Figure 19.1a).

Today, holistic imaging concepts include soft X-ray tomography, high resolution episcopic microscopy (HREM), (transmission)electron microscopy ((T)EM), computed tomography (CT), optical imaging (OI) including

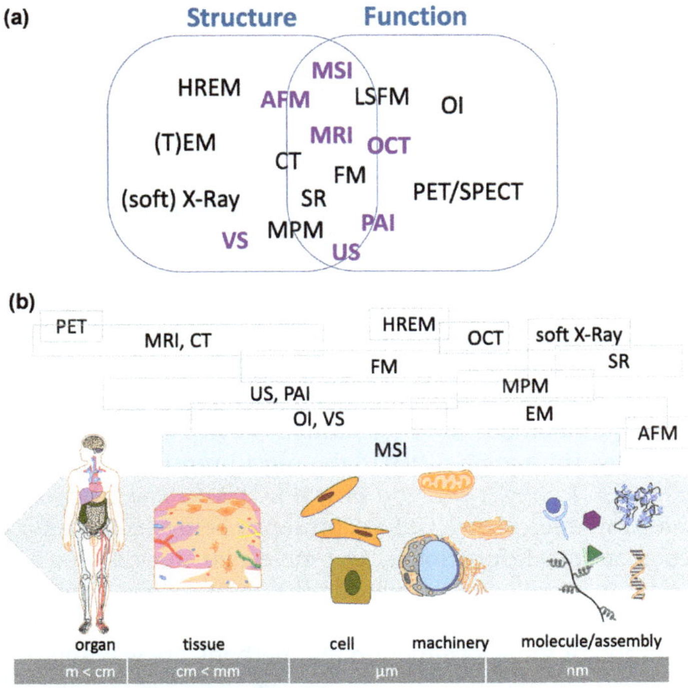

Figure 19.1 (a) Overview of imaging modalities delivering structural and functional information with and without *a priori* labeling strategies; bold/violet highlights modalities working label-free. (b) Imaging across the scales: overview of spatial resolutions achieved by different imaging technologies. X-Ray Tomography, High Resolution Episcopic Microscopy (HREM), (transmission) electron microscopy ((T)EM), Computed Tomography (CT), optical imaging (OI), Vibrational Spectroscopy (VR), Light Sheet Fluorescence Microscopy (LSFM), MultiPhoton Microscopy (MPM), Confocal Microscopy (CM), Magnetic Resonance Imaging (MRI), Ultrasound (US), Photoacoustic Imaging (PAI), Optical Coherence Tomography (OCT), Atomic Force Microscopy (AFM), Super-resolution Microscopy (SR) such as STORM, PALM, STED; Positron Emission Tomography (PET)/Single Photon Emission Computed Tomography (SPECT), Mass Spectrometry Imaging (MSI).

bioluminescence, fluorescence imaging or vibrational spectroscopy (VR), light sheet fluorescence microscopy (LSFM), multiphoton microscopy (MPM), confocal microscopy (CM), magnetic resonance imaging (MRI), ultrasound (US), photoacoustic imaging (PAI), optical coherence tomography (OCT), and atomic force microscopy (AFM); super-resolution microscopy (SR) such as stochastical optical reconstruction microscopy (STORM), photoactivated localization microscopy (PALM), stimulated emission depletion (STED), positron emission tomography (PET)/single photon emission computed tomography (SPECT) and, of course, MSI. Detailed information on the different imaging technologies can be found in a recently published compendium.[5,6] From early on, scientists were aware of the benefit of correlating data from complementary instrumental setups. Furthermore, the combination of high-resolution imaging techniques like EM working at the nm-scale with OI at the μm-level became one of the first correlative imaging approaches. Today's endeavors to improve correlative microscopy are mostly focusing on two modalities in the field of OI and EM. A recent review article on correlative multimodal imaging (CMI)[7] highlights the fact that the imaging of a specimen with two complementary modalities creates an informative, composite view of the sample, providing a holistic overview of structural, functional, temporal, and chemical information spanning over all relevant resolution ranges over a sample (Figure 19.1b).

Today, there are only few hardware-fused platforms commercially available. One multimodal imaging approach well established for preclinical imaging covers all methods used for humans *in vivo* and is often referred to as pre-clinical hybrid imaging (PHI),[8] the most prominent being μPET/μCT, PET/MR, SPECT/CT, SPECT/MR and PET/SPECT/MR. The other is correlative light and electron microscopy (CLEM), presently being explored for its ability to enhance anatomical, functional, and molecular imaging for health care. CLEM is also an excellent example for the terminology "multimodal imaging" and PHI is, as the name already suggests, the example for "hybrid imaging". Although the above-mentioned methods are very important, they will not be discussed in detail (for a review see Walter *et al.*[9]) but will be put into relation to MSI in a later section of this chapter.

The term "multisensory" imaging is mostly used in the computer science and information technology community and describes the same concept as "multimodal" imaging. More than one sensor, synonymous with mode or instrument, is used to measure a sample and data are correlated for a concise understanding of lateral and spatial orders. In a more sophisticated manner, data are not only correlated but fused for further analysis driven by artificial intelligence and deep learning approaches because it is generally agreed that no single imaging technology/mode/sensor can grasp the entire information of a biological machinery. This basically reflects the need for sensitive and specific imaging technologies with translational potential to applications in basic research, diagnosis, and disease treatment. It is noteworthy that MSI has become one of such techniques since its introduction in the 1990s.

Solely the term CMI, correlated multimodal imaging, will be used in this chapter, and will include instrumental setups and/or workflows using more than one instrument or even hardware-fused instrumentation combining completely complementary techniques.

19.3 MSI in the Context of Correlative Multimodal Imaging

MSI has become a very important tool for targeted and untargeted *in vitro* analyses of biological samples and biomaterials where most of the applications can be found in the biomedical environment. Its capability to profile and image molecular and elemental information without knowing *a priori* of the presence and distribution of biomolecules is of great value and is possible without the need for labels, or complicated sample preparation or staining protocols. With the advantage of no labeling, MSI has opened a new frontier not only for protein analysis but also for small molecule imaging or even the visualization of post-translational modifications. So far, no other imaging technology can visualize lipid, metabolite or *N*-glycan distributions, thus making it a unique tool in the field. In 2011, MSI has even been termed "mass microscopy" to emphasis the fact that MS is able to investigate morphologies in tissues at µm-resolutions, specifications usually connected to optical imaging.[10] So, it is no surprise that MSI is considered a highly valuable contributor to the field of CMI, and many studies show the benefit of MSI data to understand tissue integrity, health, and disease. Although MSI has been implemented in microscopy workflows, so far hybrid instruments on the same commercial level as PHI are not available, but can be foreseen for the next generation of MSI instruments in the very near future.

The combination of MS-based distribution maps with OI has been one of the first correlative investigations.[11,12] The high lateral resolution for morphological features as acquired from histological stains like HE was aligned with pathological features identified by MS in a vast spectrum of disease pathology.

Different available ionization techniques have their respective advantages for imaging approaches in MS and an overview is given in a recently published review.[13] Briefly, MALDI-MS produces mainly singly charged ions which simplify the complexity of mass spectra. This is an essential point for MS-based imaging considering that multiple analytes are simultaneously desorbed/ionized from one single spot. The latest developments include high lateral resolutions down to a few µm (1–5 µm)[14,15] and high ionization efficiency enhanced by post-ionization,[16] significantly increase sensitivity. However, the sample will be consumed under laser irradiation, although it has been shown for a sample rastered twice by MALDI-MS that this consumption is below 10% for a 10 µm-thin tissue section.[17] Desorption electrospray ionization (DESI) is capable of *in situ* analysis at limited lateral resolution (30–150 µm). DESI is a minimally destructive technique and imaging is performed in a continuous manner and the lateral resolution can be reduced to 10 µm in the case of a liquid junction system named nano-DESI. MALDI and DESI are soft ionization techniques allowing for the measurement of intact molecules from a few hundred Daltons (Da) up to kDa. Secondary ion mass spectrometry (SIMS) uses primary ion beams or clusters to ablate and ionize analytes of interest from a sample surface. The rather high energy levels introduced into the sample during the desorption/ionization process fragments larger molecules limiting SIMS to a m/z range <2 kDa. It was shown that in the case of supporting the SIMS process with a matrix as it is used for MALDI, intact peptides up to 2 kDa can be imaged in tissue.[18] However, the primary advantage of SIMS is its lateral resolution. Indeed, the development of nano-SIMS made imaging feasible down to a lateral resolution of 50 nm and lower, generating an exceptional potential to study biological events below cellular resolution.[19] Laser ablation-inductively coupled plasma-mass spectrometry (LA-ICP-MS), a method not so commonly used for imaging of biological samples so far, is a quick method not only to localize but also quantify metals on any surface. LA-ICP-MS employs a high-energy laser to sample a specimen prior to introducing the ablated material into a high temperature plasma to atomize and ionize the sample's elements. A large series of quantitative element maps (such as Fe, Cu, Zn, and Mn) were produced in native brain sections of a Parkinson's disease mouse model by this technique.[20] The ionization behavior of elements and molecules differs for LA-ICP, SIMS, MALDI, and DESI. Considering this, the combination of more than one ionization technique in a CMI approach becomes easily evident. Furthermore, the mentioned ionization techniques can be coupled to mass analyzers of different performance. Fast scanning ion selectors like quadrupoles, parallel multi-ion measurements in time-of-flight instruments (TOF) or high resolving Orbitraps and Fourier Transform Ion cyclotron resonance (FT-ICR) analyzers are all used in the field of MSI and the combination of different analyzers can be helpful in deciphering ambiguous species identification as demonstrated by the application of a hybrid SIMS (TOF SIMS/Orbitrap).[21] Recently ion mobility was added for its potential to separate ion species in the gas phase to simplify data interpretation by generating images of analytes exhibiting similar 3-dimensional structures (for a review see Rivera *et al.*[22]). And the increased sensitivity and mass resolution for all available mass spectrometers paved the way for *in situ* identification of analytes by MS/MS experiments. More details of the instrumentation used for (MALDI)-MSI can be found in Chapter 4 of this book.

The combination of images gathered from more than one mass analyzer or ionization technique was the first and most obvious step in pursuing the idea of CMI for MSI.

One may argue that MSI approaches are not very complementary as all setups provide the same output after all – a re-coding of *m/z* values over (potentially) the same area. So it can be stated that the combination of different MS approaches is not a correlative imaging approach strictly speaking, but rather a multisensory approach. Yet, some ionization techniques provide the unique potential of gathering molecular information of intact analytes (DESI, MALDI), some even providing elemental (LA-ICP, SIMS) and even elemental and partial molecular information *via* a characteristic fragmentation pattern (SIMS). So, combining different MSI methodologies can indeed be considered a CMI approach. Such considerations can also be made for the investigated samples. Is the co-registration and correlation of images generated by two different instruments a multimodal approach, or is only the collection of images from one sample a multimodal concept?

And on top of this, there is also a debate on the term "correlative" as some approaches only visually-matched structures and data are mostly handled separately from the different modalities. Multi-modality is especially instrumental to overcoming some of the ubiquitous limitations of individual imaging techniques, *e.g.* lack of sensitivity or specificity, limited lateral and spatial resolution or knowledge biases (*i.e.* the target of interest has to be known for being visualized), and the integration of MS technologies into other imaging workflows is not so straight forward after all as will be pointed out at the end of this chapter.

This chapter gives an overview on CMI pipelines including any kind of MSI and does not lay strong emphasis on things like sample preparation (*e.g.* consecutive tissue sections or one sample), if data are fused or handled separately or if the chosen modalities are rather similar or very complementary. In the future, there will be a common ground in the CMI community for the currently present uncertainties and linguistic discrepancies.

19.4 Combination of MS Techniques

The latest innovations for combining different mass spectrometers will be highlighted in the following sections as each ionization technique and mass analyzer has its own benefits to constantly decipher molecular complexities relevant for life sciences such as cell transformation, disease biology but also geo- or material sciences.

The least common CMI approach is the combination of MALDI and DESI. This concept enables analysis of different lipid species[23] and co-mapping of lipids and proteins.[24] However, the similar range of lateral resolutions seems to make this combination not very attractive despite the fact that ionization mechanisms are different and DESI can be "tuned" toward analyte classes just by careful adjustment of the ion source setting.[25] SIMS and MALDI imaging have been combined to detect multiple chemical classes, such as low molecular weight metabolites and lipids within individual cells and tissue. An interesting concept for correlative imaging was presented for bioactive metabolites.[26] A biofilm was first imaged by low resolution MALDI-MSI (1 mm). While under-sampling left the surface mostly unperturbed, it generated an ablation array that was used for a MALDI-guided SIMS approach at the low µm-scale. Image correlation was by this method straightforward because MALDI-MSI produced fiducial markers for finding the region of interest in the second modality and the holes in the tissue were used for image alignment during data processing.

The correlation of images produced by LA-ICP-MS to histological information was shown by Becker *et al.* who already mentioned the potential of combining their newly introduced LA-ICP-MS imaging approach with already available MALDI-MSI.[27] The introduction of a non-targeted method to exploit pathological features in histologically stained tissue can be considered one of the first CMI approaches for LA-ICP-MS.[28] Today, LA-ICP-MS has a broad field of application ranging from cancer research to elemental uptake and accumulation studies, plant bioimaging or nanomaterial tracking in environmental science, as recently reviewed by Doble *et al.*[29]

MALDI-MSI is generally compatible with LA-ICP-MS especially if images from consecutive tissue sections are collected. This has impressively been shown for neurotoxin 6-hydroxydopamine-induced brain lesions[30] or Pt-based anti-cancer drugs in human tumor biopsies.[31] The potential and accuracy of LA-ICP-MS for absolute element quantitation in tissue has been developed over the years and strategies including matrix matched[32] calibrants and normalization strategies for ionization efficiencies[33] are well accepted today. Gelatine- and agarose-based materials are reported with increasing frequency as matrix-matched standards because of their potential to be sectioned, spotted or filled into molds excellently mimicking tissue thin sections. The first multimodal imaging combining MALDI and LA-ICP-MS from one tissue section was studying Pt-concentrations in human malignant pleural mesothelioma samples.[17] Lipid distributions were acquired from a tissue section after matrix sublimation, immediately before consuming the rest of the tissue sections totally by LA-ICP MS to image Pt, P, and S (Figure 19.2). Comparing the determined Pt concentrations to the

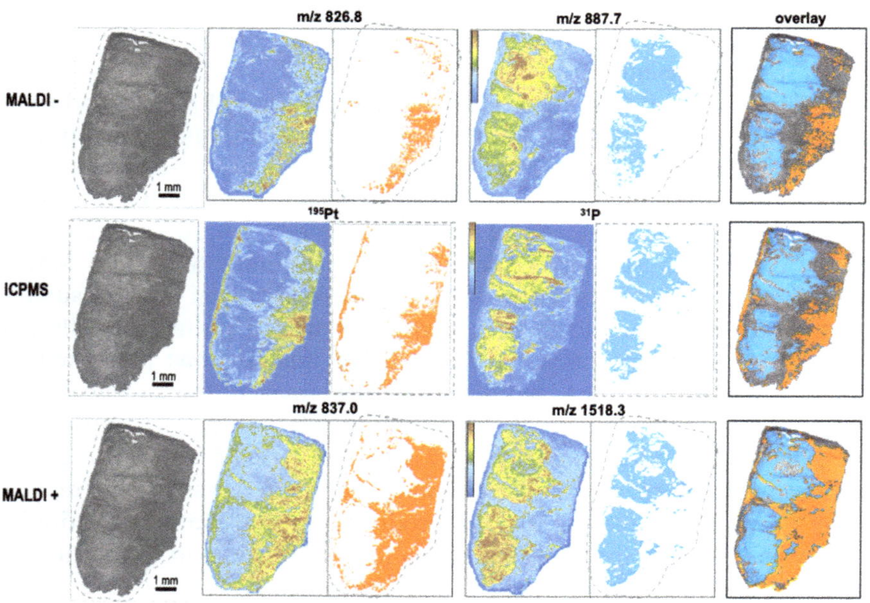

Figure 19.2 Correlative multimodal imaging of human malignant pleural mesothelioma using MALDI and LA-ICP-MS. Microscope image of a tissue section measured consecutively in positive- and negative-ion mode by MALDI-MS for lipid analysis. The same tissue section was transferred to LA-ICP-MS analysis for ^{135}Pt, and ^{34}S analysis. Correlation of high ^{31}P and low ^{135}Pt levels with tumor tissue (light gray area in microscopic image) could be identified. Certain lipid species correlated well with ^{135}Pt and were mostly observed outside the tumor region.
Adapted from ref. 17 with permission from American Chemical Society, Copyright 2018.

Pt content of a consecutive section after digestion showed only 10% relative deviation between the imaging experiment and the total analysis. Considering biological and technical variations this study showed that MALDI-MSI is not as material consuming as considered if experimental parameters are carefully adjusted.

DESI-MSI, immunohistochemistry (IHC), and conventional LA-ICP-MS imaging were combined to characterize a lung adenocarcinoma model using spheroids.[34] DESI-MSI identified major metabolites of glycolysis and the TCA cycle in the hypoxic core of the spheroid; free fatty acids were located within the outer region of the spheroid colocalized with glutathione. IHC confirmed markers of proliferation and hypoxia and LA-ICP-MS localized endogenous elements like Mg, Cu, and Zn in the outer region accordingly with cell proliferation.

IHC was also the driving force to develop an imaging system named "imaging mass cytometry" (IMC), offering commercial solutions for antibodies labeled with rare earth elements. The LA-ICP system is equipped with a fast TOF analyzer allowing comprehensive, lateral investigation in cases of known targets and availability of respective antibodies. A panel of 35 antibodies has been used to characterize the phenotype of tumor and stromal cells from 352 patients with breast cancer.[35] The development of a 40-marker panel allowed the study of the cancer immune microenvironment[36] at a lateral resolution of approx. 100 μm. A similar concept is named multiplexed ion beam imaging (MIBI) by TOF[37,38] and facilitates SIMS to measure metal-tagged antibodies at subcellular resolution in clinical tissue sections. MIBI-TOF produces ion images at a resolution comparable to light microscopy (200–300 nm), a factor of nearly three orders of magnitude lower than IMC (Figure 19.2). Although this concept can now not be considered as a multimodal imaging approach, the combination of this methodology with other modalities is so evident, that interesting studies can be expected in the very near future.

Using non-destructive methods like vibrational spectroscopy or fluorescence imaging before antibody labeling, or following *in vivo* imaging with tissue extraction, fixation, and cutting before immunohistochemical treatment are self-evident combinations for seminal CMI pipelines.

19.5 Combination of MS with Other Modalities

19.5.1 *In Vivo* and *Ex Vivo* Imaging

Connecting *in vivo* with *ex vivo* imaging provides several advantages in clinical and research applications. Low resolution *in vivo* imaging visualizes structural and partially functional dependencies. In comparison, MSI adds high resolution data of unambiguous chemical specificity.

Molecular imaging by MRI can be impaired by a lack of validation of the specific substrate responsible for the MRI signal. Furthermore, it has a rather low resolution usually hampering image co-registration and finally data correlation. However, MRI has 3-dimensional spatial information not provided by many other methods. This of course creates a problem of complex multi-dimensional data that has to be addressed by data scientists. Recently, a combination of MRI, MALDI-MSI, and LA-ICP-MS was used to visualize the distribution of Gadofluorine P in plaque tissue with high resolution, adding novel insights into molecular MR imaging of atherosclerosis.[39] High-resolution imaging at 15 μm-lateral resolution by MALDI and LA-ICP provided information for the atherosclerosis-affine MRI contrast agent Gadofluorine, yet this information could so far not be correlated to the low-resolution MRI data (mm). Section thickness (MRI 1 mm, MSI 5–10 μm) and tissue deformation (*in vivo*/*ex vivo*) lead to errors for image overlays as the simplest approach for image correlation. Improved image co-registration using deep learning may allow co-registration of all data sets.

CT and MRI can also be performed without administration of contrast agents. Quantitative iron (Fe)-sensitive MRI was used to image Fe in the frontal cortex of healthy human controls with comparisons against histochemical and quantitative LA-ICP-MS imaging in post-mortem tissues of Alzheimer's disease patients.[40] In a first step, MRI and LA-ICP-MS images were correlated only by visual inspection because of the already mentioned dimensionality mismatch. Subsequently the authors applied non-linear registration strategies, a multi-resolution registration strategy and t-distributed stochastic neighbor embedding (t-SNE) to register the low-resolution MRI images to the high-resolution images from histology and MSI. The recently introduced combination of μCT and LA-ICP-MS opens the path to quantitative elemental distribution analysis in biological specimens at high resolution. Rather small specimens (*i.e.* the water flea *Ceriodaphnia dubia*) were chosen to demonstrate that a workflow of resin embedding, μCT measurement, tissue sectioning, LA-ICP-MS analysis, and data processing can provide spatially aligned quantitative 3D-multi-isotope data on different tissue types at moderate resolution (600×600 μm).[41] Fiducial markers and the potential of acquiring rather high-resolution images by μCT (approx. 250–200 μm^2) allowed the co-registration of quantitative information of elements (Na, Mg, P, K, Ca, Mn, Fe, Cu, Zn, Rb) with anatomical features after embedding and sectioning of the specimen.[42] This study shows the high-potential CMI for quantitative 3D-mapping of elements in whole body studies at rather high spatial resolution.

The early conceptualization and development of algorithms for automatic serial image registration in the context of MALDI-MS[43] have already enabled the 3D reconstruction of MSI data of mouse heads and the co-registration with MR images[44] in 2007. This proof-of-concept study clearly pointed out the potential of generating laterally-resolved protein data to help in understanding different MRI responses. The correlation of MALDI-MSI and MRI is based on rendering along an image plane and precise co-localization of MRI and MSI data could not be achieved at that time. Yet a few years later the same research group presented 3D protein data co-localized with bone disease and presented a workflow including rigid co-registration of images from each tissue section directly after cryo-sectioning. These intermediate images were used as a reference for MRI and MALDI-MSI co-registration, respectively. The latest trend for MSI/MR imaging is the correlation of hyperpolarized [1-^{13}C] metabolites measured by (^{13}C)-MRSI in tumor models and comparing them to (^{12}C) levels in MSI data generated by DESI.[45] Ambient ionization techniques like DESI were also used to correlate autoradiography and PET images of rat brains for pharmacokinetic profiles of a 5-$HT_{2A/2C}$ receptor agonist.[46] This study shows that compound specificity is of utmost importance and can only be acquired by MS. While autoradiography and PET show the distribution of radioactive labels, DESI-MSI visualized the distribution of the unaltered antagonist besides the distribution of its demethylated metabolite. The combinatorial information of localized chemical specificity by MS, high sensitivity as given by autoradiography, and metabolic *in vivo* imaging delivered by PET is now unmatched by any other imaging method. MALDI-MSI was able to detect differences in lipid distributions between the scar and the healthy brain for a stroke investigation, thanks to its high resolving power, while PET imaging showed only minimal neurodegeneration and inflammation.[47] The proof-of-concept approach, since not directly correlating the data together, demonstrates the power of CMI. It is apparent that overcoming the hurdle of imaging technologies visualizing distributions at different scales is a goal for future CMI pipelines.

19.5.2 X-Ray Fluorescence Imaging

XRF is a nondestructive, label-free, multi-elemental mapping technique of high specificity and sensitivity for light-weight elements. Lateral resolutions for benchtop instruments are in the mid-μm range, but high-end instrumentation such as particle accelerators have much improved options for resolution and also in terms of sensitivity. Combining XRF with elemental imaging (*i.e.* LA-ICP-MS) could be considered as not beneficial, especially because LA-ICP-MS is consuming all the samples. However, the potential of absolute quantification, high resolution down to a few μm, and the potential to assess and quantify also larger elements make this combination highly attractive as shown for bone-cartilage samples.[48] In this example, LA-ICP-MS was used subsequently to μXRF on the very same sample and region and to map elements cartilaginous areas. Besides many other medical applications, the beneficial combination of μXRF and LA-ICP has also been demonstrated in areas of biomedical engineering[49] and plant research.[50,51] Correlated μCT/LA-ICP-MS imaging has also been facilitated to corroborate the distribution of MRI contrast agents in tissue.[52,53] The capabilities of SIMS and XRF for nano-chemical imaging were evaluated at 50 nm resolution for polymorphonuclear human neutrophils using nanoSIMS and synchrotron radiation-based XRF (SR-XRF).[54] Correlating isotopic maps obtained by nano-SIMS to TEM and SR-XRF images showed the crucial point of sample preparation.

An example of a novel, so far not further pursued CMI pipeline, is µXRF in combination with MALDI-MSI. The non-destructive analysis of elemental distributions can be combined with the analysis of molecules without complex sample preparation steps. The distribution of lipids and elements relevant to bone structures, nail, cartilage, muscle, and skin were visualized at the µm scales in the very same sample section of a chicken phalanx without tissue decalcification.[55] This is also one of the rarely found studies using a sample which contains soft and hard matter. This is a particular challenge for cryo-sectioning of the tissue but also MALDI-MSI because of the insulating character of the sample.

19.5.3 Optical Imaging

The potential of combining MSI with images coming from classical OI approaches (*i.e.* from histology or immunostaining), has been thoroughly reviewed.[56] The potential of MALDI, nanoSIMS, LA-ICP, and IMC is shown in impressive studies for cancer tissue. Also refer to Chapter 17 of this book for further details on MALDI in the clinical field. Some examples for CMI pipelines including OI (*i.e.* histology) are also given throughout this section. An outstanding application is the LSFM-guided MALDI-imaging of cleared tissue samples.[57] Tissue fixation and chemical clearing for 3D fluorescence analysis did not interfere with the analysis of proteins in the tissue sections (Figure 19.3).

Blending MSI with vibrational spectroscopy sheds light on spatial details by combining the benefit of a non-destructive, high resolution, optical method targeting analyte classes with the specificity and sensitivity of MSI as reviewed by Lakab *et al.*[58] A very interesting aspect of this CMI approach is the use of Fourier-transform infrared (FT-IR) microscopy to identify disease relevant tissue areas by applying tissue classification strategies and to use this information for a more detailed study by MALDI-MSI to identify molecules of interest (FTIR-guided MALDI-MSI).[59] The combination of synchrotron-FTIR, synchrotron-UV, TOF-SIMS, and histologically assessed serial tissue section of liver cirrhosis gave access to imaging the distribution of lipids, proteins, sugars, and nucleic acids.[60] In this study, a sample holder was introduced making the moving of a tissue section from one instrument to another feasible. Easy relocation of the regions of interest was achieved.

Multimodal imaging approaches for microbial interactions demand imaging modalities of high resolution. Recent improvements in SIMS design have increased sensitivity and lateral resolution down to approx. 50 nm, so that it is now possible to quantify the isotopic composition of single microbial cells in environmental samples. To separately determine microbial identity fluorescence *in situ* hybridization (FISH) using epifluorescence microscopy was used to visualize the fate of substrates labeled with ^{13}C-carbon and ^{15}N-nitrogen, while individual cells were identified simultaneously by halogen labeling *via* FISH.[61]

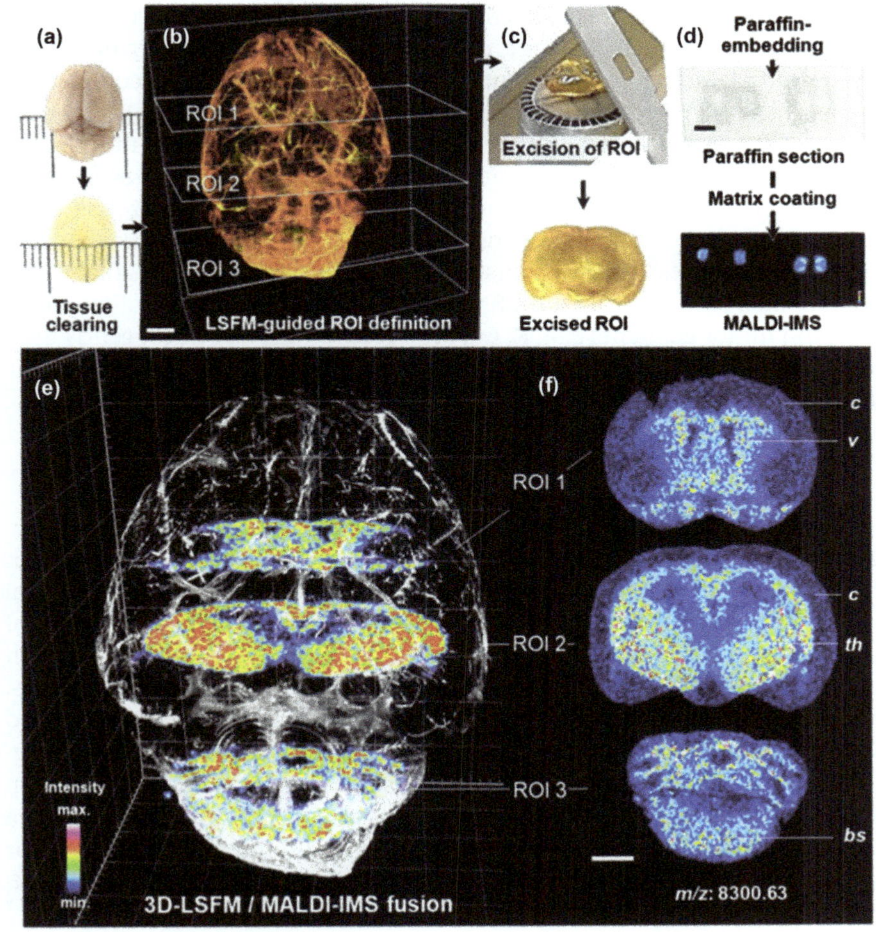

Figure 19.3 3D-Light sheet fluorescence microscopy (LSFM)-guided MALDI-imaging mass spectrometry (IMS) in an optically cleared mouse brain. (a) Optical clearing of a fluorescent-labeled mouse brain. (b) LSFM-3D reconstruction of the brain. Bar = 500 µm. (c) Tissue sections corresponding to the LSFM-defined ROIs are excised from the cleared tissue. (c) Excised ROI 2 (frontal diencephalic brain section). Bar = 1 mm. (d) The excised tissue is embedded in paraffin, sectioned, and the sections coated with matrix for MALDI-MSI. Bars = 5 mm. (e) Fused image of the 3D-LSFM reconstruction of the cleared brain and MALDI-MS images of guanine nucleotide-binding protein subunit gamma-3 (*m/z*: 8,300.63). (f) Spatial distribution of GNG3, distinct brain structures are indicated for orientation: Cerebral cortex (c), ventricles (v), thalamus (th), brain stem (bs).

Additionally, MALDI was correlated to FISH analysis.[62] Although spatial resolution was worse by a factor of 100, the benefit of detecting molecular species was evident. Detection of intact metabolites allowed a metabolic landscape for symbiotic organs to be established.

Laser-induced breakdown spectroscopy (LIBS) harnesses short-lived plasmas above the sample to excite elements for optical emission detection. Rapid multielement analyses, simplicity, and ease of operation come at the trade of low sensitivity for trace elements, matrix effects, and spectral interferences. Nevertheless, LIBS can seamlessly be integrated into LA-ICP-MS workflows for simultaneous atomic emission and isotopic mass measurement in the same experiment. Such a hybrid instrument has been demonstrated to be useful for elemental mapping of trace and bulk elements in biological tissues[63] showing that LA-ICP-MS is well suited for analysis of trace elements (Fe, Zn, Cu) and LIBS is optimal for the mapping of major (C, H, O) and minor elements (Na, K, Ca, Mg). LIBS/LA-ICP MS was recently used to analyze metals in prehistoric brown bears (*Ursus arctos*) to reconstruct the nutrition, health, and migration of the animal, and it was used to detect Pb, Mn, and Cu in plant tissues.[64,65]

19.5.4 Electron Microscopy

The correlation of MSI with EM combines inherently lower-resolution chemical images obtained from MS instrumentation with ultrastructural images from EM. Varying concentrations of elements over several orders of magnitude and the need for isotopic selectivity in a number of research areas (*e.g.* geophysics, corrosion research, the dating of fossilized materials, life science research) led to the combination of TEM and SIMS. TEM and SIMS combine high-resolution with high-sensitivity analysis and were reviewed for *ex situ* correlation in materials science and beyond.[66] The instrument development aspects are discussed in detail followed by a computational analysis of the fundamental ion-solid interaction relevant for data correlation. A concept of high potential is the combination of CLEM, TEM, and nanoSIMS as shown for an individual cell in a biological sample[67] (Figure 19.4).

In a more biology-related study, image fusion was applied to SIMS and EM images to study the potential of this CMI pipeline for histology of unlabeled tissue, nanotoxicology, and metabolic studies using isotopically-labeled tissue.[68] The integration of TOF-SIMS into a helium ion microscope showed the potential of extremely high-resolution imaging (*i.e.* 8 nm) with the potential to gather molecular information for biological samples.

The limitation of significantly different imaging scales of EM and MALDI-MSI was recently overcome by imaging of consecutive tissue sections of rat lung tissue infected by bacteria and fungi. Although image alignment was carried out manually so far, biomarker distributions could be correlated to areas infected by bacterial cells and fungal hyphae.[69] Implementing more advanced image fusion or correlation strategies in the presented workflow had high potential for future applications targeting the combination of morphological structures and molecular information at the cellular level.

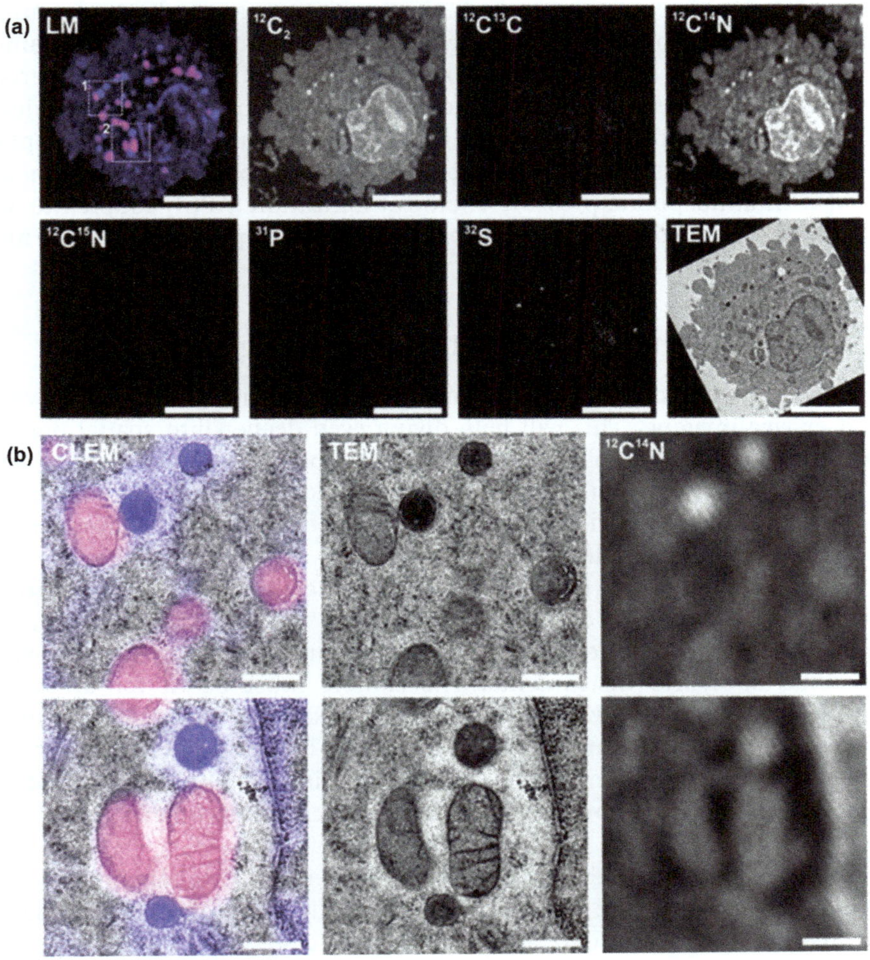

Figure 19.4 CLEM-SIMS in cell culture. (a) An individual cell was imaged using light microscopy, TEM, and nanoSIMS. The fluorescence channels show the uranyl acetate fluorescence (blue) and the MitoTracker fluorescence (magenta). The isotopes measured by nano-SIMS MS are indicated. Scale bars: 2 μm. (b) Higher zoom views of the areas marked by white squares in panel A. Scale bars: 500 nm.

19.5.5 Atomic Force Microscopy

A scanning probe module based on atomic force microscopy (AFM) was integrated into a nanoSIMS to achieve high-resolution SIMS analyses without risking artifacts induced by surface roughness, which is changing during ion bombardment.[70] Combining both modalities in one instrument led to 3D-chemical maps with the highest sensitivity and enhanced spatial accuracy for a polymer blend. Beyond topographic imaging, scanning probe techniques have found a broad range of applications for material sciences to test mechanical properties of a sample. Combining mechanical information like the spatially-resolved elastic modulus with topography and mass analysis was of interest. A newly developed combination of AFM with atmospheric pressure chemical ionization (APCI) coupled to an LTQ was designed to image the physico-chemical properties at the nm scale.[71]

19.6 Challenges for Multimodal Imaging

New combinations of imaging technologies will continue to bridge *in vivo* and *ex vivo* imaging. This will help diving even deeper into spatial organizations to understand native tissue pathology, physiology, and morphology in the context of subcellular molecular information that helps to understand biological events. Yet, bridging the gap between the different imaging scales comes with several challenges.

Sample preparation that is compatible across imaging modalities without compromising data quality is now at the forefront of many investigations. Sample preparation is also very important for future developments of hybrid instruments. The sample must be fit for different physical concepts and workflows. To give an example: tissue fixation may be well suited for one modality but absolutely incompatible with another one, *e.g.* resin embedding for EM and metabolite imaging by MALDI-MSI. Limitations in sample preparation can be indicative for potential new hybrid instruments.

Solutions to relocate the same regions of interest after changing the imaging platform must be found. This is not only indispensable for image acquisition but also for data analysis. A simple co-registration of images is too superficial to understand the relationship of structures and molecular positions, especially if images at very different lateral resolutions are combined. For this, robust markers that can be detected in different imaging techniques also have to be found. Such markers can either be sample inherent features (*e.g.* structures, analytes) or available on the sample holder (*e.g.* grids, wells, geometrically uniform sample holders fitting into multiple instruments).

The lack of high throughput and automatization for the different imaging technologies is especially critical for the sample stability, and subsequently analyte information. Considering metabolic studies this becomes obvious, because biological turn-over rates can be rather high and samples can only be measured on different instrumentations if the metabolic state is fixed properly. Yet, not every MSI method is fit for metabolic quenching and tissue embedding.

Huge steps forward are made for software solutions to correlate the imaging data and standards for data handling and storage are emerging.

19.7 Conclusions and Future Directions

CMI is a constantly developing field, relying on broad academic and industrial expertise. Setting up standard operating procedures (SOPs), quality control settings, and sharing of protocols for new developments are the basis for future success. CMI will lead to hybrid, multimodal imaging platforms that allow the entire sample to be characterized both functionally and morphologically, quickly, and at high spatial resolution to gain a 2- and 3-dimensional mechanistic understanding of the tissue context.

A real holistic view will be achieved if data processing and analysis is incorporated at a very early stage of development leading to a continuous and iterative dialog with image data analysts to communicate and understand the requirements and needs of each step in the process, from sample preparation to image acquisition, from technical potentials to data concepts.

The synergistic fusion of complementary imaging techniques provides rich information by machine-learning-assisted data analysis to ultimately find novel biomarkers, understand interface associated (bio-) chemical reactions or to instantly see hidden connections between so far independently seen events without the need for post-processing correlation software. CMI will substantially transform biomedical research diagnostics and material characterization in the future.

References

1. A. Belianinov, A. V. Ievlev, M. Lorenz, N. Borodinov, B. Doughty, S. V. Kalinin, F. M. Fernandez and O. S. Ovchinnikova, *ACS Nano*, 2018, **12**, 11798–11818.
2. A. P. James and B. V. Dasarathy, *Inf. Fusion*, 2014, **19**, 4–19.
3. L. W. McDonald and T. L. Hayes, *Exp. Mol. Pathol.*, 1969, **10**, 186–198.
4. H. D. Geissinger, *J. Microsc.*, 1974, **100**, 113–117.
5. A. Walter, J. G. Mannheim and C. J. Caruana, Part I: biological imaging, *Imaging Modalities for Biological and Preclinical Research: A Compendium*, IOP Publishing, Bristol, UK, 2021, vol. 1.
6. A. Walter, J. G. Mannheim and C. J. Caruana, Parts II–IV: preclinical imaging, multimodality imaging and outlook, *Imaging Modalities for Biological and Preclinical Research: A Compendium*, IOP Publishing, Bristol, UK, 2021, vol. 2.
7. A. Walter, P. Paul-Gilloteaux, B. Plochberger, L. Sefc, P. Verkade and J. G. Mannheim, *et al.*, *Front. Phys.*, 2020, **8**, 47.
8. T. Beyer, L. S. Freudenberg, D. W. Townsend and J. Czernin, *Insight. Imaging*, 2011, **2**, 161–169.
9. P. Verkade and L. Collinson, *Correlative Imaging: Focusing on the Future*, John Wiley & Sons, Inc., Hoboken, NJ, 2020.
10. M. Setou and N. Kurabe, *J. Electron Microsc.*, 2011, **60**, 47–56.
11. K. Chughtai and R. M. A. Heeren, *Chem. Rev.*, 2010, **110**, 3237–3277.

12. R. W. Vachet, *Nat. Nanotechnol.*, 2015, **10**, 103–104.
13. E. K. Neumann, K. V. Djambazova, R. M. Caprioli and J. M. Spraggins, *J. Am. Soc. Mass Spectrom.*, 2020, **31**, 2401–2415.
14. S. Heiles, M. Kompauer, M. A. Müller and B. Spengler, *J. Am. Soc. Mass Spectrom.*, 2020, **31**, 326–335.
15. A. Zavalin, J. Yang, K. Hayden, M. Vestal and R. M. Caprioli, *Anal. Bioanal. Chem.*, 2015, **407**, 2337–2342.
16. M. Niehaus, J. Soltwisch, M. E. Belov and K. Dreisewerd, *Nat. Methods*, 2019, **16**, 925–931.
17. M. Holzlechner, M. Bonta, H. Lohninger, A. Limbeck and M. Marchetti-Deschmann, *Anal. Chem.*, 2018, **90**, 8831–8837.
18. N. Ogrinc Potočnik, G. L. Fisher, A. Prop and R. M. A. Heeren, *Anal. Chem.*, 2017, **89**, 8223–8227.
19. F. Gyngard and M. L. Steinhauser, *J. Anal. At. Spectrom.*, 2019, **34**, 1534–1545.
20. J. S. Becker, A. Matusch, C. Palm, D. Salber, K. A. Morton and J. S. Becker, *Metallomics*, 2010, **2**, 104–111.
21. S. Van Nuffel, M. Quatredeniers, A. Pirkl, J. Zakel, J.-P. Le Caer and N. Elie, *et al.*, *Anal. Chem.*, 2020, **92**, 12079–12087.
22. E. S. Rivera, K. V. Djambazova, E. K. Neumann, R. M. Caprioli and J. M. Spraggins, *J. Mass Spectrom.*, 2020, **55**, e4614.
23. K. Škrášková, E. Claude, E. A. Jones, M. Towers, S. R. Ellis and R. M. A. Heeren, *Methods*, 2016, **104**, 69–78.
24. L. S. Eberlin, X. Liu, C. R. Ferreira, S. Santagata, N. Y. R. Agar and R. G. Cooks, *Anal. Chem.*, 2011, **83**, 8366–8371.
25. Z. Takáts, J. M. Wiseman and R. G. Cooks, *J. Mass Spectrom.*, 2005, **40**, 1261–1275.
26. E. J. Lanni, R. N. Masyuko, C. M. Driscoll, J. T. Aerts, J. D. Shrout, P. W. Bohn and J. V. Sweedler, *Anal. Chem.*, 2014, **86**, 9139–9145.
27. A. Matusch, C. Depboylu, C. Palm, B. Wu, G. U. Hoglinger, M. K. Schafer and J. S. Becker, *J. Am. Soc. Mass Spectrom.*, 2010, **21**, 161–171.
28. O. Reifschneider, C. A. Wehe, K. Diebold, C. Becker, M. Sperling and U. Karst, *J. Anal. At. Spectrom.*, 2013, **28**, 989–993.
29. P. A. Doble, R. G. de Vega, D. P. Bishop, D. J. Hare and D. Clases, *Chem. Rev.*, 2021, DOI: 10.1021/acs.chemrev.0c01219. Ahead of Print.
30. A. Matusch, L. S. Fenn, C. Depboylu, M. Klietz, S. Strohmer, J. A. McLean and J. S. Becker, *Anal. Chem.*, 2012, **84**, 3170–3178.
31. J. Bianga, A. Bouslimani, N. Bec, F. Quenet, S. Mounicou, J. Szpunar, B. Bouyssiere, R. Lobinski and C. Larroque, *Metallomics*, 2014, **6**, 1382–1386.
32. O. Reifschneider, K. S. Wentker, K. Strobel, R. Schmidt, M. Masthoff, M. Sperling, C. Faber and U. Karst, *Anal. Chem.*, 2015, **87**, 4225–4230.
33. M. Bonta, H. Lohninger, M. Marchetti-Deschmann and A. Limbeck, *Analyst*, 2014, **139**, 1521–1531.
34. L. E. Flint, G. Hamm, J. D. Ready, S. Ling, C. J. Duckett, N. A. Cross, L. M. Cole, D. P. Smith, R. J. A. Goodwin and M. R. Clench, *Anal. Chem.*, 2020, **92**, 12538–12547.

35. H. W. Jackson, J. R. Fischer, V. R. T. Zanotelli, H. R. Ali, R. Mechera and S. D. Soysal, *et al.*, *Nature*, 2020, **578**, 615–620.
36. M. E. Ijsselsteijn, R. van der Breggen, A. Farina Sarasqueta, F. Koning and N. F. C. C. de Miranda, *Front. Immunol.*, 2019, **10**, 2534.
37. L. Keren, M. Bosse, S. Thompson, T. Risom, K. Vijayaragavan and E. McCaffrey, *et al.*, *Sci. Adv.*, 2019, **5**, eaax5851.
38. M. Angelo, S. C. Bendall, R. Finck, M. B. Hale, C. Hitzman and A. D. Borowsky, *et al.*, *Nat. Med.*, 2014, **20**, 436–442.
39. F. Lohöfer, R. Buchholz, A. Glinzer, K. Huber, H. Haas and G. Kaissis, *et al.*, *Sci. Rep.*, 2020, **10**, 79.
40. M. Bulk, W. M. Abdelmoula, H. Geut, W. Wiarda, I. Ronen, J. Dijkstra and L. van der Weerd, *NeuroImage*, 2020, **215**, 116808.
41. S. J. M. Van Malderen, B. Laforce, T. Van Acker, C. Nys, M. De Rijcke and R. de Rycke, *et al.*, *Anal. Chem.*, 2017, **89**, 4161–4168.
42. E. Marillo-Sialer, J. R. Black, B. Paul, K. Kysenius, P. J. Crouch, J. M. Hergt, J. D. Woodhead and D. J. Hare, *J. Anal. At. Spectrom.*, 2020, **35**, 671–678.
43. H. Thiele, S. Heldmann, D. Trede, J. Strehlow, S. Wirtz and W. Dreher, *et al.*, *Biochim. Biophys. Acta*, 2014, **1844**, 117–137.
44. T. K. Sinha, S. Khatib-Shahidi, T. E. Yankeelov, K. Mapara, M. Ehtesham, D. S. Cornett, B. M. Dawant, R. M. Caprioli and J. C. Gore, *Nat. Methods*, 2008, **5**, 57–59.
45. M. Fala, V. Somai, A. Dannhorn, G. Hamm, K. Gibson and D.-L. Couturier, *et al.*, *Magn. Reson. Med.*, 2021, **85**, 3027–3035.
46. S. C. Jacobsen, N. R. Speth, M. Xiong, M. M. Herth, J. L. Kristensen, M. Palner and C. Janfelt, *Mol. Imaging Biol.*, 2021, **23**(5), 676–685.
47. F. Henderson, P. J. Hart, J. M. Pradillo, M. Kassiou, L. Christie, K. J. Williams, H. Boutin and A. McMahon, *Rapid Commun. Mass Spectrom.*, 2018, **32**, 721–729.
48. A. Turyanskaya, S. Smetaczek, V. Pichler, M. Rauwolf, L. Perneczky, A. Roschger, P. Roschger, P. Wobrauschek, A. Limbeck and C. Streli, *J. Anal. At. Spectrom.*, 2021, **36**, 1512–1523.
49. F. Blaske, O. Reifschneider, G. Gosheger, C. A. Wehe, M. Sperling, U. Karst, G. Hauschild and S. Höll, *Anal. Chem.*, 2014, **86**, 615–620.
50. S. Gupta, B. A. Boughton, T. Rupasinghe, S. H. A. Natera, U. Roessner, B. A. Boughton, D. L. Callahan, P. M. C. Smith and C. B. Hill, *Front. Plant Sci.*, 2019, **10**, 1139.
51. L. Feng, J. Wang, H. Li, X. Luo and J. Li, *Anal. Chim. Acta*, 2017, **984**, 66–75.
52. Y. O. Uca, D. Hallmann, B. Hesse, C. Seim, N. Stolzenburg, H. Pietsch, J. Schnorr and M. Taupitz, *Mol. Imaging Biol.*, 2021, **23**, 382–393.
53. D. E. Morrison, J. B. Aitken, M. D. de Jonge, F. Issa, H. H. Harris and L. M. Rendina, *Chem. - Eur. J.*, 2014, **20**, 16602–16612.
54. B. De Samber, R. De Rycke, M. De Bruyne, M. Kienhuis, L. Sandblad, S. Bohic, P. Cloetens, C. Urban, L. Polerecky and L. Vincze, *Anal. Chim. Acta*, 2020, **1106**, 22–32.

55. A. Svirkova, A. Turyanskaya, L. Perneczky, C. Streli and M. Marchetti-Deschmann, *Analyst*, 2018, **143**, 2587–2595.
56. T. Porta Siegel, G. Hamm, J. Bunch, J. Cappell, J. S. Fletcher and K. Schwamborn, *Mol. Imaging Biol.*, 2018, **20**, 888–901.
57. A. Blutke, N. Sun, Z. Xu, A. Buck, L. Harrison and S. C. Schriever, *et al.*, *Sci. Rep.*, 2020, **10**, 14461.
58. S. A. Iakab, P. Ràfols, X. Correig-Blanchar and M. García-Altares, *Anal. Chem.*, 2021, **93**, 6301–6310.
59. J.-H. Rabe, D. A. Sammour, S. Schulz, B. Munteanu, M. Ott and K. Ochs, *et al.*, *Sci. Rep.*, 2018, **8**, 313.
60. V. W. Petit, M. Réfrégiers, C. Guettier, F. Jamme, K. Sebanayakam, A. Brunelle, O. Laprévote, P. Dumas and F. Le Naour, *Anal. Chem.*, 2010, **82**, 3963–3968.
61. S. Behrens, T. Lösekann, J. Pett-Ridge, P. K. Weber, W.-O. Ng, B. S. Stevenson, I. D. Hutcheon, D. A. Relman, A. M. Spormann and A. M. Spormann, *Appl. Environ. Microbiol.*, 2008, **74**(10), 3143–3150.
62. B. Geier, E. M. Sogin, D. Michellod, M. Janda, M. Kompauer, B. Spengler, N. Dubilier and M. Liebeke, *Nat. Microbiol.*, 2020, **5**, 498–510.
63. M. Bonta, J. J. Gonzalez, C. D. Quarles, R. E. Russo, B. Hegedus and A. Limbeck, *J. Anal. At. Spectrom.*, 2016, **31**, 252–258.
64. M. Galiová, J. Kaiser, F. J. Fortes, K. Novotný, R. Malina and L. Prokeš, *et al.*, *Appl. Opt.*, 2010, **49**, C191–C199.
65. J. Kaiser, M. Galiová, K. Novotný, R. Červenka, L. Reale and J. Novotný, *et al.*, *Spectrochim. Acta, Part B*, 2009, **64**, 67–73.
66. S. Eswara, A. Pshenova, L. Yedra, Q. H. Hoang, J. Lovric, P. Philipp and T. Wirtz, *Appl. Phys. Rev.*, 2019, **6**, 021312.
67. F. Lange, P. Agüi-Gonzalez, D. Riedel, N. T. N. Phan, S. Jakobs and S. A.-O. Rizzoli, *PLoS One*, 2021, **16**(5), e0240768.
68. F. Vollnhals, J.-N. Audinot, T. Wirtz, M. Mercier-Bonin, I. Fourquaux, B. Schroeppel, U. Kraushaar, V. Lev-Ram, M. H. Ellisman and S. Eswara, *Anal. Chem.*, 2017, **89**, 10702–10710.
69. T. Juříková, D. Luptáková, O. Kofroňová, A. Škríba, J. Novák, H. Marešová, A. Palyzová, M. Petřík, V. Havlíček and O. Benada, *J. Fungi.*, 2020, **6**(4), 257.
70. Y. Fleming and T. Wirtz, *Beilstein J. Nanotechnol.*, 2015, **6**, 1091–1099.
71. O. S. Ovchinnikova, T. Tai, V. Bocharova, M. B. Okatan, A. Belianinov, V. Kertesz, S. Jesse and G. J. Van Berkel, *ACS Nano*, 2015, **9**, 4260–4269.

Subject Index

A2E, 435
ABCA4 gene, 442
Abcr4 gene, 442
aberrant *N*-glycosylation, 214
ablation, 4
absorption, distribution, metabolism, excretion, and toxicity (ADMET), 325
AC. *See* alternating current (AC)
accuracy, 116
acetaminophen, 372–373
acetonitrile (ACN), 241, 244
ACN. *See* acetonitrile (ACN)
acquisition, 393
actinobacteria, 294
AD. *See* Alzheimer's disease (AD)
ADMET. *See* absorption, distribution, metabolism, excretion, and toxicity (ADMET)
advanced microscopy, 59
AFM. *See* atomic force microscopy (AFM)
age-related macular degeneration (AMD), 442
age-related nuclear cataract (ARN cataract), 434–435
AH. *See* aqueous humor (AH)
alcohol, 124
alpha-cyano-4-chlorocinnamic acid (ClCCA), 9
alpha-cyano-4-hydroxycinnamic acid (CHCA), 9, 141–142, 242
4-alpha-cyano-hydroxycinnamic acid (CHCA), 193, 244
α-crystallin proteins, 440
α-cyano-hydroxycinnamic acid (α-CHCA), 22, 159, 212, 297, 364
alternating current (AC), 67
alternative desorption/ionization agents, 29–30
Alzheimer's disease (AD), 171, 219
AMD. *See* age-related macular degeneration (AMD)
amidase PNGase F, 94
amines, 122–124
amino acids, 95
9-aminoacridine (9-AA), 141, 160, 364
amyloid beta (Aβ), 197, 200, 219
analyte suppression effect (ASE), 12–15
angular illumination, 49
ANI. *See* aniline (ANI)
aniline (ANI), 240
animal disease models, MALDI-MSI of metabolites in, 148
anterior eye, 434–435
    MALDI-MSI in, 436–441
antibody panel based *N*-glycan imaging (APB *N*-glycan imaging), 213
antibody-based techniques, 197
antigen retrieval (AR), 242
AP. *See* atmospheric pressure (AP)
AP-MALDI. *See* atmospheric pressure MALDI (AP-MALDI)
APB *N*-glycan imaging. *See* antibody panel based *N*-glycan imaging (APB *N*-glycan imaging)
APCI. *See* atmospheric pressure chemical ionization (APCI)

Apis mellifera. *See* honeybees (*Apis mellifera*)
Aplysia califonica, 197
APOE. *See* apolipoprotein E (APOE)
apolipoprotein E (APOE), 219, 437
APP gene, 200
APPI. *See* atmospheric pressure photo ionization (APPI)
AQP0, 440–441
aqueous humor (AH), 432
AR. *See* antigen retrieval (AR)
area under the curve (AUC), 215
ARN cataract. *See* age-related nuclear cataract (ARN cataract)
ASE. *See* analyte suppression effect (ASE)
Aspergillus, 305
  A. fumigatus, 305
atmospheric pressure (AP), 40
  desorption ionization technologies, 21
  ion sources, 44
atmospheric pressure chemical ionization (APCI), 53, 472
atmospheric pressure MALDI (AP-MALDI), 77
atmospheric pressure photo ionization (APPI), 53
atomic force microscopy (AFM), 459, 471–472
atropine, 449–450
AUC. *See* area under the curve (AUC)
automated matrix deposition systems, 25–26
automated matrix dispensing systems, 28
axial MALDI–TOF-MS imaging, 41–42
Aβ. *See* amyloid beta (Aβ)

Bacillus
  B. anthracis, 375–376
  B. subtilis, 300, 304–305
BAK. *See* benzalkonium chloride (BAK)
baseline subtraction, 265
basepeak spectrum. *See* skyline spectrum
Bathymodiolus puteoserpentis, 306
BBB. *See* blood–brain barrier (BBB)
"beam-type" CID experiment, 65
beetle (*Paederus riparius*), 375
benzalkonium chloride (BAK), 450
betaine-like lipid (BLL), 303
bile acids, 160
1,1′-binaphthyl-2,2′ diamine (BNDM), 225
biochemical imaging techniques, 188
biological tissues, 365–366
  for MALDI-MSI
    derivatization, 94–95
    embedding and storage, 89–90
    enzymatic digestion, 93–94
    matrix selection and application methods, 95–98
    sample collection, 88–89
    sectioning, 90–91
    washing, 92–93
biomaterial analysis, 22
biomolecules, 21
4,5-(bis(dimethylamino)naphthalen-1-yl)furan-2,5-dione, 141
bisecting k-means, 279
bisphenol A (BPA), 398
bisphenol S (BPS), 398
BLL. *See* betaine-like lipid (BLL)
blood–brain barrier (BBB), 125, 332
blue crab (*Callinectes sapidus*), 197
blueprint, 59
BNDM. *See* 1,1′-binaphthyl-2,2′ diamine (BNDM)
bottom-up approach, 93, 235
botulinum neurotoxin, 375
BPA. *See* bisphenol A (BPA)
BPS. *See* bisphenol S (BPS)
brain injury and disease, 170–173
brimonidine, 449
butylamine 2,5-dihydroxybenzoate (DHB-BuN), 225

CA. *See* 4-hydroxy-3-methoxycinnamaldehyde (CA)
Callinectes sapidus. See blue crab (*Callinectes sapidus*)
Cancer borealis. See Jonah crab (*Cancer borealis*)
cancer research, 167–170
carboxyl groups, 139
carboxymethylcellulose (CMC), 89, 296, 389
Carcinus maenas. See European green crab (*Carcinus maenas*)
cataract, 434–435
ccRCC. *See* clear cell renal cell carcinoma (ccRCC)
CCS. *See* collision cross section (CCS)
CDC. *See* Center for Disease Control and Prevention (CDC)
Center for Disease Control and Prevention (CDC), 377
central nervous system (CNS), 332
Cer. *See* ceramides (Cer)
ceramide 1-phosphate (CerP), 160
ceramides (Cer), 162
CerP. *See* ceramide 1-phosphate (CerP)
CHCA. *See* alpha-cyano-4-hydroxycinnamic acid (CHCA); 4-alpha-cyano-hydroxycinnamic acid (CHCA)
chemical derivatization, 392–393
chemical environmental pollutants, 387
chemical noise, 43
chlordecone, 393–394
5-chloro-2-mercaptobenzothiazole (CMBT), 160
chlorpyrifos (CPF), 400
cholesterol sulfate, 160
chondroitin sulfates, 223
chRCC. *See* chromophobe renal cell carcinoma (chRCC)
chromophobe renal cell carcinoma (chRCC), 413
chymotrypsin, 94
CID. *See* collision-induced dissociation (CID)
CIVMs. *See* complex *in vitro* models (CIVMs)
ClCCA. *See* alpha-cyano-4-chlorocinnamic acid (ClCCA)
clear cell renal cell carcinoma (ccRCC), 413
Clostridium botulinum, 375–376
clustering techniques, 278–279
CM. *See* confocal microscopy (CM)
CMBT. *See* 5-chloro-2-mercaptobenzothiazole (CMBT)
CMC. *See* carboxymethylcellulose (CMC)
CMI. *See* correlative multimodal imaging (CMI)
CNS. *See* central nervous system (CNS)
collagen IV alpha 1, 437
collagenase, 94
collision cross section (CCS), 72, 143–144
collision-induced dissociation (CID), 65, 165, 246
colony morphology, 300
complex *in vitro* models (CIVMs), 326–327
compound identification, 306–309
computed tomography (CT), 458–459
confocal microscopy (CM), 459
Congenital Disorders of Glycosylation, 214
conjugated polymers, 141
consistency in matrix deposition, 25
continuous raster mode, 48
conventional MALDI, 16
cornea, 434–435
corneal dystrophies, 436
corneal structure, 436
correlation analysis, 274
correlative microscopy, 458
correlative multimodal imaging (CMI), 460

correlative multimodal mass spectrometry imaging, 457–460
challenges for multimodal imaging, 472–473
combination of MS techniques, 463–465
combination of MS with other modalities, 465–472
MSI in context of, 460–463
cortical GSH, 437
Coupled Chemical and Physical Dynamics model (CPCD model), 9, 11–12
CP49, 441
CPCD model. *See* Coupled Chemical and Physical Dynamics model (CPCD model)
CPF. *See* chlorpyrifos (CPF)
crime scene investigators (CSI), 360
criminal chemical profiling, 361, 363
cross-validation (CV), 276
cryoprotectants, 88
cryostats, 90
CSI. *See* crime scene investigators (CSI)
CT. *See* computed tomography (CT)
CV. *See* cross-validation (CV)
cyanobacteria, 293

DA. *See* discriminant analysis (DA)
DAN. *See* 1,5-diaminonaphthalene (DAN)
DAN-HCL. *See* 1,5-diaminonaphthalene hydrochloride (DAN-HCL)
DANI. *See* *N,N* dimethylaniline (DANI)
DAPI, 195
DAPNe. *See* direct analyte-probed nano-extraction (DAPNe)
DART–HRMS. *See* direct analysis in real time–high resolution mass spectrometry (DART–HRMS)
data analysis, 117–121
data integration, 280
data-dependent acquisition (DDA), 165, 235
data-independent acquisition (DIA), 235
DC. *See* direct current (DC)
DD preparations. *See* dried droplet preparations (DD preparations)
DDA. *See* data-dependent acquisition (DDA)
delayed extraction, 63
delocalization, 93, 126
ΔabrB mutant, 301
Δphz mutant, 301
ΔpqsH mutant, 301
Δspo0A mutant, 301
deposition method of derivatization reagents, 95
depth profiling, 119
derivatization, 94–95, 138–139
reaction deposition techniques, 97
reagents and targets, 96
dermatology, 341
desalting, 15
DESI. *See* desorption electrospray ionization (DESI)
DESI-FTICR. *See* desorption electrospray ionization Fourier transform ion cyclotron resonance (DESI-FTICR)
desorption (D), 3
desorption electrospray ionization (DESI), 105, 166, 322, 439, 461
desorption electrospray ionization Fourier transform ion cyclotron resonance (DESI-FTICR), 306
DGCC. *See* diacylglyceryl-3-*O*-carboxy-(hydroxymethyl)-choline (DGCC)
DGDG. *See* digalactosyldiacylglycerol (DGDG)
DHA. *See* dihydroxyacetophenone (DHA); 2,6-dihydroxy-acetophenone (DHA)
DHB. *See* 2,5-dihydroxy benzoic acid (DHB); dihydroxybenzoic acid (DHB)
DHB-BuN. *See* butylamine 2,5-dihydroxybenzoate (DHB-BuN)

DIA. *See* data-independent acquisition (DIA)
diacylglyceryl-3-*O*-carboxy-(hydroxymethyl)-choline (DGCC), 303
1,5-diaminonaphthalene hydrochloride (DAN-HCL), 141
1,5-diaminonaphthalene matrix (1,5-DAN), 247
1,5-diaminonaphthalene (DAN), 141, 160, 225
differential mobility spectrometry (DMS), 77
digalactosyldiacylglycerol (DGDG), 301
2,5-dihydroxy benzoic acid (DHB), 212, 224
2,5-dihydroxy-acetophenone (2,5-DHA), 160, 193
dihydroxyacetophenone (DHA), 240
2,6-dihydroxyacetophenone (DHA), 217
dihydroxybenzoic acid (DHB), 9, 217, 240
2,5-dihydroxybenzoic acid (2,5-DHB), 141, 159, 193, 242, 297
diketopiperazine, 306
dimethyl sulfoxide, 88
diode pumped solid state lasers (DPSS lasers), 47
1,2-dipalmitoyl-*sn-glycero*-3-phosphoglycerol (DPPG), 334
1,6-diphenyl-1,3,5-hexatriene (DPH), 217
2,3-diphenyl-pyranylium tetrafluoroborate (DPP-TFB), 138–139
direct analysis in real time–high resolution mass spectrometry (DART–HRMS), 375
direct analyte-probed nanoextraction (DAPNe), 380
direct current (DC), 63
discriminant analysis (DA), 275
discriminatory *m*/*z* values, 273
    discriminatory metrics, 275
    inferential statistics, 273–274
disease pathogenesis, 336–338
dispersion electric field, 77
1,2-distearoyl-*sn-glycero*-3-phosphoethanol-amine conjugated with PEG (PEG_{36}-DSPE), 334
distribution, 329–333
DMS. *See* differential mobility spectrometry (DMS)
DOA. *See* drugs of abuse (DOA)
double bond (db), 166
DPH. *See* 1,6-diphenyl-1,3,5-hexatriene (DPH)
DPP-TFB. *See* 2,3-diphenyl-pyranylium tetrafluoroborate (DPP-TFB)
DPPG. *See* 1,2-dipalmitoyl-*sn-glycero*-3-phosphoglycerol (DPPG)
DPSS lasers. *See* diode pumped solid state lasers (DPSS lasers)
dried droplet preparations (DD preparations), 247
drift tube ion mobility spectrometry (DTIMS), 72–74
drug administration and absorption, 328–329
drug detection in eye for off-target effects, 448–449
drug development, 28
drug discovery and development, 321–322
    clinical, 340–342
    early discovery, 325–327
    mass spectrometry imaging, 322–323
    pipeline, 323–325
    preclinical phase of development, 327
        disease pathogenesis, 336–338
        drug disposition and pharmacokinetics, 328–334
        pharmacodynamics, 335–336
        targeted drug delivery, 334–335
        toxicology and safety investigations, 338–340

drug disposition, 328
    distribution, 329–333
    drug administration and absorption, 328–329
    metabolism, 333
    pharmacokinetics, 333–334
drugs of abuse (DOA), 368
DTIMS. *See* drift tube ion mobility spectrometry (DTIMS)

EC tissues. *See* endometrial cancer tissues (EC tissues)
ECD. *See* electron capture dissociation (ECD)
ECM. *See* extracellular matrix (ECM)
EDDP. *See* 2-ethylidene-1,5-dimethyl-3,3-diphenylpyrrolidine (EDDP)
EFGR. *See* epidermal growth factor receptor (EFGR)
EhV. *See Emiliania huxleyi* virus (EhV)
eicosapentaenoic acid (EPA), 447
elastase, 94
electron capture dissociation (ECD), 246
electron microscopy (EM), 470
electron transfer dissociation (ETD), 246
electrospray ionization (ESI), 43, 74, 158, 235
electrospray-assisted methods, 40
ELOVL6, 170
EM. *See* electron microscopy (EM)
embedding, 89–90
Emiliania huxleyi, 303
Emiliania huxleyi virus (EhV), 303
EMR. *See* extended mass range (EMR)
endometrial cancer tissues (EC tissues), 216
energetic decoupling, 43
enhanced permeability and retention effect (EPR effect), 334
environmental pollutants, 387
    application of MALDI-MSI for study of environmental pollutant distribution, 393–397
    combination of MALDI-MSI and LC–MS/MS for identification of responses, 397–401
    sample preparation for study of environmental pollutant distribution, 389–393
enzymatic digestion, 93–94
EPA. *See* eicosapentaenoic acid (EPA)
epidermal growth factor receptor (EFGR), 327
EPR effect. *See* enhanced permeability and retention effect (EPR effect)
EPS. *See* extracellular polymeric substance (EPS)
ERα. *See* estrogen receptor α (ERα)
Escherichia coli, 133
Escovopsis fungi, 305
ESI. *See* electrospray ionization (ESI)
estrogen receptor α (ERα), 398
ETD. *See* electron transfer dissociation (ETD)
2-ethylidene-1,5-dimethyl-3,3-diphenylpyrrolidine (EDDP), 373
Eucalyptus cladocalyx, 224
Euprymna scolopes. See Hawaiian bobtail squid (*Euprymna scolopes*)
European green crab (*Carcinus maenas*), 197
evaporative ionization mass spectrometry (REIMS), 378
ex vivo
    imaging technologies, 340–341, 465–467
    integration with *ex vivo* techniques, 280–281
exact mass, 142–143
extended mass range (EMR), 71
extracellular matrix (ECM), 212
extracellular polymeric substance (EPS), 300
eye, 432–434
    diseases, 432
        MALDI-MSI for ocular therapeutics detection, 445–450

MALDI-MSI in anterior eye, 436–441
MALDI-MSI in posterior eye, 441–445
ocular structure, function, and common diseases, 432–435

FA. *See* ferulic acid (FA); formaldehyde (FA)
FANGS. *See* filter-aided *N*-glycan separation (FANGS)
fatty acids, 95, 157
FDA. *See* United States Food and Drug Administration (FDA); US Food and Drug Administration (FDA)
FDIH. *See* first dose in humans (FDIH)
ferulic acid (FA), 139, 240
FET. *See* fingermark enhancement techniques (FET)
FF. *See* fresh/frozen (FF)
FF sample. *See* formalin-fixed sample (FF sample)
FFPE. *See* formalin fixation paraffin embedding (FFPE)
field free path, 41
filensin, 441
filter-aided *N*-glycan separation (FANGS), 212
fine vacuum conditions, 40
fingermark elimination, 361
fingermark enhancement techniques (FET), 360
Fingermark Visualization Manual (FVM), 360, 364
fingermarks, 360–365
first dose in humans (FDIH), 325
FISH. *See* fluorescence *in situ* hybridization (FISH)
flash-freezing. *See* snap freezing
flash-frozen tissue samples preparation of, 135–138
flavonol glycosides, 224
flavonols, 224
fluorescence *in situ* hybridization (FISH), 376, 468
fluoromethylpyridinium based reactive matrices, 26
FMW. *See* focused microwave irradiation (FMW)
focused microwave irradiation (FMW), 135
food forensics, 377–378
forensic analysis of fibers, 378–379
forensic botany, 376–377
forensic entomology, 373–375
forensic evidence, 359
MALDI-MSI application in, 360–366
MALDI-MSI application in forensic toxicology, 366
MALDI-MSI of drugs and metabolites in hair, 366–372
potential use of MALDI-MSI of drugs and metabolites in tissues, 372–373
prospective applications of MALDI-MSI, 373–380
formaldehyde (FA), 377
formalin, 88
fixation, 89, 93, 158–159
formalin fixation paraffin embedding (FFPE), 159
formalin-fixed paraffin embedded (FFPE), 26, 87, 191, 236, 413
animal tissues, 148
material, 91
samples, 137, 241
tissues, 93, 210, 322
formalin-fixed sample (FF sample), 236, 241
Fourier transform ion cyclotron resonance (FT-ICR), 68–71, 114, 142, 163–164, 462
Fourier Transform-MS (FT-MS), 235
mass analyzers, 43
Fourier-transform infrared microscopy (FT-IR microscopy), 378, 468

fragile specimens, 88
free fatty acids, 160
fresh/frozen (FF), 210, 236
front side illumination, 48–50
frozen material, 90–91
frozen tissue sections, 92–93
FT-ICR. *See* Fourier transform ion cyclotron resonance (FT-ICR)
FT-IR microscopy. *See* Fourier-transform infrared microscopy (FT-IR microscopy)
FT-MS. *See* Fourier Transform-MS (FT-MS)
full width at half maximum (FWHM), 45, 62
FVM. *See* Fingermark Visualization Manual (FVM)
FWHM. *See* full width at half maximum (FWHM)

GAGs. *See* glycosaminoglycans (GAGs)
galactose (Gal), 208
galactosylceramide (GalCer), 166
gangliosides (GM), 161
gas chromatography (GC), 361, 394
gas cluster ion beam secondary ion mass spectrometry (GCIB-SIMS), 145
gaseous organic molecules, 8
Gaucher disease (GD), 171, 219
GC. *See* gas chromatography (GC)
GCIB-SIMS. *See* gas cluster ion beam secondary ion mass spectrometry (GCIB-SIMS)
gcnE protein, 301
GD. *See* Gaucher disease (GD)
gelatin, 389
    gelatin-embedded material, 89
Ginkgo biloba leaves, 223–224
Girard's reagent T (GirT), 139
GirT. *See* Girard's reagent T (GirT)
GlaxoSmithKline, 448
GlcNAc. *See N*-acetylglucosamine (GlcNAc)
Gleason score system (GS system), 168
glioma, 444–445
Global Natural Product Social molecular networking (GNPS), 295
globotetraosylceramide (Gb4Cer), 222
globotriaosylceramide (Gb3Cer), 222
GLP. *See* Good Laboratory Practice (GLP)
glucocerebrosidase (GCase), 171
glucose, 207
glucosylceramide (GluCer), 166
Glutamicibacter arilaitensis, 304
glutathione (GSH), 434–435, 437
glycans, 92, 94, 207
glycerides, 157
glycerol, 157
glycerophospholipids, 157
glycobiology, 207–208
glycomics, 207
    glycosaminoglycans, 222–223
    glycosides, 223–224
    glycosphingolipids, 216–222
    N-glycans, 208–216
    non-conjugated saccharides, 224–226
glycosaminoglycans (GAGs), 212, 222–223
glycosides, 223–224
glycosphingolipids (GSL), 162, 216
    biological application, 218–222
    method development and optimization, 217–218
GMD. *See* Golm metabolite database (GMD)
GNPS. *See* Global Natural Product Social molecular networking (GNPS)
Golm metabolite database (GMD), 145
Good Laboratory Practice (GLP), 106
GS system. *See* Gleason score system (GS system)
GSL. *See* glycosphingolipids (GSL)

H&E. *See* hematoxylin and eosin (H&E)
HA. *See* hydroxylamine hydrochloride (HA)

Hawaiian bobtail squid (*Euprymna scolopes*), 294, 299, 306
HCC. *See* hepatocellular carcinoma (HCC)
HCD. *See* higher collision-induced dissociation (HCD)
HD. *See* Hunter's disease (HD)
HDPE. *See* high-density polyethylene (HDPE)
heat-induced antigen retrieval (HIAR), 243
hematoxylin and eosin (H&E), 195, 236, 408
hemispheric disease models, 198
hepatocellular carcinoma (HCC), 216
HER2. *See* human epidermal growth factor receptor 2 (HER2)
herpesviridae, 91
hexosylceramide (HexCer), 173
HIAR. *See* heat-induced antigen retrieval (HIAR)
hierarchical clustering, 279
high lateral resolution – emerging applications, imaging of metabolites with, 149–150
high quality control (HQC), 115
high resolution episcopic microscopy (HREM), 458–459
high-density polyethylene (HDPE), 397
high-field asymmetric waveform ion mobility, 77
high-resolution accurate mass (HRAM), 69
high-throughput screening (HTS), 326
higher collision-induced dissociation (HCD), 246
HILIC. *See* hydrophilic interaction liquid chromatography (HILIC)
HMDB. *See* human metabolome database (HMDB)
homogenization of tissue, 248–249
honeybees (*Apis mellifera*), 197
HPMC. *See* hydroxypropyl methylcellulose (HPMC)
HQC. *See* high quality control (HQC)
HRAM. *See* high-resolution accurate mass (HRAM)
HREM. *See* high resolution episcopic microscopy (HREM)
HTS. *See* high-throughput screening (HTS)
human epidermal growth factor receptor 2 (HER2), 216, 409
human immunodeficiency virus, 91
human metabolome database (HMDB), 139
Hunter's disease (HD), 219
hydrazinonicotinic acid (HYNIC), 212
hydrogels, 250–251
hydrophilic interaction liquid chromatography (HILIC), 212
hydrophilic molecules, 93
4-hydroxy-3-methoxycinnamaldehyde (CA), 138–139
6-hydroxy-dopamine (6-OHDA), 198
hydroxylamine hydrochloride (HA), 239
hydroxypropyl methylcellulose (HPMC), 90
HYNIC. *See* hydrazinonicotinic acid (HYNIC)

ICP-MS. *See* inductively coupled plasma-mass spectrometry (ICP-MS)
IDH. *See* isocitrate dehydrogenase (IDH)
idiopathic pulmonary fibrosis (IPF), 336
IFS. *See* isotopic fine structure (IFS)
IHC. *See* immunohistochemistry (IHC)
IIR. *See* inflammation-induced optic nerve regeneration (IIR)

illumination
angular, 49
front side, 48–50
transmission mode, 50–51
IM-MS. *See* ion mobility mass spectrometry (IM-MS)
image
processing, 271
registration, 280–282
segmentation, 278
imaging, 59
stable isotope labeled lipids, 177
technologies, 59
of xenobiotics, 27–28
imaging mass cytometry (IMC), 465
imaging mass spectrometry (IMS). *See* mass spectrometry imaging (MSI)
IMC. *See* imaging mass cytometry (IMC)
immune checkpoint inhibition, 415–416
immunohistochemistry (IHC), 171, 195–196, 215, 241, 409, 465
IMS. *See* ion mobility separation (IMS)
in situ
analyte derivatization, 26
identification, 246
in situ bottom-up approach, 246–247
in situ top-down approaches, 247–248
peak annotation, 196
in situ hybridization (ISH), 245
in vivo
imaging, 465–467
integration with *in vivo* techniques, 281–282
in-solution, 111, 114
in-source decay (ISD), 41, 247
in-source fragmentation of lipids during MALDI-MSI, 162–163
"in-tissue" method, 111, 113
indium tin oxide (ITO), 42, 91, 240, 322
inductively coupled plasma-mass spectrometry (ICP-MS), 372, 396–397
inferential statistics, 273–274
inflammation-induced optic nerve regeneration (IIR), 222, 445
informatics for success of mass spectrometry imaging in biomedical research, 262–263
data analysis, 273–280
data processing, 265–272
image registration, 280–282
software for MSI data creation and handling, 263
MSI data representation and handling, 264
software for MSI data acquisition, 263–264
infrared (IR), 44, 68
IR–MALDI-MSI, 53
infrared-matrix assisted laser desorption electrospray ionization (IR-MALD-ESI), 74, 378
ink/documents, 379–380
inlet method, 8
instrumentation for MALDI-MSI
ion mobility spectrometry for improved MALDI-MSI performance, 72–77
ionization sources and design
front side illumination, 48–50
ion sources, 39–40
optical setups, 47
post-ionization, 51–54
pressure regime, 40–44
spatial resolution in MALDI-MSI, 44–47
transmission mode illumination, 50–51
mass analyzers and performance for MALDI imaging, 60–71

intermediate pressure
conditions, 40
ion sources, 42–43
internal standard (IS), 106, 110–111
intraocular lens (IOL), 434
intraocular pressure (IOP), 434
IOL. *See* intraocular lens (IOL)
ion microscope, 21, 60
ion mobility, 143–144
resolving lipids using, 166–167
spectrometry
DTIMS, 72–74
high-field asymmetric waveform ion mobility, 77
for improved MALDI-MSI performance, 72
TIMS, 75–77
TWIMS, 74–75
ion mobility mass spectrometry (IM-MS), 60
ion mobility separation (IMS), 114, 363
ion traps, 66–68
ion-suppression effects, 107
ionization (I), 4
strategies to improve ionization efficiency of lipids, 173–175
IOP. *See* intraocular pressure (IOP)
IPF. *See* idiopathic pulmonary fibrosis (IPF)
IR. *See* infrared (IR)
IR-MALD-ESI. *See* infrared-matrix assisted laser desorption electrospray ionization (IR-MALD-ESI)
iron (Fe), 466
IS. *See* internal standard (IS)
ISD. *See* in-source decay (ISD)
ISH. *See in situ* hybridization (ISH)
isocitrate dehydrogenase (IDH), 149
isomer-resolved MALDI-MSI of lipids, 166
isotope distribution, 142–143
isotopic fine structure (IFS), 69
ITO. *See* indium tin oxide (ITO)

Jonah crab (*Cancer borealis*), 197

KCNAB2 gene, 168
kdmA protein, 301
keratoconus, 436
ketone steroids, 26
knocked out (KO), 436
KO. *See* knocked out (KO)

L-DOPA induced dyskinesia (LID), 198
LA-ICP. *See* laser ablation inductively coupled plasma (LA-ICP)
LA-ICP-MS. *See* laser ablation-inductively coupled plasma-mass spectrometry (LA-ICP-MS)
laeA protein, 301
LAESI. *See* laser ablation electrospray ionization (LAESI)
laser ablation electrospray ionization (LAESI), 53, 297
laser ablation in MALDI, 4
laser pulse dependence, 5
laser spot size, 6
local thermal equilibrium, 6
plume pressure, temperature and velocity, 5
laser ablation inductively coupled plasma (LA-ICP), 62
laser ablation-inductively coupled plasma-mass spectrometry (LA-ICP-MS), 245, 462
laser desorption ionization (LDI), 22, 53, 161, 296, 301, 339, 379
laser induced ionization-MSI (LDI-MSI), 142
laser micro-dissection (LCM), 249
laser post-ionization, 174
laser pulse dependence, 5
laser spot size, 6
laser spray method, 8
laser-induced breakdown spectroscopy (LIBS), 470
LC. *See* liquid chromatography (LC)

LC–ESI-MS/MS. *See* liquid chromatography electrospray ionization tandem mass spectrometry (LC–ESI-MS/MS)
LCM. *See* laser micro-dissection (LCM)
LC–MS. *See* liquid chromatography coupled to mass spectrometry (LC–MS)
LC–MS/MS. *See* liquid chromatography–tandem mass spectrometry (LC–MS/MS)
LDA. *See* linear DA (LDA)
LDI. *See* laser desorption ionization (LDI)
LDI-MSI. *See* laser induced ionization-MSI (LDI-MSI)
lens, 434–435
  in aging and cataracts, 436–441
Lepidium meyenii, 225
LIBS. *See* laser-induced breakdown spectroscopy (LIBS)
LID. *See* L-DOPA induced dyskinesia (LID)
light sheet fluorescence microscopy (LSFM), 459
limit of detection (LOD), 109–110, 368
limit of quantification (LOQ), 109
linear DA (LDA), 275
linear ion trap (LIT), 67
linear projection techniques, 278
lipids, 92, 122, 157–158
  identification in MSI, 161
  identification of lipids detected with MALDI-MSI, 161–167
    identification of lipids using accurate mass measurements, 163–165
    identification of lipids using tandem mass spectrometry, 165–166
    in-source fragmentation of lipids during MALDI-MSI, 162–163
    isomer-resolved MALDI-MSI of lipids, 166
    lipid-related ions generated by MALDI-MSI, 162
    resolving lipids using ion mobility, 166–167
  recent developments in MALDI-MSI of, 173
    imaging stable isotope labeled lipids, 177
    lipid analysis and imaging from single cells, 175–177
    strategies to improve ionization efficiency of lipids, 173–175
  sample preparation for lipid MSI, 158–161
  selected applications of MALDI-MSI of, 167
LipostarMSI, 280
liquid chromatography (LC), 211, 361
liquid chromatography coupled to mass spectrometry (LC–MS), 328, 366
liquid chromatography electrospray ionization tandem mass spectrometry (LC–ESI-MS/MS), 412
liquid chromatography–tandem mass spectrometry (LC–MS/MS), 106, 235
liquid micro-junction (LMJ), 249
liquid microextraction, 249–250
Listeria monocytogenes, 300
LIT. *See* linear ion trap (LIT)
liver disease, 173
LLOQ. *See* lower limit of quantification (LLOQ)
LMJ. *See* liquid micro-junction (LMJ)
LMW. *See* low molecular weight (LMW)
local thermal equilibrium (LTE), 6
LOD. *See* limit of detection (LOD)

Lomatia, 224
LOQ. *See* limit of quantification (LOQ)
low molecular weight (LMW), 161
low quality control (LQC), 115
lower limit of quantification (LLOQ), 110
LPC. *See* lysophosphatidylcholines (LPC)
LPG. *See* lysyl-phosphatidylglycerol (LPG)
LQC. *See* low quality control (LQC)
LSFM. *See* light sheet fluorescence microscopy (LSFM)
LTE. *See* local thermal equilibrium (LTE)
Lucky Survivors, 7
lysophosphatidylcholines (LPC), 160
lysyl-phosphatidylglycerol (LPG), 301

m/*z* value identification, 279–280
MA. *See* methamphetamine (MA)
macular degeneration, 435, 441–445
MAF. *See* maximum autocorrelation factor analysis (MAF)
magnetic resonance, 59
magnetic resonance imaging (MRI), 459
magnetic sector mass analyzers, 60
MALDESI. *See* matrix assisted laser desorption electrospray ionization (MALDESI)
MALDI. *See* matrix-assisted laser desorption/ionization (MALDI)
MALDI-MSI. *See* Matrix Assisted Laser Desorption Ionization–Mass Spectrometry Imaging (MALDI-MSI)
MALDI-selected reaction monitoring MS (MALDI-SRM/MS), 368
4-maleicanhydrido proton sponge (MAPS), 141
mannose (Man), 208
manual micro-excision, 250
MAPS. *See* 4-maleicanhydrido proton sponge (MAPS)
mass analyzers
  FT-ICR, 68–71
  ion traps, 66–68
  merit for, 61
  Orbitrap, 71
  and performance for MALDI imaging, 60
  QTOF, 65–66
  quadrupole mass filters, 62–63
  sector analyzers, 60–62
  time-of-flight, 63–65
  triple quadrupoles, 68
mass spectrometry (MS), 235, 322
mass spectrometry imaging (MSI), 20, 60, 87, 91, 158, 189, 208, 235, 262, 321–323. *See also* correlative multimodal mass spectrometry imaging
  data acquisition and instrumentation, 24–25
  experiment for metabolic analysis, 134
    derivatization, 138–139
    MALDI-ionization process, 139–142
    metabolic quenching, 135
    preparation of flash-frozen tissue samples, 135–138
mass spectrometry-based neuropeptide identification, 196
mass-to-charge ratio values (*m*/*z* values), 295
matrix
  application, 160–161
  deposition, 241, 244, 393
  selection
    and application methods, 95–98
    based on analytes of interest, 97
  sublimation systems, 25
matrix assisted laser desorption electrospray ionization (MALDESI), 53

Matrix Assisted Laser Desorption Ionization–Mass Spectrometry Imaging (MALDI-MSI), 3, 20–21, 39–40, 235, 359, 388, 410, 434
automated matrix deposition, 25
beginning, 21–23
future challenges, 28
guided spatially-resolved proteomics, 251
identification of lipids detected with, 161–167
imaging of xenobiotics, 27–28
increasing sensitivity and specificity, 29–31
MALDI-MSI of digested proteins or bottom-up pMSI, 241
sample preparation protocols for digested proteins pMSI, 242–243
MALDI-MSI of intact proteins or top-down pMSI, 238–240
sample preparation protocols, 239–241
and metabolomics
applications, 146–150
instrumentation with strengths and current limitations, 142–144
MSI experiment for metabolic analysis, 134–142
strategies for structural analysis/identification, 144–146
MS instrumentation, 23–24
MSI data acquisition and instrumentation, 24–25
on-tissue chemistries, 25–27
opening statement, 20–21
push for higher spatial resolution, 28–29
specific MALDI of proteins using tagged probes, 244–245
standard protocols and methods, 29
matrix suppression effect (MSE), 11–12
matrix-assisted laser desorption/ionization (MALDI), 3, 20, 60, 64, 87, 95, 105, 158, 208, 262, 322
laser ablation in, 4–6
MALDI-2, 15–16, 29, 143
MALDI-based MSI, 189
MALDI-ionization process, 139–142
MALDI-Orbitrap FTMS, 142
MALDI-Time of Flight instruments, 143
MALDI–QqTOF-MS instrument, 28
matrices, 159–160, 240–241
MS-based neuropeptide imaging in biological research, 196
MALDI-MSI of neuropeptides in neurodegenerative diseases, 197–201
neuropeptide imaging in cells and animal models, 196–197
primary ionization, 7–10
secondary ionization, 10–15
maximum autocorrelation factor analysis (MAF), 194
MCP detectors. *See* multichannel plate detectors (MCP detectors)
MD. *See* method development (MD)
mechanical charge separation, 7–8
melphalan, 450
MetaboAnalyst, 280
metabolic quenching, 135
metabolic responses in plants, 400–401
metabolism, 133, 333
metabolites, 92, 134
application of metabolite imaging to animal tissues, 147–148
metal adduct ions, 10
metal surfaces, 9–10

METASPACE project, 145
methamphetamine (MA), 369
method development (MD), 106–110
instrumental and experimental conditions, 114
internal standard, 110–111
mimetic tissue, 111–113
normalization method comparison, 114
TEC, 111
method validation (MV), 106–107, 114
quality control studies in QMSI, 117
quality controls, 115–117
1-methyl-4-phenyl-1,2,3,6-tetrahydropyridine (MPTP), 198
MIBI. *See* multiplexed ion beam imaging (MIBI)
microbial communities, 295
microbial forensics, 375–376
microbial MSI, sample preparation for, 295–299
microbiology as field of study for MSI development, 293–295
applications for imaging
host–microbe interactions, 305–306
microbial co-cultures, 303–305
microbial monocultures, 299–303
compound identification, 306–309
sample preparation for microbial MSI, 295–299
micropollutants, 400–401
microprobe MALDI-MSI, 45
microscope MALDI-MSI, 45
middle quality control (MQC), 115
mimetic tissue, 111–113
MLS. *See* myxoid liposarcoma (MLS)
modern instrumentation, 59
modern mass spectrometers, 60
modern TOF instruments, 63
molecular dynamics simulations of MALDI, 5
molecular preservation, 88
molecular weight cut-off filter (MWCO filter), 196
molecular weight (MW), 21, 235
momentum analysis, 62–63
MPM. *See* multiphoton microscopy (MPM)
MPTP. *See* 1-methyl-4-phenyl-1,2,3,6-tetrahydropyridine (MPTP)
MQC. *See* middle quality control (MQC)
MRI. *See* magnetic resonance imaging (MRI)
MRM. *See* multiple reaction monitoring (MRM)
MS. *See* mass spectrometry (MS)
MSE. *See* matrix suppression effect (MSE)
MSI. *See* mass spectrometry imaging (MSI)
MSiReader, 120
multichannel plate detectors (MCP detectors), 65
multimaging, 120
multimodal imaging, 457–458, 472
multiphoton microscopy (MPM), 459
multiple matrix deposition methods, 95
multiple reaction monitoring (MRM), 235
multiplexed ion beam imaging (MIBI), 465
"multisensory" imaging, 460
multivariate statistical analysis (MVA), 194
MV. *See* method validation (MV)
MVA. *See* multivariate statistical analysis (MVA)
MW. *See* molecular weight (MW)
MWCO filter. *See* molecular weight cut-off filter (MWCO filter)
myopia, 449
myxoid liposarcoma (MLS), 214

N,*N*,*N*-trimethylammonioanilyl *N*-hydroxysuccinimidyl carbamate iodide (TAHS), 138–139
N,*N*-dimethylaniline (DANI), 240
N-(1-naphthyl)ethylenediamine dihydrochloride (NEDC), 141
N-acetylglucosamine (GlcNAc), 208
N-glycans, 95, 208
  biological applications, 213–216
  method development and optimization, 210–213
N-glycosylation, 213–214
N-hydroxysuccinimide (NHS), 245
NAFLD. *See* non-alcoholic fatty liver disease (NAFLD)
NALDI. *See* nanostructure-assisted laser desorption/ionization (NALDI)
nanoDESI. *See* nanospray desorption electrospray ionization (nanoDESI)
nanomaterials, 396
nanoparticles, 125
  MALDI-MSI of, 396–397
nanoSIMS, 467
nanospray desorption electrospray ionization (nanoDESI), 297
nanostructure-assisted laser desorption/ionization (NALDI), 142
nanostructure-initiator mass spectrometry (NIMS), 161
NBA. *See* 2-nitrobenzaldehyde (NBA)
NEDC. *See* *N*-(1-naphthyl)ethylenediamine dihydrochloride (NEDC)
nematodes, 294
neurodegenerative diseases, MALDI-MSI of neuropeptides in, 197–201
neurofibromatosis type 1 (NF1), 445
neuronal networks (NN), 275
neuropeptides, 188
  data acquisition and data analysis, 193
    data processing, 194
    MALDI-MSI instrument parameters for acquisition, 193–194
    statistical analysis, 194–195
  MALDI MS-based neuropeptide imaging in biological research, 196–201
  molecular identification, 195
  sample preparation for MALDI-MS imaging of, 189
    matrix application, 192–193
    sample cleanup and tissue derivatization strategies, 192
    tissue collection, 191–192
neurotransmitters, 95
neutral matrix molecules, 16
next-generation sequencing (NGS), 409
NF1. *See* neurofibromatosis type 1 (NF1)
NGS. *See* next-generation sequencing (NGS)
NHS. *See* *N*-hydroxysuccinimide (NHS)
nicosulfuron, 396
Niemann–Pick Disease, 219
Niemann–Pick disease type C (NPC), 171
NIMS. *See* nanostructure-initiator mass spectrometry (NIMS)
niraparib, 333
nitramide RDX, 364
2-nitrobenzaldehyde (NBA), 192
NMF. *See* non-negative matrix factorization (NMF)
NN. *See* neuronal networks (NN)
non-alcoholic fatty liver disease (NAFLD), 173
non-conjugated saccharides, 224–226
non-linear projection techniques, 278
non-negative matrix factorization (NMF), 277
non-small cell lung cancer (NSCLC), 416
non-thermal ionization, 9
norharmane, 160

normalization, 269–270
  method comparison, 114
NPC. *See* Niemann–Pick disease type C (NPC)
NSCLC. *See* non-small cell lung cancer (NSCLC)

OCT. *See* optical coherence tomography (OCT); optimal cutting temperature (OCT)
ocular therapeutics detection
  advantages and applications in pre-clinical settings, 445–448
  drug detection in eye for off-target effects, 448–449
  MALDI-MSI for, 445
  on-target ocular drug detection, 449–450
off-tissue identification, 248–251
OI. *See* optical imaging (OI)
on-target ocular drug detection, 449–450
on-tissue
  chemistries, 25–27
  derivatization, 174–175
  digestion, 244
  method, 110
on-tissue chemical derivatization (OTCD), 138, 392–393
oncology, 332, 341
optic nerve, 435, 444–445
optical coherence tomography (OCT), 459
optical imaging (OI), 458–459, 468–470
optimal cutting temperature (OCT), 88, 137, 191, 296
Orbitrap, 71, 163–164, 462
  mass analyzers, 69, 143
orthogonal extraction TOF (oTOF), 43
OTCD. *See* on-tissue chemical derivatization (OTCD)
oTOF. *See* orthogonal extraction TOF (oTOF)
oversampling methods, 48
oxidized GSH (GSSG), 437

P-glycoprotein (P-gp), 332
P-gp. *See* P-glycoprotein (P-gp)
PA. *See* phosphatidic acids (PA)
Paederus riparius. *See* beetle (*Paederus riparius*)
PAI. *See* photoacoustic imaging (PAI)
PALM. *See* photoactivated localization microscopy (PALM)
PAM. *See* pieces of M-parafilm (PAM)
pancreatic ductal adenocarcinoma (PDAC), 215
Panulirus interruptus. *See* spiny lobster (*Panulirus interruptus*)
papillary renal cell carcinoma (pRCC), 413
papovaviridae, 91
paraformaldehyde (PFA), 241
partial least squares DA (PLS-DA), 275
pathology, 408
  challenges and considerations, 416–420
  outcome and response prediction, 415–416
  possibilities and advantages of MALDI-MSI in pathology research and practice, 410–411
  tumor and disease detection, 411–412
  tumor and disease subtyping, 412–414
PAX fixation, 89
PBS. *See* phosphate-buffered saline (PBS)
PCA. *See* phenazine-1-carboxylic acid (PCA); principal component analysis (PCA)
PCA–LDA. *See* principal component analysis LDA (PCA–LDA)
PCho. *See* phosphatidylcholine (PCho)
PCLD. *See* polycaprolactone diol (PCLD)
PD. *See* pharmacodynamics (PD); photodissociation (PD)

PD-1. *See* programmed death-1 (PD-1)
PD-L1. *See* programmed cell death ligand 1 (PD-L1)
PDAC. *See* pancreatic ductal adenocarcinoma (PDAC)
PDPT. *See* phosphatidyl-*S*,*S*-dimethylpropanethiol (PDPT)
PE. *See* phosphatidylethanolamines (PE); plasmalogen glycerophosphoethanolamine (PE)
peak filtering, 270–271
peak picking, 270–271
PEG_{36}-DSPE. *See* 1,2-distearoyl-*sn-glycero*-3-phosphoethanolamine conjugated with PEG (PEG_{36}-DSPE)
Penicillium, 304
pentaerythritol tetranitrate (PETN), 364
pepsin, 94
peptide *N*-glycosidase F (PNGase F), 26, 210
peptides, 92, 122–124
perfluorooctanesulfonic acid (PFOS), 394–395
persistent organic pollutants (POP), 387
pesticides in plants, MALDI-MSI of, 396
PET. *See* positron emission tomography (PET)
PETN. *See* pentaerythritol tetranitrate (PETN)
PFA. *See* paraformaldehyde (PFA)
PFM. *See* polar fluid model (PFM)
PFOS. *See* perfluorooctanesulfonic acid (PFOS)
pharmacodynamics (PD), 325, 335–336
pharmacokinetics (PK), 325, 333–334
pharmacology, MALDI-MSI in, 148
phase explosion, 5
phenazine-1-carboxylic acid (PCA), 301
phenols, 124
PHI. *See* pre-clinical hybrid imaging (PHI)
phos-tag, 175
phosphate-buffered saline (PBS), 88
phosphatidic acids (PA), 160
phosphatidyl-*S*,*S*-dimethylpropanethiol (PDPT), 303
phosphatidylcholine (PCho), 15, 122, 159–160
phosphatidylethanolamines (PE), 15, 67, 160
phosphatidylinositols (PI), 159
phosphatidylserines (PS), 15
photoacoustic imaging (PAI), 459
photoactivated localization microscopy (PALM), 459
photodissociation (PD), 68
pHPMA. *See* poly[*N*-(2-hydroxypropyl)methacrylamide] (pHPMA)
PI. *See* phosphatidylinositols (PI)
2-picolylamine (2PA), 139
pieces of M-parafilm (PAM), 250
PK. *See* pharmacokinetics (PK)
plant metabolites, MALDI-MSI of, 146–147
plasmalogen glycerophosphoethanolamine (PE), 444
plastics, MALDI-MSI of, 396–397
PLS-DA. *See* partial least squares DA (PLS-DA)
pLSA. *See* probabilistic latent semantic analysis (pLSA)
plume, 4–5
  pressure, temperature and velocity, 5
PMIs. *See* post-mortem intervals (PMIs)
pMSI. *See* protein MSI (pMSI)
PNGase F. *See* peptide *N*-glycosidase F (PNGase F)
polar fluid model (PFM), 8
poly-lysine coated slides, 91
poly[*N*-(2-hydroxypropyl)-methacrylamide] (pHPMA), 90
polycaprolactone diol (PCLD), 397

polyimides, 124
polyketides, 157
polymers, 124
polyvinylpyrrolidone (PVP), 90
POP. *See* persistent organic pollutants (POP)
positron emission tomography (PET), 59, 336, 459
post source decay (PSD), 41
post-ionization, 15, 51
    alternative approaches, 53
    MALDESI/LAESI, 53
    MALDI-2, 52
    modalities, 40
post-mortem intervals (PMIs), 365–366
post-translational modifications (PTMs), 235
posterior eye, 435
    MALDI-MSI in, 441–445
pRCC. *See* papillary renal cell carcinoma (pRCC)
pre-clinical hybrid imaging (PHI), 460
precision, 116
prenols, 157
pressure regime, 40
    atmospheric pressure ion sources, 44
    intermediate pressure ion sources, 42–43
    plume development, 40–41
    vacuum ion sources, 41–42
primary ionization, 7. *See also* secondary ionization
    mechanical charge separation, 7–8
    metal surfaces, 9–10
    non-thermal ionization, 9
    preformed ions and "Lucky Survivors", 7
    thermal ionization, 8
    types of primary ions, 10
principal component analysis (PCA), 168, 194, 277
principal component analysis LDA (PCA–LDA), 275
probabilistic latent semantic analysis (pLSA), 160, 277
programmed cell death ligand 1 (PD-L1), 409
programmed death-1 (PD-1), 415–416
prostate cancer, 168
prostate-specific antigen (PSA), 168
protein MSI (pMSI), 235–236
    to spatially-resolved proteomics, 234–237
proteins, 92, 94, 234
    identification in MALDI-MSI and spatially-resolved proteomics, 245
        in situ identification, 246–248
        off-tissue identification and spatially-resolved proteomics, 248–251
proteoglycans, 222–223
proteolytic enzymes, 26
proteomics, 234–235
protons, 10
prunasin, 224
PS. *See* phosphatidylserines (PS)
PSA. *See* prostate-specific antigen (PSA)
PSD. *See* post source decay (PSD)
Pseudomonas
    P. aeruginosa, 294, 297, 301, 304, 306
    P. cichorii, 305
PTMs. *See* post-translational modifications (PTMs)
PVP. *See* polyvinylpyrrolidone (PVP)
pyochelin, 301

QC. *See* quality controls (QC)
QIT. *See* quadrupole ion trap (QIT)
QMFs. *See* quadrupole mass filters (QMFs)
QMSI. *See* quantitative mass spectrometry imaging (QMSI)
QQQ. *See* triple quadrupole mass spectrometry (QQQ)

QqQ. *See* triple quadrupoles (QqQ)
QqTOF platforms, 66–68
quadrupole ion trap (QIT), 67
quadrupole mass analyzers (QMAs). *See* quadrupole mass filters (QMFs)
quadrupole mass filters (QMFs), 62–63, 68
quality controls (QC), 115–117
  studies in QMSI, 117
quantification, 392
quantitative imaging, 106
quantitative MALDI-MSI, 394
  of environmental pollutants, 393–395
quantitative mass spectrometry imaging (QMSI), 105–106
  current challenges, 125–126
  history and progress, 106–107
  method workflow, 107–121
  versatility, 121–125
quantitative whole-body autoradiography (QWBA), 120, 328, 445
QWBA. *See* quantitative whole-body autoradiography (QWBA)

R&D. *See* research and development (R&D)
radio frequency (RF), 63
random forest (RF), 275
rapid acquisition, 64
RCC. *See* renal cell carcinoma (RCC)
reactome, 280
reagent selection, 95
recalibration, 269
receiver-operating-characteristics (ROC), 275
  analyses, 275
  assessment, 215
recrystallization, matrix application and, 97–98
region of interest (ROI), 110, 194
REIMS. *See* evaporative ionization mass spectrometry (REIMS)
relative standard deviation (RSD), 116
REMPI. *See* resonance-enhanced multi-photon ionization (REMPI)
renal cell carcinoma (RCC), 413
renal oncocytoma (RO), 413
reproducibility, 419
  in matrix deposition, 25
research and development (R&D), 321
resolution power, 142–143
resonance-enhanced multi-photon ionization (REMPI), 51
retigabine, 448
retina, 435, 441–445
retinal pigmented epithelium (RPE), 435
retinoblastoma, 450
RF. *See* radio frequency (RF); random forest (RF)
Rhizoctonia solani, 305
rMSIproc (R-package), 269
RO. *See* renal oncocytoma (RO)
ROC. *See* receiver-operating-characteristics (ROC)
ROI. *See* region of interest (ROI)
room temperature (RT), 239
RPE. *See* retinal pigmented epithelium (RPE)
rpoB protein, 301
RSD. *See* relative standard deviation (RSD)
RT. *See* room temperature (RT)

S1P. *See* sphingosine-1-phosphate (S1P)
SA. *See* sinapinic acid (SA)
saccharides, 207
saccharolipids, 157
SALDI. *See* surface-assisted laser desorption ionization (SALDI)
Salmonella enterica, 300
Salvia miltiorrhiza Bge, 225
sample collection, 88–89
sample preparation, 87–88
SARS-CoV-2, 91
scanning electron microscopy (SEM), 369, 374

scanning mode MALDI-MSI, 45
SCFV. *See* single chain fragment variable (SCFV)
Schistosoma, 294
Scutellaria baicalensis, 224
SDP. *See* sporulation delaying protein (SDP)
SDS. *See* sodium dodecyl sulfate (SDS)
secondary ion mass spectrometry (SIMS), 21–22, 60, 105, 235, 296, 322, 364, 461
secondary ion mass spectrometry-MSI (SIMS-MSI), 162
secondary ionization, 9, 10. *See also* primary ionization
  analyte suppression effect, 12–15
  matrix suppression effect, 11–12
sectioning, 90–91
sector analyzers, 60–62
secular frequencies, 63
self-organized map (SOM), 277
SEM. *See* scanning electron microscopy (SEM)
shotgun approach. *See* bottom-up approach
signal-to-noise ratio (*S*/*N* ratio), 43, 68, 109, 297
SIL standard. *See* stable isotopically labeled standard (SIL standard)
silicon nanospot arrays (NAPA), 161
silver nanoparticles (AgNPs), 396
SIMS. *See* secondary ion mass spectrometry (SIMS)
SIMS-MSI. *See* secondary ion mass spectrometry-MSI (SIMS-MSI)
sinapinic acid (SA), 193, 240
single chain fragment variable (SCFV), 245
single photon emission computed tomography (SPECT), 459
single photon ionization (SPI), 51
single reaction monitoring (SRM), 235
single-cell MALDI analysis, 175–177
single-focusing magnetic mass analyzer, 62
SKF. *See* sporulation killing factor (SKF)
skyline spectrum, 270
SMM. *See* Spitzoid malignant melanoma (SMM)
smoothing, 268
SN. *See* Spitz nevi (SN)
snap freezing, 88–89
"snow plough" effect, 40
sodium dodecyl sulfate (SDS), 240
"soft" ionization technique, 64
SOM. *See* self-organized map (SOM)
SOPs. *See* standard operating procedures (SOPs)
spatial analysis of metabolites in human tissues, 148–149
spatial resolution, 5
  in MALDI-MSI, 44–47
spatially-resolved proteomics, 248–251
  protein MS imaging to, 234–237
SPECT. *See* single photon emission computed tomography (SPECT)
spectral alignment, 269
spectral resampling, 268
Spectroglyph source, 143
sphingolipids, 157
sphingosine-1-phosphate (S1P), 175
SPI. *See* single photon ionization (SPI)
spiny lobster (*Panulirus interruptus*), 197
Spitz nevi (SN), 413
Spitzoid malignant melanoma (SMM), 413
sporulation delaying protein (SDP), 305
sporulation killing factor (SKF), 305
SR. *See* super-resolution microscopy (SR)
SR-XRF. *See* synchrotron radiation-based XRF (SR-XRF)

SRM. *See* single reaction monitoring (SRM)
ST. *See* sulfatides (ST)
stable isotopically labeled standard (SIL standard), 323
standard deviation, 116
standard operating procedures (SOPs), 115, 135, 473
Staphylococcus aureus, 304, 306
STED. *See* stimulated emission depletion (STED)
sterols, 157
stimulated emission depletion (STED), 459
stochastical optical reconstruction microscopy (STORM), 459
stop-motion scheme, 48
storage, 89–90
STORM. *See* stochastical optical reconstruction microscopy (STORM)
Streptomyces
    S. coelicolor, 305
    S. griseus, 303–304
STRING, 280
sucrose, 88
sulfatides (ST), 159
SunCollect, 192
super-resolution microscopy (SR), 459
supervised machine learning, 275
    algorithms in MSI, 275–276
    feature selection and extraction, 276–277
    training and testing, 276
support vector machine (SVM), 275
surface-assisted laser desorption ionization (SALDI), 29, 142
SVM. *See* support vector machine (SVM)
synchrotron radiation-based XRF (SR-XRF), 467
systems biology approach, 59

t-distributed stochastic neighborhood embedding (t-SNE), 277, 466
t-MALDI-MS. *See* transmission mode MALDI-MS (t-MALDI-MS)
t-SNE. *See* t-distributed stochastic neighborhood embedding (t-SNE)
TAG. *See* triacylglycerol (TAG)
Tag-Mass method, 244–245
TAHS. *See* *N*,*N*,*N*-trimethylammonioanilyl *N*-hydroxysuccinimidyl carbamate iodide (TAHS)
tandem mass spectrometry (MS/MS), 65, 93, 143–144, 165, 211
    identification of lipids using, 165–166
targeted drug delivery, 334–335
taurolithocholic acid (TLCA), 301
TBI. *See* traumatic brain injury (TBI)
TEC. *See* Tissue Extinction Coefficient (TEC)
technology readiness level (TRL), 359, 365
tetra-antennary *N*-glycans, 216
TFA. *See* trifluoroacetic acid (TFA)
TGs. *See* triacylglycerides (TGs)
tgSwe AD model. *See* transgenic AD model (tgSwe AD model)
thermal ionization, 8
thin layer chromatography (TLC), 22, 218
three-dimensional QIT (3-D QIT), 67
time-lag focusing. *See* delayed extraction
time-of-flight (ToF), 63–65, 114, 295
    analyzers, 41
    instruments, 462
    mass analyzers, 21, 25
TIMS. *See* trapped ion mobility spectrometry (TIMS)
tissue
    collection, 389
    fixation, 158–160
    sectioning, 91, 240, 242–243
    washing, 159, 240, 243
Tissue Extinction Coefficient (TEC), 111
tissue microarrays (TMAs), 91, 214, 242, 264, 341, 411

TLC. *See* thin layer chromatography (TLC)
TLCA. *See* taurolithocholic acid (TLCA)
TM Sprayer, 192
TMAs. *See* tissue microarrays (TMAs)
TNBC. *See* triple-negative breast cancer (TNBC)
TNT. *See* trinitrotoluene (TNT)
ToF. *See* time-of-flight (ToF)
top-down proteomics, 235
toxicology, MALDI-MSI in, 148
transgenic AD model (tgSwe AD model), 171
translucent substrates, 42
transmission mass analyzers, 70
transmission mode illumination, 50–51
transmission mode MALDI-MS (t-MALDI-MS), 50
(transmission)electron microscopy ((T) EM), 458–459
transmit-MALDI source, 143
trapped ion mobility spectrometry (TIMS), 75–77, 143, 167
trauma, 444–445
traumatic brain injury (TBI), 170
traveling wave ion mobility spectrometry (TWIMS), 74–75
triacylglycerides (TGs), 161
triacylglycerol (TAG), 303
triboelectricity, 7–8
Trichoderma atroviride, 305
trifluoroacetic acid (TFA), 212
trinitrotoluene (TNT), 364
triple quadrupole mass spectrometry (QQQ), 114
triple quadrupoles (QqQ), 68, 235
triple-negative breast cancer (TNBC), 216
TRL. *See* technology readiness level (TRL)
trypsin, 26, 94
tumor
 and disease detection, 411–412
 and disease subtyping, 412–414
turn-around problem, 64
TWIMS. *See* traveling wave ion mobility spectrometry (TWIMS)
two-dimensional LIT (2-D LIT), 67

ULOQ. *See* upper limit of quantification (ULOQ)
ultrasound (US), 459
ultraviolet (UV), 68
ultraviolet photodissociation (UVPD), 246
UMAP. *See* uniform manifold approximation and projection (UMAP)
uniform manifold approximation and projection (UMAP), 277
United States Food and Drug Administration (FDA), 214
unsupervised projection and clustering methods, 277
 clustering techniques, 278–279
 linear projection techniques, 278
 non-linear projection techniques, 278
 unsupervised algorithms in MSI, 277
upper limit of quantification (ULOQ), 110
US. *See* ultrasound (US)
US Food and Drug Administration (FDA), 325
UV. *See* ultraviolet (UV)
UVPD. *See* ultraviolet photodissociation (UVPD)

versatility of QMSI, 121
 alcohol and phenols, 124
 lipids, 122
 nanoparticles, 125
 peptides and amines, 122–124
 polymers and polyimides, 124
VH. *See* vitreous humor (VH)
vibrational spectroscopy (VR), 459
Vibrio fischeri, 294, 306
vimentin, 441
visualization, 271

vitamin metabolites, 95
vitreous humor (VH), 432
VR. *See* vibrational spectroscopy (VR)

washing, 92–93
wastewater treatment plant (WWTP), 397
WBA. *See* Whole Body Autoradiography (WBA)
WHO. *See* World Health Organization (WHO)
Whole Body Autoradiography (WBA), 366
wild type corneas (WT corneas), 436
World Health Organization (WHO), 294, 413
WT corneas. *See* wild type corneas (WT corneas)
WWTP. *See* wastewater treatment plant (WWTP)

X-ray fluorescence imaging (XRF imaging), 467–468
xenobiotics, imaging of, 27–28
XRF imaging. *See* X-ray fluorescence imaging (XRF imaging)

Yersinia enterocolitica, 30

www.ingramcontent.com/pod-product-compliance
Ingram Content Group UK Ltd.
Pitfield, Milton Keynes, MK11 3LW, UK
UKHW020731190926
8624IPUK00035B/15

9 781839 162411